## Pressure Conversion Factors

| | Pa | bar | atm | torr |
|---|---|---|---|---|
| 1 Pa = | 1 | $10^{-5}$ | $9.869\ 23 \times 10^{-6}$ | $7.500\ 62 \times 10^{-3}$ |
| 1 bar = | $10^5$ | 1 | 0.986 923 | 750.062 |
| 1 atm = | $1.013\ 25 \times 10^5$ | 1.013 25 | 1 | 760 |
| 1 torr = | 133.322 | $1.333\ 22 \times 10^{-3}$ | $1.315\ 79 \times 10^{-3}$ | 1 |

## Some Commonly Used Non-SI Units

| Unit | Quantity | Symbol | SI value |
|---|---|---|---|
| Angstrom | length | Å | $10^{-10}$ m = 100 pm |
| Micron | length | $\mu$ | $10^{-6}$ m |
| Calorie | energy | cal | 4.184 J (defined) |
| Debye | dipole moment | D | $3.3356 \times 10^{-30}$ C · m |
| Gauss | magnetic field strength | G | $10^{-4}$ T |

## Greek Alphabet

| | | | | | | | | |
|---|---|---|---|---|---|---|---|---|
| Alpha | A | $\alpha$ | Iota | I | $\iota$ | Rho | P | $\rho$ |
| Beta | B | $\beta$ | Kappa | K | $\kappa$ | Sigma | $\Sigma$ | $\sigma$ |
| Gamma | $\Gamma$ | $\gamma$ | Lambda | $\Lambda$ | $\lambda$ | Tau | T | $\tau$ |
| Delta | $\Delta$ | $\delta$ | Mu | M | $\mu$ | Upsilon | $\Upsilon$ | $\upsilon$ |
| Epsilon | E | $\epsilon$ | Nu | N | $\nu$ | Phi | $\Phi$ | $\phi$ |
| Zeta | Z | $\zeta$ | Xi | $\Xi$ | $\xi$ | Chi | X | $\chi$ |
| Eta | H | $\eta$ | Omicron | O | $o$ | Psi | $\Psi$ | $\psi$ |
| Theta | $\Theta$ | $\theta$ | Pi | $\Pi$ | $\pi$ | Omega | $\Omega$ | $\omega$ |

| $E_h$ | cm$^{-1}$ | Hz |
|---|---|---|
| $2.293\ 710 \times 10^{17}$ | $5.034\ 11 \times 10^{22}$ | $1.509\ 189 \times 10^{33}$ |
| $3.808\ 798 \times 10^{-4}$ | 83.5935 | $2.506\ 069 \times 10^{12}$ |
| $3.674\ 931 \times 10^{-2}$ | 8065.54 | $2.417\ 988 \times 10^{14}$ |
| 1 | $2.194\ 7463 \times 10^5$ | $6.579\ 684 \times 10^{15}$ |
| $4.556\ 335 \times 10^{-6}$ | 1 | $2.997\ 925 \times 10^{10}$ |
| $1.519\ 830 \times 10^{-16}$ | $3.335\ 64 \times 10^{-11}$ | 1 |

# MOLECULAR THERMODYNAMICS

# MOLECULAR THERMODYNAMICS

Donald A. McQuarrie
UNIVERSITY OF CALIFORNIA, DAVIS

John D. Simon
George B. Geller Professor of Chemistry
DUKE UNIVERSITY

University Science Books
Sausalito, California

**University Science Books**
55D Gate Five Road
Sausalito, CA 94965

Fax: (415) 332-5393
www.uscibooks.com

Production manager: *Susanna Tadlock*
Manuscript editor: *Ann McGuire*
Designer: *Robert Ishi*
Illustrator: *John Choi*
Compositor: *Eigentype*
Printer & Binder: *Edwards Brothers, Inc.*

This book is printed on acid-free paper.



Library of Congress Cataloging-in-Publication Data

McQuarrie, Donald A. (Donald Allen)
Molecular thermodynamics / Donald A. McQuarrie, John D.Simon.
p. cm.
Includes bibliographical references and index.
ISBN 1-891389-05-X
1. Thermodynamics I. Simon, John D. (John Douglas), 1957–
II. Title.
QD504.M335 1999
541.3′69—dc21 98-48543
CIP

Printed in the United States of America
10 9 8 7 6 5 4 3 2

# Contents

## CHAPTER 5 / The First Law of Thermodynamics 185



## MATHCHAPTER E / The Binomial Distribution and Stirling's Approximation 229



## CHAPTER 6 / Entropy and the Second Law of Thermodynamics 237



## CHAPTER 7 / Entropy and the Third Law of Thermodynamics 273

## CHAPTER 8 / Helmholtz and Gibbs Energies 301



## CHAPTER 9 / Phase Equilibria 349



## CHAPTER 10 / Solutions I: Liquid–Liquid Solutions 387

## CHAPTER 11 / Solutions II: Solid–Liquid Solutions 439



## CHAPTER 12 / Chemical Equilibrium 477



## CHAPTER 13 / Thermodynamics of Electrochemical Cells 529

# Preface

This book has evolved from our physical chemistry text, *Physical Chemistry, A Molecular Approach*. In that book, we emphasize a molecular approach to the teaching of physical chemistry. Consequently, unlike most other physical chemistry books, ours discusses the principles of quantum mechanics first and then uses these ideas in the subsequent development of thermodynamics. We follow that same approach in this text, *Molecular Thermodynamics*. Many of the chapters in this book are similar to those in our physical chemistry book, but we have added new material to several chapters and have three entirely new chapters. The first chapter, The Energy Levels of Atoms and Molecules, is new and presents results of quantum mechanics that are needed in later chapters. Only a few quantum-mechanical results are really needed. Students learn to think in terms of discrete electronic energy levels in general chemistry and organic chemistry. We begin our discussion of thermodynamics in Chapter 2, where we discuss the properties of gases. We introduce the Boltzmann factor and partition functions in Chapter 3 and then use the results of that chapter in Chapter 4 to calculate energies and heat capacities of ideal gases. Note that we do this before we introduce the First Law of Thermodynamics in Chapter 5. This approach works perfectly well because we treat only mechanical properties (pressure, energy, and heat capacity) that students have encountered in previous chemistry and physics courses. This approach allows us to immediately give a molecular interpretation to the three laws of thermodynamics and to many thermodynamic relations. The molecular interpretation of entropy is an obvious example, but even the concepts of work and heat in the First Law of Thermodynamics have a nice, physical, molecular interpretation in terms of energy levels and their populations. In Chapters 5 through 8, we introduce the principal thermodynamic state functions, giving them a molecular interpretation throughout. In Chapter 9, we examine one-component phase equilibria and introduce chemical potential. The next two chapters concern solutions: Chapter 10 discusses solutions of two liquids and Chapter 11, solutions of solids dissolved in liquids. Following Chapter 12 on chemical equilibrium, we present a new chapter on electrochemical cells. The final chapter serves as an introduction to nonequilibrium thermodynamics, or irreversible thermodynamics. This chapter is a unique feature of our book because few pedagogical introductions are available to this topic, although it finds extensive application in biophysics and biology.

As in our physical chemistry book, we have included a number of so-called MathChapters, which are short reviews of the mathematical topics used in subsequent chapters. The five MathChapters are Numerical Methods, Probability and Statistics, Series and Limits, Partial Differentiation, and The Binomial Distribution and Stirling's Approximation. In each one, the discussions are brief, elementary, and self-contained. After reading each MathChapter and doing the problems, a student will be able to focus on the subsequent physical chemical material rather than having to cope simultaneously with the physical chemistry and the mathematics. We believe this feature greatly enhances the pedagogy of our text.

An important feature of the text is the inclusion of about 10 to 12 worked examples in each chapter. In addition, each chapter provides some 60 problems, and solutions to the numerical problems are given in the back of the book. Some problems extend the material in the chapters and introduce new topics that are somewhat more advanced. They have been written in such a way as to lead the student step by step through the material.

Today's students are comfortable with computers. In the past few years, we have seen homework assignments turned in for which students used programs such as MathCad and Mathematica to solve problems, rather than pencil and paper. Data obtained in laboratory courses are now graphed and fit to functions using programs such as Excel, Lotus 123, and Kaleidagraph. Almost all students have access to personal computers, and a modern course in the physical sciences should encourage students to take advantage of these tremendous resources. As a result, we have written a number of our problems with the use of computers in mind. For example, the first MathChapter introduces the Newton-Raphson method for solving higher-order algebraic equations and transcedental equations numerically. We see no reason any longer to limit calculations in a physical chemistry course to solving quadratic equations and other artificial examples. Students should graph data, explore expressions that fit experimental data, and plot functions that describe physical behavior. The understanding of physical concepts is greatly enhanced by exploring the properties of real data. Such exercises remove the abstractness of many theories and enable students to appreciate the mathematics of physical chemistry so that they can describe and predict the physical behavior of chemical systems.

Keeping in mind that our purpose is to teach the next generation of chemists, the quantities, units, and symbols used in this text are those presented in the 1993 International Union of Pure and Applied Chemistry (IUPAC) publication, *Quantities, Units, and Symbols in Physical Chemistry* by Ian Mills et al. (Blackwell Scientific Publications, Oxford). Our decision to follow the IUPAC recommendations means that some of the symbols, units, and standard states presented in this book may differ from those used in the literature and older textbooks and may be unfamiliar to some instructors. In some instances, we took some time ourselves to come to grips with the new notation and units, but there is, indeed, an underlying logic to their use, and we found the effort worthwhile.

# Acknowledgments

Many people have contributed to the writing and production of this book. We thank our colleagues Paul Barbara, James T. Hynes, Veronica Vaida, John Crowell, Andy Kummel, Robert Continetti, Amit Sinha, John Weare, John Wheeler, Kim Baldridge, Jack Kyte, Bill Trogler, and Jim Ely for stimulating discussions on the topics that should be included in a modern physical chemistry course, and our students Barry Bolding, Peijun Cong, Robert Dunn, Scott Feller, Susan Forest, Jeff Greathouse, Kerry Hanson, Bulang Li, and Sunney Xie for reading portions of the manuscript and making many helpful suggestions. We are especially indebted to our superb reviewers Merv Hanson, John Frederick, Anne Meyers, George Shields, and Peter Rock; to Heather Cox, who also read the entire manuscript, made numerous insightful suggestions, and did every problem in the course of preparing the accompanying Solution Manual; to Carole McQuarrie, who spent many hours in the library and using the internet looking up experimental data and biographical data to write all the biographical sketches; and to Kenneth Pitzer and Joe Hubbard for supplying us with some critical biographical data. We also thank Susanna Tadlock for coordinating the entire project, Bob Ishi for designing what we think is a beautiful-looking book, Jane Ellis for competently dealing with many of the production details, John Choi for creatively handling all the artwork, Ann McGuire for a very helpful copyediting of the manuscript, and our publisher, Bruce Armbruster, for encouraging us to write our own book and for being an exemplary publisher and a good friend. Last, we thank our wives, Carole and Diane, both of whom are chemists, for being great colleagues as well as great wives.

# MOLECULAR THERMODYNAMICS

**Max Planck** was born in Kiel, Germany (then Prussia) on April 23, 1858, and died in 1948. He showed early talent in both music and science. He received his Ph.D. in theoretical physics in 1879 at the University of Munich for his dissertation on the second law of thermodynamics. He joined the faculty of the University of Kiel in 1885, and in 1888 he was appointed director of the Institute of Theoretical Physics, which was formed for him at the University of Berlin, where he remained until 1926. His application of thermodynamics to physical chemistry won him an early international reputation. Planck was president of the Kaiser Wilhelm Society, later renamed the Max Planck Society, from 1930 until 1937, when he was forced to retire by the Nazi government. Planck is known as the father of the quantum theory because of his theoretical work on blackbody radiation at the end of the 1890s, during which time he introduced a quantum hypothesis to achieve agreement between his theoretical equations and experimental data. He maintained his interest in thermodynamics throughout his long career in physics. Planck was awarded the Nobel Prize in physics in 1918 "in recognition of services he rendered to the advancement of physics by his discovery of energy quanta." Planck's personal life was clouded by tragedy. His two daughters died in childbirth, one son died in World War I, and another son was executed in World War II for his part in the assassination attempt on Hitler in 1944.

# CHAPTER 1

# The Energy Levels of Atoms and Molecules

Thermodynamics is the study of the various properties, particularly the relations between the various properties, of systems in equilibrium. It is primarily an experimental science that was developed in the 1800s and still is of great practical value in many fields such as chemistry, biology, geology, physics, environmental science, and engineering. We will use thermodynamics, for example, to derive a quantitative relationship between the vapor pressure of a liquid and its heat of vaporization; to show that if a gas obeys the equation of state $PV = nRT$, then its energy depends only upon its temperature; and to show that if a solution of two volatile liquids obeys Raoult's law, there is no volume change nor any heat evolved or absorbed upon mixing these two liquids. One of the most important and fruitful applications of thermodynamics is in the analysis of chemical equilibria, in which thermodynamics can be used to determine the temperature and pressure that optimize the products of a given chemical reaction. No industrial process would ever be undertaken without a thorough thermodynamic analysis of the chemical reactions involved.

All the results of thermodynamics are based on three fundamental laws. These laws summarize an enormous body of experimental data, and there are no known exceptions. In fact, Einstein said of thermodynamics:

> A theory is the more impressive the greater the simplicity of its premises is, the more different kinds of things it relates, and the more extended is its area of applicability. Therefore, the deep impression which classical thermodynamics made upon me. It is the only physical theory of universal content concerning which I am convinced that, within the framework of the applicability of its basic concepts, it will never be overthrown.[1]

Einstein's assessment is worth comment. Realize that thermodynamics was developed in the 1800s before the atomic theory of matter was generally accepted. The laws

[1]From *Albert Einstein: Philosopher-Scientist*, P.A. Schlipp, ed., La Salle, IL: Open Court Publishing Company, 1973.

and results of thermodynamics are not based on any atomic or molecular theory; they are independent of atomic and molecular models. The development of thermodynamics along these lines is called *classical thermodynamics*. This character of classical thermodynamics is both a strength and a weakness. We can be assured that classical thermodynamic results will never need to be modified as our knowledge of atomic and molecular structure improves, but classical thermodynamics gives us only a limited insight at the molecular level.

With the development of atomic and molecular theories in the late 1800s and early 1900s, thermodynamics was given a molecular interpretation, or a molecular basis. This field is called *statistical thermodynamics* because it relates averages of molecular properties to macroscopic thermodynamic properties such as temperature or pressure. The material in Chapters 3 and 4 is actually an elementary treatment of statistical thermodynamics. Many of the results of statistical thermodynamics depend upon the molecular models used, so these results are not as solidly based as are those of classical thermodynamics. Nevertheless, the intuitive advantage of having a molecular picture of certain quantities or processes is very convenient. Consequently, in our development of thermodynamics in this book, we will use a mixture of classical and statistical thermodynamics, even though this approach will cost us some of the rigor of the results.

For us to use a statistical thermodynamic approach, we must use a few quantum mechanical results for atoms and molecules. It is not at all necessary to be an expert in quantum mechanics to use these results. In this chapter, we will discuss the quantum mechanical energies of atoms and molecules and relate them to atomic and molecular spectroscopy. Every college general chemistry course treats the energy levels of atomic hydrogen and its associated spectrum in some detail, so we begin with a discussion of this topic in the first section. We then discuss multielectron atoms, the vibrational and rotational energies of diatomic molecules, and polyatomic molecules.

## 1–1. The Electronic Energy of Atomic Hydrogen Is Quantized

You may have learned in general chemistry and organic chemistry that the energies of the electrons in atoms and molecules are quantized; that is, they are restricted to only certain discrete values. For example, the energy of an electron in a hydrogen atom is given by the formula

$$\varepsilon_n = -\frac{2.17869 \times 10^{-18}\ \mathrm{J}}{n^2} \qquad n = 1, 2, \ldots \tag{1.1}$$

where the quantum number, $n$, is restricted to the integer values $1, 2, 3, \ldots$. The units of energy in the International System of Units (abbreviated SI) from the French Système Internationale d'Unites are joules, designated by J. One joule is equal to the kinetic energy of a mass of two kilograms moving with a speed of one meter per second. If you remember that kinetic energy is $\frac{1}{2}mv^2$, then you can see that $1\ \mathrm{J} = 1\ \mathrm{kg{\cdot}m^2{\cdot}s^{-2}}$. The electronic energy of a hydrogen atom given by Equation 1.1 is plotted in Figure 1.1.

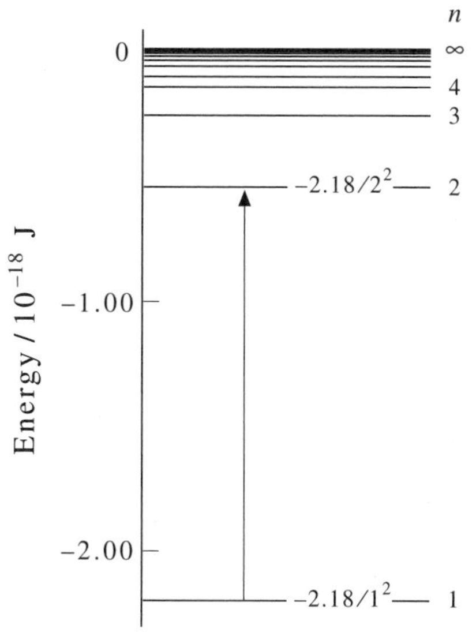


**FIGURE 1.1**
The allowed electronic energies of a hydrogen atom. The energies are given by the formula $\varepsilon_n = -2.17869 \times 10^{-18}\ \mathrm{J}/n^2$, where the quantum number $n = 1, 2, 3, \ldots$. Note that the vertical axis is label by $\varepsilon_n/10^{-18}$ J. This notation means that the dimensionless numbers on that axis are energies divided by $10^{-18}$ J. We will use this notation to label columns in tables and axes in figures because of its unambiguous nature and algebraic convenience.

Note that $\varepsilon_1 < \varepsilon_2 < \varepsilon_3 \ldots$ because of the negative sign in Equation 1.1. The state of zero energy occurs when $n \to \infty$ in Equation 1.1. In this state, the proton and the electron are so far apart that they do not attract each other at all, so we take their interaction energy to be zero. At closer distances (smaller values of $n$), the proton and the electron attract each other because of their opposite charges. A state that has negative energy is more stable than one that has zero energy.

The allowed energy states given by Equation 1.1 are called *stationary states*. The state of lowest energy ($n = 1$) is called the *ground state*. The other states are called *excited states*; the state with $n = 2$ is called the *first excited state*, that with $n = 3$ is called the *second excited state*, and so on. When an electron makes a transition from one stationary state to another, it emits or absorbs electromagnetic radiation. We picture electromagnetic radiation as consisting of packets of energy called *photons*, whose energy is equal to $h\nu$, where $\nu$ is the frequency of the radiation and $h$ is the Planck constant ($h = 6.6261 \times 10^{-34}$ J·s). Consider a transition from the state $n = 1$ to $n = 2$ (see Figure 1.1). Because $\varepsilon_2 > \varepsilon_1$, energy in the form of a photon must be absorbed, and we have $\varepsilon_2 = \varepsilon_1 + h\nu_{1\to 2}$ by conservation of energy, or $h\nu_{1\to 2} = \varepsilon_2 - \varepsilon_1$. For the $2 \to 1$ transition, conservation of energy gives $\varepsilon_2 = \varepsilon_1 + h\nu_{2\to 1}$, or $h\nu_{2\to 1} = \varepsilon_2 - \varepsilon_1$. Notice that $h\nu_{1\to 2} = h\nu_{2\to 1}$; the frequency depends only upon the magnitude of the difference between the energies of the two states. The general result

$$h\nu = \Delta\varepsilon \tag{1.2}$$

where $\Delta\varepsilon$ is the (positive) difference in the energy of the two states involved in the transition, is called the *Bohr frequency condition*.

Figure 1.2 shows the observed emission spectrum of atomic hydrogen in the visible and near ultraviolet region of the electromagnetic spectrum. Note that the lines

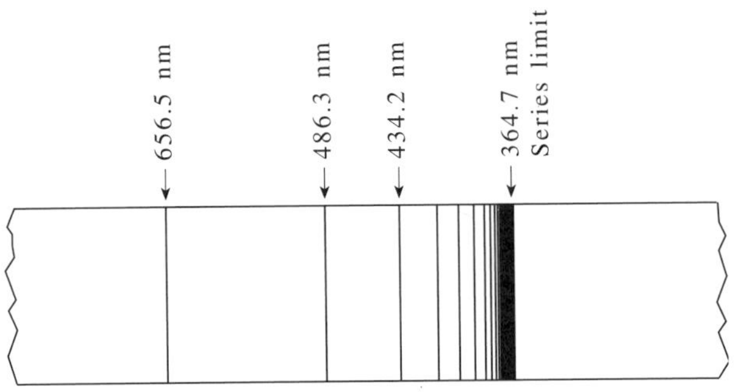


**FIGURE 1.2**
Emission spectrum of the hydrogen atom in the visible and the near ultraviolet region, showing that the emission spectrum of atomic hydrogen is a line spectrum.

in Figure 1.2 are designated by wavelength (nanometers) rather than frequency as in Equation 1.2. Recall that the relation between wavelength ($\lambda$) and frequency ($\nu$) of electromagnetic radiation is given by

$$c = \lambda\nu \tag{1.3}$$

where $c$ is the speed of light ($c = 2.9979 \times 10^8\ \text{m}\cdot\text{s}^{-1}$). Let's use Equations 1.1 through 1.3 to calculate the wavelengths shown in Figure 1.2. It turns out that the series of lines shown in Figure 1.2 are due to $n \to 2$ transitions, and so we use the relation

$$h\nu_{n\to 2} = \varepsilon_n - \varepsilon_2 = 2.17869 \times 10^{-18}\ \text{J}\left(\frac{1}{2^2} - \frac{1}{n^2}\right) \qquad n = 3, 4, \ldots$$

$$\begin{aligned}\nu_{n\to 2} &= \left(\frac{2.17869 \times 10^{-18}\ \text{J}}{6.6261 \times 10^{-34}\ \text{J}\cdot\text{s}}\right)\left(\frac{1}{4} - \frac{1}{n^2}\right)\\ &= 3.2880 \times 10^{15}\ \text{s}^{-1}\left(\frac{1}{4} - \frac{1}{n^2}\right) \qquad n = 3, 4, \ldots\end{aligned} \tag{1.4}$$

Table 1.1 shows the frequencies and the associated wavelengths calculated from Equation 1.4. The agreement between values calculated from Equation 1.4 and the experimental values (Figure 1.2) is excellent. Equation 1.4 was discovered empirically by the amateur Swiss astronomer Johann Balmer in the 1880s, and the series of lines that it predicts is now called the *Balmer series*.

**EXAMPLE 1–1**

Another series of lines in the emission of atomic hydrogen, called the *Lyman series*, occurs in the ultraviolet region. This series results from $n \to 1$ transitions. Calculate the values of the wavelengths (in nm $= 10^{-9}$ m) of the lines in the Lyman series.

SOLUTION: We use the $n \to 1$ analog of Equation 1.4

$$\nu_{n\to 1} = 3.2880 \times 10^{15}\ \text{s}^{-1}\left(\frac{1}{1^2} - \frac{1}{n^2}\right) \qquad n = 2,\ 3,\ \ldots$$

We set up the following table:

| $n$ | $\nu_{n\to 1}/10^{15}\ \text{s}^{-1}$ | $\lambda_{n\to 1}/\text{nm}$ |
|---|---|---|
| 2 | 2.4660 | 121.57 |
| 3 | 2.9227 | 102.57 |
| 4 | 3.0825 | 97.255 |
| ⋮ | ⋮ | ⋮ |
| $\infty$ | 3.2880 | 91.177 |

Once again, the agreement with the experimental values is excellent.

We can express Equation 1.2 directly in terms of wavelength by using Equation 1.3

$$\Delta\varepsilon = h\nu = \frac{hc}{\lambda}$$

This result shows that $\Delta\varepsilon$ is directly proportional to $1/\lambda$. We define a quantity called a *wavenumber*, $\tilde{\nu}$, by $\tilde{\nu} = 1/\lambda$, so that

$$\Delta\varepsilon = hc\tilde{\nu} \tag{1.5}$$

The SI units of wavenumber are $\text{m}^{-1}$, but units of $\text{cm}^{-1}$ are very widely used. Because of the simple and direct proportionality between energy and wavenumber given by Equation 1.5, it is not uncommon to express energy in terms of wavenumber.

**TABLE 1.1**
The calculated frequencies and wavelengths for the $n \to 2$ transitions in the emission spectrum of atomic hydrogen according to Equation 1.4.

| $n$ | $\nu_{n\to 2}/10^{14}\ \text{s}^{-1}$ | $\lambda_{n\to 2}/\text{nm}$ |
|---|---|---|
| 3 | 4.5667 | 656.47 |
| 4 | 6.1650 | 486.27 |
| 5 | 6.9048 | 434.18 |
| ⋮ | ⋮ | ⋮ |
| $\infty$ | 8.2200 | 364.71 |

**EXAMPLE 1–2**

Express the allowed electronic energies of atomic hydrogen given by Equation 1.1 in units of $\text{cm}^{-1}$ instead of joules.

SOLUTION: The conversion from joules to $\text{cm}^{-1}$ is given by Equation 1.5.

$$\tilde{\nu}_n = \frac{\varepsilon_n}{hc} = -\frac{2.17869 \times 10^{-18}\ \text{J}}{(6.6261 \times 10^{-34}\ \text{J}\cdot\text{s})(2.9979 \times 10^{8}\ \text{m}\cdot\text{s}^{-1})}\frac{1}{n^2}$$

$$= -\frac{1.096\,80 \times 10^{7}\ \text{m}^{-1}}{n^2} = -\frac{109\,680\ \text{cm}^{-1}}{n^2} \tag{1.6}$$

Thus, the energies are

| $n$ | $-\varepsilon_n/\text{cm}^{-1}$ |
|---|---|
| 1 | 109 680 |
| 2 | 27 420 |
| 3 | 12 187 |
| 4 | 6 855 |
| ⋮ | ⋮ |
| ∞ | 0 |

## 1–2. The Allowed Energies of a System Are Obtained from the Schrödinger Equation

The expression for the allowed electronic energies of atomic hydrogen given by Equation 1.1 as well as the allowed energies of any other atom or molecule are obtained from the Schrödinger equation, which is the central equation of quantum mechanics. We will not have to solve the Schrödinger equation nor use it at all in this book, but we will need to use some of the solutions to the Schrödinger equation for atoms and molecules. Consequently, we should discuss briefly the nature of the Schrödinger equation. For a single particle in one dimension, the Schrödinger equation takes the form

$$-\frac{\hbar^2}{2m}\frac{d^2\psi(x)}{dx^2} + V(x)\psi(x) = \varepsilon\psi(x) \tag{1.7}$$

where $\hbar$ (called h bar) $= h/2\pi$, $m$ is the mass of the particle, $x$ denotes the location of the particle along the $x$ axis, $V(x)$ is the potential energy of the particle, $\varepsilon$ is an allowed energy, and $\psi(x)$ is the wave function of the particle. When you "solve" the Schrödinger equation for a particle in a particular potential, $V(x)$, you get a set of wave functions, $\psi_n(x)$, and a set of energies, $\varepsilon_n$, associated with the wave functions, where $n$ (often an integer) labels the wave function and the energy. For example, the electronic energy of a hydrogen atom is given by Equation 1.1 and the corresponding wave functions are the hydrogen atomic orbitals, $1s, 2s, 2p, \ldots$.

A wave function has the physical interpretation that $\psi^2(x)dx$ is the probability that the particle is located between $x$ and $x + dx$. A wave function is the most complete description of a particle that is possible, and is said to specify the *state* of the particle. Quantum mechanics (and the Schrödinger equation) forces us to abandon our intuitive notion that we can locate a particle as well as we wish and replace that notion with a probabilistic interpretation instead. It turns out that for the types of particles that we work with in everyday life, the probabilities are so sharp that we can locate the particle essentially exactly, but for very small particles such as electrons, atoms, and molecules, the consequences of the probability distribution are paramount.

When we solve the Schrödinger equation, each wave function, $\psi_n(x)$, has an associated energy, $\varepsilon_n$. More than one wave function may be associated with the same energy, however. The set of energies that have the same value is called a *level*. The number of wave functions that have the same energy, $\varepsilon_n$, is called the *degeneracy* of that level and is denoted by $g_n$.

## 1–3. Atoms Have Translational Energy in Addition to Electronic Energy

In the two previous sections, we used the allowed electronic energies of atomic hydrogen to illustrate the ideas of a stationary state, the line spectrum associated with transitions between stationary states, wave functions, degeneracies, and other quantum concepts. We used atomic hydrogen as an example because atomic hydrogen, its atomic orbitals, and its emission spectrum are discussed in all general chemistry courses. Atoms and molecules, however, have other types of energies in addition to electronic energies. We discuss these other modes of energy in this and the following sections of this chapter.

Let's consider atoms first. In addition to having electronic energy, atoms have kinetic (translational) energy. The allowed translational energies of a particle are found by solving the Schrödinger equation of the particle of mass $m$ confined to some region in space. If we consider the one-dimensional case, for simplicity, then we have a particle constrained to the one-dimensional interval $0 \le x \le a$. The allowed energies in this case are given by

$$\varepsilon_n = \frac{n^2h^2}{8ma^2} \qquad n = 1,\ 2,\ \ldots \tag{1.8}$$

with an associated degeneracy, $g_n = 1$. (When $g_n = 1$, we say that the state is *nondegenerate*.) The three-dimensional version of this system is a particle confined to a three-dimensional volume, $V$, rather than a one-dimensional interval. If we take the volume to be a rectangular parallelepiped of sides $a$, $b$, and $c$, then the allowed energies are given by

$$\varepsilon_{n_xn_yn_z} = \frac{h^2}{8m}\left(\frac{n_x^2}{a^2} + \frac{n_y^2}{b^2} + \frac{n_z^2}{c^2}\right) \qquad \begin{aligned} n_x &= 1,\ 2,\ \ldots \\ n_y &= 1,\ 2,\ \ldots \\ n_z &= 1,\ 2,\ \ldots \end{aligned} \tag{1.9}$$

In Equation 1.9, the three quantum numbers independently take on the integer values 1, 2, . . . .

Equation 1.9 gives the kinetic (translational) energies of any particle of mass $m$. Atoms also have electronic energy, but there is no simple formula for the electronic energies of atoms other than hydrogen. Nevertheless, the electronic energies of atoms and their ions, which are determined by atomic spectroscopy, are well tabulated. The standard tabulation of atomic energies appears in the US Government Printing Office publication, *Atomic Energy Levels*, by Charlotte E. Moore. Chemists usually refer to these tables as "Moore's tables." Table 1.2 lists the energy-level data for the first few levels of atomic sodium, whose ground-state electron configuration is $1s^2 2s^2 2p^6 3s^1$. The first column in Table 1.2 shows the outer-shell electron configuration of the level; the second column lists the degeneracy of the level; and the third column lists the energy of the state in $\text{cm}^{-1}$. We will see that the energies of the excited electronic states usually lie so much higher than the ground-state energy that we do not have to consider them in the thermodynamic calculations that we will carry out in later chapters.

**TABLE 1.2**
Data from "Moore's tables", listing the degeneracies and energies (in $\text{cm}^{-1}$) of the first few states of atomic sodium.

| Electron configuration | Degeneracy | Energy/$\text{cm}^{-1}$ |
|---|---|---|
| $3s$ | 2 | 0.00 |
| $3p$ | 2 | 16 956.183 |
| $3p$ | 4 | 16 978.379 |
| $4s$ | 2 | 25 739.86 |
| $3d$ | 6 | 29 172.855 |
| $3d$ | 4 | 29 172.904 |
| $4p$ | 3 | 30 266.88 |
| $4p$ | 4 | 30 272.51 |
| $5s$ | 2 | 33 200.696 |
| $4d$ | 6 | 34 548.754 |
| $4d$ | 4 | 34 548.789 |

[a] From C.E. Moore, *Atomic Energy Levels*, Natl. Bur. Std. Circ. No. 467 (US Government Printing Office, Washington, DC, 1949).

**EXAMPLE 1–3**

Table 1.3 shows the first few electronic states of atomic hydrogen according to Moore's tables. Compare the values of the energies in this table with those given by the expression for $\varepsilon_n$ in Equation 1.6. Note that the energies in Tables 1.2 and 1.3 occur as sets of closely spaced values. This so-called splitting results from magnetic effects associated with the spin of the electron that are not included in Equation 1.6.

SOLUTION: The ground-state energy in Table 1.3 is taken to be 0.000 $cm^{-1}$. To compare the results of Equation 1.6 with the data in Table 1.3, we must take the difference between $\varepsilon_1$ and $\varepsilon_n$ in Equation 1.6. Thus, we have

| $n$ | $(\varepsilon_n - \varepsilon_1)/cm^{-1}$ |
|---|---|
| 1 | 0.000 |
| 2 | 82 260 |
| 3 | 97 493 |
| 4 | 102 825 |

The small discrepency between these values and those in Table 1.3 stem from the magnetic effects mentioned above.

**TABLE 1.3**
The first few electronic states of atomic hydrogen.[a]

| Electron configuration | Degeneracy | Energy/$cm^{-1}$ |
|---|---|---|
| 1*s* | 2 | 0.000 |
| 2*p* | 2 | 82 258.917 |
| 2*s* | 2 | 82 258.942 |
| 2*p* | 4 | 82 259.272 |
| 3*p* | 2 | 97 492.198 |
| 3*s* | 2 | 97 492.208 |
| 3*p*, 3*d* | 4 | 97 492.306 |
| 3*d* | 6 | 97 492.342 |
| 4*p* | 2 | 102 823.835 |
| 4*s* | 2 | 102 823.839 |
| 4*p*, 4*d* | 4 | 102 823.881 |
| 4*d*, 4*f* | 6 | 102 823.896 |
| 4*f* | 6 | 102 823.904 |

[a]From C.E. Moore, *Atomic Energy Levels*, Natl. Bur. Std. Circ. No. 467 (US Government Printing Office, Washington, DC, 1949).

## 1–4. The Vibrational Motion of a Diatomic Molecule Can Be Modeled by a Harmonic Oscillator

In the previous section, we discussed the translational and electronic energies of atoms. The expression for the translational energy of a diatomic molecule is the same as that for an atom, where the mass in this case is the mass of the diatomic molecule. The electronic energies of diatomic molecules are well tabulated, but for most of the cases we will consider, we will need only the energy and the degeneracy of the ground electronic state. We will come back to this point later. Diatomic molecules not only have translational and electronic energy, but, in addition, they have vibrational and rotational energy. In this and the next two sections, we discuss the vibrational motion of a diatomic molecule, and in Section 1–8, we discuss its rotational energy.

Figure 1.3 shows two masses connected by a spring, which is a model used to describe the vibrational motion of a diatomic molecule. Let $R_e$ be the undistorted (equilibrium) length of the spring and $R$ be its actual length. The force between the two masses will depend upon $R - R_e$, the displacement of the spring from its undistorted length. If $R - R_e$ is positive ($R > R_e$), then the two masses will be pulled toward each other. If $R - R_e$ is negative ($R < R_e$), then the two masses will be pushed away from each other.

The simplest assumption we can make about the force between the two masses as a function of the displacement is that the force is directly proportional to the displacement and to write

$$f = -k(R - R_e) = -kx \tag{1.10}$$

where the displacement $x = R - R_e$. The negative sign indicates that the force is repulsive if the spring is compressed ($R < R_e$) and is attractive if the spring is stretched ($R > R_e$). Equation 1.10 is called *Hooke's law*, and the (positive) proportionality constant $k$ is called the *force constant* of the spring. The SI units of force are newtons (N). If you remember that force is mass times acceleration, then you can see that $1\ \mathrm{N} = 1\ \mathrm{kg \cdot m \cdot s^{-2}}$. The SI units of displacement are meters, so Equation 1.10 shows that the SI units of a force constant are $\mathrm{N \cdot m^{-1}}$ or $\mathrm{kg \cdot s^{-2}}$. A small value of $k$ implies a weak or loose spring, and a large value of $k$ implies a stiff spring. If the spring is compressed or extended and then allowed to undergo its natural motion, then $x$ will vary with time according to (see Problems 1–17 through 1–20)

$$x(t) = A \cos 2\pi \nu t \tag{1.11}$$

where $A$ is called the *amplitude* of the motion and

$$\nu = \frac{1}{2\pi}\left(\frac{k}{\mu}\right)^{1/2} \tag{1.12}$$

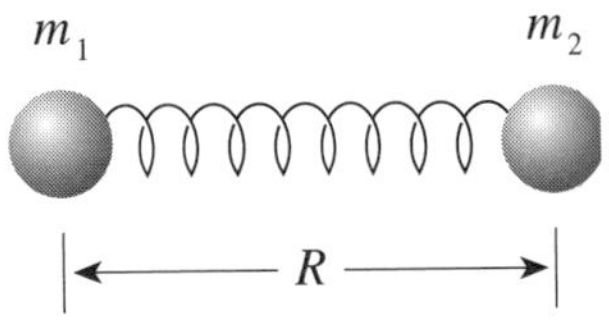


**FIGURE 1.3**
Two masses connected by a spring, which is a model used to describe the vibrational motion of a diatomic molecule.

where $\mu$ is the *reduced mass* of the molecule

$$\mu = \frac{m_1 m_2}{m_1 + m_2} \tag{1.13}$$

The units of $\nu$ are cycles per second, but cycles is usually omitted and we say that the units of $\nu$ are $s^{-1}$, which is called a hertz and is denoted by Hz.

Equation 1.11 is plotted in Figure 1.4. Motion that varies as $\cos 2\pi\nu t$ or $\sin 2\pi\nu t$ is called harmonic motion, so an oscillator that is governed by Hooke's law is called a *harmonic oscillator*. Let's look at the total energy of a harmonic oscillator. The force is given by Equation 1.10. Recall from physics that a force can be expressed as a derivative of a potential energy or that

$$f(x) = -\frac{dV(x)}{dx} \tag{1.14}$$

so that the potential energy is

$$V(x) = -\int f(x)dx + \text{constant} \tag{1.15}$$

Using Equation 1.10 for $f(x)$, we see that

$$V(x) = \frac{k}{2}x^2 + \text{constant} \tag{1.16}$$

The constant term here is an arbitrary constant that can be used to fix the zero of energy. If we choose the potential energy of the system to be zero when the spring is undistorted ($x = 0$), then we have

$$V(x) = \frac{k}{2}x^2 \tag{1.17}$$

for the potential energy associated with a simple harmonic oscillator.

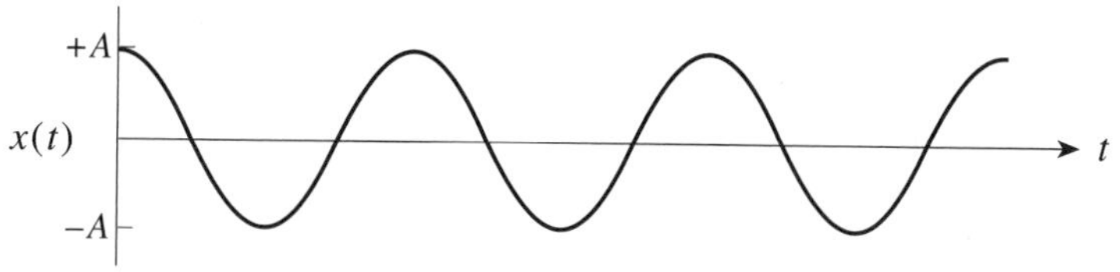


**FIGURE 1.4**
An illustration of the displacement of a harmonic oscillator versus time.

**EXAMPLE 1–4**
Show that the harmonic-oscillator potential energy, given by Equation 1.17, has units of energy.

SOLUTION: Recall that the units of $k$ are $\text{N}\cdot\text{m}^{-1}$, and that $1\ \text{N} = 1\ \text{kg}\cdot\text{m}\cdot\text{s}^{-2}$ (force = mass times acceleration). Therefore, the units of $V$ are given by

$$V = (\text{N}\cdot\text{m}^{-1})(\text{m}^2) = (\text{kg}\cdot\text{s}^{-2})(\text{m}^2) = \text{kg}\cdot\text{m}\cdot\text{s}^{-2} = \text{J}$$

Before we discuss the quantum-mechanical treatment of a harmonic oscillator, we should discuss how good an approximation it is for a vibrating diatomic molecule. The internuclear potential for a diatomic molecule is illustrated by the solid line in Figure 1.5. Notice that the curve rises steeply to the left of the minimum, indicating the difficulty of pushing the two nuclei closer together. The curve to the right side of the equilibrium position rises intially but eventually levels off. The potential energy at large separations is essentially the bond energy. The dashed line shows the potential $\frac{1}{2}k(R - R_e)^2$ associated with Hooke's law. Although the harmonic-oscillator potential may appear to be a terrible approximation to the experimental curve, note that it is, indeed, a good approximation in the region of the minimum. It turns out that this region is the physically important region for many molecules at room temperature. Although the harmonic oscillator unrealistically allows the displacement to vary from $-\infty$ to $+\infty$, these large displacements produce potential energies that are so large that they do not often occur in practice. The harmonic oscillator will be a good approximation for vibrations with small amplitudes (Problem 1–24).

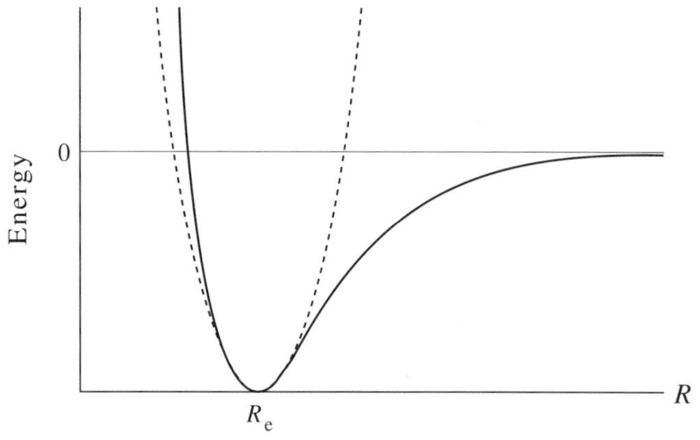


**FIGURE 1.5**
A comparison of the harmonic-oscillator potential ($kx^2/2$; dashed line) with the complete internuclear potential (solid line) of a diatomic molecule. The harmonic oscillator potential is a satisfactory approximation at small displacements.

## 1–5. The Energy Levels of a Quantum-Mechanical Harmonic Oscillator Are $\varepsilon_v = h\nu(v + \frac{1}{2})$ with $v = 0,\ 1,\ 2,\ \ldots$

When the Schrödinger equation for a one-dimensional harmonic oscillator is solved, physically well-behaved wave functions can be obtained only if the energy is restricted to the quantized values

$$\begin{aligned}\varepsilon_v &= \frac{h}{2\pi}\left(\frac{k}{\mu}\right)^{1/2}\left(v + \frac{1}{2}\right)\\ &= h\nu\left(v + \frac{1}{2}\right) \qquad v = 0,\ 1,\ 2,\ \ldots \end{aligned} \tag{1.18}$$

where

$$\nu = \frac{1}{2\pi}\left(\frac{k}{\mu}\right)^{1/2} \tag{1.19}$$

Each energy level of a harmonic oscillator is nondegenrate. In other words, the degeneracy, $g_v$, is equal to one for all values of $v$. The allowed energies of a harmonic oscillator are plotted in Figure 1.6. Note that the energy levels are equally spaced, with a separation $h\nu$. Note also that the energy of the ground state, the state with $v = 0$, is $\frac{1}{2}h\nu$ and is not zero as the lowest classical energy is. The energy of this lowest energy level is called the *zero-point energy* of the harmonic oscillator, and the fact that it is not zero is strictly a quantum-mechanical result.

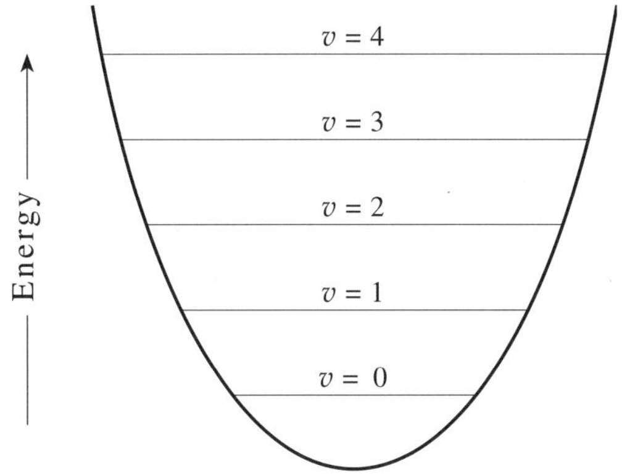


**FIGURE 1.6**
The energy levels of a quantum-mechanical harmonic oscillator.

**EXAMPLE 1–5**
Given that the value of the force constant of $H_2$ is 575 $N \cdot m^{-1}$, calculate the fundamental vibrational frequency of $H_2$.

SOLUTION: According to Equation 1.19, $\nu = (k/\mu)^{1/2}/2\pi$. The reduced mass of $H_2$ is given by (Equation 1.13)

$$\mu = \frac{m_1 m_2}{m_1 + m_2} = \frac{1.0078 \text{ amu}}{2}(1.661 \times 10^{-27} \text{ kg}\cdot\text{amu}^{-1})$$
$$= 8.370 \times 10^{-28} \text{ kg}$$

and so

$$\nu = \frac{1}{2\pi}\left(\frac{575 \text{ kg}\cdot\text{s}^{-2}}{8.370 \times 10^{-28} \text{ kg}}\right)^{1/2}$$
$$= 1.319 \times 10^{14} \text{ s}^{-1} = 4401 \text{ cm}^{-1}$$

## 1–6. The Harmonic Oscillator Accounts for the Infrared Spectrum of a Diatomic Molecule

A diatomic molecule can make a transition from one vibrational energy state to another by absorbing or emitting electromagnetic radiation whose observed frequency satisfies the Bohr frequency condition

$$\Delta\varepsilon = h\nu_{\text{obs}} \tag{1.20}$$

It turns out that the harmonic-oscillator model allows transitions only between adjacent energy states, so that we have the condition that $\Delta v = \pm 1$. Such a condition is called a *selection rule*.

For absorption to occur, $\Delta v = +1$ and so

$$\Delta\varepsilon = \varepsilon_{v+1} - \varepsilon_v = \frac{h}{2\pi}\left(\frac{k}{\mu}\right)^{1/2} = h\nu \tag{1.21}$$

Thus, the observed frequency of the radiation absorbed is

$$\nu_{\text{obs}} = \frac{1}{2\pi}\left(\frac{k}{\mu}\right)^{1/2} \tag{1.22}$$

or, in wavenumbers,

$$\tilde{\nu}_{\text{obs}} = \frac{1}{2\pi c}\left(\frac{k}{\mu}\right)^{1/2} \tag{1.23}$$

where the tilde indicates that the units are $\text{cm}^{-1}$. Furthermore, because successive energy states of a harmonic oscillator are separated by the same energy, $\Delta\varepsilon$ is the same for all allowed transitions, so this model predicts that the spectrum consists of just one line whose frequency is given by Equation 1.22. This prediction is in good accord with experiment, and this line is called the *fundamental vibrational frequency*. For diatomic molecules, these lines occur at around $10^{13}$ Hz to $10^{14}$ Hz, which is in the infrared region. Equation 1.22 or 1.23 enables us to determine force constants if the fundamental vibrational frequency is known. For example, for $H^{35}Cl(g)$, $\tilde{\nu}_{obs}$ is 2886 $cm^{-1}$ and so, according to Equation 1.23, the force constant of $H^{35}Cl(g)$ is

$$\begin{aligned} k &= (2\pi c\tilde{\nu}_{obs})^2\mu \\ &= [2\pi(2.998 \times 10^8\ m\cdot s^{-1})(2886\ cm^{-1})(100\ cm\cdot m^{-1})]^2 \\ &\qquad \times \frac{(34.97\ amu)(1.008\ amu)}{(34.97 + 1.008)\ amu}(1.661 \times 10^{-27}\ kg\cdot amu^{-1}) \\ &= 481\ kg\cdot s^{-2} = 481\ N\cdot m^{-1} \end{aligned}$$

Force constants for diatomic molecules are typically of the order of $10^2\ N\cdot m^{-1}$.

Table 1.4 lists the fundamental vibrational frequencies, rotational constants (Section 1–7), bond lengths, degeneracies, and electronic energies of the ground states of some diatomic molecules.

**TABLE 1.4**

The fundamental vibrational frequencies ($\tilde{\nu}$), rotational constants ($\tilde{B}$), bond lengths ($R_e$), degeneracies, and equilibrium energies ($D_e$) of the ground states of some diatomic molecules. These parameters were obtained from a variety of sources and do not represent the most accurate values because they were obtained under the rigid-rotator harmonic-oscillator approximation.

| Molecule | $\tilde{\nu}/cm^{-1}$ | $\tilde{B}/cm^{-1}$ | $R_e/pm$ | Degeneracy | $D_e/kJ\cdot mol^{-1}$ |
|---|---|---|---|---|---|
| $H_2$ | 4401 | 59.32 | 74.2 | 1 | 457.6 |
| $D_2$ | 2990 | 29.90 | 74.2 | 1 | 453.9 |
| $H^{35}Cl$ | 2886 | 10.44 | 127.5 | 1 | 440.2 |
| $H^{79}Br$ | 2630 | 8.348 | 141.4 | 1 | 377.7 |
| $H^{127}I$ | 2230 | 6.428 | 160.9 | 1 | 308.6 |
| $^{35}Cl^{35}Cl$ | 554 | 0.2433 | 198.8 | 1 | 242.3 |
| $^{79}Br^{79}Br$ | 323 | 0.08194 | 228.4 | 1 | 191.9 |
| $^{127}I^{127}I$ | 213 | 0.03731 | 266.7 | 1 | 150.3 |
| $^{16}O^{16}O$ | 1556 | 1.438 | 120.7 | 3 | 503.0 |
| $^{14}N^{14}N$ | 2330 | 1.990 | 109.4 | 1 | 953.0 |
| $^{12}C^{16}O$ | 2143 | 1.923 | 112.8 | 1 | 1085 |
| $^{14}N^{16}O$ | 1876 | 1.663 | 115.1 | 2 | 638.1 |
| $^{23}Na^{23}Na$ | 158 | 0.1543 | 307.8 | 1 | 72.1 |
| $^{39}K^{39}K$ | 91.9 | 0.05666 | 390.5 | 1 | 53.5 |

**EXAMPLE 1–6**
Use the data in Table 1.4 to calculate the value of the force constant of $^{39}K^{39}K(g)$.

SOLUTION: The force constant is given by

$$k = (2\pi c\tilde{\nu})^2\mu$$

The reduced mass of $^{39}K^{39}K$ is

$$\mu = \frac{(38.964 \text{ amu})(38.964 \text{ amu})}{(2)(38.964 \text{ amu})}(1.661 \times 10^{-26} \text{ kg}\cdot\text{amu}^{-1})$$
$$= 3.236 \times 10^{-26} \text{ kg}$$

and so

$$k = [2\pi(2.998 \times 10^8 \text{ m}\cdot\text{s}^{-1})(91.9 \text{ cm}^{-1})(100 \text{ cm}\cdot\text{m}^{-1})]^2(3.236 \times 10^{-26} \text{ kg})$$
$$= 9.70 \text{ kg}\cdot\text{s}^{-2} = 9.70 \text{ N}\cdot\text{m}^{-1}$$

The harmonic-oscillator selection rule says not only that $\Delta v = \pm 1$, but that the dipole moment of the molecule must change as the molecule vibrates. Thus, the harmonic-oscillator model predicts that HCl(g) absorbs in the infrared but $N_2(g)$ does not, which is in good agreement with experiment. There are, indeed, deviations from the harmonic-oscillator model, but they are fairly small, and we can systematically introduce corrections and extensions to account for them (see Problems 1–31 through 1–33).

## 1–7. The Dissociation Energy and the Ground-State Electronic Energy of a Diatomic Molecule Are Related by $D_e = D_0 + h\nu/2$

Figure 1.5 showed the potential energy of a diatomic molecule as a function of the internuclear separation, $R$. This potential energy that the two nuclei experience results from the distribution of the electrons around the two nuclei and is called the electronic energy of the molecule. The complete curve can be calculated from the Schrödinger equation by holding the nuclei a fixed distance $R$ apart and then solving for the distribution of electrons and the corresponding energy. Calculations like this must be done on a computer, but they have been carried out for a number of molecules. The internuclear separation at the minimum of the curve in Figure 1.5 is the equilibrium internuclear separation, $R_e$, or the bond length of the molecule.

Figure 1.7 shows the potential-energy curves of the ground and first excited electronic states of a typical diatomic molecule. Note that the bond length in the excited electronic state is not necessarily equal to the bond length in the ground electronic state. As for atoms, usually the energy of the first excited electronic state of a diatomic molecule lies so much higher than the ground electronic state that we need not consider it in the thermodynamic calculations that we carry out in later chapters.

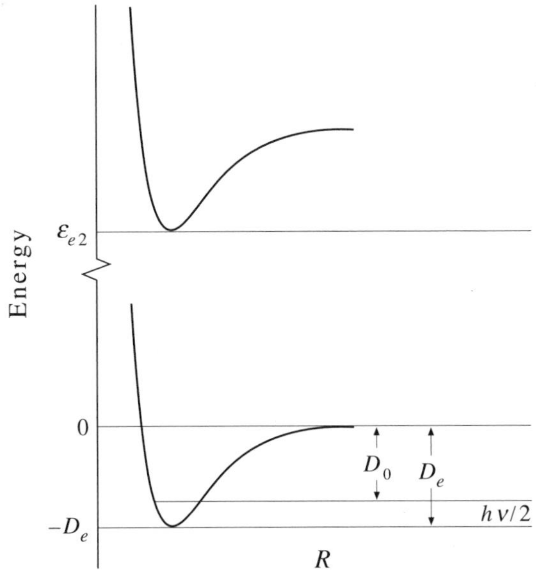


**FIGURE 1.7**
The ground-state and first excited-state electronic energy of a diatomic molecule plotted against the internuclear separation, $R$, showing that the equilibrium ground-state electronic energy, $D_e$, and the dissociation energy, $D_0$, are related by $D_e = D_0 + h\nu/2$.

According to Figures 1.5 and 1.7, the electronic energy of the ground electronic state goes to zero as the internuclear separation increases. Energies must always be reckoned relative to some (arbitrary) zero of energy. In this case, we are taking the zero of the ground-state electronic energy of a diatomic molecule to be the separated constituent atoms in their ground electronic states. Consequently, the ground-state electronic energy of the molecule at its equilibrium internuclear separation is $-D_e$, as shown in Figure 1.7. For example, the value of $D_e$ for $H_2(g)$ is 458 kJ·mol$^{-1}$, which means that the ground-state electronic energy of $H_2(g)$ at its equilibrium internuclear separation (74.1 pm) is 458 kJ·mol$^{-1}$ below that of two separated hydrogen atoms in their ground electronic state ($1s$), or that the electronic energy is $-458$ kJ·mol$^{-1}$. This is *not* the energy required to dissociate $H_2(g)$ into two ground-state hydrogen atoms, however. Recall that the lowest vibrational energy state has a zero-point energy $\varepsilon_0 = h\nu/2$. Consequently, the energy required to dissociate $H_2(g)$ from its ground state into two ground-state hydrogen atoms is $D_e - h\nu/2$, which we designate by $D_0$. Thus, we write in an equation

$$D_0 = D_e - \frac{h\nu}{2} \tag{1.24}$$

Given that $\nu = 4401$ cm$^{-1}$ for $H_2(g)$ (from Table 1.4), we see that $D_0$, the dissociation energy of $H_2(g)$ is 458 kJ·mol$^{-1}$ − 26 kJ·mol$^{-1}$ = 432 kJ·mol$^{-1}$.

**EXAMPLE 1–7**

Given that the dissociation energy of $F_2(g)$ is 154 $kJ\cdot mol^{-1}$ and that the vibrational frequency is 892 $cm^{-1}$, calculate the value of the ground-state electronic energy.

SOLUTION: We are given values of $D_0$ and $\tilde{\nu}$ and wish to calculate the value of $D_e$. From Equation 1.24, we have

$$D_e = D_0 + \frac{h\nu}{2}$$

$$= 154\ \text{kJ}\cdot\text{mol}^{-1}$$

$$+ \frac{(6.626 \times 10^{-34}\ \text{J}\cdot\text{s})(892\ \text{cm}^{-1})(2.998 \times 10^{8}\ \text{cm}\cdot\text{s}^{-1})(6.022 \times 10^{23}\ \text{mol}^{-1})}{2}$$

$$= 154\ \text{kJ}\cdot\text{mol}^{-1} + 5.33\ \text{kJ}\cdot\text{mol}^{-1}$$

$$= 159\ \text{kJ}\cdot\text{mol}^{-1}$$

## 1–8. The Energy Levels of a Rigid Rotator Are $\varepsilon_J = \hbar^2 J(J+1)/2I$

Molecules rotate as well as vibrate, and in this section we will discuss a simple model for a rotating diatomic molecule. The model consists of two point masses $m_1$ and $m_2$ at fixed distances $R_1$ and $R_2$ from their center of mass (cf. Figure 1.8). Because the distance between the two masses is fixed, this model is referred to as the *rigid-rotator model*. Even though a diatomic molecule vibrates as it rotates, the vibrational amplitude is small compared with the bond length, so considering the bond length fixed is a good approximation.

A rigid rotator rotates around its center of mass, which is given by the condition $m_1R_1 = m_2R_2$ (see Figure 1.8). If the molecule rotates about its center of mass at a

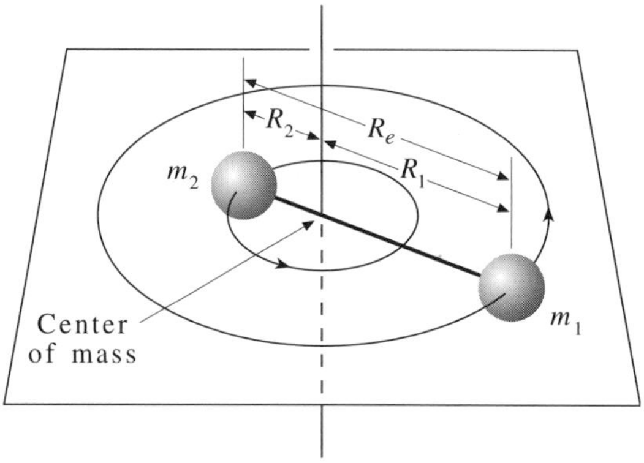


**FIGURE 1.8**
Two masses $m_1$ and $m_2$ shown rotating about their center of mass.

frequency of $\nu_{rot}$ cycles per second, then the velocities of the two masses are $v_1 = 2\pi R_1 \nu_{rot}$ and $v_2 = 2\pi R_2 \nu_{rot}$. We define the *angular velocity*, $\omega$, by $\omega$ (radians per second) $= 2\pi \nu_{rot}$, so we write $v_1$ and $v_2$ as $v_1 = R_1\omega$ and $v_2 = R_2\omega$. The kinetic energy of the rigid rotator is

$$K = \frac{1}{2}m_1 v_1^2 + \frac{1}{2}m_2 v_2^2 = \frac{1}{2}(m_1 R_1^2 + m_2 R_2^2)\omega^2$$

$$= \frac{1}{2}I\omega^2 \tag{1.25}$$

where $I$, the *moment of inertia*, is given by

$$I = m_1 R_1^2 + m_2 R_2^2 \tag{1.26}$$

The moment of inertia can be rewritten as (Problem 1–35)

$$I = \mu R_e^2 \tag{1.27}$$

where $R_e = R_1 + R_2$ (the fixed separation of the two masses) and $\mu$ is the *reduced mass*.

When we solve the Schrödinger equation for a rigid rotator, the expression for the allowed energies comes out to be

$$\varepsilon_J = \frac{\hbar^2}{2I}J(J+1) \qquad J = 0,\ 1,\ 2,\ \ldots \tag{1.28}$$

Once again, we obtain a set of discrete energy levels. In addition to the allowed energies given by Equation 1.28, we also find that each energy level has a degeneracy $g_J$ given by

$$g_J = 2J + 1 \tag{1.29}$$

The selection rule for a rigid rotator says that transitions are allowed only from adjacent states or that $\Delta J = \pm 1$. In addition to the requirement that $\Delta J = \pm 1$, the molecule must also possess a permanent dipole moment to absorb electromagnetic radiation. Thus, we say that HCl(g) has a pure rotational spectrum, but $N_2$(g) does not. In the case of absorption of electromagnetic radiation, the molecule goes from a state with a quantum number $J$ to one with $J + 1$. The energy difference, then, is

$$\Delta\varepsilon = \varepsilon_{J+1} - \varepsilon_J = \frac{\hbar^2}{2I}[(J+1)(J+2) - J(J+1)]$$

$$= \frac{\hbar^2}{I}(J+1) = \frac{h^2}{4\pi^2 I}(J+1) \tag{1.30}$$

Using the Bohr frequency condition $\Delta\varepsilon = h\nu$, the frequencies at which the absorption transitions occur are

$$\nu = \frac{h}{4\pi^2 I}(J+1) \qquad J = 0,\ 1,\ 2,\ \ldots \tag{1.31}$$

The common practice in microwave spectroscopy is to write Equation 1.31 as

$$\nu = 2B(J+1) \qquad J = 0,\ 1,\ 2,\ \ldots \tag{1.32}$$

where

$$B = \frac{h}{8\pi^2 I} \tag{1.33}$$

is called the *rotational constant* of the molecule. Also, the transition frequency is often expressed in terms of wave numbers ($\text{cm}^{-1}$) rather than hertz (Hz). If we use the relation $\tilde{\nu} = \nu/c$, then Equation 1.32 becomes

$$\tilde{\nu} = 2\tilde{B}(J+1) \qquad J = 0,\ 1,\ 2,\ \ldots \tag{1.34}$$

where $\tilde{B}$ is the rotational constant expressed in units of wave numbers

$$\tilde{B} = \frac{B}{c} = \frac{h}{8\pi^2 cI} \qquad (\text{cm}^{-1}) \tag{1.35}$$

From either Equation 1.32 or 1.34, we see that the rigid-rotator model predicts that the microwave spectrum of a diatomic molecule consists of a series of equally spaced lines with a separation of $2B$ Hz or $2\tilde{B}$ $\text{cm}^{-1}$, as shown in Figure 1.9.

Let's use Equation 1.32 to calculate the values of the absorption frequencies of $H^{35}Cl(g)$ using the equilibrium bond length given in Table 1.4 (127.5 pm). Because $I = \mu R_e^2$, we must first calculate the value of the reduced mass of $H^{35}Cl$.

$$\begin{aligned}\mu &= \frac{(1.008\ \text{amu})(34.97\ \text{amu})}{1.008\ \text{amu} + 34.97\ \text{amu}}(1.661 \times 10^{-27}\ \text{kg}\cdot\text{amu}^{-1})\\ &= 1.627 \times 10^{-27}\ \text{kg}\end{aligned}$$

and so

$$\begin{aligned}I = \mu R_e^2 &= (1.627 \times 10^{-27}\ \text{kg})(127.5 \times 10^{-12}\ \text{m})^2\\ &= 2.646 \times 10^{-47}\ \text{kg}\cdot\text{m}^2\end{aligned}$$

Therefore,

$$\begin{aligned}\nu = \frac{h}{4\pi^2 I}(J+1) &= \frac{(6.626 \times 10^{-34}\ \text{J}\cdot\text{s})(J+1)}{4\pi^2(2.646 \times 10^{-47}\ \text{kg}\cdot\text{m}^2)}\\ &= (6.343 \times 10^{11}\ \text{s}^{-1})(J+1) \qquad J = 0,\ 1,\ 2, \ldots\\ &= (6.343 \times 10^{11}\ \text{Hz})(J+1) \qquad J = 0,\ 1,\ 2, \ldots\end{aligned}$$

As $J$ takes on the values 0, 1, 2, . . ., $\nu$ takes on the values $6.343 \times 10^{11}$ Hz, $12.686 \times 10^{11}$ Hz, $19.029 \times 10^{11}$ Hz, and so on. By referring to Figure 1.11 in Problem 1–1, we see that these frequencies lie in the microwave region. Consequently, rotational

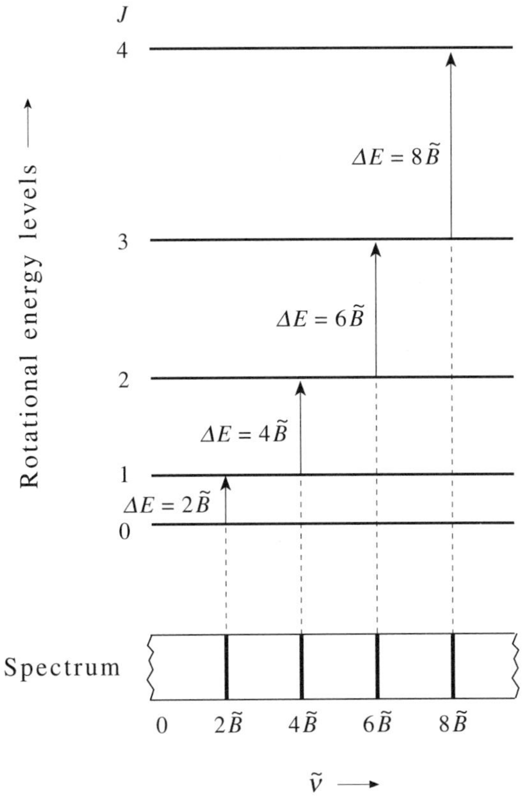


**FIGURE 1.9**
The energy levels and absorption transitions of a rigid rotator. The absorption transitions occur between adjacent levels, so the absorption spectrum shown below the energy levels consists of a series of equally spaced lines. The quantity $\tilde{B}$ is $h/8\pi^2 cI$ (Equation 1.35).

transitions of diatomic molecules occur in the microwave region, and the direct study of rotational transitions in molecules is called *microwave spectroscopy*.

From the separation between the absorption frequencies in a rotational spectrum of a diatomic molecule, we can determine the rotational constant and hence the moment of inertia of the molecule. Furthermore, because $I = \mu R_e^2$, where $R_e$ is the internuclear distance or bond length, we can determine the bond length. This procedure is illustrated in Example 1–8.

**EXAMPLE 1–8**
To a good approximation, the microwave spectrum of $^{23}Na^{35}Cl(g)$ consists of a series of equally spaced lines separated by $0.4322\ cm^{-1}$. Calculate the bond length of $^{23}Na^{35}Cl(g)$.

SOLUTION: According to Equation 1.35, the spacing of the lines in the microwave spectrum of $^{23}Na^{35}Cl(g)$ is given by

$$2\tilde{B} = \frac{h}{4\pi^2 cI}$$

and so

$$\frac{h}{4\pi^2 cI} = 0.4322 \text{ cm}^{-1}$$

Solving this equation for $I$, we have

$$I = \frac{6.626 \times 10^{-34} \text{ J}\cdot\text{s}}{4\pi^2(2.998 \times 10^8 \text{ m}\cdot\text{s}^{-1})(100 \text{ cm}\cdot\text{m}^{-1})(0.4322 \text{ cm}^{-1})}$$

$$= 1.295 \times 10^{-45} \text{ kg}\cdot\text{m}^2$$

The reduced mass of $^{23}Na^{35}Cl$ is

$$\mu = \frac{(22.99 \text{ amu})(34.97 \text{ amu})}{57.96 \text{ amu}}(1.661 \times 10^{-27} \text{ kg}\cdot\text{amu}^{-1}) = 2.304 \times 10^{-26} \text{ kg}$$

Using the fact that $I = \mu R_e^2$, we obtain

$$R_e = \left(\frac{1.295 \times 10^{-45} \text{ kg}\cdot\text{m}^2}{2.304 \times 10^{-26} \text{ kg}}\right)^{1/2} = 2.371 \times 10^{-10} \text{ m} = 237.1 \text{ pm}$$

Problems 1–36, 1–38, and 1–39 give other examples of the determination of bond lengths from microwave data.

A diatomic molecule is not truly a rigid rotator, because it simultaneously vibrates, however small the amplitude. Consequently, we might expect that although the microwave spectrum of a diatomic molecule consists of a series of equally spaced lines, their separation is not *exactly* constant. There is a straightforward procedure to correct for the fact that the bond is not exactly rigid (Problem 1–41).

## **1–9.** The Vibrations of Polyatomic Molecules Are Represented by Normal Modes

Polyatomic molecules have translational, vibrational, rotational, and electronic energies. The translational energy is the same as that of an atom or a diatomic molecule, and the electronic energies of many polyatomic molecules are well tabulated. We will discuss the vibrational energy of polyatomic molecules in this section and their rotational energy in the next section.

Consider a molecule containing $n$ nuclei. A complete specification of this molecule in space requires $3n$ coordinates, three Cartesian coordinates for each nucleus. We say that the $n$-atomic molecule has a total of $3n$ *degrees of freedom*. Of these $3n$ coordinates, three can be used to specify the center of mass of the molecule. Motion along these three coordinates corresponds to translational motion of the center of mass of the molecule, and so we call these three coordinates *translational degrees of freedom*. Two coordinates are required to specify the orientation of a linear molecule about its center of mass and three coordinates to specify the orientation of a nonlinear molecule about its center of mass. Because motion along these coordinates corresponds to rotational motion, we say that a linear molecule has two *degrees of rotational freedom* and that a nonlinear molecule has three degrees of rotational freedom. The remaining coordinates ($3n - 5$ for a linear molecule and $3n - 6$ for a nonlinear molecule) specify the relative positions of the $n$ nuclei. Because motion along these coordinates corresponds to vibrational motion, we say that a linear molecule has $3n - 5$ *vibrational degrees of freedom* and that a nonlinear molecule has $3n - 6$ vibrational degrees of freedom. These results are summarized in Table 1.5.

**EXAMPLE 1–9**

Determine the number of various degrees of freedom of HCl, $CO_2$, $H_2O$, $NH_3$, and $CH_4$.

SOLUTION:

| | Total | Translational | Rotational | Vibrational |
|---|---|---|---|---|
| HCl | 6 | 3 | 2 | 1 |
| $CO_2$ (linear) | 9 | 3 | 2 | 4 |
| $H_2O$ | 9 | 3 | 3 | 3 |
| $NH_3$ | 12 | 3 | 3 | 6 |
| $CH_4$ | 15 | 3 | 3 | 9 |

Under the harmonic-oscillator approximation, the vibrational motion of a polyatomic molecule can be pictured as the motion of $n_{\text{vib}}$ independent harmonic oscillators, where $n_{\text{vib}}$ is the number of vibrational degrees of freedom. For example, a water molecule has three degrees of vibrational freedom ($3n - 6 = 3 \times 3 - 6 = 3$), and the

**TABLE 1.5**

The number of various degrees of freedom of a polyatomic molecule containing $n$ atoms.

| Degrees of freedom | Linear | Nonlinear |
|---|---|---|
| Translational | 3 | 3 |
| Rotational | 2 | 3 |
| Vibrational | $3n - 5$ | $3n - 6$ |

vibrational motion of a water molecule can be broken down into the three characteristic vibrational motions shown below.

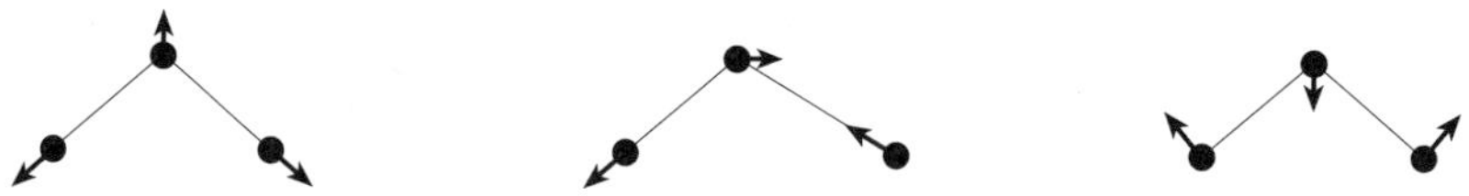

Symmetric stretch $3586\ \mathrm{cm}^{-1}$ | Asymmetric stretch $3725\ \mathrm{cm}^{-1}$ | Bending motion $1595\ \mathrm{cm}^{-1}$

These three characteristic vibrational modes of a water molecule are examples of *normal modes*. The arrows in the above drawings indicate the direction that each atom moves in the normal mode. Because the normal modes act as independent harmonic oscillators, the vibrational energy of a polyatomic molecule is given by

$$\varepsilon_{\mathrm{vib}} = \sum_{j=1}^{n_{\mathrm{vib}}} h\nu_j(v_j + \tfrac{1}{2}) \qquad \text{each } v_j = 0,\ 1,\ 2,\ \ldots \tag{1.36}$$

where $\nu_j$ is the vibrational frequency of the $j$th normal mode. These frequencies are indicated for the normal modes of a water molecule shown above. Each type of polyatomic molecule has $n_{\mathrm{vib}}$ characteristic normal modes. For $CO_2$, for example, there are four normal modes ($3n - 5 = 3 \times 3 - 5 = 4$):

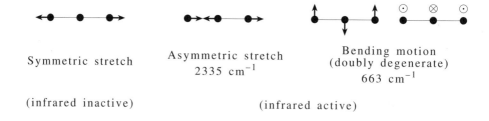

Symmetric stretch (infrared inactive) | Asymmetric stretch $2335\ \mathrm{cm}^{-1}$ | Bending motion (doubly degenerate) $663\ \mathrm{cm}^{-1}$ (infrared active)

The normal mode shown above at the far right indicates vibrational motion perpendicular to the page. Note that the bending mode is doubly degenerate; the bending motions in the plane of the page and perpendicular to it are the same motion but occur in different directions.

A selection rule for vibrational absorption spectroscopy is that the dipole moment of the molecule must vary during the motion of the normal mode. When this is so, the normal mode is said to be *infrared active*. Otherwise, it is *infrared inactive*. Note that the dipole moment changes during the motion of all three normal modes of a water molecule (shown above), so all three normal modes of $H_2O$ are infrared active. Therefore, all three lines are observed in its infrared spectrum. Note that there is no change in dipole moment during the symmetric stretch of $CO_2$, however, so this mode is infrared inactive. The other modes are infrared active, but the bending mode is doubly degenerate, so it leads to only one infrared line. Table 1.6 lists the characteristic vibrational frequencies of the normal modes of some polyatomic molecules. The numbers in parentheses indicate the degeneracies of the modes.

**TABLE 1.6**
The fundamental vibrational frequencies and the rotational constants of some polyatomic molecules. The numbers in parentheses indicate the degeneracy of the normal mode. These parameters were obtained from a variety of sources and do not represent the most accurate values because they were obtained under the rigid-rotator harmonic-oscillator approximation.

| Molecule | Shape | $\tilde{\nu}/\text{cm}^{-1}$ | $\tilde{A}/\text{cm}^{-1}$ | $\tilde{B}/\text{cm}^{-1}$ | $\tilde{C}/\text{cm}^{-1}$ |
|---|---|---|---|---|---|
| $CO_2$ | linear | 2335, 1313, 663(2) | | 0.390 | |
| $N_2O$ | linear | 2224, 1279, 590.8(2) | | 0.419 | |
| $CH_4$ | spherical top | 2898, 1515(2), 3002(3), 1300(3) | | 5.27 | |
| $CCl_4$ | spherical top | 458.7, 215.5(2), 778.4(3), 312.8(3) | | 0.057 | |
| $NH_3$ | symmetric top | 3336, 945.2, 3392(2), 1619(2) | 6.20 | 9.45 | 9.45 |
| $CH_3Cl$ | symmetric top | 2968, 1355, 729.8, 3044(2), 1487(2), 1015(2) | 5.09 | 0.443 | 0.443 |
| $H_2O$ | asymmetric top | 3725, 3586, 1595 | 27.9 | 14.6 | 9.31 |
| $SO_2$ | asymmetric top | 1153, 1362, 521.3 | 2.03 | 0.344 | 0.293 |

## 1–10. The Rotational Spectrum of a Polyatomic Molecule Depends Upon the Moments of Inertia of the Molecule

The rotational motion of a linear polyatomic molecule (such as $CO_2$ or HCN) is given by the same expression as that for a diatomic molecule. Thus,

$$\varepsilon_J = \frac{\hbar^2}{2I} J(J+1) \qquad J = 0,\ 1,\ 2, \ldots \tag{1.37}$$

with $g_J = 2J + 1$, where $I$ is the moment of inertia. In this case, however, $I$ is given by

$$I = \sum_{j=1}^{n} m_j (x_j - x_{\text{cm}})^2 \tag{1.38}$$

where $m_j$ is the mass of the $j$th atom and $x_j - x_{\text{cm}}$ is its distance from the center of mass of the molecule.

**EXAMPLE 1–10**
Given that $R_{\text{HC}} = 106.6$ pm and $R_{\text{CN}} = 115.3$ pm for $H^{12}C^{14}N$, calculate the values of the moment of inertia, $B$, and $\tilde{B}$.

SOLUTION: First we must find the center of mass, which is given by

$$x_{\text{cm}} = \frac{x_{\text{H}}m_{\text{H}} + x_{\text{C}}m_{\text{C}} + x_{\text{N}}m_{\text{N}}}{m_{\text{H}} + m_{\text{C}} + m_{\text{N}}}$$

Let's take the origin of our coordinate system to be the position of the hydrogen atom (see Figure 1.10 for an illustration of the distances in HCN), so

$$x_{\text{cm}} = \frac{(106.6 \text{ pm})(12.00 \text{ amu}) + (221.9 \text{ pm})(14.00 \text{ amu})}{(1.0079 + 12.00 + 14.00) \text{ amu}} = 162.4 \text{ pm}$$

Therefore

$$\begin{aligned} I &= m_{\text{H}}(162.4 \text{ pm})^2 + m_{\text{C}}(162.4 \text{ pm} - 106.6 \text{ pm})^2 + m_{\text{N}}(221.9 \text{ pm} - 162.4 \text{ pm})^2 \\ &= (1.135 \times 10^5 \text{ amu}\cdot\text{pm}^2)(1.661 \times 10^{-27} \text{ kg}\cdot\text{amu}^{-1})(10^{-12} \text{ m}\cdot\text{pm}^{-1})^2 \\ &= 1.885 \times 10^{-46} \text{ kg}\cdot\text{m}^2 \end{aligned}$$

and

$$\begin{aligned} B &= \frac{h}{8\pi^2 I} = \frac{6.626 \times 10^{-34} \text{ J}\cdot\text{s}}{8\pi^2(1.885 \times 10^{-46} \text{ kg}\cdot\text{m}^2)} \\ &= 4.451 \times 10^{10} \text{ Hz} \end{aligned}$$

and

$$\tilde{B} = \frac{B}{c} = 1.485 \text{ cm}^{-1}$$

Let's now consider a rigid nonlinear polyatomic molecule. Whereas the rotational properties of a linear molecule can be characterized by a single moment of inertia, three moments of inertia, $I_A$, $I_B$, and $I_C$, are needed to characterize the rotational properties of a rigid nonlinear polyatomic molecule. Each of these three moments of inertia has

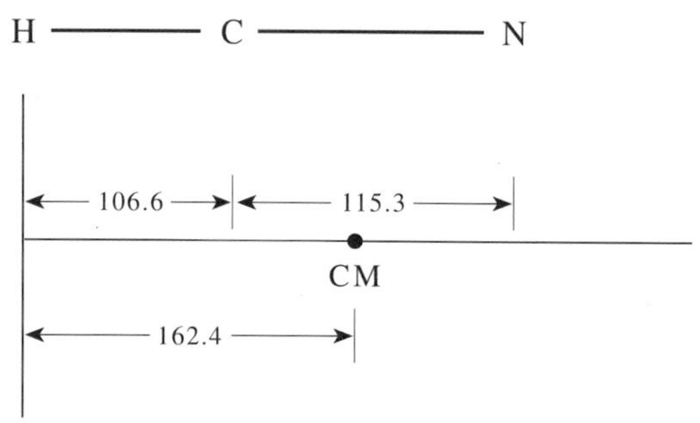


**FIGURE 1.10**
An illustration of the distances (in pm) used in Example 1–10.

a corresponding rotational constant, which is usually tabulated in units of $\text{cm}^{-1}$. The three rotational constants are denoted by $\tilde{A}$, $\tilde{B}$, and $\tilde{C}$ and are

$$\tilde{A} = \frac{h}{8\pi^2 c I_A}, \qquad \tilde{B} = \frac{h}{8\pi^2 c I_B}, \qquad \text{and} \qquad \tilde{C} = \frac{h}{8\pi^2 c I_C} \tag{1.39}$$

The rotational constants of some polyatomic molecules are given in Table 1.6.

The relative magnitudes of the three moments of inertia are used to characterize a rigid body. If all three are equal, the body is called a *spherical top*; if only two are equal, the body is called a *symmetric top*; and if all three are different, the body is called an *asymmetric top*. The molecules $CH_4$ and $SF_6$ are examples of spherical tops; $NH_3$ and $C_6H_6$ are examples of symmetric tops; and $H_2O$ is the classic example of an asymmetric top. The rotational energy levels of a polyatomic molecule depend upon whether the molecule is a spherical top, a symmetric top, or an asymmetric top.

Polyatomic molecules also have electronic energy, and as for diatomic molecules, we will need only the value of the ground-state electronic energy in later chapters.

## 1–11. The Energy of a Molecule in the Rigid-Rotator Harmonic-Oscillator Approximation Can Be Written as $\varepsilon = \varepsilon_{\text{trans}} + \varepsilon_{\text{rot}} + \varepsilon_{\text{vib}} + \varepsilon_{\text{elec}}$.

We now summarize the principal results of the chapter. An atom has translational and electronic energy, where the translational energy is given by

$$\varepsilon_{n_x n_y n_z} = \frac{h^2}{8m}\left(\frac{n_x^2}{a^2} + \frac{n_y^2}{b^2} + \frac{n_z^2}{c^2}\right) \qquad \begin{aligned} n_x &= 1, 2, 3, \ldots \\ n_y &= 1, 2, 3, \ldots \\ n_z &= 1, 2, 3, \ldots \end{aligned} \tag{1.40}$$

where $h$ is the Planck constant ($h = 6.626 \times 10^{-34}$ J·s), $m$ is the mass of the atom, $a$, $b$, and $c$ are the lengths of the container (assumed to be a rectangular parallelepiped), and $n_x$, $n_y$, and $n_z$ are the three translational quantum numbers (one for each dimension). There is no corresponding analytic expression for the electronic energies of multi-electron atoms or molecules, but they are well tabulated (see Table 1.2, for example).

A diatomic molecule has translational, vibrational, rotational, and electronic energy. In the rigid-rotator harmonic-oscillator approximation, the energy of the molecule is given by

$$\varepsilon = \varepsilon_{\text{trans}} + \varepsilon_{\text{rot}} + \varepsilon_{\text{vib}} + \varepsilon_{\text{elec}} \tag{1.41}$$

The translational energy of a diatomic molecule is the same as that of an atom, where $m$ is the mass of the molecule. The rotational motion of the molecule is modeled by a rigid rotator, whose energy is given by

$$\varepsilon_J = \frac{\hbar^2}{2I} J(J+1) \tag{1.42}$$

where $\hbar = h/2\pi$ and $I$ is the moment of inertia of the molecule. For a diatomic rigid rotator, $I = \mu R_e^2$, where $\mu = m_1 m_2/(m_1 + m_2)$ is the reduced mass, and $R_e$ is the equilibrium bond length. The energy is commonly expressed in units of $\mathrm{cm}^{-1}$ and Equation 1.42 written as

$$\tilde{\varepsilon}_J = \frac{h}{8\pi^2 Ic} J(J+1) = \tilde{B} J(J+1) \qquad J = 0,\ 1,\ 2,\ \ldots \tag{1.43}$$

where $\tilde{B} = h/8\pi^2 Ic$ is the rotational constant expressed in $\mathrm{cm}^{-1}$. Rotational transitions occur in the microwave and far infrared region of the spectrum and yield information about the moment of inertia and hence the equilibrium bond lengths of molecules (Example 1–8).

The vibrational motion of a diatomic molecule is modeled as a harmonic oscillator, whose energy is given by

$$\varepsilon_v = \hbar \left(\frac{k}{\mu}\right)^{1/2} \left(v + \frac{1}{2}\right) \qquad v = 0,\ 1,\ 2,\ \ldots \tag{1.44}$$

$$= h\nu \left(v + \frac{1}{2}\right) \qquad v = 0,\ 1,\ 2,\ \ldots \tag{1.45}$$

where $\mu$ is the reduced mass of the molecule, $k$ is the force constant, and $\nu$ is the natural vibrational frequency of the molecule. Vibrational transitions occur in the infrared region of the spectrum and can yield information about the force constants of molecules (Example 1–6).

As for atoms, there is no analytic expression for the electronic energies of diatomic molecules, but they are well tabulated. For most cases, we need only the energy and the degeneracy of the ground electronic state.

The energy of a polyatomic molecule in the rigid-rotator harmonic-oscillator approximation is also given by Equation 1.41. The translational energy is the same as that of an atom or diatomic molecule. The rotational energy depends upon the shape of the molecule. For a linear, polyatomic molecule, the rotational energy is same as that of a diatomic molecule, where the moment of inertia is given by Equation 1.38. For a nonlinear polyatomic molecule, the rotational energy depends upon whether the molecule is a spherical top (all three moments of inertia equal), a symmetric top (two of the three moments of inertia equal), or an asymmetric top (all three moments of inertia different).

In the harmonic-oscillator approximation, the vibrational motion of a polyatomic molecule decomposes into $3n - 5$ (linear) or $3n - 6$ (nonlinear) normal modes, each of which acts as an independent harmonic oscillator. The vibrational energy is given by

$$\varepsilon_{\mathrm{vib}} = \sum_{j=1}^{n_{\mathrm{vib}}} h\nu_j (v_j + \tfrac{1}{2}) \qquad v_j = 0,\ 1,\ 2,\ \ldots \tag{1.46}$$

where $n_{\mathrm{vib}}$ is the number of vibrational degrees of freedom.

To conclude this chapter, we present a summary of the energy levels of atoms and molecules that we will use in later chapters.

Atoms

$$\varepsilon_{\text{total}} = \varepsilon_{\text{trans}} + \varepsilon_{\text{elec}}$$

$$\varepsilon_{\text{trans}} = \frac{h^2}{8m}\left(\frac{n_x^2}{a^2} + \frac{n_y^2}{b^2} + \frac{n_z^2}{c^2}\right) \qquad \begin{matrix} n_x = 1,\ 2,\ \ldots \\ n_y = 1,\ 2,\ \ldots \\ n_z = 1,\ 2,\ \ldots \end{matrix} \qquad \text{(Equation 1.9)}$$

$\varepsilon_{\text{elec}} =$ see Tables 1.2 and 1.3, for example

Diatomic molecules

$$\varepsilon_{\text{total}} = \varepsilon_{\text{trans}} + \varepsilon_{\text{rot}} + \varepsilon_{\text{vib}} + \varepsilon_{\text{elec}}$$

$$\varepsilon_{\text{trans}} = \frac{h^2}{8M}\left(\frac{n_x^2}{a^2} + \frac{n_y^2}{b^2} + \frac{n_z^2}{c^2}\right) \qquad \begin{matrix} n_x = 1,\ 2,\ \ldots \\ n_y = 1,\ 2,\ \ldots \\ n_z = 1,\ 2,\ \ldots \end{matrix} \qquad \text{(Equation 1.9)}$$

$$\varepsilon_{\text{rot}} = \frac{\hbar^2 J(J+1)}{2I} \qquad J = 0,\ 1,\ 2,\ \ldots \qquad \text{(Equation 1.28)}$$

$$\varepsilon_{\text{vib}} = h\nu(v + \tfrac{1}{2}) \qquad v = 0,\ 1,\ 2,\ \ldots \qquad \text{(Equation 1.18)}$$

$$\varepsilon_{\text{elec}} = -D_{\text{e}} \qquad \text{(Table 1.4)}$$

Polyatomic molecules
linear

$$\varepsilon_{\text{total}} = \varepsilon_{\text{trans}} + \varepsilon_{\text{rot}} + \varepsilon_{\text{vib}} + \varepsilon_{\text{elec}}$$

$$\varepsilon_{\text{trans}} = \frac{h^2}{8M}\left(\frac{n_x^2}{a^2} + \frac{n_y^2}{b^2} + \frac{n_z^2}{c^2}\right) \qquad \begin{matrix} n_x = 1,\ 2,\ \ldots \\ n_y = 1,\ 2,\ \ldots \\ n_z = 1,\ 2,\ \ldots \end{matrix} \qquad \text{(Equation 1.9)}$$

$$\varepsilon_{\text{rot}} = \frac{\hbar^2 J(J+1)}{2I} \qquad J = 0,\ 1,\ 2,\ \ldots \qquad \text{(Equation 1.37)}$$

$$\varepsilon_{\text{vib}} = \sum_{j=1}^{n_{\text{vib}}} h\nu_j(v_j + \tfrac{1}{2}) \qquad \text{each } v_j = 0,\ 1,\ 2,\ \ldots \qquad \text{(Equation 1.36)}$$

$$\varepsilon_{\text{elec}} = -D_{\text{e}} \qquad \text{(Table 1.4)}$$

nonlinear

$$\varepsilon_{\text{total}} = \varepsilon_{\text{trans}} + \varepsilon_{\text{rot}} + \varepsilon_{\text{vib}} + \varepsilon_{\text{elec}}$$

$$\varepsilon_{\text{trans}} = \frac{h^2}{8M}\left(\frac{n_x^2}{a^2} + \frac{n_y^2}{b^2} + \frac{n_z^2}{c^2}\right) \qquad \begin{aligned} n_x &= 1, 2, \ldots \\ n_y &= 1, 2, \ldots \\ n_z &= 1, 2, \ldots \end{aligned} \qquad \text{(Equation 1.9)}$$

$$\varepsilon_{\text{rot}} = \text{see later chapters}$$

$$\varepsilon_{\text{vib}} = \sum_{j=1}^{n_{\text{vib}}} h\nu_j(v_j + \tfrac{1}{2}) \qquad \text{each } v_j = 0, 1, 2, \ldots \qquad \text{(Equation 1.36)}$$

$$\varepsilon_{\text{elec}} = -D_{\text{e}}$$

## Problems

**1-1.** Radiation in the ultraviolet region of the electromagnetic spectrum is usually described in terms of wavelength, $\lambda$, and is given in nanometers ($10^{-9}$ m). Calculate the values of $\nu$, $\tilde{\nu}$, and $\varepsilon$ for ultraviolet radiation with $\lambda = 200$ nm, and compare your results with those in Figure 1.11.

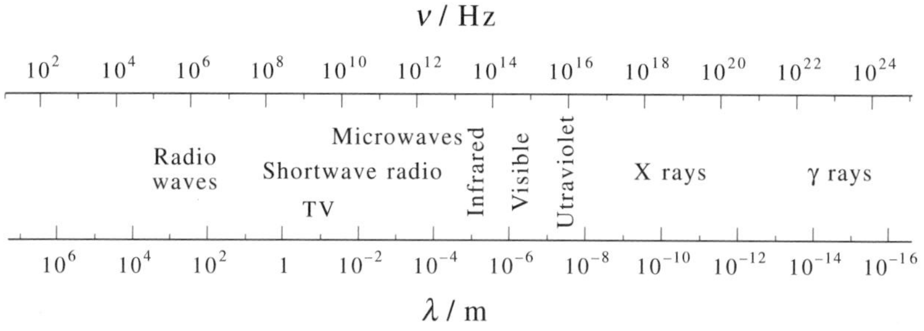


**FIGURE 1.11**
The regions of electromagnetic radiation.

**1-2.** Radiation in the infrared region is often expressed in terms of wave numbers, $\tilde{\nu} = 1/\lambda$. A typical value of $\tilde{\nu}$ in this region is $10^3$ cm$^{-1}$. Calculate the values of $\nu$, $\lambda$, and $\varepsilon$ for radiation with $\tilde{\nu} = 10^3$ cm$^{-1}$, and compare your results with those in Figure 1.11.

**1-3.** Past the infrared region, in the direction of lower energies, is the microwave region. In this region, radiation is usually characterized by its frequency, $\nu$, expressed in units of megahertz (MHz), where the unit, hertz (Hz), is one cycle per second. A typical microwave frequency is $2.0 \times 10^4$ MHz. Calculate the values of $\tilde{\nu}$, $\lambda$, and $\varepsilon$ for this radiation, and compare your results with those in Figure 1.11.

**1-4.** Calculate the value of the energy of a photon for a wavelength of 100 pm (about one atomic radius).

**1-5.** Calculate the number of photons in a 2.00-mJ light pulse at (a) 1.06 $\mu$m, (b) 537 nm, and (c) 266 nm.

**1-6.** A helium-neon laser (used in supermarket scanners) emits light at 632.8 nm. Calculate the frequency of this light. What is the energy of the photon generated by this laser?

**1-7.** The power output of a laser is measured in units of watts (W), where one watt is equal to one joule per second ($1 \text{ W} = 1 \text{ J}\cdot\text{s}^{-1}$). How many photons are emitted per second by a 1.00-mW nitrogen laser? The wavelength emitted by a nitrogen laser is 337 nm.

**1-8.** Use Equation 1.1 to calculate the value of the ionization energy of a hydrogen atom in its ground electronic state.

**1-9.** A line in the Lyman series of hydrogen has a wavelength of $1.026 \times 10^{-7}$ m. Find the original energy level of the electron.

**1-10.** A ground-state hydrogen atom absorbs a photon of light that has a wavelength of 97.2 nm. It then gives off a photon that has a wavelength of 486 nm. What is the final state of the hydrogen atom?

**1-11.** A commonly used non-SI unit of energy is an electron volt (eV), which is the energy that an electron picks up when it passes through a potential difference of one volt. Given that a joule is a coulomb times a volt, show that $1 \text{ eV} = 1.6022 \times 10^{-19}$ J.

**1-12.** Using the result of the previous problem, calculate the value of the ionization energy of a hydrogen atom in units of electron volts.

**1-13.** Using the data in Figure 1.2, show that a plot of frequency against $1/n^2$ is linear.

**1-14.** There is another series of lines in the emission spectrum of atomic hydrogen in the near infrared region called the Paschen series. This series results from $n \rightarrow 3$ transitions. Calculate the wavelength of the second line in the Paschen series, and show that this line lies in the near infrared region; that is, in the infrared region near the visible region.

**1-15.** Consider an electron in a $1s$ atomic orbital in a hydrogen atom. The average distance of this electron from the proton is given by $a_0 = 4\pi\varepsilon_0\hbar^2/m_e e^2$, where $\varepsilon_0$ is the permittivity of free space, $\hbar = h/2\pi$, where $h$ is the Planck constant, $m_e$ is the mass of an electron, and $e$ is the protonic charge. Calculate the value of $a_0$, which is called the *Bohr radius*, in units of picometers.

**1-16.** Equation 1.8 has a nice physical interpretation based upon the idea of a de Broglie wavelength. Recall that moving particles have an associated de Broglie wavelength given by $\lambda = h/mv$, where $m$ is the mass of the particle and $v$ is its speed. Show that if we assume that only standing de Broglie waves can fit in the interval 0 to $a$, then $\lambda = 2a/n$. Now use $\lambda = h/mv$ and $E = \frac{1}{2}mv^2$ to derive Equation 1.8.

**1-17.** The motion of macroscopic particles is governed by Newton's equation of motion, which can be written in the form

$$m\frac{d^2x}{dt^2} = f(x) \tag{1}$$

where $x(t)$ is the position of the mass $m$ and $f(x)$ is the force acting on the particle. Equation 1 is a differential equation whose solution gives $x(t)$, the *trajectory* of the mass. In the case of a harmonic oscillator, $m$ is the reduced mass and $f(x) = -kx$ (Hooke's law), so that Newton's equation is

$$\mu\frac{d^2x}{dt^2} = -kx$$

Show that $x(t) = A\cos 2\pi\nu t$ satisfies this equation if $\nu = (1/2\pi)(k/\mu)^{1/2}$. This result is valid only for a macroscopic oscillator, called a *classical harmonic oscillator*.

**1-18.** The kinetic energy of a classical harmonic oscillator is

$$\text{KE} = \frac{1}{2}\mu\left(\frac{dx}{dt}\right)^2$$

Using Equations 1.11 and 1.12, show that

$$\text{KE} + V(x) = \frac{kA^2}{2}$$

Interpret this result physically.

**1-19.** The solution for the classical harmonic oscillator is $x(t) = A\cos 2\pi\nu t$ (Equation 1.11). Show that the displacement oscillates between $+A$ and $-A$ with a frequency $\nu$ cycle·s$^{-1}$. What is the period of the oscillations; that is, how long is one cycle?

**1-20.** From Problem 1–19, we see that the period of a harmonic vibration is $\tau = 1/\nu$. The average of the kinetic energy over one cycle is given by

$$\langle K\rangle = \frac{1}{\tau}\int_0^{\tau}\frac{4\pi^2\nu^2 mA^2}{2}\sin^2 2\pi\nu t\,dt$$

Show that $\langle K\rangle = E/2$ where $E$ is the total energy. Show also that $\langle V\rangle = E/2$, where the instantaneous potential energy is given by

$$V = \frac{kA^2}{2}\cos^2 2\pi\nu t$$

Interpret the result $\langle K\rangle = \langle V\rangle$.

**1-21.** Calculate the value of the reduced mass of an electron in a hydrogen atom. Take the masses of the electron and proton to be $9.109390 \times 10^{-31}$ kg and $1.672623 \times 10^{-27}$ kg, respectively. What is the percent difference between this result and the rest mass of an electron?

**1-22.** Quantum mechanics gives that the electronic energy of a hydrogen atom is

$$\varepsilon_n = -\frac{\mu e^4}{8\varepsilon_0^2 h^2}\frac{1}{n^2} \qquad n = 1, 2, \ldots$$

where $\mu$ is the reduced mass of an electron and proton, $e$ is the protonic charge, $\varepsilon_0$ is the permittivity of free space, and $h$ is the Planck constant. Using the values given in the inside front cover, calculate the value of $\varepsilon_n$ in terms of $n^2$. Compare your result with Equation 1.1.

**1-23.** Show that the reduced mass of two equal masses, $m$, is $m/2$.

**1-24.** In this problem, we investigate the harmonic oscillator potential as the leading term in a Taylor expansion of the actual internuclear potential, $V(R)$, about its equilibrium position, $R_e$. According to Problem C–13, the first few terms in this expansion are

$$V(R) = V(R_e) + \left(\frac{dV}{dR}\right)_{R=R_e}(R - R_e) + \frac{1}{2!}\left(\frac{d^2V}{dR^2}\right)_{R=R_e}(R - R_e)^2 + \frac{1}{3!}\left(\frac{d^3V}{dR^3}\right)_{R=R_e}(R - R_e)^3 + \cdots \tag{1}$$

If $R$ is always close to $R_e$, then $R - R_e$ is always small. Consequently, the terms on the right side of Equation 1 get smaller and smaller. The first term in Equation 1 is a constant and depends upon where we choose the zero of energy. It is convenient to choose the zero of energy such that $V(R_e)$ equals zero and to relate $V(R)$ to this convention. Explain why the linear term in the displacement vanishes in Equation 1. (Note that $-dV/dR$ is the force acting between the two nuclei.)

Denote $R - R_e$ by $x$, $(d^2V/dR^2)_{R=R_e}$ by $k$, and $(d^3V/dR^3)_{R=R_e}$ by $\gamma$ to write Equation 1 as

$$\begin{aligned} V(x) &= \frac{1}{2}k(R - R_e)^2 + \frac{1}{6}\gamma(R - R_e)^3 + \cdots \\ &= \frac{1}{2}kx^2 + \frac{1}{6}\gamma x^3 + \cdots \end{aligned} \tag{2}$$

Argue that if we restrict ourselves to small displacements, then $x$ will be small and we can neglect the terms beyond the quadratic term in Equation 2, showing that the general potential energy function $V(R)$ can be approximated by a harmonic-oscillator potential. We can consider corrections or extensions of the harmonic-oscillator model by the higher-order terms in Equation 2. These terms are called *anharmonic terms*.

An analytic expression that is a good approximation to an intermolecular potential energy curve is a *Morse potential*

$$V(R) = D_e(1 - e^{-\beta(R-R_e)})^2$$

where $D_e$ and $\beta$ are parameters that depend upon the molecule. The parameter $D_e$ is the ground-state electronic energy of the molecule measured from the minimum of $V(R)$, and $\beta$ is a measure of the curvature of $V(R)$ at its minimum. Derive a relation between the force constant and the parameters $D_e$ and $\beta$. Given that $D_e = 7.31 \times 10^{-19}$ J·molecule$^{-1}$, $\beta = 0.0181$ pm$^{-1}$, and $R_e = 127.5$ pm for HCl(g), calculate the force constant of HCl(g). Plot the Morse potential for HCl(g), and plot the corresponding harmonic oscillator potential on the same graph (cf. Figure 1.5).

**1-25.** Use the result of Problem 1–24 and Equation 1.23 to show that

$$\beta = 2\pi c\tilde{\nu}\left(\frac{\mu}{2D_e}\right)^{1/2}$$

Given that $\tilde{\nu} = 2886\ \text{cm}^{-1}$ and $D_e = 440.2\ \text{kJ}\cdot\text{mol}^{-1}$ for $H^{35}Cl(g)$, calculate the value of $\beta$. Compare your result with that in Problem 1–24.

**1-26.** Carry out the Taylor expansion of the Morse potential in Problem 1–24 through terms in $(R - R_e)^3$. Express $\gamma$ in Equation 2 of Problem 1–24 in terms of $D_e$ and $\beta$.

**1-27.** It turns out that the solution of the Schrödinger equation for the Morse potential (Problem 1–24) can be expressed as

$$\tilde{\varepsilon}_v = \tilde{\nu}\left(v + \frac{1}{2}\right) - \tilde{\nu}\tilde{x}\left(v + \frac{1}{2}\right)^2$$

where

$$\tilde{x} = \frac{hc\tilde{\nu}}{4D_e}$$

Given that $\tilde{\nu} = 2886\ \text{cm}^{-1}$ and $D_e = 440.2\ \text{kJ}\cdot\text{mol}^{-1}$ for $H^{35}Cl(g)$, calculate the values of $\tilde{x}$ and $\tilde{\nu}\tilde{x}$. Plot the vibrational energies of $H^{35}Cl(g)$ for a Morse potential.

**1-28.** In the infrared spectrum of $H^{79}Br(g)$, there is an intense line at $2630\ \text{cm}^{-1}$. Use the harmonic-oscillator approximation to calculate the values of the force constant of $H^{79}Br(g)$ and the period of vibration of $H^{79}Br(g)$.

**1-29.** The force constant of $^{79}Br^{79}Br(g)$ is $240\ \text{N}\cdot\text{m}^{-1}$. Use the harmonic-oscillator approximation to calculate the values of the fundamental vibrational frequency and the zero-point energy (in joules per mole) of $^{79}Br^{79}Br(g)$.

**1-30.** In the far-infrared spectrum of $^{39}K^{35}Cl(g)$, there is an intense line at $278.0\ \text{cm}^{-1}$. Calculate the values of the force constant and the period of vibration of $^{39}K^{35}Cl(g)$.

**1-31.** Thus far, we have treated the vibrational motion of a diatomic molecule by means of a harmonic-oscillator model. We saw in Section 1–4, however, that the internuclear potential energy is not a simple parabola but is more like that illustrated in Figure 1.5. The harmonic-oscillator approximation consists of keeping only the quadratic term in the Taylor expansion of $V(R)$ (see Problem 1–24), and it predicts that there will be only one line in the vibrational spectrum of a diatomic molecule. Experimental data show there is, indeed, one dominant line (called the *fundamental*) but also lines of weaker intensity at almost integral multiples of the fundamental. These lines are called *overtones* (see Table 1.7).

If the anharmonic terms in Equation 1.19 are taken into account, then a quantum mechanical calculation gives

$$\tilde{\varepsilon}_v = \tilde{\nu}(v + \tfrac{1}{2}) - \tilde{x}\tilde{\nu}(v + \tfrac{1}{2})^2 + \cdots \qquad v = 0,\ 1,\ 2,\ \ldots \qquad (1)$$

where $\tilde{x}$ is called the *anharmonicity constant.* The anharmonic correction in Equation 1 is much smaller than the harmonic term because $\tilde{x} \ll 1$. Show that the levels are not equally spaced as they are for a harmonic oscillator and, in fact, that their separation decreases

**TABLE 1.7**
The vibrational spectrum of $H^{35}Cl(g)$.

| Transition | $\tilde{\nu}_{obs}/cm^{-1}$ | $\tilde{\nu}_{obs}/cm^{-1}$ Harmonic oscillator $\tilde{\nu} = 2885.90v$ |
|---|---|---|
| $0 \rightarrow 1$ (fundamental) | 2885.9 | 2885.9 |
| $0 \rightarrow 2$ (first overtone) | 5668.0 | 5771.8 |
| $0 \rightarrow 3$ (second overtone) | 8347.0 | 8657.7 |
| $0 \rightarrow 4$ (third overtone) | 10 923.1 | 11 543.6 |
| $0 \rightarrow 5$ (fourth overtone) | 13 396.5 | 14 429.5 |

with increasing $v$. The selection rule for an anharmonic oscillator is that $\Delta v$ can have any integral value, although the intensities of the $\Delta v = \pm 2, \pm 3, \ldots$ transitions are much less than for the $\Delta v = \pm 1$ transitions. Show that if we recognize that most diatomic molecules are in the ground vibrational state at room temperature, the frequencies of the observed $0 \rightarrow v$ transitions will be given by

$$\tilde{\nu}_{obs} = \tilde{\nu}v - \tilde{x}\tilde{\nu}v(v+1) \qquad v = 1, 2, \ldots \qquad (2)$$

Curve fit Equation 2 to the experimental data in Table 1.7 to find the optimum values of $\tilde{\nu}$ and $\tilde{x}\tilde{\nu}$. Use Equation 2 to calculate the values of the observed frequencies and compare your results with the experimental data.

**1-32.** Given that $\tilde{\nu} = 536.10\ cm^{-1}$ and $\tilde{x}\tilde{\nu} = 3.4\ cm^{-1}$ for $^{23}Na^{19}F(g)$, calculate the values of the frequencies of the first and second vibrational overtones (see Problem 1–31).

**1-33.** The fundamental line in the infrared spectrum of $^{12}C^{16}O(g)$ occurs at $2143.0\ cm^{-1}$, and the first overtone occurs at $4260.0\ cm^{-1}$. Calculate the values of $\tilde{\nu}$ and $\tilde{x}\tilde{\nu}$ for $^{12}C^{16}O(g)$ (see Problem 1–31).

**1-34.** The Morse potential is presented in Problem 1–24. Given that $D_e = 8.35 \times 10^{-19}\ J\cdot mol^{-1}$, $\tilde{\nu} = 1556\ cm^{-1}$, and $R_e = 120.7$ pm for $O_2$, plot a Morse potential for $O_2$. Plot the corresponding harmonic-oscillator potential on the same graph.

**1-35.** Show that the moment of inertia for a rigid diatomic rotator can be written as $I = \mu R_e^2$, where $R_e = R_1 + R_2$ (the fixed separation of the two masses), $R_1$ and $R_2$ are the distances of the two masses from the center of mass, and $\mu$ is the reduced mass.

**1-36.** In the far-infrared spectrum of $H^{79}Br(g)$, there is a series of lines separated by $16.72\ cm^{-1}$. Calculate the values of the moment of inertia and the internuclear separation in $H^{79}Br(g)$.

**1-37.** Given that $\tilde{\nu} = 2330\ cm^{-1}$ and that $D_0 = 78\ 715\ cm^{-1}$ for $N_2(g)$, calculate the value of $D_e$.

**1-38.** The $J = 0$ to $J = 1$ transition for carbon monoxide [$^{12}C^{16}O(g)$] occurs at $1.153 \times 10^5$ MHz. Calculate the value of the bond length in carbon monoxide.

**1-39.** The microwave spectrum of $^{39}K^{127}I(g)$ consists of a series of lines whose spacing is almost constant at 3634 MHz. Calculate the bond length of $^{39}K^{127}I(g)$.

**1-40.** Assuming the rotation of a diatomic molecule in the $J = 10$ state may be approximated using classical mechanics, calculate how many revolutions per second $^{23}Na^{35}Cl(g)$ makes in the $J = 10$ rotational state. The rotational constant of $^{23}Na^{35}Cl(g)$ is 6500 MHz.

**1-41.** The rigid-rotator model predicts that the lines in the rotational spectrum of a diatomic molecule should be equally spaced. The following table lists some of the observed lines in the rotational spectrum of $H^{35}Cl(g)$.

| Transition | $\tilde{\nu}_{obs}/\text{cm}^{-1}$ | $\Delta\tilde{\nu}_{obs}/\text{cm}^{-1}$ | $c\tilde{\nu}_{calc} = 2\tilde{B}(J+1)$ $\tilde{B} = 10.243\ \text{cm}^{-1}$ |
|---|---|---|---|
| $3 \rightarrow 4$ | 83.03 | | 82.72 |
| | | 21.07 | |
| $4 \rightarrow 5$ | 104.10 | | 103.40 |
| | | 20.20 | |
| $5 \rightarrow 6$ | 124.30 | | 124.08 |
| | | 20.73 | |
| $6 \rightarrow 7$ | 145.03 | | 144.76 |
| | | 20.48 | |
| $7 \rightarrow 8$ | 165.51 | | 165.44 |
| | | 20.35 | |
| $8 \rightarrow 9$ | 185.86 | | 186.12 |
| | | 20.52 | |
| $9 \rightarrow 10$ | 206.38 | | 206.80 |
| | | 20.12 | |
| $10 \rightarrow 11$ | 226.50 | | 227.48 |

The differences listed in the third column clearly show that the lines are not exactly equally spaced as the rigid-rotator approximation predicts. The discrepancy can be resolved by realizing that a chemical bond is not truly rigid. As the molecule rotates more energetically (increasing $J$), the centrifugal force causes the bond to stretch slightly. If this small effect is taken into account, then the energy is given by

$$\tilde{\varepsilon}_J = \frac{\varepsilon_J}{hc} = \tilde{B}J(J+1) - \tilde{D}J^2(J+1)^2 \tag{1}$$

where $\tilde{D}$ is called the *centrifugal distortion constant.* Show that the frequencies of the absorption due to $J \rightarrow J+1$ transitions are given by

$$\tilde{\nu} = 2\tilde{B}(J+1) - 4\tilde{D}(J+1)^3 \qquad J = 0,\ 1,\ 2,\ \ldots \tag{2}$$

Curve fit Equation 2 to the experimental data above and find the optimum values of $\tilde{B}$ and $\tilde{D}$. Compare the predictions of the resulting Equation 2 with the experimental data.

**1-42.** The following data are obtained in the microwave spectrum of $^{12}C^{16}O(g)$. Use the method of Problem 1–41 to determine the values of $\tilde{B}$ and $\tilde{D}$ from these data.

| Transition | Frequency/cm$^{-1}$ |
|---|---|
| $0 \rightarrow 1$ | 3.845 40 |
| $1 \rightarrow 2$ | 7.690 60 |
| $2 \rightarrow 3$ | 11.535 50 |
| $3 \rightarrow 4$ | 15.379 90 |
| $4 \rightarrow 5$ | 19.223 80 |
| $5 \rightarrow 6$ | 23.066 85 |

**1-43.** Given that $\tilde{B} = 8.465$ cm$^{-1}$ and $\tilde{D} = 0.000346$ cm$^{-1}$ for $H^{79}Br(g)$, calculate the frequency of the $J = 0 \rightarrow J = 1$, $J = 1 \rightarrow J = 2$, $J = 2 \rightarrow J = 3$, $\cdots$, $J = 6 \rightarrow J = 7$ transitions in the rotational spectrum of $H^{79}Br(g)$.

**1-44.** Determine the number of various degrees of freedom of $N_2$, $C_2H_2$, $C_2H_4$, $C_2H_6$, and $C_6H_6$.

**1-45.** Determine the total number of normal modes of vibration of HCN, $CD_4$, $SO_3$, $SF_6$, and $(CH_3)_2CO$.

**1-46.** Using the data in Table 1.6, calculate the value of the zero-point vibrational energy of a water molecule.

**1-47.** Using the data in Table 1.6, calculate the value of the zero-point vibrational energy of a methane molecule.

**1-48.** Given that $R_{NN} = 112.8$ pm and $R_{NO} = 118.4$ pm for $^{14}N^{14}N^{16}O$, calculate the values of the moment of inertia and $\tilde{B}$.

**1-49.** In this problem, we will see how the concept of reduced mass arises naturally when discussing the interaction of two particles. Consider two masses, $m_1$ and $m_2$, in one dimension, interacting through a potential that depends only upon their relative separation $(x_1 - x_2)$ so that $U(x_1, x_2) = U(x_1 - x_2)$. Given that the force acting upon the $j$th particle is $f_j = -(\partial U/\partial x_j)$, show that $f_1 = -f_2$. What law is this?

Newton's equations for $m_1$ and $m_2$ are

$$m_1 \frac{d^2x_1}{dt^2} = -\frac{\partial U}{\partial x_1} \qquad m_2 \frac{d^2x_2}{dt^2} = -\frac{\partial U}{\partial x_2}$$

Now introduce center-of-mass and relative coordinates by

$$X = \frac{m_1x_1 + m_2x_2}{M} \qquad x = x_1 - x_2$$

Solve for $x_1$ and $x_2$ to obtain

$$x_1 = X + \frac{m_2}{M}x \qquad x_2 = X - \frac{m_1}{M}x$$

Show that Newton's equations in these coordinates are

$$m_1 \frac{d^2X}{dt^2} + \frac{m_1 m_2}{M} \frac{d^2x}{dt^2} = -\frac{\partial U}{\partial x}$$

and

$$m_2 \frac{d^2X}{dt^2} - \frac{m_1 m_2}{M} \frac{d^2x}{dt^2} = +\frac{\partial U}{\partial x}$$

Now add these two equations to find

$$M \frac{d^2X}{dt^2} = 0$$

Interpret this result. Now divide the first equation by $m_1$ and the second by $m_2$ and subtract to obtain

$$\frac{d^2x}{dt^2} = -\left(\frac{1}{m_1} + \frac{1}{m_2}\right) \frac{\partial U}{\partial x}$$

or

$$\mu \frac{d^2x}{dt^2} = -\frac{\partial U}{\partial x}$$

where $\mu$ is the reduced mass. Interpret this result and discuss how the original two-body problem has been reduced to two one-body problems.

MATHCHAPTER A

# NUMERICAL METHODS

You learned in high school that a quadratic equation $ax^2 + bx + c = 0$ has two roots, given by the so-called quadratic formula:

$$x = \frac{-b \pm \sqrt{b^2 - 4ac}}{2a}$$

Thus, the two values of $x$ (called roots) that satisfy the equation $x^2 + 3x - 2 = 0$ are

$$x = \frac{-3 \pm \sqrt{17}}{2}$$

Although there are general formulas for the roots of cubic and quartic equations, they are very inconvenient to use, and furthermore, there are no formulas for equations of the fifth degree or higher. Unfortunately, in practice we encounter such equations frequently and must learn to deal with them. Fortunately, with the advent of hand calculators and personal computers, the numerical solution of polynomial equations and other types of equations, such as $x - \cos x = 0$, is routine. Although these and other equations can be solved by "brute force" trial and error, much more organized procedures can arrive at an answer to almost any desired degree of accuracy. Perhaps the most widely known procedure is the Newton-Raphson method, which is best illustrated by a figure. Figure A.1 shows a function $f(x)$ plotted against $x$. The solution to the equation $f(x) = 0$ is denoted by $x_*$. The idea behind the Newton-Raphson method is to guess an initial value of $x$ (call it $x_0$) "sufficiently close" to $x_*$, and draw the tangent to the curve $f(x)$ at $x_0$, as shown in Figure A.1. Very often, the extension of the tangent line through the horizontal axis will lie closer to $x_*$ than does $x_0$. We denote this value of $x$ by $x_1$ and repeat the process using $x_1$ to get a new value of $x_2$, which will lie even closer to $x_*$. By repeating this process (called iteration) we can approach $x_*$ to essentially any desired degree of accuracy.

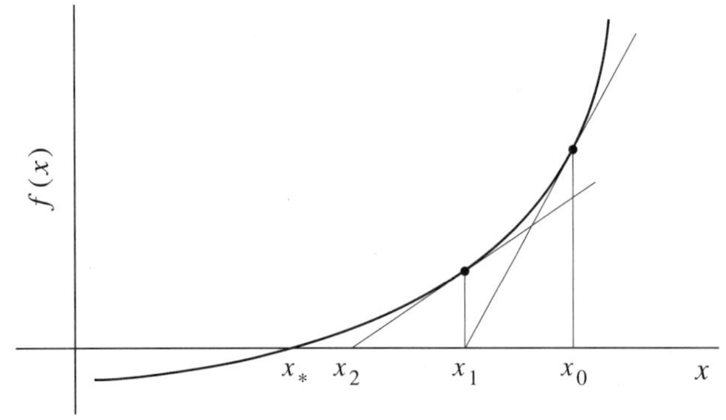


**FIGURE A.1**
A graphical illustration of the Newton-Raphson method.

We can use Figure A.1 to derive a convenient formula for the iterative values of $x$. The slope of $f(x)$ at $x_n$, $f'(x_n)$, is given by

$$f'(x_n) = \frac{f(x_n) - 0}{x_n - x_{n+1}}$$

Solving this equation for $x_{n+1}$ gives

$$x_{n+1} = x_n - \frac{f(x_n)}{f'(x_n)} \qquad \text{(A.1)}$$

which is the iterative formula for the Newton-Raphson method. As an application of this formula, consider the chemical equation

$$2\,\mathrm{NOCl(g)} \rightleftharpoons 2\,\mathrm{NO(g)} + \mathrm{Cl_2(g)}$$

whose related equilibrium constant is 2.18 at a certain temperature. (Chemical equilibrium is discussed in Chapter 12, but we're simply using the algebraic equation below as an example at this point.) If 1.00 atm of NOCl(g) is introduced into a reaction vessel, then at equilibrium $P_{\mathrm{NOCl}} = 1.00 - 2x$, $P_{\mathrm{NO}} = 2x$, and $P_{\mathrm{Cl_2}} = x$; these pressures satisfy the equilibrium-constant expression

$$\frac{P_{\mathrm{NO}}^2 P_{\mathrm{Cl_2}}}{P_{\mathrm{NOCl}}^2} = \frac{(2x)^2 x}{(1.00 - 2x)^2} = 2.18$$

which we write as

$$f(x) = 4x^3 - 8.72x^2 + 8.72x - 2.18 = 0$$

Because of the stoichiometry of the reaction equation, the value of $x$ we are seeking must be between 0 and 0.5, so let's choose 0.250 as our initial guess ($x_0$). Table A.1 shows the results of using Equation A.1. Notice that we have converged to three significant figures in just three steps.

**TABLE A.1**
The results of the application of the Newton-Raphson method to the solution of the equation $f(x) = 4x^3 - 8.72x^2 + 8.72x - 2.18 = 0$.

| $n$ | $x_n$ | $f(x_n)$ | $f'(x_n)$ |
|---|---|---|---|
| 0 | 0.2500 | $-4.825 \times 10^{-1}$ | 5.110 |
| 1 | 0.3442 | $-4.855 \times 10^{-2}$ | 4.139 |
| 2 | 0.3559 | $-6.281 \times 10^{-4}$ | 4.033 |
| 3 | 0.3561 | $-1.704 \times 10^{-5}$ | 4.031 |
| 4 | 0.3561 | | |

**EXAMPLE A–1**
In Chapter 2, we will solve the cubic equation

$$x^3 + 3x^2 + 3x - 1 = 0$$

Use the Newton-Raphson method to find the real root of this equation to five significant figures.

SOLUTION: We write the equation as

$$f(x) = x^3 + 3x^2 + 3x - 1 = 0$$

By inspection, a solution lies between 0 and 1. Using $x_0 = 0.5$ results in the following table:

| $n$ | $x_n$ | $f(x_n)$ | $f'(x_n)$ |
|---|---|---|---|
| 0 | 0.500000 | 1.37500 | 6.7500 |
| 1 | 0.296300 | 0.178294 | 5.04118 |
| 2 | 0.260930 | 0.004809 | 4.76983 |
| 3 | 0.259920 | −0.000005 | 4.76220 |
| 4 | 0.259920 | | |

The answer to five significant figures is $x = 0.25992$. Note that $f(x_n)$ is significantly smaller at each step, as it should be as we approach the value of $x$ that satisfies $f(x) = 0$, but that $f'(x_n)$ does not vary appreciably. The same behavior can be seen in Table A.1.

As powerful as it is, the Newton-Raphson method does not always work; when it does work, it is obvious the method is working, and when it doesn't work, it may be even more obvious. A spectacular failure is provided by the equation $f(x) = x^{1/3} = 0$, for which $x_* = 0$. If we begin with $x_0 = 1$, we will obtain $x_1 = -2$, $x_2 = +4$, $x_3 = -8$, and so on. Figure A.2 shows why the method is failing to converge. The message here is that you should always plot $f(x)$ first to get an idea of where the relevant roots

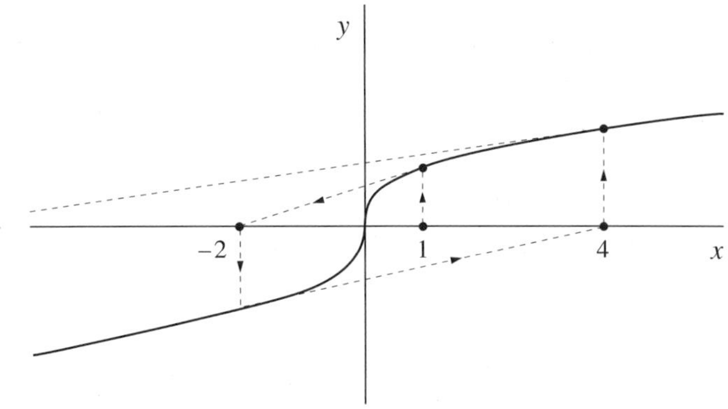


FIGURE A.2
A plot of $y = x^{1/3}$, illustrating that the Newton-Raphson method fails in this case.

lie and to see that the function does not have any peculiar properties. You should do Problems A–1 to A–8 to become proficient with the Newton-Raphson method.

There are also numerical methods to evaluate integrals. You learned in calculus that an integral is the area between a curve and the horizontal axis (area under a curve) between the integration limits, so that the value of

$$I = \int_a^b f(u)du \tag{A.2}$$

is given by the shaded area in Figure A.3. Recall a fundamental theorem of calculus, which says that if

$$F(x) = \int_a^x f(u)du$$

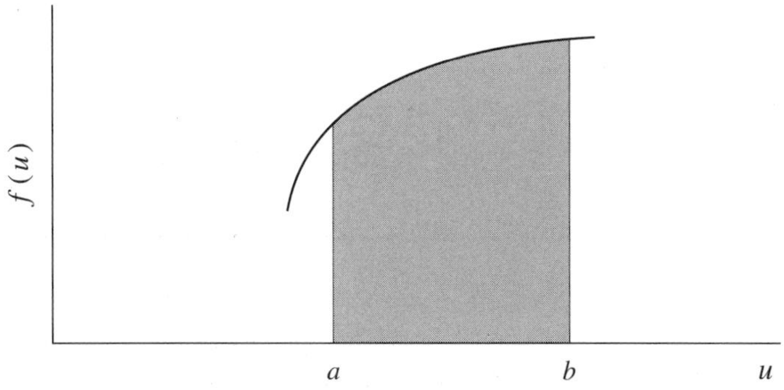


FIGURE A.3
The integral of $f(u)$ from $a$ to $b$ is given by the shaded area.

then

$$\frac{dF}{dx} = f(x)$$

The function $F(x)$ is sometimes called the antiderivative of $f(x)$. If there is no elementary function $F(x)$ whose derivative is $f(x)$, we say that the integral of $f(x)$ cannot be evaluated analytically. By elementary function, we mean a function that can be expressed as a finite combination of polynomial, trigonometric, exponential, and logarithmic functions.

It turns out that numerous integrals cannot be evaluated analytically. A particularly important example of an integral that cannot be evaluated in terms of elementary functions is

$$\phi(x) = \int_0^x e^{-u^2} du \tag{A.3}$$

Equation A.3 serves to define the (nonelementary) function $\phi(x)$. The value of $\phi(x)$ for any value of $x$ is given by the area under the curve $f(u) = e^{-u^2}$ from $u = 0$ to $u = x$.

Let's consider the more general case given by Equation A.2 or the shaded area in Figure A.3. We can approximate this area in a number of ways. First divide the interval $(a,b)$ into $n$ equally spaced subintervals $u_1 - u_0,\ u_2 - u_1,\ \ldots,\ u_n - u_{n-1}$ with $u_0 = a$ and $u_n = b$. We will let $h = u_{j+1} - u_j$ for $j = 0,\ 1,\ \ldots,\ n - 1$. Figure A.4 shows a magnification of one of the subintervals, say the $u_j, u_{j+1}$ subinterval. One way to approximate the area under the curve is to connect the points $f(u_j)$ and $f(u_{j+1})$ by a straight line as shown in Figure A.4. The area under the straight line approximation to $f(u)$ in the interval is the sum of the area of the rectangle $[hf(u_j)]$ and the area of the triangle $\{\frac{1}{2}h[f(u_{j+1}) - f(u_j)]\}$. Using this approximation for all intervals, the total area under the curve from $u = a$ to $u = b$ is given by the sum

$$\begin{aligned} I \approx I_n = {} & hf(u_0) + \frac{h}{2}[f(u_1) - f(u_0)] \\ & + hf(u_1) + \frac{h}{2}[f(u_2) - f(u_1)] \\ & \vdots \\ & + hf(u_{n-2}) + \frac{h}{2}[f(u_{n-1}) - f(u_{n-2})] \\ & + hf(u_{n-1}) + \frac{h}{2}[f(u_n) - f(u_{n-1})] \\ = {} & \frac{h}{2}[f(u_0) + 2f(u_1) + 2f(u_2) + \cdots + 2f(u_{n-1}) + f(u_n)] \end{aligned} \tag{A.4}$$

Note that the coefficients in Equation A.4 go as $1,\ 2,\ 2,\ \ldots,\ 2,\ 1$. Equation A.4 is easy to implement on a hand calculator for $n = 10$ or so and on a personal computer using a spreadsheet for larger values of $n$. The approximation to the integral given by Equation A.4 is called the *trapezoidal approximation*. [The error goes as $Ah^2$, where $A$ is a constant that depends upon the nature of the function $f(u)$. In fact, if $M$ is the

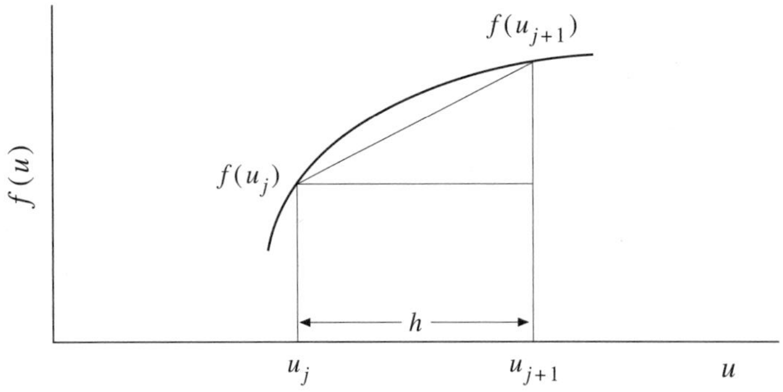


**FIGURE A.4**
An illustration of the area of the $j + 1$st subinterval for the trapezoidal approximation.

largest value of $|f''(u)|$ in the interval $(a, b)$, then the error is *at most* $M(b - a)h^2/12$.] Table A.2 shows the values of

$$\phi(1) = \int_0^1 e^{-u^2} du \tag{A.5}$$

for $n = 10$ $(h = 0.1)$, $n = 100$ $(h = 0.01)$, and $n = 1000$ $(h = 0.001)$. The "accepted" value (using more sophisticated numerical integration methods) is 0.74682413, to eight decimal places.

We can develop a more accurate numerical integration routine by approximating $f(u)$ in Figure A.4 by something other than a straight line. If we approximate $f(u)$ by a quadratic function, we have *Simpson's rule*, whose formula is

$$I_{2n} = \frac{h}{3}[f(u_0) + 4f(u_1) + 2f(u_2) + 4f(u_3) + 2f(u_4) + \cdots + 2f(u_{2n-2}) + 4f(u_{2n-1}) + f(u_{2n})] \tag{A.6}$$

Note that the coefficients go as 1, 4, 2, 4, 2, 4, ..., 4, 2, 4, 1. We write $I_{2n}$ in Equation A.6 because Simpson's rule requires that there be an even number of intervals. Table A.2 shows the values of $\phi(1)$ in Equation A.5 for $n = 10$, 100, and 1000. Note

**TABLE A.2**
The application of the trapezoidal approximation (Equation A.4) and Simpson's rule (Equation A.6) to the evaluation of $\phi(1)$ given by Equation A.5. The exact value to eight decimal places is 0.74682413.

| $n$ | $h$ | $I_n$ (trapezoidal) | $I_{2n}$ (Simpson's tule) |
|---|---|---|---|
| 10 | 0.1 | 0.74621800 | 0.74682494 |
| 100 | 0.01 | 0.74681800 | 0.74682414 |
| 1000 | 0.001 | 0.74682407 | 0.74682413 |

that with $n = 100$, the result for Simpson's rule differs from the "accepted" value by only one unit in the eighth decimal place. The error for Simpson's rule goes as $h^4$ compared with $h^2$ for the trapezoidal approximation. In fact, if $M$ is the largest value of $|f^{(4)}(u)|$ in the interval $(a, b)$, then the error is *at most* $M(b - a)h^4/180$. Problems A–9 to A–12 illustrate the use of the trapezoidal approximation and Simpson's rule.

**EXAMPLE A–2**

One theory (from Debye) of the molar heat capacity of a monatomic crystal gives

$$\overline{C}_V = 9R\left(\frac{T}{\Theta_D}\right)^3 \int_0^{\Theta_D/T} \frac{x^4 e^x}{(e^x - 1)^2}\, dx$$

where $R$ is the molar gas constant ($8.314\ \text{J}\cdot\text{K}^{-1}\cdot\text{mol}^{-1}$) and $\Theta_D$, the Debye temperature, is a parameter characteristic of the crystalline substance. Given that $\Theta_D = 309$ K for copper, calculate the molar heat capacity of copper at $T = 103$ K.

SOLUTION: At $T = 103$ K, the basic integral to evaluate numerically is

$$I = \int_0^3 \frac{x^4 e^x}{(e^x - 1)^2}\, dx$$

Using the trapezoidal approximation (Equation A.5) and Simpson's rule (Equation A.6), we find the following values of $I$:

| $n$ | $h$ | $I_n$ (trapezoidal) | $I_{2n}$ (Simpson's rule) |
|---|---|---|---|
| 10 | 0.3 | 5.9725 | 5.9648 |
| 100 | 0.03 | 5.9649 | 5.9648 |
| 1000 | 0.003 | 5.9648 | 5.9648 |

The molar heat capacity at 103 K is given by

$$\overline{C}_V = 9R\left(\frac{1}{3}\right)^3 I$$

or $\overline{C}_V = 16.5\ \text{J}\cdot\text{mol}^{-1}\cdot\text{K}^{-1}$, in agreement with the experimental value.

Although the Newton-Raphson method and Simpson's rule can be implemented easily on a spreadsheet, there are a number of easy-to-use numerical software packages such as *MathCad*, *Kaleidagraph*, *Mathematica*, or *Maple* that can be used to evaluate the roots of algebraic equations and integrals by even more sophisticated numerical methods.

## Problems

**A-1.** Solve the equation $x^5 + 2x^4 + 4x = 5$ to four significant figures for the root that lies between 0 and 1.

**A-2.** Use the Newton-Raphson method to derive the iterative formula

$$x_{n+1} = \frac{1}{2}\left(x_n + \frac{A}{x_n}\right)$$

for the value of $\sqrt{A}$. This formula was discovered by a Babylonian mathematician more than 2000 years ago. Use this formula to evaluate $\sqrt{2}$ to five significant figures.

**A-3.** Use the Newton-Raphson method to solve the equation $e^{-x} + (x/5) = 1$ to four significant figures.

**A-4.** Consider the chemical reaction described by the equation

$$CH_4(g) + H_2O(g) \rightleftharpoons CO(g) + 3\,H_2(g)$$

at 300 K. If 1.00 atm of $CH_4(g)$ and $H_2O(g)$ are introduced into a reaction vessel, the pressures at equilibrium obey the equation

$$\frac{P_{CO}P_{H_2}^3}{P_{CH_4}P_{H_2O}} = \frac{(x)(3x)^3}{(1-x)(1-x)} = 26$$

Solve this equation for $x$.

**A-5.** In Chapter 2, we will solve the cubic equation

$$64x^3 + 6x^2 + 12x - 1 = 0$$

Use the Newton-Raphson method to find the only real root of this equation to five significant figures.

**A-6.** Solve the equation $x^3 - 3x + 1 = 0$ for all three of its roots to four decimal places.

**A-7.** In Example 2–3 we will solve the cubic equation

$$\overline{V}^3 - 0.1231\overline{V}^2 + 0.02056\overline{V} - 0.001271 = 0$$

Use the Newton-Raphson method to find the root to this equation that is near $\overline{V} = 0.1$.

**A-8.** In Section 2–3 we will solve the cubic equation

$$\overline{V}^3 - 0.3664\overline{V}^2 + 0.03802\overline{V} - 0.001210 = 0$$

Use the Newton-Raphson method to show that the three roots to this equation are 0.07073, 0.07897, and 0.2167.

**A-9.** Use the trapezoidal approximation and Simpson's rule to evaluate

$$I = \int_0^1 \frac{dx}{1+x^2}$$

This integral can be evaluated analytically; it is given by $\tan^{-1}(1)$, which is equal to $\pi/4$, so $I = 0.78539816$ to eight decimal places.

**A-10.** Evaluate ln 2 to six decimal places by evaluating

$$\ln 2 = \int_1^2 \frac{dx}{x}$$

What must $n$ be to assure six-digit accuracy?

**A-11.** Use Simpson's rule to evaluate

$$I = \int_0^\infty e^{-x^2} dx$$

and compare your result with the exact value, $\sqrt{\pi}/2$.

**A-12.** Use Simpson's rule to evaluate

$$I = \int_0^\infty \frac{x^3 dx}{e^x - 1}$$

to six decimal places. The exact value is $\pi^4/15$.

**A-13.** Use a numerical software package such as *MathCad*, *Kaleidagraph*, or *Mathematica* to evaluate the integral

$$S = 4\pi^{1/2}\left(\frac{2\alpha}{\pi}\right)^{3/4} \int_0^\infty r^2 e^{-r} e^{-\alpha r^2} dr$$

for values of $\alpha$ between 0.200 and 0.300 and show that $S$ has a maximum value at $\alpha = 0.271$.

**A-14.** Use Simpson's rule to evaluate the integral (see Equation 2.31)

$$B_{2V}^*(T^*) = -3\int_0^\infty \left\{\exp\left[-\frac{4}{T^*}(x^{-12} - x^{-6})\right] - 1\right\} x^2\, dx$$

for $T^* = 2.00$ and 3.00. The accepted values are $-0.6276$ and $-0.1152$, respectively.

**Johannes Diderik van der Waals** was born in Leiden, the Netherlands, on November 23, 1837, and died in 1923. Because he had not learned Latin and Greek, he was at first not able to continue with university studies and so worked as a school teacher in a secondary school. After passage of new legislation, however, van der Waals obtained an exemption from the university requirements in classical languages and defended his doctoral dissertation at Leyden University in 1873. In his dissertation, he proposed an explanation of the continuity of the gaseous and liquid phases and the phenomenon of the critical point, as well as a derivation of a new equation of state of gases, now called the van der Waals equation. A few years later, he proposed the law of corresponding states, which reduces the properties of all gases to one common denominator. Although his dissertation was written in Dutch, his work quickly came to the attention of Maxwell, who published a review of it in English in the British journal *Nature* in 1875 and so brought the work to the attention of a much broader audience. In 1876, van der Waals was appointed the first Professor of Physics at the newly created University of Amsterdam. The University became a center for both theoretical and experimental research on fluids, largely through van der Waals' influence. Van der Waals was awarded the Nobel Prize for physics in 1910 "for the work on the equation of state for gases and liquids."

CHAPTER 2

# The Properties of Gases

We begin our study of thermodynamics with the properties of gases. First, we will discuss the ideal-gas equation and then some extensions of this equation, of which the van der Waals equation is the most famous. Although the van der Waals equation accounts in part for deviations from ideal-gas behavior, a more systematic and accurate approach is to use a so-called virial expansion, which is an expression for the pressure of a gas as a polynomial in the density. We will relate the coefficients in this polynomial to the energy of interaction between the molecules of the gas. This relation will take us into a discussion of how molecules interact with one another. We will see that deviations from ideal-gas behavior teach us a great deal about molecular interactions.

## 2–1. All Gases Behave Ideally If They Are Sufficiently Dilute

If a gas is sufficiently dilute that its constituent molecules are so far apart from each other on the average that we can ignore their interactions, it obeys the equation of state

$$PV = nRT \tag{2.1a}$$

If we divide both sides of this equation by $n$, we obtain

$$P\overline{V} = RT \tag{2.1b}$$

where $\overline{V} = V/n$ is the molar volume. We will always indicate a molar quantity by drawing a line above the symbol. Either of Equations 2.1, familiar even to high school students, is called the *ideal-gas equation of state*. Equations 2.1 are called an equation of state because they serve as a relation between the pressure, volume, and temperature of the gas for a given quantity of gas. A gas that obeys Equations 2.1 is called an ideal gas, or the gas is said to behave ideally.

The distinction between $V$ and $\overline{V}$ illustrates an important character of the quantities or the variables used to describe macroscopic systems. These quantities are of two types, called extensive quantities and intensive quantities. *Extensive quantities*, or *extensive variables*, are directly proportional to the size of a system. Volume, mass, and energy are examples of extensive quantities. *Intensive quantities*, or *intensive variables*, do not depend upon the size of the system. Pressure, temperature, and density are examples of intensive quantities. If we divide an extensive quantity by the number of particles or the number of moles in a system, we obtain an intensive quantity. For example, $V$ ($dm^3$) is an extensive quantity but $\overline{V}$ ($dm^3 \cdot mol^{-1}$) is an intensive quantity. Distinguishing between extensive and intensive quantities is often important in describing the properties of chemical systems.

The reason Equations 2.1 are encountered so frequently in chemistry courses is that *all* gases obey Equations 2.1, as long as they are sufficiently dilute. Any individual characteristics of the gas, such as the shape or size of its molecules or how the molecules interact with each other, are lost in Equations 2.1. In a sense, these equations are a common denominator for all gases. Experimentally, most gases satisfy Equations 2.1 to approximately 1% at one atm and 0°C.

Equations 2.1 require us to discuss the system of units (SI) adopted by the International Union of Pure and Applied Chemistry (IUPAC). For example, although the SI unit of volume is $m^3$ (meters cubed), the unit L (liter), which is defined as exactly 1 $dm^3$ (decimeters cubed), is an acceptable unit of volume in the IUPAC system. The SI unit of pressure is a pascal (Pa), which is equal to one newton per square meter ($\text{Pa} = \text{N}\cdot\text{m}^{-2} = \text{kg}\cdot\text{m}^{-1}\cdot\text{s}^{-2}$). Recall that a newton is the SI unit of force, so we see that pressure is a force per unit area. Pressure can be measured experimentally by observing how high a column of liquid (usually mercury) is supported by the gas. If $m$ is the mass of the liquid and $g$ is the gravitational acceleration constant, the pressure is given by

$$P = \frac{F}{A} = \frac{mg}{A} = \frac{\rho h A g}{A} = \rho h g \tag{2.2}$$

where $A$ is the base area of the column, $\rho$ is the density of the fluid, and $h$ is the height of the column. The gravitational acceleration constant is equal to $9.8067\ \text{m}\cdot\text{s}^{-2}$, or $980.67\ \text{cm}\cdot\text{s}^{-2}$. Note that the area cancels out in Equation 2.2.

**EXAMPLE 2–1**

Calculate the pressure exerted by a 76.000-cm column of mercury. Take the density of mercury to be $13.596\ \text{g}\cdot\text{cm}^{-3}$.

SOLUTION:
$$\begin{aligned} P &= (13.596\ \text{g}\cdot\text{cm}^{-3})(76.000\ \text{cm})(980.67\ \text{cm}\cdot\text{s}^{-2}) \\ &= 1.0133 \times 10^6\ \text{g}\cdot\text{cm}^{-1}\cdot\text{s}^{-2} \end{aligned}$$

A pascal is equal to $\text{N}\cdot\text{m}^{-2}$ or $\text{kg}\cdot\text{m}^{-1}\cdot\text{s}^{-2}$, so the pressure in pascals is

$$\begin{aligned} P &= (1.0133 \times 10^6\ \text{g}\cdot\text{cm}^{-1}\cdot\text{s}^{-2})(10^{-3}\ \text{kg}\cdot\text{g}^{-1})(100\ \text{cm}\cdot\text{m}^{-1}) \\ &= 1.0133 \times 10^5\ \text{Pa} = 101.33\ \text{kPa} \end{aligned}$$

Strictly speaking, new textbooks should use the IUPAC-suggested SI units, but the units of pressure are particularly problematic. Although a pascal is the SI unit of pressure and will probably see increasing use, the atmosphere will undoubtly continue to be widely used. One *atmosphere* (atm) is defined as $1.01325 \times 10^5$ Pa = 101.325 kPa. [One atmosphere used to be defined as the pressure that supports a 76.0-cm column of mercury (see Example 2.1).] Note that one kPa is approximately 1% of an atmosphere. One atmosphere used to be the standard of pressure, in the sense that tabulated properties of substances were presented at one atm. With the change to SI units, the standard is now one *bar*, which is equal to $10^5$ Pa, or 0.1 MPa. The relation between bars and atmospheres is 1 atm = 1.01325 bar. One other commonly used unit of pressure is a *torr*, which is the pressure that supports a 1.00-mm column of mercury. Thus 1 torr = (1/760) atm. Because we are experiencing a transition period between the widespread use of atm and torr on the one hand and the future use of bar and kPa on the other hand, students of physical chemistry must be proficient in both sets of pressure units. The relations between the various units of pressure are collected in Table 2.1

Of the three quantities, volume, pressure, and temperature, temperature is the most difficult to conceptualize. We will present a molecular interpretation of temperature later, but here we will give an operational definition. The fundamental temperature scale is based upon the ideal-gas law, Equations 2.1. Specifically, we define $T$ to be

$$T = \lim_{P \to 0} \frac{P\overline{V}}{R} \tag{2.3}$$

because all gases behave ideally in the limit of $P \to 0$. The unit of temperature is the kelvin, which is denoted K. Note that we do not use a degree symbol when the temperature is expressed in kelvin. Because $P$ and $\overline{V}$ cannot take on negative values, the lowest possible value of the temperature is 0 K. Temperatures as low as $1 \times 10^{-7}$ K have been achieved in the laboratory. The temperature of absolute zero (0 K) corresponds to a substance that has no thermal energy. There is no fundamental limit to the maximum value of $T$. There are, of course, practical limitations, and the highest value of $T$ achieved in the laboratory is around 100 million ($10^8$) K, which has been generated inside a magnetic confinement in nuclear fusion research facilities.

**TABLE 2.1**
Various units for expressing pressure.

| | | |
|---|---|---|
| 1 pascal (Pa) | = | $1\ \mathrm{N \cdot m^{-2}} = 1\ \mathrm{kg \cdot m^{-1} \cdot s^{-2}}$ |
| 1 atmosphere (atm) | = | $1.01325 \times 10^5$ Pa |
| | = | 1.01325 bar |
| | = | 101.325 kPa |
| | = | 1013.25 mbar |
| | = | 760 torr |
| 1 bar | = | $10^5$ Pa = 0.1 MPa |

To establish the unit of kelvin, the triple point of water has been assigned the temperature of 273.16 K. (We will learn about the properties of a "triple point" in Chapter 9. For our present purposes, it is sufficient to know that the triple point of a substance corresponds to an equilibrium system that contains gas, liquid, and solid.) We now have a definition for 0 K and 273.16 K. A kelvin is then defined as 1/273.16 of the temperature of the triple point of water. These definitions of 0 K and 273.16 K generate a linear temperature scale.

Figure 2.1 plots experimental $\overline{V}$ versus $T$ for Ar(g) at different pressures. As expected from our definition of the temperature scale, the extraplotation of these data shows that $T \rightarrow 0$ as $\overline{V} \rightarrow 0$.

The kelvin scale is related to the commonly used Celsius scale by

$$t/^\circ\text{C} = T/\text{K} - 273.15 \tag{2.4}$$

We will use the lower case $t$ for °C and the upper case $T$ for K. Note also that the degree symbol (°) is associated with values of the temperature in the Celsius scale. Equation 2.4 tells us that 0 K = −273.15°C, or that 0°C = 273.15 K. Because of the general use of °C in laboratories, a significant amount of thermodynamic data are tabulated for substances at 0°C (273.15 K) and 25°C (298.15 K); this latter value is commonly called "room temperature."

If we measure $P\overline{V}$ at 273.15 K for any gas at a sufficiently low pressure that its behavior is ideal, then

$$P\overline{V} = R(273.15\ \text{K})$$

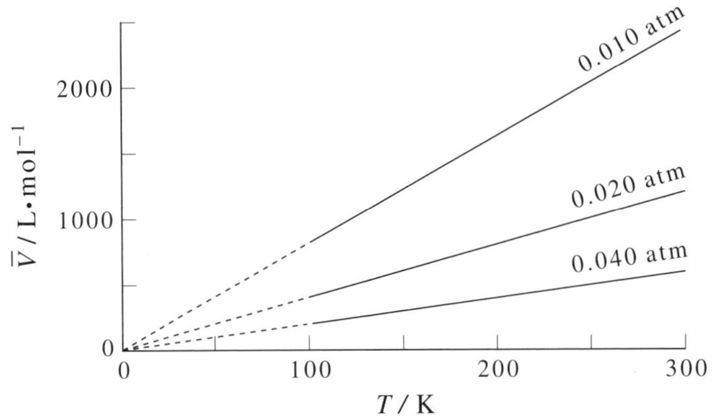


**FIGURE 2.1**
Experimental molar volumes (solid lines) of Ar(g) are plotted as a function of $T$/K at 0.040 atm, 0.020 atm, and 0.010 atm. All three pressures extrapolate to the origin (dashed lines).

Figure 2.2 shows $P\overline{V}$ data plotted against $P$ for several gases at $T = 273.15$ K. All the data plotted extrapolate to $P\overline{V} = 22.414$ L·atm as $P \to 0$, where the gases certainly behave ideally. Therefore, we can write

$$R = \frac{P\overline{V}}{T} = \frac{22.414 \text{ L·atm}}{273.15 \text{ K}} = 0.082058 \text{ L·atm·mol}^{-1}\text{·K}^{-1}$$

Using the fact that 1 atm $= 1.01325 \times 10^5$ Pa and that 1 L $= 10^{-3}$ m$^3$, we have

$$\begin{aligned} R &= (0.082058 \text{ L·atm·mol}^{-1}\text{·K}^{-1})(1.01325 \times 10^5 \text{ Pa·atm}^{-1})(10^{-3} \text{ m}^3\text{·L}^{-1}) \\ &= 8.3145 \text{ Pa·m}^3\text{·mol}^{-1}\text{·K}^{-1} \\ &= 8.3145 \text{ J·mol}^{-1}\text{·K}^{-1} \end{aligned}$$

where we have used the fact that 1 Pa·m$^3$ = 1 N·m = 1 J. Because of the change of the standard of pressure from atmospheres to bars, it is also convenient to know the value of $R$ in units of L·bar·mol$^{-1}$·K$^{-1}$. Using the fact that 1 atm = 1.01325 bar, we see that

$$\begin{aligned} R &= (0.082058 \text{ L·atm·mol}^{-1}\text{·K}^{-1})(1.01325 \text{ bar·atm}^{-1}) \\ &= 0.083145 \text{ L·bar·mol}^{-1}\text{·K}^{-1} = 0.083145 \text{ dm}^3\text{·bar·mol}^{-1}\text{·K}^{-1} \end{aligned}$$

Table 2.2 gives the value of $R$ in various units.

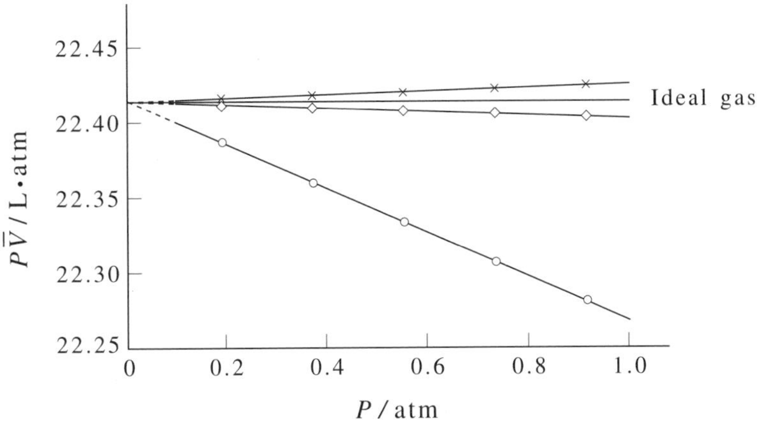


**FIGURE 2.2**
A plot of experimental values of $P\overline{V}$ versus $P$ for $H_2$(g) (crosses), $N_2$(g) (diamonds), and $CO_2$(g) (circles) at $T = 273.15$ K. The data for all three gases extrapolate to a value of $P\overline{V} = 22.414$ L·atm as $P \to 0$ (ideal behavior).

**TABLE 2.2**
The values of the molar gas constant $R$ in various units.

| | | |
|---|---|---|
| $R$ | $=$ | $8.3145\ \text{J}\cdot\text{mol}^{-1}\cdot\text{K}^{-1}$ |
| | $=$ | $0.083145\ \text{dm}^3\cdot\text{bar}\cdot\text{mol}^{-1}\cdot\text{K}^{-1}$ |
| | $=$ | $83.145\ \text{cm}^3\cdot\text{bar}\cdot\text{mol}^{-1}\cdot\text{K}^{-1}$ |
| | $=$ | $0.082058\ \text{L}\cdot\text{atm}\cdot\text{mol}^{-1}\cdot\text{K}^{-1}$ |
| | $=$ | $82.058\ \text{cm}^3\cdot\text{atm}\cdot\text{mol}^{-1}\cdot\text{K}^{-1}$ |

## 2–2. The van der Waals Equation and the Redlich–Kwong Equation Are Examples of Two-Parameter Equations of State

The ideal-gas equation is valid for all gases at sufficiently low pressures. As the pressure on a given quantity of gas is increased, however, deviations from the ideal-gas equation appear. These deviations can be displayed graphically by plotting $P\overline{V}/RT$ as a function of pressure, as shown in Figure 2.3. The quantity $P\overline{V}/RT$ is called the *compressibility factor* and is denoted by $Z$. Note that $Z = 1$ under all conditions for an ideal gas. For real gases, $Z = 1$ at low pressures, but deviations from ideal behavior ($Z \neq 1$) are seen as the pressure increases. The extent of the deviations from ideal behavior at a given pressure depends upon the temperature and the nature of the gas. The closer the gas is to the point at which it begins to liquefy, the larger the deviations from ideal behavior will be. Figure 2.4 shows $Z$ plotted against $P$ for methane at various temperatures. Note that $Z$ dips below unity at lower temperatures but lies above unity at higher temperatures. At lower temperatures the molecules are moving less rapidly, and so

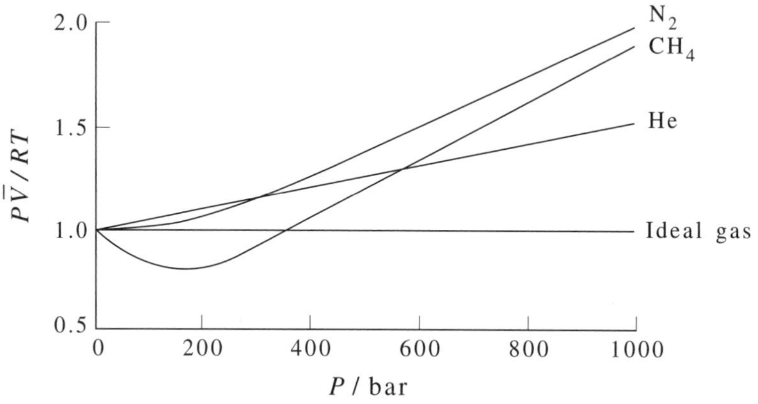


**FIGURE 2.3**
A plot of $P\overline{V}/RT$ versus $P$ for one mole of helium, nitrogen, and methane at 300 K. This figure shows that the ideal-gas equation, for which $P\overline{V}/RT = 1$, is not valid at high pressure.

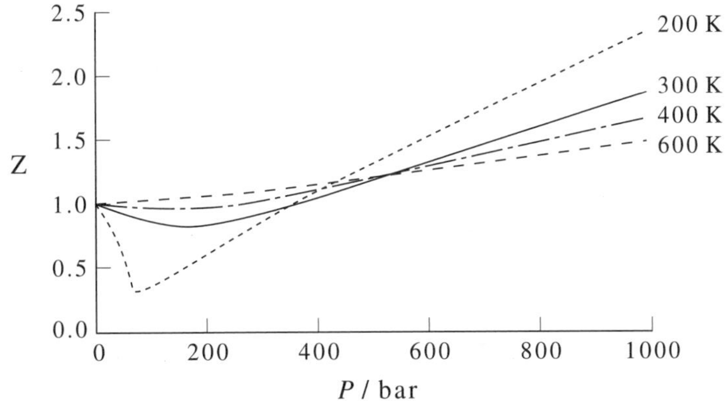


**FIGURE 2.4**
The compressibility factor of methane versus pressure at various temperatures. This figure shows that the effect of molecular attraction becomes less important at higher temperatures.

are more influenced by their attractive forces. Because of these attractive forces, the molecules are drawn together, thus making $\overline{V}_{\text{real}}$ less than $\overline{V}_{\text{ideal}}$, which in turn causes $Z$ to be less than unity. A similar effect can be seen in Figure 2.3: the order of the curves shows that the effect of molecular attractions are in the order $CH_4 > N_2 > He$ at 300 K. At higher temperatures, the molecules are moving rapidly enough that their attraction is much smaller than $k_BT$ (which we will see in Chapter 4 is a measure of their thermal energy). The molecules are influenced primarily by their repulsive forces at higher temperatures, which tend to make $\overline{V} > \overline{V}_{\text{ideal}}$, and so $Z > 1$.

Our picture of an ideal gas views the molecules as moving independently of each other, not experiencing any intermolecular interactions. Figures 2.3 and 2.4 show that this picture fails at high pressures, and that the attractive and repulsive intermolecular interactions must be taken into account. Many equations extend the ideal-gas equation to account for the intermolecular interactions. Perhaps the most well known is the *van der Waals equation*,

$$\left(P + \frac{a}{\overline{V}^2}\right)(\overline{V} - b) = RT \tag{2.5}$$

where $\overline{V}$ designates molar volume. Notice that Equation 2.5 reduces to the ideal-gas equation when $\overline{V}$ is large, as it must. The constants $a$ and $b$ in Equation 2.5 are called *van der Waals constants*, whose values depend upon the particular gas (Table 2.3). We will see in Section 2–7 that the value of $a$ reflects how strongly the molecules of a gas attract each other and the value of $b$ reflects the size of the molecules.

Let's use Equation 2.5 to calculate the pressure (in bars) exerted by 1.00 mol of $CH_4$(g) that occupies a 250-mL container at 0°C. From Table 2.3, we find that

**TABLE 2.3**
van der Waals constants for various substances.

| Species | $a/\text{dm}^6\cdot\text{bar}\cdot\text{mol}^{-2}$ | $a/\text{dm}^6\cdot\text{atm}\cdot\text{mol}^{-2}$ | $b/\text{dm}^3\cdot\text{mol}^{-1}$ |
|---|---|---|---|
| Helium | 0.034598 | 0.034145 | 0.023733 |
| Neon | 0.21666 | 0.21382 | 0.017383 |
| Argon | 1.3483 | 1.3307 | 0.031830 |
| Krypton | 2.2836 | 2.2537 | 0.038650 |
| Hydrogen | 0.24646 | 0.24324 | 0.026665 |
| Nitrogen | 1.3661 | 1.3483 | 0.038577 |
| Oxygen | 1.3820 | 1.3639 | 0.031860 |
| Carbon monoxide | 1.4734 | 1.4541 | 0.039523 |
| Carbon dioxide | 3.6551 | 3.6073 | 0.042816 |
| Ammonia | 4.3044 | 4.2481 | 0.037847 |
| Methane | 2.3026 | 2.2725 | 0.043067 |
| Ethane | 5.5818 | 5.5088 | 0.065144 |
| Ethene | 4.6112 | 4.5509 | 0.058199 |
| Propane | 9.3919 | 9.2691 | 0.090494 |
| Butane | 13.888 | 13.706 | 0.11641 |
| 2-Methyl propane | 13.328 | 13.153 | 0.11645 |
| Pentane | 19.124 | 18.874 | 0.14510 |
| Benzene | 18.876 | 18.629 | 0.11974 |

$a = 2.3026\ \text{dm}^6\cdot\text{bar}\cdot\text{mol}^{-2}$ and $b = 0.043067\ \text{dm}^3\cdot\text{mol}^{-1}$ for methane. If we divide Equation 2.5 by $\overline{V} - b$ and solve for $P$, we obtain

$$\begin{aligned} P &= \frac{RT}{\overline{V} - b} - \frac{a}{\overline{V}^2} \\ &= \frac{(0.083145\ \text{dm}^3\cdot\text{bar}\cdot\text{mol}^{-1}\cdot\text{K}^{-1})(273.15\ \text{K})}{(0.250\ \text{dm}^3\cdot\text{mol}^{-1} - 0.043067\ \text{dm}^3\cdot\text{mol}^{-1})} - \frac{2.3026\ \text{dm}^6\cdot\text{bar}\cdot\text{mol}^{-2}}{(0.250\ \text{dm}^3\cdot\text{mol}^{-1})^2} \\ &= 72.9\ \text{bar} \end{aligned}$$

By comparison, the ideal-gas equation predicts that $P = 90.8$ bar. The prediction of the van der Waals equation is in much better agreement with the experimental value of 78.6 bar than is the ideal-gas equation.

The van der Waals equation qualitatively gives the behavior shown in Figures 2.3 and 2.4. We can rewrite Equation 2.5 in the form

$$Z = \frac{P\overline{V}}{RT} = \frac{\overline{V}}{\overline{V} - b} - \frac{a}{RT\overline{V}} \tag{2.6}$$

At high pressures, the first term in Equation 2.6 dominates because $\overline{V} - b$ becomes small, and at low pressures the second term dominates.

**EXAMPLE 2–2**

Use the van der Waals equation to calculate the molar volume of ethane at 300 K and 200 atm.

SOLUTION: When we try to solve the van der Waals equation for $\overline{V}$, we obtain a cubic equation,

$$\overline{V}^3 - \left(b + \frac{RT}{P}\right)\overline{V}^2 + \frac{a}{P}\overline{V} - \frac{ab}{P} = 0$$

which we must solve numerically using the Newton-Raphson method (MathChapter G). Using the values of $a$ and $b$ from Table 2.3, we have

$$\overline{V}^3 - (0.188\ \text{L}\cdot\text{mol}^{-1})\overline{V}^2 + (0.0275\ \text{L}^2\cdot\text{mol}^{-1})\overline{V} - 0.00179\ \text{L}^3\cdot\text{mol}^{-3} = 0$$

The Newton-Raphson method gives us

$$\overline{V}_{n+1} = \overline{V}_n - \frac{\overline{V}_n^3 - 0.188\overline{V}_n^2 + 0.0275\overline{V}_n - 0.00179}{3\overline{V}_n^2 - 0.376\overline{V}_n + 0.0275}$$

where we have suppressed the units for convenience. The ideal-gas value of $\overline{V}$ is $\overline{V}_{\text{ideal}} = RT/P = 0.123\ \text{L}\cdot\text{mol}^{-1}$, so let's use $0.10\ \text{L}\cdot\text{mol}^{-1}$ as our initial guess. In this case, we obtain

| $n$ | $\overline{V}_n/\text{L}\cdot\text{mol}^{-1}$ | $f(\overline{V}_n)/\text{L}^3\cdot\text{mol}^{-3}$ | $f'(\overline{V}_n)/\text{L}^2\cdot\text{mol}^{-2}$ |
|---|---|---|---|
| 0 | 0.100 | $8.00 \times 10^{-5}$ | $2.00 \times 10^{-2}$ |
| 1 | 0.096 | $2.53 \times 10^{-6}$ | $1.90 \times 10^{-2}$ |
| 2 | 0.096 | | |

The experimental value is $0.071\ \text{L}\cdot\text{mol}^{-1}$. The calculation of pressure preceding this example and the calculation of the volume in this example show that the van der Waals equation, while more accurate than the ideal-gas equation, is not particularly accurate. We will learn shortly that there are more accurate equations of state.

Two other relatively simple equations of state that are much more accurate and hence more useful than the van der Waals equation are the *Redlich-Kwong equation*

$$P = \frac{RT}{\overline{V} - B} - \frac{A}{T^{1/2}\overline{V}(\overline{V} + B)} \tag{2.7}$$

and the *Peng-Robinson equation*

$$P = \frac{RT}{\overline{V} - \beta} - \frac{\alpha}{\overline{V}(\overline{V} + \beta) + \beta(\overline{V} - \beta)} \tag{2.8}$$

**TABLE 2.4**
The Redlich-Kwong equation parameters for various substances.

| Species | $A/\mathrm{dm^6 \cdot bar \cdot mol^{-2} \cdot K^{1/2}}$ | $A/\mathrm{dm^6 \cdot atm \cdot mol^{-2} \cdot K^{1/2}}$ | $B/\mathrm{dm^3 \cdot mol^{-1}}$ |
|---|---|---|---|
| Helium | 0.079905 | 0.078860 | 0.016450 |
| Neon | 1.4631 | 1.4439 | 0.012049 |
| Argon | 16.786 | 16.566 | 0.022062 |
| Krypton | 33.576 | 33.137 | 0.026789 |
| Hydrogen | 1.4333 | 1.4145 | 0.018482 |
| Nitrogen | 15.551 | 15.348 | 0.026738 |
| Oxygen | 17.411 | 17.183 | 0.022082 |
| Carbon monoxide | 17.208 | 16.983 | 0.027394 |
| Carbon dioxide | 64.597 | 63.752 | 0.029677 |
| Ammonia | 87.808 | 86.660 | 0.026232 |
| Methane | 32.205 | 31.784 | 0.029850 |
| Ethane | 98.831 | 97.539 | 0.045153 |
| Ethene | 78.512 | 77.486 | 0.040339 |
| Propane | 183.02 | 180.63 | 0.062723 |
| Butane | 290.16 | 286.37 | 0.08068 |
| 2-Methyl propane | 272.73 | 269.17 | 0.080715 |
| Pentane | 419.97 | 414.48 | 0.10057 |
| Benzene | 453.32 | 447.39 | 0.082996 |

where $A$, $B$, $\alpha$, and $\beta$, are parameters that depend upon the gas. The values of $A$ and $B$ in the Redlich-Kwong equation are listed in Table 2.4 for a variety of substances. The parameter $\alpha$ in the Peng-Robinson equation is a somewhat complicated function of temperature, so we will not tabulate values of $\alpha$ and $\beta$. Equations 2.7 and 2.8, like the van der Waals equation (Example 2–2), can be written as cubic equations in $\overline{V}$. For example, the Redlich-Kwong equation becomes (Problem 2–26)

$$\overline{V}^3 - \frac{RT}{P}\overline{V}^2 - \left(B^2 + \frac{BRT}{P} - \frac{A}{T^{1/2}P}\right)\overline{V} - \frac{AB}{T^{1/2}P} = 0 \tag{2.9}$$

Problem 2–28 has you show that the Peng-Robinson equation of state is also a cubic equation in $\overline{V}$.

**EXAMPLE 2–3**

Use the Redlich-Kwong equation to calculate the molar volume of ethane at 300 K and 200 atm.

SOLUTION: Substitute $T = 300\,\text{K}$, $P = 200\,\text{atm}$, $A = 97.539\,\text{dm}^6\cdot\text{atm}\cdot\text{mol}^{-1}\cdot\text{K}^{1/2}$, and $B = 0.045153\,\text{dm}^3\cdot\text{mol}^{-1}$ into Equation 2.9, to obtain

$$\overline{V}^3 - 0.1231\overline{V}^2 + 0.02056\overline{V} - 0.001271 = 0$$

where we have suppressed the units for convenience. Solving this equation by the Newton-Raphson method gives $\overline{V} = 0.0750\,\text{dm}^3\cdot\text{mol}^{-1}$, compared with the van der Waals result of $\overline{V} = 0.096\,\text{dm}^3\cdot\text{mol}^{-1}$ and the experimental result of $0.071\,\text{dm}^3\cdot\text{mol}^{-1}$ (see Example 2–2). The prediction of the Redlich-Kwong equation is nearly quantitative, unlike the van der Waals equation, which predicts a value of $\overline{V}$ that is about 30% too large.

Figure 2.5 compares experimental pressure versus density data for ethane at 400 K with the predictions of the various equations of state introduced in this chapter. Note that the Redlich-Kwong and Peng-Robinson equations are nearly quantitative, whereas the van der Waals equation fails completely at pressures greater than 200 bar. One of the impressive features of the Redlich-Kwong and Peng-Robinson equations is that they are nearly quantitative in regions where the gas liquefies. For example, Figure 2.6 shows pressure versus density data for ethane at 305.33 K, where it liquefies at around 40 bar. The horizontal region in the figure represents liquid and vapor in equilibrium with each other. Note that the Peng-Robinson equation is better in the liquid-vapor region but that the Redlich-Kwong equation is better at high pressures. The van der Waals equation is not shown because it gives negative values of the pressure under these conditions.

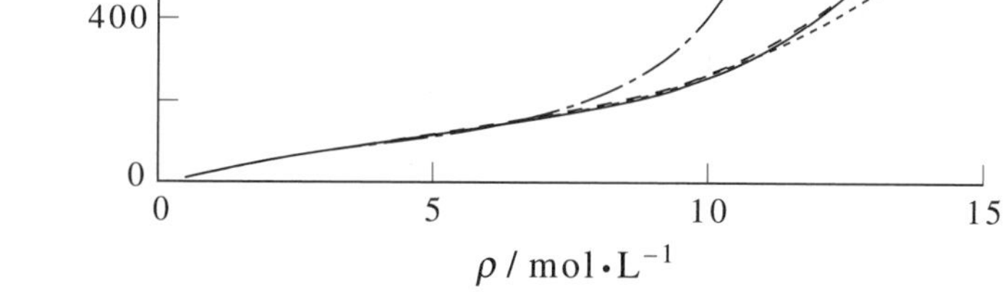

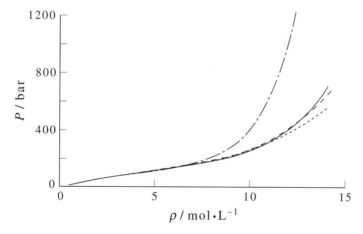

**FIGURE 2.5**
Experimental pressure versus density data for ethane at 400 K (solid line) is compared with the predictions of the van der Waals equation (dot-dashed line), the Redlich-Kwong equation (long dashed line), and the Peng-Robinson equation (short dashed line).

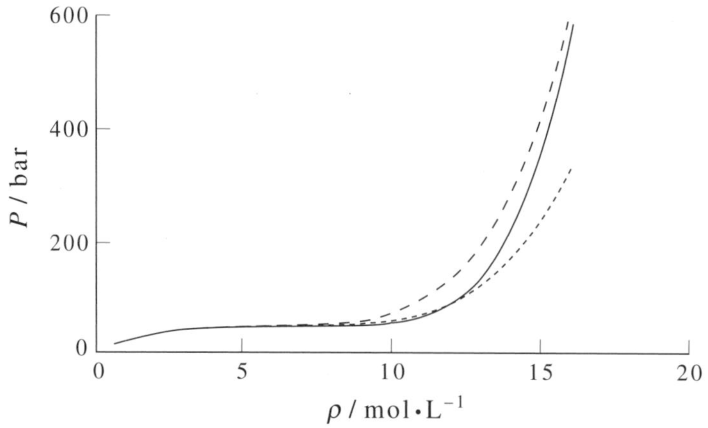


**FIGURE 2.6**
The experimental pressure versus density data (solid line) for ethane at 305.33 K is compared with the predictions of the Redlich-Kwong equation (long dashed line) and the Peng-Robinson equation (short dashed line). The liquid and vapor phases are in equilibrium in the horizontal region.

Although Figures 2.5 and 2.6 show comparisons only for ethane, the conclusions as to the relative accuracies of the equations are general. In general, the Redlich-Kwong equation is superior at high pressures, whereas the Peng-Robinson equation is superior in the liquid-vapor region. In fact, these two equations of state have been "constructed" so that this is so. There are more sophisticated equations of state (some containing more than 10 parameters!) that can reproduce the experimental data to a high degree of accuracy over a large range of pressure, density, and temperature.

## 2–3. A Cubic Equation of State Can Describe Both the Gaseous and Liquid States

A remarkable feature of equations of state that can be written as cubic equations in $\overline{V}$ is that they describe both the gaseous *and* the liquid regions of a substance. To understand this feature, we start by discussing some experimentally determined plots of $P$ as a function of $\overline{V}$ at constant $T$, which are commonly called *isotherms* (*iso* = constant). Figure 2.7 shows experimental $P$ versus $\overline{V}$ isotherms for carbon dioxide. The isotherms shown are in the neighborhood of the critical temperature, $T_c$, which is the temperature above which a gas cannot be liquefied, regardless of the pressure. The critical pressure, $P_c$, and the critical volume, $\overline{V}_c$, are the corresponding pressure and the molar volume at the *critical point.* For example, for carbon dioxide, $T_c = 304.14$ K (30.99°C), $P_c = 72.9$ atm, and $\overline{V}_c = 0.094$ L·mol$^{-1}$. Note that the isotherms in Figure 2.7 flatten out as $T \to T_c$ from above and that there are horizontal regions when $T$ is less than $T_c$. In the horizontal regions, gas and liquid coexist in equilibrium with each other. The dashed curve connecting the ends of the horizontal

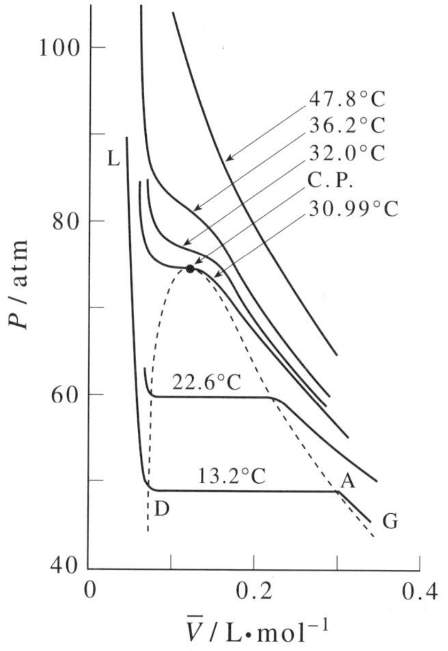


**FIGURE 2.7**
Experimental pressure-volume isotherms of carbon dioxide around its critical temperature, 30.99°C. Points G, A, D, and L are discussed in the text.

lines in Figure 2.7 is called the *coexistence curve*, because any point within this curve corresponds to liquid and gas coexisting in equilibrium with each other. At any point on or outside this curve, only one phase is present. For example, at point G in the figure we have only a gas phase. If we now start at G and compress the gas along the 13.2°C isotherm, liquid will first appear when we reach the horizontal line at point A. The pressure will remain constant as we condense the gas at molar volume 0.3 $L \cdot mol^{-1}$ (point A) to liquid at molar volume of approximately 0.07 $L \cdot mol^{-1}$ (point D). After reaching point D, the pressure increases sharply with a further decrease in volume, because we now have all liquid and the volume of a liquid changes very little with pressure.

Note that as the temperature increases toward the critical temperature, the horizontal lines shorten and disappear at the critical temperature. At this point, the meniscus between the liquid and its vapor disappears and there is no distinction between liquid and gas; the surface tension disappears and the gas and liquid phases both have the same (critical) density. We will discuss the critical point in more detail in Chapter 9.

Figure 2.8 shows similar isotherms for the van der Waals equation and the Redlich-Kwong equation. Notice that the two equations of state give fairly similar plots. The spurious loops obtained for $T < T_c$ result from the approximate nature of these equations of state. Figure 2.9 shows a single van der Waals or Redlich-Kwong isotherm for $T < T_c$. The curve GADL is the curve that would be observed experimentally upon compressing the gas. The horizontal line DA is drawn so that the areas of the loop both above and below DA are equal. (This so-called *Maxwell equal-area construction* will be justified in Chapter 9.) The line GA represents compression of the gas. Along the line AD, liquid and vapor are in equilibrium with each other. The point A represents

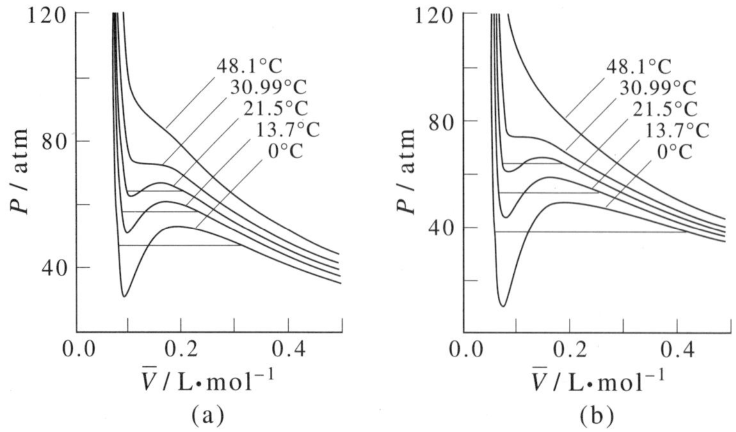


**FIGURE 2.8**
Pressure-volume isotherms of carbon dioxide around its critical temperature, as calculated from (a) the van der Waals equation (Equation 2.5) and (b) the Redlich-Kwong equation (Equation 2.7).

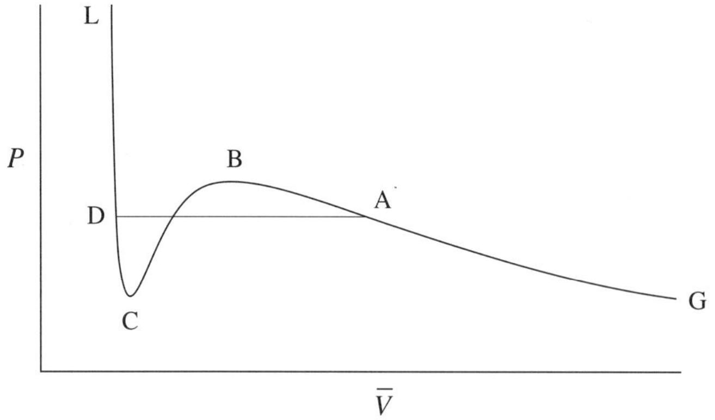


**FIGURE 2.9**
A typical van der Waals pressure-volume isotherm at a temperature less than the critical temperature. The horizontal line has been drawn so that areas of the loop above and below are equal.

the coexisting vapor and the point D represents the liquid. The line DL represents the change of volume of the liquid with increasing pressure. The steepness of this line results from the relative incompressibility of the liquid. The segment AB is a metastable region corresponding to the superheated vapor, and the segment CD corresponds to the supercooled liquid. The segment BC is a region in which $(\partial P/\partial \overline{V})_T > 0$. This condition signifies an unstable region, which is not observed for equilibrium systems.

Figure 2.9 shows that we can obtain three values of the volume along the line DA for a given pressure if the temperature is less than the critical temperature. This result is consistent with the fact that the van der Waals equation can be written as a cubic polynomial in the (molar) volume (see Example 2–2). The volume corresponding to point D is the molar volume of the liquid, the volume corresponding to point A is the molar volume of the vapor in equilibrium with the liquid, and the third root, lying between A and D is spurious.

At 142.69 K and 35.00 atm, argon exists as two phases in equilibrium with each other, and the densities of the liquid and vapor phases are 22.491 $\text{mol}\cdot\text{L}^{-1}$ and 5.291 $\text{mol}\cdot\text{L}^{-1}$, respectively. Let's see what the van der Waals equation predicts in this case. As we saw in Example 2–2, we can write the van der Waals equation as

$$\overline{V}^3 - \left(b + \frac{RT}{P}\right)\overline{V}^2 + \frac{a}{P}\overline{V} - \frac{ab}{P} = 0 \tag{2.10}$$

Using the values of $a$ and $b$ from Table 2.3, $T = 142.69$ K, and $P = 35.00$ atm, Equation 2.10 becomes

$$\overline{V}^3 - 0.3664\overline{V}^2 + 0.03802\overline{V} - 0.001210 = 0$$

where, for convenience, we have supressed the units of the coefficients. The three roots of this equation are (Problem 2–22) 0.07073 $\text{L}\cdot\text{mol}^{-1}$, 0.07897 $\text{L}\cdot\text{mol}^{-1}$, and 0.2167 $\text{L}\cdot\text{mol}^{-1}$. The smallest root represents the molar volume of liquid argon, and the largest represents the molar volume of the vapor. The corresponding densities are 14.14 $\text{mol}\cdot\text{L}^{-1}$ and 4.615 $\text{mol}\cdot\text{L}^{-1}$, which are in poor agreement with the experimental values (22.491 $\text{mol}\cdot\text{L}^{-1}$ and 5.291 $\text{mol}\cdot\text{L}^{-1}$). The Redlich-Kwong equation gives 20.13 $\text{mol}\cdot\text{L}^{-1}$ and 5.147 $\text{mol}\cdot\text{L}^{-1}$, and the Peng-Robinson equation gives 23.61 $\text{mol}\cdot\text{L}^{-1}$ and 5.564 $\text{mol}\cdot\text{L}^{-1}$ (Problem 2–23). Both the Redlich-Kwong and the Peng-Robinson equations are fairly accurate, and the Peng-Robinson equation is about 10% more accurate in this liquid region.

The point C.P. in Figure 2.7 is the critical point, where $T = T_c$, $P = P_c$, and $\overline{V} = \overline{V}_c$. The point C.P. is an inflection point, and so

$$\left(\frac{\partial P}{\partial \overline{V}}\right)_T = 0 \qquad \text{and} \qquad \left(\frac{\partial^2 P}{\partial \overline{V}^2}\right)_T = 0 \qquad \text{at C.P.}$$

We can use these two conditions to determine the critical constants in terms of $a$ and $b$ (Problem 2–25). An easier way to do this, however, is to write the van der Waals equation as a cubic equation in $\overline{V}$, Equation 2.10.

$$\overline{V}^3 - \left(b + \frac{RT}{P}\right)\overline{V}^2 + \frac{a}{P}\overline{V} - \frac{ab}{P} = 0$$

Being a cubic equation, it has three roots. For $T > T_c$, only one of these roots is real (the other two are complex), and for $T < T_c$ and $P \approx P_c$, all three roots are real. At $T = T_c$, these three roots merge into one, and so we can write Equation 2.10 as $(\overline{V} - \overline{V}_c)^3 = 0$, or

$$\overline{V}^3 - 3\overline{V}_c\overline{V}^2 + 3\overline{V}_c^2\overline{V} - \overline{V}_c^3 = 0 \tag{2.11}$$

If we compare this equation with Equation 2.10 at the critical point, we have

$$3\overline{V}_c = b + \frac{RT_c}{P_c}, \qquad 3\overline{V}_c^2 = \frac{a}{P_c}, \qquad \text{and} \qquad \overline{V}_c^3 = \frac{ab}{P_c} \tag{2.12}$$

Eliminate $P_c$ between the second two of these to obtain

$$\overline{V}_c = 3b \tag{2.13$a$}$$

and then substitute this result into the third of Equations 2.12 to obtain

$$P_c = \frac{a}{27b^2} \tag{2.13$b$}$$

and last, substitute Equations 2.13*a* and 2.13*b* into the first of Equations 2.12 to obtain

$$T_c = \frac{8a}{27bR} \tag{2.13$c$}$$

The critical constants of a number of substances are given in Table 2.5.

The values of the critical constants in terms of the parameters $A$ and $B$ of the Redlich-Kwong equation can be determined in a similar fashion. The mathematics is a bit more involved, and the results are (Problem 2–27)

$$\overline{V}_c = 3.8473B, \qquad P_c = 0.029894\frac{A^{2/3}R^{1/3}}{B^{5/3}}, \qquad \text{and} \qquad T_c = 0.34504\left(\frac{A}{BR}\right)^{2/3} \tag{2.14}$$

The following example shows that the van der Waals equation and the Redlich-Kwong equation make an interesting prediction about the value of $P_c\overline{V}_c/RT_c$.

**EXAMPLE 2–4**

Calculate the ratio $P_c\overline{V}_c/RT_c$ for the van der Waals equation and the Redlich-Kwong equation.

SOLUTION: Multiplying Equation 2.13*b* by 2.13*a* and dividing by $R$ times Equation 2.13*c* gives

$$\frac{P_c\overline{V}_c}{RT_c} = \frac{1}{R}\left(\frac{a}{27b^2}\right)(3b)\left(\frac{27bR}{8a}\right) = \frac{3}{8} = 0.375 \tag{2.15}$$

Similarly, the Redlich-Kwong equation gives

$$\frac{P_c\overline{V}_c}{RT_c} = \frac{1}{R}\left(\frac{0.029894A^{2/3}R^{1/3}}{B^{5/3}}\right)(3.8473B)\left(\frac{(BR)^{2/3}}{0.34504A^{2/3}}\right) = 0.33333 \tag{2.16}$$

Equations 2.15 and 2.2 predict that $P_c\overline{V}_c/RT_c$ should be the same value for all substances but that the numerical values differ slightly for the two approximate equations of state. The experimental values of $P_c\overline{V}_c/RT_c$ given in Table 2.5 show that neither

**TABLE 2.5**
The experimental critical constants of various substances.

| Species | $T_c$/K | $P_c$/bar | $P_c$/atm | $\overline{V}_c$/L·mol$^{-1}$ | $P_c\overline{V}_c/RT_c$ |
|---|---|---|---|---|---|
| Helium | 5.1950 | 2.2750 | 2.2452 | 0.05780 | 0.30443 |
| Neon | 44.415 | 26.555 | 26.208 | 0.04170 | 0.29986 |
| Argon | 150.95 | 49.288 | 48.643 | 0.07530 | 0.29571 |
| Krypton | 210.55 | 56.618 | 55.878 | 0.09220 | 0.29819 |
| Hydrogen | 32.938 | 12.838 | 12.670 | 0.06500 | 0.30470 |
| Nitrogen | 126.20 | 34.000 | 33.555 | 0.09010 | 0.29195 |
| Oxygen | 154.58 | 50.427 | 50.768 | 0.07640 | 0.29975 |
| Carbon monoxide | 132.85 | 34.935 | 34.478 | 0.09310 | 0.29445 |
| Chlorine | 416.9 | 79.91 | 78.87 | 0.1237 | 0.28517 |
| Carbon dioxide | 304.14 | 73.843 | 72.877 | 0.09400 | 0.27443 |
| Water | 647.126 | 220.55 | 217.66 | 0.05595 | 0.2295 |
| Ammonia | 405.30 | 111.30 | 109.84 | 0.07250 | 0.23945 |
| Methane | 190.53 | 45.980 | 45.379 | 0.09900 | 0.28735 |
| Ethane | 305.34 | 48.714 | 48.077 | 0.1480 | 0.28399 |
| Ethene | 282.35 | 50.422 | 49.763 | 0.1290 | 0.27707 |
| Propane | 369.85 | 42.477 | 41.922 | 0.2030 | 0.28041 |
| Butane | 425.16 | 37.960 | 37.464 | 0.2550 | 0.27383 |
| 2-Methylpropane | 407.85 | 36.400 | 35.924 | 0.2630 | 0.28231 |
| Pentane | 469.69 | 33.643 | 33.203 | 0.3040 | 0.26189 |
| Benzene | 561.75 | 48.758 | 48.120 | 0.2560 | 0.26724 |

equation of state is quantitative. The corresponding value for $P_c\overline{V}_c/RT_c$ for the Peng-Robinson equation is 0.30740 (Problem 2–28), which is closer to the experimental values than either of the values given by the van der Waals equation or the Redlich-Kwong equation. Note, however, that all three equations of state do predict a constant value for $P_c\overline{V}_c/RT_c$, and the experimental data in Table 2.5 show that this value is indeed fairly constant. This observation is an example of the law of corresponding states, which says that the properties of all gases are the same if we compare them under the same conditions relative to their critical point. We will discuss the law of corresponding states more thoroughly in the next section.

Although we have written $\overline{V}_c$, $P_c$, and $T_c$ in terms of $a$ and $b$ in Equations 2.13 or in terms of $A$ and $B$ in Equations 2.14, in practice these constants are usually evaluated in terms of experimental critical constants. Because there are three critical constants and only two constants for each equation of state, there is some ambiguity in doing so. For example, we could use Equations 2.13$a$ and 2.13$b$ to evaluate $a$ and $b$ in terms of $\overline{V}_c$

and $P_c$, or use another pair of equations. Because $P_c$ and $T_c$ are known more accurately, we use Equations 2.13*b* and 2.13*c* to obtain

$$a = \frac{27(RT_c)^2}{64P_c} \quad \text{and} \quad b = \frac{RT_c}{8P_c} \tag{2.17}$$

Likewise, from Equations 2.14, we obtain the Redlich-Kwong constants,

$$A = 0.42748\frac{R^2T_c^{5/2}}{P_c} \quad \text{and} \quad B = 0.086640\frac{RT_c}{P_c} \tag{2.18}$$

The van der Waals and Redlich-Kwong constants in Tables 2.3 and 2.4 have been obtained in this way.

**EXAMPLE 2–5**

Use the critical-constant data in Table 2.5 to evaluate the van der Waals constants for ethane.

SOLUTION:

$$a = \frac{27(0.083145\ \text{dm}^3\cdot\text{bar}\cdot\text{mol}^{-1}\text{K}^{-1})^2(305.34\ \text{K})^2}{64(48.714\ \text{bar})}$$
$$= 5.5817\ \text{dm}^6\cdot\text{bar}\cdot\text{mol}^{-2} = 5.5088\ \text{dm}^6\cdot\text{atm}\cdot\text{mol}^{-2}$$

and

$$b = \frac{(0.083145\ \text{dm}^3\cdot\text{bar}\cdot\text{mol}^{-1}\text{K}^{-1})(305.34\ \text{K})}{8(48.714\ \text{bar})}$$
$$= 0.065144\ \text{dm}^3\cdot\text{mol}^{-1}$$

**EXAMPLE 2–6**

Use the critical-constant data in Table 2.5 to evaluate $A$ and $B$, the Redlich-Kwong constants for ethane.

SOLUTION:

$$A = 0.42748\frac{(0.083145\ \text{dm}^3\cdot\text{bar}\cdot\text{mol}^{-1}\cdot\text{K}^{-1})^2(305.34\ \text{K})^{5/2}}{48.714\ \text{bar}}$$
$$= 98.831\ \text{dm}^6\cdot\text{bar}\cdot\text{mol}^{-2}\cdot\text{K}^{1/2} = 97.539\ \text{dm}^6\cdot\text{atm}\cdot\text{mol}^{-2}\cdot\text{K}^{1/2}$$

and

$$B = 0.086640\frac{(0.083145\ \text{dm}^3\cdot\text{bar}\cdot\text{mol}^{-1}\cdot\text{K}^{-1})(305.34\ \text{K})}{48.714\ \text{bar}}$$
$$= 0.045153\ \text{dm}^3\cdot\text{mol}^{-1}$$

## 2–4. The van der Waals Equation and the Redlich–Kwong Equation Obey the Law of Corresponding States

Let's start with the van der Waals equation, which we can write in an interesting and practical form by substituting the second of Equations 2.12 for *a* and Equation 2.13*a* for *b* into Equation 2.5:

$$\left(P + \frac{3P_c\overline{V}_c^2}{\overline{V}^2}\right)\left(\overline{V} - \frac{1}{3}\overline{V}_c\right) = RT$$

Divide through by $P_c$ and $\overline{V}_c$ to get

$$\left(\frac{P}{P_c} + \frac{3\overline{V}_c^2}{\overline{V}^2}\right)\left(\frac{\overline{V}}{\overline{V}_c} - \frac{1}{3}\right) = \frac{RT}{P_c\overline{V}_c} = \frac{RT}{\frac{3}{8}RT_c} = \frac{8}{3}\frac{T}{T_c}$$

where we have used Equation 2.15 for $P_c\overline{V}_c$. Now introduce the *reduced quantities* $P_R = P/P_c$, $\overline{V}_R = \overline{V}/\overline{V}_c$, and $T_R = T/T_c$ to obtain the van der Waals equation written in terms of reduced quantities:

$$\left(P_R + \frac{3}{\overline{V}_R^2}\right)\left(\overline{V}_R - \frac{1}{3}\right) = \frac{8}{3}T_R \qquad (2.19)$$

Equation 2.19 is remarkable in that there are no quantities in this equation that are characteristic of any particular gas; it is a universal equation for *all* gases. It says, for example, that the value of $P_R$ will be the same for all gases at the same values of $\overline{V}_R$ and $T_R$. Let's consider $CO_2$(g) and $N_2$(g) for $\overline{V}_R = 20$ and $T_R = 1.5$. According to Equation 2.19, $P_R = 0.196$ when $\overline{V}_R = 20.0$ and $T_R = 1.5$. Using the values of the critical constants given in Table 2.5, we find that the reduced quantities $P_R = 0.196$, $\overline{V}_R = 20.0$, and $T_R = 1.5$ correspond to $P_{CO_2} = 14.3$ atm $= 14.5$ bar, $\overline{V}_{CO_2} = 1.9\ \text{L·mol}^{-1}$, and $T_{CO_2} = 456$ K and to $P_{N_2} = 6.58$ atm $= 6.66$ bar, $\overline{V}_{N_2} = 1.8\ \text{L·mol}^{-1}$, and $T_{N_2} = 189$ K. These two gases under these conditions are said to be at corresponding states (same values of $P_R$, $\overline{V}_R$, and $T_R$). According to the van der Waals equation, these quantities are related by Equation 2.19, so Equation 2.19 is an example of the *law of corresponding states*, that all gases have the same properties if they are compared at corresponding conditions (same values of $P_R$, $\overline{V}_R$, and $T_R$).

---

**EXAMPLE 2–7**

Express the Redlich-Kwong equation in terms of reduced quantities.

SOLUTION: Equations 2.18 show that

$$A = 0.42748\frac{R^2T_c^{5/2}}{P_c} \quad \text{and} \quad B = 0.086640\frac{RT_c}{P_c}$$

Substituting these equivalencies into Equation 2.7 gives

$$P = \frac{RT}{\overline{V} - 0.086640\dfrac{RT_c}{P_c}} - \frac{0.42748R^2T_c^{5/2}/P_c}{T^{1/2}\overline{V}\left(\overline{V} + 0.086640\dfrac{RT_c}{P_c}\right)}$$

Divide the numerator and the denominator of the first term on the right side by $\overline{V}_c$ and the second by $\overline{V}_c^2$ to get

$$P = \frac{RT/\overline{V}_c}{\overline{V}_R - 0.086640\dfrac{RT_c}{P_c\overline{V}_c}} - \frac{0.42748R^2T_c^2/P_c\overline{V}_c^2}{T_R^{1/2}\overline{V}_R\left(\overline{V}_R + 0.086640\dfrac{RT_c}{P_c\overline{V}_c}\right)}$$

Divide both sides by $P_c$ and use the fact that $P_c\overline{V}_c/RT_c = 1/3$ in the second term to get

$$P_R = \frac{RT/P_c\overline{V}_c}{\overline{V}_R - 0.25992} - \frac{3.8473}{T_R^{1/2}\overline{V}_R(\overline{V}_R + 0.25992)}$$

Finally, multiply and divide the numerator of the first term on the right side by $T_c$ to obtain

$$P_R = \frac{3T_R}{\overline{V}_R - 0.25992} - \frac{3.8473}{T_R^{1/2}\overline{V}_R(\overline{V}_R + 0.25992)}$$

Thus, we see that the Redlich-Kwong equation also obeys a law of corresponding states.

The compressibility factor, $Z$, associated with the van der Waals equation also obeys the law of corresponding states. To demonstrate this point, we start with Equation 2.6 and substitute the second of Equations 2.12 for $a$ and Equation 2.13$b$ for $b$ to get

$$Z = \frac{P\overline{V}}{RT} = \frac{\overline{V}}{\overline{V} - \frac{1}{3}\overline{V}_c} - \frac{3P_c\overline{V}_c^2}{RT\overline{V}}$$

Now use Equation 2.15 for $P_c\overline{V}_c$ in the second term and introduce reduced variables to get

$$Z = \frac{\overline{V}_R}{\overline{V}_R - \frac{1}{3}} - \frac{9}{8\overline{V}_R T_R} \tag{2.20}$$

Similarly, the compressibility factor for the Redlich-Kwong equation is (Problem 2–30)

$$Z = \frac{\overline{V}_R}{\overline{V}_R - 0.25992} - \frac{1.2824}{T_R^{3/2}(\overline{V}_R + 0.25992)} \tag{2.21}$$

Equations 2.20 and 2.21 express $Z$ as a universal function of $\overline{V}_R$ and $T_R$, or of any other two reduced quantities, such as $P_R$ and $T_R$. Although these equations can be used to illustrate the law of corresponding states, they are based on approximate equations of state. Nevertheless, the law of corresponding states is valid for a great variety of gases. Figure 2.10 shows experimental data for $Z$ plotted against $P_R$ at various values of $T_R$ for 10 gases. Note that the data for all 10 gases fall on the same curves, thus illustrating the law of corresponding states in a more general way than either Equation 2.20 or 2.21. Much more extensive graphs are available, particularly in the engineering literature, and are of great use in practical applications.

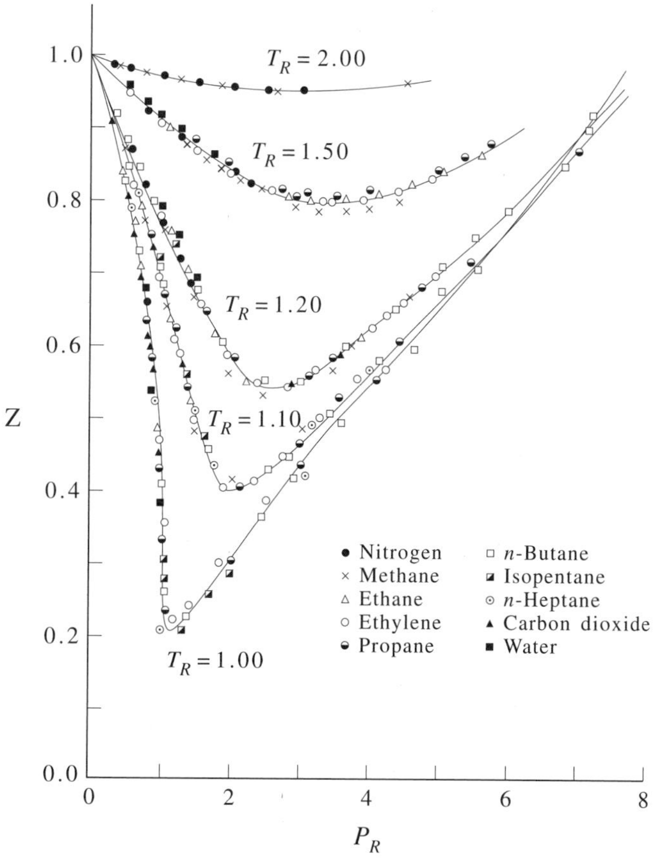


**FIGURE 2.10**
An illustration of the law of corresponding states. The compressibility factor, $Z$, is plotted against the reduced pressure, $P_R$, of each of the 10 indicated gases. Each curve represents a given reduced temperature. Note that for a given reduced temperature, all 10 gases fall on the same curve because reduced quantities are used.

**EXAMPLE 2–8**

Use Figure 2.10 to estimate the molar volume of ammonia at 215°C and 400 bar.

SOLUTION: Using the critical-constant data in Table 2.5, we find that $T_R = 1.20$ and $P_R = 3.59$. Figure 2.10 shows that $Z \approx 0.60$ under these conditions. The molar volume is

$$\begin{aligned}\overline{V} &\approx \frac{RTZ}{P} = \frac{(0.08314\ \text{L}\cdot\text{bar}\cdot\text{mol}^{-1}\cdot\text{K}^{-1})(488\ \text{K})(0.60)}{400\ \text{bar}} \\ &\approx 0.061\ \text{L}\cdot\text{mol}^{-1} = 61\ \text{cm}^3\cdot\text{mol}^{-1}\end{aligned}$$

The law of corresponding states has a nice physical interpretation. Any temperature scale we use to describe a gas is necessarily arbitrary. Even the Kelvin scale, with its fundamental zero temperature, is arbitrary in the sense that the size of a degree on the Kelvin scale is arbitrary. Thus, the numerical value we assign to the temperature is meaningless as far as the gas is concerned. A gas does "know" its critical temperature, and therefore is "aware" of its temperature *relative* to its critical temperature or its reduced temperature, $T_R = T/T_c$. Similarly, pressure and volume scales are imposed by us, but the reduced pressure and the reduced volume are quantities that are of significance to a particular gas. Thus, any gas that has a certain reduced temperature, pressure, and volume will behave in the same manner as another gas under the same conditions.

## 2–5. Second Virial Coefficients Can Be Used to Determine Intermolecular Potentials

The most fundamental equation of state, in the sense that it has the most sound theoretical foundation, is the *virial equation of state*. The virial equation of state expresses the compressibility factor as a polynomial in $1/\overline{V}$:

$$Z = \frac{P\overline{V}}{RT} = 1 + \frac{B_{2V}(T)}{\overline{V}} + \frac{B_{3V}(T)}{\overline{V}^2} + \cdots \tag{2.22}$$

The coefficients in this expression are functions of temperature only and are called *virial coefficients*. In particular, $B_{2V}(T)$ is called the *second virial coefficient*, $B_{3V}(T)$ the third, and so on. We will see later that other properties such as energy and entropy can be expressed as polynomials in $1/\overline{V}$, and generally these relations are called *virial expansions*.

We can also express the compressibility factor as a polynomial in $P$

$$Z = \frac{P\overline{V}}{RT} = 1 + B_{2P}(T)P + B_{3P}(T)P^2 + \cdots \tag{2.23}$$

Equation 2.23 is also called a virial expansion or a virial equation of state. The virial coefficients $B_{2V}(T)$ and $B_{2P}(T)$ are related by (Problem 2–36 )

$$B_{2V}(T) = RT B_{2P}(T) \tag{2.24}$$

Note in Equation 2.22 or 2.23 that $Z \to 1$ as $\overline{V}$ becomes large or as $P$ becomes small, just as it should. Table 2.6 gives an idea of the magnitudes of the terms in Equation 2.22 as a function of pressure for argon at 25°C. Notice that even at 100 bar the first three terms are sufficient for calculating $Z$.

**TABLE 2.6**
The contribution of the first few terms in the virial expansion of $Z$, Equation 2.22, for argon at 25°C.

| $P$/bar | $Z = P\overline{V}/RT$ |
|---|---|
| | $1 + \dfrac{B_{2V}(T)}{\overline{V}} + \dfrac{B_{3V}(T)}{\overline{V}^2}$ + remaining terms |
| 1 | 1 − 0.00064 + 0.00000 + (+0.00000) |
| 10 | 1 − 0.00648 + 0.00020 + (−0.00007) |
| 100 | 1 − 0.06754 + 0.02127 + (−0.00036) |
| 1000 | 1 − 0.38404 + 0.08788 + (+0.37232) |

The second virial coefficient is the most important virial coefficient because it reflects the first deviation from ideality as the pressure of the gas is increased (or the volume is decreased). As such, it is the most easily measured virial coefficient and is well tabulated for many gases. According to Equation 2.23, it can be determined experimentally from the slope of a plot of $Z$ against $P$, as shown in Figure 2.11. Figure 2.12 shows $B_{2V}(T)$ plotted against temperature for helium, nitrogen, methane, and carbon dioxide. Note that $B_{2V}(T)$ is negative at low temperatures and increases with temperature, eventually going through a shallow maximum (observable only for helium in Figure 2.12). The temperature at which $B_{2V}(T) = 0$ is called the *Boyle temperature*. At the Boyle temperature, the repulsive and attractive parts of the intermolecular interactions cancel each other, and the gas appears to behave ideally (neglecting any effect of virial coefficients beyond the second).

Not only are Equations 2.22 and 2.23 used to summarize experimental $P$–$V$–$T$ data, but they also allow us to derive exact relations between the virial coefficients and the intermolecular interactions. Consider two interacting molecules as shown in Figure 2.13. The interaction of the two molecules depends upon the distance between their centers, $r$, and upon their orientations. Because the molecules are rotating, their orientations partially average out, so for simplicity we assume that the interaction depends only upon $r$. This approximation turns out to be satisfactory for

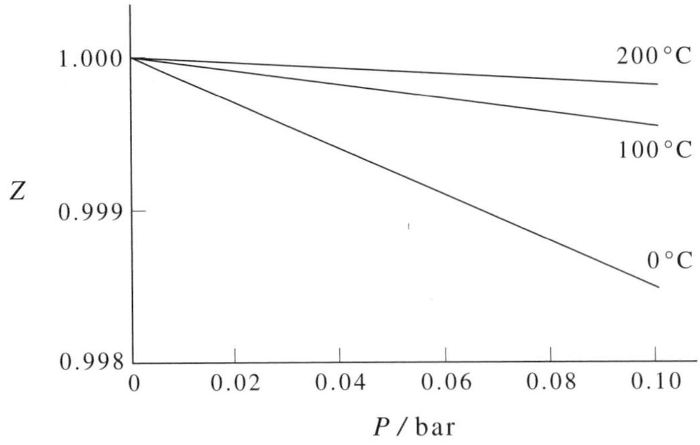


**FIGURE 2.11**
A plot of $Z$ versus $P$ at low pressures for $NH_3(g)$ at 0°C, 100°C, and 200°C. The slopes of the lines are equal to $B_{2V}(T)/RT$ according to Equations 2.23 and 2.24. The respective slopes give $B_{2V}(0°C) = -0.345\ dm^3 \cdot mol^{-1}$, $B_{2V}(100°C) = -0.142\ dm^3 \cdot mol^{-1}$, and $B_{2V}(200°C) = -0.075\ dm^3 \cdot mol^{-1}$.

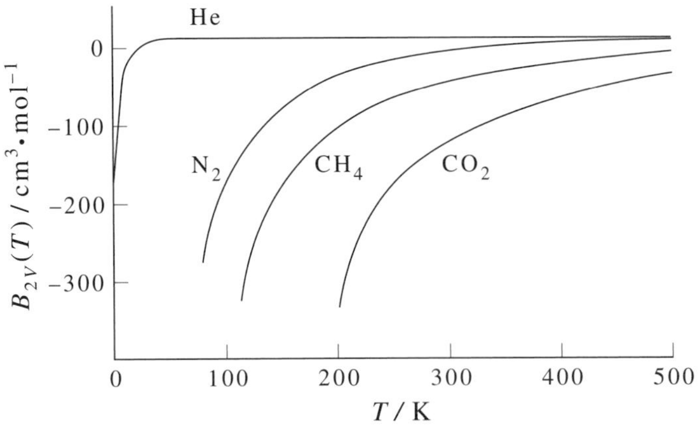


**FIGURE 2.12**
The second virial coefficient $B_{2V}(T)$ of several gases plotted against temperature. Note that $B_{2V}(T)$ is negative at low temperatures and increases with temperature up to a point, where it passes through a shallow maximum (observable here only for helium).

many molecules, especially if they are not very polar. If we let $u(r)$ be the potential energy of two molecules separated by a distance $r$, the relation between the second virial coefficient $B_{2V}(T)$ and $u(r)$ is given by

$$B_{2V}(T) = -2\pi N_A \int_0^\infty [e^{-u(r)/k_B T} - 1] r^2 dr \tag{2.25}$$

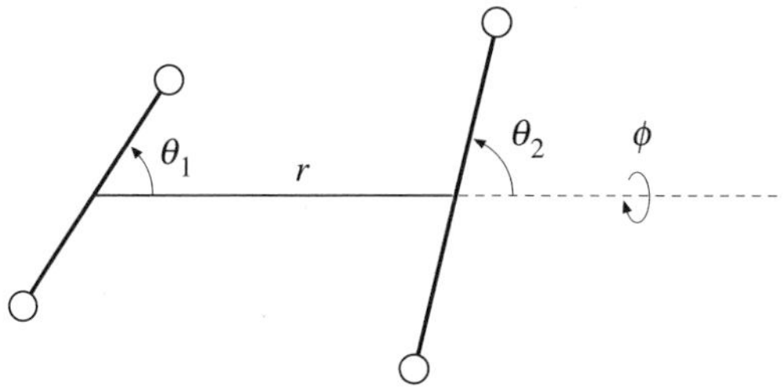


**FIGURE 2.13**
Two interacting linear molecules. Generally, the intermolecular interaction between two molecules depends upon the distance between their centers ($r$) and upon their orientations ($\theta_1$, $\theta_2$, and $\phi$) .

where $N_A$ is the Avogadro constant and $k_B$ is the Boltzmann constant, which is equal to the molar gas constant $R$ divided by the Avogadro constant. Note that $B_{2V}(T) = 0$ if $u(r) = 0$; in other words, there are no deviations from ideal behavior if there are no intermolecular interactions.

Equation 2.25 shows that once $u(r)$ is known, it is a simple matter to calculate $B_{2V}(T)$ as a function of temperature, or conversely, to determine $u(r)$ if $B_{2V}(T)$ is known. In principle, $u(r)$ can be calculated from quantum mechanics, but this is a difficult computational problem. It can be shown, however, that

$$u(r) \longrightarrow -\frac{c_6}{r^6} \tag{2.26}$$

for large values of $r$. In this expression, $c_6$ is a constant whose value depends upon the particular interacting molecules. The negative sign in Expression 2.26 indicates that the two molecules attract each other. This attraction is what causes substances to condense at sufficiently low temperatures. There is no known exact expression like 2.26 for small distances, but it must be of a form that reflects the repulsion that occurs when two molecules approach closely. Usually, we assume that

$$u(r) \longrightarrow \frac{c_n}{r^n} \tag{2.27}$$

for small values of $r$. In Equation 2.27, $n$ is an integer, often taken to be 12, and $c_n$ is a constant whose value depends upon the two molecules.

An intermolecular potential that embodies the long-range (attractive) behavior of Equation 2.26 and the short-range (repulsive) behavior of Equation 2.27 is simply the sum of the two. If we take $n$ to be 12, then

$$u(r) = \frac{c_{12}}{r^{12}} - \frac{c_6}{r^6} \tag{2.28}$$

Equation 2.28 is usually written in the form

$$u(r) = 4\varepsilon \left[ \left( \frac{\sigma}{r} \right)^{12} - \left( \frac{\sigma}{r} \right)^{6} \right] \tag{2.29}$$

where $c_{12} = 4\varepsilon\sigma^{12}$ and $c_6 = 4\varepsilon\sigma^6$. Equation 2.29, which is called the *Lennard-Jones potential*, is plotted in Figure 2.14. The two parameters in the Lennard-Jones potential have the following physical interpretation: $\varepsilon$ is the depth of the potential well and $\sigma$ is the distance at which $u(r) = 0$ (Figure 2.14). As such, $\varepsilon$ is a measure of how strongly the molecules attract each other, and $\sigma$ is a measure of the size of the molecules. These *Lennard-Jones parameters* are tabulated for a number of molecules in Table 2.7.

**EXAMPLE 2–9**

Show that the minimum of the Lennard-Jones potential occurs at $r_{\text{min}} = 2^{1/6}\sigma = 1.12\sigma$. Evaluate $u(r)$ at $r_{\text{min}}$.

SOLUTION: To find $r_{\text{min}}$, we differentiate Equation 2.29:

$$\frac{du}{dr} = 4\varepsilon\left[-\frac{12\sigma^{12}}{r^{13}} + \frac{6\sigma^6}{r^7}\right] = 0$$

which gives $r_{\text{min}}^6 = 2\sigma^6$, or $r_{\text{min}} = 2^{1/6}\sigma$. Therefore,

$$u(r_{\text{min}}) = 4\varepsilon\left[\left(\frac{\sigma}{2^{1/6}\sigma}\right)^{12} - \left(\frac{\sigma}{2^{1/6}\sigma}\right)^{6}\right] = 4\varepsilon\left(\frac{1}{4} - \frac{1}{2}\right) = -\varepsilon$$

Thus $\varepsilon$ is the depth of the potential well, relative to the infinite separation.

If we substitute the Lennard-Jones potential into Equation 2.25, we obtain

$$B_{2V}(T) = -2\pi N_{\text{A}} \int_0^{\infty} \left[\exp\left\{-\frac{4\varepsilon}{k_{\text{B}}T}\left[\left(\frac{\sigma}{r}\right)^{12} - \left(\frac{\sigma}{r}\right)^{6}\right]\right\} - 1\right] r^2 dr \quad (2.30)$$

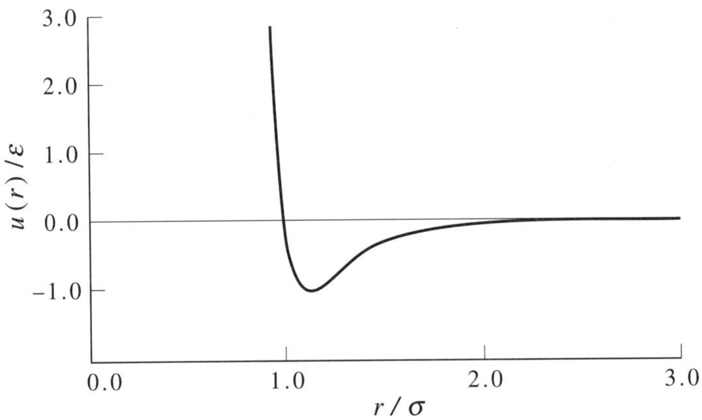


**FIGURE 2.14**
A plot of $u(r)/\varepsilon = 4\left[\left(\frac{\sigma}{r}\right)^{12} - \left(\frac{\sigma}{r}\right)^{6}\right]$ versus $r/\sigma$ for the Lennard-Jones potential. Note that the depth of the potential well is $\varepsilon$ and that $u(r) = 0$ at $r/\sigma = 1$.

**TABLE 2.7**
Lennard-Jones parameters, $\varepsilon$ and $\sigma$, for various substances.

| Species | $(\varepsilon/k_B)$/K | $\sigma$/pm | $(2\pi\sigma^3 N_A/3)/\mathrm{cm^3\cdot mol^{-1}}$ |
|---|---|---|---|
| He | 10.22 | 256 | 21.2 |
| Ne | 35.6 | 275 | 26.2 |
| Ar | 120 | 341 | 50.0 |
| Kr | 164 | 383 | 70.9 |
| Xe | 229 | 406 | 86.9 |
| $H_2$ | 37.0 | 293 | 31.7 |
| $N_2$ | 95.1 | 370 | 63.9 |
| $O_2$ | 118 | 358 | 57.9 |
| CO | 100 | 376 | 67.0 |
| $CO_2$ | 189 | 449 | 114.2 |
| $CF_4$ | 152 | 470 | 131.0 |
| $CH_4$ | 149 | 378 | 68.1 |
| $C_2H_4$ | 199 | 452 | 116.5 |
| $C_2H_6$ | 243 | 395 | 77.7 |
| $C_3H_8$ | 242 | 564 | 226.3 |
| $C(CH_3)_4$ | 232 | 744 | 519.4 |

Equation 2.30 may look complicated, but it can be simplified. We first define a reduced temperature $T^*$ by $T^* = k_B T/\varepsilon$ and let $r/\sigma = x$ to get

$$B_{2V}(T^*) = -2\pi\sigma^3 N_A \int_0^\infty \left[\exp\left\{-\frac{4}{T^*}(x^{-12} - x^{-6})\right\} - 1\right] x^2 dx$$

We then divide both sides by $2\pi\sigma^3 N_A/3$ to get

$$B^*_{2V}(T^*) = -3 \int_0^\infty \left[\exp\left\{-\frac{4}{T^*}(x^{-12} - x^{-6})\right\} - 1\right] x^2 dx \qquad (2.31)$$

where $B^*_{2V}(T^*) = B_{2V}(T^*)/(2\pi\sigma^3 N_A/3)$. Equation 2.31 shows that the reduced second virial coefficient, $B^*_{2V}(T^*)$, depends upon only the reduced temperature, $T^*$. The integral in Equation 2.31 must be evaluated numerically (MathChapter A) for each value of $T^*$. Extensive tables of $B^*_{2V}(T^*)$ versus $T^*$ are available.

Equation 2.31 is another example of the law of corresponding states. If we take experimental values of $B_{2V}(T)$, divide them by $2\pi\sigma^3 N_A/3$, and then plot the data versus $T^* = k_B T/\varepsilon$, the result for *all* gases will fall on one curve. Figure 2.15 shows such a plot for six gases. Conversely, a plot such as the one in Figure 2.15 (or better yet, numerical tables) can be used to evaluate $B_{2V}(T)$ for any gas.

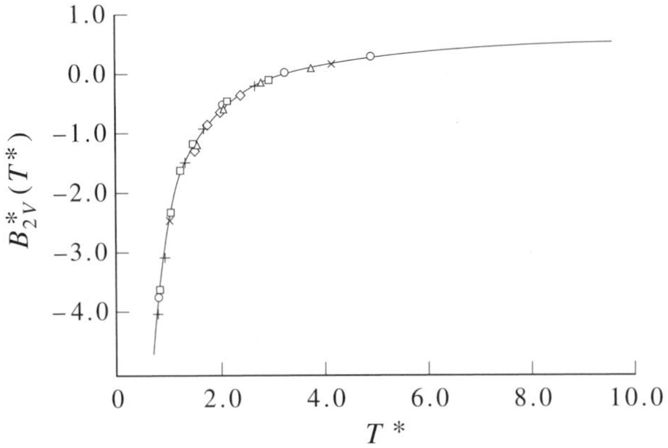


**FIGURE 2.15**
A plot of the reduced second virial coefficient $B^*_{2V}(T^*) = B_{2V}(T^*)/(2\pi\sigma^3 N_A/3)$ (solid line) against the reduced temperature $T^* = k_B T/\varepsilon$. Experimental data of six gases (argon, nitrogen, oxygen, carbon dioxide, and sulfur hexafluoride) are also plotted. This plot is another illustration of the law of corresponding states.

**EXAMPLE 2–10**
Estimate $B_{2V}(T)$ for $N_2(g)$ at 0°C.

SOLUTION: Table 2.7 gives $\varepsilon/k_B = 95.1$ K and $2\pi\sigma^3 N_A/3 = 63.9\ \text{cm}^3\cdot\text{mol}^{-1}$ for $N_2(g)$. Thus, $T^* = 2.87$, and Figure 2.15 gives $B^*_{2V}(T^*) \approx -0.2$. Therefore,

$$\begin{aligned} B_{2V}(T) &\approx (63.9\ \text{cm}^3\cdot\text{mol}^{-1})(-0.2) \\ &\approx -10\ \text{cm}^3\cdot\text{mol}^{-1} \end{aligned}$$

If we had used numerical tables for $B^*_{2V}(T^*)$ instead of Figure 2.15, we would have obtained $B^*_{2V}(T^*) = -0.16$, or $B_{2V}(T) = -10\ \text{cm}^3\cdot\text{mol}^{-1}$.

The value of $B_{2V}(T)$ has a simple interpretation. Consider Equation 2.23 under conditions where we can ignore the terms in $P^2$ and higher

$$\frac{P\overline{V}}{RT} = 1 + B_{2P}(T)P = 1 + \frac{B_{2V}(T)}{RT}P$$

By multiplying through by $RT/P$ and using $\overline{V}_{\text{ideal}} = RT/P$, we can rewrite this equation in the form

$$\overline{V} = \overline{V}_{\text{ideal}} + B_{2V}(T)$$

or

$$B_{2V}(T) = \overline{V} - \overline{V}_{\text{ideal}} \tag{2.32}$$

Thus, we see that $B_{2V}(T)$ represents the difference between the actual value of $\overline{V}$

and the ideal-gas value $\overline{V}_{\text{ideal}}$ at pressures such that the contribution of the third virial coefficient is negligible.

**EXAMPLE 2–11**

The molar volume of isobutane at 300.0 K and one bar is 24.31 $\text{dm}^3\cdot\text{mol}^{-1}$. Estimate the value of $B_{2V}$ for isobutane at 300.0 K.

SOLUTION: The ideal-gas molar volume at 300.0 K and one bar is

$$\overline{V}_{\text{ideal}} = \frac{RT}{P} = \frac{(0.083145\ \text{dm}^3\cdot\text{bar}\cdot\text{K}^{-1}\cdot\text{mol}^{-1})(300.0\ \text{K})}{1\ \text{bar}}$$
$$= 24.94\ \text{dm}^3\cdot\text{mol}^{-1}$$

Therefore, using Equation 2.32,

$$B_{2V} = \overline{V} - \overline{V}_{\text{ideal}} = 24.31\ \text{dm}^3\cdot\text{mol}^{-1} - 24.94\ \text{dm}^3\cdot\text{mol}^{-1}$$
$$= -0.63\ \text{dm}^3\cdot\text{mol}^{-1} = -630\ \text{cm}^3\cdot\text{mol}^{-1}$$

Although we have been discussing calculating $B_{2V}(T)$ in terms of the Lennard-Jones potential, in practice it's the other way around: Lennard-Jones parameters are usually determined from experimental values of $B_{2V}(T)$. This determination is usually made through trial and error using tables of $B^*_{2V}(T^*)$. The values of the Lennard-Jones parameters in Table 2.7 were determined from experimental second virial coefficient data. Because the second virial coefficient reflects the initial deviations from ideal behavior, which are caused by intermolecular interactions, experimental $P$–$V$–$T$ data turn out to be a rich source of information concerning intermolecular interactions. Once Lennard-Jones parameters have been determined, they can be used to calculate many other fluid properties such as viscosity, thermal conductivity, heats of vaporization, and various crystal properties.

## 2–6. London Dispersion Forces Are Often the Largest Contribution to the $r^{-6}$ Term in the Lennard-Jones Potential

In the previous section, we used the Lennard-Jones potential (Equation 2.29) to represent the intermolecular potential between molecules. The $r^{-12}$ term accounts for the repulsion at short distances, and the $r^{-6}$ term accounts for the attraction at larger distances. The actual form of the repulsive term is not well established, but the $r^{-6}$ dependence of the attractive term is. In this section, we will discuss three contributions to the $r^{-6}$ attraction and compare their relative importance.

Consider two dipolar molecules, whose dipole moments are $\mu_1$ and $\mu_2$. The interaction of these dipoles depends upon how they are oriented with respect to each other. The energy will vary from repulsive, when they are oriented head-to-head as shown

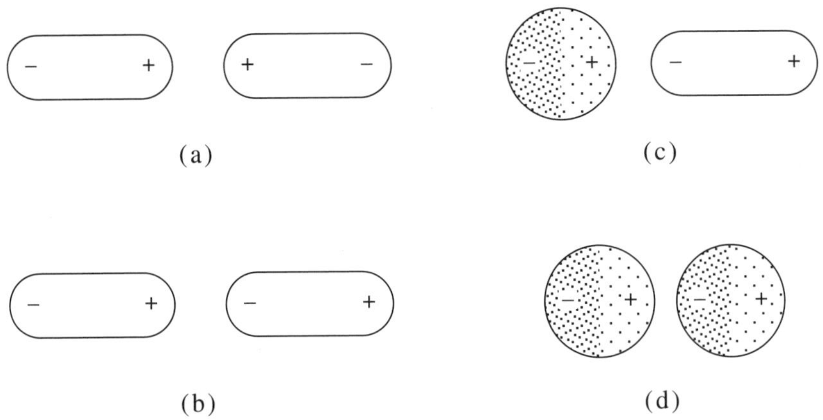


**FIGURE 2.16**
Two permanent dipoles oriented (a) head-to-head and (b) head-to-tail. The head-to-tail orientation is energetically favorable. (c) A molecule with a permanent dipole moment will induce a dipole moment in a neighboring molecule. (d) The instantaneous dipole-dipole correlation shown here is what leads to a London attraction between all atoms and molecules.

in Figure 2.16a to attractive, when they are oriented head-to-tail (Figure 2.16b). Both molecules rotate in the gas phase, and if we were to average both dipoles randomly over their orientations, the dipole-dipole interactions would average out to zero. Because different orientations have different energies, they do not occur to equal extents. Clearly, the lower-energy head-to-tail orientation is favored over the repulsive head-to-head orientation. If we take into account the energy of the orientation, then the overall average interaction between the two molecules results in an attractive $r^{-6}$ term of the form

$$u_{d.d}(r) = -\frac{2\mu_1^2\mu_2^2}{(4\pi\varepsilon_0)^2(3k_BT)}\frac{1}{r^6} \tag{2.33}$$

**EXAMPLE 2–12**
Show that the units of the right side of Equation 2.33 are energy.

SOLUTION: The units of $\mu$ are C·m (charge × separation), and so we have

$$u_{d.d}(r) \sim \frac{(\text{C}\cdot\text{m})^4}{(\text{C}^2\cdot\text{s}^2\cdot\text{kg}^{-1}\cdot\text{m}^{-3})^2\text{J m}^6}$$
$$\sim \frac{\text{kg}^2\cdot\text{m}^4\cdot\text{s}^{-4}}{\text{J}} = \text{J}$$

**EXAMPLE 2–13**

Calculate the value of the coefficient of $r^{-6}$ in Equation 2.33 at 300 K for two HCl(g) molecules. Table 2.8 lists the dipole moments of various molecules.

SOLUTION: According to Table 2.8, $\mu_1 = \mu_2 = 3.44 \times 10^{-30}$ C·m. Therefore,

$$-r^6 u_{d.d}(r) = \frac{(2)(3.44 \times 10^{-30}\ \text{C·m})^4}{(3)\left(\dfrac{8.314\ \text{J·mol}^{-1}\text{·K}^{-1}}{6.022 \times 10^{23}\ \text{mol}^{-1}}\right)(300\ \text{K})(1.113 \times 10^{-10}\ \text{C}^2\text{·s}^2\text{·kg}^{-1}\text{·m}^{-3})^2} = 1.82 \times 10^{-78}\ \text{J·m}^6$$

This numerical result may seem exceedingly small, but remember that we are calculating $-r^6 u_{d.d}(r)$. At a separation of 300 pm, $u_{d.d}(r)$ is equal to $-2.5 \times 10^{-21}$ J, compared with a thermal energy ($k_B T$) of $4.1 \times 10^{-21}$ J at 300 K.

Equation 2.33 requires that both molecules have a permanent dipole moment. Even if one molecule does not have a permanent dipole moment, the one without a permanent dipole moment will have a dipole moment induced by the other. A dipole moment can be induced in a molecule that does not have a permanent dipole moment because all atoms and molecules are *polarizable*. When an atom or a molecule interacts with an electric field, the (negative) electrons are displaced in one direction and the (positive) nuclei are displaced in the opposite direction, as illustrated in Figure 2.16c. This charge separation with its associated dipole moment, is proportional to the strength of the electric field, and if we designate the induced dipole moment by $\mu_{\text{induced}}$ and the electric field by $E$, we have that $\mu_{\text{induced}} \propto E$. The proportionality constant, which we denote by $\alpha$, is called the *polarizability*, so we have the defining expression

$$\mu_{\text{induced}} = \alpha E \tag{2.34}$$

The units of $E$ are V·m$^{-1}$, so the units of $\alpha$ in Equation 2.34 are C·m/V·m$^{-1}$ = C·m$^2$·V$^{-1}$. We can put $\alpha$ into more transparent units by using the fact that energy = (charge)$^2/4\pi\varepsilon_0$ (distance), which in SI units gives

$$\text{joule} \sim \frac{\text{C}^2}{(4\pi\varepsilon_0)\text{m}} = \text{C}^2\text{·m}^{-1}/4\pi\varepsilon_0$$

Similarly, from electrostatics, we have that

$$\text{joule} = \text{coulomb} \times \text{volt} = \text{C·V}$$

Equating these two expressions for joules gives C·V = C$^2$·m$^{-1}/4\pi\varepsilon_0$, or C·V$^{-1}$ = $(4\pi\varepsilon_0)$ m. Now we substitute this result into the above units for $\alpha$ (C·m$^2$·V$^{-1}$) to get

$$\alpha \sim (4\pi\varepsilon_0)\text{m}^3$$

Thus, we see that $\alpha/4\pi\varepsilon_0$ has units of m$^3$. The quantity $\alpha/4\pi\varepsilon_0$, which is sometimes referred to as the *polarizability volume*, has units of volume. The easier it is for the electric field to deform the atomic or molecular charge distribution, the greater is the polarizability. The polarizability of an atom or a molecule is proportional to its size (note the units of $\alpha/4\pi\varepsilon_0$), or to its number of electrons. This trend can be seen in Table 2.8, which lists the polarizability volumes of some atoms and molecules.

**TABLE 2.8**
The dipole moment ($\mu$), the polarizability volume ($\alpha/4\pi\varepsilon_0$), and the ionization energies ($I$) of various atoms and molecules.

| Species | $\mu/10^{-30}$ C·m | $(\alpha/4\pi\varepsilon_0)/10^{-30}$ m$^3$ | $I/10^{-18}$ J |
|---|---|---|---|
| He | 0 | 0.21 | 3.939 |
| Ne | 0 | 0.39 | 3.454 |
| Ar | 0 | 1.63 | 2.525 |
| Kr | 0 | 2.48 | 2.243 |
| Xe | 0 | 4.01 | 1.943 |
| $N_2$ | 0 | 1.77 | 2.496 |
| $CH_4$ | 0 | 2.60 | 2.004 |
| $C_2H_6$ | 0 | 4.43 | 1.846 |
| $C_3H_8$ | 0.03 | 6.31 | 1.754 |
| CO | 0.40 | 1.97 | 2.244 |
| $CO_2$ | 0 | 2.63 | 2.206 |
| HCl | 3.44 | 2.63 | 2.043 |
| HI | 1.47 | 5.42 | 1.664 |
| $NH_3$ | 5.00 | 2.23 | 1.628 |
| $H_2O$ | 6.14 | 1.47 | 2.020 |

We now return to the dipole-induced dipole interaction shown in Figure 2.16c. Because the induced dipole moment is always in a head-to-tail orientation with respect to the permanent dipole moment, the interaction is always attractive and is given by

$$u_{\text{induced}}(r) = -\frac{\mu_1^2\alpha_2}{(4\pi\varepsilon_0)^2 r^6} - \frac{\mu_2^2\alpha_1}{(4\pi\varepsilon_0)^2 r^6} \tag{2.35}$$

The first term represents a permanent dipole moment in molecule 1 and an induced dipole moment in molecule 2, and the second represents the opposite situation.

**EXAMPLE 2–14**
Calculate the value of the coefficient of $r^{-6}$ for $u_{\text{induced}}(r)$ for two HCl(g) molecules.

SOLUTION: The two terms in Equation 2.35 are the same for identical molecules. Using the data in Table 2.8,

$$
\begin{aligned}
-r^6 u_{\text{induced}}(r) &= \frac{2\mu^2(\alpha/4\pi\varepsilon_0)}{4\pi\varepsilon_0} \\
&= \frac{(2)(3.44 \times 10^{-30}\ \text{C}\cdot\text{m})^2(2.63 \times 10^{-30}\ \text{m}^3)}{1.113 \times 10^{-10}\ \text{C}^2\cdot\text{s}^2\cdot\text{kg}^{-1}\cdot\text{m}^{-3}} \\
&= 5.59 \times 10^{-79}\ \text{J}\cdot\text{m}^6
\end{aligned}
$$

Note that this result is about 30% of the result we obtained in Example 2–13 for $-r^6 u_{d.d}(r)$.

Both Equations 2.33 and 2.35 equal zero when neither molecule has a permanent dipole moment. The third contribution to the $r^{-6}$ term in Equation 2.29 is nonzero even if both molecules are nonpolar. This contribution was first calculated by the German scientist Fritz London in 1930 using quantum mechanics and is now called a *London dispersion attraction*. Although this attraction is a strictly quantum-mechanical effect, it lends itself to the following commonly used classical picture. Consider two atoms as shown in Figure 2.16d separated by a distance $r$. The electrons on one atom do not completely shield the high positive charge on the nucleus from the electrons on the other atom. Because the molecule is polarizable, the electronic wave function can distort a bit to further lower the interaction energy. If we average this electronic attraction quantum mechanically, we obtain an attractive term that varies as $r^{-6}$. The exact quantum-mechanical calculation is somewhat complicated, but an approximate form of the final result is

$$
u_{\text{disp}}(r) = -\frac{3}{2}\left(\frac{I_1 I_2}{I_1 + I_2}\right)\frac{\alpha_1\alpha_2}{(4\pi\varepsilon_0)^2}\frac{1}{r^6} \qquad (2.36)
$$

where $I_j$ is the ionization energy of atom or molecule $j$. Note that Equation 2.36 does not involve a permanent dipole moment and that the interaction energy is proportional to the product of the polarizability volumes. Thus, the importance of $u_{\text{disp}}(r)$ increases with the sizes of the atoms or molecules, and, in fact, is often the dominant contribution to the $r^{-6}$ term in Equation 2.29.

**EXAMPLE 2–15**

Calculate the value of the coefficient of $r^{-6}$ for $u_{\text{disp}}(r)$ for two HCl(g) molecules.

SOLUTION: Using the data in Table 2.8, we have

$$
\begin{aligned}
-r^6 u_{\text{disp}}(r) &= \frac{3}{2}\left(\frac{2.043 \times 10^{-18}\ \text{J}}{2}\right)(2.63 \times 10^{-30}\ \text{m}^3)^2 \\
&= 1.06 \times 10^{-77}\ \text{J}\cdot\text{m}^6
\end{aligned}
$$

This quantity is about six times greater than $-r^6 u_{d.d}(r)$ and 20 times greater than $-r^6 u_{\text{induced}}(r)$. Similar calculations show that the disperison term is significantly larger than either the dipole-dipole term or the dipole-induced dipole term except for very polar molecules such as $NH_3$, $H_2O$, and HCN.

The total contribution to the $r^{-6}$ term in the Lennard-Jones potential is given by the sum of Equations 2.33, 2.35, and 2.36, giving

$$u(r) = \frac{c_{12}}{r^{12}} - \frac{c_6}{r^6}$$

with (Problem 2–53)

$$c_6 = \frac{2\mu^4}{3(4\pi\varepsilon_0)^2 k_B T} + \frac{2\alpha\mu^2}{(4\pi\varepsilon_0)^2} + \frac{3}{4}\frac{I\alpha^2}{(4\pi\varepsilon_0)^2} \tag{2.37}$$

for identical atoms or molecules.

## 2–7. The van der Waals Constants Can Be Written in Terms of Molecular Parameters

Although the Lennard-Jones potential is fairly realistic, it is also difficult to use. For example, the second virial coefficient (Example 2–10) must be evaluated numerically and one must resort to numerical tables to calculate the properties of gases. Consequently, intermolecular potentials that can be evaluated analytically are often used to estimate the properties of gases. The simplest of these potentials is the so-called *hard-sphere potential* (Figure 2.17a), whose mathematical form is

$$u(r) = \begin{array}{ll} \infty & r < \sigma \\ 0 & r > \sigma \end{array} \tag{2.38}$$

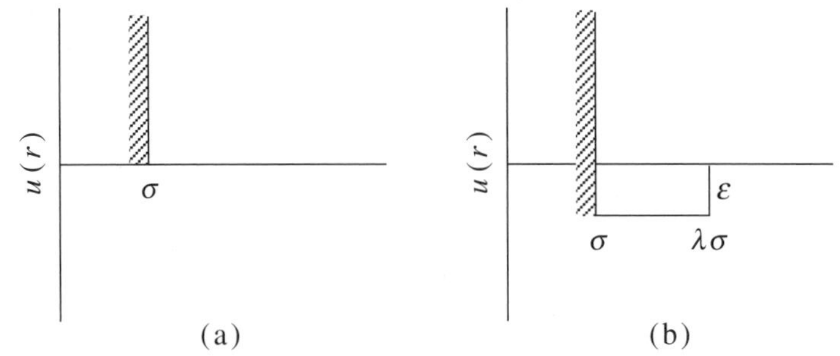


**FIGURE 2.17**
(a) A schematic illustration of a hard-sphere potential and (b) a square-well potential. The parameter $\sigma$ is the diameter of the molecules, $\varepsilon$ is the depth of the attractive well, and $(\lambda - 1)\sigma$ is the width of the well.

This potential represents hard spheres of diameter $\sigma$. Equation 2.38 depicts the repulsive region as varying infinitely steeply rather than as $r^{-12}$. As simplistic as this potential may seem, it does account for the finite size of molecules, which turns out to be the dominating feature in determining the structure of liquids and solids. Its obvious deficiency is the lack of any attractive term. At high temperatures, however, meaning high with respect to $\varepsilon/k_B$ in the Lennard-Jones potential, the molecules are traveling with enough energy that the attractive potential is significantly "washed out," so the hard-sphere potential is useful under these conditions.

The second virial coefficient is easy to evaluate for the hard sphere potential. Substituting Equation 2.38 into Equation 2.25 gives

$$\begin{aligned} B_{2V}(T) &= -2\pi N_A \int_0^\infty [e^{-u(r)/k_BT} - 1]r^2 dr \\ &= -2\pi N_A \int_0^\sigma [0 - 1]r^2 dr - 2\pi N_A \int_\sigma^\infty [e^0 - 1]r^2 dr \\ &= \frac{2\pi\sigma^3 N_A}{3} \end{aligned} \tag{2.39}$$

which is equal to four times the volume of $N_A$ spheres. (Remember that $\sigma$ is the diameter of the spheres.) Thus, the hard-sphere second virial coefficient is independent of temperature. Note that the high-temperature limit of the second virial coefficients shown in Figures 2.12 and 2.15 is fairly constant. The curves actually go through a slight maximum because molecules are not really "hard."

Another simple potential used fairly often is the *square-well potential* (Figure 2.17b):

$$u(r) = \begin{cases} \infty & r < \sigma \\ -\varepsilon & \sigma < r < \lambda\sigma \\ 0 & r > \lambda\sigma \end{cases} \tag{2.40}$$

The parameter $\varepsilon$ is the depth of the well and $(\lambda - 1)\sigma$ is its width. This potential provides an attractive region, as crude as it is. The second virial coefficient can be evaluated analytically for the square-well potential

$$\begin{aligned} B_{2V}(T) &= -2\pi N_A \int_0^\sigma [0 - 1]r^2 dr - 2\pi N_A \int_\sigma^{\lambda\sigma} [e^{\varepsilon/k_BT} - 1]r^2 dr \\ &\quad -2\pi N_A \int_{\lambda\sigma}^\infty [e^0 - 1]r^2 dr \\ &= \frac{2\pi\sigma^3 N_A}{3} - \frac{2\pi\sigma^3 N_A}{3}(\lambda^3 - 1)(e^{\varepsilon/k_BT} - 1) \\ &= \frac{2\pi\sigma^3 N_A}{3}[1 - (\lambda^3 - 1)(e^{\varepsilon/k_BT} - 1)] \end{aligned} \tag{2.41}$$

Note that Equation 2.41 reduces to Equation 2.39 when $\lambda = 1$ or $\varepsilon = 0$, there being no attractive well in either case. Figure 2.18 shows Equation 2.41 compared with

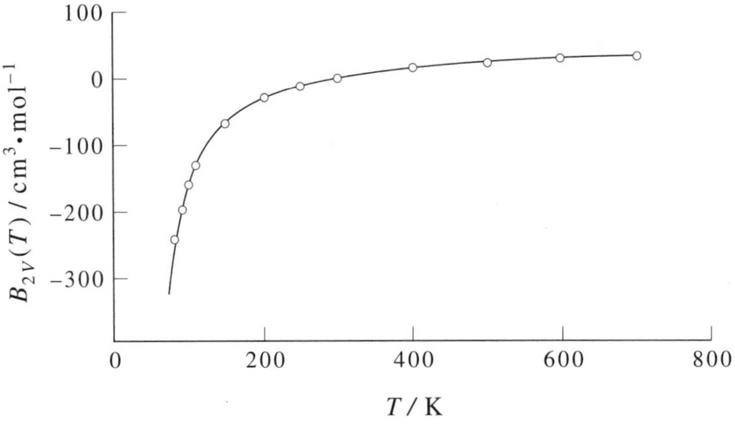


**FIGURE 2.18**
A comparison of the square-well second virial coefficient for nitrogen. The square-well parameters for nitrogen are $\sigma = 327.7$ pm, $\varepsilon/k_B = 95.2$ K, and $\lambda = 1.58$. The solid circles represent experimental data.

experimental data for nitrogen. The agreement is amazingly good, but the square-well potential does have three adjustable parameters.

We will finish this chapter with a discussion of the second virial coefficients for the three cubic equations of state introduced in Section 2–2. First, we write the van der Waals equation in the form

$$P = \frac{RT}{\overline{V} - b} - \frac{a}{\overline{V}^2}$$
$$= \frac{RT}{\overline{V}} \frac{1}{(1 - b/\overline{V})} - \frac{a}{\overline{V}^2} \tag{2.42}$$

We now use the binomial expansion of $1/(1-x)$ (MathChapter C),

$$\frac{1}{1-x} = 1 + x + x^2 + \cdots$$

to write Equation 2.42 as (letting $x = b/\overline{V}$)

$$P = \frac{RT}{\overline{V}}\left[1 + \frac{b}{\overline{V}} + \frac{b^2}{\overline{V}^2} + \cdots\right] - \frac{a}{\overline{V}^2}$$
$$= \frac{RT}{\overline{V}} + (RTb - a)\frac{1}{\overline{V}^2} + \frac{RTb^2}{\overline{V}^3} + \cdots$$

or

$$Z = \frac{P\overline{V}}{RT} = 1 + \left(b - \frac{a}{RT}\right)\frac{1}{\overline{V}} + \frac{b^2}{\overline{V}^2} + \cdots$$

Comparing this result with Equation 2.22, we see that

$$B_{2V}(T) = b - \frac{a}{RT} \tag{2.43}$$

for the van der Waals equation. We will now derive a similar result from Equation 2.25 and interpret $a$ and $b$ in terms of molecular parameters. The intermolecular potential that we will use is a hybrid of the hard-sphere potential and the Lennard-Jones potential

$$u(r) = \begin{cases} \infty & r < \sigma \\ -\dfrac{c_6}{r^6} & r > \sigma \end{cases} \tag{2.44}$$

We substitute this potential into Equation 2.25 to obtain

$$B_{2V}(T) = -2\pi N_A \int_0^\sigma (-1)r^2 dr - 2\pi N_A \int_\sigma^\infty [e^{c_6/k_B T r^6} - 1]r^2 dr$$

In the second integral, we assume that $c_6/k_B T r^6 \ll 1$ and use the expansion for $e^x$ (MathChapter C)

$$e^x = 1 + x + \frac{x^2}{2!} + \cdots$$

and keep only the first two terms to obtain

$$\begin{aligned} B_{2V}(T) &= \frac{2\pi\sigma^3 N_A}{3} - \frac{2\pi N_A c_6}{k_B T} \int_\sigma^\infty \frac{r^2 dr}{r^6} \\ &= \frac{2\pi\sigma^3 N_A}{3} - \frac{2\pi N_A c_6}{3k_B T \sigma^3} \end{aligned} \tag{2.45}$$

Comparing this result with Equation 2.43 gives

$$a = \frac{2\pi N_A^2 c_6}{3\sigma^3} \quad \text{and} \quad b = \frac{2\pi\sigma^3 N_A}{3}$$

Thus, we see that $a$ is directly proportional to $c_6$, the coefficient of $r^{-6}$ in the intermolecular potential, and that $b$ is equal to four times the volume of the molecules. From a molecular point of view, the van der Waals equation is based on an intermolecular potential that is a hard-sphere potential at small distances and a weak attractive potential (such that $c_6/k_B T r^6 \ll 1$) at larger distances.

In a similar fashion (Problem 2–55), the second virial coefficient for the Redlich-Kwong equation is

$$B_{2V}(T) = B - \frac{A}{RT^{3/2}} \tag{2.46}$$

and the second virial coefficient for the Peng-Robinson equation is (Problem 2–56)

$$B_{2V}(T) = \beta - \frac{\alpha}{RT} \tag{2.47}$$

The second virial coefficient from the van der Waals equation and the Peng-Robinson equation have the same functional form, but they have different numerical values because the values of the constants are different. Also, the parameter $\alpha$ is a function of temperature in the Peng-Robinson equation.

## Problems

**2-1.** In an issue of the journal *Science* a few years ago, a research group discussed experiments in which they determined the structure of cesium iodide crystals at a pressure of 302 gigapascals (GPa). How many atmospheres and bars is this pressure?

**2-2.** In meteorology, pressures are expressed in units of millibars (mbar). Convert 985 mbar to torr and to atmospheres.

**2-3.** Calculate the value of the pressure (in atm) exerted by a 33.9-foot column of water. Take the density of water to be 1.00 $\text{g}\cdot\text{mL}^{-1}$.

**2-4.** At which temperature are the Celsius and Farenheit temperature scales equal?

**2-5.** A travel guide says that to convert Celsius temperatures to Farenheit temperatures, double the Celsius temperature and add 30. Comment on this recipe.

**2-6.** Research in surface science is carried out using ultra-high vacuum chambers that can sustain pressures as low as $10^{-12}$ torr. How many molecules are there in a 1.00-$\text{cm}^3$ volume inside such an apparatus at 298 K? What is the corresponding molar volume $\overline{V}$ at this pressure and temperature?

**2-7.** Use the following data for an unknown gas at 300 K to determine the molecular mass of the gas.

| $P$/bar | 0.1000 | 0.5000 | 1.000 | 1.01325 | 2.000 |
|---|---|---|---|---|---|
| $\rho/\text{g}\cdot\text{L}^{-1}$ | 0.1771 | 0.8909 | 1.796 | 1.820 | 3.652 |

**2-8.** Recall from general chemistry that Dalton's law of partial pressures says that each gas in a mixture of ideal gases acts as if the other gases were not present. Use this fact to show that the partial pressure exerted by each gas is given by

$$P_j = \left( \frac{n_j}{\sum n_j} \right) P_{\text{total}} = y_j P_{\text{total}}$$

where $P_j$ is the partial pressure of the $j$th gas and $y_j$ is its mole fraction.

**2-9.** A mixture of $H_2(g)$ and $N_2(g)$ has a density of 0.216 $\text{g}\cdot\text{L}^{-1}$ at 300 K and 500 torr. What is the mole fraction composition of the mixture?

**2-10.** One liter of $N_2(g)$ at 2.1 bar and two liters of Ar(g) at 3.4 bar are mixed in a 4.0-L flask to form an ideal-gas mixture. Calculate the value of the final pressure of the mixture if the initial and final temperature of the gases are the same. Repeat this calculation if the initial temperatures of the $N_2(g)$ and Ar(g) are 304 K and 402 K, respectively, and the final temperature of the mixture is 377 K. (Assume ideal-gas behavior.)

**2-11.** It takes 0.3625 g of nitrogen to fill a glass container at 298.2 K and 0.0100 bar pressure. It takes 0.9175 g of an unknown homonuclear diatomic gas to fill the same bulb under the same conditions. What is this gas?

**2-12.** Calculate the value of the molar gas constant in units of $dm^3 \cdot torr \cdot K^{-1} \cdot mol^{-1}$.

**2-13.** Use the van der Waals equation to plot the compressibility factor, $Z$, against $P$ for methane for $T = 180$ K, 189 K, 190 K, 200 K, and 250 K. *Hint*: Calculate $Z$ as a function of $\overline{V}$ and $P$ as a function of $\overline{V}$, and then plot $Z$ versus $P$.

**2-14.** Use the Redlich-Kwong equation to plot the compressibility factor, $Z$, against $P$ for methane for $T = 180$ K, 189 K, 190 K, 200 K, and 250 K. *Hint*: Calculate $Z$ as a function of $\overline{V}$ and $P$ as a function of $\overline{V}$, and then plot $Z$ versus $P$.

**2-15.** Use both the van der Waals and the Redlich-Kwong equations to calculate the molar volume of CO at 200 K and 1000 bar. Compare your result to the result you would get using the ideal-gas equation. The experimental value is $0.04009 \text{ L} \cdot \text{mol}^{-1}$.

**2-16.** Compare the pressures given by (a) the ideal-gas equation, (b) the van der Waals equation, (c) the Redlich-Kwong equation, and (d) the Peng-Robinson equation for propane at 400 K and $\rho = 10.62 \text{ mol} \cdot \text{dm}^{-3}$. The experimental value is 400 bar. Take $\alpha = 9.6938 \text{ L}^2 \cdot \text{bar} \cdot \text{mol}^{-2}$ and $\beta = 0.05632 \text{ L} \cdot \text{mol}^{-1}$ for the Peng-Robinson equation.

**2-17.** Use the van der Waals equation and the Redlich-Kwong equation to calculate the value of the pressure of one mole of ethane at 400.0 K confined to a volume of $83.26 \text{ cm}^3$. The experimental value is 400 bar.

**2-18.** Use the van der Waals equation and the Redlich-Kwong equation to calculate the molar density of one mole of methane at 500 K and 500 bar. The experimental value is $10.06 \text{ mol} \cdot \text{L}^{-1}$.

**2-19.** Use the Redlich-Kwong equation to calculate the pressure of methane at 200 K and a density of $27.41 \text{ mol} \cdot \text{L}^{-1}$. The experimental value is 1600 bar. What does the van der Waals equation give?

**2-20.** The pressure of propane versus density at 400 K can be fit by the expression

$$
\begin{aligned}
P/\text{bar} = {} & 33.258(\rho/\text{mol} \cdot \text{L}^{-1}) - 7.5884(\rho/\text{mol} \cdot \text{L}^{-1})^2 \\
& + 1.0306(\rho/\text{mol} \cdot \text{L}^{-1})^3 - 0.058757(\rho/\text{mol} \cdot \text{L}^{-1})^4 \\
& - 0.0033566(\rho/\text{mol} \cdot \text{L}^{-1})^5 + 0.00060696(\rho/\text{mol} \cdot \text{L}^{-1})^6
\end{aligned}
$$

for $0 \leq \rho/\text{mol} \cdot \text{L}^{-1} \leq 12.3$. Use the van der Waals equation and the Redlich-Kwong equation to calculate the pressure for $\rho = 0 \text{ mol} \cdot \text{L}^{-1}$ up to $12.3 \text{ mol} \cdot \text{L}^{-1}$. Plot your results. How do they compare to the above expression?

**2-21.** The Peng-Robinson equation is often superior to the Redlich-Kwong equation for temperatures near the critical temperature. Use these two equations to calculate the pressure of $CO_2(g)$ at a density of $22.0 \text{ mol} \cdot \text{L}^{-1}$ at 280 K [the critical temperature of $CO_2(g)$ is 304.2 K]. Use $\alpha = 4.192 \text{ bar} \cdot \text{L}^2 \cdot \text{mol}^{-2}$ and $\beta = 0.02665 \text{ L} \cdot \text{mol}^{-1}$ for the Peng-Robinson equation.

**2-22.** Show that the van der Waals equation for argon at $T = 142.69$ K and $P = 35.00$ atm can be written as

$$\overline{V}^3 - 0.3664\,\overline{V}^2 + 0.03802\,\overline{V} - 0.001210 = 0$$

where, for convenience, we have supressed the units in the coefficients. Use the Newton-Raphson method (MathChapter A) to find the three roots to this equation, and calculate the values of the density of liquid and vapor in equilibrium with each other under these conditions.

**2-23.** Use the Redlich-Kwong equation and the Peng-Robinson equation to calculate the densities of the coexisting argon liquid and vapor phases at 142.69 K and 35.00 atm. Use the Redlich-Kwong constants given in Table 2.4 and take $\alpha = 1.4915\ \text{atm}\cdot\text{L}^2\cdot\text{mol}^{-2}$ and $\beta = 0.01981\ \text{L}\cdot\text{mol}^{-1}$ for the Peng-Robinson equation.

**2-24.** Butane liquid and vapor coexist at 370.0 K and 14.35 bar. The densities of the liquid and vapor phases are $8.128\ \text{mol}\cdot\text{L}^{-1}$ and $0.6313\ \text{mol}\cdot\text{L}^{-1}$, respectively. Use the van der Waals equation, the Redlich-Kwong equation, and the Peng-Robinson equation to calculate these densities. Take $\alpha = 16.44\ \text{bar}\cdot\text{L}^2\cdot\text{mol}^{-2}$ and $\beta = 0.07245\ \text{L}\cdot\text{mol}^{-1}$ for the Peng-Robinson equation.

**2-25.** Another way to obtain expressions for the van der Waals constants in terms of critical parameters is to set $(\partial P/\partial \overline{V})_T$ and $(\partial^2 P/\partial \overline{V}^2)_T$ equal to zero at the critical point. Why are these quantities equal to zero at the critical point? Show that this procedure leads to Equations 2.12 and 2.13.

**2-26.** Show that the Redlich-Kwong equation can be written in the form

$$\overline{V}^3 - \frac{RT}{P}\overline{V}^2 - \left(B^2 + \frac{BRT}{P} - \frac{A}{PT^{1/2}}\right)\overline{V} - \frac{AB}{PT^{1/2}} = 0$$

Now compare this equation with $(\overline{V} - \overline{V}_c)^3 = 0$ to get

$$3\overline{V}_c = \frac{RT_c}{P_c} \tag{1}$$

$$3\overline{V}_c^2 = \frac{A}{P_c T_c^{1/2}} - \frac{BRT_c}{P_c} - B^2 \tag{2}$$

and

$$\overline{V}_c^3 = \frac{AB}{P_c T_c^{1/2}} \tag{3}$$

Note that Equation 1 gives

$$\frac{P_c \overline{V}_c}{RT_c} = \frac{1}{3} \tag{4}$$

Now solve Equation 3 for $A$ and substitute the result and Equation 4 into Equation 2 to obtain

$$B^3 + 3\overline{V}_c B^2 + 3\overline{V}_c^2 B - \overline{V}_c^3 = 0 \tag{5}$$

Divide this equation by $\overline{V}_c^3$ and let $B/\overline{V}_c = x$ to get

$$x^3 + 3x^2 + 3x - 1 = 0$$

Solve this cubic equation by the Newton-Raphson method (MathChapter A) to obtain $x = 0.25992$, or

$$B = 0.25992\overline{V}_c \tag{6}$$

Now substitute this result and Equation 4 into Equation 3 to obtain

$$A = 0.42748\frac{R^2T_c^{5/2}}{P_c}$$

**2-27.** Use the results of the previous problem to derive Equations 2.14.

**2-28.** Write the Peng-Robinson equation as a cubic polynomial equation in $\overline{V}$ (with the coefficient of $\overline{V}^3$ equal to one), and compare it with $(\overline{V} - \overline{V}_c)^3 = 0$ at the critical point to obtain

$$\frac{RT_c}{P_c} - \beta = 3\overline{V}_c \tag{1}$$

$$\frac{\alpha_c}{P_c} - 3\beta^2 - 2\beta\frac{RT_c}{P_c} = 3\overline{V}_c^2 \tag{2}$$

and

$$\frac{\alpha_c\beta}{P_c} - \beta^2\frac{RT_c}{P_c} - \beta^3 = \overline{V}_c^3 \tag{3}$$

(We write $\alpha_c$ because $\alpha$ depends upon the temperature.) Now eliminate $\alpha_c/P_c$ between Equations 2 and 3, and then use Equation 1 for $\overline{V}_c$ to obtain

$$64\beta^3 + 6\beta^2\frac{RT_c}{P_c} + 12\beta\left(\frac{RT_c}{P_c}\right)^2 - \left(\frac{RT_c}{P_c}\right)^3 = 0$$

Let $\beta/(RT_c/P_c) = x$ and get

$$64x^3 + 6x^2 + 12x - 1 = 0$$

Solve this equation using the Newton-Raphson method to obtain

$$\beta = 0.077796\frac{RT_c}{P_c}$$

Substitute this result and Equation 1 into Equation 2 to obtain

$$\alpha_c = 0.45724\frac{(RT_c)^2}{P_c}$$

Last, use Equation 1 to show that

$$\frac{P_c\overline{V}_c}{RT_c} = 0.30740$$

**2-29.** Look up the boiling points of the gases listed in Table 2.5 and plot these values versus the critical temperatures $T_c$. Is there any correlation? Propose a reason to justify your conclusions from the plot.

**2-30.** Show that the compressibility factor $Z$ for the Redlich-Kwong equation can be written as in Equation 2.21.

**2-31.** Use the following data for ethane and argon at $T_R = 1.64$ to illustrate the law of corresponding states by plotting $Z$ against $\overline{V}_R$.

| Ethane ($T = 500$ K) | | Argon ($T = 247$ K) | |
|---|---|---|---|
| $P$/bar | $\overline{V}/\text{L}\cdot\text{mol}^{-1}$ | $P$/atm | $\overline{V}/\text{L}\cdot\text{mol}^{-1}$ |
| 0.500 | 83.076 | 0.500 | 40.506 |
| 2.00 | 20.723 | 2.00 | 10.106 |
| 10.00 | 4.105 | 10.00 | 1.999 |
| 20.00 | 2.028 | 20.00 | 0.9857 |
| 40.00 | 0.9907 | 40.00 | 0.4795 |
| 60.00 | 0.6461 | 60.00 | 0.3114 |
| 80.00 | 0.4750 | 80.00 | 0.2279 |
| 100.0 | 0.3734 | 100.0 | 0.1785 |
| 120.0 | 0.3068 | 120.0 | 0.1462 |
| 160.0 | 0.2265 | 160.0 | 0.1076 |
| 200.0 | 0.1819 | 200.0 | 0.08630 |
| 240.0 | 0.1548 | 240.0 | 0.07348 |
| 300.0 | 0.1303 | 300.0 | 0.06208 |
| 350.0 | 0.1175 | 350.0 | 0.05626 |
| 400.0 | 0.1085 | 400.0 | 0.05219 |
| 450.0 | 0.1019 | 450.0 | 0.04919 |
| 500.0 | 0.09676 | 500.0 | 0.04687 |
| 600.0 | 0.08937 | 600.0 | 0.04348 |
| 700.0 | 0.08421 | 700.0 | 0.04108 |

**2-32.** Use the data in Problem 2–31 to illustrate the law of corresponding states by plotting $Z$ against $P_R$.

**2-33.** Use the data in Problem 2.31 to test the quantitative reliability of the van der Waals equation by comparing a plot of $Z$ versus $\overline{V}_R$ from Equation 2.20 to a similar plot of the data.

**2-34.** Use the data in Problem 2.31 to test the quantitative reliability of the Redlich-Kwong equation by comparing a plot of $Z$ versus $\overline{V}_R$ from Equation 2.21 to a similar plot of the data.

**2-35.** Use Figure 2.10 to estimate the molar volume of CO at 200 K and 180 bar. An accurate experimental value is 78.3 $\text{cm}^3\cdot\text{mol}^{-1}$.

**2-36.** Show that $B_{2V}(T) = RTB_{2P}(T)$ (see Equation 2.24).

**2-37.** Use the following data for $NH_3$(g) at 273 K to determine $B_{2P}(T)$ at 273 K.

| $P$/bar | 0.10 | 0.20 | 0.30 | 0.40 | 0.50 | 0.60 | 0.70 |
|---|---|---|---|---|---|---|---|
| $(Z-1)/10^{-4}$ | 1.519 | 3.038 | 4.557 | 6.071 | 7.583 | 9.002 | 10.551 |

**2-38.** The density of oxygen as a function of pressure at 273.15 K is listed below.

| $P$/atm | 0.2500 | 0.5000 | 0.7500 | 1.0000 |
|---|---|---|---|---|
| $\rho/\text{g}\cdot\text{dm}^{-3}$ | 0.356985 | 0.714154 | 1.071485 | 1.428962 |

Use the data to determine $B_{2V}(T)$ of oxygen. Take the atomic mass of oxygen to be 15.9994 and the value of the molar gas constant to be 8.31451 $\text{J}\cdot\text{K}^{-1}\cdot\text{mol}^{-1}$ = 0.0820578 $\text{dm}^3\cdot\text{atm}\cdot\text{K}^{-1}\cdot\text{mol}^{-1}$.

**2-39.** Show that the Lennard-Jones potential can be written as

$$u(r) = \varepsilon\left(\frac{r^*}{r}\right)^{12} - 2\varepsilon\left(\frac{r^*}{r}\right)^{6}$$

where $r^*$ is the value of $r$ at which $u(r)$ is a minimum.

**2-40.** Using the Lennard-Jones parameters given in Table 2.7, compare the depth of a typical Lennard-Jones potential to the strength of a covalent bond.

**2-41.** Compare the Lennard-Jones potentials of $H_2$(g) and $O_2$(g) by plotting both on the same graph.

**2-42.** Use the data in Tables 2.5 and 2.7 to show that *roughly* $\epsilon/k = 0.75\,T_c$ and $b_0 = 0.7\,\overline{V}_c$. Thus, critical constants can be used as rough, first estimates of $\epsilon$ and $b_0$ $(= 2\pi N_0\sigma^3/3)$.

**2-43.** Prove that the second virial coefficient calculated from a general intermolecular potential of the form

$$u(r) = (\text{energy parameter}) \times f\left(\frac{r}{\text{distance parameter}}\right)$$

rigorously obeys the law of corresponding states. Does the Lennard-Jones potential satisfy this condition?

**2-44.** Use the following data for argon at 300.0 K to determine the value of $B_{2V}$. The accepted value is $-15.05\ \text{cm}^3\cdot\text{mol}^{-1}$.

| $P$/atm | $\rho$/mol·L$^{-1}$ | $P$/atm | $\rho$/mol·L$^{-1}$ |
|---|---|---|---|
| 0.01000 | 0.000406200 | 0.4000 | 0.0162535 |
| 0.02000 | 0.000812500 | 0.6000 | 0.0243833 |
| 0.04000 | 0.00162500 | 0.8000 | 0.0325150 |
| 0.06000 | 0.00243750 | 1.000 | 0.0406487 |
| 0.08000 | 0.00325000 | 1.500 | 0.0609916 |
| 0.1000 | 0.00406260 | 2.000 | 0.0813469 |
| 0.2000 | 0.00812580 | 3.000 | 0.122094 |

**2-45.** Using Figure 2.15 and the Lennard-Jones parameters given in Table 2.7, estimate $B_{2V}(T)$ for $CH_4$(g) at 0°C.

**2-46.** Show that $B_{2V}(T)$ obeys the law of corresponding states for a square-well potential with a *fixed* value of $\lambda$ (in other words, if all molecules had the same value of $\lambda$).

**2-47.** Using the Lennard-Jones parameters in Table 2.7, show that the following second virial cofficient data satisfy the law of corresponding states.

| Argon | | Nitrogen | | Ethane | |
|---|---|---|---|---|---|
| $T$/K | $B_{2V}(T)$ /$10^{-3}$ dm$^3$·mol$^{-1}$ | $T$/K | $B_{2V}(T)$ /$10^{-3}$ dm$^3$·mol$^{-1}$ | $T$/K | $B_{2V}(T)$ /$10^{-3}$ dm$^3$·mol$^{-1}$ |
| 173 | −64.3 | 143 | −79.8 | 311 | −164.9 |
| 223 | −37.8 | 173 | −51.9 | 344 | −132.5 |
| 273 | −22.1 | 223 | −26.4 | 378 | −110.0 |
| 323 | −11.0 | 273 | −10.3 | 411 | −90.4 |
| 423 | +1.2 | 323 | −0.3 | 444 | −74.2 |
| 473 | 4.7 | 373 | +6.1 | 478 | −59.9 |
| 573 | 11.2 | 423 | 11.5 | 511 | −47.4 |
| 673 | 15.3 | 473 | 15.3 | | |
| | | 573 | 20.6 | | |
| | | 673 | 23.5 | | |

**2-48.** In Section 2–4, we expressed the van der Waals equation in reduced units by dividing $P$, $\overline{V}$, and $T$ by their critical values. This suggests we can write the second virial coefficient in reduced form by dividing $B_{2V}(T)$ by $\overline{V}_c$ and $T$ by $T_c$ (instead of $2\pi N_A\sigma^3/3$ and $\varepsilon/k$ as we did in Section 2–5). Reduce the second virial coefficient data given in the previous problem by using the values of $\overline{V}_c$ and $T_c$ in Table 2.5 and show that the reduced data satisfy the law of corresponding states.

**2-49.** Listed below are experimental second virial coefficient data for argon, krypton, and xenon.

| | $B_{2V}(T)/10^{-3}\text{dm}^3\cdot\text{mol}^{-1}$ | | |
|---|---|---|---|
| $T$/K | Argon | Krypton | Xenon |
| 173.16 | −63.82 | | |
| 223.16 | −36.79 | | |
| 273.16 | −22.10 | −62.70 | −154.75 |
| 298.16 | −16.06 | | −130.12 |
| 323.16 | −11.17 | −42.78 | −110.62 |
| 348.16 | −7.37 | | − 95.04 |
| 373.16 | −4.14 | −29.28 | − 82.13 |
| 398.16 | −0.96 | | |
| 423.16 | +1.46 | −18.13 | − 62.10 |
| 473.16 | 4.99 | −10.75 | − 46.74 |
| 573.16 | 10.77 | +0.42 | − 25.06 |
| 673.16 | 15.72 | 7.42 | − 9.56 |
| 773.16 | 17.76 | 12.70 | − 0.13 |
| 873.16 | 19.48 | 17.19 | + 7.95 |
| 973.16 | | | 14.22 |

Use the Lennard-Jones parameters in Table 2.7 to plot $B^*_{2V}(T^*)$, the reduced second virial coefficient, versus $T^*$, the reduced temperature, to illustrate the law of corresponding states.

**2-50.** Use the critical temperatures and the critical molar volumes of argon, krypton, and xenon to illustrate the law of corresponding states with the data given in Problem 2–49.

**2-51.** Evaluate $B^*_{2V}(T^*)$ in Equation 2.31 numerically from $T^* = 1.00$ to 10.0 using a packaged numerical integration program such as *MathCad* or *Mathematica*. Compare the reduced second virial coefficient data from Problem 2–49 and $B^*_{2V}(T^*)$ by plotting them all on the same graph.

**2-52.** Show that the units of the right side of Equation 2.35 are energy.

**2-53.** Show that the sum of Equations 2.33, 2.35, and 2.36 gives Equation 2.37.

**2-54.** Compare the values of the coefficient of $r^{-6}$ for $N_2(g)$ using Equation 2.37 and the Lennard-Jones parameters given in Table 2.7.

**2-55.** Show that

$$B_{2V}(T) = B - \frac{A}{RT^{3/2}}$$

and

$$B_{3V}(T) = B^2 + \frac{AB}{RT^{3/2}}$$

for the Redlich-Kwong equation.

**2-56.** Show that the second and third virial coefficients of the Peng-Robinson equation are

$$B_{2V}(T) = \beta - \frac{\alpha}{RT}$$

and

$$B_{3V}(T) = \beta^2 + \frac{2\alpha\beta}{RT}$$

**2-57.** The square-well parameters for krypton are $\varepsilon/k = 136.5\text{ K}$, $\sigma = 327.8\text{ pm}$, and $\lambda = 1.68$. Plot $B_{2V}(T)$ against $T$ and compare your results with the data given in Problem 2–49.

**2-58.** The coefficient of thermal expansion $\alpha$ is defined as

$$\alpha = \frac{1}{\overline{V}}\left(\frac{\partial \overline{V}}{\partial T}\right)_P$$

Show that

$$\alpha = \frac{1}{T}$$

for an ideal gas.

**2-59.** The isothermal compressibility $\kappa$ is defined as

$$\kappa = -\frac{1}{\overline{V}}\left(\frac{\partial \overline{V}}{\partial P}\right)_T$$

Show that

$$\kappa = \frac{1}{P}$$

for an ideal gas.

MATHCHAPTER B

# PROBABILITY AND STATISTICS

In many of the following chapters, we will deal with probability distributions, average values, and standard deviations. Consequently, we take a few pages here to discuss some basic ideas of probability and show how to calculate average quantities in general.

Consider some experiment, such as the tossing of a coin or the rolling of a die, that has $n$ possible outcomes, each with probability $p_j$, where $j = 1, 2, \ldots, n$. If the experiment is repeated indefinitely, we intuitively expect that

$$p_j = \lim_{N\to\infty} \frac{N_j}{N} \qquad j = 1, 2, \ldots, n \tag{B.1}$$

where $N_j$ is the number of times that the outcome $j$ occurs and $N$ is the total number of repetitions of the experiment. Because $0 \le N_j \le N$, $p_j$ must satisfy the condition

$$0 \le p_j \le 1 \tag{B.2}$$

When $p_j = 1$, we say the event $j$ is a certainty and when $p_j = 0$, we say it is impossible. In addition, because

$$\sum_{j=1}^{n} N_j = N$$

we have the normalization condition,

$$\sum_{j=1}^{n} p_j = 1 \tag{B.3}$$

Equation B.3 means that the probability that some event occurs is a certainty. Suppose now that some number $x_j$ is associated with the outcome $j$. Then we define the *average* of $x$ or the *mean* of $x$ to be

$$\langle x \rangle = \sum_{j=1}^{n} x_j p_j = \sum_{j=1}^{n} x_j p(x_j) \tag{B.4}$$

where in the last term we have used the expanded notation $p(x_j)$, meaning the probability of realizing the number $x_j$. We will denote an average of a quantity by enclosing the quantity in angular brackets.

**EXAMPLE B–1**

Suppose we are given the following data:

| $x$ | $p(x)$ |
|---|---|
| 1 | 0.20 |
| 3 | 0.25 |
| 4 | 0.55 |

Calculate the average value of $x$.

SOLUTION: Using Equation B.4, we have

$$\langle x \rangle = (1)(0.20) + (3)(0.25) + (4)(0.55) = 3.15$$

It is helpful to interpret a probability distribution like $p_j$ as a distribution of a unit mass along the $x$ axis in a discrete manner such that $p_j$ is the fraction of mass located at the point $x_j$. This interpretation is shown in Figure B.1. According to this interpretation, the average value of $x$ is the center of mass of this system.

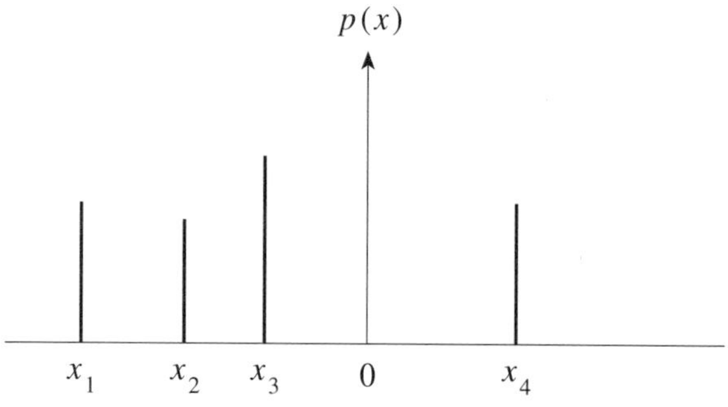


**FIGURE B.1**
The discrete probability frequency function or probability density, $p(x)$.

Another quantity of importance is

$$\langle x^2 \rangle = \sum_{j=1}^{n} x_j^2 p_j \tag{B.5}$$

The quantity $\langle x^2 \rangle$ is called the *second moment* of the distribution $\{p_j\}$ and is analogous to the moment of inertia.

**EXAMPLE B–2**

Calculate the second moment of the data given in Example B–1.

SOLUTION: Using Equation B.5, we have

$$\langle x^2 \rangle = (1)^2(0.20) + (3)^2(0.25) + (4)^2(0.55) = 11.25$$

Note from Examples B–1 and B–2 that $\langle x^2 \rangle \neq \langle x \rangle^2$. This nonequality is a general result that we will prove below.

A physically more interesting quantity than $\langle x^2 \rangle$ is the *second central moment*, or the *variance*, defined by

$$\sigma_x^2 = \langle (x - \langle x \rangle)^2 \rangle = \sum_{j=1}^{n} (x_j - \langle x \rangle)^2 p_j \tag{B.6}$$

As the notation suggests, we denote the square root of the quantity in Equation B.6 by $\sigma_x$, which is called the *standard deviation*. From the summation in Equation B.6, we can see that $\sigma_x^2$ will be large if $x_j$ is likely to differ from $\langle x \rangle$, because in that case $(x_j - \langle x \rangle)$ and so $(x_j - \langle x \rangle)^2$ will be large for the significant values of $p_j$. On the other hand, $\sigma_x^2$ will be small if $x_j$ is not likely to differ from $\langle x \rangle$, or if the $x_j$ cluster around $\langle x \rangle$, because then $(x_j - \langle x \rangle)^2$ will be small for the significant values of $p_j$. Thus, we see that either the variance or the standard deviation is a measure of the spread of the distribution about its mean.

Equation B.6 shows that $\sigma_x^2$ is a sum of positive terms, and so $\sigma_x^2 \geq 0$. Furthermore,

$$\begin{aligned}\sigma_x^2 &= \sum_{j=1}^{n} (x_j - \langle x \rangle)^2 p_j = \sum_{j=1}^{n} (x_j^2 - 2\langle x \rangle x_j + \langle x \rangle^2) p_j \\ &= \sum_{j=1}^{n} x_j^2 p_j - 2 \sum_{j=1}^{n} \langle x \rangle x_j p_j + \sum_{j=1}^{n} \langle x \rangle^2 p_j\end{aligned} \tag{B.7}$$

The first term here is just $\langle x^2 \rangle$ (cf. Equation B.5). To evaluate the second and third terms, we need to realize that $\langle x \rangle$, the average of $x_j$, is just a number and so can be factored out of the summations, leaving a summation of the form $\sum x_j p_j$ in the second term and $\sum p_j$ in the third term. The summation $\sum x_j p_j$ is $\langle x \rangle$ by definition and the summation $\sum p_j$ is unity because of normalization (Equation B.3). Putting all this together, we find that

$$\begin{aligned}\sigma_x^2 &= \langle x^2\rangle - 2\langle x\rangle^2 + \langle x\rangle^2 \\ &= \langle x^2\rangle - \langle x\rangle^2 \geq 0 \end{aligned} \tag{B.8}$$

Because $\sigma_x^2 \geq 0$, we see that $\langle x^2\rangle \geq \langle x\rangle^2$. A consideration of Equation B.6 shows that $\sigma_x^2 = 0$ or $\langle x\rangle^2 = \langle x^2\rangle$ only when $x_j = \langle x\rangle$ with a probability of one, a case that is not really probabilistic because the event $j$ occurs on every trial.

So far we have considered only discrete distributions, but continuous distributions are also important in physical chemistry. It is convenient to use the unit mass analogy. Consider a unit mass to be distributed continuously along the $x$ axis, or along some interval on the $x$ axis. We define the linear mass density $\rho(x)$ by

$$dm = \rho(x)dx$$

where $dm$ is the fraction of the mass lying between $x$ and $x + dx$. By analogy, then, we say that the probability that some quantity $x$, such as the position of a particle in a box, lies between $x$ and $x + dx$ is

$$\text{Prob}(x, x + dx) = p(x)dx \tag{B.9}$$

and that

$$\text{Prob}(a \leq x \leq b) = \int_a^b p(x)dx \tag{B.10}$$

In the mass analogy, Prob$\{a \leq x \leq b\}$ is the fraction of mass that lies in the interval $a \leq x \leq b$. The normalization condition is

$$\int_{-\infty}^{\infty} p(x)dx = 1 \tag{B.11}$$

Following Equations B.4 through B.6, we have the definitions

$$\langle x\rangle = \int_{-\infty}^{\infty} xp(x)dx \tag{B.12}$$

$$\langle x^2\rangle = \int_{-\infty}^{\infty} x^2 p(x)dx \tag{B.13}$$

and

$$\sigma_x^2 = \int_{-\infty}^{\infty} (x - \langle x\rangle)^2 p(x)dx \tag{B.14}$$

**EXAMPLE B–3**

Perhaps the simplest continuous distribution is the so-called uniform distribution, where

$$p(x) = \text{constant} = A \qquad a \le x \le b$$
$$= 0 \qquad \text{otherwise}$$

Show that $A$ must equal $1/(b-a)$. Evaluate $\langle x \rangle$, $\langle x^2 \rangle$, $\sigma_x^2$, and $\sigma_x$ for this distribution.

SOLUTION: Because $p(x)$ must be normalized,

$$\int_a^b p(x)dx = 1 = A\int_a^b dx = A(b-a)$$

Therefore, $A = 1/(b-a)$ and

$$p(x) = \frac{1}{b-a} \qquad a \le x \le b$$
$$= 0 \qquad \text{otherwise}$$

The mean of $x$ is given by

$$\langle x \rangle = \int_a^b xp(x)dx = \frac{1}{b-a}\int_a^b xdx$$
$$= \frac{b^2-a^2}{2(b-a)} = \frac{b+a}{2}$$

and the second moment of $x$ by

$$\langle x^2 \rangle = \int_a^b x^2p(x)dx = \frac{1}{b-a}\int_a^b x^2dx$$
$$= \frac{b^3-a^3}{3(b-a)} = \frac{b^2+ab+a^2}{3}$$

Last, the variance is given by Equation B.6, and so

$$\sigma_x^2 = \langle x^2 \rangle - \langle x \rangle^2 = \frac{(b-a)^2}{12}$$

and the standard deviation is

$$\sigma_x = \frac{(b-a)}{\sqrt{12}}$$

**EXAMPLE B–4**

The most commonly occurring and most important continuous probability distribution is the *Gaussian distribution*, given by

$$p(x)dx = ce^{-x^2/2a^2}dx \qquad -\infty < x < \infty$$

Find $c$, $\langle x \rangle$, $\sigma_x^2$ and $\sigma_x$.

SOLUTION: The constant $c$ is determined by normalization:

$$\int_{-\infty}^{\infty} p(x)dx = 1 = c\int_{-\infty}^{\infty} e^{-x^2/2a^2}dx \tag{B.15}$$

If you look in a table of integrals (for example, *The CRC Standard Mathematical Tables* or *The CRC Handbook of Chemistry and Physics*, CRC Press), you won't find the above integral. However, you will find the integral

$$\int_0^{\infty} e^{-\alpha x^2}dx = \left(\frac{\pi}{4\alpha}\right)^{1/2} \tag{B.16}$$

The reason that you won't find the integral with the limts $(-\infty, \infty)$ is illustrated in Figure B.2(a), where $e^{-\alpha x^2}$ is plotted against $x$. Note that the graph is symmetric about the vertical axis, so that the corresponding areas on the two sides of the axis are equal. A function that has the mathematical property that $f(x) = f(-x)$ and is called an *even function*. For an even function

$$\int_{-A}^{A} f_{\text{even}}(x)dx = 2\int_0^{A} f_{\text{even}}(x)dx \tag{B.17}$$

If we recognize that $p(x) = ce^{-x^2/2a^2}$ is an even function and use Equation B.16, then we find that

$$\begin{aligned} c\int_{-\infty}^{\infty} e^{-x^2/2a^2}dx &= 2c\int_0^{\infty} e^{-x^2/2a^2}dx \\ &= 2c\left(\frac{\pi a^2}{2}\right)^{1/2} = 1 \end{aligned}$$

or $c = 1/(2\pi a^2)^{1/2}$.

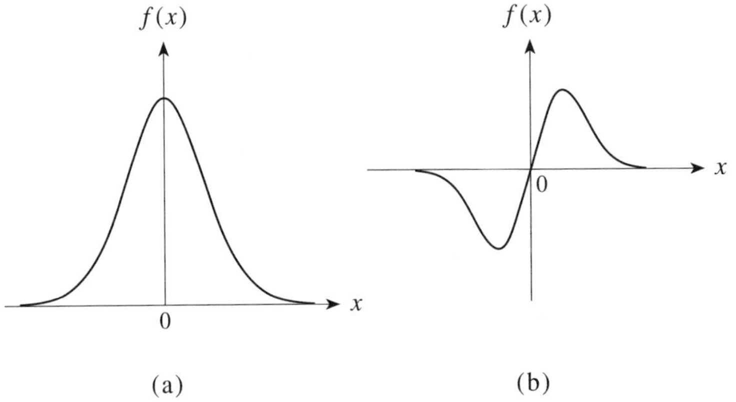


**FIGURE B.2**
(a) The function $f(x) = e^{-x^2}$ is an even function, $f(x) = f(-x)$. (b) The function $f(x) = xe^{-x^2}$ is an odd function, $f(x) = -f(-x)$.

The mean of $x$ is given by

$$\langle x \rangle = \int_{-\infty}^{\infty} xp(x)dx = (2\pi a^2)^{-1/2} \int_{-\infty}^{\infty} xe^{-x^2/2a^2} dx \tag{B.18}$$

The integrand in Equation B.18 is plotted in Figure B.2(b). Notice that this graph is antisymmetric about the vertical axis and that the area on one side of the vertical axis cancels the corresponding area on the other side. A function that has the mathematical property that $f(x) = -f(-x)$ is called an *odd function*. For an odd function,

$$\int_{-A}^{A} f_{\text{odd}}(x)dx = 0 \tag{B.19}$$

The function $xe^{-x^2/2a^2}$ is an odd function, and so

$$\langle x \rangle = (2\pi a^2)^{-1/2} \int_{-\infty}^{\infty} xe^{-x^2/2a^2} dx = 0$$

The second moment of $x$ is given by

$$\langle x^2 \rangle = (2\pi a^2)^{-1/2} \int_{-\infty}^{\infty} x^2 e^{-x^2/2a^2} dx$$

The integrand in this case is even because $y(x) = x^2 e^{-x^2/2a^2} = y(-x)$. Therefore,

$$\langle x^2 \rangle = 2(2\pi a^2)^{-1/2} \int_{0}^{\infty} x^2 e^{-x^2/2a^2} dx$$

The integral

$$\int_{0}^{\infty} x^2 e^{-\alpha x^2} dx = \frac{1}{4\alpha} \left(\frac{\pi}{\alpha}\right)^{1/2} \tag{B.20}$$

can be found in integral tables, and so

$$\langle x^2 \rangle = \frac{2}{(2\pi a^2)^{1/2}} \frac{(2\pi a^2)^{1/2} a^2}{2} = a^2$$

Because $\langle x \rangle = 0$, $\sigma_x^2 = \langle x^2 \rangle$, and so $\sigma_x$ is given by

$$\sigma_x = a$$

The standard deviation of a normal distribution is the parameter that appears in the exponential. The standard notation for a normalized Gaussian distribution function is

$$p(x)dx = (2\pi \sigma_x^2)^{-1/2} e^{-x^2/2\sigma_x^2} dx \tag{B.21}$$

Figure B.3 shows Equation B.21 for various values of $\sigma_x$. Note that the curves become narrower and taller for smaller values of $\sigma_x$.

A more general version of a Gaussian distribution is

$$p(x)dx = (2\pi\sigma_x^2)^{-1/2} e^{-(x-\langle x\rangle)^2/2\sigma_x^2} dx \tag{B.22}$$

This expression looks like those in Figure B.3 except that the curves are centered at $x = \langle x\rangle$ rather than $x = 0$. A Gaussian distribution is one of the most important and commonly used probability distributions in all of science.

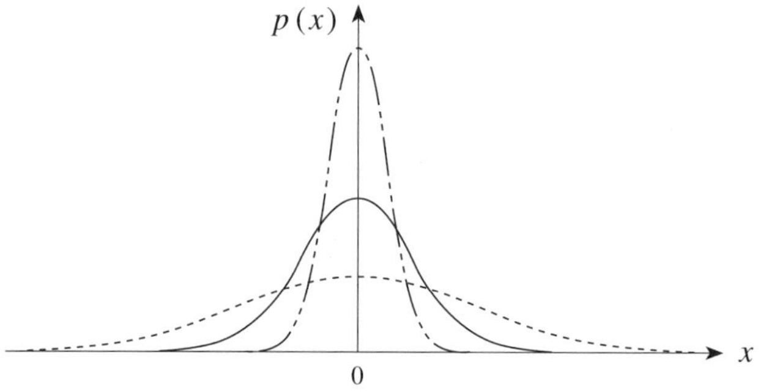


**FIGURE B.3**
A plot of a Gaussian distribution, $p(x)$, (Equation B.21) for three values of $\sigma_x$. The dotted curve corresponds to $\sigma_x = 2$, the solid curve to $\sigma_x = 1$, and the dash-dotted curve to $\sigma_x = 0.5$.

## Problems

**B-1.** Consider a particle to be constrained to lie along a one-dimensional segment 0 to $a$. Quantum mechanics tells us that the particle is found to lie between $x$ and $x + dx$ is given by

$$p(x)dx = \frac{2}{a}\sin^2\frac{n\pi x}{a}dx$$

where $n = 1,\ 2,\ 3,\ \ldots$. First show that $p(x)$ is normalized. Now calculate the average position of the particle along the line segment. The integrals that you need are (*The CRC Handbook of Chemistry and Physics* or *The CRC Standard Mathematical Tables*, CRC Press)

$$\int \sin^2 \alpha x dx = \frac{x}{2} - \frac{\sin 2\alpha x}{4\alpha}$$

and

$$\int x\sin^2 \alpha x dx = \frac{x^2}{4} - \frac{x\sin 2\alpha x}{4\alpha} - \frac{\cos 2\alpha x}{8\alpha^2}$$

**B-2.** Calculate the variance associated with the probability distribution given in Problem B–1. The necessary integral is (*CRC tables*)

$$\int x^2 \sin^2 \alpha x dx = \frac{x^3}{6} - \left(\frac{x^2}{4\alpha} - \frac{1}{8\alpha^3}\right) \sin 2\alpha x - \frac{x \cos 2\alpha x}{4\alpha^2}$$

**B-3.** Using the probability distribution given in Problem B–1, calculate the probability that the particle will be found between 0 and $a/2$. The necessary integral is given in Problem B–1.

**B-4.** Prove explicitly that

$$\int_{-\infty}^{\infty} e^{-\alpha x^2} dx = 2 \int_0^{\infty} e^{-\alpha x^2} dx$$

by breaking the integral from $-\infty$ to $\infty$ into one from $-\infty$ to 0 and another from 0 to $\infty$. Now let $z = -x$ in the first integral and $z = x$ in the second to prove the above relation.

**B-5.** By using the procedure in Problem B–4, show explicitly that

$$\int_{-\infty}^{\infty} x e^{-\alpha x^2} dx = 0$$

**B-6.** According to the kinetic theory of gases, the molecules in a gas travel at various speeds, and that the probability that a molecule has a speed between $v$ and $v + dv$ is given by

$$p(v)dv = 4\pi \left(\frac{m}{2\pi k_B T}\right)^{3/2} v^2 e^{-mv^2/2k_B T} dv \qquad 0 \le v < \infty$$

where $m$ is the mass of the particle, $k_B$ is the Boltzmann constant (the molar gas constant $R$ divided by the Avogadro constant), and $T$ is the Kelvin temperature. The probability distribution of molecular speeds is called the Maxwell-Boltzmann distribution. First show that $p(v)$ is normalized, and then determine the average speed as a function of temperature. The necessary integrals are (*CRC tables*)

$$\int_0^{\infty} x^{2n} e^{-\alpha x^2} dx = \frac{1 \cdot 3 \cdot 5 \cdots (2n-1)}{2^{n+1} \alpha^n} \left(\frac{\pi}{\alpha}\right)^{1/2} \qquad n \ge 1$$

and

$$\int_0^{\infty} x^{2n+1} e^{-\alpha x^2} dx = \frac{n!}{2\alpha^{n+1}}$$

where $n!$ is $n$ factorial, or $n! = n(n-1)(n-2)\cdots(1)$.

**B-7.** Use the Maxwell-Boltzmann distribution in Problem B–6 to determine the average kinetic energy of a gas-phase molecule as a function of temperature. The necessary integral is given in Problem B–6.

**Ludwig Boltzmann** was born in Vienna, Austria, on February 20, 1844, and died in 1906. In 1867, he received his doctorate from the University of Vienna, where he studied with Stefan (of the Stefan-Boltzmann equation). He worked on the kinetic theory of gases and did experimental work on gases and radiation during his stay there. Although known for his theoretical work, he was an able experimentalist but was handicapped by poor vision. He was an early proponent of the atomic theory, and much of his work involved a study of the atomic theory of matter. In 1869, Boltzmann extended Maxwell's theory of the distribution of energy among colliding gas molecules and gave a new expression for this distribution, now known as the Boltzmann factor. In addition, the distribution of the speeds and the energies of gas molecules is now called the Maxwell-Boltzmann distribution. In 1877, he published his famous equation, $S = k_B \ln W$, which expresses the relation between entropy and probability. At the time, the atomic nature of matter was not generally accepted, and Boltzmann's work was criticized by a number of eminent scientists. Unfortunately, Boltzmann did not live to see the atomic theory and his work corroborated. He had always suffered from depression and committed suicide in 1906 by drowning.

CHAPTER 3

# The Boltzmann Factor and Partition Functions

In previous chapters, we learned that the energy states of atoms and molecules, and for all systems in fact, are quantized. These allowed energy states are found by solving the Schrödinger equation. A practical question that arises is how the molecules are distributed over these energy states at a given temperature. For example, we may ask what fraction of the molecules are to be found in the ground vibrational state, the first excited vibrational state, and so on. You may have an intuitive feel that the populations of excited states increase with increasing temperature, and we will see in this chapter that this is the case. Two central themes of this chapter are the Boltzmann factor and the partition function. The Boltzmann factor is one of the most fundamental and useful quantities of physical chemistry. The Boltzmann factor tells us that if a system has states with energies $E_1$, $E_2$, $E_3$, ..., the probability $p_j$ that the system will be in the state with energy $E_j$ depends exponentially on the energy of that state, or

$$p_j \propto e^{-E_j/k_B T}$$

where $k_B$ is the Boltzmann constant and $T$ is the kelvin temperature. We will derive this result in Section 3–2 and then discuss its implications and applications in the remainder of the chapter.

The sum of the probabilities must equal 1, so the normalization constant for the above probability is $1/Q$ where

$$Q = \sum_j e^{-E_j/k_B T}$$

The quantity $Q$ is called a partition function, and we will see that partition functions play a central role in calculating the properties of any system. For example, we will show that we can calculate the energy, heat capacity, and pressure of a system in terms of $Q$. In Chapter 4, we will use partition functions to calculate the heat capacities of monatomic and polyatomic ideal gases.

## 3–1. The Boltzmann Factor Is One of the Most Important Quantities in the Physical Sciences

Consider some macroscopic system such as a liter of gas, a liter of water, or a kilogram of some solid. From a mechanical point of view, such a system can be described by specifying the number of particles, $N$, the volume, $V$, and the forces between the particles. Even though the system contains on the order of Avogadro's number of particles, we can still consider its Hamiltonian operator and its associated wave functions, which will depend upon the coordinates of all the particles. The Schrödinger equation for this $N$-body system is

$$\hat{H}_N \Psi_j = E_j \Psi_j \qquad j = 1, 2, 3, \ldots \tag{3.1}$$

where the energies depend upon both $N$ and $V$, which we will emphasize by writing $E_j(N, V)$.

For the special case of an ideal gas, the total energy $E_j(N, V)$ will simply be a sum of the individual molecular energies,

$$E_j(N, V) = \epsilon_1 + \epsilon_2 + \cdots + \epsilon_N \tag{3.2}$$

because the molecules of an ideal gas are independent of each other. For example, for a monatomic ideal gas in a cubic container with sides of length $a$, if we ignore the electronic states and focus only on the translational states, then the $\epsilon_j$s are just the translational energies given by (Equation 1.45)

$$\epsilon_{n_x n_y n_z} = \frac{h^2}{8ma^2}\left(n_x^2 + n_y^2 + n_z^2\right) \tag{3.3}$$

Note that $E_j(N, V)$ depends upon $N$ through the number of terms in Equation 3.2 and upon $V$ through the fact that $a = V^{1/3}$ in Equation 3.3.

For a more general system in which the particles interact with each other, the $E_j(N, V)$ cannot be written as a sum of individual particle energies, but we can still consider the set of allowed macroscopic energies $\{E_j(N, V)\}$, at least in principle.

What we want to do now is determine the probability that a system will be in the state $j$ with energy $E_j(N, V)$. To do this, we consider a huge collection of such systems in thermal contact with an essentially infinite heat bath (called a heat reservoir) at a temperature $T$. Each system has the same values of $N$, $V$, and $T$ but is likely to be in a different quantum state, consistent with the values of $N$ and $V$. Such a collection of systems is called an *ensemble* (Figure 3.1). We will denote the number of systems in the state $j$ with energy $E_j(N, V)$ by $a_j$ and the total number of systems in the ensemble by $\mathcal{A}$.

We now ask for the relative number of systems of the ensemble that would be found in each state. As an example, let's focus on two particular states, 1 and 2, with

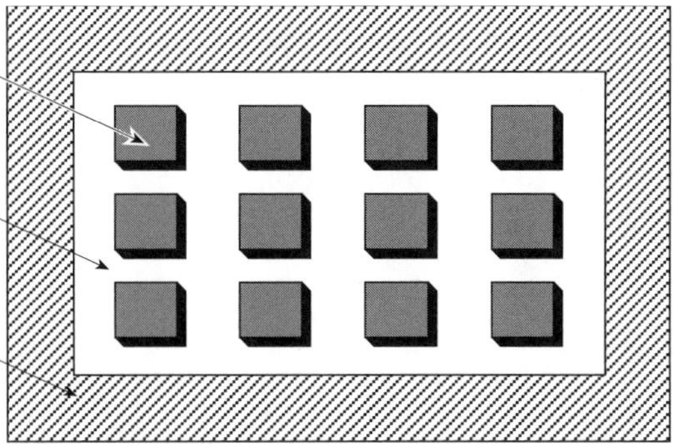


**FIGURE 3.1**
An ensemble, or collection, of (macroscopic) systems in thermal equilibrium with a heat reservoir. The number of systems in the state $j$ [with energy $E_j(N, V)$] is $a_j$, and the total number of systems in the ensemble is $\mathcal{A}$. Because the ensemble is a conceptual construction, we may consider $\mathcal{A}$ to be as large as we want.

energies $E_1(N, V)$ and $E_2(N, V)$. The relative number of systems in the states with energies $E_1$ and $E_2$ must depend upon $E_1$ and $E_2$, so we write

$$\frac{a_2}{a_1} = f(E_1, E_2) \tag{3.4}$$

where $a_1$ and $a_2$ are the number of systems in the ensemble in states 1 and 2 and where the functional form of $f$ is to be determined. Now, because energy is a quantity that must always be referred to a zero of energy, the dependence on $E_1$ and $E_2$ in Equation 3.4 must be of the form

$$f\left(E_1, E_2\right) = f\left(E_1 - E_2\right) \tag{3.5}$$

In this way, any arbitrary zero of energy associated with $E_1$ and $E_2$ will cancel. Thus, we have so far

$$\frac{a_2}{a_1} = f\left(E_1 - E_2\right) \tag{3.6}$$

Equation 3.6 must be true for any two energy states, so we can also write

$$\frac{a_3}{a_2} = f\left(E_2 - E_3\right) \quad \text{and} \quad \frac{a_3}{a_1} = f\left(E_1 - E_3\right) \tag{3.7}$$

But

$$\frac{a_3}{a_1} = \frac{a_2}{a_1} \cdot \frac{a_3}{a_2}$$

so using Equations 3.6 and 3.7, we find that the function $f$ must satisfy

$$f\left(E_1 - E_3\right) = f\left(E_1 - E_2\right) f\left(E_2 - E_3\right) \tag{3.8}$$

The form of the function $f$ that satisfies this equation may not be obvious as first sight, but if you remember that

$$e^{x+y} = e^x e^y$$

then we can see that

$$f(E) = e^{\beta E}$$

where $\beta$ is an arbitrary constant (see also Problem 3–2). To verify that this form for $f$ does indeed satisfy Equation 3.8, we substitute this functional form of $f(E)$ into Equation 3.8:

$$e^{\beta(E_1-E_3)} = e^{\beta(E_1-E_2)}e^{\beta(E_2-E_3)} = e^{\beta(E_1-E_3)}$$

Thus, we find from Equation 3.6 that

$$\frac{a_2}{a_1} = e^{\beta(E_1-E_2)} \tag{3.9}$$

There is nothing special about the states 1 and 2, so we can write Equation 3.9 more generally as

$$\frac{a_n}{a_m} = e^{\beta(E_m-E_n)} \tag{3.10}$$

The form of this equation implies that both $a_m$ and $a_n$ are given by

$$a_j = Ce^{-\beta E_j} \tag{3.11}$$

where $j$ represents either state $m$ or $n$ and $C$ is a constant.

## 3–2. The Probability That a System in an Ensemble Is in the State $j$ with Energy $E_j(N, V)$ Is Proportional to $e^{-E_j(N,V)/k_BT}$

Equation 3.11 has two quantities, $C$ and $\beta$, that we must determine. Determining $C$ is fairly easy. We sum both sides of Equation 3.11 over $j$ to obtain

$$\sum_j a_j = C\sum_j e^{-\beta E_j}$$

But the summation over $a_j$ must equal $\mathcal{A}$, the total number of systems in the ensemble. Therefore, we have

$$C = \frac{\sum_j a_j}{\sum_j e^{-\beta E_j}} = \frac{\mathcal{A}}{\sum_j e^{-\beta E_j}}$$

If we substitute this result back into Equation 3.11, we obtain

$$\frac{a_j}{\mathcal{A}} = \frac{e^{-\beta E_j}}{\sum_j e^{-\beta E_j}} \tag{3.12}$$

The ratio $a_j/\mathcal{A}$ is the fraction of systems in our ensemble that will be found in the state $j$ with energy $E_j$. In the limit of large $\mathcal{A}$, which we are certainly able to take because we can make our ensemble as large as we want, $a_j/\mathcal{A}$ becomes a probability (MathChapter B), so Equation 3.12 can be written as

$$p_j = \frac{e^{-\beta E_j}}{\sum_i e^{-\beta E_i}} \tag{3.13}$$

where $p_j$ is the probability that a randomly chosen system will be in state $j$ with energy $E_j(N, V)$.

Equation 3.13 is a central result of physical chemistry. We customarily let the denominator in this expression be denoted by $Q$, and if we specifically include the dependence of $E_j$ on $N$ and $V$, then we write

$$Q(N, V, \beta) = \sum_i e^{-\beta E_i(N,V)} \tag{3.14}$$

Equation 3.13 becomes

$$p_j(N, V, \beta) = \frac{e^{-\beta E_j(N,V)}}{Q(N, V, \beta)} \tag{3.15}$$

We are not quite ready to determine $\beta$ at this point, but later we will present several different arguments to show that

$$\beta = \frac{1}{k_\mathrm{B}T} \tag{3.16}$$

where $k_\mathrm{B}$ is the Boltzmann constant and $T$ is the kelvin temperature. Thus, we can write Equation 3.15 as

$$p_j(N, V, T) = \frac{e^{-E_j(N,V)/k_\mathrm{B}T}}{Q(N, V, T)} \tag{3.17}$$

We will use Equations 3.15 and 3.17 interchangeably. Equation 3.15 is just as acceptable as Equation 3.17. From a theoretical point of view, $\beta$, or $1/k_\mathrm{B}T$, often happens to be a more convenient quantity to use than $T$ itself.

The quantity $Q(N, V, \beta)$, or $Q(N, V, T)$, is called *the partition function* of the system, and we will see in the next few chapters that we can express all the macroscopic properties of a system in terms of $Q(N, V, \beta)$. At this point, it may not seem possible to determine all the energy states $\{E_j(N, V)\}$ never mind $Q(N, V, \beta)$, but you will learn that we can determine $Q(N, V, \beta)$ for a number of interesting and important systems.

## 3–3. We Postulate That the Average Ensemble Energy Is Equal to the Observed Energy of a System

Using Equation 3.15, we can calculate the average energy of a system in an ensemble of systems. If we denote the average energy by $\langle E \rangle$, then (see MathChapter B)

$$\langle E \rangle = \sum_j p_j(N, V, \beta) E_j(N, V) = \sum_j \frac{E_j(N, V) e^{-\beta E_j(N,V)}}{Q(N, V, \beta)} \tag{3.18}$$

Note that $\langle E \rangle$ is a function of $N$, $V$, and $\beta$. We can express Equation 3.18 entirely in terms of $Q(N, V, \beta)$. First we differentiate $\ln Q(N, V, \beta)$ with respect to $\beta$, with $N$ and $V$ held constant:

$$\begin{aligned}
\left( \frac{\partial \ln Q(N, V, \beta)}{\partial \beta} \right)_{N,V} &= \frac{1}{Q(N, V, \beta)} \left( \frac{\partial \sum e^{-\beta E_j(N,V)}}{\partial \beta} \right)_{N,V} \\
&= \frac{1}{Q(N, V, \beta)} \sum_j [-E_j(N, V)] e^{-\beta E_j(N,V)} \\
&= -\sum_j \frac{E_j(N, V) e^{-\beta E_j(N,V)}}{Q(N, V, \beta)}
\end{aligned} \tag{3.19}$$

If we compare Equation 3.19 with 3.18, we see that

$$\langle E \rangle = -\left( \frac{\partial \ln Q}{\partial \beta} \right)_{N,V} \tag{3.20}$$

We can also express Equation 3.20 as a temperature derivative rather than a $\beta$ derivative. If we use the chain rule of differentiation, we can write for any function $f$ that

$$\frac{\partial f}{\partial T} = \frac{\partial f}{\partial \beta} \cdot \frac{\partial \beta}{\partial T} = \frac{\partial f}{\partial \beta} \cdot \frac{d(1/k_{\text{B}}T)}{dT} = -\frac{1}{k_{\text{B}}T^2} \frac{\partial f}{\partial \beta}$$

or

$$\frac{\partial f}{\partial \beta} = -k_{\text{B}}T^2 \frac{\partial f}{\partial T}$$

Applying this result to Equation 3.20 with $f = \ln Q$ gives us the alternate form

$$\langle E \rangle = k_{\text{B}}T^2 \left( \frac{\partial \ln Q}{\partial T} \right)_{N,V} \tag{3.21}$$

Equation 3.20 is often the easier one to use, however.

**EXAMPLE 3–1**

Derive an equation for $\langle E \rangle$ for the simple system of a (bare) proton in a magnetic field $B_z$.

SOLUTION: The energy of a bare proton in a magnetic field $B_z$ is given by $E_{\pm\frac{1}{2}} = \mp\frac{1}{2}\hbar\gamma B_z$ where $\gamma$ is a quantity called the magnetogyric ratio. The partition function consists of just two terms:

$$Q(T, B_z) = e^{\beta\hbar\gamma B_z/2} + e^{-\beta\hbar\gamma B_z/2}$$
$$= e^{\hbar\gamma B_z/2k_\mathrm{B}T} + e^{-\hbar\gamma B_z/2k_\mathrm{B}T}$$

The average energy is obtained from either Equation 3.20 or 3.21:

$$\langle E \rangle = -\left(\frac{\partial \ln Q}{\partial \beta}\right)_{B_z} = -\frac{1}{Q(\beta, B_z)}\left(\frac{\partial Q}{\partial \beta}\right)_{B_z}$$
$$= -\frac{\hbar\gamma B_z}{2}\left(\frac{e^{\beta\hbar\gamma B_z/2} - e^{-\beta\hbar\gamma B_z/2}}{e^{\beta\hbar\gamma B_z/2} + e^{-\beta\hbar\gamma B_z/2}}\right)$$
$$= -\frac{\hbar\gamma B_z}{2}\left(\frac{e^{\hbar\gamma B_z/2k_\mathrm{B}T} - e^{-\hbar\gamma B_z/2k_\mathrm{B}T}}{e^{\hbar\gamma B_z/2k_\mathrm{B}T} + e^{-\hbar\gamma B_z/2k_\mathrm{B}T}}\right)$$

This expression for $\langle E \rangle$ (in units of $\hbar\gamma B_z/2$) is plotted against $T$ (in units of $\hbar\gamma B_z/2k_\mathrm{B}$) in Figure 3.2. Note that $\langle E \rangle \to -\hbar\gamma B_z/2$ as $T \to 0$ and that $\langle E \rangle \to 0$ as $T \to \infty$. As $T \to 0$, there is no thermal energy, so the proton orients itself parallel to the magnetic field with certainty. As $T \to \infty$, however, the thermal energy of the proton increases to such an extent that the proton is equally likely to point in either direction.

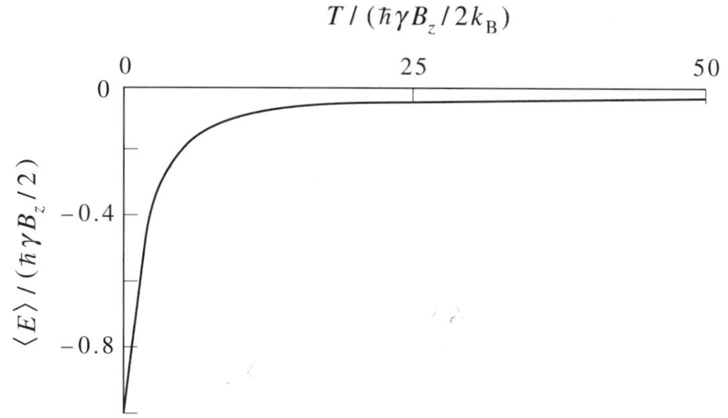


**FIGURE 3.2**
The average energy of a (bare) proton in a magnetic field plotted against the temperature (see Example 3–1).

We will learn in Chapter 4 that for a monatomic ideal gas,

$$Q(N, V, \beta) = \frac{[q(V, \beta)]^N}{N!} \tag{3.22}$$

where

$$q(V, \beta) = \left(\frac{2\pi m}{h^2\beta}\right)^{3/2} V \tag{3.23}$$

For a monatomic ideal gas in its electronic ground state, the energy of the system is only in the translational degrees of freedom. Before we substitute Equation 3.22 into Equation 3.20, we write $\ln Q$ for convenience as a sum of terms that involve $\beta$ and terms that are independent of $\beta$:

$$\begin{aligned}\ln Q &= N \ln q - \ln N! \\ &= -\frac{3N}{2}\ln\beta + \frac{3N}{2}\ln\left(\frac{2\pi m}{h^2}\right) + N\ln V - \ln N! \\ &= -\frac{3N}{2}\ln\beta + \text{terms involving only } N \text{ and } V\end{aligned}$$

Now we can see more easily that

$$\left(\frac{\partial \ln Q}{\partial\beta}\right)_{N,V} = -\frac{3N}{2}\frac{d\ln\beta}{d\beta} = -\frac{3N}{2\beta} = -\frac{3}{2}Nk_BT$$

and that (Equation 3.20)

$$\langle E\rangle = \tfrac{3}{2}Nk_BT$$

For $n$ moles, $N = nN_A$ and $k_BN_A = R$, so

$$\langle E\rangle = \tfrac{3}{2}nRT$$

This observation leads us to a fundamental postulate of physical chemistry that the ensemble average of any quantity, as calculated using the probability distribution of Equation 3.17, is the same as the experimentally observed value of that quantity. If we let the experimentally observed energy of a system be denoted by $U$, then we have

$$\overline{U} = \langle\overline{E}\rangle = \tfrac{3}{2}RT$$

for one mole of a monatomic ideal gas. (We indicate a molar quantity by an overbar.)

**EXAMPLE 3–2**

We will learn in the next chapter that for the rigid rotator-harmonic oscillator model of an ideal diatomic gas, the partition function is given by

$$Q(N, V, \beta) = \frac{[q(V, \beta)]^N}{N!}$$

where

$$q(V, \beta) = \left(\frac{2\pi m}{h^2\beta}\right)^{3/2} V \cdot \frac{8\pi^2 I}{h^2\beta} \cdot \frac{e^{-\beta h\nu/2}}{1 - e^{-\beta h\nu}}$$

In this expression, $I$ is the moment of inertia and $\nu$ is the fundamental vibrational frequency of the diatomic molecule. Note that $q(V, \beta)$ for a diatomic molecule is the same as the expression for $q(V, \beta)$ for a monatomic gas (Equation 3.23, a translational term), except that it is multiplied by a rotational term, $8\pi^2 I/h^2\beta$, and a vibrational term, $e^{-\beta h\nu/2}/(1 - e^{-\beta h\nu})$. The reason for this difference will become apparent in Section 3–8. Use this partition function to calculate the average energy of one mole of a diatomic ideal gas.

SOLUTION: Once again, for convenience we write $\ln Q$ as the sum of terms that involve $\beta$ and terms that are independent of $\beta$:

$$\begin{aligned}\ln Q &= N \ln q - \ln N! \\ &= -\frac{3N}{2}\ln\beta - N\ln\beta - \frac{N\beta h\nu}{2} - N\ln(1 - e^{-\beta h\nu}) \\ &\qquad + \text{terms not involving } \beta\end{aligned}$$

Now

$$\begin{aligned}\left(\frac{\partial \ln Q}{\partial \beta}\right)_{N,V} &= -\frac{3N}{2}\frac{d\ln\beta}{d\beta} - N\frac{d\ln\beta}{d\beta} - \frac{Nh\nu}{2} - N\frac{d\ln(1 - e^{-\beta h\nu})}{d\beta} \\ &= -\frac{3N}{2\beta} - \frac{N}{\beta} - \frac{Nh\nu}{2} - \frac{Nh\nu e^{-\beta h\nu}}{1 - e^{-\beta h\nu}}\end{aligned}$$

or

$$U = \langle E\rangle = \frac{3}{2}Nk_BT + Nk_BT + \frac{Nh\nu}{2} + \frac{Nh\nu e^{-\beta h\nu}}{1 - e^{-\beta h\nu}}$$

For one mole, $N = N_A$ and $N_A k_B = R$, so

$$\overline{U} = \frac{3}{2}RT + RT + \frac{N_A h\nu}{2} + \frac{N_A h\nu e^{-\beta h\nu}}{1 - e^{-\beta h\nu}} \qquad (3.24)$$

Equation 3.24 has a nice physical interpretation. The first term represents the average translational energy, the second term represents the average rotational energy, the third term represents the zero-point vibrational energy, and the fourth term represents the average vibrational energy. The fourth term is negligible at low temperatures for most gases but increases with increasing temperature as the excited vibrational states become populated.

## 3–4. The Heat Capacity at Constant Volume Is the Temperature Derivative of the Average Energy

The constant-volume heat capacity, $C_V$, of a system is defined as

$$C_V = \left(\frac{\partial \langle E \rangle}{\partial T}\right)_{N,V} = \left(\frac{\partial U}{\partial T}\right)_{N,V} \tag{3.25}$$

The heat capacity $C_V$ is then a measure of how the energy of the system changes with temperature at constant amount and volume. Consequently, $C_V$ can be expressed in terms of $Q(N, V, T)$ through Equation 3.21. We have seen that $\overline{U} = 3RT/2$ for one mole of a monatomic ideal gas, so

$$\overline{C}_V = \tfrac{3}{2}R \qquad \begin{pmatrix}\text{monatomic}\\ \text{ideal gas}\end{pmatrix} \tag{3.26}$$

For a diatomic ideal gas, we obtain from Equation 3.24

$$\begin{aligned}\overline{C}_V &= \frac{5}{2}R + N_A h\nu \frac{\partial}{\partial T}\left(\frac{e^{-\beta h\nu}}{1-e^{-\beta h\nu}}\right)\\ &= \frac{5}{2}R - \frac{N_A h\nu}{k_B T^2}\frac{\partial}{\partial \beta}\left(\frac{e^{-\beta h\nu}}{1-e^{-\beta h\nu}}\right)\\ &= \frac{5}{2}R + R\left(\frac{h\nu}{k_B T}\right)^2 \frac{e^{-h\nu/k_B T}}{(1-e^{-h\nu/k_B T})^2} \qquad \begin{pmatrix}\text{diatomic}\\ \text{ideal gas}\end{pmatrix}\end{aligned} \tag{3.27}$$

Figure 3.3 shows the theoretical (Equation 3.27) versus the experimental molar heat capacity of $O_2$(g) as a function of temperature. The agreement between the two is seen to be excellent.

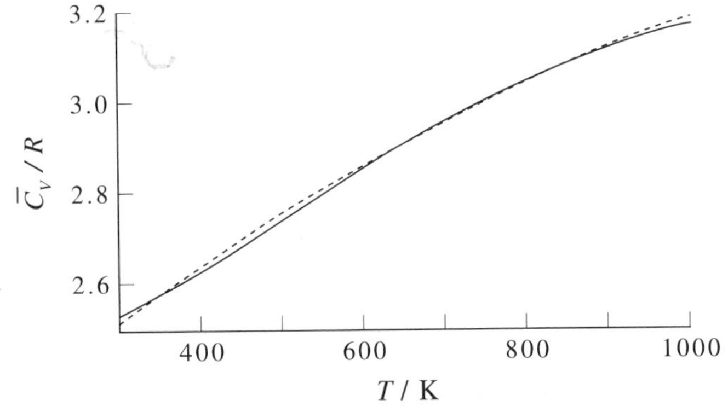


**FIGURE 3.3**
The experimental and theoretical (Equation 3.27) molar heat capacity of $O_2$(g) from 300 K to 1000 K. The theoretical curve (solid curve) is calculated using $h\nu/k = 2240$ K.

**EXAMPLE 3–3**

In 1905, Einstein proposed a simple model for an atomic crystal that can be used to calculate the molar heat capacity. He pictured an atomic crystal as $N$ atoms situated at lattice sites, with each atom vibrating as a three-dimensional harmonic oscillator. Because all the lattice sites are identical, he further assumed that each atom vibrated with the same frequency. The partition function associated with this model is (Problem 3–20)

$$Q = e^{-\beta U_0} \left( \frac{e^{-\beta h\nu/2}}{1 - e^{-\beta h\nu}} \right)^{3N} \tag{3.28}$$

where $\nu$, which is characteristic of the particular crystal, is the frequency with which the atoms vibrate about their lattice positions and $U_0$ is the sublimation energy at 0 K, or the energy needed to separate all the atoms from one another at 0 K. Calculate the molar heat capacity of an atomic crystal from this partition function.

SOLUTION: The average energy is given by (Equation 3.20)

$$\begin{aligned} U &= -\left( \frac{\partial \ln Q}{\partial \beta} \right)_{N,V} \\ &= -\left( \frac{\partial}{\partial \beta} \left[ -\beta U_0 - \frac{3N}{2}\beta h\nu - 3N \ln(1 - e^{-\beta h\nu}) \right] \right)_{N,V} \\ &= U_0 + \frac{3Nh\nu}{2} + \frac{3Nh\nu e^{-\beta h\nu}}{1 - e^{-\beta h\nu}} \end{aligned}$$

Note that $U$ consists of three terms: $U_0$, the sublimation energy at 0 K; $3Nh\nu/2$, the zero-point energy of $N$ three-dimensional harmonic oscillators; and a term that represents the increase in vibrational energy as the temperature increases.

The heat capacity at constant volume is given by

$$\begin{aligned} C_V &= \left( \frac{\partial U}{\partial T} \right)_{N,V} = -\frac{1}{k_B T^2} \left( \frac{\partial U}{\partial \beta} \right)_{N,V} \\ &= -\frac{3Nh\nu}{k_B T^2} \left[ -\frac{h\nu e^{-\beta h\nu}}{1 - e^{-\beta h\nu}} - \frac{h\nu e^{-2\beta h\nu}}{(1 - e^{-\beta h\nu})^2} \right] \end{aligned}$$

or

$$\overline{C}_V = 3R \left( \frac{h\nu}{k_B T} \right)^2 \frac{e^{-h\nu/k_B T}}{(1 - e^{-h\nu/k_B T})^2} \tag{3.29}$$

where we have used the fact that $N = N_A$ and $N_A k_B = R$ for one mole.

Equation 3.29 contains one adjustable parameter, the vibrational frequency $\nu$. Figure 3.4 shows the molar heat capacity of diamond as a function of temperature calculated with $\nu = 2.75 \times 10^{13}\ \text{s}^{-1}$. The agreement with experiment is seen to be fairly good considering the simplicity of the model.

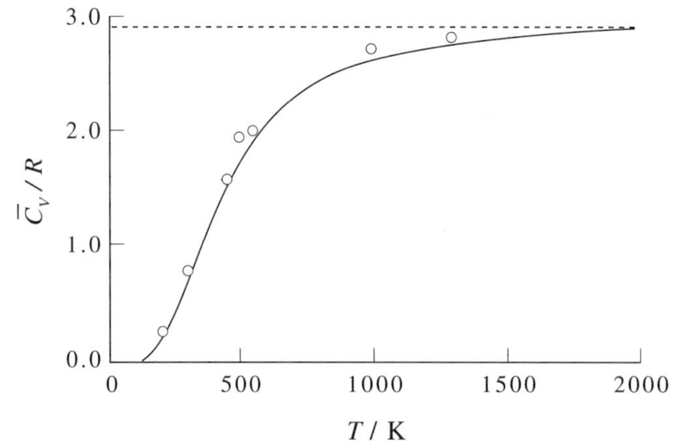


**FIGURE 3.4**
The observed and theoretical (Einstein model) molar heat capacity of diamond as a function of temperaure. The solid curve is calculated using Equation 3.29, and the circles represent experimental data.

It is interesting to look at the high-temperature limit of Equation 3.29. At high temperatures, $h\nu/k_B T$ is small, so we can use the fact that $e^x \approx 1 + x$ for small $x$ (MathChapter C). Thus, Equation 3.29 becomes

$$\overline{C}_V \approx 3R\left(\frac{h\nu}{k_B T}\right)^2 \frac{1 - \dfrac{h\nu}{k_B T} + \cdots}{\left(1 - 1 + \dfrac{h\nu}{k_B T} + \cdots\right)^2}$$

$$\approx 3R\left(\frac{h\nu}{k_B T}\right)^2 \frac{1}{\left(\dfrac{h\nu}{k_B T}\right)^2} = 3R$$

This result predicts that the molar heat capacities of atomic crystals should level off at a value of $3R = 24.9 \text{ J}\cdot\text{K}^{-1}\cdot\text{mol}^{-1}$ at high temperatures. This prediction is known as the law of Dulong and Petit, which played an important role in the determination of atomic masses in the 1800s. This prediction is in good agreement with the data shown in Figure 3.4.

## 3–5. We Can Express the Pressure in Terms of a Partition Function

We will show in Section 5–6 that the pressure of a macroscopic system is given by

$$P_j(N, V) = -\left(\frac{\partial E_j}{\partial V}\right)_N \tag{3.30}$$

Using the fact that the average pressure is given by

$$\langle P\rangle = \sum_j p_j(N, V, \beta)P_j(N, V)$$

we can write

$$\langle P\rangle = \sum_j p_j(N, V, \beta)\left(-\frac{\partial E_j}{\partial V}\right)_N = \sum_j \left(-\frac{\partial E_j}{\partial V}\right)_N \frac{e^{-\beta E_j(N,V)}}{Q(N, V, \beta)} \tag{3.31}$$

This expression can be written in a more compact form. Let's start with

$$Q(N, V, \beta) = \sum_j e^{-\beta E_j(N,V)}$$

and differentiate it with respect to $V$ keeping $N$ and $\beta$ fixed:

$$\left(\frac{\partial Q}{\partial V}\right)_{N,\beta} = -\beta \sum_j \left(\frac{\partial E_j}{\partial V}\right)_N e^{-\beta E_j(N,V)}$$

Comparing this result with the second equality of Equation 3.31 shows that

$$\langle P\rangle = \frac{k_B T}{Q(N, V, \beta)}\left(\frac{\partial Q}{\partial V}\right)_{N,\beta}$$

or equivalently,

$$\langle P\rangle = k_B T\left(\frac{\partial \ln Q}{\partial V}\right)_{N,\beta} \tag{3.32}$$

Just as we equated the ensemble average of the energy with the observed energy, we equate the ensemble average pressure with the observed pressure, $P = \langle P\rangle$. Thus, we see that we can calculate the observed pressure if we know $Q(N, V, \beta)$.

We can use this result to derive the ideal-gas equation of state. First, consider a monatomic ideal gas. Recall from Equation 3.22 that $Q(N, V, \beta)$ for a monatomic ideal gas is given by

$$Q(N, V, \beta) = \frac{[q(V, \beta)]^N}{N!}$$

where

$$q(V, \beta) = \left(\frac{2\pi m}{h^2\beta}\right)^{3/2} V$$

Let's use this result to calculate the pressure of a monatomic ideal gas. To evaluate Equation 3.32, we write out $\ln Q$ first for convenience:

$$\begin{aligned}\ln Q &= N \ln q - \ln N! \\ &= \frac{3N}{2}\ln\left(\frac{2\pi m}{h^2\beta}\right) + N\ln V - \ln N!\end{aligned}$$

Because $N$ and $\beta$ are fixed in Equation 3.32, we write $\ln Q$ as

$$\ln Q = N \ln V + \text{terms in } N \text{ and } \beta \text{ only}$$

Therefore,

$$\left(\frac{\partial \ln Q}{\partial V}\right)_{N,\beta} = \frac{N}{V}$$

and substituting this result into Equation 3.32 gives us

$$P = \frac{Nk_{\text{B}}T}{V}$$

as you might have expected.

Notice that the ideal-gas equation results from the fact that $\ln Q = N \ln V +$ terms in $N$ and $\beta$, which comes from the fact that $q(V, T)$ is directly proportional to $V$ in Equation 3.22. Example 3–2 shows that $q(V, T)$ is directly proportional to $V$ for a diatomic ideal gas also, and so $PV = Nk_{\text{B}}T$ for a diatomic ideal gas. This is the case for a polyatomic ideal gas as well, so the ideal-gas equation of state results for any ideal gas, monatomic, diatomic, or polyatomic.

**EXAMPLE 3–4**

Calculate the equation of state associated with the partition function

$$Q(N, V, \beta) = \frac{1}{N!}\left(\frac{2\pi m}{h^2\beta}\right)^{3N/2}(V - Nb)^N e^{\beta a N^2/V}$$

where $a$ and $b$ are constants. Can you identify the resulting equation of state?

SOLUTION: We use Equation 3.32 to calculate the equation of state. First, we evaluate $\ln Q$, which gives

$$\ln Q = N \ln(V - Nb) + \frac{\beta a N^2}{V} + \text{terms in } N \text{ and } \beta \text{ only}$$

We now differentiate with respect to $V$, keeping $N$ and $\beta$ constant, to get

$$\left(\frac{\partial \ln Q}{\partial V}\right)_{N,\beta} = \frac{N}{V - Nb} - \frac{\beta a N^2}{V^2}$$

and so

$$P = \frac{Nk_{\text{B}}T}{V - Nb} - \frac{aN^2}{V^2}$$

Bringing the last term to the left side and multiplying by $V - Nb$ gives us

$$\left(P + \frac{aN^2}{V^2}\right)(V - Nb) = Nk_{\text{B}}T$$

which is the van der Waals equation.

## 3–6. The Partition Function of a System of Independent, Distinguishable Molecules Is the Product of Molecular Partition Functions

The general results we have derived up to now are valid for arbitrary systems. To apply these equations, we need to have the set of eigenvalues $\{E_j(N, V)\}$ for the $N$-body Schrödinger equation. In general, this is an impossible task. For many important physical systems, however, writing the total energy of the system as a sum of individual energies is a good approximation. This procedure leads to a great simplification of the partition function and allows us to apply the results with relative ease.

First, let's consider a system that consists of independent, distinquishable particles. Although atoms and molecules are certainly not distinguishable in general, they can be treated as such in a number of cases. An excellent example is that of a perfect crystal. In a perfect crystal, each atom is confined to one and only one lattice site, which we could, at least in principle, identify by a set of three coordinates. Because each particle, then, is confined to a lattice site and the lattice sites are distinguishable, the particles themselves are distinguishable. We can treat the vibration of each particle about its lattice site as independent to a fairly good approximation, just as we did for normal modes of polyatomic molecules.

We will denote the individual particle energies by $\{\varepsilon_j^a\}$, where the superscript denotes the particle (they are distinguishable), and the subscript denotes the energy state of the particle. In this case, the total energy of the system $E_l(N, V)$ can be written as

$$E_l(N, V) = \underbrace{\varepsilon_i^a(V) + \varepsilon_j^b(V) + \varepsilon_k^c(V) + \cdots}_{N \text{ terms}}$$

and the system partition function becomes

$$Q(N, V, T) = \sum_l e^{-\beta E_l} = \sum_{i,j,k,\ldots} e^{-\beta(\varepsilon_i^a + \varepsilon_j^b + \varepsilon_k^c + \cdots)}$$

Because the particles are distinguishable and independent, we can sum over $i$, $j$, $k$, ... independently, in which case $Q(N, V, T)$ can be written as a product of individual summations (Problem 3–21):

$$\begin{aligned} Q(N, V, T) &= \sum_i e^{-\beta\varepsilon_i^a} \sum_j e^{-\beta\varepsilon_j^b} \sum_k e^{-\beta\varepsilon_k^c} \cdots \\ &= q_a(V, T) q_b(V, T) q_c(V, T) \cdots \end{aligned} \tag{3.33}$$

where each of the $q(V, T)$ is given by

$$q(V, T) = \sum_i e^{-\beta\varepsilon_i} = \sum_i e^{-\varepsilon_i/k_\mathrm{B}T} \tag{3.34}$$

In many cases, the $\{\varepsilon_i\}$ is a set of molecular energies; thus $q(V, T)$ is called a *molecular partition function.*

Equation 3.33 is an important result. It shows that if we can write the total energy as a sum of individual, independent terms, and if the atoms or molecules are *distinguishable*, then the system partition function $Q(N, V, T)$ reduces to a product of molecular partition functions $q(V, T)$. Because $q(V, T)$ requires a knowledge of the allowed energies of only individual atoms or molecules, its evaluation is often feasible, as we will see for a number of cases in Chapter 4.

If the energy states of all the atoms or molecules are the same (as for a monatomic crystal), then Equation 3.33 becomes

$$Q(N, V, T) = [q(V, T)]^N \qquad \begin{pmatrix}\text{independent, distinguishable} \\ \text{atoms or molecules}\end{pmatrix} \tag{3.35}$$

where

$$q(V, T) = \sum_j e^{-\varepsilon_j/k_\mathrm{B}T}$$

The Einstein model of atomic crystals (Example 3–3) considers the atoms to be fixed at lattice sites, so Equation 3.35 should be applicable to that model. Notice that the partition function of that model (Equation 3.28) can be written in the form of Equation 3.35 if we let $u_0 = U_0/N$ be the sublimation energy per atom at 0 K, in which case we have

$$Q = \left[ e^{-\beta u_0} \left( \frac{e^{-\beta h\nu/2}}{1 - e^{-\beta h\nu}} \right)^3 \right]^N \tag{3.36}$$

## 3–7. The Partition Function of a System of Independent, Indistinguishable Atoms or Molecules Can Usually Be Written as $[q(V, T)]^N/N!$

Equation 3.35 is an attractive result, but atoms and molecules are, in general, not distinguishable; thus the utility of Equation 3.35 is severely limited. The reduction of a system partition function $Q(N, V, T)$ to molecular partition functions $q(V, T)$ becomes somewhat more complicated when the inherent indistinguishability of atoms and molecules cannot be ignored. For indistinguishable particles, the total energy is

$$E_{ijk\ldots} = \underbrace{\varepsilon_i + \varepsilon_j + \varepsilon_k + \cdots}_{N \text{ terms}}$$

(note the lack of distinguishing superscripts, as in Equation 3.33) and the system partition function is

$$Q(N, V, T) = \sum_{i,j,k,\ldots} e^{-\beta(\varepsilon_i+\varepsilon_j+\varepsilon_k+\cdots)} \tag{3.37}$$

Because the particles are indistinguishable, we cannot sum over $i, j, k, \ldots$ separately as we did in Equation 3.33. To see why, we must consider a fundamental property of all particles.

You learned in general chemistry that the Pauli Exclusion Principle says that no two electrons in an atom can have the same set of four quantum numbers. Another way of saying this is that no two electrons in an atom can be in the same quantum state. The Pauli Exclusion Principle is actually more general than the above statement, and applies to all particles of spin 1/2, 3/2, 5/2, and so on. Such particles, called *fermions*, have the restriction that no two fermions in a system can occupy the same quantum state. Examples of fermions are electrons (spin 1/2), protons (spin 1/2), and neutrons (spin 1/2). Particles that have spin of 0, 1, 2,. . . , called *bosons*, do not have any restriction regarding the occupancy of individual quantum states. Examples of bosons are alpha particles (spin 0) and photons (spin 1). It turns out that fermions and bosons constitute all the known particles in nature. We must recognize the occupancy requirements of fermions and bosons when we attempt to carry out the summation in Equation 3.37.

Let's go back now to the summation in Equation 3.37 for the case of fermions. Because no two identical fermions can occupy the same single-particle energy state, terms in which two or more indices are the same cannot be included in the summation. Therefore, the indices $i, j, k, \ldots$ are *not* independent of one another, so a direct evaluation of $Q(N, V, T)$ by means of Equation 3.37 poses problems for fermions.

**EXAMPLE 3–5**

Consider a system of two noninteracting identical fermions, each of which has states with energies $\varepsilon_1$, $\varepsilon_2$, $\varepsilon_3$, and $\varepsilon_4$. Enumerate the allowed total energies in the summation in Equation 3.37.

SOLUTION: For this system

$$Q(2, V, T) = \sum_{i,j=1}^{4} e^{-\beta(\varepsilon_i+\varepsilon_j)}$$

Of the 16 terms that would occur in an unrestricted evaluation of $Q$, only six are allowed for two identical fermions; these are the terms with energies

$$\begin{array}{ll} \varepsilon_1+\varepsilon_2 & \varepsilon_2+\varepsilon_3 \\ \varepsilon_1+\varepsilon_3 & \varepsilon_2+\varepsilon_4 \\ \varepsilon_1+\varepsilon_4 & \varepsilon_3+\varepsilon_4 \end{array}$$

The six terms in which the $\varepsilon_j$ are written in reverse order are the same as those above (because the particles are indistinguishable), and the four terms in which the $\varepsilon_j$ are the same are not allowed (because the particles are fermions).

Bosons do not have the restriction that no two of the same type can occupy the same single-particle state, but the summation in Equation 3.37 is still complicated. To see why, consider a term in Equation 3.37 in which all the indices are the same except for one; for example, a term like

$$E = \underbrace{\varepsilon_2 + \varepsilon_{10} + \varepsilon_{10} + \varepsilon_{10} + \cdots + \varepsilon_{10}}_{N \text{ particles, } N \text{ terms}}$$

(in reality, these indices might be enormous numbers). Because the particles are indistinguishable, the position of the term $\varepsilon_2$ is not important, and we could just as easily have $\varepsilon_{10} + \varepsilon_2 + \varepsilon_{10} + \varepsilon_{10} + \cdots + \varepsilon_{10}$ or $\varepsilon_{10} + \varepsilon_{10} + \varepsilon_2 + \varepsilon_{10} + \cdots + \varepsilon_{10}$ and so on. Because these terms all represent the same state, such a state should be included only once in Equation 3.37, but an unrestricted summation over all the indices (summing over $i, j, k, \ldots$ independently) in Equation 3.37 would produce $N$ terms of this type (the $\varepsilon_2$ can be located in any of the $N$ positions).

Now consider the other extreme in which all the $N$ particles are in different molecular states; that is, for example, a system energy of $\varepsilon_1 + \varepsilon_2 + \varepsilon_3 + \varepsilon_4 + \cdots + \varepsilon_N$. Because the particles are indistinguishable, all $N!$ arrangements obtained by permuting these $N$ terms are identical and should occur only once in Equation 3.37. Yet such terms will appear $N!$ times in an unrestricted summation. Consequently, a direct evaluation of $Q(N, V, T)$ by means of Equation 3.37 poses problems for bosons as well as fermions.

**EXAMPLE 3–6**

Redo Example 3–5 for bosons instead of fermions.

SOLUTION: In this case there are 10 allowed terms: the six that are allowed in Example 3–5 and the four in which the $\varepsilon_j$ are the same (bosons do not have the restriction that no two can occupy the same state).

Note that in every case, the terms in Equation 3.37 that cause difficulty are those in which two or more indices are the same. If it were not for such terms, we could carry out the summation in Equation 3.37 in an unrestricted manner (obtaining $[q(V, T)]^N$ as in Section 3–6) and then divide by $N!$ (to obtain $[q(V, T)]^N/N!$) to account for the over-counting. For example, if we could ignore terms like $\varepsilon_1 + \varepsilon_1$, $\varepsilon_2 + \varepsilon_2$, etc. in the evaluation of $Q(2, V, T)$, there would be a total of 12 terms, the six enumerated in Example 3–5 and the six in which the energies are written in reverse order. By dividing by 2!, we would obtain the correct number allowed terms.

Certainly, if the number of quantum states available to any particle is much greater than the number of particles, it would be unlikely for any two particles to be in the same state. Although most of the quantum-mechanical systems we have studied have an infinite number of energy states, at any given temperature many of these will not be readily accessible because the energies of these states are much larger than $k_B T$, which is roughly the average energy of a molecule. If, however, the number of quantum states with energies less than roughly $k_B T$ is much larger than the number of particles, then essentially all the terms in Equation 3.37 will contain $\varepsilon$'s with different indices, and so we can evaluate $Q(N, V, T)$ to a good approximation by summing over $i, \; j, \; k, \; \ldots$ independently in Equation 3.37 and then dividing by $N!$ to get

$$Q(N, V, T) = \frac{[q(V, T)]^N}{N!} \quad \left(\begin{array}{c}\text{independent, indistinguishable} \\ \text{atoms or molecules}\end{array}\right) \tag{3.38}$$

where

$$q(V, T) = \sum_j e^{-\varepsilon_j / k_B T} \tag{3.39}$$

The number of translational states alone is usually sufficient to guarantee that the number of energy states available to any atom or molecule is greater than the number of particles in the system. Therefore, this procedure yields an excellent approximation in many cases. The criterion that the number of available states exceeds the number of particles so that Equation 3.38 can be used is

$$\frac{N}{V}\left(\frac{h^2}{8mk_B T}\right)^{3/2} \ll 1 \tag{3.40}$$

Notice that this criterion is favored by large particle mass, high temperature, and low density.

Although our discussion at this point is limited to ideal gases (independent, indistinguishable particles), we show the values of $(N/V)(h^2/8mk_B T)^{3/2}$ in Table 3.1 even for some liquids at their boiling points, just to show that Inequality 3.40 is easily satisfied in most cases. Note that the exceptional systems include liquid helium and liquid hydrogen (because of their small masses and low temperatures) and electrons in metals (because of their very small mass). These systems are the prototype examples of quantum systems that must be treated by special methods (which we will not discuss).

When Equation 3.38 is valid, that is, when the number of available molecular states is much greater than the number of particles, we say that the particles obey *Boltzmann statistics*. As Inequality 3.40 indicates, Boltzmann statistics becomes increasingly valid with increasing temperature. Let's test Inequality 3.40 for $N_2(g)$ at 20°C and one bar. Under these conditions,

$$\begin{aligned}\frac{N}{V} = \frac{P}{k_B T} &= \frac{10^5 \text{ Pa}}{(1.381 \times 10^{-23} \text{ J}\cdot\text{K}^{-1})(293.2 \text{ K})} \\ &= 2.470 \times 10^{25} \text{ m}^{-3}\end{aligned}$$

**TABLE 3.1**

The quantity $(N/V)(h^2/8mk_BT)^{3/2}$ at a pressure of one bar for a number of simple systems.

| System | $T$/K | $\frac{N}{V}\left(\frac{h^2}{8mk_BT}\right)^{3/2}$ |
|---|---|---|
| Liquid helium | 4 | 1.5 |
| Gaseous helium | 4 | 0.11 |
| Gaseous helium | 20 | $1.8 \times 10^{-3}$ |
| Gaseous helium | 100 | $3.3 \times 10^{-5}$ |
| Liquid hydrogen | 20 | 0.29 |
| Gaseous hydrogen | 20 | $5.1 \times 10^{-3}$ |
| Gaseous hydrogen | 100 | $9.4 \times 10^{-5}$ |
| Liquid neon | 27 | $1.0 \times 10^{-2}$ |
| Gaseous neon | 27 | $7.8 \times 10^{-5}$ |
| Liquid krypton | 127 | $5.1 \times 10^{-5}$ |
| Electrons in metals (Na) | 300 | 1400 |

and

$$\frac{h^2}{8mk_BT} = \frac{(6.626 \times 10^{-34}\ \text{J}\cdot\text{s})^2}{(8)(4.653 \times 10^{-26}\ \text{kg})(1.381 \times 10^{-23}\ \text{J}\cdot\text{K}^{-1})(293.2\ \text{K})}$$
$$= 2.913 \times 10^{-22}\ \text{m}^2$$

and so

$$\frac{N}{V}\left(\frac{h^2}{8mk_BT}\right)^{3/2} = (2.470 \times 10^{25}\ \text{m}^{-3})(2.913 \times 10^{-22}\ \text{m}^2)^{3/2}$$
$$= 1.23 \times 10^{-7}$$

which is much less than unity.

Let's test Inequality 3.40 for liquid nitrogen at its boiling point, $-195.8°$C. Experimentally, the density of $N_2$(l) is 0.808 g·mL$^{-1}$ at its boiling point. Therefore,

$$\frac{N}{V} = (0.808\ \text{g}\cdot\text{mL}^{-1})\left(\frac{1\ \text{mol N}_2}{28.02\ \text{g N}_2}\right)\left(\frac{6.022 \times 10^{23}}{1\ \text{mol}}\right)\left(\frac{10^6\ \text{mL}}{1\ \text{m}^3}\right)$$
$$= 1.737 \times 10^{28}\ \text{m}^{-3}$$

and

$$\frac{N}{V}\left(\frac{h^2}{8mk_BT}\right)^{3/2} = (1.737 \times 10^{28}\ \text{m}^{-3})(1.104 \times 10^{-21}\ \text{m}^2)^{3/2}$$
$$= 6.37 \times 10^{-4}$$

Thus, Equation 3.38 is valid, even for liquid nitrogen at its boiling point.

## 3–8. A Molecular Partition Function Can Be Decomposed into Partition Functions for Each Degree of Freedom

In this section, we will explore the similarity between a system partition function, Equation 3.14, and a molecular partition function, Equation 3.39. We will start by substituting Equation 3.38 into Equation 3.21:

$$\begin{aligned}\langle E\rangle &= k_{\mathrm{B}}T^2\left(\frac{\partial \ln Q}{\partial T}\right)_{N,V} \\ &= Nk_{\mathrm{B}}T^2\left(\frac{\partial \ln q}{\partial T}\right)_{V} \\ &= N\sum_j \varepsilon_j \frac{e^{-\varepsilon_j/k_{\mathrm{B}}T}}{q(V,T)}\end{aligned} \tag{3.41}$$

But Equation 3.38 is valid only for independent particles, so

$$\langle E\rangle = N\langle\varepsilon\rangle \tag{3.42}$$

where $\langle\varepsilon\rangle$ is the average energy of any one molecule. If we compare Equations 3.41 and 3.42, we see that

$$\langle\varepsilon\rangle = \sum_j \varepsilon_j \frac{e^{-\varepsilon_j/k_{\mathrm{B}}T}}{q(V,T)} \tag{3.43}$$

We can conclude from this equation that the probability that a molecule is in its $j$th molecular energy state, $\pi_j$, is given by

$$\pi_j = \frac{e^{-\varepsilon_j/k_{\mathrm{B}}T}}{q(V,T)} = \frac{e^{-\varepsilon_j/k_{\mathrm{B}}T}}{\sum_j e^{-\varepsilon_j/k_{\mathrm{B}}T}} \tag{3.44}$$

Note how similar this equation is to Equation 3.13.

Equation 3.44 can be reduced even further if we assume that the energy of a molecule can be written as

$$\varepsilon = \varepsilon_i^{\mathrm{trans}} + \varepsilon_j^{\mathrm{rot}} + \varepsilon_k^{\mathrm{vib}} + \varepsilon_l^{\mathrm{elec}} \tag{3.45}$$

Because the various energy terms are distinguishable here, we can apply the reasoning behind Equation 3.33 and write

$$q(V,T) = q_{\mathrm{trans}}q_{\mathrm{rot}}q_{\mathrm{vib}}q_{\mathrm{elec}} \tag{3.46}$$

where, for example

$$q_{\mathrm{trans}} = \sum_j e^{-\varepsilon_j^{\mathrm{trans}}/k_{\mathrm{B}}T} \tag{3.47}$$

Note that the partition function for a diatomic molecule we used in Example 3–2 was expressed as

$$q(V,\beta) = q_{\text{trans}}(V,T)q_{\text{rot}}(T)q_{\text{vib}}(T)$$

where

$$q_{\text{trans}}(V,T) = \left(\frac{2\pi m}{h^2\beta}\right)^{3/2} V$$

$$q_{\text{rot}}(T) = \frac{8\pi^2 I}{h^2\beta}$$

and

$$q_{\text{vib}}(T) = \frac{e^{-\beta h\nu/2}}{1 - e^{-\beta h\nu}}$$

If we substitute Equations 3.45 and 3.46 into Equation 3.44, we obtain

$$\pi_{ijkl} = \frac{e^{-\varepsilon_i^{\text{trans}}/k_{\text{B}}T} e^{-\varepsilon_j^{\text{rot}}/k_{\text{B}}T} e^{-\varepsilon_k^{\text{vib}}/k_{\text{B}}T} e^{-\varepsilon_l^{\text{elec}}/k_{\text{B}}T}}{q_{\text{trans}}q_{\text{rot}}q_{\text{vib}}q_{\text{elec}}} \tag{3.48}$$

where $\pi_{ijkl}$ is the probability that a molecule is in the $i$th translational state, the $j$th rotational state, the $k$th vibrational state, and the $l$th electronic state. Now if we sum Equation 3.48 over $i$ (all translational states), $j$ (all rotational states), and $l$ (all electronic states), we obtain

$$\begin{aligned} \pi_k^{\text{vib}} &= \sum_{i,j,l} \pi_{ijkl} = \frac{\left(\sum_i e^{-\varepsilon_i^{\text{trans}}/k_{\text{B}}T}\right)\left(\sum_j e^{-\varepsilon_j^{\text{rot}}/k_{\text{B}}T}\right)\left(\sum_l e^{-\varepsilon_l^{\text{elec}}/k_{\text{B}}T}\right) e^{-\varepsilon_k^{\text{vib}}/k_{\text{B}}T}}{q_{\text{trans}}q_{\text{rot}}q_{\text{vib}}q_{\text{elec}}} \\ &= \frac{e^{-\varepsilon_k^{\text{vib}}/k_{\text{B}}T}}{q_{\text{vib}}} = \frac{e^{-\varepsilon_k^{\text{vib}}/k_{\text{B}}T}}{\sum_k e^{-\varepsilon_k^{\text{vib}}/k_{\text{B}}T}} \end{aligned} \tag{3.49}$$

where, as the notation suggests, $\pi_k^{\text{vib}}$ is the probability that a molecule is in its $k$th vibrational state. Furthermore, the average vibrational energy of a molecule is given by

$$\begin{aligned} \langle\varepsilon^{\text{vib}}\rangle &= \sum_k \varepsilon_k^{\text{vib}} \frac{e^{-\varepsilon_k^{\text{vib}}/k_{\text{B}}T}}{q_{\text{vib}}} \\ &= k_{\text{B}}T^2 \frac{\partial \ln q_{\text{vib}}}{\partial T} = -\frac{\partial \ln q_{\text{vib}}}{\partial \beta} \end{aligned} \tag{3.50}$$

Again, note the similarity with Equation 3.21. Of course, we also have the relations

$$\langle\varepsilon^{\text{trans}}\rangle = k_{\text{B}}T^2 \left(\frac{\partial \ln q_{\text{trans}}}{\partial T}\right)_V = -\left(\frac{\partial \ln q_{\text{trans}}}{\partial \beta}\right)_V \tag{3.51}$$

and

$$\langle \varepsilon^{\text{rot}} \rangle = k_{\text{B}} T^2 \frac{\partial \ln q_{\text{rot}}}{\partial T} = -\frac{\partial \ln q_{\text{rot}}}{\partial \beta} \tag{3.52}$$

**EXAMPLE 3–7**

Use the partition function for a diatomic molecule given in Example 3–2 to calculate $\langle \varepsilon^{\text{vib}} \rangle$.

SOLUTION: According to Example 3–2, we can write

$$q_{\text{vib}}(T) = \frac{e^{-\beta h\nu/2}}{1 - e^{-\beta h\nu}}$$

and so

$$\begin{aligned}\langle \varepsilon^{\text{vib}} \rangle &= -\left( \frac{\partial \ln q_{\text{vib}}}{\partial \beta} \right) \\ &= \frac{h\nu}{2} + \frac{h\nu e^{-\beta h\nu}}{1 - e^{-\beta h\nu}}\end{aligned}$$

in agreement with Equation 3.24.

To this point, we have written partition functions as summations over energy *states*. Each state is represented by a wave function with an associated energy. Thus, we write

$$q(V, T) = \sum_{\substack{j \\ \text{(states)}}} e^{-\varepsilon_j / k_{\text{B}} T} \tag{3.53}$$

We will call sets of states that have the same energy, *levels*. We can write $q(V, T)$ as a summation over levels by including the degeneracy, $g_j$, of the level:

$$q(V, T) = \sum_{\substack{j \\ \text{(levels)}}} g_j e^{-\varepsilon_j / k_{\text{B}} T} \tag{3.54}$$

In the notation of Equation 3.53, the terms representing a degenerate level are repeated $g_j$ times, whereas in Equation 3.54, they are written once and multiplied by $g_j$. For example, we learned in Section 1–8 (Equations 1.28 and 1.29) that the energy and degeneracy for a linear rigid rotator are

$$\varepsilon_J = \frac{\hbar^2}{2I} J(J+1)$$

and

$$g_J = 2J + 1$$

Thus, we can write the rotational partition function by summing over levels:

$$q_{\text{rot}}(T) = \sum_{J=0}^{\infty}(2J+1)e^{-\hbar^2 J(J+1)/2Ik_{\text{B}}T} \tag{3.55}$$

Including degeneracies explicitly as in Equation 3.54 is usually more convenient, so we will use Equation 3.54 rather than Equation 3.53 in later chapters.

## Problems

**3-1.** How would you describe an ensemble whose systems are one-liter containers of water at 25°C?

**3-2.** Show that Equation 3.8 is equivalent to $f(x+y) = f(x)f(y)$. In this problem, we will prove that $f(x) \propto e^{ax}$. First, take the logarithm of the above equation to obtain

$$\ln f(x+y) = \ln f(x) + \ln f(y)$$

Differentiate both sides with respect to $x$ (keeping $y$ fixed) to get

$$\left[\frac{\partial \ln f(x+y)}{\partial x}\right]_y = \frac{d\ln f(x+y)}{d(x+y)}\left[\frac{\partial (x+y)}{\partial x}\right]_y = \frac{d\ln f(x+y)}{d(x+y)}$$
$$= \frac{d\ln f(x)}{dx}$$

Now differentiate with respect to $y$ (keeping $x$ fixed) and show that

$$\frac{d\ln f(x)}{dx} = \frac{d\ln f(y)}{dy}$$

For this relation to be true for all $x$ and $y$, each side must equal a constant, say $a$. Show that

$$f(x) \propto e^{ax} \qquad \text{and} \qquad f(y) \propto e^{ay}$$

**3-3.** Show that $a_l/a_i = e^{\beta(E_i - E_l)}$ implies that $a_j = Ce^{-\beta E_j}$.

**3-4.** Prove to yourself that $\sum_i e^{-\beta E_i} = \sum_j e^{-\beta E_j}$.

**3-5.** Show that the partition function in Example 3–1 can be written as

$$Q(\beta, B_z) = 2\cosh\left(\frac{\beta\hbar\gamma B_z}{2}\right) = 2\cosh\left(\frac{\hbar\gamma B_z}{2k_{\text{B}}T}\right)$$

Use the fact that $d\cosh x/dx = \sinh x$ to show that

$$\langle E\rangle = -\frac{\hbar\gamma B_z}{2}\tanh\frac{\beta\hbar\gamma B_z}{2} = -\frac{\hbar\gamma B_z}{2}\tanh\frac{\hbar\gamma B_z}{2k_{\text{B}}T}$$

**3-6.** Use either the expression for $\langle E\rangle$ in Example 3–1 or the one in Problem 3–5 to show that

$$\langle E\rangle \longrightarrow -\frac{\hbar\gamma B_z}{2} \qquad \text{as} \qquad T \longrightarrow 0$$

and that

$$\langle E \rangle \longrightarrow 0 \quad \text{as} \quad T \longrightarrow \infty$$

**3-7.** Generalize the results of Example 3–1 to the case of a spin-1 nucleus. Determine the low-temperature and high-temperature limits of $\langle E \rangle$.

**3-8.** If $N_w$ is the number of protons aligned with a magnetic field $B_z$ and $N_o$ is the number of protons opposed to the field, show that

$$\frac{N_o}{N_w} = e^{-\hbar \gamma B_z / k_B T}$$

Given that $\gamma = 26.7522 \times 10^7 \text{ rad}\cdot\text{T}^{-1}\cdot\text{s}^{-1}$ for a proton, calculate $N_o/N_w$ as a function of temperature for a field strength of 5.0 T. At what temperature is $N_o = N_w$? Interpret this result physically.

**3-9.** In Section 3–3, we derived an expression for $\langle E \rangle$ for a monatomic ideal gas by applying Equation 3.20 to $Q(N, V, T)$ given by Equation 3.22. Apply Equation 3.21 to

$$Q(N, V, T) = \frac{1}{N!} \left( \frac{2\pi m k_B T}{h^2} \right)^{3N/2} V^N$$

to derive the same result. Note that this expression for $Q(N, V, T)$ is simply Equation 3.22 with $\beta$ replaced by $1/k_B T$.

**3-10.** A gas absorbed on a surface can sometimes be modelled as a two-dimensional ideal gas. We will learn in Chapter 4 that the partition function of a two-dimensional ideal gas is

$$Q(N, A, T) = \frac{1}{N!} \left( \frac{2\pi m k_B T}{h^2} \right)^{N} A^N$$

where $A$ is the area of the surface. Derive an expression for $\langle E \rangle$ and compare your result with the three-dimensional result.

**3-11.** Although we will not do so in this book, it is possible to derive the partition function for a monatomic van der Waals gas.

$$Q(N, V, T) = \frac{1}{N!} \left( \frac{2\pi m k_B T}{h^2} \right)^{3N/2} (V - Nb)^N e^{aN^2/V k_B T}$$

where $a$ and $b$ are the van der Waals constants. Derive an expression for the energy of a monatomic van der Waals gas.

**3-12.** An approximate partition function for a gas of hard spheres can be obtained from the partition function of a monatomic gas by replacing $V$ in Equation 3.22 (and the following equation) by $V - b$, where $b$ is related to the volume of the $N$ hard spheres. Derive expressions for the energy and the pressure of this system.

**3-13.** Use the partition function in Problem 3–10 to calculate the heat capacity of a two-dimensional ideal gas.

**3-14.** Use the partition function for a monatomic van der Waals gas given in Problem 3–11 to calculate the heat capacity of a monatomic van der Waals gas. Compare your result with that of a monatomic ideal gas.

**3-15.** Using the partition function given in Example 3–2, show that the pressure of an ideal diatomic gas obeys $PV = Nk_BT$, just as it does for a monatomic ideal gas.

**3-16.** Show that if a partition function is of the form

$$Q(N, V, T) = \frac{[q(V, T)]^N}{N!}$$

and if $q(V, T) = f(T)V$ [as it does for a monatomic ideal gas (Equation 3.22) and a diatomic ideal gas (Example 3–2)], then the ideal-gas equation of state results.

**3-17.** Use Equation 3.27 and the value of $\tilde{\nu}$ for $O_2$ given in Table 1.4 to calculate the value of the molar heat capacity of $O_2$(g) from 300 K to 1000 K (see Figure 3.3).

**3-18.** Show that the heat capacity given by Equation 3.29 in Example 3–3 obeys a law of corresponding states.

**3-19.** Consider a system of independent, distinguishable particles that have only two quantum states with energy $\varepsilon_0$ (let $\varepsilon_0 = 0$) and $\varepsilon_1$. Show that the molar heat capacity of such a system is given by

$$\overline{C}_V = R(\beta\varepsilon)^2 \frac{e^{-\beta\varepsilon}}{(1 + e^{-\beta\varepsilon})^2}$$

and that $\overline{C}_V$ plotted against $\beta\varepsilon$ passes through a maximum value at $\beta\varepsilon$, given by the solution to $\beta\varepsilon/2 = \coth \beta\varepsilon/2$. Use a table of values of $\coth x$ (for example, the *CRC Standard Mathematical Tables*) to show that $\beta\varepsilon = 2.40$.

**3-20.** Deriving the partition function for an Einstein crystal is not difficult (see Example 3–3). Each of the $N$ atoms of the crystal is assumed to vibrate independently about its lattice position, so that the crystal is pictured as $N$ independent harmonic oscillators, each vibrating in three directions. The partition function of a harmonic oscillator is

$$q_{\text{ho}}(T) = \sum_{v=0}^{\infty} e^{-\beta(v+\frac{1}{2})h\nu}$$
$$= e^{-\beta h\nu/2} \sum_{v=0}^{\infty} e^{-\beta v h\nu}$$

This summation is easy to evaluate if you recognize it as the so-called geometric series (MathChapter C)

$$\sum_{v=0}^{\infty} x^v = \frac{1}{1 - x}$$

Show that

$$q_{\text{ho}}(T) = \frac{e^{-\beta h\nu/2}}{1 - e^{-\beta h\nu}}$$

and that

$$Q = e^{-\beta U_0} \left( \frac{e^{-\beta h\nu/2}}{1 - e^{-\beta h\nu}} \right)^{3N}$$

where $U_0$ simply represents the zero-of-energy, where all $N$ atoms are infinitely separated.

**3-21.** Show that

$$S = \sum_{i=1}^{2} \sum_{j=0}^{1} x^i y^j = x(1+y) + x^2(1+y) = (x+x^2)(1+y)$$

by summing over $j$ first and then over $i$. Now obtain the same result by writing $S$ as a product of two separate summations.

**3-22.** Evaluate

$$S = \sum_{i=0}^{2} \sum_{j=0}^{1} x^{i+j}$$

by summing over $j$ first and then over $i$. Now obtain the same result by writing $S$ as a product of two separate summations.

**3-23.** How many terms are there in the following summations?

**a.** $S = \sum_{i=1}^{3} \sum_{j=1}^{2} x^i y^j$ **b.** $S = \sum_{i=1}^{3} \sum_{j=0}^{2} x^i y^j$ **c.** $S = \sum_{i=1}^{3} \sum_{j=1}^{2} \sum_{k=1}^{2} x^i y^j z^k$

**3-24.** Consider a system of two noninteracting identical fermions, each of which has states with energies $\varepsilon_1$, $\varepsilon_2$, and $\varepsilon_3$. How many terms are there in the unrestricted evaluation of $Q(2, V, T)$? Enumerate the allowed total energies in the summation in Equation 3.37 (see Example 3–5). How many terms occur in $Q(2, V, T)$ when the fermion restriction is taken into account?

**3-25.** Redo Problem 3–24 for the case of bosons instead of fermions.

**3-26.** Consider a system of three noninteracting identical fermions, each of which has states with energies $\varepsilon_1$, $\varepsilon_2$, and $\varepsilon_3$. How many terms are there in the unrestricted evaluation of $Q(3, V, T)$? Enumerate the allowed total energies in the summation of Equation 3.37 (see Example 3–5). How many terms occur in $Q(3, V, T)$ when the fermion restriction is taken into account?

**3-27.** Redo Problem 3–26 for the case of bosons instead of fermions.

**3-28.** Evaluate $(N/V)(h^2/8mk_BT)^{3/2}$ (see Table 3.1) for $O_2(g)$ at its normal boiling point, 90.20 K. Use the ideal-gas equation of state to calculate the density of $O_2(g)$ at 90.20 K.

**3-29.** Evaluate $(N/V)(h^2/8mk_BT)^{3/2}$ (see Table 3.1) for He(g) at its normal boiling point 4.22 K. Use the ideal-gas equation of state to calculate the density of He(g) at 4.22 K.

**3-30.** Evaluate $(N/V)(h^2/8mk_BT)^{3/2}$ for the electrons in sodium metal at 298 K. Take the density of sodium to 0.97 $g \cdot mL^{-1}$. Compare your result with the value given in Table 3.1.

**3-31.** Evaluate $(N/V)(h^2/8mk_BT)^{3/2}$ (see Table 3.1) for liquid hydrogen at its normal boiling point 20.3 K. The density of $H_2$(l) at its boiling point is 0.067 g·mL$^{-1}$.

**3-32.** Because the molecules in an ideal gas are independent, the partition function of a mixture of monatomic ideal gases is of the form

$$Q(N_1, N_2, V, T) = \frac{[q_1(V,T)]^{N_1}}{N_1!}\frac{[q_2(V,T)]^{N_2}}{N_2!}$$

where

$$q_j(V,T) = \left(\frac{2\pi m_j k_B T}{h^2}\right)^{3/2} V \qquad j = 1, 2$$

Show that

$$\langle E\rangle = \frac{3}{2}(N_1 + N_2)k_BT$$

and that

$$PV = (N_1 + N_2)k_BT$$

for a mixture of monatomic ideal gases.

**3-33.** We will learn in Chapter 4 that the rotational partition function of an asymmetric top molecule is given by

$$q_{rot}(T) = \frac{\pi^{1/2}}{\sigma}\left(\frac{8\pi^2 I_A k_B T}{h^2}\right)^{1/2}\left(\frac{8\pi^2 I_B k_B T}{h^2}\right)^{1/2}\left(\frac{8\pi^2 I_C k_B T}{h^2}\right)^{1/2}$$

where $\sigma$ is a constant and $I_A$, $I_B$, and $I_C$ are the three (distinct) moments of inertia. Show that the rotational contribution to the molar heat capacity is $\overline{C}_{V,rot} = \frac{3}{2}R$.

**3-34.** The allowed energies of a harmonic oscillator are given by $\varepsilon_v = (v + \frac{1}{2})h\nu$. The corresponding partition function is given by

$$q_{vib}(T) = \sum_{v=0}^{\infty} e^{-(v+\frac{1}{2})h\nu/k_BT}$$

Let $x = e^{-h\nu/k_BT}$ and use the formula for the summation of a geometric series (Problem 3–20) to show that

$$q_{vib}(T) = \frac{e^{-h\nu/2k_BT}}{1 - e^{-h\nu/k_BT}}$$

**3-35.** Derive an expression for the probability that a harmonic oscillator will be found in the $v$th state. Calculate the probability that the first few vibrational states are occupied for HCl(g) at 300 K. (See Table 1–4 and Problem 3–34.)

**3-36.** Show that the fraction of harmonic oscillators in the ground vibrational state is given by

$$f_0 = 1 - e^{-h\nu/k_BT}$$

Calculate $f_0$ for $N_2$(g) at 300 K, 600 K, and 1000 K (see Table 1.4).

**3-37.** Use Equation 3.55 to show that the fraction of rigid rotators in the $J$th rotational level is given by

$$f_J = \frac{(2J+1)e^{-\hbar^2 J(J+1)/2Ik_BT}}{q_{rot}(T)}$$

Plot the fraction in the $J$th level relative to the $J = 0$ level ($f_J/f_0$) against $J$ for HCl(g) at 300 K. Take $\tilde{B} = 10.44\ \text{cm}^{-1}$.

**3-38.** Equations 3.20 and 3.21 give the ensemble average of $E$, which we assert is the same as the experimentally observed value. In this problem, we will explore the standard deviation about $\langle E \rangle$ (MathChapter B). We start with either Equation 3.20 or 3.21:

$$\langle E \rangle = U = -\left(\frac{\partial \ln Q}{\partial \beta}\right)_{N,V} = k_BT^2\left(\frac{\partial \ln Q}{\partial T}\right)_{N,V}$$

Differentiate again with respect to $\beta$ or $T$ to show that (MathChapter B)

$$\sigma_E^2 = \langle E^2 \rangle - \langle E \rangle^2 = k_BT^2C_V$$

where $C_V$ is the heat capacity. To explore the relative magnitude of the spread about $\langle E \rangle$, consider

$$\frac{\sigma_E}{\langle E \rangle} = \frac{(k_BT^2C_V)^{1/2}}{\langle E \rangle}$$

To get an idea of the size of this ratio, use the values of $\langle E \rangle$ and $C_V$ for a (monatomic) ideal gas, namely, $\frac{3}{2}Nk_BT$ and $\frac{3}{2}Nk_B$, respectively, and show that $\sigma_E/\langle E \rangle$ goes as $N^{-1/2}$. What does this trend say about the likely observed deviations from the average macroscopic energy?

**3-39.** Following Problem 3–38, show that the variance about the average values of a *molecular* energy is given by

$$\sigma_\varepsilon^2 = \langle \varepsilon^2 \rangle - \langle \varepsilon \rangle^2 = \frac{k_BT^2C_V}{N}$$

and that $\sigma_\varepsilon/\langle \varepsilon \rangle$ goes as order unity. What does this result say about the deviations from the average molecular energy?

**3-40.** Use the result of Problem 3–38 to show that $C_V$ is never negative.

**3-41.** The energies and degeneracies of the four lowest electronic states of Na(g) are tabulated below.

| Energy/cm$^{-1}$ | Degeneracy |
|---|---|
| 0.000 | 2 |
| 16 956.183 | 2 |
| 16 973.379 | 4 |
| 25 739.86 | 2 |

Calculate the fraction of the atoms in each of these electronic states in a sample of Na(g) at 1000 K. Repeat this calculation for a temperature of 2500 K.

**3-42.** The vibrational frequency of NaCl(g) is 159.23 $\text{cm}^{-1}$. Calculate the molar heat capacity, $\overline{C}_V$, at 1000 K. (See Equation 3.27.)

**3-43.** The energies and degeneracies of the two lowest electronic states of atomic iodine are listed below.

| Energy/$\text{cm}^{-1}$ | Degeneracy |
|---|---|
| 0 | 4 |
| 7603.2 | 2 |

What temperature is required so that 2% of the atoms are in the excited state?

# MATHCHAPTER C

# SERIES AND LIMITS

Frequently, we need to investigate the behavior of an equation for small values (or perhaps large values) of one of the variables in the equation. For example, we shall show in Chapter 4 that the temperature dependence of the molar heat capacity of a diatomic gas is given by

$$\overline{C}_V = \frac{5R}{2} + R\left(\frac{\Theta_{\text{vib}}}{T}\right)^2 \frac{e^{-\Theta_{\text{vib}}/T}}{(1 - e^{-\Theta_{\text{vib}}/T})^2} \tag{C.1}$$

where $R$ is the molar gas constant and $\Theta_{\text{vib}}$ is a constant that is characteristic of the gas. Suppose now that we wish to determine $\overline{C}_V$ at high temperatures, where $\Theta_{\text{vib}}/T$ is small. To do this, we first have to use the fact that $e^x$ can be written as the infinite series (i.e., a series containing an unending number of terms)

$$e^x = \sum_{n=0}^{\infty} \frac{x^n}{n!} = 1 + x + \frac{x^2}{2!} + \frac{x^3}{3!} + \cdots \tag{C.2}$$

and then realize that if $x$ is small, then $x^2$, $x^3$, etc. are even smaller. We can express this result by writing

$$e^x = 1 + x + O(x^2)$$

where $O(x^2)$ is a bookkeeping symbol that reminds us we are neglecting terms involving $x^2$ and higher powers of $x$. If we apply this result to Equation C.1, we have

$$\begin{aligned}\overline{C} &= \frac{5R}{2} + R\left(\frac{\Theta_{\text{vib}}}{T}\right)^2 \frac{1 - \frac{\Theta_{\text{vib}}}{T} + \cdots}{\left(1 - 1 - \frac{\Theta_{\text{vib}}}{T} + \cdots\right)^2} \\ &= \frac{5R}{2} + R\left(\frac{\Theta_{\text{vib}}}{T}\right)^2 \frac{1}{(\Theta_{\text{vib}}/T)^2} = \frac{7R}{2}\end{aligned}$$

Thus, we see that $\overline{C}_V$ takes on a limiting value of $7R/2$ at high temperatures. In this MathChapter, we will review some useful series and apply them to some physical problems.

One of the most useful series we will use is the geometric series:

$$\frac{1}{1-x} = \sum_{n=0}^{\infty} x^n = 1 + x + x^2 + x^3 + \cdots \qquad |x| < 1 \tag{C.3}$$

This result can be derived by algebraically dividing 1 by $1 - x$, or by the following trick. Consider the finite series (i.e., a series with a finite number of terms)

$$S_N = 1 + x + x^2 + \cdots + x^N$$

Now multiply $S_N$ by $x$:

$$xS_N = x + x^2 + \cdots + x^{N+1}$$

Now notice that

$$S_N - xS_N = 1 - x^{N+1}$$

or that

$$S_N = \frac{1 - x^{N+1}}{1-x} \tag{C.4}$$

If $|x| < 1$, then $x^{N+1} \to 0$ as $N \to \infty$, so we recover Equation C.3.

Recovering Equation C.3 from Equation C.4 brings us to an important point regarding infinite series: Equation C.3 is valid only if $|x| < 1$. It makes no sense at all if $|x| \geq 1$. We say that the infinite series in Equation C.3 converges for $|x| < 1$ and diverges for $|x| \geq 1$. How can we tell whether a given infinite series converges or diverges? There are a number of so-called convergence tests, but one simple and useful one is the *ratio test*. To apply the ratio test, we form the ratio of the $(n + 1)$th term, $u_{n+1}$, to the $n$th term, $u_n$, and then let $n$ become very large:

$$r = \lim_{n\to\infty} \left| \frac{u_{n+1}}{u_n} \right| \tag{C.5}$$

If $r < 1$, the series converges; if $r > 1$, the series diverges; and if $r = 1$, the test is inconclusive. Let's apply this test to the geometric series (Equation C.3). In this case, $u_{n+1} = x^{n+1}$ and $u_n = x^n$, so

$$r = \lim_{n\to\infty} \left| \frac{x^{n+1}}{x^n} \right| = |x|$$

Thus, we see that the series converges if $|x| < 1$ and diverges if $|x| > 1$. It actually diverges at $x = 1$, but the ratio test does not tell us that. We would have to use a more sophisticated convergence test to determine the behavior at $x = 1$.

For the exponential series (Equation C.2), we have

$$r = \lim_{n\to\infty} \left| \frac{x^{n+1}/(n+1)!}{x^n/n!} \right| = \lim_{n\to\infty} \left| \frac{x}{n+1} \right|$$

Thus, we conclude that the exponential series converges for all values of $x$.

In Chapter 4, we encounter the summation

$$S = \sum_{v=0}^{\infty} e^{-vh\nu/k_\mathrm{B}T} \tag{C.6}$$

where $\nu$ represents the vibrational frequency of a diatomic molecule and the other symbols have their usual meanings. We can sum this series by letting

$$x = e^{-h\nu/k_\mathrm{B}T}$$

in which case we have

$$S = \sum_{v=0}^{\infty} x^v$$

The quantity $x$ is less than 1, and according to Equation C.3, $S = 1/(1-x)$, or

$$S = \frac{1}{1 - e^{-h\nu/k_\mathrm{B}T}} \tag{C.7}$$

We say that $S$ has been evaluated in closed form because its numerical evaluation requires only a finite number of steps, in contrast to Equation C.6, which would require an infinite number of steps.

A practical question that arises is how do we find the infinite series that corresponds to a given function. For example, how do we derive Equation C.2? First, assume that the function $f(x)$ can be expressed as a power series in $x$:

$$f(x) = c_0 + c_1x + c_2x^2 + c_3x^3 + \cdots$$

where the $c_j$ are to be determined. Then let $x = 0$ and find that $c_0 = f(0)$. Now differentiate once with respect to $x$

$$\frac{df}{dx} = c_1 + 2c_2x + 3c_3x_2 + \cdots$$

and let $x = 0$ to find that $c_1 = (df/dx)_{x=0}$. Differentiate again,

$$\frac{d^2f}{dx^2} = 2c_2 + 3 \cdot 2c_3x + \cdots$$

and let $x = 0$ to get $c_2 = (d^2f/dx^2)_{x=0}/2$. Differentiate once more,

$$\frac{d^3f}{dx^3} = 3 \cdot 2c_3 + 4 \cdot 3 \cdot 2x + \cdots$$

and let $x = 0$ to get $c_3 = (d^3 f/dx^3)_{x=0}/3!$. The general result is

$$c_n = \frac{1}{n!}\left(\frac{d^n f}{dx^n}\right)_{x=0} \tag{C.8}$$

so we can write

$$f(x) = f(0) + \left(\frac{df}{dx}\right)_{x=0} x + \frac{1}{2!}\left(\frac{d^2 f}{dx^2}\right)_{x=0} x^2 + \frac{1}{3!}\left(\frac{d^3 f}{dx^3}\right)_{x=0} x^3 + \cdots \tag{C.9}$$

Equation C.9 is called the Maclaurin series of $f(x)$. If we apply Equation C.9 to $f(x) = e^x$, we find that

$$\left(\frac{d^n e^x}{dx^n}\right)_{x=0} = 1$$

so

$$e^x = 1 + x + \frac{x^2}{2!} + \frac{x^3}{3!} + \cdots$$

Some other important Maclaurin series, which can be obtained from a straightforward application of Equation C.9 (Problem C–7) are

$$\sin x = x - \frac{x^3}{3!} + \frac{x^5}{5!} - \frac{x^7}{7!} + \cdots \tag{C.10}$$

$$\cos x = 1 - \frac{x^2}{2!} + \frac{x^4}{4!} - \frac{x^6}{6!} + \cdots \tag{C.11}$$

$$\ln(1 + x) = x - \frac{x^2}{2} + \frac{x^3}{3} - \frac{x^4}{4} + \cdots \qquad -1 < x \le 1 \tag{C.12}$$

and

$$(1 + x)^n = 1 + nx + \frac{n(n-1)}{2!}x^2 + \frac{n(n-1)(n-2)}{3!}x^3 + \cdots \qquad x^2 < 1 \tag{C.13}$$

Series C.10 and C.11 converge for all values of $x$, but as indicated, Series C.12 converges only for $-1 < x \le 1$ and Series C.13 converges only for $x^2 < 1$. Note that if $n$ is a positive integer in Series C.13, the series truncates. For example, if $n = 2$ or 3, we have

$$(1 + x)^2 = 1 + 2x + x^2$$

and

$$(1 + x)^3 = 1 + 3x + 3x^2 + x^3$$

Equation C.13 for a positive integer is called the binomial expansion. If $n$ is not a positive integer, the series continues indefinitely, and Equation C.13 is called the binomial series. Any handbook of mathematical tables will have the Maclaurin series for many functions. Problem C–13 discusses a Taylor series, which is an extension of a Maclaurin series.

We can use the series presented here to derive a number of results used throughout the book. For example, the limit

$$\lim_{x \to 0} \frac{\sin x}{x}$$

occurs several times. Because this limit gives 0/0, we could use l'Hôpital's rule, which tells us that

$$\lim_{x \to 0} \frac{\sin x}{x} = \lim_{x \to 0} \frac{\dfrac{d \sin x}{dx}}{\dfrac{dx}{dx}} = \lim_{x \to 0} \cos x = 1$$

We could derive the same result by dividing Equation C.10 by $x$ and then letting $x \to 0$. (These two methods are really equivalent. See Problem C–14.)

We will do one final example involving series and limits. According to a theory by Debye, the temperature dependence of the molar heat capacity of a crystal is given by

$$\overline{C}_V(T) = 9R\left(\frac{T}{\Theta_D}\right)^3 \int_0^{\Theta_D/T} \frac{x^4 e^x dx}{(e^x - 1)^2} \tag{C.14}$$

In this equation, $T$ is the kelvin temperature, $R$ is the molar gas constant, and $\Theta_D$ is a parameter characteristic of the particular crystal. The parameter $\Theta_D$ has units of temperature and is called the Debye temperature of the crystal. We want to determine both the low-temperature and the high-temperature limits of $\overline{C}_V(T)$. In the low-temperature limit, the upper limit of the integral becomes very large. For large values of $x$, we can neglect 1 compared with $e^x$ in the denominator of the integrand, showing that the integrand goes as $x^4 e^{-x}$ for large $x$. But $x^4 e^{-x} \to 0$ as $x \to \infty$, so the upper limit of the integral can safely be set to $\infty$, giving

$$\lim_{T \to 0} \overline{C}_V(T) = 9R\left(\frac{T}{\Theta_D}\right)^3 \int_0^{\infty} \frac{x^4 e^x dx}{(e^x - 1)^2}$$

Whatever the value of the integral here, it is just a constant, so we see that

$$\overline{C}_V(T) \to \text{constant} \times T^3 \quad \text{as} \quad T \to 0$$

This famous result for the low-temperature heat capacity of a crystal is called the $T^3$ law. The low-temperature heat capacity goes to zero as $T^3$. We will use the $T^3$ law in Chapter 7.

Now let's look at the high-temperature limit. For high temperatures, the upper limit of the integral in Equation C.14 becomes very small. Consequently, during the integration from 0 to $\Theta_D/T$, $x$ is always small. Therefore, we can use Equation C.2 for $e^x$, giving

$$
\begin{aligned}
\lim_{T\to\infty} \overline{C}_V(T) &= 9R\left(\frac{T}{\Theta_D}\right)^3 \int_0^{\Theta_D/T} \frac{x^4[1+x+O(x^2)]dx}{[1+x+O(x^2)-1]^2} \\
&= 9R\left(\frac{T}{\Theta_D}\right)^3 \int_0^{\Theta_D/T} x^2 dx \\
&= 9R\left(\frac{T}{\Theta_D}\right)^3 \cdot \frac{1}{3}\left(\frac{\Theta_D}{T}\right)^3 = 3R
\end{aligned}
$$

This result is called the Law of Dulong and Petit; the molar heat capacity of a crystal becomes $3R = 24.9\ \text{J}\cdot\text{K}^{-1}\cdot\text{mol}^{-1}$ for monatomic crystals at high temperatures. By "high temperatures", we actually mean that $T \gg \Theta_D$, which for many substances is less than 1000 K.

## Problems

**C-1.** Calculate the percentage difference between $e^x$ and $1+x$ for $x = 0.0050, 0.0100, 0.0150, \ldots, 0.1000$.

**C-2.** Calculate the percentage difference between $\ln(1+x)$ and $x$ for $x = 0.0050, 0.0100, 0.0150, \ldots, 0.1000$.

**C-3.** Write out the expansion of $(1+x)^{1/2}$ through the quadratic term.

**C-4.** Evaluate the series

$$
S = \sum_{v=0}^{\infty} e^{-(v+\frac{1}{2})\beta h\nu}
$$

**C-5.** Show that

$$
\frac{1}{(1-x)^2} = 1 + 2x + 3x^2 + 4x^3 + \cdots
$$

**C-6.** Evaluate the series

$$
S = \frac{1}{2} + \frac{1}{4} + \frac{1}{8} + \frac{1}{16} + \cdots
$$

**C-7.** Use Equation C.9 to derive Equations C.10 and C.11.

**C-8.** Show that Equations C.2, C.10, and C.11 are consistent with the relation $e^{ix} = \cos x + i \sin x$.

**C-9.** In Example 3–3, we derived a simple formula for the molar heat capacity of a solid based on a model by Einstein:

$$
\overline{C}_V = 3R\left(\frac{\Theta_E}{T}\right)^2 \frac{e^{-\Theta_E/T}}{(1-e^{-\Theta_E/T})^2}
$$

where $R$ is the molar gas constant and $\Theta_E = h\nu/k_B$ is a constant, called the Einstein temperature, that is characteristic of the solid. Show that this equation gives the Dulong and Petit limit ($\overline{C}_V \to 3R$) at high temperatures.

**C-10.** Evaluate the limit of

$$f(x) = \frac{e^{-x}\sin^2 x}{x^2}$$

as $x \to 0$.

**C-11.** Evaluate the integral

$$I = \int_0^a x^2 e^{-x}\cos^2 x dx$$

for small values of $a$ by expanding $I$ in powers of $a$ through quadratic terms.

**C-12.** Prove that the series for $\sin x$ converges for all values of $x$.

**C-13.** A Maclaurin series is an expansion about the point $x = 0$. A series of the form

$$f(x) = c_0 + c_1(x - x_0) + c_2(x - x_0)^2 + \cdots$$

is an expansion about the point $x_0$ and is called a Taylor series. First show that $c_0 = f(x_0)$. Now differentiate both sides of the above expansion with respect to $x$ and then let $x = x_0$ to show that $c_1 = (df/dx)_{x=x_0}$. Now show that

$$c_n = \frac{1}{n!}\left(\frac{d^n f}{dx^n}\right)_{x=x_0}$$

and so

$$f(x) = f(x_0) + \left(\frac{df}{dx}\right)_{x_0}(x - x_0) + \frac{1}{2}\left(\frac{d^2 f}{dx^2}\right)_{x_0}(x - x_0)^2 + \cdots$$

**C-14.** Later on, we will need to sum the series

$$s_1 = \sum_{v=0}^{\infty} v x^v$$

and

$$s_2 = \sum_{v=0}^{\infty} v^2 x^v$$

To sum the first one, start with (Equation C.3)

$$s_0 = \sum_{v=0}^{\infty} x^v = \frac{1}{1-x}$$

Differentiate with respect to $x$ and then multiply by $x$ to obtain

$$s_1 = \sum_{v=0}^{\infty} v x^v = x\frac{ds_0}{dx} = x\frac{d}{dx}\left(\frac{1}{1-x}\right) = \frac{x}{(1-x)^2}$$

Using the same approach, show that

$$s_2 = \sum_{v=0}^{\infty} v^2 x^v = \frac{x + x^2}{(1-x)^3}$$

**William Francis Giauque** was born on May 12, 1895, in Niagara Falls, Ontario, Canada, to American parents and died in 1982. After working for two years in the laboratory at Hooker Electro-Chemical Company in Niagara Falls, he entered the University of California at Berkeley with the intent of becoming a chemical engineer. He decided to study chemistry, however, and remained at Berkeley to receive his Ph.D. in chemistry with a minor in physics in 1922. His dissertation was on the behavior of materials at very low temperatures. Upon receiving his Ph.D., Giauque accepted a faculty position in the College of Chemistry at Berkeley and remained there for the rest of his life. He made exhaustive and meticulous thermochemical studies that explored the Third Law of Thermodynamics. In particular, his very low temperature studies of the entropies of substances validated the Third Law. Giauque developed the technique of adiabatic demagnetization to achieve low temperatures. He achieved a temperature of 0.25 K, and other research groups subsequently reached temperatures as low as 0.0014 K using Giauque's technique. Together with his graduate student Herrick Johnston, he spectroscopically identified the two hitherto unknown oxygen isotopes 17 and 18 in 1929. He was awarded the Nobel Prize for chemistry in 1949 "for his contributions in the field of chemical thermodynamics, particularly concerning the behavior of substances at extremely low temperatures."

CHAPTER 4

# Partition Functions and Ideal Gases

In this chapter, we will apply the general results of the preceding chapter to calculate the partition functions and heat capacities of ideal gases. We have shown in Section 3–7 that if the number of available quantum states is much greater than the number of particles, we can write the partition function of the entire system in terms of the individual atomic or molecular partition functions:

$$Q(N, V, T) = \frac{[q(V, T)]^N}{N!}$$

This equation is particularly applicable to ideal gases because the molecules are independent and the densities of gases that behave ideally are low enough that the inequality given by Equation 3.40 is satisfied. We will discuss a monatomic ideal gas first and then diatomic and polyatomic ideal gases.

## 4–1. The Translational Partition Function of an Atom in a Monatomic Ideal Gas is $(2\pi m k_B T/h^2)^{3/2} V$

The energy of an atom in an ideal monatomic gas can be written as the sum of its translational energy and its electronic energy

$$\varepsilon_{\text{atomic}} = \varepsilon_{\text{trans}} + \varepsilon_{\text{elec}}$$

so the atomic partition function can be written as

$$q(V, T) = q_{\text{trans}}(V, T)q_{\text{elec}}(T) \tag{4.1}$$

We will evaluate the translational partition function first.

The translational energy states in a cubic container are given by (Equation 1–45)

$$\varepsilon_{n_x n_y n_z} = \frac{h^2}{8ma^2}\left(n_x^2 + n_y^2 + n_z^2\right) \qquad n_x, n_y, n_z = 1, 2, \ldots \tag{4.2}$$

We substitute Equation 4.2 into $q_{\text{trans}}$ (Equation 3.47) to get

$$q_{\text{trans}} = \sum_{n_x, n_y, n_z=1}^{\infty} e^{-\beta\varepsilon_{n_x n_y n_z}} = \sum_{n_x=1}^{\infty}\sum_{n_y=1}^{\infty}\sum_{n_z=1}^{\infty} \exp\left[-\frac{\beta h^2}{8ma^2}\left(n_x^2 + n_y^2 + n_z^2\right)\right] \tag{4.3}$$

Because $e^{a+b+c} = e^a e^b e^c$, we can write the triple summation as a product of three single summations:

$$q_{\text{trans}} = \sum_{n_x=1}^{\infty} \exp\left(-\frac{\beta h^2 n_x^2}{8ma^2}\right) \sum_{n_y=1}^{\infty} \exp\left(-\frac{\beta h^2 n_y^2}{8ma^2}\right) \sum_{n_z=1}^{\infty} \exp\left(-\frac{\beta h^2 n_z^2}{8ma^2}\right)$$

Now, each of these three single summations is alike, because each one is simply

$$\sum_{n=1}^{\infty} \exp\left(-\frac{\beta h^2 n^2}{8ma^2}\right) = e^{-\beta h^2/8ma^2} + e^{-4\beta h^2/8ma^2} + e^{-9\beta h^2/8ma^2} + \cdots$$

Thus, we can write Equation 4.3 as

$$q_{\text{trans}}(V, T) = \left[\sum_{n=1}^{\infty} \exp\left(-\frac{\beta h^2 n^2}{8ma^2}\right)\right]^3 \tag{4.4}$$

This summation cannot be expressed in terms of any simple analytic function. This situation does not present any difficulty, however, for the following reason. Graphically, a summation such as $\sum_{n=1}^{\infty} f_n$ is equal to the sum of the areas under rectangles of unit width centered at $1, 2, 3, \ldots$ and of height $f_1, f_2, f_3, \ldots$ as shown in Figure 4.1. If the heights of successive rectangles differ by a very small amount, the area of the rectangles is essentially equal to the area under the continuous curve obtained by letting the summation index $n$ be a continuous variable (Figure 4.1). Problem 4–2 helps you prove that the successive terms in the summation in Equation 4.4 do indeed differ very little from each other under most conditions.

Thus, it is an excellent approximation to replace the summation in Equation 4.4 by an integration:

$$q_{\text{trans}}(V, T) = \left(\int_0^{\infty} e^{-\beta h^2 n^2/8ma^2}\, dn\right)^3 \tag{4.5}$$

Note that the integral starts at $n = 0$, whereas the summation in Equation 4.4 starts at $n = 1$. For the small values of $\beta h^2/8ma^2$ we are considering here, the difference is negligible (Problem 4–41). If we denote $\beta h^2/8ma^2$ by $\alpha$, the above integral becomes (see MathChapter B)

$$\int_0^{\infty} e^{-\alpha n^2}\, dn = \left(\frac{\pi}{4\alpha}\right)^{1/2}$$

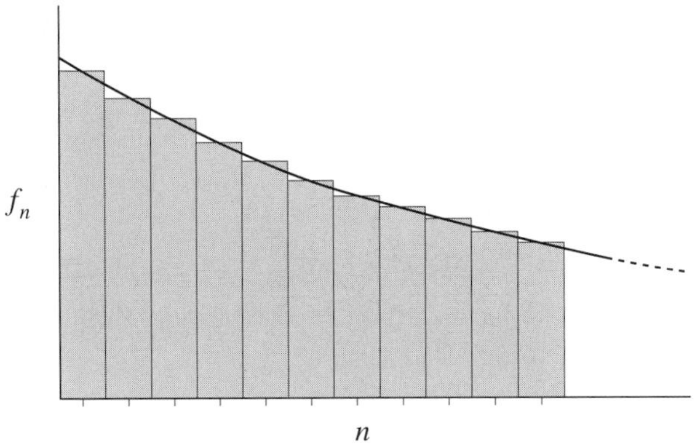


**FIGURE 4.1**
An illustration of the approximation of a summation $\sum_{n=1}^{\infty} f_n$ by an integral. The summation is equal to the areas of the rectangles and the integral is equal to the area under the curve obtained by letting $n$ be a continuous variable.

so we have that

$$q_{\text{trans}}(V, T) = \left(\frac{2\pi m k_{\text{B}} T}{h^2}\right)^{3/2} V \tag{4.6}$$

where we have written $V$ for $a^3$. Note that $q_{\text{trans}}$ is a function of $V$ and $T$.

We can calculate the average translational energy of an ideal-gas atom from this partition function by using Equation 3.51:

$$\begin{aligned}\langle \varepsilon_{\text{trans}} \rangle &= k_{\text{B}} T^2 \left(\frac{\partial \ln q_{\text{trans}}}{\partial T}\right)_V \\ &= k_{\text{B}} T^2 \left(\frac{\partial}{\partial T}\left[\frac{3}{2} \ln T + \text{terms independent of } T\right]\right)_V \\ &= \tfrac{3}{2} k_{\text{B}} T \end{aligned} \tag{4.7}$$

in agreement with what we found in Section 3–3.

## 4–2. Most Atoms Are in the Ground Electronic State at Room Temperature

In this section, we will investigate the electronic contributions to $q(V, T)$. It is more convenient to write the electronic partition function as a sum over levels rather than a sum over states (Section 3–8), so we write

$$q_{\text{elec}} = \sum_i g_{ei} e^{-\beta \varepsilon_{ei}} \tag{4.8}$$

where $g_{ei}$ is the degeneracy, and $\varepsilon_{ei}$ the energy of the $i$th electronic level. We first fix the arbitrary zero of energy such that $\varepsilon_{e1} = 0$; that is, we will measure all electronic energies relative to the ground electronic state. The electronic contribution to $q$ can then be written as

$$q_{\text{elec}}(T) = g_{e1} + g_{e2}e^{-\beta\varepsilon_{e2}} + \cdots \tag{4.9}$$

where $\varepsilon_{ej}$ is the energy of the $j$th electronic level relative to the ground state. Note that $q_{\text{elec}}$ is a function of $T$ but not of $V$.

As we have seen in Chapter 1, these $\varepsilon$'s are typically of the order of tens of thousands of wave numbers. Using the fact that $1.986 \times 10^{-23}\ \text{J} = 1\ \text{cm}^{-1}$, the Boltzmann constant in wave numbers is $k_{\text{B}} = 0.6950\ \text{cm}^{-1}\cdot\text{K}^{-1}$. Thus, we see that typically

$$\beta\varepsilon_{\text{elec}} \approx \frac{40\,000\ \text{cm}^{-1}}{0.6950\ \text{cm}^{-1}\cdot\text{K}^{-1}}\frac{1}{T} \approx \frac{10^4\ \text{K}}{T}$$

which is equal to 10 even for $T = 1000$ K. Therefore, $e^{-\beta\varepsilon_{e2}}$ in Equation 4.9 typically is around $10^{-5}$ for most atoms at ordinary temperatures, so only the first term in the summation for $q_{\text{elec}}$ is significantly different from zero. There are some cases, however, such as the halogen atoms, for which the first excited state lies only a few hundred wave numbers above the ground state, so that several terms in $q_{\text{elec}}$ are necessary. Even in these cases, the sum in Equation 4.9 converges very rapidly.

As we learned in Chapter 1, the electronic energies of atoms and ions are determined by atomic spectroscopy and are well tabulated. The standard reference, "Moore's tables," lists the energy levels and energies of many atoms and ions. Table 4.1 lists the first few levels for H, He, Li, and F. We can make some general observations from tables like Table 4.1. The first excited states of the noble gas atoms are of order of $10^5\ \text{cm}^{-1}$ or higher than the ground states; the first excited states at the alkali metal atoms are of order of $10^4\ \text{cm}^{-1}$ or higher than the ground states; and the first excited states of the halogen atoms are only of order of $10^2\ \text{cm}^{-1}$ higher than the ground states. Thus, at ordinary temperatures, the electronic partition function of noble gas atoms is essentially unity and that of alkali metal atoms is two, while those for halogen atoms consist of two terms.

Using the data in Table 4.1, we can now calculate the fraction of helium atoms in the first excited state. This fraction is given by

$$\begin{aligned} f_2 &= \frac{g_{e2}e^{-\beta\varepsilon_{e2}}}{q_{\text{elec}}(T)} \\ &= \frac{g_{e2}e^{-\beta\varepsilon_{e2}}}{g_{e1} + g_{e2}e^{-\beta\varepsilon_{e2}} + g_{e3}e^{-\beta\varepsilon_{e3}} + \cdots} \\ &= \frac{3e^{-\beta\varepsilon_{e2}}}{1 + 3e^{-\beta\varepsilon_{e2}} + e^{-\beta\varepsilon_{e3}} + \cdots} \end{aligned} \tag{4.10}$$

At 300 K, $\beta\varepsilon_{e2} = 770$, so $f_2 \approx 10^{-334}$. Even at 3000 K, $f_2 \approx 10^{-33}$. This is typical of the noble gases. The energy separation between the ground and excited levels must

**TABLE 4.1**
Some atomic energy levels.[a]

| Atom | Electron configuration | Degeneracy $g_e = 2J + 1$ | Energy/cm$^{-1}$ |
|---|---|---|---|
| H | $1s$ | 2 | 0. |
| | $2p$ | 2 | 82 258.907 |
| | $2s$ | 2 | 82 258.942 |
| | $2p$ | 4 | 82 259.272 |
| He | $1s^2$ | 1 | 0. |
| | $1s2p$ | 3 | 159 850.318 |
| | | 1 | 166 271.70 |
| Li | $1s^22s$ | 2 | 0. |
| | $1s^22p$ | 2 | 14 903.66 |
| | | 4 | 14 904.00 |
| | $1s^23s$ | 2 | 27 206.12 |
| F | $1s^22s^22p^5$ | 4 | 0. |
| | | 2 | 404.0 |
| | $1s^22s^22p^43s$ | 6 | 102 406.50 |
| | | 4 | 102 681.24 |
| | | 2 | 102 841.20 |
| | | 4 | 104 731.86 |
| | | 2 | 105 057.10 |

[a]From C.E. Moore, "Atomic Energy Levels" *Natl. Bur. Std, Circ.* 1 467, U.S. Government Printing Office, Washington D.C., 1949

be less than a few hundred cm$^{-1}$ or so before any population of the excited level is significant.

**EXAMPLE 4–1**

Using the data in Table 4.1, calculate the fraction of fluorine atoms in the first excited state at 300 K, 1000 K, and 2000 K.

SOLUTION: Using the second line of Equation 4.10 with $g_{e1} = 4$, $g_{e2} = 2$, and $g_{e3} = 6$, we have

$$f_2 = \frac{2e^{-\beta\varepsilon_{e2}}}{4 + 2e^{-\beta\varepsilon_{e2}} + 6e^{-\beta\varepsilon_{e3}} + \cdots}$$

with $\varepsilon_{e2} = 404.0\ \text{cm}^{-1}$ and $\varepsilon_{e3} = 102\,406.50\ \text{cm}^{-1}$. We also have

$$\beta\varepsilon_{e2} = \frac{404.0\ \text{cm}^{-1}}{(0.6950\ \text{cm}^{-1}\cdot\text{K}^{-1})T} = \frac{581.3\ \text{K}}{T}$$

and

$$\beta\varepsilon_{e3} = \frac{147\,300\ \text{K}}{T}$$

Clearly, we can neglect the third term in the denominator of $f_2$.

The value of $f_2$ for the various temperatures is

$$f_2(T = 300\ \text{K}) = \frac{2e^{-581/300}}{4 + 2e^{-581/300}} = 0.0672$$
$$f_2(T = 1000\ \text{K}) = \frac{2e^{-581/1000}}{4 + 2e^{-581/1000}} = 0.219$$
$$f_2(T = 2000\ \text{K}) = 0.272$$

Thus, the population of the first excited state is significant at these temperatures and so the first two terms of the summation in Equation 4.9 must be evaluated in determining $q_{\text{elec}}(T)$.

For most atoms and molecules, the first two terms of the electronic partition function are sufficient, or

$$q_{\text{elec}}(T) \approx g_{e1} + g_{e2}e^{-\beta\varepsilon_{e2}} \tag{4.11}$$

At temperatures at which the second term is not negligible with respect to the first term, we must check the possible contribution of higher terms as well.

This completes our discussion of the partition function of monatomic ideal gases. In summary, we have

$$Q(N, V, T) = \frac{(q_{\text{trans}}q_{\text{elec}})^N}{N!} \tag{4.12}$$

where

$$q_{\text{trans}}(V, T) = \left(\frac{2\pi mk_BT}{h^2}\right)^{3/2} V$$
$$q_{\text{elec}}(T) = g_{e1} + g_{e2}e^{-\beta\varepsilon_{e2}} + \cdots \tag{4.13}$$

We can now calculate some of the properties of a monatomic ideal gas. The average energy is

$$U = k_BT^2\left(\frac{\partial \ln Q}{\partial T}\right)_{N,V} = Nk_BT^2\left(\frac{\partial \ln q}{\partial T}\right)_V = \frac{3}{2}Nk_BT + \frac{Ng_{e2}\varepsilon_{e2}e^{-\beta\varepsilon_{e2}}}{q_{\text{elec}}} + \cdots \tag{4.14}$$

The first term represents the average kinetic energy, and the second term represents the average electronic energy (in excess of the ground-state energy). The contribution of the electronic degrees of freedom to the average energy is small at ordinary temperatures. If we ignore the very small contribution from the electronic degrees of freedom, the molar heat capacity at constant volume is given by

$$\overline{C}_V = \left(\frac{d\overline{U}}{dT}\right)_{N,V} = \frac{3}{2}R$$

The pressure is

$$\begin{aligned} P &= k_B T \left(\frac{\partial \ln Q}{\partial V}\right)_{N,T} = N k_B T \left(\frac{\partial \ln q}{\partial V}\right)_T \\ &= N k_B T \left[\frac{\partial}{\partial V}(\ln V + \text{terms not involving } V)\right]_T \\ &= \frac{N k_B T}{V} \end{aligned} \tag{4.15}$$

which is the ideal gas equation of state. Note that Equation 4.15 results because $q(V, T)$ is of the form $f(T)V$, and only the translational energy of the atoms contributes to the pressure. This is expected intuitively, because the pressure is due to bombardment of the walls of the container by the atoms and molecules of the gas.

In the next few sections, we will treat a diatomic ideal gas. In addition to translational and electronic degrees of freedom, diatomic molecules also possess vibrational and rotational degrees of freedom. The general procedure would be to set up the Schrödinger equation for two nuclei and $n$ electrons and to solve this equation for the set of eigenvalues of the diatomic molecule. Fortunately, a series of very good approximations can be used to reduce this complicated two-nuclei, $n$-electron problem to a set of simpler problems. The simplest of these approximations is the rigid rotator-harmonic oscillator approximation, which we described in Chapter 1. We will set up this approximation in the next section and then discuss the vibrational and rotational partition functions within this approximation in Sections 4–4 and 4–5.

## 4–3. The Energy of a Diatomic Molecule Can Be Approximated as a Sum of Separate Terms

When treating diatomic or polyatomic molecules, we use the rigid rotator-harmonic oscillator approximation (Chapter 1). In this case, we can write the total energy of the molecule as a sum of its translational, rotational, vibrational, and electronic energies:

$$\varepsilon = \varepsilon_{\text{trans}} + \varepsilon_{\text{rot}} + \varepsilon_{\text{vib}} + \varepsilon_{\text{elec}} \tag{4.16}$$

As for a monatomic ideal gas, the inequality given by Equation 3.40 is easily satisfied at normal temperatures, and so we can write

$$Q(N, V, T) = \frac{[q(V, T)]^N}{N!} \tag{4.17}$$

Furthermore, Equation 4.16 allows us to write

$$q(V, T) = q_{\text{trans}} q_{\text{rot}} q_{\text{vib}} q_{\text{elec}} \tag{4.18}$$

so the partition function of a molecular ideal gas is given by

$$Q(N, V, T) = \frac{\left(q_{\text{trans}} q_{\text{rot}} q_{\text{vib}} q_{\text{elec}}\right)^N}{N!} \tag{4.19}$$

The translational partition function of a diatomic molecule is similar to the result we found in Section 4–1 for an atom:

$$q_{\text{trans}}(V, T) = \left[\frac{2\pi(m_1 + m_2)k_B T}{h^2}\right]^{3/2} V \tag{4.20}$$

Note that Equation 4.20 is essentially the same as Equation 4.6. The electronic partition function will be similar to Equation 4.9. We will discuss the vibrational and rotational contributions to the partition function in the next two sections. Although Equation 4.19 is not exact, it is often a good approximation, particularly for small molecules.

Before we consider $q_{\text{rot}}$ and $q_{\text{vib}}$, we must choose a zero of energy for the rotational, vibrational, and electronic states. The natural choice for the zero of rotational energy is the $J = 0$ state, where the rotational energy is zero. In the vibrational case, however, we have two sensible choices. One is to take the zero of vibrational energy to be that of the ground state, and the other is take the zero to be the bottom of the internuclear potential well. In the first case, the energy of the ground vibrational state is zero, and in the second case it is $h\nu/2$. We will choose the zero of vibrational energy to be the bottom of the internuclear potential well of the lowest electronic state, so the energy of the ground vibrational state will be $h\nu/2$.

Last, we take the zero of the electronic energy to be the separated atoms at rest in their ground electronic states (see Figure 4.2). Recall that the depth of the ground electronic state potential well is denoted by $D_e$ ($D_e$ is a positive number; see Section 1–7), and so the energy of the ground electronic state is $\varepsilon_{e1} = -D_e$, and the electronic partition function is

$$q_{\text{elec}} = g_{e1} e^{D_e/k_B T} + g_{e2} e^{-\varepsilon_{e2}/k_B T} \tag{4.21}$$

where $D_e$ and $\varepsilon_{e2}$ are shown in Figure 4.2. We also introduced in Section 1–7 a quantity $D_0$ that is equal to $D_e - \frac{1}{2}h\nu$. As Figure 4.2 shows, $D_0$ is the energy difference between the lowest vibrational state and the dissociated molecule. The quantity $D_0$ can be measured spectroscopically, and values of $D_0$ and $D_e$ for several diatomic molecules are given in Table 4.2.

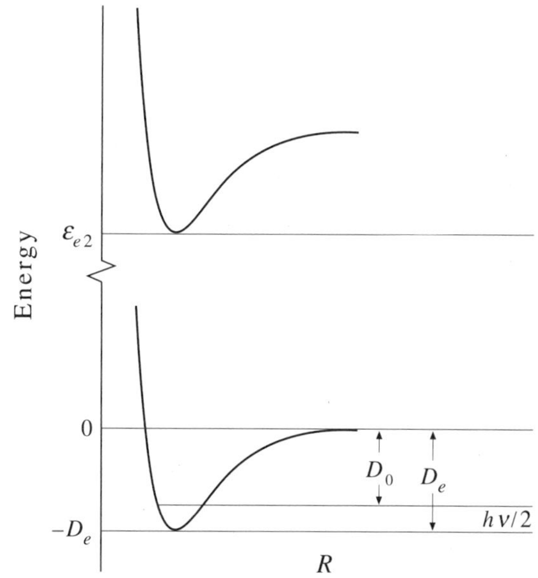


**FIGURE 4.2**
The ground and first excited electronic states as a function of the internuclear separation, illustrating the quantities $D_e$ and $D_0$ of the ground state and $\varepsilon_{e2}$. The quantities $D_e$ and $D_0$ are related by $D_e = D_0 + h\nu/2$ as shown in the figure.

**TABLE 4.2**
Molecular constants for several diatomic molecules. These parameters were obtained from a variety of sources and do not represent the most accurate values because they were obtained under the rigid rotator-harmonic oscillator approximation.

| Molecule | $\Theta_{vib}$/K | $\Theta_{rot}$/K | $D_0$/kJ·mol$^{-1}$ | $D_e$/kJ·mol$^{-1}$ | Degeneracy of the ground electronic state |
|---|---|---|---|---|---|
| $H_2$ | 6332 | 85.3 | 432.1 | 457.6 | 1 |
| $D_2$ | 4480 | 42.7 | 435.6 | 453.9 | 1 |
| $Cl_2$ | 805 | 0.351 | 239.2 | 242.3 | 1 |
| $Br_2$ | 463 | 0.116 | 190.1 | 191.9 | 1 |
| $I_2$ | 308 | 0.0537 | 148.8 | 150.3 | 1 |
| $O_2$ | 2256 | 2.07 | 493.6 | 503.0 | 3 |
| $N_2$ | 3374 | 2.88 | 941.6 | 953.0 | 1 |
| CO | 3103 | 2.77 | 1070 | 1085 | 1 |
| NO | 2719 | 2.39 | 626.8 | 638.1 | 2 |
| HCl | 4227 | 15.02 | 427.8 | 445.2 | 1 |
| HBr | 3787 | 12.02 | 362.6 | 377.7 | 1 |
| HI | 3266 | 9.25 | 294.7 | 308.6 | 1 |
| $Na_2$ | 229 | 0.221 | 71.1 | 72.1 | 1 |
| $K_2$ | 133 | 0.081 | 53.5 | 54.1 | 1 |

## 4–4. Most Molecules Are in the Ground Vibrational State at Room Temperature

In this section, we will evaluate the vibrational part of the partition function of a diatomic molecule under the harmonic-oscillator approximation. If we measure the vibrational energy levels relative to the bottom of the internuclear potential well, the energies are given by (Equation 1–22)

$$\varepsilon_v = \left(v + \tfrac{1}{2}\right) h\nu \qquad v = 0,\ 1,\ 2,\ \ldots \tag{4.22}$$

with $\nu = (k/\mu)^{1/2}/2\pi$, where $k$ is the force constant of the molecule and $\mu$ is its reduced mass. The vibrational partition function $q_{\text{vib}}$ becomes

$$q_{\text{vib}}(T) = \sum_v e^{-\beta\varepsilon_v} = \sum_{v=0}^{\infty} e^{-\beta(v+\frac{1}{2})h\nu}$$
$$= e^{-\beta h\nu/2} \sum_{v=0}^{\infty} e^{-\beta h\nu v}$$

This summation can be evaluated easily by recognizing it to be a geometric series (MathChapter C):

$$\sum_{n=0}^{\infty} x^n = \frac{1}{1-x}$$

with $x = e^{-\beta h\nu} < 1$. Thus we can write

$$\sum_{v=0}^{\infty} e^{-\beta h\nu v} = \sum_{v=0}^{\infty} \left(e^{-\beta h\nu}\right)^v = \frac{1}{1 - e^{-\beta h\nu}}$$

so $q_{\text{vib}}(T)$ becomes

$$q_{\text{vib}}(T) = \frac{e^{-\beta h\nu/2}}{1 - e^{-\beta h\nu}} \tag{4.23}$$

Note that this is the vibrational term encountered in Example 3–2, which presented the partition function for the rigid rotator-harmonic oscillator model of an ideal diatomic gas. If we introduce a quantity, $\Theta_{\text{vib}} = h\nu/k_{\text{B}}$, called the *vibrational temperature*, $q_{\text{vib}}(T)$ can be written as

$$q_{\text{vib}}(T) = \frac{e^{-\Theta_{\text{vib}}/2T}}{1 - e^{-\Theta_{\text{vib}}/T}} \tag{4.24}$$

This is one of the rare cases in which $q$ can be summed directly without having to approximate it by an integral, as we did for the translational case in Section 4–1 and will do shortly for the rotational case in Section 4–5.

We can calculate the average vibrational energy from $q_{vib}(T)$

$$\langle E_{vib}\rangle = Nk_B T^2 \frac{d\ln q_{vib}}{dT} = Nk_B\left(\frac{\Theta_{vib}}{2} + \frac{\Theta_{vib}}{e^{\Theta_{vib}/T}-1}\right) \tag{4.25}$$

Table 4.2 gives $\Theta_{vib}$ for several diatomic molecules. The vibrational contribution to the molar heat capacity is

$$\overline{C}_{V,vib} = \frac{d\langle \overline{E}_{vib}\rangle}{dT} = R\left(\frac{\Theta_{vib}}{T}\right)^2 \frac{e^{-\Theta_{vib}/T}}{\left(1-e^{-\Theta_{vib}/T}\right)^2} \tag{4.26}$$

Figure 4.3 shows the vibrational contribution of an ideal diatomic gas to the molar heat capacity as a function of temperature. The high temperature limit of $\overline{C}_{V,vib}$ is $R$, and $\overline{C}_{V,vib}$ is one-half of this value at $T/\Theta_{vib} = 0.34$.

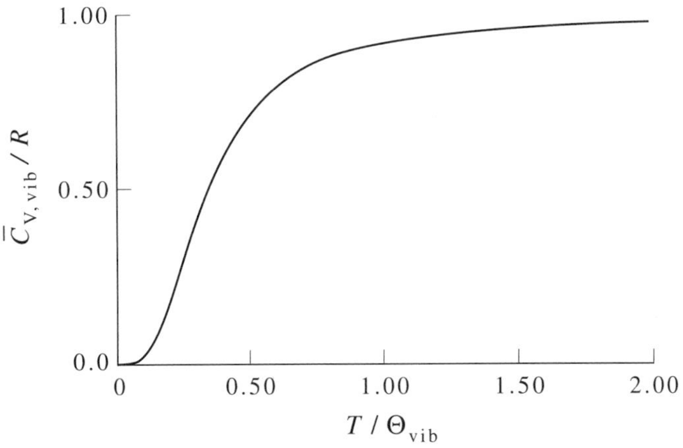


**FIGURE 4.3**
The vibrational contribution to the molar heat capacity of an ideal diatomic gas as a function of reduced temperature, $T/\Theta_{vib}$.

**EXAMPLE 4–2**
Calculate the vibrational contribution to the molar heat capacity of $N_2(g)$ at 1000 K. The experimental value is $3.43\ J\cdot K^{-1}\cdot mol^{-1}$.

SOLUTION: We use Equation 4.26 with $\Theta_{vib} = 3374$ (Table 4.2). Thus, $\Theta_{vib}/T = 3.374$ and

$$\frac{\overline{C}_{V,vib}}{R} = (3.374)^2 \frac{e^{-3.374}}{(1-e^{-3.374})^2} = 0.418$$

or

$$\overline{C}_{V,vib} = (0.418)(8.314\ J\cdot K^{-1}\cdot mol^{-1}) = 3.48\ J\cdot K^{-1}\cdot mol^{-1}$$

The agreement with the experimental value is quite good.

An interesting quantity to calculate is the fraction of molecules in various vibrational states. The fraction of molecules in the $v$th vibrational state is

$$f_v = \frac{e^{-\beta h\nu(v+\frac{1}{2})}}{q_{\text{vib}}} \tag{4.27}$$

If we substitute Equation 4.23 into this equation, we obtain

$$f_v = \left(1 - e^{-\beta h\nu}\right) e^{-\beta h\nu v} = \left(1 - e^{-\Theta_{\text{vib}}/T}\right) e^{-v\Theta_{\text{vib}}/T} \tag{4.28}$$

The following example illustrates the use of this equation.

**EXAMPLE 4–3**

Use Equation 4.28 to calculate the fraction of $N_2(g)$ molecules in the $v = 0$ and $v = 1$ vibrational states at 300 K.

SOLUTION: We first calculate $\exp(-\Theta_{\text{vib}}/T)$ for 300 K:

$$e^{-\Theta_{\text{vib}}/T} = e^{-3374\text{ K}/300\text{ K}} = e^{-11.25} = 1.31 \times 10^{-5}$$

Therefore,

$$f_0 = 1 - e^{-\Theta_{\text{vib}}/T} \approx 1$$

and

$$f_1 = (1 - e^{-\Theta_{\text{vib}}/T})e^{-\Theta_{\text{vib}}/T} \approx 1.31 \times 10^{-5}$$

Notice that essentially all the nitrogen molecules are in the ground vibrational state at 300 K.

Figure 4.4 shows the population of vibrational levels of $Br_2(g)$ at 300 K. Notice that most molecules are in the ground vibrational state and that the population of the higher vibrational states decreases exponentially. Bromine has a smaller force constant and a larger mass (and hence a smaller value of $\Theta_{\text{vib}}$) than most diatomic molecules, however (*cf.* Table 4.2), so the population of excited vibrational states of $Br_2(g)$ at a given temperature is greater than most other molecules.

We can use Equation 4.28 to calculate the fraction of molecules in all excited vibrational states. This quantity is given by $\sum_{v=1}^{\infty} f_v$ but because $\sum_{v=0}^{\infty} f_v = 1$, we can write

$$f_{v>0} = \sum_{v=1}^{\infty} f_v = 1 - f_0 = 1 - (1 - e^{-\Theta_{\text{vib}}/T})$$

or simply

$$f_{v>0} = e^{-\Theta_{\text{vib}}/T} = e^{-\beta h\nu} \tag{4.29}$$

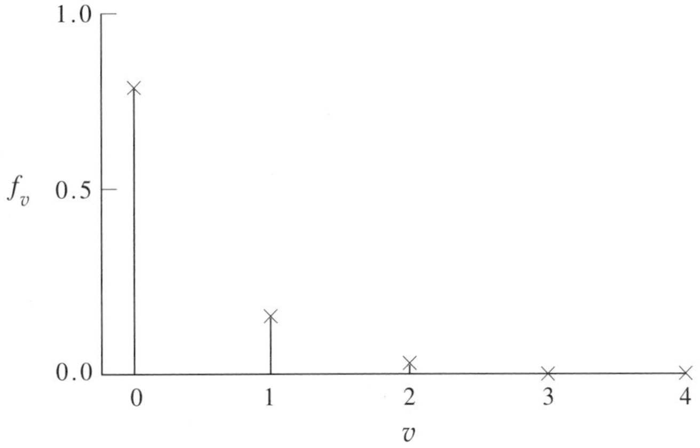


**FIGURE 4.4**
The population of the vibrational levels of $Br_2(g)$ at 300 K.

Table 4.3 gives the fraction of molecules in excited vibrational states for several diatomic molecules.

**TABLE 4.3**
The fraction of molecules in excited vibrational states at 300 K and 1000 K.

| Gas | $\Theta_{vib}$/K | $f_{v>0}$ (T = 300 K) | $f_{v>0}$ (T = 1000 K) |
|---|---|---|---|
| $H_2$ | 6332 | $1.01 \times 10^{-9}$ | $2.00 \times 10^{-3}$ |
| HCl | 4227 | $7.59 \times 10^{-7}$ | $1.46 \times 10^{-2}$ |
| $N_2$ | 3374 | $1.30 \times 10^{-5}$ | $3.43 \times 10^{-2}$ |
| CO | 3103 | $3.22 \times 10^{-5}$ | $4.49 \times 10^{-2}$ |
| $Cl_2$ | 805 | $6.82 \times 10^{-2}$ | $4.47 \times 10^{-1}$ |
| $I_2$ | 308 | $3.58 \times 10^{-1}$ | $7.35 \times 10^{-1}$ |

## 4–5. Most Molecules Are in Excited Rotational States at Ordinary Temperatures

The energy levels of a rigid rotator are given by (Equation 1–28)

$$\varepsilon_J = \frac{\hbar^2 J(J+1)}{2I} \qquad J = 0,\ 1,\ 2,\ \ldots \tag{4.30a}$$

where $I$ is the moment of inertia of the rotator. Each energy level has a degeneracy of

$$g_J = 2J + 1 \tag{4.30b}$$

Using Equations 4.30*a* and 4.30*b*, we can write the rotational partition function of a rigid rotator as

$$q_{\text{rot}}(T) = \sum_{J=0}^{\infty}(2J+1)e^{-\beta\hbar^2 J(J+1)/2I} \tag{4.31}$$

where we sum over levels rather than states by including the degeneracy explicitly. For convenience, we introduce a quantity that has units of temperature and is called the *rotational temperature*, $\Theta_{\text{rot}}$:

$$\Theta_{\text{rot}} = \frac{\hbar^2}{2Ik_{\text{B}}} = \frac{hB}{k_{\text{B}}} \tag{4.32}$$

where $B = h/8\pi^2 I$ (Equation 1.33). Substituting Equation 4.32 into Equation 4.31 gives

$$q_{\text{rot}}(T) = \sum_{J=0}^{\infty}(2J+1)e^{-\Theta_{\text{rot}}J(J+1)/T} \tag{4.33}$$

Unlike the harmonic-oscillator partition function, the summation in Equation 4.33 cannot be written in closed form. However, as the data in Table 4.2 will verify, the value of $\Theta_{\text{rot}}/T$ is quite small at ordinary temperatures for diatomic molecules that do not contain hydrogen atoms. For example, $\Theta_{\text{rot}}$ for CO(g) is 2.77 K, so $\Theta_{\text{rot}}/T$ is about $10^{-2}$ at room temperature. Just as we were able to approximate the summation in Equation 4.4 very well by an integral because $\alpha = \beta h^2/8ma^2$ is typically small at normal temperatures, we are able to approximate the summation in Equation 4.33 by an integral because $\Theta_{\text{rot}}/T$ is small for most molecules at ordinary temperatures. Therefore, it is an excellent approximation to write $q_{\text{rot}}(T)$ as

$$q_{\text{rot}}(T) = \int_0^{\infty}(2J+1)e^{-\Theta_{\text{rot}}J(J+1)/T}dJ$$

This integral is easy to evaluate because if we let $x = J(J+1)$, then $dx = (2J+1)dJ$ and $q_{\text{rot}}(T)$ becomes

$$\begin{aligned} q_{\text{rot}}(T) &= \int_0^{\infty} e^{-\Theta_{\text{rot}}x/T}dx \\ &= \frac{T}{\Theta_{\text{rot}}} = \frac{8\pi^2 Ik_{\text{B}}T}{h^2} \qquad \Theta_{\text{rot}} \ll T \end{aligned} \tag{4.34}$$

Note that this is the rotational term encountered in Example 3–2, which presented the partition function for the rigid rotator-harmonic oscillator model of an ideal diatomic gas. This approximation improves as the temperature increases and is called the high-temperature limit. For low temperatures or for molecules with large values of $\Theta_{\text{rot}}$, say $H_2$(g) with $\Theta_{\text{rot}} = 85.3$ K, we can use Equation 4.33 directly. For example, the first four terms of Equation 4.33 are sufficient to calculate $q_{\text{rot}}(T)$ to within 0.1% for $T < 3\Theta_{\text{rot}}$.

For simplicity, we will use only the high-temperature limit, because $\Theta_{\text{rot}} \ll T$ for most molecules at room temperature. (See Table 4.2.)

The average rotational energy is

$$\langle E_{\text{rot}} \rangle = Nk_{\text{B}}T^2 \left( \frac{d \ln q_{\text{rot}}}{dT} \right) = Nk_{\text{B}}T$$

and the rotational contribution to the molar heat capacity is

$$\overline{C}_{V,\text{rot}} = R$$

A diatomic molecule has two rotational degrees of freedom, and each one contributes $R/2$ to $\overline{C}_{V,\text{rot}}$.

We can also calculate the fraction of molecules in the $J^{\text{th}}$ rotational level:

$$\begin{aligned} f_J &= \frac{(2J+1)e^{-\Theta_{\text{rot}}J(J+1)/T}}{q_{\text{rot}}} \\ &= (2J+1)(\Theta_{\text{rot}}/T)e^{-\Theta_{\text{rot}}J(J+1)/T} \end{aligned} \tag{4.35}$$

**EXAMPLE 4–4**

Use Equation 4.35 to calculate the population of the rotational levels of CO at 300 K.

SOLUTION: Using $\Theta_{\text{rot}} = 2.77$ K from Table 4.2, we have that $\Theta_{\text{rot}}/T = 0.00923$ at 300 K. Therefore,

$$f_J = (2J+1)(0.00923)e^{-0.00923J(J+1)}$$

We can present our results in the form of a table:

| $J$ | $f_J$ |
|---|---|
| 0 | 0.00923 |
| 2 | 0.0437 |
| 4 | 0.0691 |
| 6 | 0.0814 |
| 8 | 0.0807 |
| 10 | 0.0702 |
| 12 | 0.0547 |
| 16 | 0.0247 |
| 18 | 0.0145 |

These results are plotted in Figure 4.5.

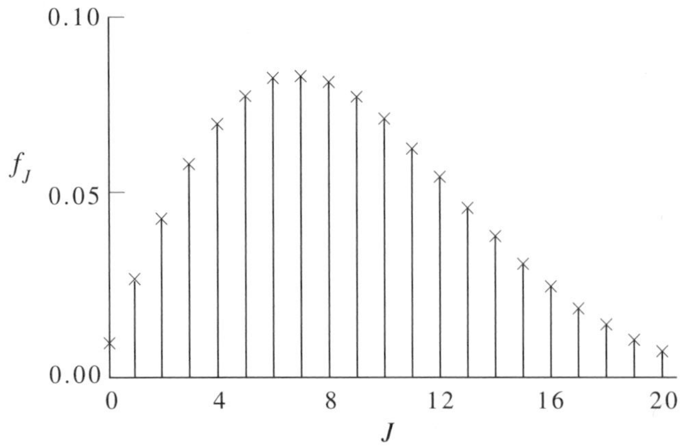


**FIGURE 4.5**
The fraction of molecules in the $J$th rotational level for CO at 300 K.

Contrary to the case for vibrational levels, most molecules are in the excited rotational levels at ordinary temperatures. We can estimate the most probable value of $J$ by treating Equation 4.35 as if $J$ were continuous and by setting the derivative with respect to $J$ equal to zero to obtain (Problem 4–18)

$$J_{\text{mp}} \approx \left(\frac{T}{2\Theta_{\text{rot}}}\right)^{1/2} - \frac{1}{2} \tag{4.36}$$

This equation gives a value of 7 for CO at 300 K (in agreement with Figure 4.5).

## 4–6. Rotational Partition Functions Contain a Symmetry Number

Although it is not apparent from our derivation of $q_{\text{rot}}(T)$, Equations 4.33 and 4.34 apply only to heteronuclear diatomic molecules. The underlying reason is that the wave function of a homonuclear diatomic molecule must possess a certain symmetry with respect to the interchange of the two identical nuclei in the molecule. In particular, if the two nuclei have integral spins (bosons), the molecular wave function must remain unchanged under interchange of the two nuclei; if the nuclei have half odd integer spin (fermions), the molecular wave function must change sign. This symmetry requirement has a profound effect on the population of the rotational energy levels of a homonuclear diatomic molecule, which can be understood only by a careful analysis of the general symmetry properties of the wave function of a diatomic molecule. This analysis is somewhat involved and will not be done here, but we need the final result. At temperatures such that $\Theta_{\text{rot}} \ll T$, which we have seen applies to most molecules at ordinary temperatures, $q_{\text{rot}}$ for a homonuclear diatomic molecule is

$$q_{\text{rot}}(T) = \frac{T}{2\Theta_{\text{rot}}} \tag{4.37}$$

Note that this equation is the same as Equation 4.34 for a heteronuclear diatomic molecule except for the factor of 2 in the denominator. This factor comes from the additional symmetry of the homonuclear diatomic molecule; in particular, a homonuclear diatomic molecule has two indistinguishable orientations. There is a two-fold axis of symmetry perpendicular to the internuclear axis.

Equations 4.34 and 4.37 can be written as one equation by writing $q_{\text{rot}}$ as

$$q_{\text{rot}}(T) = \frac{T}{\sigma \Theta_{\text{rot}}} \tag{4.38}$$

where $\sigma = 1$ for a heteronuclear diatomic molecule and 2 for a homonuclear diatomic molecule. The factor $\sigma$ is called the *symmetry number* of the molecule and represents the number of indistinguishable orientations of the molecule.

Having studied each contribution to the molecular partition function of a diatomic molecule, we can now include the rigid rotator-harmonic oscillator approximation in the partition function of a diatomic molecule to obtain

$$\begin{aligned} q(V,T) &= q_{\text{trans}} q_{\text{rot}} q_{\text{vib}} q_{\text{elec}} \\ &= \left(\frac{2\pi M k_{\text{B}} T}{h^2}\right)^{3/2} V \cdot \frac{T}{\sigma \Theta_{\text{rot}}} \cdot \frac{e^{-\Theta_{\text{vib}}/2T}}{1 - e^{-\Theta_{\text{vib}}/T}} \cdot g_{e1} e^{D_e/k_{\text{B}}T} \end{aligned} \tag{4.39}$$

Remember that this expression requires that $\Theta_{\text{rot}} \ll T$, that only the ground electronic state is populated, that the zero of the electronic energy is taken to be the separated atoms at rest in their ground electronic states, and that the zero of energy for the vibrational energy is that at the bottom of the internuclear potential well of the lowest electronic state. Note that only $q_{\text{trans}}$ is a function of $V$, and that this function is of the form $f(T)V$, which, as we have seen before, is responsible for the ideal-gas equation of state.

**EXAMPLE 4–5**

Derive an expression for the molar energy $\overline{U}$ of a diatomic ideal gas from Equation 4.39. Identify each of the terms.

SOLUTION: We start with

$$Q(N,V,T) = \frac{[q(V,T)]^N}{N!}$$

and

$$U = k_{\text{B}} T^2 \left(\frac{\partial \ln Q}{\partial T}\right)_{N,V} = N k_{\text{B}} T^2 \left(\frac{\partial \ln q}{\partial T}\right)_V$$

Using Equation 4.39 for $q(V,T)$, we have

$$\ln q = \frac{3}{2} \ln T + \ln T - \frac{\Theta_{\text{vib}}}{2T} - \ln(1 - e^{-\Theta_{\text{vib}}/T}) + \frac{D_e}{k_{\text{B}} T}$$
$$+ \text{ terms not containing } T$$

Therefore,

$$\left(\frac{\partial \ln q}{\partial T}\right)_V = \frac{3}{2T} + \frac{1}{T} + \frac{\Theta_{\text{vib}}}{2T^2} + \frac{(\Theta_{\text{vib}}/T^2)e^{-\Theta_{\text{vib}}/T}}{1 - e^{-\Theta_{\text{vib}}/T}} - \frac{D_e}{k_B T^2}$$

and letting $N = N_A$ and $N_A k_B = R$ for one mole,

$$\overline{U} = \frac{3}{2}RT + RT + R\frac{\Theta_{\text{vib}}}{2} + R\frac{\Theta_{\text{vib}}e^{-\Theta_{\text{vib}}/T}}{1 - e^{-\Theta_{\text{vib}}/T}} - N_A D_e \qquad (4.40)$$

The first term represents the average translational energy ($RT/2$ for each of the three translational degrees of freedom), the second term represents the average rotational energy ($RT/2$ for each of the two rotational degrees of freedom), the third term represents the zero-point vibrational energy, the fourth term represents the average vibrational energy in excess of the zero-point vibrational energy, and the last term reflects the electronic energy relative to the zero of electronic energy that we have chosen, namely the two separated atoms at rest in their ground electronic states.

The heat capacity is obtained by differentiating Equation 4.40 with respect to $T$:

$$\frac{\overline{C}_V}{R} = \frac{5}{2} + \left(\frac{\Theta_{\text{vib}}}{T}\right)^2 \frac{e^{-\Theta_{\text{vib}}T}}{(1 - e^{-\Theta_{\text{vib}}T})^2} \qquad (4.41)$$

Figure 3.3 presents a comparison of Equation 4.41 with experimental data for oxygen. The agreement is good and is typical of that found for other properties. The agreement can be improved considerably by including the first corrections to the rigid rotator-harmonic oscillator model. These include effects such as centrifugal distortion and anharmonicity. The consideration of these effects introduces a new set of molecular constants, all of which can be determined spectroscopically and are well tabulated. The use of such additional parameters from spectroscopic data can give calculated values of the heat capacity that are actually more accurate than calorimetric ones.

## 4–7. The Vibrational Partition Function of a Polyatomic Molecule Is a Product of Harmonic Oscillator Partition Functions for Each Normal Coordinate

The discussion in Section 4–3 for diatomic molecules applies equally well to polyatomic molecules, and so

$$Q(N, V, T) = \frac{[q(V, T)]^N}{N!}$$

As before, the number of translational energy states alone is sufficient to guarantee that the number of energy states available to any molecule is much greater than the number of molecules in the system.

As for diatomic molecules, we use a rigid rotator-harmonic oscillator approximation. This allows us to separate the rotational motion from the vibrational motion of the molecule, so that we can treat each one separately. Both problems are somewhat more complicated for polyatomic molecules than for diatomic molecules. Nevertheless, we can write the polyatomic analog of Equation 4.19:

$$Q(N, V, T) = \frac{(q_{\text{trans}} q_{\text{rot}} q_{\text{vib}} q_{\text{elec}})^N}{N!} \tag{4.42}$$

In Equation 4.42, $q_{\text{trans}}$ is given by

$$q_{\text{trans}}(V, T) = \left[\frac{2\pi M k_{\text{B}} T}{h^2}\right]^{3/2} V \tag{4.43}$$

where $M$ is the total mass of the molecule. We choose as the zero of energy the $n$ atoms completely separated in their ground electronic states. Thus, the energy of the ground electronic state is $-D_e$, and then the electronic partition function is

$$q_{\text{elec}} = g_{e1} e^{D_e/k_{\text{B}}T} + \cdots \tag{4.44}$$

To calculate $Q(N, V, T)$ we must investigate $q_{\text{rot}}$ and $q_{\text{vib}}$.

We learned in Section 1–9 that the vibrational motion of a polyatomic molecule can be expressed in terms of normal coordinates. By introducing normal coordinates, the vibrational motion of a polyatomic molecule can be expressed as a set of *independent* harmonic oscillators. Consequently, the vibrational energy of a polyatomic molecule can be written as

$$\varepsilon_{\text{vib}} = \sum_{j=1}^{\alpha} \left(v_j + \tfrac{1}{2}\right) h\nu_j \qquad v_j = 0,\ 1,\ 2,\ \ldots \tag{4.45}$$

where $\nu_j$ is the vibrational frequency associated with the $j$th normal mode and $\alpha$ is the number of vibrational degrees of freedom ($3n - 5$ for a linear molecule and $3n - 6$ for a nonlinear molecule, where $n$ is the number of atoms in the molecule). Because the normal modes are independent,

$$q_{\text{vib}} = \prod_{j=1}^{\alpha} \frac{e^{-\Theta_{\text{vib},j}/2T}}{(1 - e^{-\Theta_{\text{vib},j}/T})} \tag{4.46}$$

$$E_{\text{vib}} = N k_{\text{B}} \sum_{j=1}^{\alpha} \left(\frac{\Theta_{\text{vib},j}}{2} + \frac{\Theta_{\text{vib},j} e^{-\Theta_{\text{vib},j}/T}}{1 - e^{-\Theta_{\text{vib},j}/T}}\right) \tag{4.47}$$

and

$$C_{V,\text{vib}} = N k_{\text{B}} \sum_{j=1}^{\alpha} \left[\left(\frac{\Theta_{\text{vib},j}}{T}\right)^2 \frac{e^{-\Theta_{\text{vib},j}/T}}{(1 - e^{-\Theta_{\text{vib},j}/T})^2}\right] \tag{4.48}$$

where $\Theta_{\text{vib},j}$ is a characteristic vibrational temperature defined by

$$\Theta_{\text{vib},j} = \frac{h\nu_j}{k_B} \tag{4.49}$$

Table 4.4 contains values of $\Theta_{\text{vib},j}$ for several polyatomic molecules.

**TABLE 4.4**

Values of the characteristic rotational temperatures, the characteristic vibrational temperatures, $D_0$ for the ground state, and the symmetry number, $\sigma$, for some polyatomic molecules. The numbers in parentheses indicate the degeneracy of that mode.

| Molecule | $\Theta_{\text{rot}}$/K | $\Theta_{\text{vib},j}$/K | $D_0$/kJ·mol$^{-1}$ | $\sigma$ |
|---|---|---|---|---|
| $CO_2$ | 0.561 | 3360, 954(2), 1890 | 1596 | 2 |
| $H_2O$ | 40.1, 20.9, 13.4 | 5360, 5160, 2290 | 917.6 | 2 |
| $NH_3$ | 13.6, 13.6, 8.92 | 4800, 1360, 4880(2), 2330(2) | 1158 | 3 |
| $ClO_2$ | 2.50, 0.478, 0.400 | 1360, 640, 1600 | 378 | 2 |
| $SO_2$ | 2.92, 0.495, 0.422 | 1660, 750, 1960 | 1063 | 2 |
| $N_2O$ | 0.603 | 3200, 850(2), 1840 | 1104 | 2 |
| $NO_2$ | 11.5, 0.624, 0.590 | 1900, 1080, 2330 | 928.0 | 2 |
| $CH_4$ | 7.54, 7.54, 7.54 | 4170, 2180(2), 4320(3), 1870(3) | 1642 | 12 |
| $CH_3Cl$ | 7.32, 0.637, 0.637 | 4270, 1950, 1050, 4380(2) 2140(2), 1460(2) | 1551 | 3 |
| $CCl_4$ | 0.0823, 0.0823, 0.0823 | 660, 310(2), 1120(3), 450(3) | 1292 | 12 |

**EXAMPLE 4–6**

Calculate the contribution of each normal mode to the vibrational heat capacity of $CO_2$ at 400 K.

SOLUTION: The values of $\Theta_{\text{vib},j}$ are given in Table 4.4. Note that the $\Theta_{\text{vib}} = 954$ K mode (bending mode) is doubly degenerate. For $\Theta_{\text{vib},j} = 954$ K (the doubly degerate bending mode),

$$\frac{\overline{C}_{V,j}}{R} = \left(\frac{954}{400}\right)^2 \frac{e^{-954/400}}{(1 - e^{-954/400})^2} = 0.635$$

For $\Theta_{\text{vib},j} = 1890$ K (the asymmetric stretch),

$$\frac{\overline{C}_{V,j}}{R} = \left(\frac{1890}{400}\right)^2 \frac{e^{-1890/400}}{(1 - e^{-1890/400})^2} = 0.202$$

For $\Theta_{\text{vib},j} = 3360$ K (the symmetric stretch),

$$\frac{\overline{C}_{V,j}}{R} = \left(\frac{3360}{400}\right)^2 \frac{e^{-3360/400}}{(1 - e^{-3360/400})^2} = 0.016$$

The total vibrational heat capacity at 400 K is

$$\frac{\overline{C}_{V,\text{vib}}}{R} = 2(0.635) + 0.202 + 0.016 = 1.488$$

Note that the contribution from each mode decreases as $\Theta_{\text{vib},j}$ increases. Because $\Theta_{\text{vib},j}$ is proportional to the frequency of the mode, it requires higher temperatures to excite modes with larger values of $\Theta_{\text{vib},j}$. The molar vibrational heat capacity from 200 K to 2000 K contributed by each mode is shown in Figure 4.6.

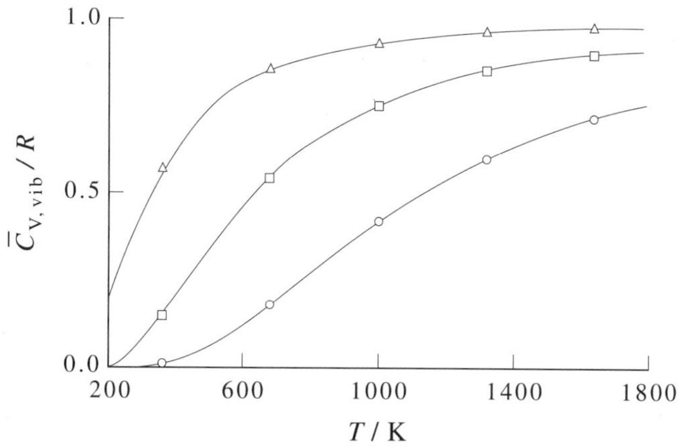


**FIGURE 4.6**
The contribution of each normal mode to the molar vibrational heat capacity of $CO_2$. The curve indicated by triangles corresponds to $\Theta_{\text{vib},j} = 954$ K; the curve indicated by squares to $\Theta_{\text{vib},j} = 1890$ K; and the curve indicated by circles to $\Theta_{\text{vib},j} = 3360$ K. Note that modes with smaller values of $\Theta_{\text{vib},j}$, or $\nu_j$, contribute more at a given temperature.

## 4–8. The Form of the Rotational Partition Function of a Polyatomic Molecule Depends Upon the Shape of the Molecule

In this section, we will discuss the rotational partition functions of polyatomic molecules. Let's consider a linear polyatomic molecule first. In the rigid-rotator approximation, the energies and degeneracies of a linear polyatomic molecule are the same as for a diatomic molecule, $\varepsilon_J = J(J+1)h^2/8\pi^2 I$ with $J = 0, 1, 2, \ldots$ and $g_J = 2J+1$. In this case, the moment of inertia $I$ is

$$I = \sum_{j=1}^{n} m_j d_j^2$$

where $d_j$ is the distance of the $j$th nucleus from the center of mass of the molecule. Consequently, the rotational partition function of a linear polyatomic molecule is the same as that of a diatomic molecule, namely,

$$q_{\text{rot}} = \frac{8\pi^2 I k_B T}{\sigma h^2} = \frac{T}{\sigma \Theta_{\text{rot}}} \tag{4.50}$$

As before, we have introduced a symmetry number, which is unity for unsymmetrical molecules such as $N_2O$ and COS and equal to two for symmetrical molecules such as $CO_2$ and $C_2H_2$. Recall that the symmetry number is the number of different ways the molecule can be rotated into a configuration indistinguishable from the original.

**EXAMPLE 4–7**

What is the symmetry number of ammonia, $NH_3$?

SOLUTION: Ammonia is a trigonal pyramidal molecule and has the three indistinguishable orientations shown below looking down the three-fold axis of symmetry.

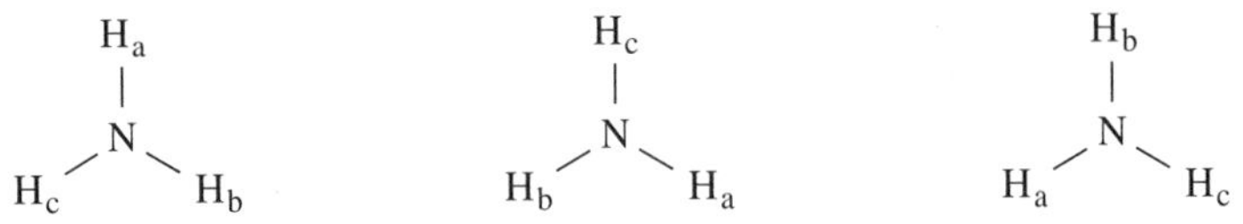

Therefore, the symmetry number is three.

In Section 1–10, we learned that the rotational properties of nonlinear polyatomic molecules depend upon the relative magnitudes of their moments of inertia. If all three moments of inertia are equal, the molecule is called a *spherical top*. If two of the three are equal, the molecule is called a *symmetric top*. If all three are different, the molecule is called an *asymmetric top*. Just as we defined a characteristic rotational temperature of a diatomic molecule by Equation 4.32, $\Theta_{\text{rot}} = \hbar^2/2Ik_B$, we define three characteristic rotational temperatures in terms of the three moments of inertia according to

$$\Theta_{\text{rot},j} = \frac{\hbar^2}{2I_j k_B} \qquad j = \text{A, B, C} \tag{4.51}$$

Thus, we have the various cases

$$\Theta_{\text{rot,A}} = \Theta_{\text{rot,B}} = \Theta_{\text{rot,C}} \qquad \text{spherical top}$$

$$\Theta_{\text{rot,A}} = \Theta_{\text{rot,B}} \neq \Theta_{\text{rot,C}} \qquad \text{symmetric top}$$

$$\Theta_{\text{rot,A}} \neq \Theta_{\text{rot,B}} \neq \Theta_{\text{rot,C}} \qquad \text{asymmetric top}$$

The quantum-mechanical problem of a spherical top can be solved exactly to give

$$\begin{aligned} \varepsilon_J &= \frac{J(J+1)\hbar^2}{2I} \\ g_J &= (2J+1)^2 \qquad J = 0,\ 1,\ 2,\ \ldots \end{aligned} \tag{4.52}$$

The rotational partition function is

$$q_{\text{rot}}(T) = \sum_{J=0}^{\infty} (2J+1)^2 e^{-\hbar^2 J(J+1)/2Ik_{\text{B}}T} \tag{4.53}$$

For almost all spherical top molecules $\Theta_{\text{rot}} \ll T$ at ordinary temperatures, so we convert the sum in Equation 4.53 to an integral:

$$q_{\text{rot}}(T) = \frac{1}{\sigma} \int_0^{\infty} (2J+1)^2 e^{-\Theta_{\text{rot}} J(J+1)/T} dJ$$

Note that we have included the symmetry number $\sigma$. For $\Theta_{\text{rot}} \ll T$, the most important values of $J$ are large (Problem 4–26), and so we may neglect 1 compared with $J$ in the integrand of the above expression for $q_{\text{rot}}$ to obtain

$$q_{\text{rot}}(T) = \frac{1}{\sigma} \int_0^{\infty} 4J^2 e^{-\Theta_{\text{rot}} J^2/T} dJ$$

If we let $\Theta_{\text{rot}}/T = a$, we can write

$$\begin{aligned} q_{\text{rot}}(T) &= \frac{4}{\sigma} \int_0^{\infty} x^2 e^{-ax^2} dx \\ &= \frac{4}{\sigma} \cdot \frac{1}{4a} \left(\frac{\pi}{a}\right)^{1/2} \end{aligned}$$

or, upon substituting $\Theta_{\text{rot}}/T$ for $a$,

$$q_{\text{rot}}(T) = \frac{\pi^{1/2}}{\sigma} \left(\frac{T}{\Theta_{\text{rot}}}\right)^{3/2} \qquad \text{spherical top} \tag{4.54}$$

The corresponding expressions for a symmetric top and an asymmetric top are

$$q_{\text{rot}}(T) = \frac{\pi^{1/2}}{\sigma} \left(\frac{T}{\Theta_{\text{rot,A}}}\right) \left(\frac{T}{\Theta_{\text{rot,C}}}\right)^{1/2} \qquad \text{symmetric top} \tag{4.55}$$

and

$$q_{\text{rot}}(T) = \frac{\pi^{1/2}}{\sigma}\left(\frac{T^3}{\Theta_{\text{rot,A}}\Theta_{\text{rot,B}}\Theta_{\text{rot,C}}}\right)^{1/2} \quad \text{asymmetric top} \tag{4.56}$$

Notice how Equation 4.56 reduces to Equation 4.55 when $\Theta_{\text{rot,A}} = \Theta_{\text{rot,B}}$ and how both Equations 4.55 and 4.56 reduce to Equation 4.54 when $\Theta_{\text{rot,A}} = \Theta_{\text{rot,B}} = \Theta_{\text{rot,C}}$. Table 4.4 contains values of $\Theta_{\text{rot,A}}$, $\Theta_{\text{rot,B}}$, and $\Theta_{\text{rot,C}}$ for several polyatomic molecules.

The average molar rotational energy of a nonlinear polyatomic molecule is

$$\begin{aligned}\overline{U}_{\text{rot}} &= N_A k_B T^2\left(\frac{d\ln q_{\text{rot}}(T)}{dT}\right)\\ &= RT^2\left(\frac{d\ln T^{3/2}}{dT}\right) = \frac{3RT}{2}\end{aligned}$$

or $RT/2$ for each rotational degree of freedom, and $\overline{C}_{V,\text{rot}} = 3R/2$.

## 4–9. Calculated Molar Heat Capacities Are in Very Good Agreement with Experimental Data

We can now use the results of Sections 4–7 and 4–8 to construct $q(V,T)$ for polyatomic molecules. For an ideal gas of linear polyatomic molecules, $q(V,T)$ is the product of Equations 4.43, 4.44, 4.46, and 4.50:

$$q(V,T) = \left(\frac{2\pi M k_B T}{h^2}\right)^{3/2} V \cdot \frac{T}{\sigma\Theta_{\text{rot}}} \cdot \left(\prod_{j=1}^{3n-5}\frac{e^{-\Theta_{\text{vib},j}/2T}}{1-e^{-\Theta_{\text{vib},j}/T}}\right) \cdot g_{e1}e^{D_e/k_BT} \tag{4.57}$$

The energy is

$$\frac{U}{Nk_BT} = \frac{3}{2} + \frac{2}{2} + \sum_{j=1}^{3n-5}\left(\frac{\Theta_{\text{vib},j}}{2T} + \frac{\Theta_{\text{vib},j}/T}{e^{\Theta_{\text{vib},j}/T}-1}\right) - \frac{D_e}{k_BT} \tag{4.58}$$

and the heat capacity is

$$\frac{C_V}{Nk_B} = \frac{3}{2} + \frac{2}{2} + \sum_{j=1}^{3n-5}\left(\frac{\Theta_{\text{vib},j}}{T}\right)^2 \frac{e^{-\Theta_{\text{vib},j}/T}}{(1-e^{-\Theta_{\text{vib},j}/T})^2} \tag{4.59}$$

For an ideal gas of nonlinear polyatomic molecules,

$$\begin{aligned}q(V,T) &= \left(\frac{2\pi M k_B T}{h^2}\right)^{3/2} V \cdot \frac{\pi^{1/2}}{\sigma}\left(\frac{T^3}{\Theta_{\text{rot,A}}\Theta_{\text{rot,B}}\Theta_{\text{rot,C}}}\right)^{1/2}\\ &\times \left[\prod_{j=1}^{3n-6}\frac{e^{-\Theta_{\text{vib},j}/2T}}{(1-e^{-\Theta_{\text{vib},j}/T})}\right] \cdot g_{e1}e^{D_e/k_BT}\end{aligned} \tag{4.60}$$

$$\frac{U}{Nk_BT} = \frac{3}{2} + \frac{3}{2} + \sum_{j=1}^{3n-6}\left(\frac{\Theta_{\text{vib},j}}{2T} + \frac{\Theta_{\text{vib},j}/T}{e^{\Theta_{\text{vib},j}/T} - 1}\right) - \frac{D_e}{k_BT} \tag{4.61}$$

and

$$\frac{C_V}{Nk_B} = \frac{3}{2} + \frac{3}{2} + \sum_{j=1}^{3n-6}\left(\frac{\Theta_{\text{vib},j}}{T}\right)^2 \frac{e^{-\Theta_{\text{vib},j}/T}}{(1 - e^{-\Theta_{\text{vib},j}/T})^2} \tag{4.62}$$

**EXAMPLE 4–8**

Calculate the molar heat capacity of gaseous water at 300 K.

SOLUTION: We use Equation 4.62 with $\Theta_{\text{vib},j} = 2290$ K, 5160 K, and 5360 K (Table 4.4). For $\Theta_{\text{vib},j} = 2290$ K,

$$\frac{\overline{C}_{V,j}}{R} = \left(\frac{2290}{300}\right)^2 \frac{e^{2290/300}}{(e^{2290/300} - 1)^2} = 0.0282$$

Similarly $\overline{C}_{V,j}/R = 1.00 \times 10^{-5}$ for $\Theta_{\text{vib},j} = 5160$ K and $5.56 \times 10^{-6}$ for $\Theta_{\text{vib},j} =$ 5360 K. The total molar heat capacity of water at 300 K is

$$\frac{\overline{C}_V}{R} = 3.000 + 0.0282 + 1.00 \times 10^{-5} + 5.56 \times 10^{-6} = 3.028$$

The experimental value is 3.011. Notice that the vibrational degrees of freedom contribute very little to the heat capacity of water at 300 K. The calculated and experimental values at 1000 K are 3.948 and 3.952, respectively. Figure 4.7 shows the molar heat capacity of water from 300 K to 1200 K.

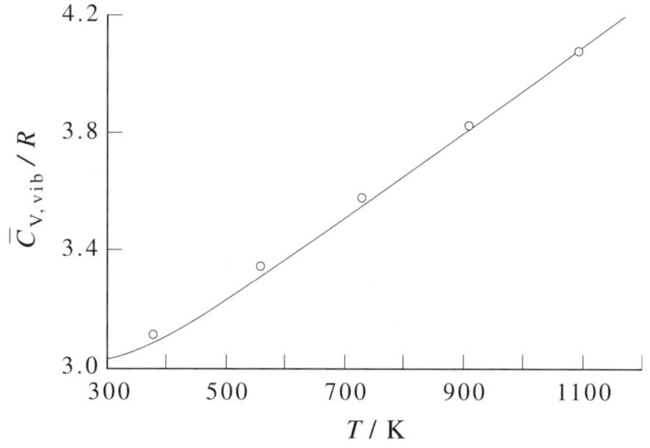


**FIGURE 4.7**
A comparison of the molar heat capacity of water vapor calculated from Equation 4.62 and the experimental value. The experimental data are indicated by the circles.

Table 4.5 gives the vibrational contribution to the molar heat capacity at 300 K for a variety of molecules of different shapes. It can be seen that the vibrational contributions are far from their high-temperature limits and that the agreement between the calculated and experimental values of $\overline{C}_V/R$ is good. A calculation for more complicated molecules would show similar agreement between the calculated values and the experimental data.

**TABLE 4.5**
Vibrational contributions to the molar heat capacity of some polyatomic molecules at 300 K.

| Molecule | $\Theta_{\text{vib}}$/K | Degeneracy | Vibrational Contribution to $\overline{C}_V$ | $\overline{C}_{V,\text{vib}}/R$ | Total $\overline{C}_V/R$ (calc) | Total $\overline{C}_V/R$ (exptl) |
|---|---|---|---|---|---|---|
| $CO_2$ | 1890 | 1 | 0.073 | | | |
| | 3360 | 1 | 0.000 | | | |
| | 954 | 2 | 0.458 | 0.99 | 3.49 | 3.46 |
| $N_2O$ | 1840 | 1 | 0.082 | | | |
| | 3200 | 1 | 0.003 | | | |
| | 850 | 2 | 0.533 | 1.15 | 2.65 | |
| $NH_3$ | 4800 | 1 | 0.000 | | | |
| | 1360 | 1 | 0.226 | | | |
| | 4880 | 2 | 0.000 | | | |
| | 2330 | 2 | 0.026 | 0.28 | 3.28 | |
| $CH_4$ | 4170 | 1 | 0.000 | | | |
| | 2180 | 2 | 0.037 | | | |
| | 4320 | 3 | 0.000 | | | |
| | 1870 | 3 | 0.077 | 0.30 | 3.30 | 3.29 |
| $H_2O$ | 2290 | 1 | 0.028 | | | |
| | 5160 | 1 | 0.000 | | | |
| | 5360 | 1 | 0.000 | 0.03 | 3.03 | 3.01 |

# Problems

**4-1.** Equation 4.7 shows that $\langle \varepsilon_{\text{trans}} \rangle = \frac{3}{2} k_{\text{B}} T$ in three dimensions, and Problem 4–3 shows that $\langle \varepsilon_{\text{trans}} \rangle = \frac{1}{2} k_{\text{B}} T$ in one dimension and $\frac{2}{2} k_{\text{B}} T$ in two dimensions. Show that typical values of translational quantum numbers at room temperature are $O(10^9)$ for $m = 10^{-26}$ kg, $a = 1$ dm, and $T = 300$ K.

**4-2.** Show that the difference between the successive terms in the summation in Equation 4.4 is very small for $m = 10^{-26}$ kg, $a = 1$ dm, and $T = 300$ K. Recall from Problem 4–1 that typical values of $n$ are $O(10^9)$.

**4-3.** Show that

$$q_{\text{trans}}(a, T) = \left( \frac{2\pi m k_{\text{B}} T}{h^2} \right)^{1/2} a$$

in one dimension and that

$$q_{\text{trans}}(a, T) = \left( \frac{2\pi m k_{\text{B}} T}{h^2} \right) a^2$$

in two dimensions. Use these results to show that $\langle \varepsilon_{\text{trans}} \rangle$ has a contribution of $k_{\text{B}} T/2$ to its total value for each dimension.

**4-4.** Using the data in Table 1.2, calculate the fraction of sodium atoms in the first excited state at 300 K, 1000 K, and 2000 K.

**4-5.** Using the data in Table 4.1, evaluate the fraction of lithium atoms in the first excited state at 300 K, 1000 K, and 2000 K.

**4-6.** Show that each dimension contributes $R/2$ to the molar translational heat capacity.

**4-7.** Using the values of $\Theta_{\text{vib}}$ and $D_0$ in Table 4.2, calculate the vaues of $D_e$ for CO, NO, and $K_2$.

**4-8.** Calculate the characteristic vibrational temperature $\Theta_{\text{vib}}$ for $H_2$(g) and $D_2$(g) ($\tilde{\nu}_{H_2} = 4401\ \text{cm}^{-1}$ and $\tilde{\nu}_{D_2} = 3112\ \text{cm}^{-1}$).

**4-9.** Plot the vibrational contribution to the molar heat capacity of $Cl_2$(g) from 250 K to 1000 K.

**4-10.** Plot the fraction of HCl(g) molecules in the first few vibrational states at 300 K and 1000 K.

**4-11.** Calculate the fraction of molecules in the ground vibrational state and in all the excited states at 300 K for each of the molecules in Table 4.2.

**4-12.** Calculate the value of the characteristic rotational temperature $\Theta_{\text{rot}}$ for $H_2$(g) and $D_2$(g). (The bond lengths of $H_2$ and $D_2$ are 74.16 pm.) The atomic mass of deuterium is 2.014.

**4-13.** The average molar rotational energy of a diatomic molecule is $RT$. Show that typical values of $J$ are given by $J(J+1) = T/\Theta_{\text{rot}}$. What are typical values of $J$ for $N_2$(g) at 300 K?

**4-14.** There is a mathematical procedure to calculate the error in replacing a summation by an integral as we do for the translational and rotational partition functions. The formula is called the Euler-Maclaurin summation formula and goes as follows:

$$\sum_{n=a}^{b} f(n) = \int_a^b f(n)dn + \frac{1}{2}\{f(b) + f(a)\} - \frac{1}{12}\left\{ \left.\frac{df}{dn}\right|_{n=a} - \left.\frac{df}{dn}\right|_{n=b} \right\}$$
$$+ \frac{1}{720}\left\{ \left.\frac{d^3 f}{dn^3}\right|_{n=a} - \left.\frac{d^3 f}{dn^3}\right|_{n=b} \right\} + \cdots$$

Apply this formula to Equation 4.33 to obtain

$$q_{\mathrm{rot}}(T) = \frac{T}{\Theta_{\mathrm{rot}}}\left\{ 1 + \frac{1}{3}\left(\frac{\Theta_{\mathrm{rot}}}{T}\right) + \frac{1}{15}\left(\frac{\Theta_{\mathrm{rot}}}{T}\right)^2 + O\left[\left(\frac{\Theta_{\mathrm{rot}}}{T}\right)^3\right]\right\}$$

Calculate the correction to replacing Equation 4.33 by an integral for $N_2(g)$ at 300 K; $H_2(g)$ at 300 K (being so light, $H_2$ is an extreme example).

**4-15.** Apply the Euler-Maclaurin summation formula (Problem 4–14) to the one-dimensional version of Equation 4.4 to obtain

$$q_{\mathrm{trans}}(a, T) = \left(\frac{2\pi m k_{\mathrm{B}} T}{h^2}\right)^{1/2} a + \left[\frac{1}{2} + \frac{h^2}{48 m a^2 k_{\mathrm{B}} T}\right] e^{-h^2/8ma^2 k_{\mathrm{B}} T}$$

Show that the correction amounts to about $10^{-8}\%$ for $m = 10^{-26}$ kg, $a = 1$ dm, and $T = 300$ K.

**4-16.** We were able to evaluate the vibrational partition function for a harmonic oscillator exactly by recognizing the summation as a geometric series. Apply the Euler-Maclaurin summation formula (Problem 4–14) to this case and show that

$$\sum_{v=0}^{\infty} e^{-\beta(v+\frac{1}{2})h\nu} = e^{-\Theta_{\mathrm{vib}}/2T} \sum_{v=0}^{\infty} e^{-v\Theta_{\mathrm{vib}}/T}$$
$$= e^{-\Theta_{\mathrm{vib}}/2T}\left[\frac{T}{\Theta_{\mathrm{vib}}} + \frac{1}{2} + \frac{\Theta_{\mathrm{vib}}}{12T} + \cdots\right]$$

Show that the corrections to replacing the summation by an integration are very large for $O_2(g)$ at 300 K. Fortunately, we don't need to replace the summation by an integration in this case.

**4-17.** Plot the fraction of NO(g) molecules in the various rotational levels at 300 K and at 1000 K.

**4-18.** Show that the values of $J$ at the maximum of a plot of $f_J$ versus $J$ (Equation 4.35) is given by

$$J_{\max} \approx \left(\frac{T}{2\Theta_{\mathrm{rot}}}\right)^{1/2} - \frac{1}{2}$$

*Hint*: Treat $J$ as a continuous variable. Use this result to verify the values of $J$ at the maxima in the plots in Problem 4–17.

**4-19.** The experimental heat capacity of $N_2(g)$ can be fit to the empirical formula

$$\overline{C}_V(T)/R = 2.283 + (6.291 \times 10^{-4}\ \mathrm{K}^{-1})T - (5.0 \times 10^{-10}\ \mathrm{K}^{-2})T^2$$

over the temperature range $300\ \mathrm{K} < T < 1500\ \mathrm{K}$. Plot $\overline{C}_V(T)/R$ versus $T$ over this range using Equation 4.41, and compare your results with the experimental curve.

**4-20.** The experimental heat capacity of CO(g) can be fit to the empirical formula

$$\overline{C}_V(T)/R = 2.192 + (9.240 \times 10^{-4}\ \mathrm{K}^{-1})T - (1.41 \times 10^{-7}\ \mathrm{K}^{-2})T^2$$

over the temperature range $300\ \mathrm{K} < T < 1500\ \mathrm{K}$. Plot $\overline{C}_V(T)/R$ versus $T$ over this range using Equation 4.41, and compare your results with the experimental curve.

**4-21.** Calculate the contribution of each normal mode to the molar vibrational heat capacity of $H_2O(g)$ at 600 K.

**4-22.** In analogy to the characteristic vibrational temperature, we can define a characteristic electronic temperature by

$$\Theta_{\mathrm{elec},j} = \frac{\varepsilon_{ej}}{k_\mathrm{B}}$$

where $\varepsilon_{ej}$ is the energy of the $j$th excited electronic state relative to the ground state. Show that if we define the ground state to be the zero of energy, then

$$q_{\mathrm{elec}} = g_0 + g_1 e^{-\Theta_{\mathrm{elec},1}/T} + g_2 e^{-\Theta_{\mathrm{elec},2}/T} + \cdots$$

The first and second excited electronic states of O(g) lie 158.2 $\mathrm{cm}^{-1}$ and 226.5 $\mathrm{cm}^{-1}$ above the ground electronic state. Given $g_0 = 5$, $g_1 = 3$, and $g_2 = 1$, calculate the values of $\Theta_{\mathrm{elec},1}$, $\Theta_{\mathrm{elec},2}$, and $q_{\mathrm{elec}}$ (ignoring any higher states) for O(g) at 5000 K.

**4-23.** Determine the symmetry numbers for $H_2O$, HOD, $CH_4$, $SF_6$, $C_2H_2$, and $C_2H_4$.

**4-24.** The HCN(g) molecule is a linear molecule, and the following constants determined spectroscopically are $I = 18.816 \times 10^{-47}\ \mathrm{kg{\cdot}m^2}$, $\tilde{\nu}_1 = 2096.7\ \mathrm{cm}^{-1}$ (HC–N stretch), $\tilde{\nu}_2 = 713.46\ \mathrm{cm}^{-1}$ (H–C–N bend, two-fold degeneracy), and $\tilde{\nu}_3 = 3311.47\ \mathrm{cm}^{-1}$ (H–C stretch). Calculate the values of $\Theta_{\mathrm{rot}}$ and $\Theta_{\mathrm{vib}}$ and $\overline{C}_V$ at 3000 K.

**4-25.** The acetylene molecule is linear, the C≡C bond length is 120.3 pm, and the C–H bond length is 106.0 pm. What is the symmetry number of acetylene? Determine the moment of inertia (Section 1–8) of acetylene and calculate the value of $\Theta_{\mathrm{rot}}$. The fundamental frequencies of the normal modes are $\tilde{\nu}_1 = 1975\ \mathrm{cm}^{-1}$, $\tilde{\nu}_2 = 3370\ \mathrm{cm}^{-1}$, $\tilde{\nu}_3 = 3277\ \mathrm{cm}^{-1}$, $\tilde{\nu}_4 = 729\ \mathrm{cm}^{-1}$, and $\tilde{\nu}_5 = 600\ \mathrm{cm}^{-1}$. The normal modes $\tilde{\nu}_4$ and $\tilde{\nu}_5$ are doubly degenerate. All the other modes are nondegenerate. Calculate $\Theta_{\mathrm{vib},j}$ and $\overline{C}_V$ at 300 K.

**4-26.** Plot the summand in Equation 4.53 versus $J$, and show that the most important values of $J$ are large for $T \gg \Theta_{\mathrm{rot}}$. We use this fact in going from Equation 4.53 to Equation 4.54.

**4-27.** Use the Euler-Maclaurin summation formula (Problem 4–14) to show that

$$q_{\mathrm{rot}}(T) = \frac{\pi^{1/2}}{\sigma}\left(\frac{T}{\Theta_{\mathrm{rot}}}\right)^{3/2} + \frac{1}{6} + O\left(\frac{\Theta_{\mathrm{rot}}}{T}\right)$$

for a spherical top molecule. Show that the correction to replacing Equation 4.53 by an integral is about 1% for $CH_4$ and 0.001% for $CCl_4$ at 300 K.

**4-28.** The N–N and N–O bond lengths in the (linear) molecule $N_2O$ are 109.8 pm and 121.8 pm, respectively. Calculate the center of mass and the moment of inertia of $^{14}N^{14}N^{16}O$. Compare your answer with the value obtained from $\Theta_{rot}$ in Table 4.4.

**4-29.** $NO_2(g)$ is a bent triatomic molecule. The following data determined from spectroscopic measurements are $\tilde{\nu}_1 = 1319.7\ \text{cm}^{-1}$, $\tilde{\nu}_2 = 749.8\ \text{cm}^{-1}$, $\tilde{\nu}_3 = 1617.75\ \text{cm}^{-1}$, $\tilde{A}_0 = 8.0012\ \text{cm}^{-1}$, $\tilde{B}_0 = 0.43304\ \text{cm}^{-1}$, and $\tilde{C}_0 = 0.41040\ \text{cm}^{-1}$. Determine the three characteristic vibrational temperatures and the characteristic rotational temperatures for each of the principle axes of $NO_2(g)$ at 1000 K. Calculate the value of $\overline{C}_V$ at 1000 K.

**4-30.** The experimental heat capacity of $NH_3(g)$ can be fit to the empirical formula

$$\overline{C}_V(T)/R = 2.115 + (3.919 \times 10^{-3}\ \text{K}^{-1})T - (3.66 \times 10^{-7}\ \text{K}^{-2})T^2$$

over the temperature range 300 K $< T <$ 1500 K. Plot $\overline{C}_V(T)/R$ versus $T$ over this range using Equation 4.62 and the molecular parameters in Table 4.4, and compare your results with the experimental curve.

**4-31.** The experimental heat capacity of $SO_2(g)$ can be fit to the empirical formula

$$\overline{C}_V(T)/R = 6.8711 - \frac{1454.62\ \text{K}}{T} + \frac{160\,351\ \text{K}^2}{T^2}$$

over the temperature range 300 K $< T <$ 1500 K. Plot $\overline{C}_V(T)/R$ versus $T$ over this range using Equation 4.62 and the molecular parameters in Table 4.4, and compare your results with the experimental curve.

**4-32.** The experimental heat capacity of $CH_4(g)$ can be fit to the empirical formula

$$\overline{C}_V(T)/R = 1.099 + (7.27 \times 10^{-3}\ \text{K}^{-1})T + (1.34 \times 10^{-7}\ \text{K}^{-2})T^2 \\ -(8.67 \times 10^{-10}\ \text{K}^{-3})T^3$$

over the temperature range 300 K $< T <$ 1500 K. Plot $\overline{C}_V(T)/R$ versus $T$ over this range using Equation 4.62 and the molecular parameters in Table 4.4, and compare your results with the experimental curve.

**4-33.** Show that the moment of inertia of a diatomic molecule is $\mu R_e^2$, where $\mu$ is the reduced mass, and $R_e$ is the equilibrium bond length.

**4-34.** Given that the values of $\Theta_{rot}$ and $\Theta_{vib}$ for $H_2$ are 85.3 K and 6332 K, respectively, calculate these quantities for HD and $D_2$. *Hint*: Use the Born-Oppenheimer approximation.

**4-35.** Using the result for $q_{rot}(T)$ obtained in Problem 4–14, derive corrections to the expressions $\langle E_{rot}\rangle = RT$ and $C_{V,rot} = R$ given in Section 4–5. Express your result in terms of powers of $\Theta_{rot}/T$.

**4-36.** Show that the thermodynamic quantities $P$ and $C_V$ are independent of the choice of a zero of energy.

**4-37.** Molecular nitrogen is heated in an electric arc. The spectroscopically determined relative populations of excited vibrational levels are listed below.

| $v$ | 0 | 1 | 2 | 3 | 4 | ... |
|---|---|---|---|---|---|---|
| $\frac{f_v}{f_0}$ | 1.000 | 0.200 | 0.040 | 0.008 | 0.002 | ... |

Is the nitrogen in thermodynamic equilibrium with respect to vibrational energy? What is the vibrational temperature of the gas? Is this value necessarily the same as the translational temperature? Why or why not?

**4-38.** Consider a system of independent diatomic molecules constrained to move in a plane, that is, a two-dimensional ideal diatomic gas. How many degrees of freedom does a two-dimensional diatomic molecule have? Given that the energy eigenvalues of a two-dimensional rigid rotator are

$$\varepsilon_J = \frac{\hbar^2 J^2}{2I} \qquad J = 0,\ 1,\ 2,\ \ldots$$

(where $I$ is the moment of inertia of the molecule) with a degeneracy $g_J = 2$ for all $J$ except $J = 0$, derive an expression for the rotational partition function. The vibrational partition function is the same as for a three-dimensional diatomic gas. Write out

$$q(T) = q_{\text{trans}}(T)q_{\text{rot}}(T)q_{\text{vib}}(T)$$

and derive an expression for the average energy of this two-dimensional ideal diatomic gas.

**4-39.** What molar constant-volume heat capacities would you expect under classical conditions for the following gases: (a) Ne, (b) $O_2$, (c) $H_2O$, (d) $CO_2$, and (e) $CHCl_3$?

**4-40.** In this problem, we will derive an expression for the number of translational energy states with (translational) energy between $\varepsilon$ and $\varepsilon + d\varepsilon$. This expression is essentially the degeneracy of the state whose energy is

$$\varepsilon_{n_x n_y n_z} = \frac{h^2}{8ma^2}(n_x^2 + n_y^2 + n_z^2) \qquad n_x,\ n_y,\ n_z = 1,\ 2,\ 3,\ \ldots \tag{1}$$

The degeneracy is given by the number of ways the integer $M = 8ma^2\varepsilon/h^2$ can be written as the sum of the squares of three positive integers. In general, this is an erratic and discontinuous function of $M$ (the number of ways will be zero for many values of $M$), but it becomes smooth for large $M$, and we can derive a simple expression for it. Consider a three-dimensional space spanned by $n_x$, $n_y$, and $n_z$. There is a one-to-one correspondence between energy states given by Equation 1 and the points in this $n_x$, $n_y$, $n_z$ space with coordinates given by positive integers. Figure 4.8 shows a two-dimensional version of this space. Equation 1 is an equation for a sphere of radius $R = (8ma^2\varepsilon/h^2)^{1/2}$ in this space

$$n_x^2 + n_y^2 + n_z^2 = \frac{8ma^2\varepsilon}{h^2} = R^2$$

We want to calculate the number of lattice points that lie at some fixed distance from the origin in this space. In general, this is very difficult, but for large $R$ we can proceed as follows. We treat $R$, or $\varepsilon$, as a continuous variable and ask for the number of lattice points

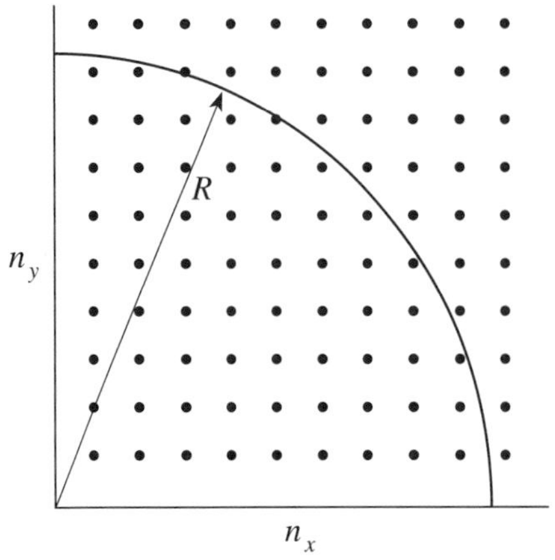


**FIGURE 4.8**
A two-dimensional version of the $(n_x, n_y, n_z)$ space, the space with the quantum numbers $n_x$, $n_y$, and $n_z$ as axes. Each point corresponds to an energy of a particle in a (two-dimensional) box.

between $\varepsilon$ and $\varepsilon + \Delta\varepsilon$. To calculate this quantity, it is convenient to first calculate the number of lattice points consistent with an energy $\leq \varepsilon$. For large $\varepsilon$, an excellent approximation can be made by equating the number of lattice points consistent with an energy $\leq \varepsilon$ with the volume of one octant of a sphere of radius $R$. We take only one octant because $n_x$, $n_y$, and $n_z$ are restricted to be positive integers. If we denote the number of such states by $\Phi(\varepsilon)$, we can write

$$\Phi(\varepsilon) = \frac{1}{8}\left(\frac{4\pi R^3}{3}\right) = \frac{\pi}{6}\left(\frac{8ma^2\varepsilon}{h^2}\right)^{3/2}$$

The number of states with energy between $\varepsilon$ and $\varepsilon + \Delta\varepsilon$ ($\Delta\varepsilon/\varepsilon \ll 1$) is

$$\omega(\varepsilon, \Delta\varepsilon) = \Phi(\varepsilon + \Delta\varepsilon) - \Phi(\varepsilon)$$

Show that

$$\omega(\varepsilon, \Delta\varepsilon) = \omega(\varepsilon)\Delta\varepsilon = \frac{\pi}{4}\left(\frac{8ma^2}{h^2}\right)^{3/2} \varepsilon^{1/2}\Delta\varepsilon + O[(\Delta\varepsilon)^2]$$

Show that if we take $\varepsilon = 3k_BT/2$, $T = 300$ K, $m = 10^{-25}$ kg, $a = 1$ dm, and $\Delta\varepsilon$ to be $0.010\varepsilon$ (in other words 1% of $\varepsilon$), then $\omega(\varepsilon, \Delta\varepsilon)$ is $O(10^{28})$. So, even for a system as simple as a single particle in a box, the degeneracy can be very large at room temperature.

**4-41.** The translational partition function can be written as a single integral over the energy $\varepsilon$ if we include the degeneracy

$$q_{\text{trans}}(V, T) = \int_0^\infty \omega(\varepsilon)e^{-\varepsilon/k_BT}d\varepsilon$$

where $\omega(\varepsilon)d\varepsilon$ is the number of states with energy between $\varepsilon$ and $\varepsilon + d\varepsilon$. Using the result from the previous problem, show that $q_{\text{trans}}(V, T)$ is the same as that given by Equation 4.6.

# MATHCHAPTER D

# PARTIAL DIFFERENTIATION

Recall from your course in calculus that the derivative of a function $y(x)$ at some point $x$ is defined as

$$\frac{dy}{dx} = \lim_{\Delta x \to 0} \frac{y(x + \Delta x) - y(x)}{\Delta x} \tag{D.1}$$

Physically, $dy/dx$ expresses the variation of $y$ when $x$ is varied. Much of your calculus course was spent in starting with Equation D.1 to derive formulas for the derivatives of the commonly occurring functions. The function $y$ in Equation D.1 depends upon only one variable, $x$. For the function $y(x)$, $x$ is called the independent variable and $y$, whose value depends upon the value of $x$, is called the dependent variable.

Functions can depend upon more than one variable. For example, we know that the pressure of an ideal gas depends upon the temperature, volume, and number of moles through the equation

$$P = \frac{nRT}{V} \tag{D.2}$$

In this case, there are three independent variables; the temperature, volume, and amount of gas can be varied independently. The pressure is the dependent variable. We can emphasize this dependency by writing

$$P = P(n, T, V)$$

Experimentally, we may wish to vary only one of the independent variables at a time (say the temperature) to produce a change in pressure with two of the independent variables fixed (fixed volume and fixed number of moles). To form the derivative of $P$ with respect to $T$ with $n$ and $V$ held constant, we simply refer to Equation D.1 and write

$$\left(\frac{\partial P}{\partial T}\right)_{n,V} = \lim_{\Delta T \to 0} \frac{P(n, T + \Delta T, V) - P(n, T, V)}{\Delta T} \tag{D.3}$$

We call $(\partial P/\partial T)_{n,V}$ the partial derivative of $P$ with respect to $T$, with $n$ and $V$ held constant. To actually evaluate this partial derivative, we simply differentiate $P$ with respect to $T$ in Equation D.2, treating $n$ and $V$ as if they were constants. Thus, for an ideal gas

$$\left(\frac{\partial P}{\partial T}\right)_{n,V} = \frac{nR}{V}$$

We can also have

$$\left(\frac{\partial P}{\partial n}\right)_{T,V} = \frac{RT}{V}$$

and

$$\left(\frac{\partial P}{\partial V}\right)_{n,T} = -\frac{nRT}{V^2}$$

**EXAMPLE D–1**

Evaluate the two first partial derivatives of $P$ for the van der Waals equation

$$P = \frac{RT}{\overline{V} - b} - \frac{a}{\overline{V}^2} \tag{D.4}$$

SOLUTION: In this case, $P$ depends upon $T$ and $\overline{V}$, so we have $P = P(T, \overline{V})$. The two first partial derivatives of $P$ are

$$\left(\frac{\partial P}{\partial T}\right)_{\overline{V}} = \frac{R}{\overline{V} - b} \tag{D.5}$$

and

$$\left(\frac{\partial P}{\partial \overline{V}}\right)_T = -\frac{RT}{(\overline{V} - b)^2} + \frac{2a}{\overline{V}^3} \tag{D.6}$$

The partial derivatives given by Equations D.5 and D.6 are themselves functions of $T$ and $\overline{V}$, so we can form second partial derivatives by differentiating Equations D.5 and D.6:

$$\left(\frac{\partial^2 P}{\partial T^2}\right)_{\overline{V}} = 0$$

and

$$\left(\frac{\partial^2 P}{\partial \overline{V}^2}\right)_T = \frac{2RT}{(\overline{V} - b)^3} - \frac{6a}{\overline{V}^4}$$

We can also form another type of second derivative, however. For example, we can form

$$\left[\frac{\partial}{\partial \overline{V}}\left(\frac{\partial P}{\partial T}\right)_{\overline{V}}\right]_T = \left[\frac{\partial}{\partial \overline{V}}\left(\frac{R}{\overline{V}-b}\right)\right]_T$$
$$= -\frac{R}{(\overline{V}-b)^2} \qquad \text{(D.7)}$$

and we can also form

$$\left[\frac{\partial}{\partial T}\left(\frac{\partial P}{\partial \overline{V}}\right)_T\right]_{\overline{V}} = \left[\frac{\partial}{\partial T}\left(-\frac{RT}{(\overline{V}-b)^2}+\frac{2a}{\overline{V}^3}\right)\right]_{\overline{V}}$$
$$= -\frac{R}{(\overline{V}-b)^2} \qquad \text{(D.8)}$$

The above two second derivatives are called cross derivatives, mixed derivatives, or second cross partial derivatives. These derivatives are commonly written as

$$\left(\frac{\partial^2 P}{\partial \overline{V}\partial T}\right) \quad \text{or} \quad \left(\frac{\partial^2 P}{\partial T\partial \overline{V}}\right)$$

We don't indicate which variable is held constant because they differ with each differentiation. Notice that these two cross derivatives are equal (see Equations D.7 and D.8), so that

$$\left(\frac{\partial^2 P}{\partial \overline{V}\partial T}\right) = \left(\frac{\partial^2 P}{\partial T\partial \overline{V}}\right) \qquad \text{(D.9)}$$

Thus, the order in which we take the two partial derivatives of $P$ makes no difference in this case. It turns out that cross derivatives are generally equal.

**EXAMPLE D–2**

Suppose that

$$S = -\left(\frac{\partial A}{\partial T}\right)_V \quad \text{and} \quad P = -\left(\frac{\partial A}{\partial V}\right)_T$$

where $A$, $S$, and $P$ are functions of $T$ and $V$. Prove that

$$\left(\frac{\partial S}{\partial V}\right)_T = \left(\frac{\partial P}{\partial T}\right)_V$$

SOLUTION: Take the partial derivative of $S$ with respect to $V$ at constant $T$:

$$\left(\frac{\partial S}{\partial V}\right)_T = -\left(\frac{\partial^2 A}{\partial V\partial T}\right)$$

and the partial derivative of $P$ with respect to $T$ at constant $V$:

$$\left(\frac{\partial P}{\partial T}\right)_V = -\left(\frac{\partial^2 A}{\partial T \partial V}\right)$$

and equate the two cross derivatives of $A$ to obtain

$$\left(\frac{\partial S}{\partial V}\right)_T = \left(\frac{\partial P}{\partial T}\right)_V$$

The partial derivatives given in Equations D.5 and D.6 indicate how $P$ changes with one independent variable, keeping the other one fixed. We often want to know how a dependent variable changes with a change in the values of both (or more) of its independent variables. Using the example $P = P(T, \overline{V})$ (for one mole), we write

$$\Delta P = P(T + \Delta T, \overline{V} + \Delta\overline{V}) - P(T, \overline{V})$$

If we add and subtract $P(T, \overline{V} + \Delta\overline{V})$ to this equation, we obtain

$$\begin{aligned}\Delta P = &[P(T + \Delta T, \overline{V} + \Delta\overline{V}) - P(T, \overline{V} + \Delta\overline{V})] \\ &+ [P(T, \overline{V} + \Delta\overline{V}) - P(T, \overline{V})]\end{aligned}$$

Multiply the first two terms in brackets by $\Delta T/\Delta T$ and the second two terms by $\Delta\overline{V}/\Delta\overline{V}$ to get

$$\begin{aligned}\Delta P = &\left[\frac{P(T + \Delta T, \overline{V} + \Delta\overline{V}) - P(T, \overline{V} + \Delta\overline{V})}{\Delta T}\right]\Delta T \\ &+ \left[\frac{P(T, \overline{V} + \Delta\overline{V}) - P(T, \overline{V})}{\Delta\overline{V}}\right]\Delta\overline{V}\end{aligned}$$

Now let $\Delta T \to 0$ and $\Delta\overline{V} \to 0$, in which case we have

$$\begin{aligned}dP = &\lim_{\Delta T \to 0}\left[\frac{P(T + \Delta T, \overline{V}) - P(T, \overline{V})}{\Delta T}\right]\Delta T \\ &+ \lim_{\Delta\overline{V} \to 0}\left[\frac{P(T, \overline{V} + \Delta\overline{V}) - P(T, \overline{V})}{\Delta\overline{V}}\right]\Delta\overline{V}\end{aligned} \tag{D.10}$$

The first limit gives $(\partial P/\partial T)_{\overline{V}}$ (by definition) and the second gives $(\partial P/\partial\overline{V})_T$, so that Equation D.10 gives our desired result:

$$dP = \left(\frac{\partial P}{\partial T}\right)_{\overline{V}} dT + \left(\frac{\partial P}{\partial\overline{V}}\right)_T d\overline{V} \tag{D.11}$$

Equation D.11 is called the total derivative of $P$. It simply says that the change in $P$ is given by how $P$ changes with $T$ (keeping $\overline{V}$ constant) times the infinitesimal change in $T$ plus how $P$ changes with $\overline{V}$ (at constant $T$) times the infinitesimal change in $\overline{V}$.

**EXAMPLE D–3**

We can use Equation D.11 to estimate the change in pressure when both the temperature and the molar volume are changed slightly. To this end, for finite $\Delta T$ and $\Delta\overline{V}$, we write Equation D.11 as

$$\Delta P \approx \left(\frac{\partial P}{\partial T}\right)_{\overline{V}} \Delta T + \left(\frac{\partial P}{\partial \overline{V}}\right)_T \Delta\overline{V}$$

Use this equation to estimate the change in pressure of one mole of an ideal gas if the temperature is changed from 273.15 K to 274.00 K and the volume is changed from 10.00 L to 9.90 L.

SOLUTION: We first need

$$\left(\frac{\partial P}{\partial T}\right)_{\overline{V}} = \left[\frac{\partial}{\partial T}\left(\frac{RT}{\overline{V}}\right)\right]_{\overline{V}} = \frac{R}{\overline{V}}$$

and

$$\left(\frac{\partial P}{\partial \overline{V}}\right)_T = \left[\frac{\partial}{\partial \overline{V}}\left(\frac{RT}{\overline{V}}\right)\right]_{\overline{V}} = -\frac{RT}{\overline{V}^2}$$

so that

$$\begin{aligned}
\Delta P &\approx \frac{R}{\overline{V}}\Delta T - \frac{RT}{\overline{V}^2}\Delta\overline{V} \\
&\approx \frac{(8.314\ \mathrm{J\cdot K^{-1}\cdot mol^{-1}})}{(10.00\ \mathrm{L\cdot mol^{-1}})}(0.85\ \mathrm{K}) \\
&\qquad - \frac{(8.314\ \mathrm{J\cdot K^{-1}\cdot mol^{-1}})(273.15\ \mathrm{K})}{(10.00\ \mathrm{L\cdot mol^{-1}})^2}(-0.10\ \mathrm{L\cdot mol^{-1}}) \\
&\approx 3.0\ \mathrm{J\cdot L^{-1}} \\
&\approx 3.0 \times 10^3\ \mathrm{J\cdot m^{-3}} = 3.0 \times 10^3\ \mathrm{Pa} = 0.030\ \mathrm{bar}
\end{aligned}$$

Incidently, in this particularly simple case, we calculate the exact change in $P$ from

$$\begin{aligned}
\Delta P &= \frac{RT_2}{\overline{V}_2} - \frac{RT_1}{\overline{V}_1} \\
&= (8.314\ \mathrm{J\cdot K^{-1}\cdot mol^{-1}})\left(\frac{274.00\ \mathrm{K}}{9.90\ \mathrm{L\cdot mol^{-1}}} - \frac{273.15\ \mathrm{K}}{10.00\ \mathrm{L\cdot mol^{-1}}}\right) \\
&= 3.0\ \mathrm{J\cdot L^{-1}} = 3.0\ \mathrm{J\cdot dm^{-3}} = 0.030\ \mathrm{bar}
\end{aligned}$$

Equation D.4 gives $P$ as a function of $T$ and $\overline{V}$, or $P = P(T, \overline{V})$. We can form the total derivative of $P$ by differentiating the right side of Equation D.4 with respect to $T$ and $\overline{V}$ to obtain

$$\begin{aligned} dP &= \frac{R}{\overline{V} - b} dT - \frac{RT}{(\overline{V} - b)^2} d\overline{V} + \frac{2a}{\overline{V}^3} d\overline{V} \\ &= \frac{R}{\overline{V} - b} dT + \left[ -\frac{RT}{(\overline{V} - b)^2} + \frac{2a}{\overline{V}^3} \right] d\overline{V} \end{aligned} \tag{D.12}$$

We can see from Example D–1 that Equation D.12 is just Equation D.11 written for the van der Waals equation. Suppose, however, that we are given an arbitrary expression for $dP$, say

$$dP = \frac{RT}{\overline{V} - b} dT + \left[ \frac{RT}{(\overline{V} - b)^2} - \frac{a}{T\overline{V}^2} \right] d\overline{V} \tag{D.13}$$

and are asked to determine the equation of state $P = P(T, \overline{V})$ that leads to Equation D.13. In fact, a simpler question is to ask if there even is a function $P(T, \overline{V})$ whose total derivative is given by Equation D.13. How can we tell? If there is such a function $P(T, \overline{V})$, then its total derivative is (Equation D.11)

$$dP = \left( \frac{\partial P}{\partial T} \right)_{\overline{V}} dT + \left( \frac{\partial P}{\partial \overline{V}} \right)_T d\overline{V}$$

Furthermore, according to Equation D.9, the cross derivatives of a function $P(T, \overline{V})$,

$$\left( \frac{\partial^2 P}{\partial \overline{V} \partial T} \right) = \left[ \frac{\partial}{\partial \overline{V}} \left( \frac{\partial P}{\partial T} \right)_{\overline{V}} \right]_T$$

and

$$\left( \frac{\partial^2 P}{\partial T \partial \overline{V}} \right) = \left[ \frac{\partial}{\partial T} \left( \frac{\partial P}{\partial \overline{V}} \right)_T \right]_{\overline{V}}$$

must be equal. If we apply this requirement to Equation D.13, we find that

$$\frac{\partial}{\partial T} \left[ \frac{RT}{(\overline{V} - b)^2} - \frac{a}{T\overline{V}^2} \right] = \frac{R}{(\overline{V} - b)^2} + \frac{a}{T^2\overline{V}^2}$$

and

$$\frac{\partial}{\partial \overline{V}} \left( \frac{RT}{\overline{V} - b} \right) = -\frac{RT}{(\overline{V} - b)^2}$$

Thus, we see that the cross-derivatives are not equal, so the expression given by Equation D.13 is not the derivative of any function $P(T, \overline{V})$. The differential given by Equation D.13 is called an *inexact differential.*

We can obtain an example of an *exact differential* simply by explicitly differentiating any function $P(T, \overline{V})$, such as we did for the van der Waals equation to obtain

Equation D.12. Equations D.7 and D.8 show that the cross derivatives are equal, as they must be for an exact differential.

**EXAMPLE D–4**

Is

$$dP = \left[\frac{R}{\overline{V}-B} + \frac{A}{2T^{3/2}\overline{V}(\overline{V}+B)}\right] dT + \left[-\frac{RT}{(\overline{V}-B)^2} + \frac{A(2\overline{V}+B)}{T^{1/2}\overline{V}^2(\overline{V}+B)^2}\right] d\overline{V} \qquad \text{(D.14)}$$

an exact differential?

SOLUTION: We evaluate the two derivatives

$$\left[\frac{\partial}{\partial \overline{V}}\left\{\frac{R}{\overline{V}-B} + \frac{A}{2T^{3/2}\overline{V}(\overline{V}+B)}\right\}\right]_T = -\frac{R}{(\overline{V}-B)^2} - \frac{A(2\overline{V}+B)}{2T^{3/2}\overline{V}^2(\overline{V}+B)^2}$$

and

$$\left[\frac{\partial}{\partial T}\left\{-\frac{RT}{(\overline{V}-B)^2} + \frac{A(2\overline{V}+B)}{T^{1/2}\overline{V}^2(\overline{V}+B)^2}\right\}\right]_{\overline{V}} = -\frac{R}{(\overline{V}-B)^2} - \frac{A(2\overline{V}+B)}{2T^{3/2}\overline{V}^2(\overline{V}+B)^2}$$

These derivatives are equal and so Equation D.14 represents an exact differential. Equation D.14 is the total derivative of $P$ for the Redlich-Kwong equation of state.

Exact and inexact differentials play a significant role in physical chemistry. If $dy$ is an exact differential, then

$$\int_1^2 dy = y_2 - y_1 \qquad \text{(exact differential)}$$

so the integral depends only upon the end points (1 and 2) and not upon the path from 1 to 2. This statement is not true for an inexact differential, however, so

$$\int_1^2 dy \neq y_2 - y_1 \qquad \text{(inexact differential)}$$

The integral in this case depends not only upon the end points but also upon the path from 1 to 2.

## Problems

**D-1.** The isothermal compressibility, $\kappa_T$, of a substance is defined as

$$\kappa_T = -\frac{1}{V}\left(\frac{\partial V}{\partial P}\right)_T$$

Obtain an expression for the isothermal compressibility of an ideal gas.

**D-2.** The coefficient of thermal expansion, $\alpha$, of a substance is defined as

$$\alpha = \frac{1}{V}\left(\frac{\partial V}{\partial T}\right)_P$$

Obtain an expression for the coefficient of thermal expansion of an ideal gas.

**D-3.** Prove that

$$\left(\frac{\partial P}{\partial V}\right)_{n,T} = \frac{1}{\left(\dfrac{\partial V}{\partial P}\right)_{n,T}}$$

for an ideal gas and for a gas whose equation of state is $P = nRT/(V - nb)$, where $b$ is a constant. This relation is generally true and is called the reciprocal identity. Notice that the same variables must be held fixed on both sides of the identity.

**D-4.** Given that

$$U = kT^2\left(\frac{\partial \ln Q}{\partial T}\right)_{N,V}$$

where

$$Q(N, V, T) = \frac{1}{N!}\left(\frac{2\pi m k_B T}{h^2}\right)^{3N/2} V^N$$

and $k_B$, $m$, and $h$ are constants, determine $U$ as a function of $T$.

**D-5.** Show that the total derivative of $P$ for the Redlich-Kwong equation,

$$P = \frac{RT}{\overline{V} - B} - \frac{A}{T^{1/2}\overline{V}(\overline{V} + B)}$$

is given by Equation D.14.

**D-6.** Show explicitly that

$$\left(\frac{\partial^2 P}{\partial \overline{V}\partial T}\right) = \left(\frac{\partial^2 P}{\partial T \partial \overline{V}}\right)$$

for the Redlich-Kwong equation (Problem D–5).

**D-7.** We will derive the following equation in Chapter 5:

$$\left(\frac{\partial U}{\partial V}\right)_T = T\left(\frac{\partial P}{\partial T}\right)_V - P$$

Evaluate $(\partial U/\partial V)_T$ for an ideal gas, for a van der Waals gas (Equation D.4), and for a Redlich-Kwong gas (Problem D–5).

**D-8.** Given that the heat capacity at constant volume is defined by

$$C_V = \left(\frac{\partial U}{\partial T}\right)_V$$

and given the expression in Problem D–7, derive the equation

$$\left(\frac{\partial C_V}{\partial V}\right)_T = T\left(\frac{\partial^2 P}{\partial T^2}\right)_V$$

**D-9.** Use the expression in Problem D–8 to determine $(\partial C_V/\partial V)_T$ for an ideal gas, a van der Waals gas (Equation D.4), and a Redlich-Kwong gas (see Problem D–5).

**D-10.** Is

$$dV = \pi r^2 dh + 2\pi r h dr$$

an exact or inexact differential?

**D-11.** Is

$$dx = C_V(T)dT + \frac{nRT}{V}dV$$

an exact or inexact differential? The quantity $C_V(T)$ is simply an arbitrary function of $T$. What about $dx/T$?

**D-12.** Prove that

$$\frac{1}{Y}\left(\frac{\partial Y}{\partial P}\right)_{T,n} = \frac{1}{\overline{Y}}\left(\frac{\partial \overline{Y}}{\partial P}\right)_T$$

and that

$$\left(\frac{\partial P}{\partial \overline{Y}}\right)_T = n\left(\frac{\partial P}{\partial Y}\right)_{T,n}$$

where $Y = Y(P, T, n)$ is an extensive variable.

**D-13.** Equation 2.5 gives $P$ for the van der Waals equation as a function of $\overline{V}$ and $T$. Show that $P$ expressed as a function of $V$, $T$, and $n$ is

$$P = \frac{nRT}{V - nb} - \frac{n^2 a}{V^2} \qquad (1)$$

Now evaluate $(\partial P/\partial \overline{V})_T$ from Equation 2.5 and $(\partial P/\partial V)_{T,n}$ from Equation 1 above and show that (see Problem D–12)

$$\left(\frac{\partial P}{\partial \overline{V}}\right)_T = n\left(\frac{\partial P}{\partial V}\right)_{T,n}$$

**D-14.** Referring to Problem D–13, show that

$$\left(\frac{\partial P}{\partial T}\right)_{\overline{V}} = \left(\frac{\partial P}{\partial T}\right)_{V,n}$$

and generally that

$$\left[\frac{\partial y(x, \overline{V})}{\partial x}\right]_{\overline{V}} = \left[\frac{\partial y(x, n, V)}{\partial x}\right]_{V,n}$$

where $y$ and $x$ are intensive variables and $y(x, n, V)$ can be written as $y(x, V/n)$.

**James Prescott Joule** was born in Salford, near Manchester, England, on December 24, 1818, and died in 1889. He and his elder brother were tutored at home by John Dalton, then in his 70s. Joule's father was a wealthy brewer, which allowed Joule freedom from having to seek employment. Joule conducted his pioneering experiments in laboratories he built at his own expense in his home or in his father's brewery. From 1837 to 1847, he carried out a series of experiments that led to the general law of energy conservation and to the mechanical equivalent of heat. Joule announced all his measurements in a public lecture at St. Ann's Church in Manchester, England and, because his earlier reports had been rejected by the British Association, later had his lecture published in the *Manchester Courier*, a newspaper for which his brother wrote musical critiques. In 1847, he presented his results to the British Association meeting in Oxford, where the 22-year-old William Thomson (later Lord Kelvin) immediately appreciated the importance of Joule's work. Thomson later asked Joule to carry out experiments on the expansion of gases. This work led to the discovery of the Joule-Thomson effect, which demonstrated that a nonideal gas cools when undergoing a free expansion. Joule was elected to the Royal Society in 1850. Later in life, he suffered severe financial losses, and in 1878 friends obtained a pension for him from the government. The SI unit of energy is named in his honor.

CHAPTER 5

# The First Law of Thermodynamics

As we said in the introduction to Chapter 1, thermodynamics is the study of the various properties and, particularly, the relations between the various properties of macroscopic systems in equilbrium. For example, thermodynamics tells us the conditions under which a gas will cool upon expansion (an important relation for liquefying gases) and how the vapor pressure of a droplet depends upon the radius of the droplet (an important relation in meteorology). Thermodynamics is primarily an experimental science and is extensively applied to chemistry, biology, geology, physics, environmental science, and engineering. All the results of thermodynamics rest upon three fundamental laws. These laws, appropriately known as the First, Second, and Third Laws of Thermodynamics, summarize a vast body of experimental data on macroscopic systems, and there are absolutely no known exceptions. We shall take up each one of these laws in this and the next two chapters. The First Law of Thermodynamics is essentially a statement of the law of conservation of energy applied to macroscopic systems. To present the First Law, we must introduce the concepts of work and heat as they are used in thermodynamics. As we will see in the next section, work and heat are modes of energy transfer between a system and its surroundings. We will also see that although the amount of work and heat involved in a process depend upon how the process is carried out, the energy depends only upon the initial and final states and does not depend upon how the final state is reached from the initial state. Later in the chapter we shall introduce another important thermodynamic function, the enthalpy, which we will see arises naturally for processes that are carried out at constant pressure. Finally we will introduce standard molar enthalpies of formation, from which we can calculate energy and enthalpy changes associated with chemical reactions.

## 5–1. A Common Type of Work Is Pressure–Volume Work

The concepts of work and heat play important roles in thermodynamics. Both work and heat refer to the manner in which energy is transferred between some system of interest and its surroundings. By *system* we mean that part of the world we are investigating and by *surroundings* we mean everything else. We define *heat*, $q$, to be the manner of energy transfer that results from a temperature difference between the system and its surroundings. Heat input to a system is considered a positive quantity; heat evolved by a system is considered a negative quantity. We define *work*, $w$, to be the transfer of energy between the system of interest and its surroundings as a result of the existence of unbalanced forces between the two. If the energy of the system is increased by the work, we say that work is done *on* the system by the surroundings, and we take it to be a positive quantity. On the other hand, if the energy of the system is decreased by the work, we say that the system does work on the surroundings, or that work is done *by* the system, and we take it to be a negative quantity. A common example of work in physical chemistry occurs during the expansion or compression of a gas as a result of the difference in pressures exerted by the gas and on the gas.

An important aspect of work is that it can always be related to the raising or lowering of a mass in the surroundings. To see the consequences of this statement, consider the situation in Figure 5.1, where a gas is confined to a cylinder that exerts a force $Mg$ on the gas. In Figure 5.1a, the initial pressure of the gas, $P_i$, is sufficient to push the piston upward, so there are pins holding it in position. Now, remove the pins and allow the gas to lift the mass upward to the new position shown, and let the pressure of the gas now be $P_f$. In this process, the mass $M$ has been raised a distance $h$, so the work done by the system is

$$w = -Mgh$$

The negative sign here is in accord with our convention that work done *by* a system is taken to be a negative quantity. If we divide $Mg$ by $A$, the area of the piston, and multiply $h$ by $A$, then we have

$$w = -\frac{Mg}{A} \cdot Ah$$

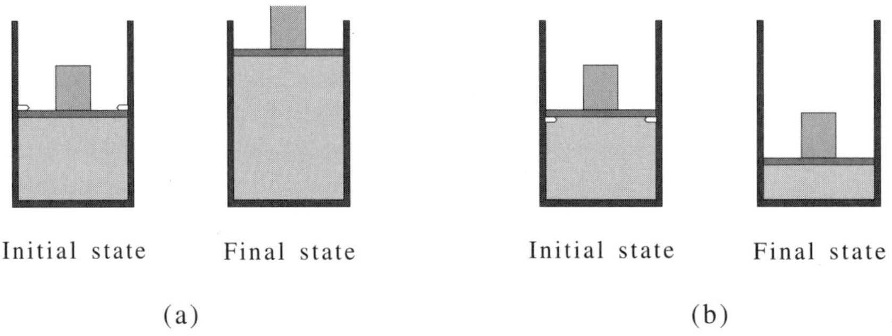


**FIGURE 5.1**
The effect of work is equivalent to the raising or lowering of a mass in the surroundings. In (a) work is done *by* the system because the mass is raised, and in (b) work is done *on* the system because the mass is lowered. (The system is defined as the gas inside the piston.)

But $Mg/A$ is the external pressure exerted on the gas and $Ah$ is the change in volume experienced by the gas, so we have

$$w = -P_{\text{ext}}\Delta V \tag{5.1}$$

Note that $\Delta V > 0$ in an expansion, so $w < 0$. Clearly, the external pressure must be less than the pressure of the inital state of the gas in order that the expansion occur. After the expansion, $P_{\text{ext}} = P_{\text{f}}$.

Now consider the situation in Figure 5.1b, where the initial pressure of the gas is less than the external pressure $P_{\text{ext}} = Mg/A$, so the gas is compressed when the pins are removed. In this case, the mass $M$ is lowered a distance $h$, and the work is given by

$$w = -Mgh = -\frac{Mg}{A}(Ah) = -P_{\text{ext}}\Delta V$$

But now $\Delta V < 0$, so $w > 0$. After the compression, we have $P_{\text{ext}} = P_{\text{f}}$. The work is positive because work is done *on* the gas when it is compressed.

If $P_{\text{ext}}$ is not constant during the expansion, the work is given by

$$w = -\int_{V_{\text{i}}}^{V_{\text{f}}} P_{\text{ext}} dV \tag{5.2}$$

where the limits on the integral indicate an initial state and a final state; we must have knowledge of how $P_{\text{ext}}$ varies with $V$ along the path connecting these two states so we can carry out the integration in Equation 5.2. Equation 5.2 is applicable to either expansion or compression. If $P_{\text{ext}}$ is constant, Equation 5.2 gives Equation 5.1

$$w = -P_{\text{ext}}\left(V_{\text{f}} - V_{\text{i}}\right) = -P_{\text{ext}}\,\Delta V$$

**EXAMPLE 5–1**

Consider an ideal gas that occupies 1.00 $\text{dm}^3$ at a pressure of 2.00 bar. If the gas is compressed isothermally at a constant external pressure, $P_{\text{ext}}$, so that the final volume is 0.500 $\text{dm}^3$, what is the smallest value $P_{\text{ext}}$ can have? Calculate the work involved using this value of $P_{\text{ext}}$.

SOLUTION: For a compression to occur, the value of $P_{\text{ext}}$ must be at least as large as the final pressure of the gas. Given the inital pressure and volume, and the final volume, we can determine the final pressure. The final pressure of the gas is

$$P_{\text{f}} = \frac{P_{\text{i}}V_{\text{i}}}{V_{\text{f}}} = \frac{(2.00\text{ bar})(1.00\text{ dm}^3)}{0.500\text{ dm}^3} = 4.00\text{ bar}$$

This is the smallest value $P_{\text{ext}}$ can be to compress the gas isothermally from 1.00 $\text{dm}^3$ to 0.500 $\text{dm}^3$. The work involved using this value of $P_{\text{ext}}$ is

$$\begin{aligned} w = -P_{\text{ext}}\,\Delta V &= -(4.00\text{ bar})(-0.500\text{ dm}^3) = 2.00\text{ dm}^3\cdot\text{bar} \\ &= (2.00\text{ dm}^3\cdot\text{bar})(10^{-3}\text{ m}^3\cdot\text{dm}^{-3})(10^5\text{ Pa}\cdot\text{bar}^{-1}) = 200\text{ Pa}\cdot\text{m}^3 = 200\text{ J} \end{aligned}$$

Of course, $P_{ext}$ can be any value greater than 4.00 bar, so 200 J represents the smallest value of $w$ for the isothermal compression at constant pressure from a volume of $1.00\ dm^3$ to $0.500\ dm^3$.

Figure 5.2 illustrates the work involved in Example 5–1. As Equation 5.2 implies, the work is the area under the curve of $P_{ext}$ versus $V$. The smooth curve is an isotherm ($P$ versus $V$ at constant $T$) of an ideal gas; Figure 5.2a shows a constant-pressure compression at an external pressure equal to $P_f$, the final pressure of the gas; and Figure 5.2b shows one at an external pressure greater than $P_f$. We see that the work is different for different values of $P_{ext}$.

## 5–2. Work and Heat Are Not State Functions, but Energy Is a State Function

Work and heat have a property that makes them quite different from energy. To appreciate this difference, we must first discuss what we mean by the state of a system. We say that a system is in a definite state when all the variables needed to describe the system completely are defined. For example, the state of one mole of an ideal gas can be described completely by specifying $P$, $\overline{V}$, and $T$. In fact, because $P$, $\overline{V}$, and $T$ are related by $P\overline{V} = RT$, any two of these three variables will suffice to specify the state of the gas. Other systems may require more variables, but usually only a few will

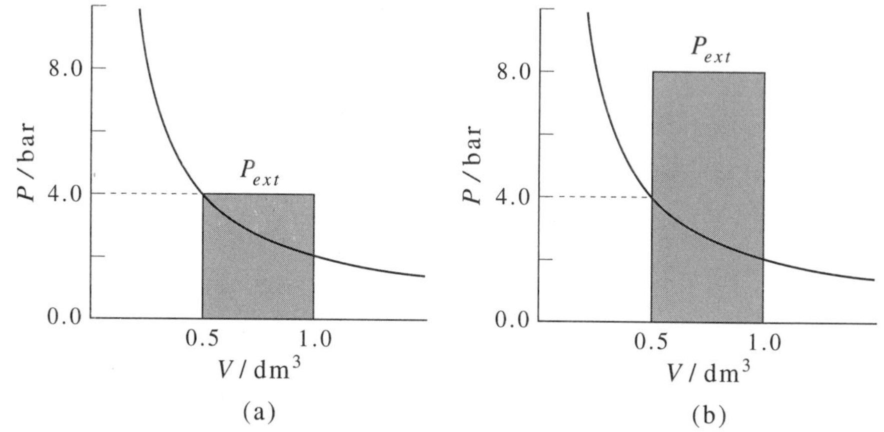


**FIGURE 5.2**
An illustration of the work involved in an isothermal constant-pressure compression from $V_i = 1.00\ dm^3$ to $V_f = 0.500\ dm^3$ at different values of $P_{ext}$. The smooth curve is an isotherm ($P$ vs. $V$ at constant $T$) for an ideal gas. In (a) $P_{ext}$ is equal to $P_f$, the final pressure of the gas, and in (b) $P_{ext}$ is larger than $P_f = 4.00$ bar, and pins must be used to stop the compression at $V_f = 0.500\ dm^3$. Otherwise the gas would be compressed further, until it reaches the volume that corresponds to $P_{ext}$ on the isotherm. The work is equal to the area of the $P_{ext}$-$V$ rectangles.

suffice. A *state function* is a property that depends only upon the state of the system, and not upon how the system was brought to that state, or upon the history of the system. Energy is an example of a state function. An important mathematical property of a state function is that its differential can be integrated in a normal way:

$$\int_1^2 dU = U_2 - U_1 = \Delta U \tag{5.3}$$

As the notation suggests, the value of $\Delta U$ is *independent* of the path taken between the initial and final states 1 and 2; it depends only upon the initial and final states through $\Delta U = U_2 - U_1$.

Work and heat are *not* state functions. For example, the external pressure used to compress a gas can have any value as long as it is large enough to compress the gas. Consequently, the work done on the gas,

$$w = -\int_1^2 P_{\text{ext}} dV$$

will depend upon the pressure used to compress the gas. The value of $P_{\text{ext}}$ must exceed the pressure of the gas to compress it. The minimum work required occurs when $P_{\text{ext}}$ is just infinitesimally greater than the pressure of the gas at every stage of the compression, which means that the gas is essentially in equilibrium during the entire compression. In this special but important case, we can replace $P_{\text{ext}}$ by the pressure of the gas ($P$) in Equation 5.2. When $P_{\text{ext}}$ and $P$ differ only infinitesimally, the process is called a *reversible process* because the process could be reversed (from compression to expansion) by decreasing the external pressure infinitesimally. Necessarily, a strictly reversible process would require an infinite time to carry out because the process must be adjusted by an infinitesimal amount at each stage. Nevertheless, a reversible process serves as a useful idealized limit.

Figure 5.3 shows that a reversible, isothermal compression of a gas requires the minimum possible amount of work. Let $w_{\text{rev}}$ denote the reversible work. To calculate $w_{\text{rev}}$ for the compression of an ideal gas isothermally from $V_1$ to $V_2$, we use Equation 5.2 with $P_{\text{ext}}$ replaced by the equilibrium value of the pressure of the gas, which is $nRT/V$ for an ideal gas. Therefore,

$$\begin{aligned} w_{\text{rev}} &= -\int_1^2 P_{\text{gas}} dV = -\int_1^2 \frac{nRT}{V} dV = -nRT \int_1^2 \frac{dV}{V} \\ &= -nRT \ln \frac{V_2}{V_1} \end{aligned} \tag{5.4}$$

Because $V_2 < V_1$ for compression, we see that $w_{\text{rev}} > 0$ as it should be; in other words, we have done work on the gas.

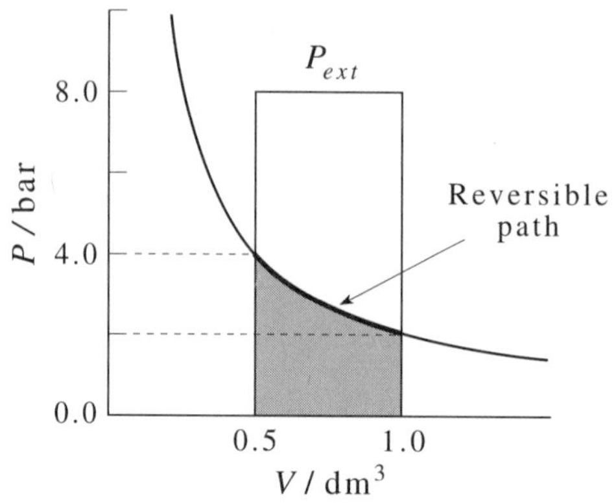


**FIGURE 5.3**
The work of isothermal compression is the area under the $P_{ext}$ versus $V$ curves shown in the figure. The external pressure must exceed the pressure of the gas in order to compress it. The minimum amount of work occurs when the expansion is carried out reversibly; that is, when $P_{ext}$ is just infinitesimally greater than the pressure of the gas at every stage of the compression. The gray area is the minimum work needed to compress the gas from $V_1 = 1.00$ dm$^3$ to $V_2 = 0.500$ dm$^3$. The constant-pressure compression curves are the same as those in Figure 5.2.

**EXAMPLE 5–2**

Consider an ideal gas that occupies 1.00 dm$^3$ at 2.00 bar. Calculate the work required to compress the gas isothermally to a volume of 0.667 dm$^3$ at a constant pressure of 3.00 bar followed by another isothermal compression to 0.500 dm$^3$ at a constant pressure of 4.00 bar (Figure 5.4). Compare the result with the work of compressing

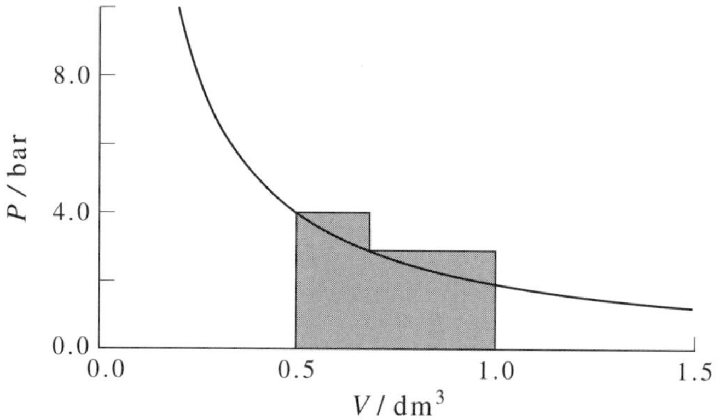


**FIGURE 5.4**
An illustration of the constant-pressure compression of a gas as described in Example 5–2. The work required is given by the areas under the two rectangles.

the gas isothermally and reversibly from 1.00 $dm^3$ to 0.500 $dm^3$. Compare both results to the one obtained in Example 5–1.

SOLUTION: In the two-stage compression, $\Delta V = -(1.00 - 0.667)$ $dm^3$ in the first step and $-(0.667 - 0.500)$ $dm^3$ in the second step. Therefore,

$$\begin{aligned} w &= -(3.00 \text{ bar})(-0.333 \text{ dm}^3) - (4.00 \text{ bar})(-0.167 \text{ dm}^3) \\ &= 1.67 \text{ dm}^3\cdot\text{bar} = 167 \text{ J} \end{aligned}$$

We use Equation 5.4 for the reversible process

$$w_{\text{rev}} = -nRT \ln \frac{V_2}{V_1} = -nRT \ln \frac{0.500 \text{ dm}^3}{1.00 \text{ dm}^3}$$

Because the gas is ideal and the process is isothermal, $nRT$ is equal to either $P_1V_1$ or $P_2V_2$, both of which equal 2.00 $dm^3\cdot$bar, and so

$$w_{\text{rev}} = -(2.00 \text{ dm}^3\cdot\text{bar}) \ln 0.500 = 1.39 \text{ dm}^3\cdot\text{bar} = 139 \text{ J}$$

Note that $w_{\text{rev}}$ is less than that for the two-stage process and that the work for that process is less than the work required in Example 5–1 (200 J). (Compare Figures 5.2, 5.3, and 5.4.)

Just as the reversible isothermal compression of a gas requires the minimum amount of work to be done on the gas, a reversible isothermal expansion requires the gas to do a maximum amount of work in the process. In a reversible expansion, the external pressure is infinitesimally less than the pressure of the gas at each stage. If $P_{\text{ext}}$ were any larger, the expansion would not occur. The work involved in the reversible isothermal expansion of an ideal gas is also given by Equation 5.4. Because $V_2 > V_1$ for expansion, we see that $w_{\text{rev}} < 0$; the gas has done work on the surroundings, in fact, the maximum possible.

**EXAMPLE 5–3**
Derive an expression for the reversible isothermal work of an expansion of a van der Waals gas.

SOLUTION: The expression for the reversible work is

$$w_{\text{rev}} = -\int_1^2 P dV$$

where

$$P = \frac{nRT}{V - nb} - \frac{an^2}{V^2}$$

We substitute this expression for $P$ into $w_{\text{rev}}$ to obtain

$$\begin{aligned} w_{\text{rev}} &= -nRT \int_1^2 \frac{dV}{V - nb} - an^2 \int_1^2 \frac{dV}{V^2} \\ &= -nRT \ln \frac{V_2 - nb}{V_1 - nb} + an^2 \left( \frac{1}{V_2} - \frac{1}{V_1} \right) \end{aligned}$$

Note that this equation reduces to Equation 5.4 when $a = b = 0$.

## 5–3. The First Law of Thermodynamics Says the Energy Is a State Function

Because the work involved in a process depends upon how the process is carried out, work is *not* a state function. Thus, we write

$$\int_1^2 \delta w = w \qquad (\text{not } \Delta w \text{ or } w_2 - w_1) \tag{5.5}$$

It makes no sense at all to write $w_2$, $w_1$, $w_2 - w_1$, or $\Delta w$. The value of $w$ obtained in Equation 5.5 depends upon the *path* from state 1 to 2, so work is called a *path function*. Mathematically, $\delta w$ in Equation 5.5 is called an *inexact differential*, as opposed to an *exact differential* like $dU$, which can be integrated in the normal way to obtain $U_2 - U_1$ (see MathChapter D).

Work and heat are defined only for processes in which energy is transferred between a system and its surroundings. Both work and heat are path functions. Although a system in a given state has a certain amount of energy, it does not possess work or heat. The difference between energy and work and heat can be summarized by writing

$$\int_1^2 dU = U_2 - U_1 = \Delta U \qquad (U \text{ is a state function}) \tag{5.6}$$

$$\int_1^2 \delta w = w \qquad (\text{not } w_2 - w_1) \qquad (\text{path function}) \tag{5.7}$$

and

$$\int_1^2 \delta q = q \qquad (\text{not } q_2 - q_1) \qquad (\text{path function}) \tag{5.8}$$

For a process in which energy is transferred both as work and heat, the law of conservation of energy says that the energy of the system obeys the equation

$$dU = \delta q + \delta w \tag{5.9}$$

in differential form, or

$$\Delta U = q + w \tag{5.10}$$

in integrated form. Equations 5.9 and 5.10 are statements of the *First Law of Thermodynamics*. The First Law of Thermodynamics, which is essentially a statement of the law of conservation of energy, also says that even though $\delta q$ and $\delta w$ are separately path functions or inexact differentials, their sum is a state function or an exact differential. All state functions are exact differentials.

## 5–4. An Adiabatic Process Is a Process in Which No Energy as Heat Is Transferred

Not only are work and heat not state functions, but we can prove that even reversible work and reversible heat are not state functions by a direct calculation. Consider the three paths, depicted in Figure 5.5, that occur between the same initial and final states, $P_1, V_1, T_1$ and $P_2, V_2, T_1$. Path A involves a reversible isothermal expansion of an ideal gas from $P_1, V_1, T_1$ to $P_2, V_2, T_1$. Because the energy of an ideal gas depends upon only the temperature (see Equation 4.40, for example),

$$\Delta U_A = 0 \tag{5.11}$$

so

$$\delta w_{\text{rev,A}} = -\delta q_{\text{rev,A}}$$

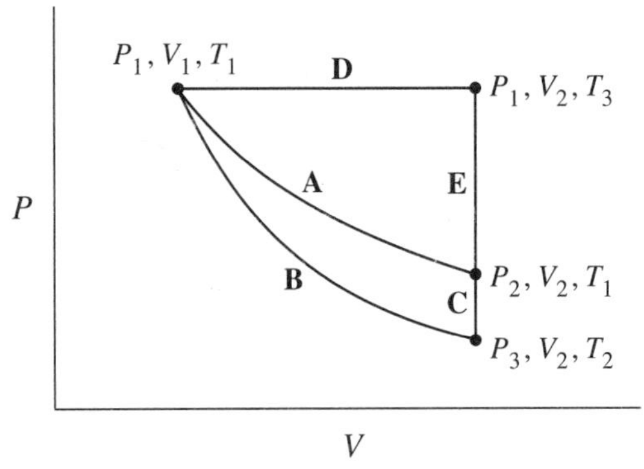


**FIGURE 5.5**
An illustration of three different pathways (A, B + C, and D + E) to take an ideal gas from $P_1, V_1, T_1$ to $P_2, V_2, T_1$. In each case, the value of $\Delta U$ is the same ($\Delta U$ is a state function), but the values of $q$ and $w$ are different ($q$ and $w$ are path functions).

for an isothermal process involving an ideal gas. Furthermore, because the process is reversible,

$$\delta w_{\text{rev,A}} = -\delta q_{\text{rev,A}} = -\frac{RT_1}{V}dV \tag{5.12}$$

so

$$w_{\text{rev,A}} = -q_{\text{rev,A}} = -RT_1 \int_{V_1}^{V_2} \frac{dV}{V} = -RT_1 \ln \frac{V_2}{V_1} \tag{5.13}$$

Note that $w_{\text{rev}}$ is negative ($V_2 > V_1$) because work is done by the gas. Furthermore, $q_{\text{rev}}$ is positive because energy as heat entered the system to maintain the temperature constant as the system used its energy to do the work.

Another path (B + C) in Figure 5.5 consists of two parts. The first part (B) involves a reversible expansion from $P_1, V_1, T_1$ to $P_3, V_2, T_2$ and is carried out such that no energy as heat is transferred between the system and its surroundings. A process in which no energy as heat is transferred is called an *adiabatic process*. For an adiabatic process, $q = 0$, so

$$dU = \delta w \tag{5.14}$$

Path C of the path B + C involves heating the gas reversibly at constant volume from $P_3, V_2, T_2$ to $P_2, V_2, T_1$. As stated above, $\Delta U$ depends upon only temperature; it is independent of $P$ and $V$ for an ideal gas. To calculate $\Delta U$ for a change from state 1 of temperature $T_1$ to state 2 of temperature $T_2$, recall that the constant volume heat capacity is defined as (Equation 3.25)

$$C_V(T) = \left(\frac{\partial U}{\partial T}\right)_V$$

and therefore, for an ideal gas

$$\frac{dU}{dT} = \left(\frac{\partial U}{\partial T}\right)_V = C_V(T)$$

or $dU = C_V(T)dT$, which can be integrated to give

$$\Delta U = \int_{T_1}^{T_2} C_V(T)dT$$

We can now calculate the total work involved in the path B + C. Because process B is adiabatic,

$$q_{\text{rev,B}} = 0 \tag{5.15}$$

so

$$w_{\text{rev,B}} = \Delta U_{\text{B}} = \int_{T_1}^{T_2} \left(\frac{\partial U}{\partial T}\right)_V dT = \int_{T_1}^{T_2} C_V(T)dT \tag{5.16}$$

For process C, no pressure-volume work is involved (it is a constant-volume process), so

$$q_{\text{rev,C}} = \Delta U_{\text{C}} = \int_{T_2}^{T_1} C_V(T)dT \tag{5.17}$$

For the total path B + C, then

$$\begin{aligned} q_{\text{rev,B+C}} &= q_{\text{rev,B}} + q_{\text{rev,C}} = 0 + \int_{T_2}^{T_1} C_V(T)dT \\ &= \int_{T_2}^{T_1} C_V(T)dT \end{aligned} \tag{5.18}$$

and

$$\begin{aligned} w_{\text{rev,B+C}} &= w_{\text{rev,B}} + w_{\text{rev,C}} = \int_{T_1}^{T_2} C_V(T)dT + 0 \\ &= \int_{T_1}^{T_2} C_V(T)dT \end{aligned} \tag{5.19}$$

Note that

$$\Delta U_{\text{B+C}} = \Delta U_{\text{B}} + \Delta U_{\text{C}} = \int_{T_1}^{T_2} C_V(T)dT + \int_{T_2}^{T_1} C_V(T)dT = 0$$

which is the same as in path A, because the energy $U$ is a state function. However, $w_{\text{rev,A}} \neq w_{\text{rev,B+C}}$ and $q_{\text{rev,A}} \neq q_{\text{rev,B+C}}$, because both work and heat are path functions.

**EXAMPLE 5–4**

Calculate $\Delta U$, $w_{\text{rev}}$, and $q_{\text{rev}}$ for the paths D + E in Figure 5.5, where D represents a reversible constant-pressure ($P_1$) expansion of an ideal gas from $V_1, T_1$ to $V_2, T_3$ and E represents cooling the gas reversibly from $T_3$ to $T_1$ at a constant volume $V_2$.

SOLUTION: For path D,

$$\Delta U_{\text{D}} = \int_{T_1}^{T_3} C_V(T)dT$$

$$w_{\text{rev,D}} = -P_1(V_2 - V_1)$$

and

$$q_{\text{rev,D}} = \Delta U_{\text{D}} - w_{\text{rev,D}} = \int_{T_1}^{T_3} C_V(T)dT + P_1(V_2 - V_1)$$

For path E,

$$\Delta U_{\text{E}} = \int_{T_3}^{T_1} C_V(T)dT$$

$$w_{\text{rev,E}} = 0$$

and

$$q_{\text{rev,E}} = \Delta U_{\text{E}} = \int_{T_3}^{T_1} C_V(T)dT$$

Therefore, for the overall process,

$$\Delta U_{\text{D+E}} = \Delta U_{\text{D}} + \Delta U_{\text{E}} = \int_{T_1}^{T_3} C_V(T)dT + \int_{T_3}^{T_1} C_V(T)dT = 0$$

$$w_{\text{rev,D+E}} = w_{\text{rev,D}} + w_{\text{rev,E}} = -P_1(V_2 - V_1)$$

and

$$q_{\text{rev,D+E}} = q_{\text{rev,D}} + q_{\text{rev,E}} = P_1(V_2 - V_1)$$

Note that $\Delta U = 0$ for all three processes indicated in Figure 5.5, but that $w_{\text{rev}}$ and $q_{\text{rev}}$ are different for each one.

## 5–5. The Temperature of a Gas Decreases in a Reversible Adiabatic Expansion

Path B in Figure 5.5 represents the reversible adiabatic expansion of an ideal gas from $T_1, V_1$ to $T_2, V_2$. As the figure suggests, $T_2 < T_1$, which means that the gas cools during a (reversible) adiabatic expansion. We can determine the final temperature $T_2$ for this process. For an adiabatic process, $q = 0$, and so

$$dU = \delta w = dw$$

Note that the above expression tells us that $\delta w = dw$ is an exact differential when $\delta q = 0$. Likewise, $\delta q = dq$ is an exact differential if $\delta w = 0$. The work done by the gas (the system) in the expansion is "paid for" by a decrease in the energy of the gas, which amounts to a decrease in the temperature of the gas. Because the work involved in a reversible expansion is maximum, the gas must suffer a maximum drop in temperature in a reversible adiabatic expansion. Recall that for an ideal gas, $U$ depends only upon the temperature and $dU = C_V(T)dT = n\overline{C}_V(T)dT$, where $\overline{C}_V(T)$ is the molar constant-volume heat capacity. Using the fact that $dw = -PdV = -nRTdV/V$ for a reversible expansion, the relation $dU = dw$ gives

$$C_V(T)dT = -\frac{nRT}{V}dV \tag{5.20}$$

We divide both sides by $T$ and $n$ and integrate to obtain

$$\int_{T_1}^{T_2} \frac{\overline{C}_V(T)}{T}dT = -R\int_{V_1}^{V_2} \frac{dV}{V} = -R\ln\frac{V_2}{V_1} \tag{5.21}$$

We learned in Section 4–2 that $\overline{C}_V = 3R/2$ for a monatomic ideal gas, so Equation 5.21 becomes

$$\frac{3R}{2}\int_{T_1}^{T_2}\frac{dT}{T} = \frac{3R}{2}\ln\frac{T_2}{T_1} = -R\ln\frac{V_2}{V_1}$$

or

$$\frac{3}{2}\ln\frac{T_2}{T_1} = -\ln\frac{V_2}{V_1} = \ln\frac{V_1}{V_2}$$

or

$$\left(\frac{T_2}{T_1}\right)^{3/2} = \frac{V_1}{V_2} \qquad \left(\begin{array}{c}\text{monatomic}\\ \text{ideal gas}\end{array}\right) \tag{5.22}$$

Thus, the gas cools in a reversible adiabatic expansion ($V_2 > V_1$).

**EXAMPLE 5–5**

Calculate the final temperature if argon (assumed to be ideal) at an initial temperature of 300 K expands reversibly and adiabatically from a volume of 50.0 L to 200 L.

SOLUTION: First solve Equation 5.22 for $T_2/T_1$,

$$\frac{T_2}{T_1} = \left(\frac{V_1}{V_2}\right)^{2/3}$$

and then let $T_1 = 300$ K, $V_1 = 50.0$ L, and $V_2 = 200$ L to obtain

$$T_2 = (300\text{K})\left(\frac{50.0\text{ L}}{200\text{ L}}\right)^{2/3} = 119\text{ K}$$

We can express Equation 5.22 in terms of pressure and volume by using $PV = nRT$ to eliminate $T_1$ and $T_2$:

$$\left(\frac{P_2V_2}{P_1V_1}\right)^{3/2} = \frac{V_1}{V_2}$$

Upon taking both sides to the 2/3 power and rearranging, we obtain

$$P_1V_1^{5/3} = P_2V_2^{5/3} \qquad \text{(monatomic ideal gas)} \tag{5.23}$$

This equation shows how the pressure and volume are related in a reversible, adiabatic process for an ideal monatomic gas. Compare this result to Boyle's law, which says that

$$P_1V_1 = P_2V_2$$

for an isothermal process.

**EXAMPLE 5–6**

Derive the analogs of Equations 5.22 and 5.23 for an ideal diatomic gas. Assume the temperature is such that the vibrational contribution to the heat capacity can be ignored.

SOLUTION: Assuming that $\overline{C}_{V,\text{vib}} \approx 0$, we have from Equation 4.41 that $\overline{C}_V = 5R/2$. Equation 5.20 for a diatomic ideal gas becomes

$$\frac{5R}{2}\int_{T_1}^{T_2}\frac{dT}{T} = \frac{5R}{2}\ln\frac{T_2}{T_1} = -R\ln\frac{V_2}{V_1}$$

so

$$\left(\frac{T_2}{T_1}\right)^{5/2} = \frac{V_1}{V_2} \qquad \text{(diatomic ideal gas)}$$

Substituting $T = PV/nR$ into the above equation gives

$$\left(\frac{P_2V_2}{P_1V_1}\right)^{5/2} = \frac{V_1}{V_2}$$

or

$$P_1V_1^{7/5} = P_2V_2^{7/5} \qquad \text{(diatomic ideal gas)}$$

## 5–6. Work and Heat Have a Simple Molecular Interpretation

Let's go back to Equation 3.18 for the average energy of a macroscopic system,

$$U = \sum_j p_j(N, V, \beta)E_j(N, V) \tag{5.24}$$

with

$$p_j(N, V, \beta) = \frac{e^{-\beta E_j(N,V)}}{Q(N, V, \beta)} \tag{5.25}$$

Equation 5.24 represents the average energy of an equilibrium system that has the variables $N$, $V$, and $T$ fixed. If we differentiate Equation 5.24, we obtain

$$dU = \sum_j p_j dE_j + \sum_j E_j dp_j \tag{5.26}$$

Because $E_j = E_j(N, V)$, we can view $dE_j$ as the change in $E_j$ due to a small change in the volume, $dV$, keeping $N$ fixed. Therefore, substituting $dE_j = (\partial E_j/\partial V)_N dV$ into Equation 5.26 gives

$$dU = \sum_j p_j \left(\frac{\partial E_j}{\partial V}\right)_N dV + \sum_j E_j dp_j$$

This result suggests we can interpret the first term in Equation 5.26 to be the average change in energy of the system caused by a small change in its volume, in other words, the average work.

Furthermore, if this change is done reversibly, so that the system remains essentially in equilibrium at each stage, then the $p_j$ in Equation 5.26 will be given by Equation 5.25 thoughout the entire process. We can emphasize this by writing

$$dU = \sum_j p_j(N, V, \beta) \left(\frac{\partial E_j}{\partial V}\right)_N dV + \sum_j E_j(N, V) dp_j(N, V, \beta) \tag{5.27}$$

If we compare this result with the macroscopic equation (Equation 5.9)

$$dU = \delta w_{\text{rev}} + \delta q_{\text{rev}} \tag{5.28}$$

we see that

$$\delta w_{\text{rev}} = \sum_j p_j(N, V, \beta) \left(\frac{\partial E_j}{\partial V}\right)_N dV \tag{5.29}$$

and

$$\delta q_{\text{rev}} = \sum_j E_j(N, V) dp_j(N, V, \beta) \tag{5.30}$$

Thus, we see that reversible work, $\delta w_{\text{rev}}$, results from an infinitesimal change in the allowed energies of a system, without changing the probability distribution of its states. Reversible heat, on the other hand, results from a change in the probability distribution of the states of a system, without changing the allowed energies.

If we compare Equation 5.29 with

$$\delta w_{\text{rev}} = -PdV$$

we see that we can identify the pressure of the gas with

$$P = -\sum_j p_j(N, V, \beta) \left(\frac{\partial E_j}{\partial V}\right)_N = -\left\langle \left(\frac{\partial E}{\partial V}\right)_N \right\rangle \tag{5.31}$$

Recall that we used this equation without proof in Section 3–5 to show that $PV = RT$ for one mole of an ideal gas.

## 5–7. The Enthalpy Change Is Equal to the Energy Transferred as Heat in a Constant-Pressure Process Involving Only $P$–$V$ Work

For a reversible process in which the only work involved is pressure-volume work, the first law tells us that

$$\Delta U = q + w = q - \int_{V_1}^{V_2} P dV \tag{5.32}$$

If the process is carried out at constant volume, then $V_1 = V_2$ and

$$\Delta U = q_V \tag{5.33}$$

where the subscript $V$ on $q$ emphasizes that Equation 5.33 applies to a constant-volume process. Thus, we see that $\Delta U$ can be measured experimentally by measuring the energy as heat (by means of a calorimeter) associated with a constant-volume process (in a rigid closed container).

Many processes, particularly chemical reactions, are carried out at constant pressure (open to the atmosphere). The energy as heat associated with a constant-pressure process, $q_P$, is not equal to $\Delta U$. It would be convenient to have a state function analogous to $U$ so that we could write an expression like that in Equation 5.33. To this end, let $P$ be constant in Equation 5.32 so that

$$q_P = \Delta U + P_{\text{ext}} \int_{V_1}^{V_2} dV = \Delta U + P\Delta V \tag{5.34}$$

where we have used the subscript $P$ on $q_P$ to emphasize that this is a constant-pressure process. This equation suggests that we define a new state function by

$$H = U + PV \tag{5.35}$$

At constant pressure,

$$\Delta H = \Delta U + P\Delta V \qquad \text{(constant pressure)} \tag{5.36}$$

Equation 5.34 shows that

$$q_P = \Delta H \tag{5.37}$$

Thus, this new state function $H$, called the *enthalpy*, plays the same role in a constant-pressure process that $U$ plays in a constant-volume process. The value of $\Delta H$ can be determined experimentally by measuring the energy as heat associated with a constant-pressure process, or conversely, $q_P$ can be determined from $\Delta H$. Because most chemical reactions take place at constant pressure, the enthalpy is a practical and important thermodynamic function.

Let's apply these results to the melting of ice at 0°C and one atm. For this process, $q_P = 6.01\ \text{kJ}\cdot\text{mol}^{-1}$. Using Equation 5.37, we find that

$$\Delta\overline{H} = q_P = 6.01\ \text{kJ}\cdot\text{mol}^{-1}$$

where the overbar on $H$ signifies that $\Delta\overline{H}$ is a molar quantity. We can also calculate the value of $\Delta\overline{U}$ using Equation 5.36 and the fact that the molar volume of ice ($\overline{V}_s$) is $0.0196\ \text{L}\cdot\text{mol}^{-1}$ and that of water ($\overline{V}_l$) is $0.0180\ \text{L}\cdot\text{mol}^{-1}$:

$$\begin{aligned}\Delta\overline{U} &= \Delta\overline{H} - P\Delta\overline{V} \\ &= 6.01\ \text{kJ}\cdot\text{mol}^{-1} - (1\ \text{atm})(0.0180\ \text{L}\cdot\text{mol}^{-1} - 0.0196\ \text{L}\cdot\text{mol}^{-1}) \\ &= 6.01\ \text{kJ}\cdot\text{mol}^{-1} - (1.60\times10^{-3}\ \text{L}\cdot\text{atm}\cdot\text{mol}^{-1})\left(\frac{8.314\ \text{J}}{0.08206\ \text{L}\cdot\text{atm}}\right)\left(\frac{1\ \text{kJ}}{10^3\ \text{J}}\right) \\ &\approx 6.01\ \text{kJ}\cdot\text{mol}^{-1}\end{aligned}$$

Thus, in this case, there is essentially no difference between $\Delta\overline{H}$ and $\Delta\overline{U}$.

Let's look at the vaporization of water at 100°C and one atm. For this process, $q_P = 40.7\ \text{kJ}\cdot\text{mol}^{-1}$, $\overline{V}_l = 0.0180\ \text{L}\cdot\text{mol}^{-1}$, and $\overline{V}_g = 30.6\ \text{L}\cdot\text{mol}^{-1}$. Therefore,

$$\Delta\overline{H} = q_P = 40.7\ \text{kJ}\cdot\text{mol}^{-1}$$

But

$$\Delta\overline{V} = 30.6\ \text{L}\cdot\text{mol}^{-1} - 0.0180\ \text{L}\cdot\text{mol}^{-1} = 30.6\ \text{L}\cdot\text{mol}^{-1}$$

so

$$\begin{aligned}\Delta\overline{U} &= \Delta\overline{H} - P\Delta\overline{V} \\ &= 40.7\ \text{kJ}\cdot\text{mol}^{-1} - (1\ \text{atm})(30.6\ \text{L}\cdot\text{mol}^{-1})\left(\frac{8.314\ \text{J}}{0.08206\ \text{L}\cdot\text{atm}}\right) \\ &= 37.6\ \text{kJ}\cdot\text{mol}^{-1}\end{aligned}$$

Notice that the numerical values of $\Delta\overline{H}$ and $\Delta\overline{U}$ are significantly different ($\approx 8\%$) in this case because $\Delta\overline{V}$ for this process is fairly large. We can give a physical interpretation of these results. Of the 40.7 kJ that are absorbed at constant pressure, 37.6 kJ ($q_V = \Delta\overline{U}$) are used to overcome the intermolecular forces holding the water molecules in the liquid state (hydrogen bonds) and 3.1 kJ (40.7 kJ − 37.6 kJ) are used to increase the volume of the system against the atmospheric pressure.

**EXAMPLE 5–7**

The value of $\Delta H$ at 298 K and one bar for the reaction described by

$$2\,H_2(g) + O_2(g) \longrightarrow 2\,H_2O(l)$$

is −572 kJ. Calculate $\Delta U$ for this reaction as written.

SOLUTION: Because the reaction is carried out at a constant pressure of 1.00 bar, $\Delta H = q_P = -572$ kJ. To calculate $\Delta U$, we must first calculate $\Delta V$. Initially, we have three moles of gas at 298 K and 1.00 bar, and so

$$V = \frac{nRT}{P} = \frac{(3\text{ mol})(0.08314\text{ L}\cdot\text{bar}\cdot\text{K}^{-1}\cdot\text{mol}^{-1})(298\text{ K})}{1.00\text{ bar}}$$
$$= 74.3\text{ L}$$

Afterward, we have two moles of liquid water, whose volume is about 36 mL, which is negligible compared with 74.3 L. Thus, $\Delta V = -74.3$ L and

$$\Delta U = \Delta H - P\Delta V$$
$$= -572\text{ kJ} + (1.00\text{ bar})(73.4\text{ L})\left(\frac{1\text{ kJ}}{10\text{ bar}\cdot\text{L}}\right) = -572\text{ kJ} + 7.43\text{ kJ}$$
$$= -565\text{ kJ}$$

The numerical difference between $\Delta H$ and $\Delta U$ in this case is about 1%.

Example 5–7 is a special case of a general result for reactions or processes that involve ideal gases, which says that

$$\Delta H = \Delta U + RT\Delta n_{\text{gas}} \tag{5.38}$$

where

$$\Delta n_{\text{gas}} = \begin{pmatrix}\text{number of moles of}\\ \text{gaseous products}\end{pmatrix} - \begin{pmatrix}\text{number of moles of}\\ \text{gaseous reactants}\end{pmatrix}$$

As Example 5–7 implies, the numerical difference between $\Delta H$ and $\Delta U$ is usually small.

## 5–8. Heat Capacity Is a Path Function

Recall that heat capacity is defined as the energy as heat required to raise the temperature of a substance by one degree. The heat capacity also depends upon the temperature $T$. Because the energy required to raise the temperature of a substance by one kelvin depends upon the amount of substance, heat capacity is an *extensive quantity*. Heat capacity is also a path function; for example, its value depends upon whether we heat the substance at constant volume or at constant pressure. If the substance is heated at constant volume, the added energy as heat is $q_V$ and the heat capacity is denoted by $C_V$. Because $\Delta U = q_V$, $C_V$ is given by

$$C_V = \left(\frac{\partial U}{\partial T}\right)_V \approx \frac{\Delta U}{\Delta T} = \frac{q_V}{\Delta T} \tag{5.39}$$

If the substance is heated at constant pressure, the added energy as heat is $q_P$ and the heat capacity is denoted by $C_P$. Because $\Delta H = q_P$, $C_P$ is given by

$$C_P = \left(\frac{\partial H}{\partial T}\right)_P \approx \frac{\Delta H}{\Delta T} = \frac{q_P}{\Delta T} \tag{5.40}$$

We expect that $C_P$ is larger than $C_V$ because not only do we increase the temperature when we add energy as heat in a constant-pressure process, but we also do work against atmospheric pressure as the substance expands as it is heated. Calculating the difference between $C_P$ and $C_V$ for an ideal gas is easy. We start with $H = U + PV$ and replace $PV$ by $nRT$ to obtain

$$H = U + nRT \qquad \text{(ideal gas)} \tag{5.41}$$

Notice that because $U$ depends only upon the temperature (at constant $n$) for an ideal gas, $H$ also depends only upon temperature. Thus, we can differentiate Equation 5.41 with respect to temperature to obtain

$$\frac{dH}{dT} = \frac{dU}{dT} + nR \tag{5.42}$$

But

$$\frac{dH}{dT} = \left(\frac{\partial H}{\partial T}\right)_P = C_P \qquad \text{(ideal gas)}$$

and

$$\frac{dU}{dT} = \left(\frac{\partial U}{\partial T}\right)_V = C_V \qquad \text{(ideal gas)}$$

so Equation 5.42 becomes

$$C_P - C_V = nR \qquad \text{(ideal gas)} \tag{5.43}$$

Recall from Chapter 3 that $C_V$ is $3R/2$ for one mole of a monatomic ideal gas and is approximately $3R$ for one mole of a nonlinear polyatomic ideal gas at room temperature. Therefore, the difference between $\overline{C}_P$ and $\overline{C}_V$ is significant for gases. For solids and liquids, however, the difference is small.

**EXAMPLE 5–8**

We will prove generally that (Section 8–3)

$$\overline{C}_P - \overline{C}_{\overline{V}} = T\left(\frac{\partial P}{\partial T}\right)_{\overline{V}}\left(\frac{\partial \overline{V}}{\partial T}\right)_P$$

First, use this result to show that $\overline{C}_P - \overline{C}_V = R$ for an ideal gas and then derive an expression for $\overline{C}_P - \overline{C}_V$ for a gas that obeys the equation of state

$$P\overline{V} = RT + B(T)P$$

SOLUTION: For an ideal gas, $P\overline{V} = RT$, so

$$\left(\frac{\partial P}{\partial T}\right)_{\overline{V}} = \frac{R}{\overline{V}} \qquad \text{and} \qquad \left(\frac{\partial \overline{V}}{\partial T}\right)_P = \frac{R}{P}$$

and so

$$\overline{C}_P - \overline{C}_V = T\left(\frac{R}{\overline{V}}\right)\left(\frac{R}{P}\right) = R\left(\frac{RT}{P\overline{V}}\right) = R$$

To determine $(\partial P/\partial T)_{\overline{V}}$ for a gas that obeys the equation of state, $P\overline{V} = RT + B(T)P$, we first solve for $P$.

$$P = \frac{RT}{\overline{V} - B(T)}$$

and then differentiate with respect to temperature:

$$\left(\frac{\partial P}{\partial T}\right)_{\overline{V}} = \frac{R}{\overline{V} - B(T)} + \frac{RT}{[\overline{V} - B(T)]^2}\frac{dB}{dT}$$

$$= \frac{P}{T} + \frac{P}{\overline{V} - B(T)}\frac{dB}{dT}$$

Similarly,

$$\overline{V} = \frac{RT}{P} + B(T)$$

and

$$\left(\frac{\partial \overline{V}}{\partial T}\right)_P = \frac{R}{P} + \frac{dB}{dT}$$

Therefore, using the equation for $\overline{C}_P - \overline{C}_V$ given in the statement of this example,

$$\overline{C}_P - \overline{C}_V = T\left(\frac{\partial P}{\partial T}\right)_{\overline{V}}\left(\frac{\partial \overline{V}}{\partial T}\right)_P$$

$$= T\left[\frac{P}{T} + \frac{P}{\overline{V} - B(T)}\frac{dB}{dT}\right]\left[\frac{R}{P} + \frac{dB}{dT}\right]$$

$$= R + \left[\frac{RT}{\overline{V} - B(T)} + P\right]\frac{dB}{dT} + \frac{PT}{\overline{V} - B(T)}\left(\frac{dB}{dT}\right)^2$$

$$= R + 2\left(\frac{dB}{dT}\right)P + \frac{1}{R}\left(\frac{dB}{dT}\right)^2 P^2$$

where we have used the fact that $P = RT/[\overline{V} - B(T)]$ in going from the third line to the last line. Notice that this expression is the same as that for an ideal gas if $B(T)$ is a constant.

## 5–9. Relative Enthalpies Can Be Determined from Heat Capacity Data and Heats of Transition

By integrating Equation 5.40, we can calculate the difference in the enthalpy of a substance that does not change phase between two temperatures:

$$H(T_2) - H(T_1) = \int_{T_1}^{T_2} C_P(T)dT \tag{5.44}$$

If we let $T = 0$ K, we have

$$H(T) - H(0) = \int_0^T C_P(T')dT' \tag{5.45}$$

[Notice that we have written the integration variable in Equation 5.45 with a prime, which is standard mathematical notation used to distinguish an integration limit ($T$ in this case) from the integration variable, $T'$.] It would appear from Equation 5.44 that if we had heat-capacity data from 0 K to any other temperature, $T$, we could calculate $H(T)$ relative to $H(0)$. That is not entirely true, however. Equation 5.45 is applicable to a temperature range in which no phase transitions occur. If there is a phase transition, we must add the enthalpy change for that transition because heat is absorbed without a change in $T$ for a phase transition. For example, if $T$ in Equation 5.45 is in the liquid region of a substance and the only phase change between 0 K and $T$ is a solid-liquid transition, then

$$H(T) - H(0) = \int_0^{T_{\text{fus}}} C_P^{\text{s}}(T)dT + \Delta_{\text{fus}}H + \int_{T_{\text{fus}}}^{T} C_P^{\text{l}}(T')dT' \tag{5.46}$$

where $C_P^{\text{s}}(T)$ and $C_P^{\text{l}}(T)$ stand for the heat capacity of the solid and liquid phases, respectively, $T_{\text{fus}}$ stands for the melting temperature, and $\Delta_{\text{fus}}H$ is the enthalpy change upon melting (the heat of fusion):

$$\Delta_{\text{fus}}H = H^{\text{l}}(T_{\text{fus}}) - H^{\text{s}}(T_{\text{fus}})$$

Figure 5.6 shows the molar heat capacity of benzene as a function of temperature. Notice that the plot of $C_P$ versus $T$ is not continuous, but has jump discontinuities at the temperatures corresponding to phase transitions. The melting point and boiling point of benzene at one atm are 278.7 K and 353.2 K, respectively. As Equation 5.45 implies, the area under the curve in Figure 5.6 from 0 K to $T \leq 278.7$ K gives the molar enthalpy of solid benzene [relative to $\overline{H}(0)$]. To calculate the molar enthalpy of liquid

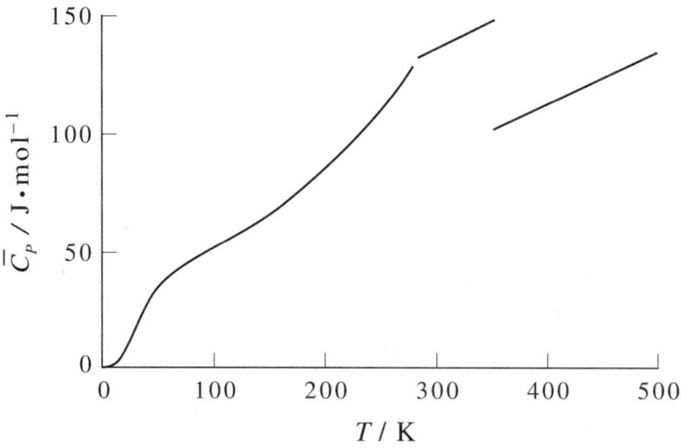


**FIGURE 5.6**
The constant-pressure molar heat capacity of benzene from 0 K to 500 K. The melting point and boiling point of benzene at one atm are 278.7 K and 353.2 K, respectively

benzene, say, at 300 K and one atm, we take the area under the curve in Figure 5.6 from 0 K to 300 K and add the molar enthalpy of fusion, which is 9.95 kJ·mol$^{-1}$. Figure 5.7 shows the molar enthalpy of benzene as a function of temperature. Notice that $\overline{H}(T) - \overline{H}(0)$ increases smoothly within a phase and that there is a jump at a phase transition.

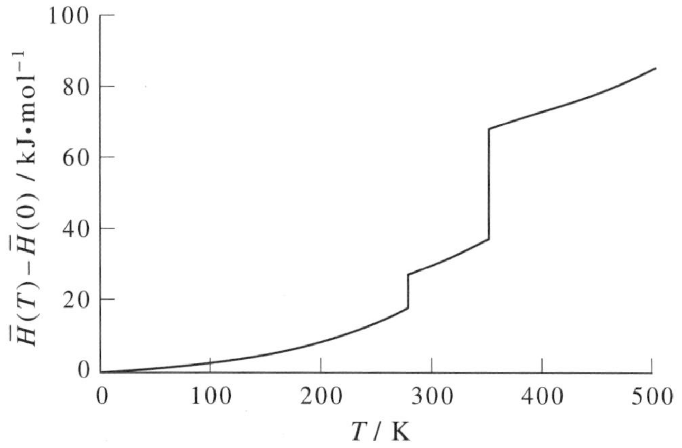


**FIGURE 5.7**
The molar enthalpy of benzene [relative to $\overline{H}(0)$] from 0 K to 500 K.

## 5–10. Enthalpy Changes for Chemical Equations Are Additive

Because most chemical reactions take place at constant pressure (open to the atmosphere), the enthalpy change associated with chemical reactions, $\Delta_r H$, (the subscript r indicates that the enthalpy change is for a chemical reaction) plays a central role in *thermochemistry*, which is the branch of thermodynamics that concerns the measurement of the evolution or absorption of energy as heat associated with chemical reactions. For example, the combustion of methane,

$$CH_4(g) + 2\,O_2(g) \longrightarrow CO_2(g) + 2\,H_2O(l)$$

releases energy as heat and is called an exothermic reaction (*exo* = out). Most combustion reactions are highly exothermic. The heat evolved in a combustion reaction is called the *heat of combustion*. Chemical reactions that absorb energy as heat are called endothermic reactions (*endo* = in). Exothermic and endothermic reactions are illustrated schematically in Figure 5.8.

The enthalpy change for a chemical reaction can be viewed as the total enthalpy of the products minus the total enthalpy of the reactants:

$$\Delta_r H = H_{prod} - H_{react} \tag{5.47}$$

For an exothermic reaction, $H_{prod}$ is less than $H_{react}$, so $\Delta_r H < 0$. Figure 5.8a represents an exothermic reaction; the enthalpy of the reactants is greater than the enthalpy of the products, so $q_P = \Delta_r H < 0$, and energy as heat is evolved as the reaction proceeds. For an endothermic reaction, $H_{prod}$ is greater than $H_{react}$, so $\Delta_r H > 0$. Figure 5.8b represents an endothermic reaction; the enthalpy of the reactants is less than the enthalpy of the products, so $q_P = \Delta_r H > 0$, and energy as heat must be supplied to drive the reaction up the enthalpy "hill."

Let's consider several examples of chemical reactions carried out at one bar. For the combustion of one mole of methane to form one mole of $CO_2(g)$ and two moles of

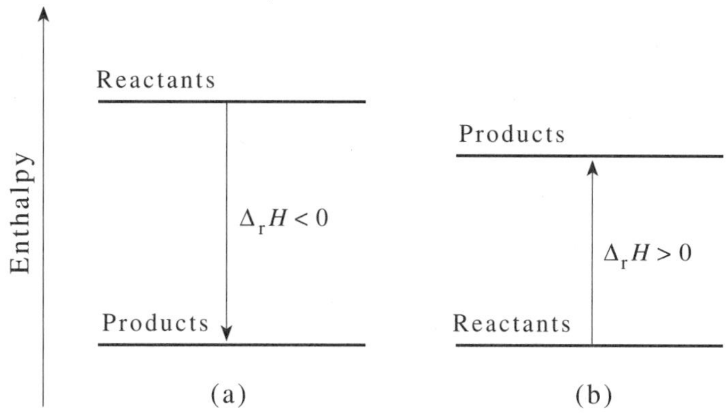


**FIGURE 5.8**
An enthalpy diagram for (a) an exothermic reaction and (b) and endothermic reaction.

$H_2O(l)$, the value of $\Delta_r H$ is $-890.36$ kJ at 298 K. The negative value of $\Delta_r H$ tells us that the reaction gives off energy as heat and is therefore exothermic.

An example of an endothermic reaction is the water-gas reaction:

$$C(s) + H_2O(g) \longrightarrow CO(g) + H_2(g)$$

For this reaction, $\Delta_r H = +131$ kJ at 298 K, so energy as heat must be supplied to drive the reaction from left to right.

An important and useful property of $\Delta_r H$ for chemical equations is additivity. This property of $\Delta_r H$ follows directly from the fact that the enthalpy is a state function. If we add two chemical equations to obtain a third chemical equation, the value of $\Delta_r H$ for the resulting equation is equal to the sum of the $\Delta_r H$ for the two equations being added together. The additivity of $\Delta_r H$ is best illustrated by example. Consider the following two chemical equations.

$$C(s) + \tfrac{1}{2}O_2(g) \longrightarrow CO(g) \qquad \Delta_r H(1) = -110.5 \text{ kJ} \tag{1}$$

$$CO(g) + \tfrac{1}{2}O_2(g) \longrightarrow CO_2(g) \qquad \Delta_r H(2) = -283.0 \text{ kJ} \tag{2}$$

If we add these two chemical equations as if they were algebraic equations, we get

$$C(s) + O_2(g) \longrightarrow CO_2(g) \tag{3}$$

The additive property of $\Delta_r H$ tells us that $\Delta_r H$ for Equation 3 is simply

$$\begin{aligned} \Delta_r H(3) &= \Delta_r H(1) + \Delta_r H(2) \\ &= -110.5 \text{ kJ} + (-283.0 \text{ kJ}) = -393.5 \text{ kJ} \end{aligned}$$

In effect, we can imagine Equations 1 and 2 as representing a two-step process with the same initial and final states as Equation 3. The total enthalpy change for the two equations together must, therefore, be the same as if the reaction proceeded in a single step.

The additivity property of $\Delta_r H$ values is known as *Hess's Law*. Thus, if the values of $\Delta_r H(1)$ and $\Delta_r H(2)$ are known, we need not independently determine the experimental value of $\Delta_r H(3)$ because its value is equal to the sum $\Delta_r H(1) + \Delta_r H(2)$.

Now let's consider the following combination of chemical equations.

$$SO_2(g) \longrightarrow S(s) + O_2(g) \tag{1}$$

$$S(s) + O_2(g) \longrightarrow SO_2(g) \tag{2}$$

Because Equation 2 is simply the reverse of Equation 1, we conclude from Hess's Law that

$$\Delta_r H(\text{reverse}) = -\Delta_r H(\text{forward}) \tag{5.48}$$

As an example of the application of Hess's Law, consider the use of

$$2\,\mathrm{P(s)} + 3\,\mathrm{Cl_2(g)} \longrightarrow 2\,\mathrm{PCl_3(l)} \qquad \Delta_r H(1) = -640\ \mathrm{kJ} \qquad (1)$$

and

$$2\,\mathrm{P(s)} + 5\,\mathrm{Cl_2(g)} \longrightarrow 2\,\mathrm{PCl_5(s)} \qquad \Delta_r H(2) = -887\ \mathrm{kJ} \qquad (2)$$

to calculate the value of $\Delta_r H$ for the equation

$$\mathrm{PCl_3(l)} + \mathrm{Cl_2(g)} \longrightarrow \mathrm{PCl_5(s)} \qquad (3)$$

In this case, we add Equation 2 to the reverse of Equation 1 to obtain Equation 4:

$$2\,\mathrm{PCl_3(l)} + 2\,\mathrm{Cl_2(g)} \longrightarrow 2\,\mathrm{PCl_5(s)} \qquad (4)$$

Thus, from Hess's law, we obtain

$$\begin{aligned}\Delta_r H(4) &= \Delta_r H(2) - \Delta_r H(1)\\ &= -887\ \mathrm{kJ} + 640\ \mathrm{kJ} = -247\ \mathrm{kJ}\end{aligned}$$

We now multiply Equation 4 through by 1/2 to obtain Equation 3:

$$\mathrm{PCl_3(l)} + \mathrm{Cl_2(g)} \longrightarrow \mathrm{PCl_5(s)} \qquad (3)$$

and so

$$\Delta_r H(3) = \frac{1}{2}\Delta_r H(4) = \frac{-247\ \mathrm{kJ}}{2} = -124\ \mathrm{kJ}$$

**EXAMPLE 5–9**

The molar enthalpies of combustion of isobutane and $n$-butane are $-2871\ \mathrm{kJ\cdot mol^{-1}}$ and $-2878\ \mathrm{kJ\cdot mol^{-1}}$, respectively at 298K and one atm. Calculate $\Delta_r H$ for the conversion of one mole of $n$-butane to one mole of isobutane.

SOLUTION: The equations for the two combustion reactions are

$$n\text{–}\mathrm{C_4H_{10}(g)} + \tfrac{13}{2}\,\mathrm{O_2(g)} \longrightarrow 4\,\mathrm{CO_2(g)} + 5\,\mathrm{H_2O(l)} \qquad (1)$$

$$\Delta_r H(1) = -2877\ \mathrm{kJ\cdot mol^{-1}}$$

and

$$i\text{–}\mathrm{C_4H_{10}(g)} + \tfrac{13}{2}\,\mathrm{O_2(g)} \longrightarrow 4\,\mathrm{CO_2(g)} + 5\,\mathrm{H_2O(l)} \qquad (2)$$

$$\Delta_r H(2) = -2869\ \mathrm{kJ\cdot mol^{-1}}$$

If we reverse the second equation and add the result to the first equation, then we obtain the desired equation

$$n\text{–}C_4H_{10}(g) \longrightarrow i\text{–}C_4H_{10}(g) \qquad (3)$$

$$\begin{aligned}\Delta_r H(3) &= \Delta_r H(1) - \Delta_r H(2) \\ &= -2877\ \text{kJ}\cdot\text{mol}^{-1} - (-2869\ \text{kJ}\cdot\text{mol}^{-1}) = -8\ \text{kJ}\cdot\text{mol}^{-1}\end{aligned}$$

The heat of this reaction cannot be measured directly because competing reactions occur.

## 5–11. Heats of Reactions Can Be Calculated from Tabulated Heats of Formation

The enthalpy change of a chemical reaction, $\Delta_r H$, depends upon the number of moles of the reactants. Recently, the physical chemistry division of the International Union of Pure and Applied Chemistry (IUPAC) has proposed a systematic procedure for tabulating reaction enthalpies. The *standard reaction enthalpy* of a chemical reaction is denoted by $\Delta_r H^\circ$ and refers to the enthalpy change associated with one mole of a specified reagent when all reactants and products are in their standard states, which for a gas is the equivalent hypothetical ideal gas at a pressure of one bar at the temperature of interest.

For example, consider the combustion of carbon to form carbon dioxide $CO_2(g)$. (The standard state of a solid is the pure crystalline substance at one bar pressure at the temperature of interest.) The balanced reaction can be written in many ways, including

$$C(s) + O_2(g) \longrightarrow CO_2(g) \qquad (5.49)$$

and

$$2\,C(s) + 2\,O_2(g) \longrightarrow 2\,CO_2(g) \qquad (5.50)$$

The quantity $\Delta_r H^\circ$ implies Equation 5.49 because only one mole of the (specified) reactant C(s) is combusted. The value of $\Delta_r H^\circ$ for this reaction at 298 K is $\Delta_r H^\circ = -393.5\ \text{kJ}\cdot\text{mol}^{-1}$. The corresponding reaction enthalpy for Equation 5.50 is

$$\Delta_r H = 2\Delta_r H^\circ = -787.0\ \text{kJ}$$

We see that $\Delta_r H$ is an extensive quantity, whereas $\Delta_r H^\circ$ is an intensive quantity. The advantage of the terminology is that it removes the ambiguity of how the balanced reaction corresponding to an enthalpy change is written.

Certain subscripts are used in place of r to indicate specific types of processes. For example, the subscript "c" is used for a combustion reaction and "vap" is used for vaporization [e.g., $H_2O(l) \rightarrow H_2O(g)$]. Table 5.1 lists many of the subscripts you will encounter.

The *standard molar enthalpy of formation*, $\Delta_f H^\circ$, is a particularly useful quantity. This intensive quantity is the standard reaction enthalpy for the formation of one mole of a molecule from its constituent elements. The degree superscript tells us that all reactants and products are in their standard states. The value of $\Delta_f H^\circ$ of $H_2O(l)$ is $-285.8\ \text{kJ}\cdot\text{mol}^{-1}$ at 298.15 K. This quantity implies that the balanced reaction is written as

$$H_2(g) + \tfrac{1}{2}O_2(g) \longrightarrow H_2O(l)$$

because $\Delta_f H^\circ$ refers to the heat of formation of one mole of $H_2O(l)$. (The standard state for a liquid is the normal state of the liquid at one bar at the temperature of interest.) A value of $\Delta_f H^\circ$ for $H_2O(l)$ equal to $-285.8\ \text{kJ}\cdot\text{mol}^{-1}$ tells us that one mole of $H_2O(l)$ lies 285.8 kJ "downhill" on the enthalpy scale relative to its constituent elements (Figure 5.9b) when the reactants and products are in their standard states.

Most compounds cannot be formed directly from their elements. For example, an attempt to make the hydrocarbon acetylene ($C_2H_2$) by the direct reaction of carbon with hydrogen

$$2\,C(s) + H_2(g) \longrightarrow C_2H_2(g) \tag{5.51}$$

yields not just $C_2H_2$ but a complex mixture of various hydrocarbons such as $C_2H_4$ and $C_2H_6$, among others. Nevertheless, we can determine the value of $\Delta_f H^\circ$ for acetylene by using Hess's Law, together with the available $\Delta_c H^\circ$ data on combustion reactions. All three species in Equation 5.51 burn in oxygen, and at 298 K we have

$$C(s) + O_2(g) \longrightarrow CO_2(g) \qquad \Delta_c H^\circ(1) = -393.5\ \text{kJ}\cdot\text{mol}^{-1} \tag{1}$$

$$H_2(g) + \tfrac{1}{2}O_2(g) \longrightarrow H_2O(l) \qquad \Delta_c H^\circ(2) = -285.8\ \text{kJ}\cdot\text{mol}^{-1} \tag{2}$$

$$C_2H_2(g) + \tfrac{5}{2}O_2(g) \longrightarrow 2\,CO_2(g) + H_2O(l) \qquad \Delta_c H^\circ(3) = -1299.6\ \text{kJ}\cdot\text{mol}^{-1} \tag{3}$$

If we multiply Equation 1 by 2, reverse Equation 3, and add the results to Equation 2, we obtain

$$2\,C(s) + H_2(g) \longrightarrow C_2H_2(g) \tag{4}$$

with

$$\begin{aligned}
\Delta_r H^\circ(4) &= 2\,\Delta_c H^\circ(1) + \Delta_c H^\circ(2) - \Delta_c H^\circ(3) \\
&= (2)(-393.5\ \text{kJ}\cdot\text{mol}^{-1}) + (-285.8\ \text{kJ}\cdot\text{mol}^{-1}) - (-1299.5\ \text{kJ}\cdot\text{mol}^{-1}) \\
&= +226.7\ \text{kJ}\cdot\text{mol}^{-1}
\end{aligned}$$

**TABLE 5.1**
Common subscripts for the enthalpy changes of processes.

| Subscript | Reaction |
|---|---|
| vap | Vaporization, evaporation |
| sub | Sublimation |
| fus | Melting, fusion |
| trs | Transition between phases in general |
| mix | Mixing of fluids |
| ads | Adsorption |
| c | Combustion |
| f | Formation |

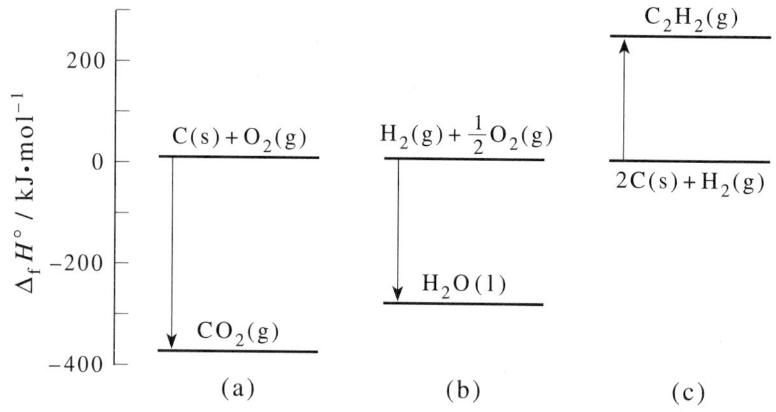


**FIGURE 5.9**
Standard enthalpy changes involved in the formation of $CO_2$(g), $H_2O$(l), and $C_2H_2$(g) from their elements, based upon the convention that $\Delta_f H^\circ = 0$ for a pure element in its stable form at one bar and at the temperature of interest.

Note that the stoichiometric coefficients have no units in the IUPAC convention. Because Equation 4 represents the formation of one mole of $C_2H_2$(g) from its elements, $\Delta_f H^\circ[C_2H_2(g)] = +226.7\ \text{kJ}\cdot\text{mol}^{-1}$ at 298 K (Figure 5.9c). Thus, we see that we can obtain values of $\Delta_f H^\circ$ even if the compound cannot be formed directly from its elements.

**EXAMPLE 5–10**
Given that the standard enthalpies of combustion of C(s), $H_2$(g), and $CH_4$(g) are $-393.5\ \text{kJ}\cdot\text{mol}^{-1}$, $-285.8\ \text{kJ}\cdot\text{mol}^{-1}$, and $-890.4\ \text{kJ}\cdot\text{mol}^{-1}$, respectively, at 298 K, calculate the standard enthalpy of formation of methane, $CH_4$(g).

SOLUTION: The chemical equations for the three combustion reactions are as follows:

$$C(s) + O_2(g) \longrightarrow CO_2(g) \qquad \Delta_c H^\circ(1) = -393.51 \text{ kJ}\cdot\text{mol}^{-1} \tag{1}$$

$$H_2(g) + \tfrac{1}{2}O_2(g) \longrightarrow H_2O(l) \qquad \Delta_c H^\circ(2) = -285.83 \text{ kJ}\cdot\text{mol}^{-1} \tag{2}$$

$$CH_4(g) + 2O_2(g) \longrightarrow CO_2(g) + 2H_2O(l) \;\; \Delta_c H^\circ(3) = -890.36 \text{ kJ}\cdot\text{mol}^{-1} \tag{3}$$

If we reverse Equation 3, multiply Equation 2 by 2, and add the results to Equation 1, we obtain the equation for the formation of $CH_4(g)$ from its elements.

$$C(s) + 2H_2(g) \longrightarrow CH_4(g) \tag{4}$$

along with

$$\begin{aligned}\Delta_r H^\circ(4) &= \Delta_c H^\circ(1) + 2\,\Delta_c H^\circ(2) - \Delta_c H^\circ(3)\\ &= (-393.51 \text{ kJ}\cdot\text{mol}^{-1}) + (2)(-285.83 \text{ kJ}\cdot\text{mol}^{-1}) - (-890.36 \text{ kJ}\cdot\text{mol}^{-1})\\ &= -74.81 \text{ kJ}\cdot\text{mol}^{-1}\end{aligned}$$

Because Equation (4) represents the formation of one mole of $CH_4(g)$ directly from its elements, we have $\Delta_f H^\circ[CH_4(g)] = -74.81 \text{ kJ}\cdot\text{mol}^{-1}$ at 298 K.

As suggested by Figure 5.9, we can set up a table of $\Delta_f H^\circ$ values for compounds by setting the values of $\Delta_f H^\circ$ for the elements equal to zero. That is, for each pure element in its stable form at one bar at the temperature of interest, we set $\Delta_f H^\circ$ equal to zero. Thus, standard enthalpies of formation of compounds are given relative to the elements in their normal physical states at one bar. Table 5.2 lists values of $\Delta_f H^\circ$ at 25°C for a number of substances. If you look at Table 5.2, you will see that $\Delta_f H^\circ[\text{C(diamond)}] = +1.897 \text{ kJ}\cdot\text{mol}^{-1}$, $\Delta_f H^\circ[Br_2(g)] = +30.907 \text{ kJ}\cdot\text{mol}^{-1}$, and $\Delta_f H^\circ[I_2(g)] = +62.438 \text{ kJ}\cdot\text{mol}^{-1}$. The values of $\Delta_f H^\circ$ for these forms of the elements are not equal to zero because C(diamond), $Br_2(g)$, and $I_2(g)$ are not the normal physical states of these elements at 25°C and one bar. The normal physical states of these elements at 25°C and one bar are C(graphite), $Br_2(l)$, and $I_2(s)$.

**EXAMPLE 5–11**

Use Table 5.2 to calculate the molar enthalpy of vaporization $\Delta_{vap}H^\circ$ of bromine at 25°C.

SOLUTION: The equation that represents the vaporization of one mole of bromine is

$$Br_2(l) \longrightarrow Br_2(g)$$

**TABLE 5.2**
Standard molar enthalpies of formation, $\Delta_f H^\circ$, for various substances at 25°C and one bar.

| Substance | Formula | $\Delta_f H^\circ/\text{kJ}\cdot\text{mol}^{-1}$ |
|---|---|---|
| Acetylene | $C_2H_2(g)$ | +226.73 |
| Ammonia | $NH_3(g)$ | −46.11 |
| Benzene | $C_6H_6(l)$ | +49.03 |
| Bromine | $Br_2(g)$ | +30.907 |
| Butane | $C_4H_{10}(g)$ | −125.6 |
| Carbon(diamond) | $C(s)$ | +1.897 |
| Carbon(graphite) | $C(s)$ | 0 |
| Carbon dioxide | $CO_2(g)$ | −393.509 |
| Carbon monoxide | $CO(g)$ | −110.5 |
| Cyclohexane | $C_6H_{12}(l)$ | −156.4 |
| Ethane | $C_2H_6(g)$ | −84.68 |
| Ethanol | $C_2H_5OH(l)$ | −277.69 |
| Ethene | $C_2H_4(g)$ | +52.28 |
| Glucose | $C_6H_{12}O_6(s)$ | −1260 |
| Hexane | $C_6H_{14}(l)$ | −198.7 |
| Hydrazine | $N_2H_4(l)$ | +50.6 |
| | $N_2H_4(g)$ | +95.40 |
| Hydrogen bromide | $HBr(g)$ | −36.3 |
| Hydrogen chloride | $HCl(g)$ | −92.31 |
| Hydrogen fluoride | $HF(g)$ | −273.3 |
| Hydrogen iodide | $HI(g)$ | +26.5 |
| Hydrogen peroxide | $H_2O_2(l)$ | −187.8 |
| Iodine | $I_2(g)$ | +62.438 |
| Methane | $CH_4(g)$ | −74.81 |
| Methanol | $CH_3OH(l)$ | −239.1 |
| | $CH_3OH(g)$ | −201.5 |
| Nitrogen oxide | $NO(g)$ | +90.37 |
| Nitrogen dioxide | $NO_2(g)$ | +33.85 |
| Dinitrogen tetraoxide | $N_2O_4(g)$ | +9.66 |
| | $N_2O_4(l)$ | −19.5 |
| Octane | $C_8H_{18}(l)$ | −250.1 |
| Pentane | $C_5H_{12}(l)$ | −173.5 |
| Propane | $C_3H_8(g)$ | −103.8 |
| Sucrose | $C_{12}H_{22}O_{11}(s)$ | −2220 |
| Sulfur dioxide | $SO_2(g)$ | −296.8 |
| Sulfur trioxide | $SO_3(g)$ | −395.7 |
| Tetrachloromethane | $CCl_4(l)$ | −135.44 |
| | $CCl_4(g)$ | −102.9 |
| Water | $H_2O(l)$ | −285.83 |
| | $H_2O(g)$ | −241.8 |

Therefore,

$$\begin{aligned}\Delta_{\text{vap}}H^\circ &= \Delta_f H^\circ[\text{Br}_2(\text{g})] - \Delta_f H^\circ[\text{Br}_2(\text{l})] \\ &= 30.907 \text{ kJ}\cdot\text{mol}^{-1}\end{aligned}$$

Note that this result is not the value of $\Delta_{\text{vap}}H^\circ$ at its normal boiling point of 58.8°C. The value of $\Delta_{\text{vap}}H^\circ$ at 58.8°C is 29.96 $\text{kJ}\cdot\text{mol}^{-1}$. (We will learn how to calculate the temperature variation of $\Delta H$ in the next section.)

We can use Hess's law to understand how enthaplies of formation are used to calculate enthalpy changes. Consider the general chemical equation

$$a\text{A} + b\text{B} \longrightarrow y\text{Y} + z\text{Z}$$

where $a$, $b$, $y$, and $z$ are the number of moles of the respective species. We can calculate $\Delta_r H$ in two steps, as shown in the following diagram:

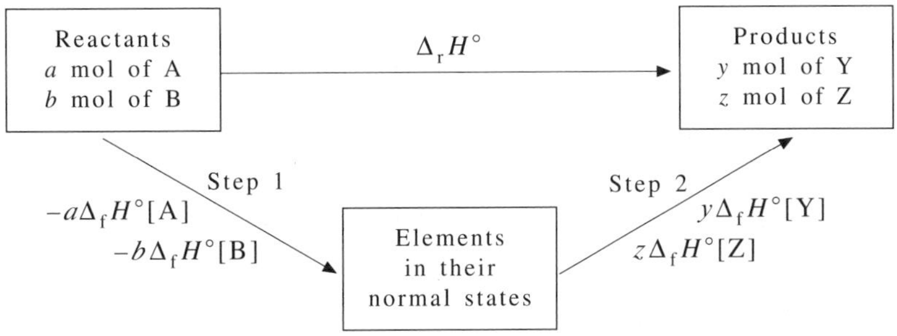


First, we decompose compounds A and B into their constituent elements (step 1); and then we combine the elements to form the compounds Y and Z (step 2). In the first step, we have

$$\Delta_r H(1) = -a\Delta_f H^\circ[\text{A}] - b\Delta H_f^\circ[\text{B}]$$

We have omitted the degree superscript on the $\Delta_r H$ because this value is not necessarily referenced to one mole of a particular reagent. The minus signs occur here because the reaction involved is the reverse of the formation of the compounds from their elements; we are forming the elements from the compounds. In the second step, we have

$$\Delta_r H(2) = y\Delta_f H^\circ[\text{Y}] + z\Delta_f H^\circ[\text{Z}]$$

The sum of $\Delta_r H(1)$ and $\Delta_r H(2)$ gives $\Delta_r H$ for the general equation:

$$\Delta_r H = y\Delta_f H^\circ[\text{Y}] + z\Delta_f H^\circ[\text{Z}] - a\Delta_f H^\circ[\text{A}] - b\Delta_f H^\circ[\text{B}] \tag{5.52}$$

Note that the right side of Equation 5.52 is the total enthalpy of the products minus the total enthalpy of the reactants (see Equation 5.47).

When using Equation 5.52, you need to specify whether each substance is a gas, liquid, or solid because the value of $\Delta_f H^\circ$ depends upon the physical state of the substance. Using Equation 5.52, we determine $\Delta_r H$ for the reaction

$$C_2H_2(g) + \tfrac{5}{2}\,O_2(g) \longrightarrow 2\,CO_2(g) + H_2O(l)$$

at 298 K to be

$$\Delta_r H = (2)\Delta_f H^\circ[CO_2(g)] + (1)\Delta_f H^\circ[H_2O(l)]$$
$$-(1)\Delta_f H^\circ[C_2H_2(g)] - (\tfrac{5}{2})\Delta_f H^\circ[O_2(g)]$$

Using the data in Table 5.2, we obtain

$$\Delta_r H = (2)(-393.509\ \text{kJ}\cdot\text{mol}^{-1}) + (1)(-285.83\ \text{kJ}\cdot\text{mol}^{-1})$$
$$-(1)(+226.73\ \text{kJ}\cdot\text{mol}^{-1}) - (\tfrac{5}{2})(0\ \text{kJ}\cdot\text{mol}^{-1})$$
$$= -1299.58\ \text{kJ}\cdot\text{mol}^{-1}$$

Note that $\Delta_f H^\circ[O_2(g)] = 0$ because the $\Delta_f H^\circ$ value for any element in its stable state at 298 K and one bar is zero. To determine $\Delta_r H$ for

$$2\,C_2H_2(g) + 5\,O_2(g) \longrightarrow 4\,CO_2(g) + 2\,H_2O(l)$$

we multiply $\Delta_r H = -1299.58\ \text{kJ}\cdot\text{mol}^{-1}$ by 2 mol to obtain $\Delta_r H = -2599.16$ kJ.

**EXAMPLE 5–12**

Use the $\Delta_f H^\circ$ data in Table 5.2 to calculate the value of $\Delta_r H^\circ$ for the combustion of liquid ethanol, $C_2H_5OH(l)$, at 25°C:

$$C_2H_5OH(l) + 3\,O_2(g) \longrightarrow 2\,CO_2(g) + 3\,H_2O(l)$$

SOLUTION: Referring to Table 5.2, we find that $\Delta_f H^\circ[CO_2(g)] = -393.509$ $\text{kJ}\cdot\text{mol}^{-1}$; $\Delta_f H^\circ[H_2O(l)] = -285.83\text{kJ}\cdot\text{mol}^{-1}$; $\Delta_f H^\circ[O_2(g)] = 0$; and $\Delta_f H^\circ[C_2H_5OH(l)] = -277.69\ \text{kJ}\cdot\text{mol}^{-1}$. Application of Equation 5.52 yields

$$\Delta_c H = (2)\Delta_f H^\circ[CO_2(g)] + (3)\Delta_f H^\circ[H_2O(l)]$$
$$-(1)\Delta_f H^\circ[C_2H_5OH(l)] - (3)\Delta_f H^\circ[O_2(g)]$$
$$= (2)(-393.509\ \text{kJ}\cdot\text{mol}^{-1}) + (3)(-285.83\ \text{kJ}\cdot\text{mol}^{-1})$$
$$-(1)(-277.69\ \text{kJ}\cdot\text{mol}^{-1}) - (3)(0)$$
$$= -1366.82\ \text{kJ}\cdot\text{mol}^{-1}$$

## 5–12. The Temperature Dependence of $\Delta_r H$ Is Given in Terms of the Heat Capacities of the Reactants and Products

Up to now, we have calculated reaction enthalpies at 25°C. We will see in this section that we can calculate $\Delta_r H$ at other temperatures if we have sufficient heat-capacity data. Consider the general reaction

$$a\text{A} + b\text{B} \longrightarrow y\text{Y} + z\text{Z}$$

We can express $\Delta_r H$ at a temperature $T_2$ in the form

$$\Delta_r H(T_2) = y[H_Y(T_2) - H_Y(0)] + z[H_Z(T_2) - H_Z(0)]$$
$$-a[H_A(T_2) - H_A(0)] - b[H_B(T_2) - H_B(0)] \qquad (5.53)$$

where, from Equation 5.45,

$$H_Y(T_2) - H_Y(0) = \int_0^{T_2} C_{P,Y}(T)dT \qquad (5.54)$$

etc. Similarly, $\Delta_r H(T_1)$ is given by

$$\Delta_r H(T_1) = y[H_Y(T_1) - H_Y(0)] + z[H_Z(T_1) - H_Z(0)]$$
$$-a[H_A(T_1) - H_A(0)] - b[H_B(T_1) - H_B(0)] \qquad (5.55)$$

with

$$H_Y(T_1) - H_Y(0) = \int_0^{T_1} C_{P,Y}(T)dT \qquad (5.56)$$

etc. If we substitute Equation 5.54 into Equation 5.53 and Equation 5.56 into Equation 5.55, and then subtract the resultant $\Delta_r H(T_1)$ from $\Delta_r H(T_2)$, we obtain

$$\Delta_r H(T_2) = \Delta_r H(T_1) + \int_{T_1}^{T_2} \Delta C_P(T)dT \qquad (5.57)$$

where, as the notation suggests,

$$\Delta C_P(T) = yC_{P,\text{Y}}(T) + zC_{P,\text{Z}}(T) - aC_{P,\text{A}}(T) - bC_{P,\text{B}}(T) \qquad (5.58)$$

Thus, if we know $\Delta_r H$ at $T_1$, say 25°C, we can calculate $\Delta_r H$ at any other temperature using Equation 5.57. In writing Equation 5.57 we have assumed there are no phase transitions between $T_1$ and $T_2$.

Equation 5.57 has a simple physical interpretation given by Figure 5.10. To calculate the value of $\Delta_r H$ at some temperature $T_2$ given the value of $\Delta_r H$ at $T_1$, we can follow the path 1—2—3 in Figure 5.10. This pathway involves taking the reactants

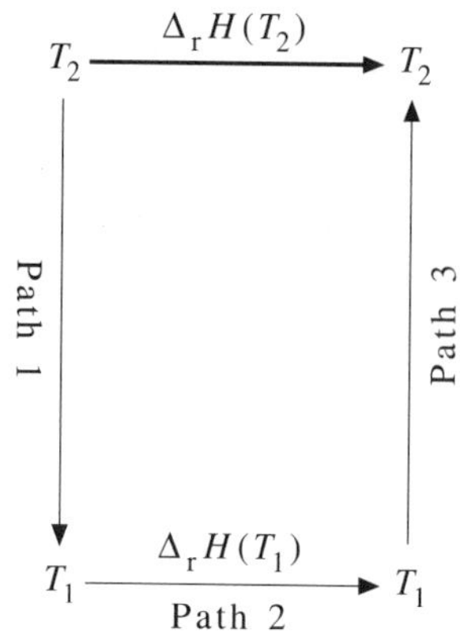


**FIGURE 5.10**
An illustration of Equation 5.57. Along path 1 we take the reactants from $T_2$ to $T_1$. Along path 2 we let the reaction occur at $T_1$. Then along path 3, we bring the products from $T_1$ back to $T_2$. Because $\Delta H$ is a state function, we have that $\Delta H(T_2) = \Delta H_1 + \Delta H_2 + \Delta H_3$.

from temperature $T_2$ to $T_1$, letting the reaction occur at $T_1$, and then taking the products from $T_1$ back to $T_2$. The mathematical expressions for $\Delta H$ for each step are

$$\Delta H_1 = \int_{T_2}^{T_1} C_P(\text{reactants})dT = -\int_{T_1}^{T_2} C_P(\text{reactants})dT$$

$$\Delta H_2 = \Delta_r H(T_1)$$

$$\Delta H_3 = \int_{T_1}^{T_2} C_P(\text{products})dT$$

and so

$$\begin{aligned}\Delta H(T_2) &= \Delta H_1 + \Delta H_2 + \Delta H_3 \\ &= \Delta_r H(T_1) + \int_{T_1}^{T_2} \left[C_P(\text{products}) - C_P(\text{reactants})\right] dT\end{aligned}$$

As a simple application of Equation 5.57, consider

$$H_2O(s) \longrightarrow H_2O(l)$$

Let's calculate $\Delta_{\text{fus}}H^\circ$ of water at $-10^\circ$C and one bar given that $\Delta_{\text{fus}}H^\circ(0^\circ\text{C}) = 6.01\ \text{kJ}\cdot\text{mol}^{-1}$, $C_P^\circ(\text{s}) = 37.7\ \text{J}\cdot\text{K}^{-1}\cdot\text{mol}^{-1}$ and $C_P^\circ(\text{l}) = 75.3\ \text{J}\cdot\text{K}^{-1}\cdot\text{mol}^{-1}$. Because the equation is written in terms of one mole of reactant and the reactants and products are in their standard states, we use a superscript ∘ on calculated thermodynamic quantities. Therefore,

$$\Delta C_P^\circ = C_P^\circ(\text{l}) - C_P^\circ(\text{s}) = 37.6\ \text{J}\cdot\text{K}^{-1}\cdot\text{mol}^{-1}$$

and

$$\begin{aligned}\Delta_{\text{fus}}H^\circ(-10^\circ\text{C}) &= \Delta_{\text{fus}}H^\circ(0^\circ\text{C}) + \int_{0^\circ\text{C}}^{-10^\circ\text{C}} (37.6\ \text{J}\cdot\text{K}^{-1}\cdot\text{mol}^{-1})\, dT \\ &= 6.01\ \text{kJ}\cdot\text{mol}^{-1} - 376\ \text{J}\cdot\text{mol}^{-1} = 5.64\ \text{kJ}\cdot\text{mol}^{-1}\end{aligned}$$

**EXAMPLE 5–13**

The standard molar enthaply of formation, $\Delta_f H^\circ$, of $NH_3(g)$ is $-46.11\ \text{kJ}\cdot\text{mol}^{-1}$ at 25°C. Using the heat capacity data given below, calculate the standard molar heat of formation of $NH_3(g)$ at 1000 K.

$$C_P^\circ(\text{H}_2)/\text{J}\cdot\text{K}^{-1}\cdot\text{mol}^{-1} = 29.07 - (0.837 \times 10^{-3}\ \text{K}^{-1})T + (2.012 \times 10^{-6}\ \text{K}^{-2})T^2$$

$$C_P^\circ(\text{N}_2)/\text{J}\cdot\text{K}^{-1}\cdot\text{mol}^{-1} = 26.98 + (5.912 \times 10^{-3}\ \text{K}^{-1})T - (0.3376 \times 10^{-6}\ \text{K}^{-2})T^2$$

$$C_P^\circ(\text{NH}_3)/\text{J}\cdot\text{K}^{-1}\cdot\text{mol}^{-1} = 25.89 + (32.58 \times 10^{-3}\ \text{K}^{-1})T - (3.046 \times 10^{-6}\ \text{K}^{-2})T^2$$

where $298\ \text{K} < T < 1500\ \text{K}$

SOLUTION: We use the equation

$$\Delta_f H^\circ(1000\ \text{K}) = \Delta_f H^\circ(298\ \text{K}) + \int_{298\ \text{K}}^{1000\ \text{K}} \Delta C_P^\circ(T)dT$$

The relevant chemical equation for the formation of one mole of $NH_3(g)$ from its elements is

$$\tfrac{1}{2}\,\text{N}_2(\text{g}) + \tfrac{3}{2}\,\text{H}_2(\text{g}) \longrightarrow \text{NH}_3(\text{g})$$

and so

$$\begin{aligned}\Delta C_P^\circ(T)/\text{J}\cdot\text{K}^{-1}\cdot\text{mol}^{-1} &= (1)\,C_P^\circ(\text{NH}_3) - (\tfrac{1}{2})\,C_P^\circ(\text{N}_2) - (\tfrac{3}{2})\,C_P^\circ(\text{H}_2)\\ &= -31.21 + (30.88 \times 10^{-3}\ \text{K}^{-1})T - (5.895 \times 10^{-6}\ \text{K}^{-2})T^2\end{aligned}$$

The integral of $\Delta C_P(T)$ is

$$\int_{298\ \text{K}}^{1000\ \text{K}} \left[-31.21 + (30.88 \times 10^{-3}\ \text{K}^{-1})T - (5.895 \times 10^{-6}\ \text{K}^{-2})T^2\right] dT$$

$$\begin{aligned}&= (-21.91 + 14.07 - 1.913)\ \text{kJ}\cdot\text{mol}^{-1}\\ &= -9.75\ \text{kJ}\cdot\text{mol}^{-1}\end{aligned}$$

so

$$\begin{aligned}\Delta_f H^\circ(1000\ \text{K}) &= \Delta_f H^\circ(25^\circ\text{C}) - 9.75\ \text{kJ}\cdot\text{mol}^{-1}\\ &= -46.11\ \text{kJ}\cdot\text{mol}^{-1} - 9.75\ \text{kJ}\cdot\text{mol}^{-1}\\ &= -55.86\ \text{kJ}\cdot\text{mol}^{-1}\end{aligned}$$

The pressure dependence of $\Delta_r H$ (which we will study in Chapter 8) is usually much smaller than its temperature dependence.

## Problems

**5-1.** Suppose that a 10-kg mass of iron at 20°C is dropped from a height of 100 meters. What is the kinetic energy of the mass just before it hits the ground? What is its speed? What would be the final temperature of the mass if all its kinetic energy at impact is transformed into internal energy? Take the molar heat capacity of iron to be $\overline{C}_P = 25.1\ \mathrm{J \cdot mol^{-1} \cdot K^{-1}}$ and the gravitational acceleration constant to be $9.80\ \mathrm{m \cdot s^{-2}}$.

**5-2.** Consider an ideal gas that occupies 2.50 $\mathrm{dm^3}$ at a pressure of 3.00 bar. If the gas is compressed isothermally at a constant external pressure, $P_{\mathrm{ext}}$, so that the final volume is 0.500 $\mathrm{dm^3}$, calculate the smallest value $P_{\mathrm{ext}}$ can have. Calculate the work involved using this value of $P_{\mathrm{ext}}$.

**5-3.** A one-mole sample of $CO_2(g)$ occupies 2.00 $\mathrm{dm^3}$ at a temperature of 300 K. If the gas is compressed isothermally at a constant external pressure, $P_{\mathrm{ext}}$, so that the final volume is 0.750 $\mathrm{dm^3}$, calculate the smallest value $P_{\mathrm{ext}}$ can have, assuming that $CO_2(g)$ satisfies the van der Waals equation of state under these conditions. Calculate the work involved using this value of $P_{\mathrm{ext}}$.

**5-4.** Calculate the work involved when one mole of an ideal gas is compressed reversibly from 1.00 bar to 5.00 bar at a constant temperature of 300 K.

**5-5.** Calculate the work involved when one mole of an ideal gas is expanded reversibly from 20.0 $\mathrm{dm^3}$ to 40.0 $\mathrm{dm^3}$ at a constant temperature of 300 K.

**5-6.** Calculate the minimum amount of work required to compress 5.00 moles of an ideal gas isothermally at 300 K from a volume of 100 $\mathrm{dm^3}$ to 40.0 $\mathrm{dm^3}$.

**5-7.** Consider an ideal gas that occupies 2.25 L at 1.33 bar. Calculate the work required to compress the gas isothermally to a volume of 1.50 L at a constant pressure of 2.00 bar followed by another isothermal compression to 0.800 L at a constant pressure of 2.50 bar (Figure 5.4). Compare the result with the work of compressing the gas isothermally and reversibly from 2.25 L to 0.800 L.

**5-8.** Show that for an isothermal reversible expansion from a molar volume $\overline{V}_1$ to a final molar volume $\overline{V}_2$, the work is given by

$$w = -RT \ln\left(\frac{\overline{V}_2 - B}{\overline{V}_1 - B}\right) - \frac{A}{BT^{1/2}} \ln\left[\frac{(\overline{V}_2 + B)\overline{V}_1}{(\overline{V}_1 + B)\overline{V}_2}\right]$$

for the Redlich-Kwong equation.

**5-9.** Use the result of Problem 5–8 to calculate the work involved in the isothermal reversible expansion of one mole of $CH_4(g)$ from a volume of $1.00\ \mathrm{dm^3 \cdot mol^{-1}}$ to $5.00\ \mathrm{dm^3 \cdot mol^{-1}}$ at 300 K. (See Table 2.4 for the values of $A$ and $B$.)

**5-10.** Repeat the calculation in Problem 5-9 for a van der Waals gas.

**5-11.** Derive an expression for the reversible isothermal work of an expansion of a gas that obeys the Peng-Robinson equation of state.

**5-12.** One mole of a monatomic ideal gas initially at a pressure of 2.00 bar and a temperature of 273 K is taken to a final pressure of 4.00 bar by the reversible path defined by $P/V =$ constant. Calculate the values of $\Delta U$, $\Delta H$, $q$, and $w$ for this process. Take $\overline{C}_V$ to be equal to $12.5\ \text{J}\cdot\text{mol}^{-1}\cdot\text{K}^{-1}$.

**5-13.** The isothermal compressibility of a substance is given by

$$\beta = -\frac{1}{V}\left(\frac{\partial V}{\partial P}\right)_T \tag{1}$$

For an ideal gas, $\beta = 1/P$, but for a liquid, $\beta$ is fairly constant over a moderate pressure range. If $\beta$ is constant, show that

$$\frac{V}{V_0} = e^{-\beta(P-P_0)} \tag{2}$$

where $V_0$ is the volume at a pressure $P_0$. Use this result to show that the reversible isothermal work of compressing a liquid from a volume $V_0$ (at a pressure $P_0$) to a volume $V$ (at a pressure $P$) is given by

$$\begin{aligned} w &= -P_0(V - V_0) + \beta^{-1}V_0\left(\frac{V}{V_0}\ln\frac{V}{V_0} - \frac{V}{V_0} + 1\right) \\ &= -P_0V_0[e^{-\beta(P-P_0)} - 1] + \beta^{-1}V_0\{1 - [1 + \beta(P - P_0)]e^{-\beta(P-P_0)}\} \end{aligned} \tag{3}$$

(You need to use the fact that $\int \ln x dx = x\ln x - x$.)

The fact that liquids are incompressible is reflected by $\beta$ being small, so that $\beta(P - P_0) \ll 1$ for moderate pressures. Show that

$$\begin{aligned} w &= \beta P_0V_0(P - P_0) + \frac{\beta V_0(P - P_0)^2}{2} + O(\beta^2) \\ &= \frac{\beta V_0}{2}(P^2 - P_0^2) + O(\beta^2) \end{aligned} \tag{4}$$

Calculate the work required to compress one mole of toluene reversibly and isothermally from 10 bar to 100 bar at 20°C. Take the value of $\beta$ to be $8.95 \times 10^{-5}\ \text{bar}^{-1}$ and the molar volume to be $0.106\ \text{mol}\cdot\text{L}^{-1}$ at 20°C.

**5-14.** In the previous problem, you derived an expression for the reversible, isothermal work done when a liquid is compressed. Given that $\beta$ is typically $O(10^{-4})\ \text{bar}^{-1}$, show that $V/V_0 \approx 1$ for pressures up to about 100 bar. This result, of course, reflects the fact that liquids are not very compressible. We can exploit this result by substituting $dV = -\beta V dP$ from the defining equation of $\beta$ into $w = -\int P dV$ and then treating $V$ as a constant. Show that this approximation gives Equation 4 of Problem 5–13.

**5-15.** Show that

$$\frac{T_2}{T_1} = \left(\frac{V_1}{V_2}\right)^{R/\overline{C}_V}$$

for a reversible adiabatic expansion of an ideal gas.

**5-16.** Show that

$$\left(\frac{T_2}{T_1}\right)^{3/2} = \frac{\overline{V}_1 - b}{\overline{V}_2 - b}$$

for a reversible, adiabatic expansion of a monatomic gas that obeys the equation of state $P(\overline{V} - b) = RT$. Extend this result to the case of a diatomic gas.

**5-17.** Show that

$$\frac{T_2}{T_1} = \left(\frac{P_2}{P_1}\right)^{R/\overline{C}_P}$$

for a reversible adiabatic expansion of an ideal gas.

**5-18.** Show that

$$P_1 V_1^{(\overline{C}_V + R)/\overline{C}_V} = P_2 V_2^{(\overline{C}_V + R)/\overline{C}_V}$$

for an adiabatic expansion of an ideal gas. Show that this formula reduces to Equation 5.23 for a monatomic gas.

**5-19.** Calculate the work involved when one mole of a monatomic ideal gas at 298 K expands reversibly and adiabatically from a pressure of 10.00 bar to a pressure of 5.00 bar.

**5-20.** A quantity of $N_2(g)$ at 298 K is compressed reversibly and adiabatically from a volume of 20.0 $dm^3$ to 5.00 $dm^3$. Assuming ideal behavior, calculate the final temperature of the $N_2(g)$. Take $\overline{C}_V = 5R/2$.

**5-21.** A quantity of $CH_4(g)$ at 298 K is compressed reversibly and adiabatically from 50.0 bar to 200 bar. Assuming ideal behavior, calculate the final temperature of the $CH_4(g)$. Take $\overline{C}_V = 3R$.

**5-22.** One mole of ethane at 25°C and one atm is heated to 1200°C at constant pressure. Assuming ideal behavior, calculate the values of $w$, $q$, $\Delta U$, and $\Delta H$ given that the molar heat capacity of ethane is given by

$$\overline{C}_P/R = 0.06436 + (2.137 \times 10^{-2}\ \text{K}^{-1})T - (8.263 \times 10^{-6}\ \text{K}^{-2})T^2 + (1.024 \times 10^{-9}\ \text{K}^{-3})T^3$$

over the above temperature range. Repeat the calculation for a constant-volume process.

**5-23.** The value of $\Delta_r H^\circ$ at 25°C and one bar is +290.8 kJ for the reaction

$$2\,\text{ZnO(s)} + 2\,\text{S(s)} \longrightarrow 2\,\text{ZnS(s)} + \text{O}_2\text{(g)}$$

Assuming ideal behavior, calculate the value of $\Delta_r U^\circ$ for this reaction.

**5-24.** Liquid sodium is being considered as an engine coolant. How many grams of sodium are needed to absorb 1.0 MJ of heat if the temperature of the sodium is not to increase by more than 10°C. Take $\overline{C}_P = 30.8\ \text{J}\cdot\text{K}^{-1}\cdot\text{mol}^{-1}$ for Na(l) and $75.2\ \text{J}\cdot\text{K}^{-1}\cdot\text{mol}^{-1}$ for $H_2O(l)$.

**5-25.** A 25.0-g sample of copper at 363 K is placed in 100.0 g of water at 293 K. The copper and water quickly come to the same temperature by the process of heat transfer from copper to water. Calculate the final temperature of the water. The molar heat capacity of copper is $24.5\ \mathrm{J\cdot K^{-1}\cdot mol^{-1}}$.

**5-26.** A 10.0-kg sample of liquid water is used to cool an engine. Calculate the heat removed (in joules) from the engine when the temperature of the water is raised from 293 K to 373 K. Take $\overline{C}_P = 75.2\ \mathrm{J\cdot K^{-1}\cdot mol^{-1}}$ for $H_2O(l)$.

**5-27.** In this problem, we will derive a general relation between $C_P$ and $C_V$. Start with $U = U(P, T)$ and write

$$dU = \left(\frac{\partial U}{\partial P}\right)_T dP + \left(\frac{\partial U}{\partial T}\right)_P dT \tag{1}$$

We could also consider $V$ and $T$ to be the independent variables of $U$ and write

$$dU = \left(\frac{\partial U}{\partial V}\right)_T dV + \left(\frac{\partial U}{\partial T}\right)_V dT \tag{2}$$

Now take $V = V(P, T)$ and substitute its expression for $dV$ into Equation 2 to obtain

$$dU = \left(\frac{\partial U}{\partial V}\right)_T \left(\frac{\partial V}{\partial P}\right)_T dP + \left[\left(\frac{\partial U}{\partial V}\right)_T \left(\frac{\partial V}{\partial T}\right)_P + \left(\frac{\partial U}{\partial T}\right)_V\right] dT$$

Compare this result with Equation 1 to obtain

$$\left(\frac{\partial U}{\partial P}\right)_T = \left(\frac{\partial U}{\partial V}\right)_T \left(\frac{\partial V}{\partial P}\right)_T \tag{3}$$

and

$$\left(\frac{\partial U}{\partial T}\right)_P = \left(\frac{\partial U}{\partial V}\right)_T \left(\frac{\partial V}{\partial T}\right)_P + \left(\frac{\partial U}{\partial T}\right)_V \tag{4}$$

Last, substitute $U = H - PV$ into the left side of Equation 4 and use the definitions of $C_P$ and $C_V$ to obtain

$$C_P - C_V = \left[P + \left(\frac{\partial U}{\partial V}\right)_T\right] \left(\frac{\partial V}{\partial T}\right)_P$$

Show that $C_P - C_V = nR$ if $(\partial U/\partial V)_T = 0$, as it is for an ideal gas.

**5-28.** Following Problem 5–27, show that

$$C_P - C_V = \left[V - \left(\frac{\partial H}{\partial P}\right)_T\right] \left(\frac{\partial P}{\partial T}\right)_V$$

**5-29.** Starting with $H = U + PV$, show that

$$\left(\frac{\partial U}{\partial T}\right)_P = C_P - P\left(\frac{\partial V}{\partial T}\right)_P$$

Interpret this result physically.

**5-30.** Given that $(\partial U/\partial V)_T = 0$ for an ideal gas, prove that $(\partial H/\partial V)_T = 0$ for an ideal gas.

**5-31.** Given that $(\partial U/\partial V)_T = 0$ for an ideal gas, prove that $(\partial C_V/\partial V)_T = 0$ for an ideal gas.

**5-32.** Show that $C_P - C_V = nR$ if $(\partial H/\partial P)_T = 0$, as is true for an ideal gas.

**5-33.** Differentiate $H = U + PV$ with respect to $V$ at constant temperature to show that $(\partial H/\partial V)_T = 0$ for an ideal gas.

**5-34.** Given the following data for sodium, plot $\overline{H}(T) - \overline{H}(0)$ against $T$ for sodium: melting point, 361 K; boiling point, 1156 K; $\Delta_{\text{fus}}H^\circ = 2.60\ \text{kJ}\cdot\text{mol}^{-1}$; $\Delta_{\text{vap}}H^\circ = 97.4\ \text{kJ}\cdot\text{mol}^{-1}$; $\overline{C}_P(\text{s}) = 28.2\ \text{J}\cdot\text{mol}^{-1}\cdot\text{K}^{-1}$; $\overline{C}_P(\text{l}) = 32.7\ \text{J}\cdot\text{mol}^{-1}\cdot\text{K}^{-1}$; $\overline{C}_P(\text{g}) = 20.8\ \text{J}\cdot\text{mol}^{-1}\cdot\text{K}^{-1}$.

**5-35.** The $\Delta_r H^\circ$ values for the following equations are

$$2\,\text{Fe(s)} + \tfrac{3}{2}\text{O}_2(\text{g}) \rightarrow \text{Fe}_2\text{O}_3(\text{s}) \qquad \Delta_r H^\circ = -206\ \text{kJ}\cdot\text{mol}^{-1}$$

$$3\,\text{Fe(s)} + 2\,\text{O}_2(\text{g}) \rightarrow \text{Fe}_3\text{O}_4(\text{s}) \qquad \Delta_r H^\circ = -136\ \text{kJ}\cdot\text{mol}^{-1}$$

Use these data to calculate the value of $\Delta_r H$ for the reaction described by

$$4\,\text{Fe}_2\text{O}_3(\text{s}) + \text{Fe(s)} \longrightarrow 3\,\text{Fe}_3\text{O}_4(\text{s})$$

**5-36.** Given the following data,

$$\tfrac{1}{2}\text{H}_2(\text{g}) + \tfrac{1}{2}\text{F}_2(\text{g}) \rightarrow \text{HF(g)} \qquad \Delta_r H^\circ = -273.3\ \text{kJ}\cdot\text{mol}^{-1}$$

$$\text{H}_2(\text{g}) + \tfrac{1}{2}\text{O}_2(\text{g}) \rightarrow \text{H}_2\text{O(l)} \qquad \Delta_r H^\circ = -285.8\ \text{kJ}\cdot\text{mol}^{-1}$$

calculate the value of $\Delta_r H$ for the reaction described by

$$2\,\text{F}_2(\text{g}) + 2\,\text{H}_2\text{O(l)} \longrightarrow 4\,\text{HF(g)} + \text{O}_2(\text{g})$$

**5-37.** The standard molar heats of combustion of the isomers *m*-xylene and *p*-xylene are $-4553.9\ \text{kJ}\cdot\text{mol}^{-1}$ and $-4556.8\ \text{kJ}\cdot\text{mol}^{-1}$, respectively. Use these data, together with Hess's Law, to calculate the value of $\Delta_r H^\circ$ for the reaction described by

$$m\text{-xylene} \longrightarrow p\text{-xylene}$$

**5-38.** Given that $\Delta_r H^\circ = -2826.7$ kJ for the combustion of 1.00 mol of fructose at 298.15 K,

$$\text{C}_6\text{H}_{12}\text{O}_6(\text{s}) + 6\,\text{O}_2(\text{g}) \longrightarrow 6\,\text{CO}_2(\text{g}) + 6\,\text{H}_2\text{O(l)}$$

and the $\Delta_f H^\circ$ data in Table 5.2, calculate the value of $\Delta_f H^\circ$ for fructose at 298.15 K.

**5-39.** Use the $\Delta_f H^\circ$ data in Table 5.2 to calculate the value of $\Delta_c H^\circ$ for the combustion reactions described by the equations:

**a.** $\text{CH}_3\text{OH(l)} + \tfrac{3}{2}\,\text{O}_2(\text{g}) \longrightarrow \text{CO}_2(\text{g}) + 2\,\text{H}_2\text{O(l)}$
**b.** $\text{N}_2\text{H}_4(\text{l}) + \text{O}_2(\text{g}) \longrightarrow \text{N}_2(\text{g}) + 2\,\text{H}_2\text{O(l)}$

Compare the heat of combustion per gram of the fuels $\text{CH}_3\text{OH(l)}$ and $\text{N}_2\text{H}_4(\text{l})$.

**5-40.** Using Table 5.2, calculate the heat required to vaporize 1.00 mol of $\text{CCl}_4(\text{l})$ at 298 K.

**5-41.** Using the $\Delta_f H^\circ$ data in Table 5.2, calculate the values of $\Delta_r H^\circ$ for the following:

**a.** $\text{C}_2\text{H}_4(\text{g}) + \text{H}_2\text{O(l)} \longrightarrow \text{C}_2\text{H}_5\text{OH(l)}$
**b.** $\text{CH}_4(\text{g}) + 4\,\text{Cl}_2(\text{g}) \longrightarrow \text{CCl}_4(\text{l}) + 4\,\text{HCl(g)}$

In each case, state whether the reaction is endothermic or exothermic.

**5-42.** Use the following data to calculate the value of $\Delta_{\text{vap}}H^\circ$ of water at 298 K and compare your answer to the one you obtain from Table 5.2: $\Delta_{\text{vap}}H^\circ$ at 373 K $= 40.7\ \text{kJ}\cdot\text{mol}^{-1}$; $\overline{C}_P(\text{l}) = 75.2\ \text{J}\cdot\text{mol}^{-1}\cdot\text{K}^{-1}$; $\overline{C}_P(\text{g}) = 33.6\ \text{J}\cdot\text{mol}^{-1}\cdot\text{K}^{-1}$.

**5-43.** Use the following data and the data in Table 5.2 to calculate the standard reaction enthalpy of the water-gas reaction at 1273 K. Assume that the gases behave ideally under these conditions.

$$\text{C(s)} + \text{H}_2\text{O(g)} \longrightarrow \text{CO(g)} + \text{H}_2\text{(g)}$$

$$C_P^\circ[\text{CO(g)}]/R = 3.231 + (8.379 \times 10^{-4}\ \text{K}^{-1})T - (9.86 \times 10^{-8}\ \text{K}^{-2})T^2$$

$$C_P^\circ[\text{H}_2\text{(g)}]/R = 3.496 + (1.006 \times 10^{-4}\ \text{K}^{-1})T + (2.42 \times 10^{-7}\ \text{K}^{-2})T^2$$

$$C_P^\circ[\text{H}_2\text{O(g)}]/R = 3.652 + (1.156 \times 10^{-3}\ \text{K}^{-1})T + (1.42 \times 10^{-7}\ \text{K}^{-2})T^2$$

$$C_P^\circ[\text{C(s)}]/R = -0.6366 + (7.049 \times 10^{-3}\ \text{K}^{-1})T - (5.20 \times 10^{-6}\ \text{K}^{-2})T^2 + (1.38 \times 10^{-9}\ \text{K}^{-3})T^3$$

**5-44.** The standard molar enthalpy of formation of $\text{CO}_2\text{(g)}$ at 298 K is $-393.509\ \text{kJ}\cdot\text{mol}^{-1}$. Use the following data to calculate the value of $\Delta_\text{f}H^\circ$ at 1000 K. Assume the gases behave ideally under these conditions.

$$C_P^\circ[\text{CO}_2\text{(g)}]/R = 2.593 + (7.661 \times 10^{-3}\ \text{K}^{-1})T - (4.78 \times 10^{-6}\ \text{K}^{-2})T^2 + (1.16 \times 10^{-9}\ \text{K}^{-3})T^3$$

$$C_P^\circ[\text{O}_2\text{(g)}]/R = 3.094 + (1.561 \times 10^{-3}\ \text{K}^{-1})T - (4.65 \times 10^{-7}\ \text{K}^{-2})T^2$$

$$C_P^\circ[\text{C(s)}]/R = -0.6366 + (7.049 \times 10^{-3}\ \text{K}^{-1})T - (5.20 \times 10^{-6}\ \text{K}^{-2})T^2 + (1.38 \times 10^{-9}\ \text{K}^{-3})T^3$$

**5-45.** The value of the standard molar reaction enthalpy for

$$\text{CH}_4\text{(g)} + 2\,\text{O}_2\text{(g)} \longrightarrow \text{CO}_2\text{(g)} + 2\,\text{H}_2\text{O(g)}$$

is $-802.2\ \text{kJ}\cdot\text{mol}^{-1}$ at 298 K. Using the heat-capacity data in Problems 5–43 and 5–44 in addition to

$$C_P^\circ[\text{CH}_4\text{(g)}]/R = 2.099 + (7.272 \times 10^{-3}\ \text{K}^{-1})T + (1.34 \times 10^{-7}\ \text{K}^{-2})T^2 - (8.66 \times 10^{-10}\ \text{K}^{-3})T^3$$

to derive a general equation for the value of $\Delta_\text{r}H^\circ$ at any temperature between 300 K and 1500 K. Plot $\Delta_\text{r}H^\circ$ versus $T$. Assume that the gases behave ideally under these conditions.

**5-46.** In all the calculations thus far, we have assumed the reaction takes place at constant temperature, so that any energy evolved as heat is absorbed by the surroundings. Suppose, however, that the reaction takes place under adiabatic conditions, so that all the energy released as heat stays within the system. In this case, the temperature of the system will increase, and the final temperature is called the *adiabatic flame temperature*. One relatively simple way to estimate this temperature is to suppose the reaction occurs at the initial temperature of the reactants and then determine to what temperature the products can be raised by the quantity $\Delta_\text{r}H^\circ$. Calculate the adiabatic flame temperature if one mole of

$CH_4(g)$ is burned in two moles of $O_2(g)$ at an initial temperature of 298 K. Use the results of the previous problem.

**5-47.** Explain why the adiabatic flame temperature defined in the previous problem is also called the maximum flame temperature.

**5-48.** How much energy as heat is required to raise the temperature of 2.00 moles of $O_2(g)$ from 298 K to 1273 K at 1.00 bar? Take

$$\overline{C}_P[O_2(g)]/R = 3.094 + (1.561 \times 10^{-3}\ K^{-1})T - (4.65 \times 10^{-7}\ K^{-2})T^2$$

**5-49.** When one mole of an ideal gas is compressed adiabatically to one-half of its original volume, the temperature of the gas increases from 273 K to 433 K. Assuming that $\overline{C}_V$ is independent of temperature, calculate the value of $\overline{C}_V$ for this gas.

**5-50.** Use the van der Waals equation to calculate the minimum work required to expand one mole of $CO_2(g)$ isothermally from a volume of 0.100 $dm^3$ to a volume of 100 $dm^3$ at 273 K. Compare your result with that which you calculate assuming ideal behavior.

**5-51.** Show that the work involved in a reversible, adiabatic pressure change of one mole of an ideal gas is given by

$$w = \overline{C}_V T_1 \left[ \left( \frac{P_2}{P_1} \right)^{R/\overline{C}_P} - 1 \right]$$

where $T_1$ is the initial temperature and $P_1$ and $P_2$ are the initial and final pressures, respectively.

**5-52.** In this problem, we will discuss a famous experiment called the *Joule-Thomson experiment*. In the first half of the 19th century, Joule tried to measure the temperature change when a gas is expanded into a vacuum. The experimental setup was not sensitive enough, however, and he found that there was no temperature change, within the limits of his error. Soon afterward, Joule and Thomson devised a much more sensitive method for measuring the temperature change upon expansion. In their experiments (see Figure 5.11), a constant applied pressure $P_1$ causes a quantity of gas to flow slowly from one chamber to another through a porous plug of silk or cotton. If a volume, $V_1$, of gas is pushed through the

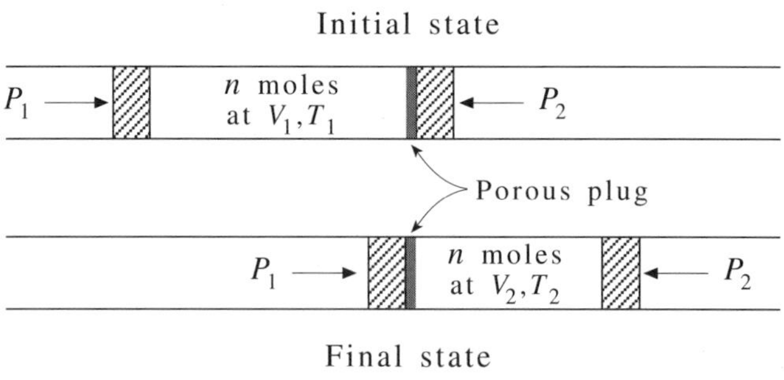


**FIGURE 5.11**
A schematic description of the Joule-Thomson experiment.

porous plug, the work done on the gas is $P_1V_1$. The pressure on the other side of the plug is maintained at $P_2$, so if a volume $V_2$ enters the right-side chamber, then the net work is given by

$$w = P_1V_1 - P_2V_2$$

The apparatus is constructed so that the entire process is adiabatic, so $q = 0$. Use the First Law of Thermodynamics to show that

$$U_2 + P_2V_2 = U_1 + P_1V_1$$

or that $\Delta H = 0$ for a Joule-Thomson expansion.

Starting with

$$dH = \left(\frac{\partial H}{\partial P}\right)_T dP + \left(\frac{\partial H}{\partial T}\right)_P dT$$

show that

$$\left(\frac{\partial T}{\partial P}\right)_H = -\frac{1}{C_P}\left(\frac{\partial H}{\partial P}\right)_T$$

Interpret physically the derivative on the left side of this equation. This quantity is called the *Joule-Thomson coefficient* and is denoted by $\mu_{\text{JT}}$. In Problem 5–54 you will show that it equals zero for an ideal gas. Nonzero values of $(\partial T/\partial P)_H$ directly reflect intermolecular interactions. Most gases cool upon expansion [a positive value of $(\partial T/\partial P)_H$] and a Joule-Thomson expansion is used to liquefy gases.

**5-53.** The Joule-Thomson coefficient (Problem 5–52) depends upon the temperature and pressure, but assuming an average constant value of $0.15\ \text{K}\cdot\text{bar}^{-1}$ for $N_2(g)$, calculate the drop in temperature if $N_2(g)$ undergoes a drop in pressure of 200 bar.

**5-54.** Show that the Joule-Thomson coefficient (Problem 5–52) can be written as

$$\mu_{\text{JT}} = \left(\frac{\partial T}{\partial P}\right)_H = -\frac{1}{C_P}\left[\left(\frac{\partial U}{\partial V}\right)_T\left(\frac{\partial V}{\partial P}\right)_T + \left(\frac{\partial (PV)}{\partial P}\right)_T\right]$$

Show that $(\partial T/\partial P)_H = 0$ for an ideal gas.

**5-55.** In this problem, we will investigate the pressure dependence of the heat capacity, $C_P$. Because $C_P = (\partial H/\partial T)_P$, we see that

$$\left(\frac{\partial C_P}{\partial P}\right)_T = \frac{\partial^2 H}{\partial P \partial T}$$

Now show that

$$\left(\frac{\partial C_P}{\partial P}\right)_T = -\mu_{\text{JT}}\left(\frac{\partial C_P}{\partial T}\right)_P - C_P\left(\frac{\partial \mu_{\text{JT}}}{\partial T}\right)_P$$

where $\mu_{\text{JT}}$ is the Joule-Thomson coefficient. Show that $C_P$ is independent of pressure for an ideal gas.

**5-56.** Given that $C_P(T)$ for $N_2(g)$ can be represented by

$$C_P(T)/\text{J}\cdot\text{mol}^{-1}\cdot\text{K}^{-1} = 27.296 + (5.230 \times 10^{-3}\ \text{K}^{-1})T - (4.0 \times 10^{-9}\ \text{K}^{-2})T^2$$

for 298 K $< T <$ 1500 K and that $\mu_{\text{JT}}$ can be represented by

$$\mu_{\text{JT}}/\text{K}\cdot\text{atm}^{-1} = 3.722 - (0.02935\,\text{K}^{-1})T + (8.106 \times 10^{-5}\,\text{K}^{-2})T^2 - (7.496 \times 10^{-8}\,\text{K}^{-3})T^3$$

for 150 K $< T <$ 500 K around one atmosphere, estimate the pressure dependence of $C_P(T)$ for $N_2(g)$ at 300 K and one atmosphere pressure.

**5-57.** In Chapter 8, we will derive the formula

$$\left(\frac{\partial U}{\partial V}\right)_T = T\left(\frac{\partial P}{\partial T}\right)_V - P \qquad (8.22)$$

Show that $U$ is independent of volume for an ideal gas. Show that

$$\Delta U = a\left(\frac{1}{V_1} - \frac{1}{V_2}\right)$$

for an isothermal expansion of a van der Waals gas and that

$$\Delta U = \frac{A}{2T^{1/2}B}\ln\left[\frac{(V_2 + B)\ V_1}{(V_1 + B)\ V_2}\right]$$

for a Redlich-Kwong gas.

**5-58.** Use the rigid rotator-harmonic oscillator model and the data in Table 4.2 to plot $\overline{C}_P(T)$ for CO(g) from 300 K to 1000 K. Compare your result with the expression given in Problem 5–43.

**5-59.** Use the rigid rotator-harmonic oscillator model and the data in Table 4.4 to plot $\overline{C}_P(T)$ for $CH_4(g)$ from 300 K to 1000 K. Compare your result with the expression given in Problem 5–45.

**5-60.** Why do you think the equations for the dependence of temperature on volume for a reversible adiabatic process (see Equation 5.22 and Example 5.6) depend upon whether the gas is a monatomic gas or a polyatomic gas?

MATHCHAPTER E

# THE BINOMIAL DISTRIBUTION AND STIRLING'S APPROXIMATION

In the next chapter, we will learn about entropy, a thermodynamic state function that has a molecular interpretation of being a measure of the disorder of a system. In doing so, we will have to put the idea of the disorder of a system on a quantitative basis. A problem we will encounter is that of determining how many ways we can arrange $N$ distinguishable objects such that there are $n_1$ objects in the first group, $n_2$ objects in the second group, and so on, such that

$$n_1 + n_2 + n_3 + \cdots = N$$

that is, such that all the objects are accounted for. This problem is actually a fairly standard one in statistics.

Let's solve the problem of dividing the $N$ distinguishable objects into two groups first and then generalize our results to any number of groups. First, we calculate the number of permutations of $N$ distinguishable objects, that is, the number of possible different arrangements or ways to order $N$ distinguishable objects. Let's choose one of the $N$ objects and place it in the first position, one of the $N-1$ remaining objects and place it in the second postion, and so on until all $N$ objects are ordered. Clearly, there are $N$ choices for the first position, $N-1$ choices for the second position, and so on until finally there is only one object left for the $N$th position. The total number of ways of doing this ordering is the product of all the choices:

$$N(N-1)(N-2)\ldots(2)(1) = N!$$

Next, we calculate the number of ways of dividing $N$ distinguishable objects into two groups, one containing $N_1$ objects and the other containing the $N - N_1 = N_2$ remaining objects. There are

$$\underbrace{N(N-1)\ldots(N-N_1+1)}_{N_1 \text{ terms}}$$

ways to form the first group. This product can be written more conveniently as

$$N(N-1)(N-2)\ldots(N-N_1+1) = \frac{N!}{(N-N_1)!} \tag{E.1}$$

as can seen by noting that

$$N! = (N)(N-1)\ldots(N-N_1+1) \times (N-N_1)!$$

The number of ways of forming the second group is $N_2! = (N - N_1)!$. You might think that the total number of arrangements is the product of the two factors, $N!/(N - N_1)!$ and $N_2!$, but this product drastically overcounts the situation because the order in which we arrange the $N_1$ objects in the first group and the $N_2$ objects in the second group is immaterial to the problem stated. All $N_1!$ orders of the first group and $N_2!$ orders of the second group correspond to just one division of $N$ distinguishable objects into two groups containing $N_1$ and $N_2$ objects. Therefore, we divide the product of $N!/(N - N_1)!$ and $N_2!$ by $N_1!$ and $N_2!$ to obtain

$$W(N_1, N_2) = \frac{N!}{(N-N_1)!N_1!} = \frac{N!}{N_1!N_2!} \tag{E.2}$$

where we let $W(N_1, N_2)$ denote the result. (Problem E–12 shows that $0! = 1$.)

**EXAMPLE E–1**

Use Equation E.2 to calculate the number of ways of arranging four distinguishable objects into two groups, containing three objects and one object. Verify your result with an explicit enumeration.

SOLUTION: We have $N = 4$, $N_1 = 3$, and $N_2 = 1$, and so Equation E.2 gives

$$W(3, 1) = \frac{4!}{3!1!} = 4$$

If we let $a$, $b$, $c$, and $d$ be the four distinguishable objects, the four arrangements are $abc : d$, $abd : c$, $acd : b$, and $bcd : a$. There are no others.

The combinatorial factor in Equation E.2 is called a binomial coefficient because the expansion of the binomial $(x + y)^N$ is given by

$$(x+y)^N = \sum_{N_1=0}^{N} \frac{N!}{N_1!(N-N_1)!} x^{N_1} y^{N-N_1} \tag{E.3}$$

For example,

$$(x+y)^2 = x^2 + 2xy + y^2 = \sum_{N_1=0}^{2} \frac{2!}{N_1!(2-N_1)!} x^{N_1} y^{2-N_1}$$

and

$$(x+y)^3 = x^3 + 3x^2y + 3xy^2 + y^3 = \sum_{N_1=0}^{3} \frac{3!}{N_1!(3-N_1)!} x^{N_1} y^{3-N_1}$$

Equation E.3 may be written in a more symmetric form:

$$(x+y)^N = \sum_{N_1=0}^{N} \sum_{N_2=0}^{N}{}^{*} \frac{N!}{N_1!N_2!} x^{N_1} y^{N_2} \tag{E.4}$$

where the asterisk on the summation signs indicates that only terms with $N_1 + N_2 = N$ are included. This symmetric form of the binomial expansion suggests the form of the multinomial expansion given below in Equation E.6. Simple numerical examples verify that Equations E.3 and E.4 are equivalent.

The generalization of Equation E.2 to the division of $N$ distinguishable objects into $r$ groups, the first containing $N_1$, the second containing $N_2$, and so on, is

$$W(N_1, N_2, \ldots, N_r) = \frac{N!}{N_1!N_2! \cdots N_r!} \tag{E.5}$$

with $N_1 + N_2 + \cdots + N_r = N$. This quantity is called a multinomial coefficient because it occurs in the multinomial expansion:

$$(x_1 + x_2 + \cdots + x_r)^N = \sum_{N_1=0}^{N} \sum_{N_2=0}^{N} \cdots \sum_{N_r=0}^{N}{}^{*} \frac{N!}{N_1!N_2! \cdots N_r!} x_1^{N_1} x_2^{N_2} \ldots x_r^{N_r} \tag{E.6}$$

where the asterisk indicates that only terms such that $N_1 + N_2 + \cdots + N_r = N$ are included. Note how Equation E.6 is a straightforward generalization of Equation E.4.

**EXAMPLE E–2**

Calculate the number of ways of dividing 10 distinguishable objects into three groups containing 2, 5, and 3 objects.

SOLUTION: We use Equation E.5:

$$W(2, 5, 3) = \frac{10!}{2!5!3!} = 2520$$

If we use Equation E.5 to calculate something like the number of ways of distributing Avogadro's number of particles over their energy states, then we are forced to deal with factorials of huge numbers. Even the evaluation of 100! would be a chore, never mind $10^{23}!$, unless we have a good approximation for $N!$. We shall see that there is an approximation for $N!$ that actually improves as $N$ gets larger. Such an approximation is called an asymptotic approximation, that is, an approximation to a function that gets better as the argument of the function increases.

Because $N!$ is a product, it is convenient to deal with $\ln N!$ because the latter is a sum. The asymptotic expansion to $\ln N!$ is called Stirling's approximation and is given by

$$\ln N! = N \ln N - N \tag{E.7}$$

which is surely a lot easier to use than calculating $N!$ and then taking its logarithm. Table E.1 shows the value of $\ln N!$ versus Stirling's approximation for a number of values of $N$. Note that the agreement, which we express in terms of relative error, improves markedly with increasing $N$.

**EXAMPLE E–3**

A more refined version of Stirling's approximation (one we will *not* have to use in the next chapter) says that

$$\ln N! = N \ln N - N + \ln(2\pi N)^{1/2}$$

Use this version of Stirling's approximation to calculate $\ln N!$ for $N = 10$ and compare the relative error with that in Table E.1.

SOLUTION: For $N = 10$,

$$\ln N! = N \ln N - N + \ln(2\pi N)^{1/2} = 15.096$$

and using the value of $\ln 10!$ from Table E.1, we see that

$$\text{relative error} = \frac{15.104 - 15.096}{15.104} = 0.0005$$

The relative error is significantly smaller than that in Table E.1. The relative errors for the other entries in Table E.1 are essentially zero for this extended version of Stirling's approximation.

**TABLE E.1**

A numerical comparison of $\ln N!$ with Stirling's approximation.

| $N$ | $\ln N!$ | $N \ln N - N$ | Relative error[a] |
|---|---|---|---|
| 10 | 15.104 | 13.026 | 0.1376 |
| 50 | 148.48 | 145.60 | 0.0194 |
| 100 | 363.74 | 360.52 | 0.0089 |
| 500 | 2611.3 | 2607.3 | 0.0015 |
| 1000 | 5912.1 | 5907.7 | 0.0007 |

[a]relative error $= (\ln N! - N \ln N + N)/\ln N!$

The proof of Stirling's approximation is not difficult. Because $N!$ is given by $N! = N(N-1)(N-2)\dots(2)(1)$, $\ln N!$ is given by

$$\ln N! = \sum_{n=1}^{N} \ln n \tag{E.8}$$

Figure E.1 shows $\ln x$ plotted versus $x$ for integer values of $x$. According to Equation E.8, the sum of the areas under the rectangles up to $N$ in Figure E.1 is $\ln N!$. Figure E.1 also shows the continuous curve $\ln x$ plotted on the same graph. Thus, $\ln x$ is seen to form an envelope to the rectangles, and this envelope becomes a steadily smoother approximation to the rectangles as $x$ increases. Therefore, we can approximate the area under these rectangles by the integral of $\ln x$. The area under $\ln x$ will poorly approximate the rectangles only in the beginning. If $N$ is large enough (we are deriving an asymptotic expansion), this area will make a negligible contribution to the total area. We may write, then,

$$\ln N! = \sum_{n=1}^{N} \ln n \approx \int_1^N \ln x dx = N \ln N - N \quad (N \text{ large}) \tag{E.9}$$

which is Stirling's approximation to $\ln N!$. The lower limit could just as well have been taken as 0 in Equation E.9, because $N$ is large. (Remember that $x \ln x \to 0$ as $x \to 0$.) We will use Stirling's approximation frequently in the next few chapters.

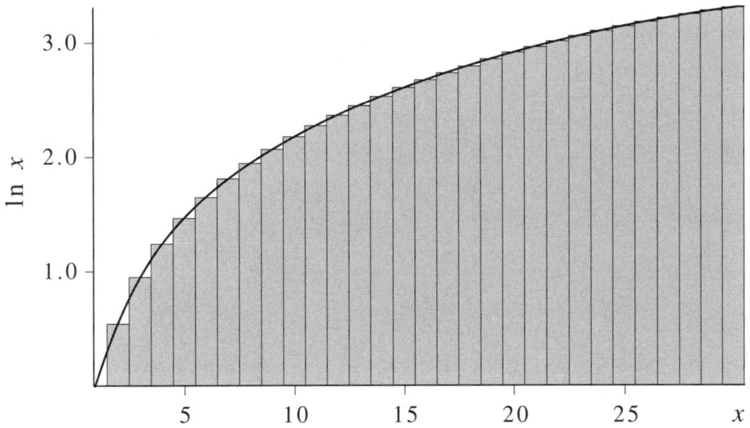


**FIGURE E.1**
A plot of $\ln x$ versus $x$. The sum of the areas under the rectangles up to $N$ is $\ln N!$.

## Problems

**E-1.** Use Equation E.3 to write the expansion of $(1+x)^5$. Use Equation E.4 to do the same thing.

**E-2.** Use Equation E.6 to write out the expression for $(x+y+z)^2$. Compare your result to the one that you obtain by multiplying $(x+y+z)$ by $(x+y+z)$.

**E-3.** Use Equation E.6 to write out the expression for $(x+y+z)^4$. Compare your result to the one that you obtain by multiplying $(x+y+z)^2$ from Problem E–2 by itself.

**E-4.** How many permutations of the letters $a$, $b$, $c$ are there?

**E-5.** The coefficients of the expansion of $(1+x)^n$ can be arranged in the following form:

| $n$ | | | | | | | | | |
|---|---|---|---|---|---|---|---|---|---|
| 0 | | | | | 1 | | | | |
| 1 | | | | 1 | | 1 | | | |
| 2 | | | 1 | | 2 | | 1 | | |
| 3 | | 1 | | 3 | | 3 | | 1 | |
| 4 | 1 | | 4 | | 6 | | 4 | | 1 |

Do you see a pattern in going from one row to the next? The triangular arrangement here is called Pascal's triangle.

**E-6.** In how many ways can a committee of three be chosen from nine people?

**E-7.** Calculate the relative error for $N = 50$ using the formula for Stirling's approximation given in Example E–3, and compare your result with that given in Table E.1 using Equation E.7. Take $\ln N!$ to be 148.47776 (*CRC Handbook of Chemistry and Physics*).

**E-8.** Prove that $x \ln x \to 0$ as $x \to 0$.

**E-9.** Prove that the maximum value of $W(N, N_1) = N!/(N-N_1)!N_1!$ is given by $N_1 = N/2$. (*Hint*: Treat $N_1$ as a continuous variable.)

**E-10.** Prove that the maximum value of $W(N_1, N_2, \ldots, N_r)$ in Equation E.5 is given by $N_1 = N_2 = \cdots = N_r = N/r$.

**E-11.** Prove that

$$\sum_{k=0}^{N} \frac{N!}{k!(N-k)!} = 2^N$$

**E-12.** The quantity $n!$ as we have defined it is defined only for positive integer values of $n$. Consider now the function of $x$ *defined* by

$$\Gamma(x) = \int_0^\infty t^{x-1} e^{-t} dt \tag{1}$$

Integrate by parts (letting $u = t^{x-1}$ and $dv = e^{-t}dt$) to get

$$\Gamma(x) = (x-1)\int_0^{\infty} t^{x-2}e^{-t}dt = (x-1)\Gamma(x-1) \qquad (2)$$

Now use Equation 2 to show that $\Gamma(x) = (x-1)!$ if $x$ is a positive integer. Although Equation 2 provides us with a general function that is equal to $(n-1)!$ when $x$ takes on integer values, it is defined just as well for non-integer values. For example, show that $\Gamma(3/2)$, which in a sense is $(\frac{1}{2})!$, is equal to $\pi^{1/2}/2$. Equation 1 can also be used to explain why $0! = 1$. Let $x = 1$ in Equation 1 to show that $\Gamma(1)$, which we can write as $0!$, is equal to 1. The function $\Gamma(x)$ defined by Equation 1 is called the *gamma function* and was introduced by Euler to generalize the idea of a factorial to general values of $n$. The gamma function arises in many problems in chemistry and physics.

**Rudolf Clausius** was born in Köslin, Prussia (now Koszalin, Poland), on January 2, 1822, and died in 1888. Although Clausius was initially attracted to history, he eventually received his Ph.D. in mathematical physics from the University of Halle in 1847. He held a position for several years at the University of Zurich but returned to Germany and in 1871 settled at the University of Bonn, where he remained for the rest of his life. Clausius is credited with creating the early foundations of thermodynamics. In 1850, he published his first great paper on the theory of heat, in which he rejected the then-current caloric theory and argued that the energy of a system is a thermodynamic state function. In 1865, he published his second landmark paper, in which he introduced another new thermodynamic state function, which he called entropy, and expressed the Second Law of Thermodynamics in terms of the entropy. Clausius also studied the kinetic theory of gases and made important contributions to it. He was chauvinistic and strongly defended German achievements against what he considered the infringements of others. Most of Clausius' work was done before 1870 because of two events in his life. In 1870, he was wounded while serving in an ambulance corps in the Franco-Prussian War and suffered life-long pain from his injury. More tragically, his wife died in childbirth, and he assumed the responsibility of raising six young children.

CHAPTER 6

# Entropy and the Second Law of Thermodynamics

In this chapter, we will introduce and develop the concept of entropy. We will see that energy considerations alone are not sufficient to predict in which direction a process or a chemical reaction can occur spontaneously. We will demonstrate that isolated systems that are not in equilibrium will evolve in a direction that increases their disorder, and then we will introduce a thermodynamic state function called entropy that gives a quantitative measure of the disorder of a system. One statement of the Second Law of Thermodynamics, which governs the direction in which systems evolve to their equilibrium states, is that the entropy of an isolated system always increases as a result of any spontaneous (irreversible) process. In the second half of this chapter, we will give a quantitative molecular definition of entropy in terms of partition function.

## 6–1. The Change of Energy Alone Is Not Sufficient to Determine the Direction of a Spontaneous Process

For years, scientists wondered why some reactions or processes proceed spontaneously and others do not. We all know that under the right conditions iron rusts, and that objects do not spontaneously unrust. We all know that hydrogen and oxygen react explosively to form water but that an input of energy by means of electrolysis is required to decompose water into hydrogen and oxygen. At one time scientists believed that a criterion for a reaction or a process to proceed spontaneously was that it should be exothermic, or evolve energy. This belief was motivated by the fact that the products of an exothermic reaction lie at a lower energy or enthalpy than the reactants. After all, balls *do* roll downhill and opposite charges *do* attract each other. Mechanical systems evolve in such a way as to minimize their energy.

Now consider the situation in Figure 6.1, however, where one bulb contains a gas at some low pressure at which it may be considered to behave ideally, and the other bulb is evacuated. When the two bulbs are connected by opening the stopcock between them, the gas will expand into the evacuated bulb until the pressures in

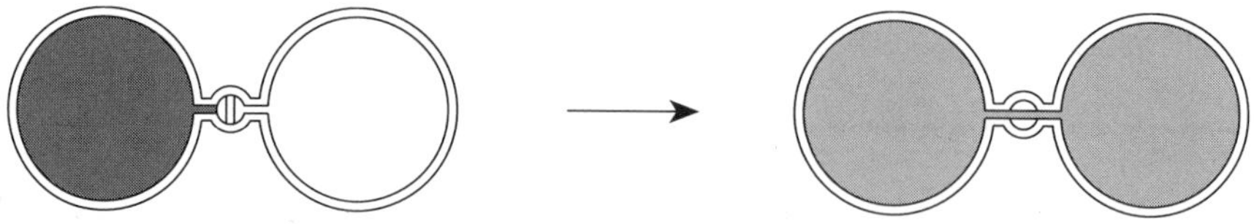

**FIGURE 6.1**
Two bulbs connected by a stopcock. Initially, one bulb contains a colored gas such as bromine and the other one is evacuated. When the two bulbs are connected by opening the stopcock, the bromine occupies both bulbs at a uniform pressure as seen by the uniform color.

the two bulbs are equal, at which time the system will be in equilibrium. Yet a careful determination of the thermal processes of this experiment shows that both $\Delta U$ and $\Delta H$ are essentially zero. Furthermore, the unaided reverse process has never been observed. Gases do not spontaneously occupy only part of a container, leaving the other part as a vacuum.

Another example of a spontaneous process that is not exothermic is depicted in Figure 6.2, where two pure gases are separated by a stopcock. When the stopcock is opened, the two gases will mix, and both will eventually become evenly distributed between the two bulbs, in which case the system will be in equilibrium. Yet once again, the value of $\Delta U$ or $\Delta H$ for this process is essentially zero. Furthermore, the reverse process has never been observed. Mixtures of gases do not spontaneously unmix.

There are many spontaneous endothermic processes. A simple example of a spontaneous endothermic reaction is the melting of ice at a temperature above 0°C. This spontaneous process has a value of $\Delta_{\text{fus}}H^\circ$ equal to $+6.0\ \text{kJ}\cdot\text{mol}^{-1}$ when the temperature is around 0°C. An especially interesting endothermic chemical reaction is the reation of solid barium hydroxide, $Ba(OH)_2(s)$, with solid ammonium nitrate, $NH_4NO_3(s)$:

$$Ba(OH)_2(s) + 2\ NH_4NO_3(s) \longrightarrow Ba(NO_3)_2(s) + 2\ H_2O(l) + 2\ NH_3(aq)$$

The energy absorbed by mixing stoichiometric amounts of these two reagents in a test tube can cool the system to below −20°C.

These and numerous other examples indicate that spontaneous processes have a direction that cannot be explained by the First Law of Thermodynamics. Of course,

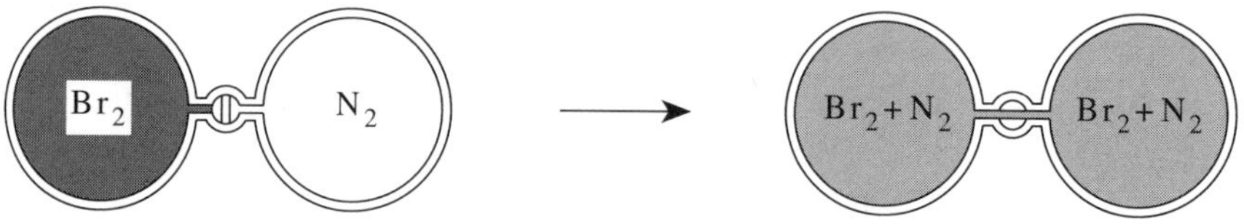


**FIGURE 6.2**
Two bulbs connected by a stopcock. Initially, each bulb is occupied by a pure gas, say bromine and nitrogen. When the two bulbs are connected by opening the stopcock, the two gases mix uniformly, so each bulb contains the same uniform mixture.

each of these processes obeys the First Law of Thermodynamics, but using this law, we cannot tell why one direction occurs spontaneously and its reverse does not. Although mechanical systems tend to achieve their state of lowest energy, clearly some other factor is involved that we have not yet discussed.

## 6–2. Nonequilibrium Isolated Systems Evolve in a Direction That Increases Their Disorder

If we examine the above processes from a microscopic or molecular point of view, we see that each one involves an increase in disorder or randomness of the system. For example, in Figure 6.1, the gas molecules in the final state are able to move over a volume that is twice as large as in the initial state. In a sense, locating any gas molecule in the final state is twice as difficult as in the initial state. Recall that we found that the number of accessible translational states increases with the volume of the container, Problem 4–42. A similar argument applies to the mixing of two gases. Not only is each gas spread over a larger volume, but they are also mixed together. Clearly the final (mixed) state is more disordered than the initial (separated) state. The melting of ice at a temperature greater than 0°C also involves an increase in disorder. Our molecular picture of a solid being an ordered lattice array of its constituent particles and a liquid being a more random arrangement directly implies that the melting of ice involves an increase in disorder.

These examples suggest that not only do systems evolve spontaneously in a direction that lowers their energy but that they also seek to increase their disorder. There is a competition between the tendency to minimize energy and to maximize disorder. If disorder is not a factor, as is the case for a simple mechanical system, then energy is the key factor and the direction of any spontaneous process is that which minimizes the energy. If energy is not a factor, however, as is the case when mixing two gases, then disorder is the key factor and the direction of any spontaneous process is that which maximizes the disorder. In general, some compromise between decreasing energy and increasing disorder must be met.

What we need is to devise some particular property that puts this idea of disorder on a useful, quantitative basis. Like energy, we want this property to be a state function because then it will be a property of the state of the system, and not of its previous history. Thus, we will rule out heat, although the transfer of energy as heat to a system certainly does increase its disorder. To try to get an idea of what an appropriate function might be, let's consider, for simplicity, the heat transfer associated with a reversible, small change in the temperature and volume of an ideal gas. From the First Law (Equation 5.9), we have

$$\begin{aligned}\delta q_{\text{rev}} &= dU - \delta w_{\text{rev}} = C_V(T)dT + PdV \\ &= C_V(T)dT + \frac{nRT}{V}dV \end{aligned} \tag{6.1}$$

Example 5–4 showed us that $\delta q_{\text{rev}}$ is not a state function. In mathematical terms, this means that the right side of Equation 6.1 is not an exact differential; in other words, it can not be written as the derivative of some function of $T$ and $V$ (see MathChapter D). The first term, however, can be written as the derivative of a function of $T$ because $C_V$ is a function of only temperature for an ideal gas , so $C_V(T)dT$ can be written as

$$C_V(T)dT = d\left[\int C_V(T)dT + \text{constant}\right]$$

The fact that the second term cannot be written as a derivative means that

$$\frac{nRT}{V}dV \neq d\left(\int \frac{nRT}{V}dV + \text{constant}\right)$$

because $T$ depends upon $V$. It is really a work term, so the evaluation of $w_{\text{rev}}$ depends upon the path. If we divide Equation 6.1 by $T$, however, we get a very interesting result:

$$\frac{\delta q_{\text{rev}}}{T} = \frac{C_V(T)dT}{T} + \frac{nR}{V}dV \tag{6.2}$$

Notice now that $\delta q_{\text{rev}}/T$ is an exact differential. The right side can be written in the form

$$d\left[\int \frac{C_V(T)}{T}dT + nR\int \frac{dV}{V} + \text{constant}\right]$$

so $\delta q_{\text{rev}}/T$ is the derivative of a state function that is a function of $T$ and $V$ (see also MathChapter D). If we let this state function be denoted by $S$, Equation 6.2 reads

$$dS = \frac{\delta q_{\text{rev}}}{T} \tag{6.3}$$

Notice that the inexact differential $\delta q_{\text{rev}}$ has been converted to an exact differential by multiplying it by $1/T$. In mathematical terms, we say that $1/T$ is an *integrating factor* of $\delta q_{\text{rev}}$.

The state function $S$ that we have described here is called the *entropy*. Because entropy is a state function, $\Delta S = 0$ for a cyclic process; that is, a process in which the final state is the same as the initial state. We can indicate this concept mathematically by writing

$$\oint dS = 0 \tag{6.4}$$

where the circle on the integral sign indicates a cyclic process. From Equation 6.3, we can also write

$$\oint \frac{\delta q_{\text{rev}}}{T} = 0 \tag{6.5}$$

Equation 6.5 is a statement of the fact that $\delta q_{rev}/T$ is the derivative of a state function. Although we proved Equation 6.5 only for the case of an ideal gas, it is generally true (Problem 6–5).

## 6–3. Unlike $q_{rev}$, Entropy Is a State Function

In the previous chapter, we calculated the reversible work and reversible heat for two processes that take place between the same initial and final states (Figure 6.3). The first process involved a reversible isothermal expansion of an ideal gas from $P_1, V_1, T_1$ to $P_2, V_2, T_1$ (path A). For this process (cf. Equations 5.12 and 5.13),

$$\delta q_{rev,A} = \frac{nRT_1}{V} dV \tag{6.6}$$

and so

$$q_{rev,A} = nRT_1 \ln \frac{V_2}{V_1}$$

The other process involved a reversible adiabatic expansion of an ideal gas from $P_1, V_1, T_1$ to $P_3, V_2, T_2$ (path B), followed by heating the gas reversibly at constant volume from $P_3, V_2, T_2$ to $P_2, V_2, T_1$ (path C). For this process (cf. Equations 5.15 and 5.17),

$$\delta q_{rev,B} = 0$$
$$\delta q_{rev,C} = C_V(T)dT \tag{6.7}$$

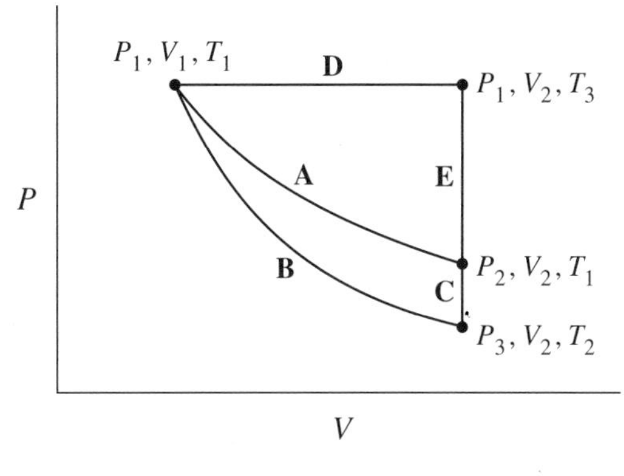


**FIGURE 6.3**
An illustration of three different paths (A, B+C, and D+E) from an initial state $P_1, V_1, T_1$ to a final state $P_2, V_2, T_1$ of an ideal gas. Path A represents a reversible isothermal expansion from $P_1, V_1$ to $P_2, V_2$. Path B+C represents a reversible adiabatic expansion (B) from $P_1, V_1, T_1$ to $P_3, V_2, T_2$ followed by reversibly heating the gas at constant volume (C) from $P_3, V_2, T_2$ to $P_2, V_2, T_1$. Path D+E represents a reversible expansion at constant pressure $P_1$ (D) from $P_1, V_1, T_1$ to $P_1, V_2, T_3$ followed by a reversible cooling at constant volume $V_2$ (E) from $P_1, V_2, T_3$ to $P_2, V_2, T_1$.

and

$$q_{\text{rev,B+C}} = \int_{T_2}^{T_1} C_V(T)dT$$

where $T_2$ is given by (cf. Equation 5.21)

$$\int_{T_1}^{T_2} \frac{C_V(T)}{T}dT = -nR\ln\frac{V_2}{V_1} \tag{6.8}$$

The point here is that $q_{\text{rev}}$ differs for the two paths, A and B + C, indicating that $q_{\text{rev}}$ is not a state function.

Now let's evaluate

$$\Delta S = \int_1^2 \frac{\delta q_{\text{rev}}}{T}$$

for these two paths. For path A from $P_1, T_1, V_1$ to $P_2, V_2, T_1$, we have, using Equation 6.6,

$$\Delta S_A = \int_1^2 \frac{\delta q_{\text{rev,A}}}{T_1} = \int_{V_1}^{V_2} \frac{1}{T_1}\frac{nRT_1}{V}dV$$

$$= nR\int_{V_1}^{V_2} \frac{dV}{V} = nR\ln\frac{V_2}{V_1} \tag{6.9}$$

For the reversible adiabatic expansion from $P_1, V_1, T_1$ to $P_3, V_2, T_2$ (path B) followed by a reversible heating at constant volume from $P_3, V_2, T_2$ to $P_2, V_2, T_1$ (path C), we have, using Equation 6.7,

$$\Delta S_B = \int_1^2 \frac{\delta q_{\text{rev,B}}}{T} = 0$$

and

$$\Delta S_C = \int_2^1 \frac{\delta q_{\text{rev,}C}}{T} = \int_{T_2}^{T_1} \frac{C_V(T)}{T}dT = -\int_{T_1}^{T_2} \frac{C_V(T)}{T}dT$$

But using Equation 6.8, $\Delta S_C$ turns out to be

$$\Delta S_C = nR\ln\frac{V_2}{V_1}$$

and so

$$\Delta S_{B+C} = \Delta S_B + \Delta S_C = 0 + nR\ln\frac{V_2}{V_1} = nR\ln\frac{V_2}{V_1} \tag{6.10}$$

Thus, we see that the $\Delta S_A$ (Equation 6.9) is equal to $\Delta S_{B+C}$ (Equation 6.10) and that the value of $\Delta S$ is independent of the path.

**EXAMPLE 6–1**

Calculate $q_{rev}$ and $\Delta S$ for a reversible expansion of an ideal gas at constant pressure $P_1$ from $T_1$, $V_1$ to $T_3$, $V_2$ (path D in Figure 6.3) followed by a reversible cooling of the gas at constant volume $V_2$ from $P_1$, $T_3$ to $P_2$, $T_1$ (path E).

SOLUTION: For path D (cf. Example 5–4),

$$\delta q_{\text{rev},D} = dU_D - \delta w_{\text{rev},D} = C_V(T)dT + P_1 dV \tag{6.11}$$

and so

$$q_{\text{rev},D} = \int_{T_1}^{T_3} C_V(T)dT + P_1(V_2 - V_1)$$

For path E, $\delta w_{\text{rev}} = 0$, and so

$$\delta q_{\text{rev},E} = dU_E = C_V(T)dT \tag{6.12}$$

and

$$q_{\text{rev},E} = \int_{T_3}^{T_1} C_V(T)dT$$

For the complete process (paths D + E),

$$q_{\text{rev},D+E} = q_{\text{rev},D} + q_{\text{rev},E} = P_1(V_2 - V_1)$$

To calculate $\Delta S$ for path D, we use Equation 6.11 to write

$$\begin{aligned}\Delta S_D &= \int \frac{\delta q_{\text{rev,D}}}{T} \\ &= \int_{T_1}^{T_3} \frac{C_V(T)}{T} dT + P_1 \int_{V_1}^{V_2} \frac{dV}{T}\end{aligned}$$

To evaluate the second integral here, we must know how $T$ varies with $V$ for this process. But this is given by $P_1 V = nRT$, so

$$\begin{aligned}\Delta S_D &= \int_{T_1}^{T_3} \frac{C_V(T)}{T} dT + nR \int_{V_1}^{V_2} \frac{dV}{V} \\ &= \int_{T_1}^{T_3} \frac{C_V(T)}{T} dT + nR \ln \frac{V_2}{V_1}\end{aligned}$$

For path E, $\delta w_{\text{rev}} = 0$, and using Equation 6.12 for $\delta q_{\text{rev},E}$ gives

$$\Delta S_E = \int \frac{\delta q_{\text{rev,E}}}{T} = \int_{T_3}^{T_1} \frac{C_V(T)}{T} dT$$

The value of $\Delta S$ for the complete process (paths D + E) is

$$\Delta S_{D+E} = \Delta S_D + \Delta S_E = nR \ln \frac{V_2}{V_1}$$

Notice that this is the very same result we obtained for paths A and B + C, once again suggesting that $S$ is a state function.

### EXAMPLE 6–2

We shall prove in Example 8–4 that similar to that found for an ideal gas, $U$ is a function of only the temperature for a gas that obeys the equation of state

$$P = \frac{RT}{\overline{V} - b}$$

where $b$ is a constant that reflects the size of the molecules. Calculate $q_{\text{rev}}$ and $\Delta S$ for both the paths A and B + C in Figure 6.3 for one mole of such a gas.

SOLUTION: Path A represents an isothermal expansion, so $dU_A = 0$ because $U$ depends only upon the temperature. Therefore,

$$\delta q_{\text{rev},A} = -\delta w_{\text{rev},A} = P d\overline{V} = \frac{RT}{\overline{V} - b} d\overline{V}$$

and

$$q_{\text{rev},A} = \int_{\overline{V}_1}^{\overline{V}_2} \frac{RT d\overline{V}}{\overline{V} - b} = RT \int_{\overline{V}_1}^{\overline{V}_2} \frac{d\overline{V}}{\overline{V} - b} = RT \ln \frac{\overline{V}_2 - b}{\overline{V}_1 - b}$$

The entropy change is given by

$$\Delta S_A = \int_1^2 \frac{\delta q_{\text{rev,A}}}{T} = R \int_{\overline{V}_1}^{\overline{V}_2} \frac{d\overline{V}}{\overline{V} - b} = R \ln \frac{\overline{V}_2 - b}{\overline{V}_1 - b}$$

For path B, a reversible adiabatic expansion, $q_{\text{rev},B} = 0$, so

$$\Delta S_B = 0$$

For path C, $\delta w_{\text{rev},C} = 0$, and

$$\delta q_{\text{rev},C} = dU_C = C_V(T) dT$$

and

$$q_{\text{rev},C} = \int_{T_2}^{T_1} C_V(T) dT$$

The molar entropy change is given by

$$\Delta \overline{S}_C = \int_{T_2}^{T_1} \frac{\delta q_{\text{rev},C}}{T} = \int_{T_2}^{T_1} \frac{\overline{C}_V(T)}{T} dT = -\int_{T_1}^{T_2} \frac{\overline{C}_V(T)}{T} dT$$

and so

$$\Delta \overline{S}_{B+C} = \Delta \overline{S}_B + \Delta \overline{S}_C = -\int_{T_1}^{T_2} \frac{\overline{C}_V(T)}{T} dT$$

But $T_2$, the temperature at the end of the reversible adiabatic expansion, can be found from

$$dU = \delta q_{\text{rev}} + \delta w_{\text{rev}}$$

Using the fact that $d\overline{U} = \overline{C}_V(T)dT$ and $\delta q_{\text{rev}} = 0$ gives

$$\overline{C}_V(T)dT = -Pd\overline{V} = -\frac{RT}{\overline{V} - b}d\overline{V}$$

Divide through by $T$ and integrate from the initial state to the final state to get

$$\int_{T_1}^{T_2} \frac{\overline{C}_V(T)}{T} dT = -R \int_{\overline{V}_1}^{\overline{V}_2} \frac{d\overline{V}}{\overline{V} - b} = -R \ln \frac{\overline{V}_2 - b}{\overline{V}_1 - b}$$

Substituting this result into the above expression for $\Delta\overline{S}_{B+C}$ gives

$$\Delta\overline{S}_{B+C} = R \ln \frac{\overline{V}_2 - b}{\overline{V}_1 - b}$$

Therefore, we see that even though $q_{\text{rev},A} \neq q_{\text{rev},B+C}$, nevertheless,

$$\Delta\overline{S}_A = \Delta\overline{S}_{B+C}$$

We will show several times in the following sections that the entropy is related to the disorder of a system, but for now, notice that if we add energy as heat to a system, then its entropy increases because its thermal disorder increases. Furthermore, notice that because $dS = \delta q_{\text{rev}}/T$, energy delivered as heat at a lower temperature contributes more to an entropy (disorder) increase than at a higher temperature. The lower the temperature, the lower the disorder, so the energy added as heat has proportionally more "order" to convert to "disorder."

## 6–4. The Second Law of Thermodynamics States That the Entropy of an Isolated System Increases as a Result of a Spontaneous Process

We all know that energy as heat will flow spontaneously from a region of high temperature to a region of low temperature. Let's investigate the role entropy plays in this process. Consider the two-compartment system shown in Figure 6.4, where parts A and B are large one-component systems. Both systems are at equilibrium, but they are not at equilibrium with each other. Let the temperatures of these two systems be $T_{\text{A}}$ and $T_{\text{B}}$. The two systems are separated from each other by a rigid, heat-conducting wall so that energy as heat can flow from one system to the other, but the two-compartment system itself is isolated. When we call a system *isolated*, we mean that the system is separated from its surroundings by rigid walls that do not allow matter or energy to pass through them. We may picture the walls as rigid, totally non-heat conducting,

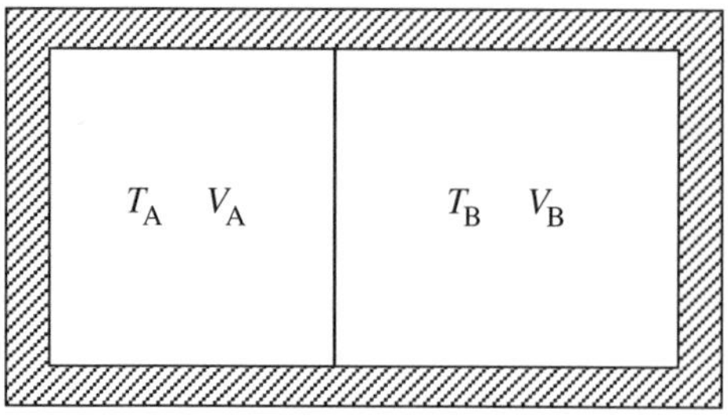


**FIGURE 6.4**
A two-compartment system in which A and B are large, one-component systems. Each system is at equilibrium, but they are not at equilibrium with each other. The two systems are separated from each other by a rigid, heat-conducting wall. The total two-compartment system itself is isolated.

and impervious to matter. Consequently, the system can do no work nor can work be done on the system, nor can it exhange energy as heat with the surroundings. The two-compartment system is described by the equations

$$\begin{gathered} U_A + U_B = \text{constant} \\ V_A = \text{constant} \qquad V_B = \text{constant} \\ S = S_A + S_B \end{gathered} \tag{6.13}$$

Because $V_A$ and $V_B$ are fixed, we have for each separate system

$$\begin{aligned} dU_A &= \delta q_{rev} + \delta w_{rev} = T_A dS_A \quad (dV_A = 0) \\ dU_B &= \delta q_{rev} + \delta w_{rev} = T_B dS_B \quad (dV_B = 0) \end{aligned} \tag{6.14}$$

The entropy change of the two-compartment system is given by

$$\begin{aligned} dS &= dS_A + dS_B \\ &= \frac{dU_A}{T_A} + \frac{dU_B}{T_B} \end{aligned} \tag{6.15}$$

But $dU_A = -dU_B$ because the two-compartment system is isolated, so we have

$$dS = dU_B \left( \frac{1}{T_B} - \frac{1}{T_A} \right) \tag{6.16}$$

Experimentally, we know that if $T_B > T_A$, then $dU_B < 0$ (energy as heat flows from system B to system A), in which case $dS > 0$. Similarly, $dS > 0$ if $T_B < T_A$ because $dU_B > 0$ in this case (energy as heat flows from system B to system A). We may interpret this result by saying that the spontaneous flow of energy as heat from a body at a higher temperature to a body at a lower temperature is governed by the condition $dS > 0$. If $T_A = T_B$, then the two-compartment system is in equilibrium and $dS = 0$.

We can generalize this result by investigating the role entropy plays in governing the direction of any spontaneous process. To be able to focus on the entropy alone, we will consider an infinitesimal spontaneous change in an isolated system. We choose an isolated system because the energy remains constant in an isolated system, and we wish to separate the effect due to a change in energy from the effect due to a change in entropy. Because the energy remains constant, the driving force for any spontaneous process in an isolated system must be due to an increase in entropy, which we can express mathematically by $dS > 0$. Because the system is isolated, this increase in entropy must be created within the system itself. Unlike energy, entropy is not necessarily conserved; it increases whenever a spontaneous process takes place. In fact, the entropy of an isolated system will continue to increase until no more spontaneous processes occur, in which case the system will be in equilibrium (Figure 6.5). Thus, we conclude that the entropy of an isolated system is a maximum when the system is in equilibrium. Consequently, $dS = 0$ at equilibrium. Furthermore, not only is $dS = 0$ in an isolated system at equilibrium, but $dS = 0$ for any reversible process in an isolated system because, by definition, a reversible process is one in which the system remains essentially in equilibrium during the entire process. To summarize our conclusions thus far, then, we write

$$\begin{aligned} dS &> 0 \quad \text{(spontaneous process in an isolated system)} \\ dS &= 0 \quad \text{(reversible process in an isolated system)} \end{aligned} \tag{6.17}$$

Because we have considered an isolated system, no energy as heat can flow in or out of the system. For other types of systems, however, energy as heat can flow in or out, and it is convenient to view $dS$ in any spontaneous infinitesimal process as consisting of two parts. One part of $dS$ is the entropy created by the irreversible process itself, and the other part is the entropy due to the energy as heat exchanged between the system and its surroundings. These two contributions account for the entire change in entropy. We will denote the part of $dS$ that is created by the irreversible process by $dS_{\text{prod}}$ because it is *produced* by the system. This quantity is always positive. We will denote the part of $dS$ that is due to the exchange of energy as heat with the surroundings by $dS_{\text{exch}}$ because it is due to *exchange*. This quantity is given by $\delta q/T$, and it can be positive, negative, or zero. Note that $\delta q$ need not be $\delta q_{\text{rev}}$. The quantity $\delta q$ will be $\delta q_{\text{rev}}$

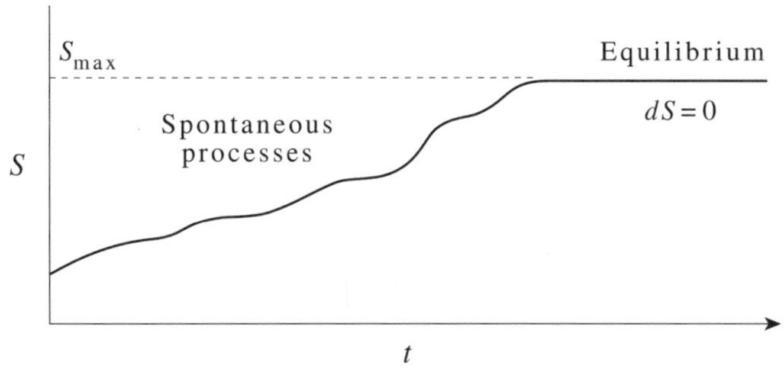


**FIGURE 6.5**
A schematic plot of entropy versus time for an isolated system. The entropy increases ($dS > 0$) until no more spontaneous processes occur, in which case the system is in equilibrium, and $dS = 0$.

if the exchange is reversible and $\delta q_{\text{irr}}$ if the exchange is irreversible. Thus, we write for *any* process

$$\begin{aligned} dS &= dS_{\text{prod}} + dS_{\text{exch}} \\ &= dS_{\text{prod}} + \frac{\delta q}{T} \end{aligned} \tag{6.18}$$

For a reversible process, $\delta q = \delta q_{\text{rev}}$, $dS_{\text{prod}} = 0$, so

$$dS = \frac{\delta q_{\text{rev}}}{T} \tag{6.19}$$

in agreement with Equation 6.3. For an irreversible or spontaneous process, $dS_{\text{prod}} > 0$, $dS_{\text{exch}} = \delta q_{\text{irr}}/T$, and so

$$dS > \frac{\delta q_{\text{irr}}}{T} \tag{6.20}$$

Equations 6.19 and 6.20 can be written as one equation,

$$dS \geq \frac{\delta q}{T} \tag{6.21}$$

or

$$\Delta S \geq \int \frac{\delta q}{T} \tag{6.22}$$

where the equality sign holds for a reversible process and the inequality sign holds for an irreversible process. Equation 6.22 is one of a number of ways of expressing the Second Law of Thermodynamics and is called the *Inequality of Clausius.*

A formal statement of the Second Law of Thermodynamics is as follows:

> There is a thermodynamic state function of a system called the entropy, $S$, such that for any change in the thermodynamic state of the system,
>
> $$dS \geq \frac{\delta q}{T}$$
>
> where the equality sign applies if the change is carried out reversibly and the inequality sign applies if the change is carried out irreversibly at any stage.

We can use Equation 6.22 to prove quite generally that the entropy of an isolated system always increases during a spontaneous (irreversible) process or that $\Delta S > 0$. Consider a cyclic process (Figure 6.6) in which a system is first isolated and undergoes an irreversible process from state 1 to state 2. Now let the system interact with its surroundings and return it to state 1 by any reversible path. Because $S$ is a state function, $\Delta S = 0$ for this cyclic process, so according to Equation 6.22,

$$\Delta S = 0 > \int_1^2 \frac{\delta q_{\text{irr}}}{T} + \int_2^1 \frac{\delta q_{\text{rev}}}{T}$$

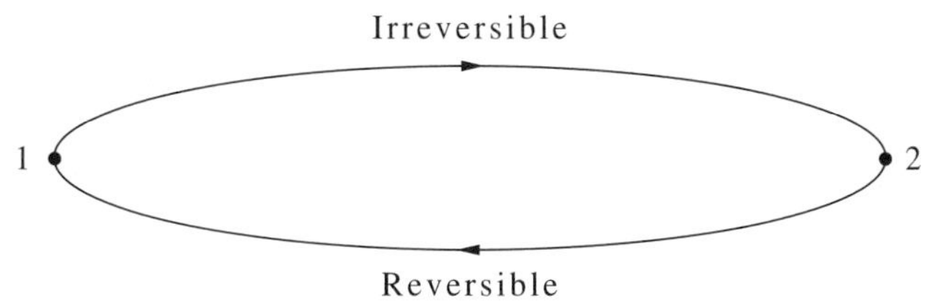


**FIGURE 6.6**
A cyclic process in which the system is first isolated and undergoes an irreversible process from state 1 to state 2. Then the system is allowed to interact with its surroundings and is brought back to state 1 by some reversible path. Because entropy is a state function, $\Delta S = 0$ for a cyclic process.

The inequality applies because the cyclic process is irreversible from 1 to 2. The first integral here equals zero because the system is isolated, i.e., $\delta q_{\text{irr}} = 0$. The second integral is by definition equal to $S_1 - S_2$, so we have $0 > S_1 - S_2$. Because the final state is state 2 and the initial state is state 1,

$$\Delta S = S_2 - S_1 > 0$$

Thus we see that the entropy increases when the isolated system goes from state 1 to state 2 by a general irreversible process.

Because the universe itself may be considered to be an isolated system and all naturally occurring processes are irreversible, one statement of the Second Law of Thermodynamics says that the entropy of the universe is constantly increasing. In fact, Clausius summarized the first two laws of thermodynamics by

> The energy of the Universe is constant;
>
> the entropy is tending to a maximum.

## 6–5. The Most Famous Equation of Statistical Thermodynamics Is $S = k_\text{B} \ln W$

In this section, we will discuss the molecular interpretation of entropy more quantitatively than we have up to now. We have shown that entropy is a state function that is related to the disorder of a system. Disorder can be expressed in a number of ways, but the way that has turned out to be the most useful is the following. Consider an ensemble of $\mathcal{A}$ isolated systems, each with energy $E$, volume $V$, and number of particles $N$. Realize that whatever the value of $E$, it must be an eigenvalue of the Schrödinger equation for the system. As we discussed in Chapter 3, the energy is a function of $N$ and $V$, so we can write $E = E(N, V)$ (see, for example, Equations 3.2 and 3.3). Although all the systems have the same energy, they may be in different quantum states because of degeneracy. Let the degeneracy associated with the energy $E$ be $\Omega(E)$, so that we can label the $\Omega(E)$ degenerate quantum states by $j = 1, 2, \ldots, \Omega(E)$. (The degeneracies of systems that consist of $N$ particles turn out to be enormous; they are numbers of the order of $e^N$ for energies not too close to the ground-state energy.) Now, let $a_j$ be the number of systems in the ensemble that are in the state $j$. Because the $\mathcal{A}$ systems of the ensemble are distinguishable, the number of ways of having $a_1$ systems in state 1, $a_2$ systems in state 2, etc. is given by (MathChapter E)

$$W(a_1, a_2, a_3, \ldots) = \frac{\mathcal{A}!}{a_1!a_2!a_3!\ldots} = \frac{\mathcal{A}!}{\prod_j a_j!} \tag{6.23}$$

with

$$\sum_j a_j = \mathcal{A}$$

If all $\mathcal{A}$ systems are in one particular state (a totally ordered arrangement), say state 1, then $a_1 = \mathcal{A}$, $a_2 = a_3 = \cdots = 0$ and $W = 1$, which is the smallest value $W$ can have. In the other extreme, when all the $a_j$ are equal (a disordered arrangement), $W$ takes on its largest value (Problem E–10). Therefore, $W$ can be taken to be a quantitative measure of the disorder of a system. We will not set the entropy proportional to $W$, however, but to $\ln W$ according to

$$S = k_B \ln W \tag{6.24}$$

where $k_B$ is the Boltzmann constant. Note that $S = 0$ for a completely ordered system ($a_1 = 1,\ a_2 = a_3 = \cdots 0$) and achieves a maximum value for a completely disordered system ($a_1 = a_2 = a_3 = \cdots$). Equation 6.24 was formulated by Boltzmann and is the most famous equation of statistical thermodynamics. In fact, this equation is the only inscription on a monument to Boltzmann in the central cemetary in Vienna. It gives us a quantitative relation between the thermodynamic quantity, entropy, and the statistical quantity, $W$.

We set $S$ equal to $\ln W$ rather than $W$ for the following reason. We want $S$ to be such that the total entropy of a system that is made up of two parts (say A and B) is given by

$$S_{\text{total}} = S_A + S_B$$

In other words, we want $S$ to be an extensive state function. Now if $W_A$ is the value of $W$ for system A and $W_B$ is the value of $W$ for system B, $W_{AB}$ for the composite system is given by

$$W_{AB} = W_A W_B$$

The entropy of the composite system is

$$\begin{aligned} S_{AB} &= k_B \ln W_{AB} = k_B \ln W_A W_B = k_B \ln W_A + k_B \ln W_B \\ &= S_A + S_B \end{aligned}$$

An alternate form of Equation 6.24 expresses $S$ in terms of the degeneracy $\Omega$. We can determine this expression in the following way. Given no other information, there is no reason to choose one of the $\Omega$ degenerate quantum states over any other; each one should occur in an ensemble with equal probability (this concept is actually one of the postulates of statistical thermodynamics). Consequently, we expect that the ensemble of isolated systems should contain equal numbers of systems in each quantum state.

Because $S$ is a maximum for an isolated system at equilibrium, $W$ must also be a maximum. The value of $W$ is maximized when all the $a_j$ are equal (Problem E–10). Let the total number of systems in the ensemble be $\mathcal{A} = n\Omega$ and let each $a_j = n$, so that the set of $\Omega$ degenerate quantum states is replicated $n$ times in the ensemble. (We will never need the value of $n$.) Using Stirling's approximation (MathChapter E) in Equation 6.23, we get

$$\begin{aligned} S_{\text{ensemble}} &= k_B \ln W = k_{\text{B}}[\mathcal{A} \ln \mathcal{A} - \sum_{j=1}^{\Omega} a_j \ln a_j] \\ &= k_{\text{B}}[n\Omega \ln(n\Omega) - \sum_{j=1}^{\Omega} (n \ln n)] = k_B[n\Omega \ln(n\Omega) - \Omega(n \ln n)] \\ &= k_{\text{B}}(n\Omega \ln \Omega) \end{aligned}$$

The entropy of a typical system in the ensemble is given by $S_{\text{ensemble}} = \mathcal{A}S_{\text{system}} = n\Omega S_{\text{system}}$, and so

$$S = k_{\text{B}} \ln \Omega \tag{6.25}$$

where we have dropped the subscript, system. Equation 6.25 is an alternate form of Equation 6.24 and relates entropy to disorder. As a concrete example, consider a system of $N$ (distinguishable) spins (or dipoles) that can be oriented in one of two possible directions with equal probability. Then, each spin has a degeneracy of 2 associated with it, and the degeneracy of the $N$ spins is $2^N$. The entropy of this system is $Nk_{\text{B}} \ln 2$. We will use this result when we discuss the entropy of carbon monoxide at 0 K in Section 7–8.

As another example of the use of Equation 6.25, Problem 6–23 has you show that

$$\Omega(E) = c(N) f(E) V^N$$

for an ideal gas of $N$ particles, where $c(N)$ is a function of $N$ and $f(E)$ is a function of the energy. Now let's determine $\Delta S$ for an isothermal expansion of one mole of an ideal gas from a volume $V_1$ to $V_2$.

$$\begin{aligned} \Delta S &= k_{\text{B}} \ln \Omega_2 - k_{\text{B}} \ln \Omega_1 \\ &= k_{\text{B}} \ln \frac{\Omega_2}{\Omega_1} = k_{\text{B}} \ln \frac{c(N) f(E_2) V_2^N}{c(N) f(E_1) V_1^N} \end{aligned}$$

But $f(E_1) = f(E_2)$ in this case because we are considering an isothermal expansion of an ideal gas, so $E_2 = E_1$. Therefore, we have for one mole

$$\Delta\overline{S} = N_A k_B \ln \frac{V_2}{V_1} = R \ln \frac{V_2}{V_1}$$

in agreement with Equation 6.9.

**EXAMPLE 6–3**

Use the fact that

$$\Omega(E) = c(N) f(E) V^N$$

for an ideal gas to show that the change in entropy (per mole) when two gases are mixed isothermally is given by

$$\Delta_{\text{mix}}\overline{S}/R = -y_1 \ln y_1 - y_2 \ln y_2 \tag{6.26}$$

where $y_1$ and $y_2$ are the mole fractions of the two gases.

SOLUTION: Consider the process depicted in Figure 6.2. Then $\Delta_{\text{mix}} S$ is given by

$$\begin{aligned}\Delta_{\text{mix}} S &= S_{\text{mixture}} - S_1 - S_2 \\ &= k_B \ln \frac{\Omega_{\text{mixture}}}{\Omega_1 \Omega_2}\end{aligned}$$

where 1 and 2 refer to $N_2(g)$ and $Br_2(g)$, respectively. The quantities $\Omega_1$ and $\Omega_2$ are given by

$$\Omega_1 = c(N_1) f(E_1) V_1^{N_1} \quad \text{and} \quad \Omega_2 = c(N_2) f(E_2) V_2^{N_2}$$

Because the molecules in a mixture of ideal gases are independent of each other,

$$\Omega_{\text{mixture}} = c(N_1) f(E_1)(V_1 + V_2)^{N_1} \times c(N_2) f(E_2)(V_1 + V_2)^{N_2}$$

Substitute these expressions for $\Omega_{N_2}$, $\Omega_{Br_2}$, and $\Omega_{\text{mixture}}$ into the above equation for $\Delta_{\text{mix}} S$ to get

$$\begin{aligned}\Delta_{\text{mix}} S &= k_B \ln \frac{(V_1 + V_2)^{N_1}}{V_1^{N_1}} \cdot \frac{(V_1 + V_2)^{N_2}}{V_2^{N_2}} \\ &= -k_B N_1 \ln \left( \frac{V_1}{V_1 + V_2} \right) - k_B N_2 \ln \left( \frac{V_2}{V_1 + V_2} \right)\end{aligned}$$

Because $V$ is proportional to $n$ for an ideal gas,

$$\frac{V_1}{V_1 + V_2} = \frac{n_1}{n_1 + n_2} = y_1 \quad \text{and} \quad \frac{V_2}{V_1 + V_2} = \frac{n_2}{n_1 + n_2} = y_2$$

so we have

$$\Delta_{\text{mix}} S = -k_B N_1 \ln y_1 - k_B N_2 \ln y_2$$
$$= -Rn_1 \ln y_1 - Rn_2 \ln y_2$$

Now, finally divide by $n_1 + n_2$ and $R$ to get

$$\Delta_{\text{mix}} \overline{S}/R = -y_1 \ln y_1 - y_2 \ln y_2$$

Note that $\Delta_{\text{mix}} \overline{S}$ is always a positive quantity because $y_1$ and $y_2$, being mole fractions, are always less than one. Thus, the isothermal mixing of two (ideal) gases is a spontaneous process. We will derive Equation 6.26 using classical thermodynamics in the next section.

## 6–6. We Must Always Devise a Reversible Process to Calculate Entropy Changes

The discussion so far has been fairly abstract, and it will be helpful at this point to illustrate the change of entropy in a spontaneous process by means of some calculations involving an ideal gas for simplicity. First, let's consider the situation in Figure 6.1, in which an ideal gas at $T$ and $V_1$ is allowed to expand into a vacuum to a total volume of $V_2$. We use Equation 6.19 even though this is *not* a reversible process. Remember that because the entropy is a state function, it depends only upon the initial and final states and not upon the path between them. Equation 6.19 tells us that we can calculate $\Delta S$ by integrating $\delta q_{\text{rev}}/T$ over a reversible path,

$$\Delta S = \int_1^2 \frac{\delta q_{\text{rev}}}{T} \tag{6.27}$$

regardless of whether the process is reversible or not. Even though the irreversible process occurs adiabatically, we use a reversible path to calculate the entropy change from the state $T$, $V_1$ to $T$, $V_2$. This path will not represent the actual adiabatic process, which does not matter because we are interested in only the entropy change between the initial state and the final state. To calculate $\Delta S$, then, we start with

$$\delta q_{\text{rev}} = dU - \delta w_{\text{rev}}$$

But $dU = 0$ for the expansion of an ideal gas into a vacuum because $U$ depends upon only temperature and is independent of volume for an ideal gas. Therefore, we have $\delta q_{\text{rev}} = -\delta w_{\text{rev}}$. The reversible work is given by

$$\delta w_{\text{rev}} = -PdV = -\frac{nRT}{V}dV$$

so

$$\Delta S = \int_1^2 \frac{\delta q_{\text{rev}}}{T} = -\int_1^2 \frac{\delta w_{\text{rev}}}{T} = nR \int_{V_1}^{V_2} \frac{dV}{V} = nR \ln \frac{V_2}{V_1} \tag{6.28}$$

Note that $\Delta S > 0$ because $V_2 > V_1$. Thus, we see that the entropy increases in the expansion of an ideal gas into a vacuum.

Because Equation 6.19 tells us to calculate $\Delta S$ by expanding the gas reversibly and isothermally from $V_1$ to $V_2$, Equation 6.28 holds for the reversible isothermal expansion. Because $S$ is a state function, however, the value of $\Delta S$ obtained from Equation 6.28 is the same as the value of $\Delta S$ for the irreversible isothermal expansion from $V_1$ to $V_2$. How, then, do a reversible and an irreversible isothermal expansion differ? The answer lies in the value of $\Delta S$ for the surroundings. (Remember that the condition $\Delta S \geq 0$ applies to an isolated system. If the system is not isolated, then the condition $\Delta S \geq 0$ applies to the sum of the entropy changes in the system and its surroundings, in other words, the entire universe.)

Let's look at the entropy change of the surroundings, $\Delta S_{\text{surr}}$, for both a reversible and an irreversible isothermal expansion. During the reversible expansion, $\Delta U = 0$ (the process is isothermal and the gas is ideal) and the gas absorbs a quantity of energy as heat, $q_{\text{rev}} = -w_{\text{rev}} = nRT \ln V_2/V_1$, from its surroundings. The entropy of the surroundings, therefore, decreases according to

$$\Delta S_{\text{surr}} = -\frac{q_{\text{rev}}}{T} = -nRT \ln \frac{V_2}{V_1}$$

The total entropy change is given by

$$\Delta S_{\text{total}} = \Delta S_{\text{sys}} + \Delta S_{\text{surr}} = nR \ln \frac{V_2}{V_1} - nR \ln \frac{V_2}{V_1} = 0$$

as it should be because the entire process is carried out reversibly.

In the irreversible expansion, $\Delta U = 0$ (the process is isothermal and the gas is ideal). The value of $P_{\text{ext}}$ is also zero, so $w_{\text{irr}} = 0$ and therefore, $q_{\text{irr}} = 0$. No energy as heat is delivered to the system by the surroundings and so

$$\Delta S_{\text{surr}} = 0$$

Thus, the total entropy change is given by

$$\Delta S_{\text{total}} = \Delta S_{\text{sys}} + \Delta S_{\text{surr}} = nR \ln \frac{V_2}{V_1} + 0 = nR \ln \frac{V_2}{V_1}$$

and so $\Delta S > 0$ as we expect for an irreversible process.

Did we use $q_{\text{irr}} = 0$ to calculate $\Delta S_{\text{surr}}$ in this process? We actually did because no work was done by the process. In the general case of an isothermal process in which no work is done ($\delta w = 0$), the process is one of pure heat transfer and $dU = \delta q = dq$, where $dq$ is an exact differential because $U$ is a state function. Therefore, $q$ is path independent and so we can use $q_{\text{irr}}$ to calculate the entropy is this particular case.

**EXAMPLE 6–4**

In Example 6–2 we stated that $U$ is a function of only the temperature for a gas that obeys the equation of state

$$P = \frac{RT}{\overline{V} - b}$$

where $b$ is a constant that reflects the size of the molecules. Calculate $\Delta\overline{S}$ when one mole of such a gas at $T$ and $\overline{V}_1$ is allowed to expand into a vacuum to a total volume of $\overline{V}_2$.

SOLUTION: We start with

$$\delta q_{\text{rev}} = dU - \delta w_{\text{rev}}$$

Because $U$ is a function of only the temperature, and hence is independent of the volume, $dU = 0$ for the expansion. Therefore,

$$\delta q_{\text{rev}} = -\delta w_{\text{rev}} = P d\overline{V} = \frac{RT}{\overline{V} - b} d\overline{V}$$

and

$$\Delta\overline{S} = \int_1^2 \frac{\delta q_{\text{rev}}}{T} = R \int_{\overline{V}_1}^{\overline{V}_2} \frac{d\overline{V}}{\overline{V} - b} = R \ln \frac{\overline{V}_2 - b}{\overline{V}_1 - b}$$

Once again, the entropy increases when a gas expands into a vacuum.

Let's look at the mixing of two ideal gases, as depicted in Figure 6.2. Because the two gases are ideal, each acts independently of the other. Thus, we can consider each gas separately to expand from $V_{\text{initial}}$ to $V_{\text{final}}$. For nitrogen, we have (using Equation 6.28)

$$\Delta S_{N_2} = n_{N_2} R \ln \frac{V_{N_2} + V_{Br_2}}{V_{N_2}} = -n_{N_2} R \ln \frac{V_{N_2}}{V_{N_2} + V_{Br_2}}$$

and for bromine,

$$\Delta S_{Br_2} = n_{Br_2} R \ln \frac{V_{N_2} + V_{Br_2}}{V_{Br_2}} = -n_{Br_2} R \ln \frac{V_{Br_2}}{V_{N_2} + V_{Br_2}}$$

The total entropy change is

$$\begin{aligned}\Delta S &= \Delta S_{N_2} + \Delta S_{Br_2} \\ &= -n_{N_2} R \ln \frac{V_{N_2}}{V_{N_2} + V_{Br_2}} - n_{Br_2} R \ln \frac{V_{Br_2}}{V_{N_2} + V_{Br_2}}\end{aligned}$$

Because $V$ is proportional to $n$ for an ideal gas, we can write the above equation as

$$\Delta S = -n_{N_2} R \ln \frac{n_{N_2}}{n_{N_2} + n_{Br_2}} - n_{Br_2} R \ln \frac{n_{Br_2}}{n_{N_2} + n_{Br_2}} \tag{6.29}$$

If we divide both sides by the total number of moles, $n_{\text{total}} = n_{N_2} + n_{Br_2}$ and introduce mole fractions

$$y_{N_2} = \frac{n_{N_2}}{n_{\text{total}}} \quad \text{and} \quad y_{Br_2} = \frac{n_{Br_2}}{n_{\text{total}}}$$

then Equation 6.29 becomes

$$\Delta_{\text{mix}}\overline{S}/R = -y_{N_2} \ln y_{N_2} - y_{Br_2} \ln y_{Br_2}$$

More generally, $\Delta_{\text{mix}}\overline{S}$ for the isothermal mixing of $N$ ideal gases is given by

$$\Delta_{\text{mix}}\overline{S} = -R \sum_{j=1}^{N} y_j \ln y_j \tag{6.30}$$

in agreement with Equation 6.26. Equation 6.30 says that $\Delta_{\text{mix}}\overline{S} > 0$ because the arguments of the logarithms are less than unity. Thus, Equation 6.30 shows that there is an increase in entropy whenever ideal gases mix isothermally.

Last, let's consider $\Delta S$ when two equal sized pieces of the same metal at different temperatures, $T_h$ and $T_c$, are brought into thermal contact and then isolated from their surroundings. Clearly, the two pieces of metal will come to the same final temperature, $T$, which can be calculated by

$$\text{heat lost by hotter piece} = \text{heat gained by colder piece}$$

$$C_V(T_h - T) = C_V(T - T_c)$$

Solving for $T$ gives

$$T = \frac{T_h + T_c}{2}$$

We now will calculate the entropy change for each piece of metal. Remember that we must calculate $\Delta S$ along a reversible path, even though the actual process is irreversible. As usual, we use Equation 6.19,

$$dS = \frac{\delta q_{\text{rev}}}{T}$$

There is essentially no work done, so $\delta q_{\text{rev}} = dU = C_V dT$. Therefore,

$$\Delta S = \int_{T_1}^{T_2} \frac{C_V dT}{T}$$

If we take $C_V$ to be constant from $T_1$ to $T_2$, then

$$\Delta S = C_V \ln \frac{T_2}{T_1} \tag{6.31}$$

Now, for the initially hotter piece, $T_1 = T_{\text{h}}$ and $T_2 = (T_{\text{h}} + T_{\text{c}})/2$, and so

$$\Delta S_{\text{h}} = C_V \ln \frac{T_{\text{h}} + T_{\text{c}}}{2T_{\text{h}}}$$

Similarly,

$$\Delta S_{\text{c}} = C_V \ln \frac{T_{\text{h}} + T_{\text{c}}}{2T_{\text{c}}}$$

The total change in entropy is given by

$$\begin{aligned} \Delta S &= \Delta S_{\text{h}} + \Delta S_{\text{c}} \\ &= C_V \ln \frac{(T_{\text{h}} + T_{\text{c}})^2}{4T_{\text{h}}T_{\text{c}}} \end{aligned} \tag{6.32}$$

We will now prove that $(T_{\text{h}} + T_{\text{c}})^2 > 4T_{\text{h}}T_{\text{c}}$, and that $\Delta S > 0$. Start with

$$(T_{\text{h}} - T_{\text{c}})^2 = T_{\text{h}}^2 - 2T_{\text{h}}T_{\text{c}} + T_{\text{c}}^2 > 0$$

Add $4T_{\text{h}}T_{\text{c}}$ to both sides and obtain

$$T_{\text{h}}^2 + 2T_{\text{h}}T_{\text{c}} + T_{\text{c}}^2 = (T_{\text{h}} + T_{\text{c}})^2 > 4T_{\text{h}}T_{\text{c}}$$

Therefore, the value of the argument of the logarithm in Equation 6.32 is greater than one, so we see that $\Delta S > 0$ in this irreversible process.

**EXAMPLE 6–5**

The constant-pressure molar heat capacity of $O_2(g)$ from 300 K to 1200 K is given by

$$\overline{C}_P(T)/\text{J}\cdot\text{K}^{-1}\cdot\text{mol}^{-1} = 25.72 + (12.98 \times 10^{-3}\ \text{K}^{-1})T - (38.62 \times 10^{-7}\ \text{K}^{-2})T^2$$

where $T$ is in kelvins. Calculate the value of $\Delta\overline{S}$ when one mole of $O_2(g)$ is heated at constant pressure from 300 K to 1200 K.

SOLUTION: As usual, we start with Equation 6.19

$$dS = \frac{\delta q_{\text{rev}}}{T}$$

In this case, $\delta q_{\text{rev}} = \overline{C}_P(T)dT$, so

$$\Delta \overline{S} = \int_{T_1}^{T_2} \frac{\overline{C}_P(T)}{T} dT$$

Using the given expression for $\overline{C}_P(T)$, we have

$$\begin{aligned}\Delta \overline{S}/\text{J}\cdot\text{K}^{-1}\cdot\text{mol}^{-1} &= \int_{300\text{ K}}^{1200\text{ K}} \frac{25.72}{T} dT + \int_{300\text{ K}}^{1200\text{ K}} (12.98 \times 10^{-3}\text{ K}^{-1})dT \\ &\quad - \int_{300\text{ K}}^{1200\text{ K}} (38.62 \times 10^{-7}\text{ K}^{-2})TdT \\ &= 25.72 \ln \frac{1200\text{ K}}{300\text{ K}} + (12.98 \times 10^{-3}\text{ K}^{-1})(900\text{ K}) \\ &\quad - (38.62 \times 10^{-7}\text{ K}^{-2})\left[(1200\text{ K})^2 - (300\text{ K})^2\right]/2 \\ &= 35.66 + 11.68 - 2.61 = 44.73\end{aligned}$$

Note the increase in entropy due to the increased thermal disorder.

## 6–7. Thermodynamics Gives Us Insight into the Conversion of Heat into Work

The concept of entropy and the Second Law of Thermodynamics was first developed by a French engineer named Sadi Carnot in the 1820s in a study of the efficiency of the newly developed steam engines and other types of heat engines. Although primarily of historical interest to chemists, the result of Carnot's analysis is still worth knowing. Basically, a steam engine works in a cyclic manner; in each cycle, it withdraws energy as heat from some high-temperature thermal reservoir, uses some of this energy to do work, and then discharges the rest of the energy as heat to a lower-temperature thermal reservoir. A schematic representation of a heat engine is shown in Figure 6.7. The maximum amount of work will be obtained if the cyclic process is carried out reversibly. Of course, the maximum amount of work cannot be acheived in practice because the reversible path is an idealized process, but the results will give us a measure of the maximum efficiency that can be expected. Because the process is cyclic and reversible,

$$\Delta U_{\text{engine}} = w + q_{\text{rev,h}} + q_{\text{rev,c}} = 0 \tag{6.33}$$

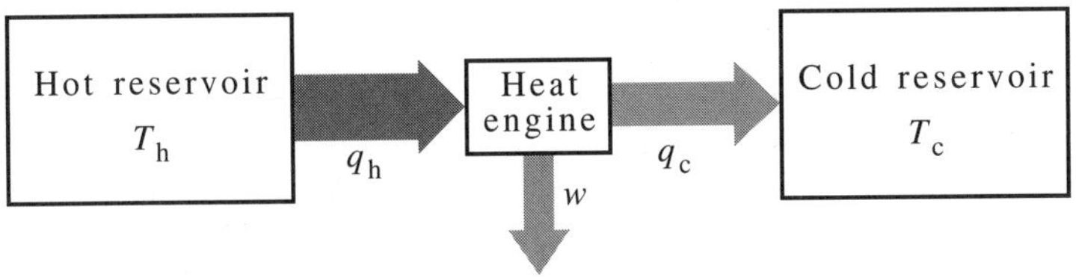


**FIGURE 6.7**
A highly schematic illustration of a heat engine. Energy as heat ($q_h$) is withdrawn from a high-temperature thermal reservoir at temperature $T_h$. The engine does work ($w$) and delivers an amount of energy as heat ($q_c$) to the lower-temperature reservoir at temperature $T_c$.

and

$$\Delta S_{\text{engine}} = \frac{\delta q_{\text{rev,h}}}{T_h} + \frac{\delta q_{\text{rev,c}}}{T_c} = 0 \tag{6.34}$$

where $\delta q_{\text{rev,h}}$ is the energy withdrawn reversibly as heat from the high-temperature reservoir at temperature $T_h$, and $\delta q_{\text{rev,c}}$ is the energy discharged reversibly as heat to the lower-temperature reservoir at temperature $T_c$. Note that the sign convention for energy transferred as heat means that $\delta q_{\text{rev,h}}$ is a positive quantity and that $\delta q_{\text{rev,c}}$ is a negative quantity. From Equation 6.33, we have that the work done by the engine is

$$-w = q_{\text{rev,h}} + q_{\text{rev,c}}$$

The work done by the engine is a negative quantity, so $-w$ is a positive quantity. We can define the efficiency of the process by the ratio of the work done by the engine divided by the amount of energy withdrawn as heat from the hot reservoir, or

$$\text{maximum efficiency} = \frac{-w}{q_{\text{rev,h}}} = \frac{q_{\text{rev,h}} + q_{\text{rev,c}}}{q_{\text{rev,h}}}$$

Equation 6.34 says that $q_{\text{rev,c}} = -q_{\text{rev,h}}(T_c/T_h)$, so the efficiency can be written as

$$\text{maximum efficiency} = 1 - \frac{T_c}{T_h} = \frac{T_h - T_c}{T_h} \tag{6.35}$$

Equation 6.35 is really a remarkable result because it is independent of the specific design of the engine or of the working substance. For a heat engine working between 373 K and 573 K, the maximum possible efficiency is

$$\text{maximum efficiency} = \frac{200}{573} = 35\%$$

In practice, the efficiency would be less due to factors such as friction. Equation 6.35 indicates that a greater efficiency is obtained by engines working with a higher value of $T_h$ or a lower value of $T_c$.

Note that the efficiency equals zero if $T_h = T_c$, which says that no net work can be obtained from an isothermal cyclic process. This conclusion is known as Kelvin's statement of the Second Law. A closed system operating in an isothermal cyclic manner cannot convert heat into work without some accompanying change in the surroundings.

## 6–8. Entropy Can Be Expressed in Terms of a Partition Function

We presented the equation $S = k_B \ln W$ in Section 6–5. This equation can be used as the starting point to derive most of the important results of statistical thermodynamics. For example, we can use it to derive an expression for the entropy in terms of the system partition function, $Q(N, V, \beta)$, as we have for the energy and the pressure:

$$U = k_B T^2 \left(\frac{\partial \ln Q}{\partial T}\right)_{N,V} = -\left(\frac{\partial \ln Q}{\partial \beta}\right)_{N,V} \tag{6.36}$$

and

$$P = k_B T \left(\frac{\partial \ln Q}{\partial V}\right)_{N,T} \tag{6.37}$$

Substitute Equation 6.23 into Equation 6.24 and then use Stirling's approximation for the factorials (MathChapter E) to get

$$\begin{aligned} S_{\text{ensemble}} &= k_B \ln \frac{\mathcal{A}!}{\prod_j a_j!} = k_B \ln \mathcal{A}! - k_B \sum_j \ln a_j! \\ &= k_B \mathcal{A} \ln \mathcal{A} - k_B \mathcal{A} - k_B \sum_j a_j \ln a_j + k_B \sum_j a_j \\ &= k_B \mathcal{A} \ln \mathcal{A} - k_B \sum_j a_j \ln a_j \end{aligned} \tag{6.38}$$

where we have used the fact that $\sum a_j = \mathcal{A}$ and have subscripted $S$ with "ensemble" to emphasize that it is the entropy of the entire ensemble of $\mathcal{A}$ systems. The entropy of a typical system is given by $S_{\text{system}} = S_{\text{ensemble}}/\mathcal{A}$. If we use the fact that the probability of finding a system in the $j$th quantum state is given by

$$p_j = \frac{a_j}{\mathcal{A}}$$

and then substitute $a_j = \mathcal{A} p_j$ into Equation 6.38, we obtain

$$\begin{aligned} S_{\text{ensemble}} &= k_B \mathcal{A} \ln \mathcal{A} - k_B \sum_j p_j \mathcal{A} \ln p_j \mathcal{A} \\ &= k_B \mathcal{A} \ln \mathcal{A} - k_B \sum_j p_j \mathcal{A} \ln p_j - k_B \sum_j p_j \mathcal{A} \ln \mathcal{A} \end{aligned} \tag{6.39}$$

But the last term here cancels with the first because

$$\sum_j p_j \mathcal{A} \ln \mathcal{A} = \mathcal{A} \ln \mathcal{A} \sum_j p_j = \mathcal{A} \ln \mathcal{A}$$

where we have used the facts that $\mathcal{A} \ln \mathcal{A}$ is a constant and $\sum_j p_j = 1$. If we furthermore divide Equation 6.39 through by $\mathcal{A}$, we obtain

$$S_{\text{system}} = -k_\text{B} \sum_j p_j \ln p_j \tag{6.40}$$

Note that if all the $p_j$'s are zero except for one (which must equal unity because $\sum_j p_j = 1$), the system is completely ordered and $S = 0$. Therefore, we see that according to our molecular picture of entropy, $S = 0$ for a perfectly ordered system. Problem 6–39 asks you to show that $S$ is a maximum when all the $p_j$'s are equal, in which case the system is maximally disordered.

To derive an expression for $S$ in terms of $Q(N, V, T)$, we substitute

$$p_j(N, V, \beta) = \frac{e^{-\beta E_j(N,V)}}{Q(N, V, \beta)} \tag{6.41}$$

into Equation 6.40 to obtain

$$\begin{aligned}
S &= -k_\text{B} \sum_j p_j \ln p_j \\
&= -k_\text{B} \sum_j \frac{e^{-\beta E_j}}{Q} \left(-\beta E_j - \ln Q\right) \\
&= \beta k_\text{B} \sum_j \frac{E_j e^{-\beta E_j}}{Q} + \frac{k_\text{B} \ln Q}{Q} \sum_j e^{-\beta E_j} \\
&= \frac{U}{T} + k_\text{B} \ln Q
\end{aligned} \tag{6.42}$$

We used the fact that $\beta k_\text{B} = 1/T$ to go from the third line to the last line. Using Equation 6.36 for $U$ gives $S$ in terms of the partition function, $Q(N, V, T)$.

$$S = k_\text{B} T \left(\frac{\partial \ln Q}{\partial T}\right)_{N,V} + k_\text{B} \ln Q \tag{6.43}$$

Recall from Chapter 4 that

$$Q(N, V, T) = \frac{1}{N!} \left(\frac{2\pi m k_\text{B} T}{h^2}\right)^{3N/2} V^N g_{e1}$$

for a monatomic ideal gas where all the atoms are in their ground electronic state. Using Equation 6.43, we obtain for the molar entropy of one mole of a monatomic ideal gas,

$$\overline{S} = \frac{3}{2}R + R\ln\left[\left(\frac{2\pi m k_{\mathrm{B}} T}{h^2}\right)^{3/2} \overline{V} g_{e1}\right] - k_{\mathrm{B}} \ln N_{\mathrm{A}}! \tag{6.44}$$

Applying Stirling's approximation to the last term gives

$$-k_{\mathrm{B}} \ln N_{\mathrm{A}}! = -k_{\mathrm{B}} N_{\mathrm{A}} \ln N_{\mathrm{A}} + k_{\mathrm{B}} N_{\mathrm{A}} = -R \ln N_{\mathrm{A}} + R$$

Therefore,

$$\overline{S} = \frac{5}{2}R + R\ln\left[\left(\frac{2\pi m k_{\mathrm{B}} T}{h^2}\right)^{3/2} \frac{\overline{V} g_{e1}}{N_{\mathrm{A}}}\right] \tag{6.45}$$

**EXAMPLE 6–6**

Use Equation 6.45 to calculate the molar entropy of argon at 298.2 K and one bar, and compare your result with the experimental value of 154.8 $\mathrm{J\cdot K^{-1}\cdot mol^{-1}}$.

SOLUTION: At 298.2 K and one bar,

$$\begin{aligned}
\frac{N_{\mathrm{A}}}{\overline{V}} &= \frac{N_{\mathrm{A}} P}{RT} \\
&= \frac{(6.022 \times 10^{23}\ \mathrm{mol^{-1}})(1\ \mathrm{bar})}{(0.08314\ \mathrm{L\cdot bar\cdot K^{-1}\cdot mol^{-1}})(298.2\ \mathrm{K})} \\
&= 2.429 \times 10^{22}\ \mathrm{L^{-1}} = 2.431 \times 10^{25}\ \mathrm{m^{-3}}
\end{aligned}$$

and

$$\begin{aligned}
\left(\frac{2\pi m k_{\mathrm{B}} T}{h^2}\right)^{3/2} &= \left[\frac{2\pi(0.03995\ \mathrm{kg\cdot mol^{-1}})(1.3806 \times 10^{-23}\ \mathrm{J\cdot K^{-1}})(298.2\ \mathrm{K})}{(6.022 \times 10^{23}\ \mathrm{mol^{-1}})(6.626 \times 10^{-34}\ \mathrm{J\cdot s})^2}\right]^{3/2} \\
&= (3.909 \times 10^{21}\ \mathrm{m^{-2}})^{3/2} \\
&= 2.444 \times 10^{32}\ \mathrm{m^{-3}}
\end{aligned}$$

Therefore

$$\begin{aligned}
\frac{\overline{S}}{R} &= \frac{5}{2} + \ln\left[\frac{2.444 \times 10^{32}\ \mathrm{m^{-3}}}{2.429 \times 10^{25}\ \mathrm{m^{-3}}}\right] \\
&= 18.62
\end{aligned}$$

or

$$\overline{S} = (18.62)(8.314\ \mathrm{J\cdot K^{-1}\cdot mol^{-1}}) = 154.8\ \mathrm{J\cdot K^{-1}\cdot mol^{-1}}$$

This value of $\overline{S}$ agrees exactly with the experimentally determined value.

**EXAMPLE 6–7**

Show that Equation 6.45 gives Equation 6.26 for the molar entropy of mixing nitrogen and bromine as ideal gases.

SOLUTION: First we write Equation 6.45 as

$$S = Nk_B \ln V + \text{terms not involving } V$$

The initial state is given by

$$\begin{aligned} S_1 &= S_{1,N_2} + S_{1,Br_2} \\ &= n_{N_2} R \ln V_{N_2} + n_{Br_2} R \ln V_{Br_2} + \text{terms not involving } V \end{aligned}$$

where we have written $Nk_B = nR$. The final state is given by

$$\begin{aligned} S_2 &= S_{2,N_2} + S_{2,Br_2} \\ &= n_{N_2} R \ln(V_{N_2} + V_{Br_2}) + n_{Br_2} R \ln(V_{N_2} + V_{Br_2}) + \text{terms not involving } V \end{aligned}$$

Therefore

$$\Delta_{mix} S = S_2 - S_1 = n_{N_2} R \ln \frac{V_{N_2} + V_{Br_2}}{V_{N_2}} + n_{Br_2} R \ln \frac{V_{N_2} + V_{Br_2}}{V_{Br_2}}$$

Because $V$ is proportional to $n$ for an ideal gas, we have

$$\Delta_{mix} S = -n_{N_2} R \ln \frac{n_{N_2}}{n_{N_2} + n_{Br_2}} - n_{Br_2} R \ln \frac{n_{Br_2}}{n_{N_2} + n_{Br_2}}$$

If we divide this result through by $n_{N_2} + n_{Br_2}$, then we obtain Equation 6.26.

## 6–9. The Molecular Formula $S = k_B \ln W$ Is Analogous to the Thermodynamic Formula $dS = \delta q_{rev}/T$

In this last section, we will show that Equation 6.24, or its equivalent, Equation 6.40, is consistent with our thermodynamic definition of the entropy. As a bonus, we will finally prove that $\beta = 1/k_B T$.

If we differentiate Equation 6.40 with respect to $p_j$, we get

$$dS = -k_B \sum_j \left(dp_j + \ln p_j dp_j\right)$$

But $\sum dp_j = 0$ because $\sum p_j = 1$, so

$$dS = -k_B \sum_j \ln p_j dp_j \tag{6.46}$$

Now substitute Equation 6.41 into the ln $p_j$ term in Equation 6.46 to obtain

$$dS = -k_B \sum_j \left[ -\beta E_j(N, V) - \ln Q \right] dp_j$$

The term involving ln $Q$ drops out because

$$\sum_j \ln Q \, dp_j = \ln Q \sum_j dp_j = 0$$

and so

$$dS = \beta k_B \sum_j E_j(N, V) dp_j(N, V, \beta) \tag{6.47}$$

But we showed in Section 5–4 that $\sum_j E_j(N, V)dp_j(N, V, \beta)$ is the energy as heat that a system gains or loses in a reversible process, so Equation 6.47 becomes

$$dS = \beta k_B \delta q_{rev} \tag{6.48}$$

Equation 6.48 shows, furthermore, that $\beta k_B$ is an integrating factor of $\delta q_{rev}$, or $\beta k_B = 1/T$, or $\beta = 1/k_B T$. Thus, we have finally proved that $\beta = 1/k_B T$.

In the next chapter, we will discuss the experimental determination of the entropies of substances.

## Problems

**6-1.** Show that

$$\oint dY = 0$$

if $Y$ is a state function.

**6-2.** Let $z = z(x, y)$ and $dz = xydx + y^2 dy$. Although $dz$ is not an exact differential (why not?), what combination of $dz$ and $x$ and/or $y$ is an exact differential?

**6-3.** Use the criterion developed in MathChapter D to prove that $\delta q_{rev}$ in Equation 6.1 is not an exact differential (see also Problem D–11).

**6-4.** Use the criterion developed in MathChapter D to prove that $\delta q_{rev}/T$ in Equation 6.1 is an exact differential.

**6-5.** In this problem, we will prove that Equation 6.5 is valid for an arbitrary system. To do this, consider an isolated system made up of two equilibrium subsystems, A and B, which are in thermal contact with each other; in other words, they can exchange energy as heat between themselves. Let subsystem A be an ideal gas and let subsystem B be arbitrary. Suppose now that an infinitesimal reversible process occurs in A accompanied by an exchange of energy as heat $\delta q_{rev}$(ideal). Simultaneously, another infinitesimal reversible process takes place in B accompanied by an exchange of energy as heat $\delta q_{\text{rev}}(\text{arbitrary})$. Because the composite system is isolated, the First Law requires that

$$\delta q_{\text{rev}}(\text{ideal}) = -\delta q_{\text{rev}}(\text{arbitrary})$$

Now use Equation 6.4 to prove that

$$\oint \frac{\delta q_{\text{rev}}(\text{arbitrary})}{T} = 0$$

Therefore, we can say that the definition given by Equation 6.4 holds for any system.

**6-6.** Calculate $q_{\text{rev}}$ and $\Delta S$ for a reversible cooling of one mole of an ideal gas at a constant volume $V_1$ from $P_1, V_1, T_1$ to $P_2, V_1, T_4$ followed by a reversible expansion at constant pressure $P_2$ from $P_2, V_1, T_4$ to $P_2, V_2, T_1$ (the final state for all the processes shown in Figure 6.3). Compare your result for $\Delta S$ with those for paths A, B + C, and D + E in Figure 6.3.

**6-7.** Derive Equation 6.8 without referring to Chapter 5.

**6-8.** Calculate the value of $\Delta S$ if one mole of an ideal gas is expanded reversibly and isothermally from 10.0 $\text{dm}^3$ to 20.0 $\text{dm}^3$. Explain the sign of $\Delta S$.

**6-9.** Calculate the value of $\Delta S$ if one mole of an ideal gas is expanded reversibly and isothermally from 1.00 bar to 0.100 bar. Explain the sign of $\Delta S$.

**6-10.** Calculate the values of $q_{\text{rev}}$ and $\Delta S$ along the path D + E in Figure 6.3 for one mole of a gas whose equation of state is given in Example 6–2. Compare your result with that obtained in Example 6–2.

**6-11.** Show that $\Delta S_{\text{D+E}}$ is equal to $\Delta S_{\text{A}}$ and $\Delta S_{\text{B+C}}$ for the equation of state given in Example 6–2.

**6-12.** Calculate the values of $q_{\text{rev}}$ and $\Delta S$ along the path described in Problem 6–6 for one mole of a gas whose equation of state is given in Example 6–2. Compare your result with that obtained in Example 6–2.

**6-13.** Show that

$$\Delta S = C_P \ln \frac{T_2}{T_1}$$

for a constant-pressure process if $C_P$ is independent of temperature. Calculate the change in entropy of 2.00 moles of $H_2O(l)$ ($\overline{C}_P = 75.2\ \text{J}\cdot\text{K}^{-1}\cdot\text{mol}^{-1}$) if it is heated from 10°C to 90°C.

**6-14.** Show that

$$\Delta \overline{S} = \overline{C}_V \ln \frac{T_2}{T_1} + R \ln \frac{V_2}{V_1}$$

if one mole of an ideal gas is taken from $T_1, V_1$ to $T_2, V_2$, assuming that $\overline{C}_V$ is independent of temperature. Calculate the value of $\Delta \overline{S}$ if one mole of $N_2(g)$ is expanded from 20.0 $\text{dm}^3$ at 273 K to 300 $\text{dm}^3$ at 400 K. Take $\overline{C}_P = 29.4\ \text{J}\cdot\text{K}^{-1}\cdot\text{mol}^{-1}$.

**6-15.** In this problem, we will consider a two-compartment system like that in Figure 6.4, except that the two subsystems have the same temperature but different pressures and the wall that separates them is flexible rather than rigid. Show that in this case,

$$dS = \frac{dV_B}{T}(P_B - P_A)$$

Interpret this result with regard to the sign of $dV_B$ when $P_B > P_A$ and when $P_B < P_A$.

**6-16.** In this problem, we will illustrate the condition $dS_{prod} \geq 0$ with a concrete example. Consider the two-component system shown in Figure 6.8. Each compartment is in equilibrium with a heat reservoir at different temperatures $T_1$ and $T_2$, and the two compartments are separated by a rigid heat-conducting wall. The total change of energy as heat of compartment 1 is

$$dq_1 = d_eq_1 + d_iq_1$$

where $d_eq_1$ is the energy as heat exchanged with the reservoir and $d_iq_1$ is the energy as heat exchanged with compartment 2. Similarly,

$$dq_2 = d_eq_2 + d_iq_2$$

Clearly,

$$d_iq_1 = -d_iq_2$$

Show that the entropy change for the two-compartment system is given by

$$\begin{aligned} dS &= \frac{d_eq_1}{T_1} + \frac{d_eq_2}{T_2} + d_iq_1\left(\frac{1}{T_1} - \frac{1}{T_2}\right) \\ &= dS_{exchange} + dS_{prod} \end{aligned}$$

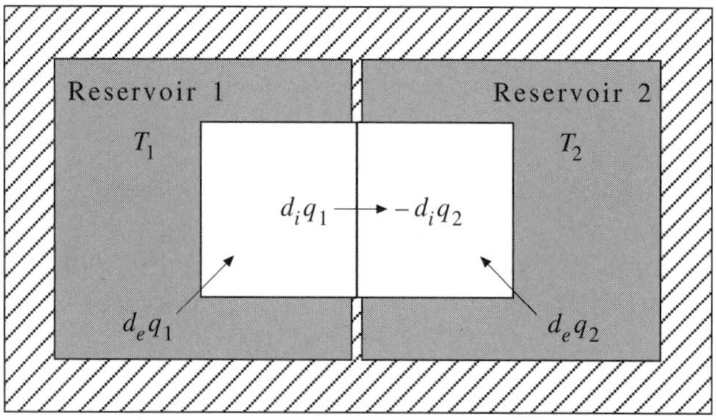


**FIGURE 6.8**
A two-compartment system with each compartment in contact with an (essentially infinite) heat reservoir, one at temperature $T_1$ and the other at temperature $T_2$. The two compartments are separated by a rigid heat-conducting wall.

where

$$dS_{\text{exchange}} = \frac{d_e q_1}{T_1} + \frac{d_e q_2}{T_2}$$

is the entropy *exchanged* with the reservoirs (surroundings) and

$$dS_{\text{prod}} = d_i q_1 \left( \frac{1}{T_1} - \frac{1}{T_2} \right)$$

is the entropy *produced* within the two-compartment system. Now show that the condition $dS_{\text{prod}} \geq 0$ implies that energy as heat flows spontaneously from a higher temperature to a lower temperature. The value of $dS_{\text{exchange}}$, however, has no restriction and can be positive, negative, or zero.

**6-17.** Show that

$$\Delta S \geq \frac{q}{T}$$

for an isothermal process. What does this equation say about the sign of $\Delta S$? Can $\Delta S$ decrease in a reversible isothermal process? Calculate the entropy change when one mole of an ideal gas is compressed reversibly and isothermally from a volume of 100 dm$^3$ to 50.0 dm$^3$ at 300 K.

**6-18.** Vaporization at the normal boiling point ($T_{\text{vap}}$) of a substance (the boiling point at one atm) can be regarded as a reversible process because if the temperature is decreased infinitesimally below $T_{\text{vap}}$, all the vapor will condense to liquid, whereas if it is increased infinitesimally above $T_{\text{vap}}$, all the liquid will vaporize. Calculate the entropy change when two moles of water vaporize at 100.0°C. The value of $\Delta_{\text{vap}}\overline{H}$ is 40.65 kJ·mol$^{-1}$. Comment on the sign of $\Delta_{\text{vap}}S$.

**6-19.** Melting at the normal melting point ($T_{\text{fus}}$) of a substance (the melting point at one atm) can be regarded as a reversible process because if the temperature is changed infinitesimally from exactly $T_{\text{fus}}$, then the substance will either melt or freeze. Calculate the change in entropy when two moles of water melt at 0°C. The value of $\Delta_{\text{fus}}\overline{H}$ is 6.01 kJ·mol$^{-1}$. Compare your answer with the one you obtained in Problem 6–18. Why is $\Delta_{\text{vap}}S$ much larger than $\Delta_{\text{fus}}S$?

**6-20.** Consider a simple example of Equation 6.23 in which there are only two states, 1 and 2. Show that $W(a_1, a_2)$ is a maximum when $a_1 = a_2$. *Hint*: Consider $\ln W$, use Stirling's approximation, and treat $a_1$ and $a_2$ as continuous variables.

**6-21.** Extend Problem 6–20 to the case of three states. Do you see how to generalize it to any number of states?

**6-22.** Show that the system partition function can be written as a summation over levels by writing

$$Q(N, V, T) = \sum_E \Omega(N, V, E) e^{-E/k_B T}$$

Now consider the case of an isolated system, for which there is only one term in $Q(N, V, T)$. Now substitute this special case for $Q$ into Equation 6.43 to derive the equation $S = k_B \ln \Omega$.

**6-23.** In this problem, we will show that $\Omega = c(N)f(E)V^N$ for an ideal gas (Example 6–3). In Problem 4–40 we showed that the number of translational energy states between $\varepsilon$ and $\varepsilon + \Delta\varepsilon$ for a particle in a box can be calculated by considering a sphere in $n_x$, $n_y$, $n_z$ space,

$$n_x^2 + n_y^2 + n_z^2 = \frac{8ma^2\varepsilon}{h^2} = R^2$$

Show that for an $N$-particle system, the analogous expression is

$$\sum_{j=1}^{N}(n_{xj}^2 + n_{yj}^2 + n_{zj}^2) = \frac{8ma^2E}{h^2} = R^2$$

or, in more convenient notation

$$\sum_{j=1}^{3N} n_j^2 = \frac{8ma^2E}{h^2} = R^2$$

Thus, instead of dealing with a three-dimensional sphere as we did in Problem 4–40, here we must deal with a $3N$-dimensional sphere. Whatever the formula for the volume of a $3N$-dimensional sphere is (it is known), we can at least say that it is proportional to $R^{3N}$. Show that this proportionality leads to the following expression for $\Phi(E)$, the number of states with energy $\leq E$,

$$\Phi(E) \propto \left(\frac{8ma^2E}{h^2}\right)^{3N/2} = c(N)E^{3N/2}V^N$$

where $c(N)$ is a constant whose value depends upon $N$ and $V = a^3$. Now, following the argument developed in Problem 4–40, show that the number of states between $E$ and $E + \Delta E$ (which is essentially $\Omega$) is given by

$$\Omega = c(N)f(E)V^N\Delta E$$

where $f(E) = E^{\frac{3N}{2}-1}$.

**6-24.** Show that if a process involves only an isothermal transfer of energy as heat (*pure heat transfer*), then

$$dS_{\text{sys}} = \frac{dq}{T} \qquad \text{(pure heat transfer)}$$

**6-25.** Calculate the change in entropy of the system and of the surroundings and the total change in entropy if one mole of an ideal gas is expanded isothermally and reversibly from a pressure of 10.0 bar to 2.00 bar at 300 K.

**6-26.** Redo Problem 6–25 for an expansion into a vacuum, with an initial pressure of 10.0 bar and a final pressure of 2.00 bar.

**6-27.** The molar heat capacity of 1-butene can be expressed as

$$\overline{C}_P(T)/R = 0.05641 + (0.04635\ \text{K}^{-1})T - (2.392 \times 10^{-5}\ \text{K}^{-2})T^2 + (4.80 \times 10^{-9}\ \text{K}^{-3})T^3$$

over the temperature range $300\ \text{K} < T < 1500\ \text{K}$. Calculate the change in entropy when one mole of 1-butene is heated from 300 K to 1000 K at constant pressure.

**6-28.** Plot $\Delta_{\text{mix}}\overline{S}$ against $y_1$ for the mixing of two ideal gases. At what value of $y_1$ is $\Delta_{\text{mix}}\overline{S}$ a maximum? Can you give a physical interpretation of this result?

**6-29.** Calculate the entropy of mixing if two moles of $N_2(g)$ are mixed with one mole $O_2(g)$ at the same temperature and pressure. Assume ideal behavior.

**6-30.** Show that $\Delta_{\text{mix}}\overline{S} = R \ln 2$ if equal volumes of any two ideal gases under the same conditions are mixed.

**6-31.** Derive the equation $dU = TdS - PdV$. Show that

$$d\overline{S} = \overline{C}_V \frac{dT}{T} + R\frac{d\overline{V}}{\overline{V}}$$

for one mole of an ideal gas. Assuming that $\overline{C}_V$ is independent of temperature, show that

$$\Delta\overline{S} = \overline{C}_V \ln \frac{T_2}{T_1} + R \ln \frac{\overline{V}_2}{\overline{V}_1}$$

for the change from $T_1, \overline{V}_1$ to $T_2, \overline{V}_2$. Note that this equation is a combination of Equations 6.28 and 6.31.

**6-32.** Derive the equation $dH = TdS + VdP$. Show that

$$\Delta\overline{S} = \overline{C}_P \ln \frac{T_2}{T_1} - R \ln \frac{P_2}{P_1}$$

for the change of one mole of an ideal gas from $T_1, P_1$ to $T_2, P_2$, assuming that $\overline{C}_P$ is independent of temperature.

**6-33.** Calculate the change in entropy if one mole of $SO_2(g)$ at 300 K and 1.00 bar is heated to 1000 K and its pressure is decreased to 0.010 bar. Take the molar heat capacity of $SO_2(g)$ to be

$$\overline{C}_P(T)/R = 7.871 - \frac{1454.6\ \text{K}}{T} + \frac{160\,351\ \text{K}^2}{T^2}$$

**6-34.** In the derivation of Equation 6.32, argue that $\Delta S_c > 0$ and $\Delta S_h < 0$. Now show that

$$\Delta S = \Delta S_c + \Delta S_h > 0$$

by showing that

$$\Delta S_c - |\Delta S_h| > 0$$

**6-35.** We can use the equation $S = k \ln W$ to derive Equation 6.28. First, argue that the probability that an ideal-gas molecule is found in a subvolume $V_s$ of some larger volume $V$ is $V_s/V$. Because the molecules of an ideal gas are independent, the probability that $N$ ideal-gas molecules are found in $V_s$ is $(V_s/V)^N$. Now show that the change in entropy when the volume of one mole of an ideal gas changes isothermally from $V_1$ to $V_2$ is

$$\Delta S = R \ln \frac{V_2}{V_1}$$

**6-36.** The relation $n_j \propto e^{-\varepsilon_j/k_B T}$ can be derived by starting with $S = k_B \ln W$. Consider a gas with $n_0$ molecules in the ground state and $n_j$ in the $j$th state. Now add an energy $\varepsilon_j - \varepsilon_0$ to this system so that a molecule is promoted from the ground state to the $j$th state. If the volume of the gas is kept constant, then no work is done, so $dU = dq$,

$$dS = \frac{dq}{T} = \frac{dU}{T} = \frac{\varepsilon_j - \varepsilon_0}{T}$$

Now, assuming that $n_0$ and $n_j$ are large, show that

$$dS = k_B \ln \left\{ \frac{N!}{(n_0 - 1)! n_1! \cdots (n_j + 1)! \cdots} \right\} - k_B \ln \left\{ \frac{N!}{(n_0! n_1! \cdots n_j! \cdots} \right\}$$

$$= k_B \ln \left\{ \frac{n_j!}{(n_j + 1)!} \frac{n_0!}{(n_0 - 1)!} \right\} = k_B \ln \frac{n_0}{n_j}$$

Equating the two expressions for $dS$, show that

$$\frac{n_j}{n_0} = e^{-(\varepsilon_j - \varepsilon_0)/k_B T}$$

**6-37.** We can use Equation 6.24 to calculate the probability of observing fluctuations from the equilibrium state. Show that

$$\frac{W}{W_{\text{eq}}} = e^{-\Delta S/k_B}$$

where $W$ represents the nonequilibrium state and $\Delta S$ is the entropy difference between the two states. We can interpret the ratio $W/W_{\text{eq}}$ as the probability of observing the nonequilibrium state. Given that the entropy of one mole of oxygen is 205.0 $\text{J}\cdot\text{K}^{-1}\cdot\text{mol}^{-1}$ at 25°C and one bar, calculate the probability of observing a decrease in entropy that is one millionth of a percent of this amount.

**6-38.** Consider one mole of an ideal gas confined to a volume $V$. Calculate the probability that all the $N_A$ molecules of this ideal gas will be found to occupy one half of this volume, leaving the other half empty.

**6-39.** Show that $S_{\text{system}}$ given by Equation 6.40 is a maximum when all the $p_j$ are equal. Remember that $\sum p_j = 1$, so that

$$\sum_j p_j \ln p_j = p_1 \ln p_1 + p_2 \ln p_2 + \cdots + p_{n-1} \ln p_{n-1}$$
$$+ (1 - p_1 - p_2 - \cdots - p_{n-1}) \ln(1 - p_1 - p_2 - \cdots - p_{n-1})$$

See also Problem E–10.

**6-40.** Use Equation 6.45 to calculate the molar entropy of krypton at 298.2 K and one bar, and compare your result with the experimental value of 164.1 $\text{J}\cdot\text{K}^{-1}\cdot\text{mol}^{-1}$.

**6-41.** Use Equation 4.39 and the data in Table 4.2 to calculate the entropy of nitrogen at 298.2 K and one bar. Compare your result with the experimental value of 191.6 $\text{J}\cdot\text{K}^{-1}\cdot\text{mol}^{-1}$.

**6-42.** Use Equation 4.57 and the data in Table 4.4 to calculate the entropy of $CO_2$(g) at 298.2 K and one bar. Compare your result with the experimental value of 213.8 $\text{J}\cdot\text{K}^{-1}\cdot\text{mol}^{-1}$.

**6-43.** Use Equation 4.60 and the data in Table 4.4 to calculate the entropy of $NH_3(g)$ at 298.2 K and one bar. Compare your result with the experimental value of 192.8 $J{\cdot}K^{-1}{\cdot}mol^{-1}$.

**6-44.** Derive Equation 6.35.

**6-45.** The boiling point of water at a pressure of 25 atm is 223°C. Compare the theoretical efficiencies of a steam engine operating between 20°C and the boiling point of water at 1 atm and at 25 atm.

**Walther Nernst** was born in Briessen, Prussia (now Wabrzezno, Poland), on June 25, 1864, and died in 1941. Although he aspired to be a poet, his chemistry teacher kindled his interest in science. Between 1883 and 1887, Nernst studied physics with von Helmholtz, Boltzmann, and Kohlrausch. He received his doctorate in physics at the University of Würzburg in 1887. Nernst was Ostwald's assistant at the University of Leibzig from 1887 to 1891, after which he went to the University of Göttingen, where he established the Kaiser Wilhelm Institute for Physical Chemistry and Electrochemistry in 1894. Upon moving to the University of Berlin in 1905, Nernst began his studies of the behavior of substances at very low temperatures. He proposed one of the early versions of the Third Law of Thermodynamics, which says that the physical activities of substances tend to vanish as the temperature approaches absolute zero. The Third Law made it possible to calculate thermodynamic quantities such as equilibrium constants from thermal data. He was awarded the Nobel Prize for chemistry in 1920 "in recognition of his work in thermochemistry." He was an early automoblile enthusiast and served during World War I as a driver. Nernst lost both of his sons in World War I. His anti-Nazi stance in the 1930s led to increasing isolation, so he retired to his country home, where he died in 1941.

CHAPTER 7

# Entropy and the Third Law of Thermodynamics

In the previous chapter, we introduced the concept of entropy. We showed that entropy is created or generated whenever a spontaneous or irreversible process occurs in an isolated system. We also showed that the entropy of an isolated system that is not in equilibrium will increase until the system reaches equilibrium, from which time the entropy will remain constant. We expressed this condition mathematically by writing $dS \geq 0$ for a process that occurs at constant $U$ and $V$. Although we calculated the *change* in entropy for a few processes, we did not attempt to calculate absolute values of the entropy of substances. (See Example 6–6 and Problems 6–41 through 6–43, however.) In this chapter, we will introduce the Third Law of Thermodynamics, so that we can calculate absolute values of the entropy of substances.

## 7–1. Entropy Increases with Increasing Temperature

We start with the First Law of Thermodynamics for a reversible process:

$$dU = \delta q_{\text{rev}} + \delta w_{\text{rev}}$$

Using the fact that $\delta q_{\text{rev}} = TdS$ and $\delta w_{\text{rev}} = -PdV$, we obtain a combination of the First and Second Laws of Thermodynamics:

$$dU = TdS - PdV \tag{7.1}$$

We can derive a number of relationships between thermodynamic quantities using the laws of thermodynamics and the fact that state functions are exact differentials. Example 7–1 derives the following two important relationships

$$\left(\frac{\partial S}{\partial T}\right)_V = \frac{C_V}{T} \tag{7.2}$$

and

$$\left(\frac{\partial S}{\partial V}\right)_T = \frac{1}{T}\left[P + \left(\frac{\partial U}{\partial V}\right)_T\right] \tag{7.3}$$

**EXAMPLE 7–1**

Express $U$ as a function of $V$ and $T$ and then use this result and Equation 7.1 to derive Equations 7.2 and 7.3.

SOLUTION: If we treat $U$ as a function of $V$ and $T$, its total derivative is (MathChapter H)

$$dU = \left(\frac{\partial U}{\partial T}\right)_V dT + \left(\frac{\partial U}{\partial V}\right)_T dV \tag{7.4}$$

We substitute Equation 7.4 into Equation 7.1 and solve for $dS$ to obtain

$$dS = \frac{1}{T}\left(\frac{\partial U}{\partial T}\right)_V dT + \frac{1}{T}\left[P + \left(\frac{\partial U}{\partial V}\right)_T\right] dV$$

Using the definition that $(\partial U/\partial T)_V = C_V$, we obtain

$$dS = \frac{C_V dT}{T} + \frac{1}{T}\left[P + \left(\frac{\partial U}{\partial V}\right)_T\right] dV$$

If we compare this equation for $dS$ with the total derivative of $S = S(T, V)$,

$$dS = \left(\frac{\partial S}{\partial T}\right)_V dT + \left(\frac{\partial S}{\partial V}\right)_T dV$$

we see that

$$\left(\frac{\partial S}{\partial T}\right)_V = \frac{C_V}{T} \quad \text{and} \quad \left(\frac{\partial S}{\partial V}\right)_T = \frac{1}{T}\left[P + \left(\frac{\partial U}{\partial V}\right)_T\right]$$

Equation 7.2 tells us how $S$ varies with temperature at constant volume. If we integrate with respect to $T$ (keeping $V$ constant), we obtain

$$\Delta S = S(T_2) - S(T_1) = \int_{T_1}^{T_2} \frac{C_V(T)dT}{T} \qquad \text{(constant } V\text{)} \tag{7.5}$$

Thus, if we know $C_V(T)$ as a function of $T$, we can calculate $\Delta S$. Note that because $C_V$ is always positive, the entropy increases with increasing temperature.

Equation 7.5 is restricted to constant volume. To derive a similar equation for constant pressure, we start with

$$dH = d(U + PV) = dU + PdV + VdP$$

and substitute Equation 7.1 for $dU$ to obtain

$$dH = TdS + VdP \tag{7.6}$$

Proceeding in a similar manner as in Example 7–1 (Problem 7–1), we obtain

$$\left(\frac{\partial S}{\partial T}\right)_P = \frac{C_P(T)}{T} \tag{7.7}$$

and

$$\left(\frac{\partial S}{\partial P}\right)_T = \frac{1}{T}\left[\left(\frac{\partial H}{\partial P}\right)_T - V\right] \tag{7.8}$$

From Equation 7.7, we get

$$\Delta S = S(T_2) - S(T_1) = \int_{T_1}^{T_2} \frac{C_P(T)dT}{T} \qquad \text{(constant } P\text{)} \tag{7.9}$$

Thus, if we know $C_P$ as a function of $T$, we can calculate $\Delta S$. Most processes we will consider occur at constant pressure, so we will usually use Equation 7.9 to calculate $\Delta S$.

If we let $T_1 = 0$ K in Equation 7.9, then we have

$$S(T) = S(0\text{ K}) + \int_0^T \frac{C_P(T')dT'}{T'} \qquad \text{(constant } P\text{)} \tag{7.10}$$

Equation 7.10 tells us that we can calculate the entropy of a substance if we know $S(0\text{ K})$ and $C_P(T)$ from $T = 0$ K to the temperature of interest. (Notice once again that we use a prime on the variable of integration to distinguish it from an integration limit.)

## 7–2. The Third Law of Thermodynamics Says That the Entropy of a Perfect Crystal Is Zero at 0 K

Let's discuss $S(0\text{ K})$ first. Around the turn of the century, the German chemist Walther Nernst, after studying numerous chemical reactions, postulated that $\Delta_r S \rightarrow 0$ as $T \rightarrow 0$. Nernst did not make any statement concerning the entropy of any particular substance at 0 K, only that all pure crystalline substances have the same entropy at 0 K. We have added the "pure crystalline" condition here to avoid some apparent exceptions to Nernst's postulate that we will resolve later. In 1911, Planck, who incidentally did a great deal of research in thermodynamics (including his doctoral thesis), extended Nernst's postulate by postulating that the entropy of a pure substance approaches zero at 0 K. Planck's postulate is consistent with Nernst's but takes it further. There are several equivalent statements of what is now called the *Third Law of Thermodynamics*, but the one we will use is

> Every substance has a finite positive entropy, but at zero kelvin the entropy may become zero, and does so in the case of a perfectly crystalline substance.

The Third Law of Thermodynamics is unlike the first two laws in that it introduces no new state function. The first law gives us the energy and the second law gives us the entropy; the third law provides a numerical scale for entropy.

Although the Third Law was formulated before the full development of the quantum theory, it is much more plausible and intuitive if we think of it in terms of molecular quantum states or levels. One of our molecular formulas for the entropy is (Equation 6.24)

$$S = k_B \ln W \tag{7.11}$$

where $W$ is the number of ways the total energy of a system may be distributed over its various energy states. At 0 K, we expect that the system will be in its lowest energy state. Therefore, $W = 1$ and $S = 0$. Another way to see this result is to start with Equation 6.40 for $S$:

$$S = -k_B \sum_j p_j \ln p_j \tag{7.12}$$

where $p_j$ is the probability of finding the system in the $j$th quantum state with energy $E_j$. At 0 K, there is no thermal energy, so we expect the system to be in the ground state; thus, $p_0 = 1$ and all the other $p_j$'s equal zero. Therefore, $S$ in Equation 7.12 equals zero. Even if the ground state has a degeneracy of $n$, say, then each of the $n$ quantum states with energy $E_0$ would have a probability of $1/n$, and $S$ in Equation 7.12 would be

$$S(0\text{ K}) = -k_B \sum_{j=1}^{n} \frac{1}{n} \ln \frac{1}{n} = k_B \ln n \tag{7.13}$$

Even if the degeneracy of the ground state were as large as the Avogadro constant, $\overline{S}$ would be equal to only $7.56 \times 10^{-22}\ \text{J}\cdot\text{K}^{-1}\cdot\text{mol}^{-1}$, which is well below a measurable value of $\overline{S}$.

Because the Third Law of Thermodynamics asserts that $S(0\text{ K}) = 0$, we can write Equation 7.10 as

$$S(T) = \int_0^T \frac{C_P(T')dT'}{T'} \tag{7.14}$$

## 7–3. $\Delta_{\rm trs}S = \Delta_{\rm trs}H/T_{\rm trs}$ at a Phase Transition

We made a tacit assumption when we wrote Equation 7.14; we assumed that there is no phase transition between 0 and $T$. Suppose there is such a transition at $T_{\rm trs}$ between 0 and $T$. We can calculate the entropy change upon the phase transition, $\Delta_{\rm trs}S$, by using the equation

$$\Delta_{\rm trs}S = \frac{q_{\rm rev}}{T_{\rm trs}} \tag{7.15}$$

A phase transition is a good example of a reversible process. A phase transition can be reversed by changing the temperature ever so slightly. In the melting of ice, for example, at one atm, the system will be all ice if $T$ is just slightly less than 273.15 K and all liquid if $T$ is just slightly greater than 273.15 K. Furthermore, a phase transition takes place at a fixed temperature, so Equation 7.15 becomes (recall that $\Delta H = q_P$ for a phase transition)

$$\Delta_{\rm trs}S = \frac{\Delta_{\rm trs}H}{T_{\rm trs}} \tag{7.16}$$

**EXAMPLE 7–2**

Calculate the molar entropy change upon melting and upon vaporization at one atm for $H_2O$. Use $\Delta_{\rm fus}\overline{H} = 6.01\ \text{kJ}\cdot\text{mol}^{-1}$ at 273.15 K and $\Delta_{\rm vap}\overline{H} = 40.7\ \text{kJ}\cdot\text{mol}^{-1}$ at 373.15 K.

SOLUTION: Using Equation 7.16, we have

$$\Delta_{\rm fus}\overline{S} = \frac{6.01\ \text{kJ}\cdot\text{mol}^{-1}}{273.15\ \text{K}} = 22.0\ \text{J}\cdot\text{K}^{-1}\cdot\text{mol}^{-1}$$

and

$$\Delta_{\rm vap}\overline{S} = \frac{40.7\ \text{kJ}\cdot\text{mol}^{-1}}{373.15\ \text{K}} = 109\ \text{J}\cdot\text{K}^{-1}\cdot\text{mol}^{-1}$$

Note that $\Delta_{\rm vap}\overline{S}$ is much larger than $\Delta_{\rm fus}\overline{S}$. This makes sense molecularly because the difference in disorder between a gas and a liquid phase is much greater than the difference in disorder between a liquid and a solid phase.

To calculate $S(T)$, we integrate $C_P(T)/T$ up to the first phase transition temperature, add a $\Delta_{\rm trs}H/T_{\rm trs}$ term for the phase transition, and then integrate $C_P(T)/T$ from the first phase transition temperature to the second, and so on. For example, if the substance has no solid-solid phase transition, we would have, for $T$ greater than the boiling point,

$$S(T) = \int_0^{T_{\rm fus}} \frac{C_P^{\rm s}(T)dT}{T} + \frac{\Delta_{\rm fus}H}{T_{\rm fus}} + \int_{T_{\rm fus}}^{T_{\rm vap}} \frac{C_P^{\rm l}(T)dT}{T} + \frac{\Delta H_{\rm vap}}{T_{\rm vap}} + \int_{T_{\rm vap}}^{T} \frac{C_P^{\rm g}(T')dT'}{T'} \tag{7.17}$$

where $T_{\text{fus}}$ is the melting point, $C_P^{\text{s}}(T)$ is the heat capacity of the solid phase, $T_{\text{vap}}$ is the boiling point, $C_P^{\text{l}}(T)$ is the heat capacity of the liquid phase, $C_P^{\text{g}}(T)$ is the heat capacity of the gaseous phase, and $\Delta_{\text{fus}}H$ and $\Delta H_{\text{vap}}$ are the enthalpies of fusion and vaporization, respectively.

## 7–4. The Third Law of Thermodynamics Asserts That $C_P \to 0$ as $T \to 0$

It has been shown experimentally and theoretically that $C_P^{\text{s}}(T) \to T^3$ as $T \to 0$ for most nonmetallic crystals ($C_P^{\text{s}}$ for metallic crystals goes as $aT + bT^3$ as $T \to 0$, where $a$ and $b$ are constants). This $T^3$ temperature dependence is valid from 0 K to about 15 K and is called the *Debye $T^3$ law*, after the Dutch chemist Peter Debye, who first showed theoretically that $C_P^{\text{s}}(T) \to T^3$ as $T \to 0$ for nonmetallic solids.

**EXAMPLE 7–3**

According to the Debye theory, the low-temperature molar heat capacity of nonmetallic solids goes as

$$\overline{C}_P(T) = \frac{12\pi^4}{5} R \left(\frac{T}{\Theta_{\text{D}}}\right)^3 \qquad 0 < T \le T_{\text{low}}$$

where $T_{\text{low}}$ depends upon the particular solid, but is about 10 K to 20 K for most solids, and $\Theta_{\text{D}}$ is a constant characteristic of the solid. The parameter $\Theta_{\text{D}}$ has units of temperature and is called the *Debye temperature* of the solid. Show that if $\overline{C}_P$ is given by the above expression, the low-temperature contribution to the molar entropy is given by

$$\overline{S}(T) = \frac{\overline{C}_P(T)}{3} \qquad 0 < T \le T_{\text{low}}$$

SOLUTION: Substitute the given expression for $\overline{C}_P(T)$ into Equation 7.14 to get

$$\overline{S}(T) = \int_0^T \frac{\overline{C}_P(T')dT'}{T'} = \frac{12\pi^4 R}{5\Theta_{\text{D}}^3} \int_0^T T'^2 dT'$$

$$= \frac{12\pi^4 R}{5\Theta_{\text{D}}^3} \frac{T^3}{3} = \frac{\overline{C}_P(T)}{3} \tag{7.18}$$

**EXAMPLE 7–4**

Given that the molar heat capacity of solid chlorine is 3.39 $\text{J}\cdot\text{K}^{-1}\cdot\text{mol}^{-1}$ at 14 K and obeys the Debye $T^3$ law below 14 K, calculate the molar entropy of solid chlorine at 14 K.

SOLUTION: We use Equation 7.18 and get

$$\overline{S}(\text{at 14 K}) = \frac{\overline{C}_P(\text{at 14 K})}{3}$$

$$= \frac{3.39\ \text{J}\cdot\text{K}^{-1}\cdot\text{mol}^{-1}}{3} = 1.13\ \text{J}\cdot\text{K}^{-1}\cdot\text{mol}^{-1}$$

## 7–5. Practical Absolute Entropies Can Be Determined Calorimetrically

Given suitable heat capacity data and enthalpies of transition and transition temperatures, we can use Equation 7.17 to calculate entropies based on the convention of setting $S(0\ \text{K}) = 0$. Such entropies are called third-law entropies, or practical absolute entropies. Table 7.1 gives the entropy of $N_2(g)$ at 298.15 K. The entropy at 10.00 K was determined by using Equation 7.18 with $\overline{C}_P = 6.15\ \text{J}\cdot\text{K}^{-1}\cdot\text{mol}^{-1}$. At 35.61 K, the solid undergoes a phase change in crystalline structure with $\Delta_{\text{trs}}\overline{H} = 0.2289\ \text{kJ}\cdot\text{mol}^{-1}$, so $\Delta_{\text{trs}}\overline{S} = 6.43\ \text{J}\cdot\text{K}^{-1}\cdot\text{mol}^{-1}$. At 63.15 K, $N_2(s)$ melts with $\Delta_{\text{fus}}\overline{H} = 0.71\ \text{kJ}\cdot\text{mol}^{-1}$, so $\Delta_{\text{fus}}\overline{S} = 11.2\ \text{J}\cdot\text{K}^{-1}\cdot\text{mol}^{-1}$. Finally, $N_2(l)$ at one atm boils at 77.36 K with $\Delta_{\text{vap}}\overline{H} = 5.57\ \text{kJ}\cdot\text{mol}^{-1}$, giving $\Delta_{\text{vap}}\overline{S} = 72.0\ \text{J}\cdot\text{K}^{-1}\cdot\text{mol}^{-1}$. For the regions between the phase transitions, $\overline{C}_P(T)/T$ data were integrated numerically (Problem 7–14). According to Equation 7.17, the molar entropy is given by the area under the curve of $\overline{C}_P(T)/T$ plotted against the temperature.

The small correction at the end of Table 7.1 needs explaining. The values of entropies of gases presented in the literature are called *standard entropies*, which by convention are corrected for the nonideality of the gas at one bar. We will learn how

**TABLE 7.1**
The standard molar entropy of nitrogen at 298.15 K.

| Process | $\overline{S}/\text{J}\cdot\text{K}^{-1}\cdot\text{mol}^{-1}$ |
|---|---|
| 0 to 10.00 K | 2.05 |
| 10.00 to 35.61 K | 25.79 |
| Transition | 6.43 |
| 35.61 to 63.15 K | 23.41 |
| Fusion | 11.2 |
| 63.15 to 77.36 K | 11.46 |
| Vaporization | 72.0 |
| 77.36 K to 298.15 K | 39.25 |
| Correction for nonideality | 0.02 |
| Total | 191.6 |

to make this correction in Section 8–6. Recall that the standard state of a (real) gas at any temperature is that of the corresponding (hypothetical) ideal gas at one bar.

Figure 7.1 shows the molar entropy of nitrogen plotted against temperature from 0 K to 400 K. Note that the molar entropy increases smoothly with temperature between phase transitions and that there are discontinuous jumps at each phase transition. Note also that the jump at the vaporization transition is much larger than the jump at the melting point. Figure 7.2 shows a similar plot for benzene. Note that benzene does not undergo any solid-solid phase transitions.

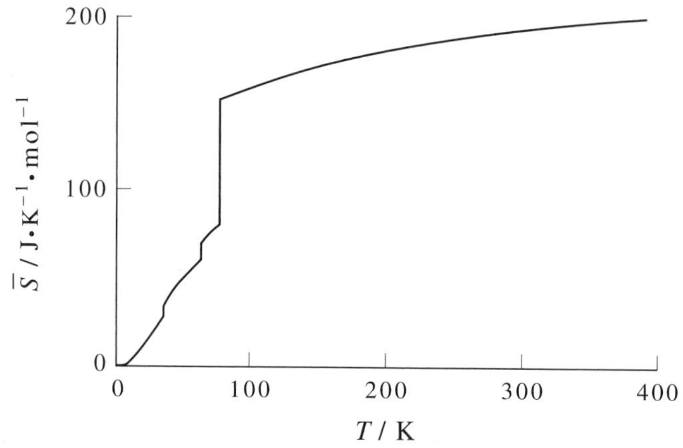


**FIGURE 7.1**
The molar entropy of nitrogen plotted against temperature from 0 K to 400 K.

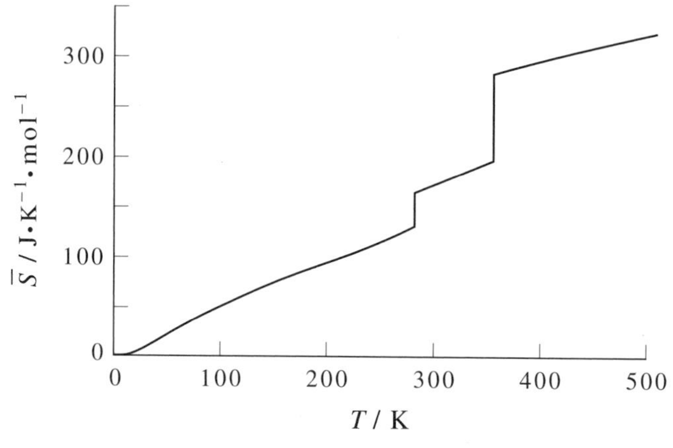


**FIGURE 7.2**
The molar entropy of benzene plotted against temperature from 0 K to 500 K.

## 7–6. Practical Absolute Entropies of Gases Can Be Calculated from Partition Functions

Recall from Section 6–8 that the entropy can be written as (Equation 6.43)

$$S = k_{\mathrm{B}} \ln Q + k_{\mathrm{B}} T \left( \frac{\partial \ln Q}{\partial T} \right)_{N,V} \tag{7.19}$$

where $Q(N, V, T)$ is the system partition function

$$Q(N, V, T) = \sum_j e^{-E_j(N,V)/k_{\mathrm{B}}T} \tag{7.20}$$

Equation 7.19 is consistent with the Third Law of Thermodynamics. Let's write Equation 7.19 for $S$ more explicitly by substituting Equation 7.20 into it:

$$S = k_{\mathrm{B}} \ln \sum_j e^{-E_j/k_{\mathrm{B}}T} + \frac{1}{T} \frac{\sum_j E_j e^{-E_j/k_{\mathrm{B}}T}}{\sum_j e^{-E_j/k_{\mathrm{B}}T}} \tag{7.21}$$

We want to study the behavior of this equation as $T \to 0$. Assume for generality that the first $n$ states have the same energy $E_1 = E_2 = \cdots = E_n$ (in other words, that the ground state is $n$-fold degenerate) and that the next $m$ states have the same energy $E_{n+1} = E_{n+2} = \cdots = E_{n+m}$ (the first excited state is $m$-fold degenerate), and so on.

Let's look at the summations in Equations 7.21 as $T \to 0$. Writing out Equation 7.20 explicitly gives

$$\sum_j e^{-E_j/k_{\mathrm{B}}T} = ne^{-E_1/k_{\mathrm{B}}T} + me^{-E_{n+1}/k_{\mathrm{B}}T} + \cdots$$

If we factor out $e^{-E_1/k_{\mathrm{B}}T}$, then

$$\sum_j e^{-E_j/k_{\mathrm{B}}T} = e^{-E_1/k_{\mathrm{B}}T} \left[ n + me^{-(E_{n+1}-E_1)/k_{\mathrm{B}}T} + \cdots \right]$$

But $E_{n+1} - E_1 > 0$ (essentially by definition), so

$$e^{-(E_{n+1}-E_1)/k_{\mathrm{B}}T} \to 0 \quad \text{as} \quad T \to 0$$

Therefore, as $T \to 0$,

$$\sum_j e^{-E_j/k_{\mathrm{B}}T} \to ne^{-E_1/k_{\mathrm{B}}T}$$

In the limit of small $T$, then, the first terms in each summation in Equation 7.21 dominate, and we have

$$S = k_B \ln(ne^{-E_1/k_BT}) + \frac{1}{T}\frac{nE_1e^{-E_1/k_BT}}{ne^{-E_1/k_BT}}$$
$$= k_B \ln n - \frac{E_1}{T} + \frac{E_1}{T} = k_B \ln n$$

Thus, as $T \to 0$, $S$ is proportional to the logarithm of the degeneracy of the ground state (see Equation 7.13). As we argued in Section 7–2, even if $n$ were as large as the Avogadro constant, $S$ itself would be completely negligible.

We learned in Chapter 3 (Equation 3.38) that

$$Q(N, V, T) = \frac{[q(V, T)]^N}{N!} \tag{7.22}$$

for an ideal gas. Furthermore, we learned in Chapter 4 that for a
(1) monatomic ideal gas (Equation 4.13):

$$q(V, T) = \left(\frac{2\pi m k_B T}{h^2}\right)^{3/2} V \cdot g_{e1} \tag{7.23}$$

(2) diatomic ideal gas (Equation 4.39):

$$q(V, T) = \left(\frac{2\pi M k_B T}{h^2}\right)^{3/2} V \cdot \frac{T}{\sigma \Theta_{\text{rot}}} \cdot \frac{e^{-\Theta_{\text{vib}}/2T}}{1 - e^{-\Theta_{\text{vib}}/T}} \cdot g_{e1} e^{D_e/k_BT} \tag{7.24}$$

(3) linear polyatomic ideal gas (Equation 4.57):

$$q(V, T) = \left(\frac{2\pi M k_B T}{h^2}\right)^{3/2} V \cdot \frac{T}{\sigma \Theta_{\text{rot}}} \cdot \left[\prod_{j=1}^{3n-5} \frac{e^{-\Theta_{\text{vib},j}/2T}}{1 - e^{-\Theta_{\text{vib},j}/T}}\right] \cdot g_{e1} e^{D_e/k_BT} \tag{7.25}$$

(4) nonlinear polyatomic ideal gas (Equation 4.60):

$$q(V, T) = \left(\frac{2\pi M k_B T}{h^2}\right)^{3/2} V \cdot \frac{\pi^{1/2}}{\sigma} \left(\frac{T^3}{\Theta_A \Theta_B \Theta_C}\right)^{1/2} \left[\prod_{j=1}^{3n-6} \frac{e^{-\Theta_{\text{vib},j}/2T}}{1 - e^{-\Theta_{\text{vib},j}/T}}\right] g_{e1} e^{D_e/k_BT} \tag{7.26}$$

The various quantities in these equations are defined and discussed in Chapter 4.

If we substitute Equation 7.22 into Equation 7.19, then we obtain

$$S = Nk_B \ln q - k_B \ln N! + Nk_B T \left(\frac{\partial \ln q}{\partial T}\right)_V$$

If we use Stirling's approximation for $\ln N!$ ($= N \ln N - N$), then (Problem 7-27)

$$S = Nk_B + Nk_B \ln\left[\frac{q(V, T)}{N}\right] + Nk_B T \left(\frac{\partial \ln q}{\partial T}\right)_V \tag{7.27}$$

Let's use Equations 7.27 and 7.24 to calculate the standard molar entropy of $N_2(g)$ at 298.15 K and compare the result with the value in Table 7.1 obtained from heat capacity data. If we substitute Equation 7.24 into Equation 7.27, we obtain

$$\frac{\overline{S}}{R} = \ln\left[\left(\frac{2\pi M k_B T}{h^2}\right)^{3/2} \frac{\overline{V} e^{5/2}}{N_A}\right] + \ln\frac{Te}{2\Theta_{\text{rot}}} - \ln(1 - e^{-\Theta_{\text{vib}}/T}) + \frac{\Theta_{\text{vib}}/T}{e^{\Theta_{\text{vib}}/T} - 1} + \ln g_{e1} \qquad (7.28)$$

The first term represents the translational contribution to $S$, the second represents the rotational contribution, the third and fourth represent the vibrational contribution, and the last term represents the electronic contribution to $S$. The necessary parameters are $\Theta_{\text{rot}} = 2.88$ K, $\Theta_{\text{vib}} = 3374$ K, and $g_{e1} = 1$. At 298.15 K and one bar, the various factors are

$$\left(\frac{2\pi M k_B T}{h^2}\right)^{3/2} = \left[\frac{2\pi(4.653\times10^{-26}\ \text{kg})(1.3807\times10^{-23}\ \text{J}\cdot\text{K}^{-1})(298.15\ \text{K})}{(6.626\times10^{-34}\ \text{J}\cdot\text{s})^2}\right]^{3/2}$$
$$= 1.436\times10^{32}\ \text{m}^{-3}$$

$$\frac{\overline{V}}{N_A} = \frac{RT}{N_A P} = \frac{(0.08314\ \text{L}\cdot\text{bar}\cdot\text{mol}^{-1}\cdot\text{K}^{-1})(298.15\ \text{K})}{(6.022\times10^{23}\ \text{mol}^{-1})(1\ \text{bar})}$$
$$= 4.117\times10^{-23}\ \text{L} = 4.117\times10^{-26}\ \text{m}^{-3}$$

$$\frac{Te}{2\Theta_{\text{rot}}} = \frac{(298.15\ \text{K})(2.71828)}{2(2.88\ \text{K})} = 140.7$$

$$1 - e^{-\Theta_{\text{vib}}/T} = 1 - e^{-11.31} \approx 1.000$$

$$\frac{\Theta_{\text{vib}}/T}{e^{\Theta_{\text{vib}}/T} - 1} = \frac{11.31}{e^{11.31} - 1} = 1.380\times10^{-4}$$

Thus, the standard molar entropy $S^\circ$ at 298.15 K is

$$S^\circ = S^\circ_{\text{trans}} + S^\circ_{\text{rot}} + S^\circ_{\text{vib}} + S^\circ_{\text{elec}}$$
$$= (150.4 + 41.13 + 1.15\times10^{-3} + 0)\ \text{J}\cdot\text{K}^{-1}\cdot\text{mol}^{-1}$$
$$= 191.5\ \text{J}\cdot\text{K}^{-1}\cdot\text{mol}^{-1}$$

compared with the value of 191.6 $\text{J}\cdot\text{K}^{-1}\cdot\text{mol}^{-1}$ given in Table 7.1. The two values agree essentially exactly. This type of agreement is quite common, and in many cases the statistical thermodynamic value is more accurate than the calorimetric value. Table 7.2 gives standard molar entropies for several substances. The accepted literature values are often a combination of statistical thermodynamic and calorimetric values.

**EXAMPLE 7–5**

Use the equations of this section to calculate the standard molar entropy of carbon dioxide at 298.15 K and compare the result with the value in Table 7.2.

**TABLE 7.2**
Standard molar entropies ($S^\circ$) of various substances at 298.15 K.

| Substance | $S^o/\mathrm{J \cdot K^{-1} \cdot mol^{-1}}$ | Substance | $S^o/\mathrm{J \cdot K^{-1} \cdot mol^{-1}}$ |
|---|---|---|---|
| Ag(s) | 42.55 | HCl(g) | 186.9 |
| Ar(g) | 154.8 | HCN(g) | 201.8 |
| $Br_2$(g) | 245.5 | HI(g) | 206.6 |
| $Br_2$(l) | 152.2 | $H_2O$(g) | 188.8 |
| C(s)(diamond) | 2.38 | $H_2O$(l) | 70.0 |
| C(s)(graphite) | 5.74 | Hg(l) | 75.9 |
| $CH_4$(g) | 186.3 | $I_2$(s) | 116.1 |
| $C_2H_2$(g) | 200.9 | $I_2$(g) | 260.7 |
| $C_2H_4$(g) | 219.6 | K(s) | 64.7 |
| $C_2H_6$(g) | 229.6 | $N_2$(g) | 191.6 |
| $CH_3OH$(l) | 126.8 | Na(s) | 51.3 |
| $CH_3Cl$(g) | 234.6 | $NH_3$(g) | 192.8 |
| CO(g) | 197.7 | NO(g) | 210.8 |
| $CO_2$(g) | 213.8 | $NO_2$(g) | 240.1 |
| $Cl_2$(g) | 223.1 | $O_2$(g) | 205.2 |
| $H_2$(g) | 130.7 | $O_3$(g) | 238.9 |
| HBr(g) | 198.7 | $SO_2$(g) | 248.2 |

SOLUTION: Carbon dioxide is a symmetric linear molecule with four vibrational degrees of freedom. We substitute Equation 7.25 into Equation 7.27 to obtain

$$\frac{S^\circ}{R} = 1 + \ln\left[\left(\frac{2\pi M k_B T}{h^2}\right)^{3/2} \frac{\overline{V}}{N_A}\right] + \ln\left(\frac{T}{\sigma\Theta_{\text{rot}}}\right)$$
$$-\sum_{j=1}^{4} \frac{\Theta_{\text{vib},j}}{2T} - \sum_{j=1}^{4} \ln\left(1 - e^{-\Theta_{\text{vib},j}/T}\right) + \ln g_{e1} + \frac{D_e}{k_B T}$$
$$+T\left[\frac{3}{2T} + \frac{1}{T} + \sum_{j=1}^{4} \frac{\Theta_{\text{vib},j}}{2T^2} + \sum_{j=1}^{4} \frac{(\Theta_{\text{vib},j}/T^2)e^{-\Theta_{\text{vib},j}/T}}{1 - e^{-\Theta_{\text{vib},j}/T}} - \frac{D_e}{k_B T^2}\right]$$

or

$$\frac{S^\circ}{R} = \frac{7}{2} + \ln\left[\left(\frac{2\pi M k_B T}{h^2}\right)^{3/2} \frac{\overline{V}}{N_A}\right] + \ln\left(\frac{T}{\sigma\Theta_{\text{rot}}}\right)$$
$$+\sum_{j=1}^{4}\left[\frac{(\Theta_{\text{vib},j}/T)e^{-\Theta_{\text{vib},j}/T}}{1 - e^{-\Theta_{\text{vib},j}/T}} - \ln(1 - e^{-\Theta_{\text{vib},j}/T})\right] + \ln g_{e1}$$

Paralleling the calculation for $N_2$(g), we find that $(2\pi M k_B T/h^2)^{3/2} = 2.826 \times 10^{32}\ \mathrm{m^{-3}}$ and $\overline{V}/N_A = 4.117 \times 10^{-26}\ \mathrm{m^{-3}}$. Using the value of $\Theta_{\text{rot}} = 0.561$ K from Table 4.4, we find that $T/2\Theta_{\text{rot}} = 265.8$. Similarly, we use Table 4.4 to show that the four values of $\Theta_{\text{vib},j}/T$ are 3.199 (twice), 6.338, and 11.27. Last, $g_{e1} = 1$. Putting all this together, we find that

$$\begin{aligned}\frac{S^\circ}{R} &= \frac{7}{2} + \ln\left[(2.826 \times 10^{32}\ \text{m}^{-3})(4.117 \times 10^{-26}\ \text{m}^{-3})\right] + \ln 265.8 \\ &\quad + 2\left[\frac{3.199e^{-3.199}}{1 - e^{-3.199}} - \ln(1 - e^{-3.199})\right] + \left[\frac{6.338e^{-6.338}}{1 - e^{-6.338}} - \ln(1 - e^{-6.338})\right] \\ &\quad + \left[\frac{11.27e^{-11.27}}{1 - e^{-11.27}} - \ln(1 - e^{-11.27})\right] \\ &= 3.5 + 16.27 + 5.58 + 2(0.178) + 0.01 + O(10^{-4}) \\ &= 25.71\end{aligned}$$

or

$$S^\circ = 25.71R = 213.8\ \text{J}\cdot\text{K}^{-1}\cdot\text{mol}^{-1}$$

which is in excellent agreement with the value in Table 7.2.

## 7–7. The Values of Standard Molar Entropies Depend Upon Molecular Mass and Molecular Structure

Let's look at the standard molar entropy values in Table 7.2 and try to determine some trends. First, notice that the standard molar entropies of the gaseous substances are the largest, and the standard molar entropies of the solid substances are the smallest. These values reflect the fact that solids are more ordered than liquids and gases.

Now consider the standard molar entropies of the noble gases given in Table 7.3. The increase in standard molar entropy of the noble gases is a consequence of their increasing mass as we move down the periodic table. Thus, an increase in mass leads to an increase in thermal disorder (more translational energy levels are available) and a greater entropy. We know from quantum theory that the greater the molecular mass, the

**TABLE 7.3**

Standard molar entropies ($S^\circ$) for the noble gases, the gaseous halogens, and the hydrogen halides at 298.15 K.

| Noble gas | $S^\circ/\text{J}\cdot\text{K}^{-1}\cdot\text{mol}^{-1}$ | Halogen | $S^\circ/\text{J}\cdot\text{K}^{-1}\cdot\text{mol}^{-1}$ | Hydrogen halide | $S^\circ/\text{J}\cdot\text{K}^{-1}\cdot\text{mol}^{-1}$ |
|---|---|---|---|---|---|
| He(g) | 126.2 | $F_2$(g) | 202.8 | HF(g) | 173.8 |
| Ne(g) | 146.3 | $Cl_2$(g) | 223.1 | HCl(g) | 186.9 |
| Ar(g) | 154.8 | $Br_2$(g) | 245.5 | HBr(g) | 198.7 |
| Kr(g) | 164.1 | $I_2$(g) | 260.7 | HI(g) | 206.6 |
| Xe(g) | 169.7 | | | | |

more closely spaced are the energy levels. The same trend can be seen by comparing the standard molar entropies at 298.15 K of the gaseous halogens and hydrogen halides (see Table 7.3 and Figure 7.3).

Generally speaking, the more atoms of a given type in a molecule, the greater is the capacity of the molecule to take up energy and thus the greater is its entropy (the greater the number of atoms, the more different ways in which the molecules can vibrate.) This trend is illustrated by the series $C_2H_2$(g), $C_2H_4$(g), and $C_2H_6$(g), whose standard molar entropies in joules per kelvin per mole at 298.15 K are 201, 220, and 230, respectively. For molecules with the same geometry and number of atoms, the standard molar entropy increases with increasing molecular mass.

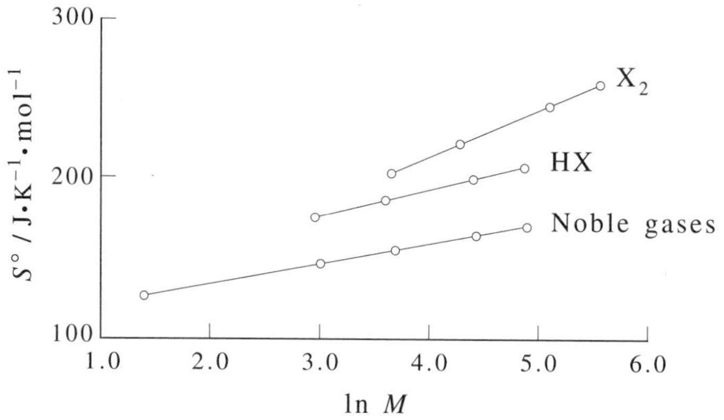


**FIGURE 7.3**
Standard molar entropies ($S°$) for the noble gases, the gaseous halogens, and the hydrogen halides at 298.15 K plotted against ln $M$, where $M$ is the molecular mass.

**EXAMPLE 7–6**

Arrange the following molecules in order of increasing standard molar entropy: $CH_2Cl_2$(g); $CHCl_3$(g); $CH_3Cl$(g).

SOLUTION: The number of atoms is the same in each case, but chlorine has a greater mass than hydrogen. Thus, we predict that

$$S°[CH_3Cl(g)] < S°[CH_2Cl_2(g)] < S°[CHCl_3(g)]$$

This ordering is in agreement with the values of the standard molar entropies at 298.15 K, which are in units of joules per kelvin per mole, 234.6, 270.2, and 295.7, respectively.

An interesting comparison is given by the isomers acetone and trimethylene oxide (whose molecular structures are shown below), whose standard molar entropies for the gaseous forms at 298.15 K are 298 $J \cdot K^{-1} \cdot mol^{-1}$ and 274 $J \cdot K^{-1} \cdot mol^{-1}$, respectively. The entropy of acetone is higher than the entropy of trimethylene oxide because of the free rotation of the methyl groups about the carbon-carbon bonds in the acetone molecule. The relatively rigid ring structure of the trimethylene oxide molecule restricts the movement of the ring atoms. This restriction gives rise to a lower molar entropy because the capacity of the rigid isomer to take up energy is less than that of the more flexible acetone molecule, which has more possibilities for intermolecular motion. For molecules with approximately the same molecular masses, the more compact the molecule is, the smaller is its entropy.

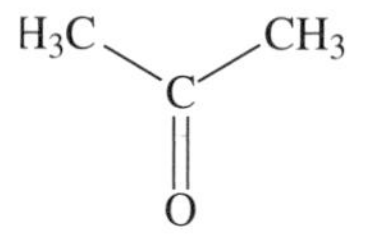


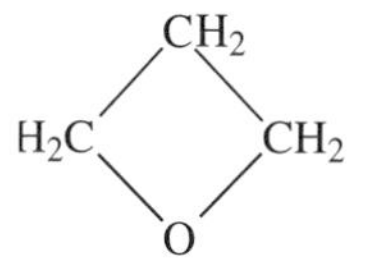


Note that Table 7.2 gives $S^\circ = 245.5\ J \cdot K^{-1} \cdot mol^{-1}$ for $Br_2(g)$ at 298.15 K and one bar. But bromine is a liquid at 298.15 K and one bar, so where does such a value come from? Even though bromine is a liquid under these conditions, we can calculate $S^\circ[Br_2(g)]$ according to the scheme outlined in Figure 7.4. Therefore, we need the values of the molar heat capacity of $Br_2(l)$ (75.69 $J \cdot K^{-1} \cdot mol^{-1}$), the molar heat capacity of $Br_2(g)$ (36.02 $J \cdot K^{-1} \cdot mol^{-1}$), the normal boiling point of $Br_2(l)$ (332.0 K), and the molar

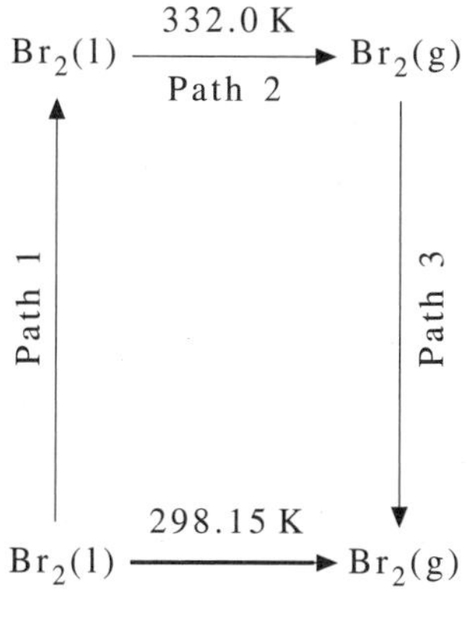


**FIGURE 7.4**
The scheme used to calculate $S^\circ[Br_2(g)]$ at 298.15 K. In Path 1, $Br_2(l)$ is heated to its boiling point, 332.0 K. Then $Br_2(l)$ is vaporized to $Br_2(g)$ at 332.0 K (Path 2), and finally $Br_2(g)$ is cooled from 332.0 K back to 298.15 K (Path 3).

enthalpy of vaporization at 332.0 K (29.54 $kJ\cdot mol^{-1}$). We start with $Br_2(l)$ at 298.15 K and heat it to its boiling point. The value of $\Delta\overline{S}$ for this first step is (Equation 7.7)

$$\Delta\overline{S}_1 = \overline{S}^{l}(332.0\text{ K}) - \overline{S}^{l}(298.15\text{ K}) = \overline{C}^{l}_{P} \ln \frac{T_2}{T_1}$$

$$= (75.69\text{ J}\cdot\text{K}^{-1}\cdot\text{mol}^{-1}) \ln \frac{332.0\text{ K}}{298.15\text{ K}} = 8.140\text{ J}\cdot\text{K}^{-1}\cdot\text{mol}^{-1}$$

Now vaporize the bromine at its normal boiling point (step 2 in Figure 7.4):

$$\Delta\overline{S}_2 = \overline{S}^{g}(332.0\text{ K}) - \overline{S}^{l}(332.0\text{ K}) = \frac{\Delta_{\text{vap}}\overline{H}}{T_{\text{vap}}} = \frac{29.54\text{ kJ}\cdot\text{mol}^{-1}}{332.0\text{ K}}$$

$$= 88.98\text{ J}\cdot\text{K}^{-1}\cdot\text{mol}^{-1}$$

Last, cool the gas from 332.0 K back to 298.15 K (step 3):

$$\Delta\overline{S}_3 = \overline{S}^{g}(298.15\text{ K}) - \overline{S}^{g}(332.0\text{ K}) = \overline{C}^{g}_{P} \ln \frac{298.15\text{ K}}{332.0\text{ K}}$$

$$= (36.02\text{ J}\cdot\text{K}^{-1}\cdot\text{mol}^{-1}) \ln \frac{298.15}{332.0} = -3.87\text{ J}\cdot\text{K}^{-1}\cdot\text{mol}^{-1}$$

If we add these three steps and then add the results to $S^{\circ}_{298}[Br_2(l)] = 152.2\text{ J}\cdot\text{K}^{-1}\cdot\text{mol}^{-1}$ (Table 7.2), we obtain

$$\begin{aligned} S^{\circ}_{298}[\text{Br}_2(\text{g})] &= S^{\circ}_{298}[\text{Br}_2(\text{l})] + \Delta\overline{S}_1 + \Delta\overline{S}_2 + \Delta\overline{S}_3 \\ &= 152.2\text{ J}\cdot\text{K}^{-1}\cdot\text{mol}^{-1} + 8.14\text{ J}\cdot\text{K}^{-1}\cdot\text{mol}^{-1} \\ &\quad + 88.98\text{ J}\cdot\text{K}^{-1}\cdot\text{mol}^{-1} - 3.87\text{ J}\cdot\text{K}^{-1}\cdot\text{mol}^{-1} \\ &= 245.5\text{ J}\cdot\text{K}^{-1}\cdot\text{mol}^{-1} \end{aligned}$$

in agreement with the value of $Br_2(g)$ in Table 7.2. Incidentally, the spectroscopic value of $S^{\circ}[Br_2(g)]$, using Equation 7.24 and the data in Chapter 4 is 245.5 $J\cdot K^{-1}\cdot mol^{-1}$ (Problem 7–33).

## 7–8. The Spectroscopic Entropies of a Few Substances Do Not Agree with the Calorimetric Entropies

Table 7.4 compares calculated values of the molar entropies of several polyatomic gases with those measured calorimetrically. Note again, that the agreement with experiment is quite good. In fact, calculated values of the entropy are often more accurate than measured values, provided sophisticated enough spectroscopic models are used.

There is, however, a class of molecules for which the type of agreement in Table 7.4 is not found. For example, for carbon monoxide, $\overline{S}_{\text{calc}} = 160.3\text{ J}\cdot\text{K}^{-1}\cdot\text{mol}^{-1}$ and $\overline{S}_{\text{exp}} = 155.6\text{ J}\cdot\text{K}^{-1}\cdot\text{mol}^{-1}$ at its boiling point (81.6 K), for a discrepancy of 4.7 $J\cdot K^{-1}\cdot mol^{-1}$. Other such discrepancies are found, and in all cases $\overline{S}_{\text{calc}} > \overline{S}_{\text{exp}}$. The difference

**TABLE 7.4**
The standard molar entropies of several polyatomic gases at 298.15 K and one bar.

| Gas | $S^\circ(\text{calc})/\text{J}\cdot\text{K}^{-1}\cdot\text{mol}^{-1}$ | $S^\circ(\text{exp})/\text{J}\cdot\text{K}^{-1}\cdot\text{mol}^{-1}$ |
|---|---|---|
| $CO_2$ | 213.8 | 213.7 |
| $NH_3$ | 192.8 | 192.6 |
| $NO_2$ | 240.1 | 240.2 |
| $CH_4$ | 186.3 | 186.3 |
| $C_2H_2$ | 200.9 | 200.8 |
| $C_2H_4$ | 219.6 | 219.6 |
| $C_2H_6$ | 229.6 | 229.5 |

$\overline{S}_{\text{calc}} - \overline{S}_{\text{exp}}$ is often referred to as *residual entropy*. The explanation of these cases is the following. Carbon monoxide has a very small dipole moment ($\approx 4 \times 10^{-31}$ C·m), so when carbon monoxide is crystallized, the molecules do not have a strong tendency to line up in an energetically favorable way. The resultant crystal, then, is a random mixture of the two possible orientations, CO and OC. As the crystal is cooled down toward 0 K, each molecule gets locked into its orientation and cannot realize the state of lowest energy with $W = 1$, that is, all the molecules oriented in the same direction. Instead, the number of configurations $W$ of the crystal is $2^N$, because each of the $N$ molecules exists equally likely (almost equally likely because the dipole moment is so small) in two states. Thus, the molar entropy of the crystal at 0 K is $S = R\ln 2$ instead of zero. If $R\ln 2 = 5.7\ \text{J}\cdot\text{K}^{-1}\cdot\text{mol}^{-1}$ is added to the experimental entropy, the agreement in the case of carbon monoxide becomes satisfactory. If it were possible to obtain carbon monoxide in its true equilibrium state at $T = 0$ K, this discrepancy would not occur. A similar situation occurs with nitrous oxide, which is a linear molecule with the structure NNO. For $H_3CD$, the residual entropy is $11.7\ \text{J}\cdot\text{K}^{-1}\cdot\text{mol}^{-1}$, which is explained by realizing that each molecule of monodeuterated methane can assume four different orientations in the low-temperature crystal, so $\overline{S}_{\text{residual}} = R\ln 4 = 11.5\ \text{J}\cdot\text{K}^{-1}\cdot\text{mol}^{-1}$, in very close agreement with the experimental value.

## 7–9. Standard Entropies Can Be Used to Calculate Entropy Changes of Chemical Reactions

One of the most important uses of tables of standard molar entropies is for the calculation of entropy changes of chemical reactions. These changes are calculated in much the same way we calculated standard enthalpy changes of reactions from standard molar enthalpies of formation in Chapter 5. For the general reaction

$$a\,\text{A} + b\,\text{B} \longrightarrow y\,\text{Y} + z\,\text{Z}$$

the standard entropy change is given by

$$\Delta_r S^\circ = yS^\circ[\mathrm{Y}] + zS^\circ[\mathrm{Z}] - aS^\circ[\mathrm{A}] - bS^\circ[\mathrm{B}]$$

For example, using the values of $S^\circ$ given in Table 7.2 for the substances in the reaction described by the chemical equation

$$\mathrm{H_2(g)} + \tfrac{1}{2}\,\mathrm{O_2(g)} \rightleftharpoons \mathrm{H_2O(l)}$$

$$\begin{aligned}\Delta_r S^\circ &= (1)S^\circ[\mathrm{H_2O(l)}] - (1)S^\circ[\mathrm{H_2(g)}] - (\tfrac{1}{2})S^\circ[\mathrm{O_2(g)}] \\ &= (1)(70.0\ \mathrm{J{\cdot}K^{-1}{\cdot}mol^{-1}}) - (1)(130.7\ \mathrm{J{\cdot}K^{-1}{\cdot}mol^{-1}}) - (\tfrac{1}{2})(205.2\ \mathrm{J{\cdot}K^{-1}{\cdot}mol^{-1}}) \\ &= -163.3\ \mathrm{J{\cdot}K^{-1}{\cdot}mol^{-1}}\end{aligned}$$

This value of $\Delta_r S^\circ$ represents the value of $\Delta_r S$ for the combustion of one mole of $H_2(g)$ or the formation of one mole of $H_2O(l)$, when all the reactants and products are in their standard states. The large negative value of $\Delta_r S^\circ$ reflects the loss of gaseous reactants to produce a condensed phase, an ordering process.

We will use tables of standard enthalpies of formation and standard entropies to calculate equilibrium constants of chemical reactions in Chapter 12.

## Problems

**7-1.** Form the total derivative of $H$ as a function of $T$ and $P$ and equate the result to $dH$ in Equation 7.6 to derive Equations 7.7 and 7.8.

**7-2.** The molar heat capacity of $H_2O(l)$ has an approximately constant value of $\overline{C}_P = 75.4\ \mathrm{J{\cdot}K^{-1}{\cdot}mol^{-1}}$ from 0°C to 100°C. Calculate $\Delta S$ if two moles of $H_2O(l)$ are heated from 10°C to 90°C at constant pressure.

**7-3.** The molar heat capacity of butane can be expressed by

$$\overline{C}_P/R = 0.05641 + (0.04631\ \mathrm{K^{-1}})T - (2.392 \times 10^{-5}\ \mathrm{K^{-2}})T^2 + (4.807 \times 10^{-9}\ \mathrm{K^{-3}})T^3$$

over the temperature range $300\ \mathrm{K} \le T \le 1500\ \mathrm{K}$. Calculate $\Delta S$ if one mole of butane is heated from 300 K to 1000 K at constant pressure.

**7-4.** The molar heat capacity of $C_2H_4(g)$ can be expressed by

$$\overline{C}_V(T)/R = 16.4105 - \frac{6085.929\ \mathrm{K}}{T} + \frac{822\,826\ \mathrm{K}^2}{T^2}$$

over the temperature range $300\ \mathrm{K} < T < 1000\ \mathrm{K}$. Calculate $\Delta S$ if one mole of ethene is heated from 300 K to 600 K at constant volume.

**7-5.** Use the data in Problem 7–4 to calculate $\Delta S$ if one mole of ethene is heated from 300 K to 600 K at constant pressure. Assume ethene behaves ideally.

**7-6.** We can calculate the difference in the results of Problems 7–4 and 7–5 in the following way. First, show that because $\overline{C}_P - \overline{C}_V = R$ for an ideal gas,

$$\Delta \overline{S}_P = \Delta \overline{S}_V + R \ln \frac{T_2}{T_1}$$

Check to see numerically that your answers to Problems 7–4 and 7–5 differ by $R \ln 2 = 0.693R = 5.76\ \mathrm{J \cdot K^{-1} \cdot mol^{-1}}$.

**7-7.** The results of Problems 7–4 and 7–5 must be connected in the following way. Show that the two processes can be represented by the diagram

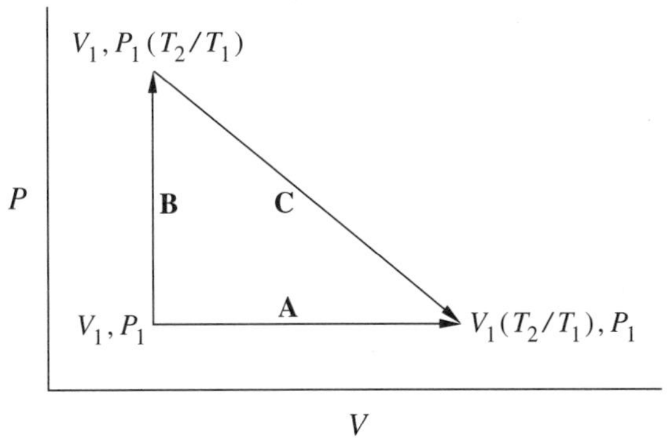


where paths A and B represent the processes in Problems 7–5 and 7–4, respectively. Now, path A is equivalent to the sum of paths B and C. Show that $\Delta S_C$ is given by

$$\Delta S_C = nR \ln \frac{V_1\left(\frac{T_2}{T_1}\right)}{V_1} = nR \ln \frac{P_1\left(\frac{T_2}{T_1}\right)}{P_1} = nR \ln \frac{T_2}{T_1}$$

and that the result given in Problem 7–6 follows.

**7-8.** Use Equations 6.23 and 6.24 to show that $S = 0$ at 0 K, where every system will be in its ground state.

**7-9.** Prove that $S = -k_B \sum p_j \ln p_j = 0$ when $p_1 = 1$ and all the other $p_j = 0$. In other words, prove that $x \ln x \to 0$ as $x \to 0$.

**7-10.** It has been found experimentally that $\Delta_{\text{vap}}\overline{S} \approx 88\ \text{J}\cdot\text{K}^{-1}\cdot\text{mol}^{-1}$ for many nonassociated liquids. This rough rule of thumb is called *Trouton's rule*. Use the following data to test the validity of Trouton's rule.

| Substance | $t_{\text{fus}}/^\circ\text{C}$ | $t_{\text{vap}}/^\circ\text{C}$ | $\Delta_{\text{fus}}\overline{H}/\text{kJ}\cdot\text{mol}^{-1}$ | $\Delta_{\text{vap}}\overline{H}/\text{kJ}\cdot\text{mol}^{-1}$ |
|---|---|---|---|---|
| Pentane | −129.7 | 36.06 | 8.42 | 25.79 |
| Hexane | −95.3 | 68.73 | 13.08 | 28.85 |
| Heptane | −90.6 | 98.5 | 14.16 | 31.77 |
| Ethylene oxide | −111.7 | 10.6 | 5.17 | 25.52 |
| Benzene | 5.53 | 80.09 | 9.95 | 30.72 |
| Diethyl ether | −116.3 | 34.5 | 7.27 | 26.52 |
| Tetrachloromethane | −23 | 76.8 | 3.28 | 29.82 |
| Mercury | −38.83 | 356.7 | 2.29 | 59.11 |
| Bromine | −7.2 | 58.8 | 10.57 | 29.96 |

**7-11.** Use the data in Problem 7–10 to calculate the value of $\Delta_{\text{fus}}\overline{S}$ for each substance.

**7-12.** Why is $\Delta_{\text{vap}}\overline{S} > \Delta_{\text{fus}}\overline{S}$?

**7-13.** Show that if $C_P^{\text{s}}(T) \rightarrow T^{\alpha}$ as $T \rightarrow 0$, where $\alpha$ is a positive constant, then $S(T) \rightarrow 0$ as $T \rightarrow 0$.

**7-14.** Use the following data to calculate the standard molar entropy of $N_2(g)$ at 298.15 K.

$$C_P^\circ[\text{N}_2(\text{s}_1)]/R = -0.03165 + (0.05460\ \text{K}^{-1})T + (3.520 \times 10^{-3}\ \text{K}^{-2})T^2 - (2.064 \times 10^{-5}\ \text{K}^{-3})T^3$$
$$10\ \text{K} \le T \le 35.61\ \text{K}$$

$$C_P^\circ[\text{N}_2(\text{s}_2)]/R = -0.1696 + (0.2379\ \text{K}^{-1})T - (4.214 \times 10^{-3}\ \text{K}^{-2})T^2 + (3.036 \times 10^{-5}\ \text{K}^{-3})T^3$$
$$35.61\ \text{K} \le T \le 63.15\ \text{K}$$

$$C_P^\circ[\text{N}_2(\text{l})]/R = -18.44 + (1.053\ \text{K}^{-1})T - (0.0148\ \text{K}^{-2})T^2 + (7.064 \times 10^{-5}\ \text{K}^{-3})T^3$$
$$63.15\ \text{K} \le T \le 77.36\ \text{K}$$

$C_P^\circ[\text{N}_2(\text{g})]/R = 3.500$ from $77.36\ \text{K} \le T \le 1000\ \text{K}$, $\overline{C}_P(T = 10.0\ \text{K}) = 6.15\ \text{J}\cdot\text{K}^{-1}\cdot\text{mol}^{-1}$, $T_{\text{trs}} = 35.61$ K, $\Delta_{\text{trs}}\overline{H} = 0.2289\ \text{kJ}\cdot\text{mol}^{-1}$, $T_{\text{fus}} = 63.15$ K, $\Delta_{\text{fus}}\overline{H} = 0.71\ \text{kJ}\cdot\text{mol}^{-1}$, $T_{\text{vap}} =$ 77.36 K, and $\Delta_{\text{vap}}\overline{H} = 5.57\ \text{kJ}\cdot\text{mol}^{-1}$. The correction for nonideality (Problem 8–20) $= 0.02\ \text{J}\cdot\text{K}^{-1}\cdot\text{mol}^{-1}$.

**7-15.** Use the data in Problem 7–14 and $\overline{C}_P[N_2(g)]/R = 3.307 + (6.29 \times 10^{-4}\ K^{-1})T$ for $T \geq 77.36$ K to plot the standard molar entropy of nitrogen as a function of temperature from 0 K to 1000 K.

**7-16.** The molar heat capacities of solid, liquid, and gaseous chlorine can be expressed as

$$C_P^\circ[Cl_2(s)]/R = -1.545 + (0.1502\ K^{-1})T - (1.179 \times 10^{-3}\ K^{-2})T^2 + (3.441 \times 10^{-6}\ K^{-3})T^3$$
$$15\ K \leq T \leq 172.12\ K$$

$$C_P^\circ[Cl_2(l)]/R = 7.689 + (5.582 \times 10^{-3}\ K^{-1})T - (1.954 \times 10^{-5}\ K^{-2})T^2$$
$$172.12\ K \leq T \leq 239.0\ K$$

$$C_P^\circ[Cl_2(g)]/R = 3.812 + (1.220 \times 10^{-3}\ K^{-1})T - (4.856 \times 10^{-7}\ K^{-2})T^2$$
$$239.0\ K \leq T \leq 1000\ K$$

Use the above molar heat capacities and $T_{fus} = 172.12$ K, $\Delta_{fus}\overline{H} = 6.406\ kJ\cdot mol^{-1}$, $T_{vap} = 239.0$ K, $\Delta_{vap}\overline{H} = 20.40\ kJ\cdot mol^{-1}$, $\Theta_D = 116$ K and the correction for nonideality $= 0.502\ J\cdot K^{-1}\cdot mol^{-1}$ to calculate the standard molar entropy of chlorine at 298.15 K. Compare your result with the value given in Table 7.2.

**7-17.** Use the data in Problem 7–16 to plot the standard molar entropy of chlorine as a function of temperature from 0 K to 1000 K.

**7-18.** Use the following data to calculate the standard molar entropy of cyclopropane at 298.15 K.

$$C_P^\circ[C_3H_6(s)]/R = -1.921 + (0.1508\ K^{-1})T - (9.670 \times 10^{-4}\ K^{-2})T^2 + (2.694 \times 10^{-6}\ K^{-3})T^3$$
$$15\ K \leq T \leq 145.5\ K$$

$$C_P^\circ[C_3H_6(l)]/R = 5.624 + (4.493 \times 10^{-2}\ K^{-1})T - (1.340 \times 10^{-4}\ K^{-2})T^2$$
$$145.5\ K \leq T \leq 240.3\ K$$

$$C_P^\circ[C_3H_6(g)]/R = -1.793 + (3.277 \times 10^{-2}\ K^{-1})T - (1.326 \times 10^{-5}\ K^{-2})T^2$$
$$240.3\ K \leq T \leq 1000\ K$$

$T_{fus} = 145.5$ K, $T_{vap} = 240.3$ K, $\Delta_{fus}\overline{H} = 5.44\ kJ\cdot mol^{-1}$, $\Delta_{vap}\overline{H} = 20.05\ kJ\cdot mol^{-1}$, and $\Theta_D = 130$ K. The correction for nonideality $= 0.54\ J\cdot K^{-1}\cdot mol^{-1}$.

**7-19.** Use the data in Problem 7–18 to plot the standard molar entropy of cyclopropane from 0 K to 1000 K.

**7-20.** The constant-pressure molar heat capacity of $N_2O$ as a function of temperature is tabulated below. Dinitrogen oxide melts at 182.26 K with $\Delta_{fus}\overline{H} = 6.54\ kJ\cdot mol^{-1}$, and boils at 184.67 K with $\Delta_{vap}\overline{H} = 16.53\ kJ\cdot mol^{-1}$ at one bar. Assuming the heat capacity of solid dinitrogen oxide can be described by the Debye theory up to 15 K, calculate the molar entropy of $N_2O(g)$ at its boiling point.

| $T$/K | $\overline{C}_P$/J·K$^{-1}$·mol$^{-1}$ | $T$/K | $\overline{C}_P$/J·K$^{-1}$·mol$^{-1}$ |
|---|---|---|---|
| 15.17 | 2.90 | 120.29 | 45.10 |
| 19.95 | 6.19 | 130.44 | 47.32 |
| 25.81 | 10.89 | 141.07 | 48.91 |
| 33.38 | 16.98 | 154.71 | 52.17 |
| 42.61 | 23.13 | 164.82 | 54.02 |
| 52.02 | 28.56 | 174.90 | 56.99 |
| 57.35 | 30.75 | 180.75 | 58.83 |
| 68.05 | 34.18 | 182.26 | Melting point |
| 76.67 | 36.57 | 183.55 | 77.70 |
| 87.06 | 38.87 | 183.71 | 77.45 |
| 98.34 | 41.13 | 184.67 | Boiling point |
| 109.12 | 42.84 | | |

**7-21.** Methylammonium chloride occurs as three crystalline forms, called $\beta$, $\gamma$, and $\alpha$, between 0 K and 298.15 K. The constant-pressure molar heat capacity of methylammonium chloride as a function of temperature is tabulated below. The $\beta \rightarrow \gamma$ transition occurs at 220.4 K with $\Delta_{\text{trs}}\overline{H} = 1.779\ \text{kJ}\cdot\text{mol}^{-1}$ and the $\gamma \rightarrow \alpha$ transition occurs at 264.5 K with $\Delta_{\text{trs}}\overline{H} = 2.818\ \text{kJ}\cdot\text{mol}^{-1}$. Assuming the heat capacity of solid methylammonium chloride can be described by the Debye theory up to 12 K, calculate the molar entropy of methylammonium chloride at 298.15 K.

| $T$/K | $\overline{C}_P$/J·K$^{-1}$·mol$^{-1}$ | $T$/K | $\overline{C}_P$/J·K$^{-1}$·mol$^{-1}$ |
|---|---|---|---|
| 12 | 0.837 | 180 | 73.72 |
| 15 | 1.59 | 200 | 77.95 |
| 20 | 3.92 | 210 | 79.71 |
| 30 | 10.53 | 220.4 | $\beta \rightarrow \gamma$ transition |
| 40 | 18.28 | 222 | 82.01 |
| 50 | 25.92 | 230 | 82.84 |
| 60 | 32.76 | 240 | 84.27 |
| 70 | 38.95 | 260 | 87.03 |
| 80 | 44.35 | 264.5 | $\gamma \rightarrow \alpha$ transition |
| 90 | 49.08 | 270 | 88.16 |
| 100 | 53.18 | 280 | 89.20 |
| 120 | 59.50 | 290 | 90.16 |
| 140 | 64.81 | 295 | 90.63 |
| 160 | 69.45 | | |

**7-22.** The constant-pressure molar heat capacity of chloroethane as a function of temperature is tabulated below. Chloroethane melts at 134.4 K with $\Delta_{\text{fus}}\overline{H} = 4.45\ \text{kJ}\cdot\text{mol}^{-1}$, and boils at 286.2 K with $\Delta_{\text{vap}}\overline{H} = 24.65\ \text{kJ}\cdot\text{mol}^{-1}$ at one bar. Furthermore, the heat capacity of solid chloroethane can be described by the Debye theory up to 15 K. Use these data to calculate the molar entropy of chloroethane at its boiling point.

| $T$/K | $\overline{C}_P$/J·K$^{-1}$·mol$^{-1}$ | $T$/K | $\overline{C}_P$/J·K$^{-1}$·mol$^{-1}$ |
|---|---|---|---|
| 15 | 5.65 | 130 | 84.60 |
| 20 | 11.42 | 134.4 | 90.83 (solid) |
| 25 | 16.53 | | 97.19 (liquid) |
| 30 | 21.21 | 140 | 96.86 |
| 35 | 25.52 | 150 | 96.40 |
| 40 | 29.62 | 160 | 96.02 |
| 50 | 36.53 | 180 | 95.65 |
| 60 | 42.47 | 200 | 95.77 |
| 70 | 47.53 | 220 | 96.04 |
| 80 | 52.63 | 240 | 97.78 |
| 90 | 55.23 | 260 | 99.79 |
| 100 | 59.66 | 280 | 102.09 |
| 110 | 65.48 | 286.2 | 102.13 |
| 120 | 73.55 | | |

**7-23.** The constant-pressure molar heat capacity of nitromethane as a function of temperature is tabulated below. Nitromethane melts at 244.60 K with $\Delta_{\text{fus}}\overline{H} = 9.70\ \text{kJ}\cdot\text{mol}^{-1}$, and boils at 374.34 K at one bar with $\Delta_{\text{vap}}\overline{H} = 38.27\ \text{kJ}\cdot\text{mol}^{-1}$. Furthermore, the heat capacity of solid nitromethane can be described by the Debye theory up to 15 K. Use these data to calculate the molar entropy of nitromethane at 298.15 K and one bar. The vapor pressure of nitromethane is 36.66 torr at 298.15 K. (Be sure to take into account $\Delta S$ for the isothermal compression of nitromethane from its vapor pressure to one bar at 298.15 K).

| $T$/K | $\overline{C}_P$/J·K$^{-1}$·mol$^{-1}$ | $T$/K | $\overline{C}_P$/J·K$^{-1}$·mol$^{-1}$ |
|---|---|---|---|
| 15 | 3.72 | 200 | 71.46 |
| 20 | 8.66 | 220 | 75.23 |
| 30 | 19.20 | 240 | 78.99 |
| 40 | 28.87 | 244.60 | melting point |
| 60 | 40.84 | 250 | 104.43 |
| 80 | 47.99 | 260 | 104.64 |
| 100 | 52.80 | 270 | 104.93 |
| 120 | 56.74 | 280 | 105.31 |
| 140 | 60.46 | 290 | 105.69 |
| 160 | 64.06 | 300 | 106.06 |
| 180 | 67.74 | | |

**7-24.** Use the following data to calculate the standard molar entropy of CO(g) at its normal boiling point. Carbon monoxide undergoes a solid-solid phase transition at 61.6 K. Compare your result with the calculated value of 160.3 $\text{J·K}^{-1}\text{·mol}^{-1}$. Why is there a discrepancy between the calculated value and the experimental value?

$$\overline{C}_P[\text{CO(s}_1)]/R = -2.820 + (0.3317\ \text{K}^{-1})T - (6.408\times10^{-3}\ \text{K}^{-2})T^2 + (6.002\times10^{-5}\ \text{K}^{-3})T^3$$
$$10\ \text{K} \le T \le 61.6\ \text{K}$$

$$\overline{C}_P[\text{CO(s}_2)]/R = 2.436 + (0.05694\ \text{K}^{-1})T$$
$$61.6\ \text{K} \le T \le 68.1\ \text{K}$$

$$\overline{C}_P[\text{CO(l)}]/R = 5.967 + (0.0330\ \text{K}^{-1})T - (2.088\times10^{-4}\ \text{K}^{-2})T^2$$
$$68.1\ \text{K} \le T \le 81.6\ \text{K}$$

and $T_{\text{trs}}(\text{s}_1 \rightarrow \text{s}_2) = 61.6$ K, $T_{\text{fus}} = 68.1$ K, $T_{\text{vap}} = 81.6$ K, $\Delta_{\text{fus}}\overline{H} = 0.836\ \text{kJ·mol}^{-1}$, $\Delta_{\text{trs}}\overline{H} = 0.633\ \text{kJ·mol}^{-1}$, $\Delta_{\text{vap}}\overline{H} = 6.04\ \text{kJ·mol}^{-1}$, $\Theta_{\text{D}} = 79.5$ K, and the correction for nonideality $= 0.879\ \text{J·K}^{-1}\text{·mol}^{-1}$.

**7-25.** The molar heat capacities of solid and liquid water can be expressed by

$$\overline{C}_P[\text{H}_2\text{O(s)}]/R = -0.2985 + (2.896\times10^{-2}\ \text{K}^{-1})T - (8.6714\times10^{-5}\ \text{K}^{-2})T^2 + (1.703\times10^{-7}\ \text{K}^{-3})T^3$$
$$10\ \text{K} \le T \le 273.15\ \text{K}$$

$$\overline{C}_P[\text{H}_2\text{O(l)}]/R = 22.447 - (0.11639\ \text{K}^{-1})T + (3.3312\times10^{-4}\ \text{K}^{-2})T^2 - (3.1314\times10^{-7}\ \text{K}^{-3})T^3$$
$$273.15\ \text{K} \le T \le 298.15\ \text{K}$$

and $T_{\text{fus}} = 273.15$ K, $\Delta_{\text{fus}}\overline{H} = 6.007\ \text{kJ·mol}^{-1}$, $\Delta_{\text{vap}}\overline{H}(T = 298.15\ \text{K}) = 43.93\ \text{kJ·mol}^{-1}$, $\Theta_{\text{D}} = 192$ K, the correction for nonideality $= 0.32\ \text{J·K}^{-1}\text{·mol}^{-1}$, and the vapor pressure of $H_2O$ at 298.15 K = 23.8 torr. Use these data to calculate the standard molar entropy of $H_2O$(g) at 298.15 K. You need the vapor pressure of water at 298.15 K because that is the equilibrium pressure of $H_2O$(g) when it is vaporized at 298.15 K. You must include the value of $\Delta S$ that results when you compress the $H_2O$(g) from 23.8 torr to its standard value of one bar. Your answer should come out to be 185.6 $\text{J·K}^{-1}\text{·mol}^{-1}$, which does not agree exactly with the value in Table 7.2. There is a residual entropy associated with ice, which a detailed analysis of the structure of ice gives as $\Delta S_{\text{residual}} = R\ln(3/2) = 3.4\ \text{J·K}^{-1}\text{·mol}^{-1}$, which is in good agreement with $\overline{S}_{\text{calc}} - \overline{S}_{\text{exp}}$.

**7-26.** Use the data in Problem 7–25 and the empirical expression

$$\overline{C}_P[\text{H}_2\text{O(g)}]/R = 3.652 + (1.156\times10^{-3}\ \text{K}^{-1})T - (1.424\times10^{-7}\ \text{K}^{-2})T^2$$
$$300\ \text{K} \le T \le 1000\ \text{K}$$

to plot the standard molar entropy of water from 0 K to 500 K.

**7-27.** Show for an ideal gas that

$$\overline{S} = R\ln\frac{qe}{N_{\text{A}}} + RT\left(\frac{\partial \ln q}{\partial T}\right)_V$$

**7-28.** Show that Equations 3.21 and 7.19 are consistent with Equations 7.2 and 7.3.

**7-29.** Substitute Equation 7.23 into Equation 7.19 and derive the equation (Problem 6–31)

$$\Delta\overline{S} = \overline{C}_V \ln \frac{T_2}{T_1} + R \ln \frac{V_2}{V_1}$$

for one mole of a monatomic ideal gas.

**7-30.** Use Equation 7.24 and the data in Chapter 4 to calculate the standard molar entropy of $Cl_2(g)$ at 298.15 K. Compare your answer with the experimental value of $223.1\ J\cdot K^{-1}\cdot mol^{-1}$.

**7-31.** Use Equation 7.24 and the data in Chapter 4 to calculate the standard molar entropy of CO(g) at its standard boiling point, 81.6 K. Compare your answer with the experimental value of $155.6\ J\cdot K^{-1}\cdot mol^{-1}$. Why is there a discrepancy of about $5\ J\cdot K^{-1}\cdot mol^{-1}$?

**7-32.** Use Equation 7.26 and the data in Chapter 4 to calculate the standard molar entropy of $NH_3(g)$ at 298.15 K. Compare your answer with the experimental value of $192.8\ J\cdot K^{-1}\cdot mol^{-1}$.

**7-33.** Use Equation 7.24 and the data in Chapter 4 to calculate the standard molar entropy of $Br_2(g)$ at 298.15 K. Compare your answer with the experimental value of $245.5\ J\cdot K^{-1}\cdot mol^{-1}$.

**7-34.** The vibrational and rotational constants for HF(g) within the harmonic oscillator-rigid rotator model are $\tilde{\nu} = 3959\ cm^{-1}$ and $\tilde{B} = 20.56\ cm^{-1}$. Calculate the standard molar entropy of HF(g) at 298.15 K. How does this value compare with that in Table 7.3?

**7-35.** Calculate the standard molar entropy of $H_2(g)$ and $D_2(g)$ at 298.15 K given that the bond length of both diatomic molecules is 74.16 pm and the vibrational temperatures of $H_2(g)$ and $D_2(g)$ are 6332 K and 4480 K, respectively. Calculate the standard molar entropy of HD(g) at 298.15 K ($R_e = 74.13$ pm and $\Theta_{vib} = 5496$ K).

**7-36.** Calculate the standard molar entropy of HCN(g) at 1000 K given that $I = 1.8816 \times 10^{-46}\ kg\cdot m^2$, $\tilde{\nu}_1 = 2096.70\ cm^{-1}$, $\tilde{\nu}_2 = 713.46\ cm^{-1}$, and $\tilde{\nu}_3 = 3311.47\ cm^{-1}$. Recall that HCN(g) is a linear triatomic molecule and therefore the bending mode, $\nu_2$, is doubly degenerate.

**7-37.** Given that $\tilde{\nu}_1 = 1321.3\ cm^{-1}$, $\tilde{\nu}_2 = 750.8\ cm^{-1}$, $\tilde{\nu}_3 = 1620.3\ cm^{-1}$, $\tilde{A} = 7.9971\ cm^{-1}$, $\tilde{B} = 0.4339\ cm^{-1}$, and $\tilde{C} = 0.4103\ cm^{-1}$, calculate the standard molar entropy of $NO_2(g)$ at 298.15 K. (Note that $NO_2(g)$ is a bent triatomic molecule.) How does your value compare with that in Table 7.2?

**7-38.** In Problem 7–48, you are asked to calculate the value of $\Delta_r S^\circ$ at 298.15 K using the data in Table 7.2 for the reaction described by

$$2\,CO(g) + O_2(g) \longrightarrow 2\,CO_2(g)$$

Use the data in Table 4.2 to calculate the standard molar entropy of each of the reagents in this reaction [see Example 7–5 for the calculation of the standard molar entropy of $CO_2(g)$]. Then use these results to calculate the standard entropy change for the above reaction. How does your answer compare with what you obtained in Problem 7–48?

**7-39.** Calculate the value of $\Delta_r S^\circ$ for the reaction described by

$$H_2(g) + \tfrac{1}{2}\,O_2(g) \longrightarrow H_2O(g)$$

at 500 K using the data in Tables 4.2 and 4.4.

**7-40.** In each case below, predict which molecule of the pair has the greater molar entropy under the same conditions (assume gaseous species).

**a.** $CO$ $CO_2$

**b.** $CH_3CH_2CH_3$ 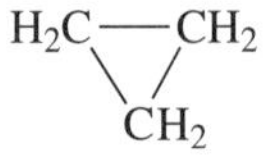


**c.** $CH_3CH_2CH_2CH_2CH_3$ 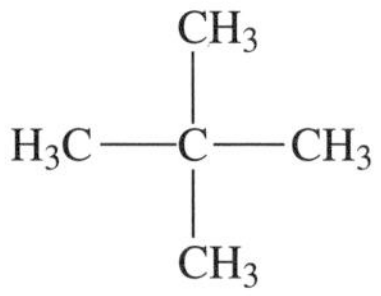


**7-41.** In each case below, predict which molecule of the pair has the greater molar entropy under the same conditions (assume gaseous species).

**a.** $H_2O$ $D_2O$

**b.** $CH_3CH_2OH$ 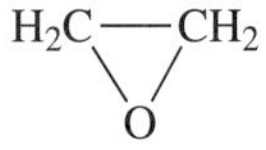


**c.** $CH_3CH_2CH_2CH_2NH_2$ 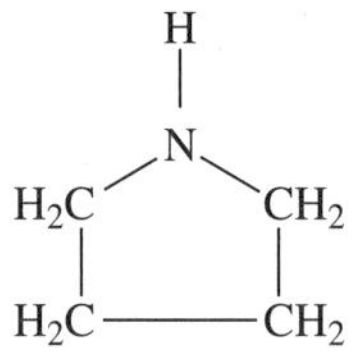


**7-42.** Arrange the following reactions according to increasing values of $\Delta_r S^\circ$ (do not consult any references).

**a.** $S(s) + O_2(g) \longrightarrow SO_2(g)$
**b.** $H_2(g) + O_2(g) \longrightarrow H_2O_2(l)$
**c.** $CO(g) + 3\,H_2(g) \longrightarrow CH_4(g) + H_2O(l)$
**d.** $C(s) + H_2O(g) \longrightarrow CO(g) + H_2(g)$

**7-43.** Arrange the following reactions according to increasing values of $\Delta_r S^\circ$ (do not consult any references).

**a.** $2\,H_2(g) + O_2(g) \longrightarrow 2\,H_2O(l)$
**b.** $NH_3(g) + HCl(g) \longrightarrow NH_4Cl(s)$
**c.** $K(s) + O_2(g) \longrightarrow KO_2(s)$
**d.** $N_2(g) + 3\,H_2(g) \longrightarrow 2\,NH_3(g)$

**7-44.** In Problem 7–40, you are asked to predict which molecule, $CO(g)$ or $CO_2(g)$, has the greater molar entropy. Use the data in Tables 4.2 and 4.4 to calculate the standard molar entropy of $CO(g)$ and $CO_2(g)$ at 298.15 K. Does this calculation confirm your intuition? Which degree of freedom makes the dominant contribution to the molar entropy of CO? Of $CO_2$?

**7-45.** Table 7.2 gives $S^\circ[\mathrm{CH_3OH(l)}] = 126.8\ \mathrm{J\cdot K^{-1}\cdot mol^{-1}}$ at 298.15 K. Given that $T_{\mathrm{vap}} = 337.7$ K, $\Delta_{\mathrm{vap}}\overline{H}(T_{\mathrm{b}}) = 36.5\mathrm{kJ\cdot mol^{-1}}$, $\overline{C}_P[\mathrm{CH_3OH(l)}] = 81.12\mathrm{J\cdot K^{-1}\cdot mol^{-1}}$, and $\overline{C}_P[\mathrm{CH_3OH(g)}] = 43.8\ \mathrm{J\cdot K^{-1}\cdot mol^{-1}}$, calculate the value of $S^\circ[\mathrm{CH_3OH(g)}]$ at 298.15 K and compare your answer with the experimental value of $239.8\ \mathrm{J\cdot K^{-1}\cdot mol^{-1}}$.

**7-46.** Given the following data, $T_{\mathrm{fus}} = 373.15$ K, $\Delta\overline{H}_{\mathrm{vap}}(T_{\mathrm{vap}}) = 40.65\ \mathrm{kJ\cdot mol^{-1}}$, $\overline{C}_P[\mathrm{H_2O(l)}] = 75.3\ \mathrm{J\cdot K^{-1}\cdot mol^{-1}}$, and $\overline{C}_P[\mathrm{H_2O(g)}] = 33.8\ \mathrm{J\cdot K^{-1}\cdot mol^{-1}}$, show that the values of $S^\circ[\mathrm{H_2O(l)}]$ and $S^\circ[\mathrm{H_2O(g)}]$ in Table 7.2 are consistent.

**7-47.** Use the data in Table 7.2 to calculate the value of $\Delta_r S^\circ$ for the following reactions at 25°C and one bar.

**a.** $\mathrm{C(s, graphite) + O_2(g) \longrightarrow CO_2(g)}$
**b.** $\mathrm{CH_4(g) + 2\,O_2(g) \longrightarrow CO_2(g) + 2\,H_2O(l)}$
**c.** $\mathrm{C_2H_2(g) + H_2(g) \longrightarrow C_2H_4(g)}$

**7-48.** Use the data in Table 7.2 to calculate the value of $\Delta_r S^\circ$ for the following reactions at 25°C and one bar.

**a.** $\mathrm{CO(g) + 2\,H_2(g) \longrightarrow CH_3OH(l)}$
**b.** $\mathrm{C(s, graphite) + H_2O(l) \longrightarrow CO(g) + H_2(g)}$
**c.** $\mathrm{2\,CO(g) + O_2(g) \longrightarrow 2\,CO_2(g)}$

**Hermann von Helmholtz** was born in Potsdam, Germany, on August 31, 1821, and died in 1894. Although he wanted to study physics, his family could not afford to send him to the University, so he studied medicine in Berlin because he could obtain state financial aid. He was, however, required to repay his stipend by service as a surgeon in the army for eight years. He was later appointed professor at the University of Königsberg, and he also held positions at Bonn, Heidelberg, and Berlin. In 1885, in recognition of his position as the foremost scientist in Germany, he was appointed president of the newly founded Physico-Technical Institute in Berlin, an institution devoted to purely scientific research. Helmholtz was one of the greatest scientists of the 19th century, making important discoveries in physiology, optics, acoustics, electromagnetic theory, and thermodynamics. His work in physiology showed that physiological phenomena are based upon the laws of physics and not on some vague "vital force." In thermodynamics, he derived the equation now known as the Gibbs-Helmholtz equation, which we will discuss in this chapter. Helmholtz was always generous with his students and other scientists, but unfortunately he was a barely intelligible lecturer, even to the likes of Planck, who was a student in several of his classes. Helmholtz's great influence in German science was recognized by the Kaiser, who bestowed him with the title "von."

CHAPTER 8

# Helmholtz and Gibbs Energies

The criterion that $dS > 0$ for a spontaneous process applies only to an isolated system. Consequently, in the various processes we discussed in Chapter 6, we had to consider the entropy change of both the system *and* its surroundings to determine the sign of $\Delta S_{\text{total}}$ and establish whether a process is spontaneous or not. Although of great fundamental and theoretical importance, the criterion that $dS \geq 0$ for a spontaneous process in an isolated system is too restrictive for practical applications. In this chapter, we will introduce two new state functions that can be used to determine the direction of a spontaneous process in systems that are not isolated.

## 8–1. The Sign of the Helmholtz Energy Change Determines the Direction of a Spontaneous Process in a System at Constant Volume and Temperature

Let's consider a system with its volume and temperature held constant. The criterion that $dS \geq 0$ for a spontaneous process does not apply to a system at constant temperature and volume because the system is not isolated; a system must be in thermal contact with a thermal reservoir to be at constant temperature. If the criterion $dS \geq 0$ does not apply, then what is the criterion for a spontaneous process that we can use for a system at constant temperature and volume? Let's start with the expression of the First Law of Thermodynamics, Equation 5.9,

$$dU = \delta q + \delta w \tag{8.1}$$

Because $\delta w = -P_{\text{ext}}dV$ and $dV = 0$ (constant volume), then $\delta w = 0$. If we substitute Equation 6.3, $dS \geq \delta q/T$, and $\delta w = 0$ into Equation 8.1, we obtain

$$dU \leq TdS \qquad \text{(constant } V\text{)} \tag{8.2}$$

The equality holds for a reversible process and the inequality for an irreversible process. Note that if the system is isolated, then $dU = 0$ and we have $dS \geq 0$ as in Chapter 6. We can write Equation 8.2 as

$$dU - TdS \leq 0$$

If $T$ and $V$ are held constant, we can write this expression as

$$d(U - TS) \leq 0 \qquad \text{(constant } T \text{ and } V) \tag{8.3}$$

Equation 8.3 prompts us to define a new thermodynamic state function by

$$A = U - TS \tag{8.4}$$

so Equation 8.3 becomes

$$dA \leq 0 \qquad \text{(constant } T \text{ and } V) \tag{8.5}$$

The quantity $A$ is called the *Helmholtz energy*. In a system held at constant $T$ and $V$, the Helmholtz energy will decrease until all the possible spontaneous processes have occurred, at which time the system will be in equilibrium and $A$ will be a minimum. At equilibrium, $dA = 0$ (see Figure 8.1). Note that Equation 8.5 is the analog of the criterion that $dS \geq 0$ for a spontaneous process to occur in an isolated system (cf. Figures 6.5 and 8.1).

For an isothermal change from one state to another, Equation 8.4 gives

$$\Delta A = \Delta U - T\Delta S \tag{8.6}$$

Using Equation 8.5, we see that

$$\Delta A = \Delta U - T\Delta S \leq 0 \qquad \text{(constant } T \text{ and } V) \tag{8.7}$$

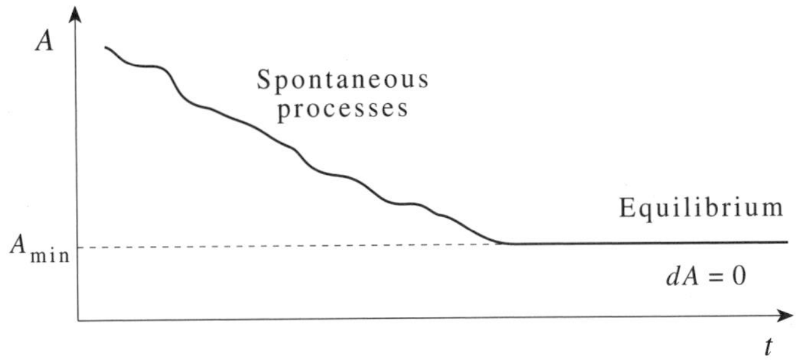


**FIGURE 8.1**
The Helmholtz energy, $A$, of a system will decrease during any spontaneous processes that occur at constant $T$ and $V$ and will achieve its minimum value at equilibrium.

where the equality holds for a reversible change and the inequality holds for an irreversible, spontaneous change. A process for which $\Delta A > 0$ cannot take place spontaneously in a system at constant $T$ and $V$. Consequently, something, such as work, must be done on the system to effect the change.

Notice that if $\Delta U < 0$ and $\Delta S > 0$ in Equation 8.6, then both the energy change and the entropy change contribute to $\Delta A$ being negative. But if they differ in sign, some sort of compromise must be reached and the value of $\Delta A$ is a quantitative measure of whether a process is spontaneous or not. The Helmholtz energy represents this compromise between the tendency of a system to decrease its energy and to increase its entropy. Because $\Delta S$ is multiplied by $T$, we see that the sign of $\Delta U$ is more important at low temperatures but the sign of $\Delta S$ is more important at high temperatures.

We can apply the criterion that $\Delta A < 0$ for an irreversible (spontaneous) process in a system at constant $T$ and $V$ to the mixing of two ideal gases, which we discussed in Section 6–6. For that process, $\Delta U = 0$ and $\Delta \overline{S} = -y_1 R \ln y_1 - y_2 R \ln y_2$. Therefore, for the mixing of two ideal gases at constant $T$ and $V$, $\Delta \overline{A} = RT(y_1 \ln y_1 + y_2 \ln y_2)$, which is a negative quantity because $y_1$ and $y_2$ are less than one. Thus, we see once again that the isothermal mixing of two ideal gases is a spontaneous process.

In addition to serving as our criterion for spontaneity in a system at constant temperature and volume, the Helmholtz energy has an important physical interpretation. Let's start with Equation 8.6

$$\Delta A = \Delta U - T\Delta S \tag{8.8}$$

for a spontaneous (irreversible) process, so that $\Delta A < 0$. In this process, the initial and final states are well-defined equilibrium states, and there is no fundamental reason we have to follow an irreversible path to get from one state to the other. In fact, we can gain some considerable insight into the process if we look at any reversible path connecting these two states. For a reversible path we can replace $\Delta S$ by $q_{\text{rev}}/T$, giving

$$\Delta A = \Delta U - q_{\text{rev}}$$

But according to the first law, $\Delta U - q_{\text{rev}}$ is equal to $w_{\text{rev}}$, so we have

$$\Delta A = w_{\text{rev}} \qquad \text{(isothermal, reversible)} \tag{8.9}$$

If $\Delta A < 0$, the process will occur spontaneously and $w_{\text{rev}}$ represents the work that can be done *by* the system if this change is carried out reversibly. This quantity is the maximum work that could be obtained. If any irreversible process such as friction occurs, then the quantity of work that can be obtained will be less than $w_{\text{rev}}$. If $\Delta A > 0$, the process will not occur spontaneously and $w_{\text{rev}}$ represents the work that must be done *on* the system to produce the change in a reversible manner. If there is any irreversibility in the process, the quantity of work required will be even greater than $w_{\text{rev}}$.

## 8–2. The Gibbs Energy Determines the Direction of a Spontaneous Process for a System at Constant Pressure and Temperature

Most reactions occur at constant pressure rather than at constant volume because they are open to the atmosphere. Let's see what the criterion of spontaneity is for a system at constant temperature and pressure. Once again, we start with Equation 8.1, but now we substitute $dS \geq \delta q/T$ and $\delta w = -PdV$ to obtain

$$dU \leq TdS - PdV$$

or

$$dU - TdS - PdV \leq 0$$

Because both $T$ and $P$ are constant, we can write this expression as

$$d(U - TS + PV) \leq 0 \qquad \text{(constant } T \text{ and } P\text{)} \tag{8.10}$$

We now define a new thermodynamic state function by

$$G = U - TS + PV \tag{8.11}$$

so Equation 8.5 becomes

$$dG \leq 0 \qquad \text{(constant } T \text{ and } P\text{)} \tag{8.12}$$

Note that Equation 8.11 is the analog of Equation 8.4.

The quantity $G$ is called the *Gibbs energy*. In a system at constant $T$ and $P$, the Gibbs energy will decrease as the result of any spontaneous processes until the system reaches equilibrium, where $dG = 0$. A plot of $G$ versus time for a system at constant $T$ and $P$ would be similar to the plot of $A$ versus time for a system at constant $T$ and $V$ (Figure 8.1). Thus, we see that the Gibbs energy, $G$, is the analog of the Helmoltz energy, $A$, for a process that takes place at constant temperature and pressure.

Equation 8.11 can also be written as

$$G = H - TS \tag{8.13}$$

where $H = U + PV$ is the enthalpy. Note that the enthalpy plays the same role in a constant $T$ and $P$ process that the energy $U$ plays in a constant $T$ and $V$ process (cf. Equation 8.4). Note also that $G$ can be written as

$$G = A + PV \tag{8.14}$$

thus relating the Gibbs energy and the Helmholtz energy in the same manner that $H$ and $U$ are related.

The analog of Equation 8.7 is

$$\Delta G = \Delta H - T\Delta S \leq 0 \qquad \text{(constant } T \text{ and } P) \qquad (8.15)$$

The equality holds for a reversible process, whereas the inequality holds for an irreversible (spontaneous) process. If $\Delta H < 0$ and $\Delta S > 0$ in Equation 8.15, both terms in Equation 8.15 contribute to $\Delta G$ being negative. But if $\Delta H$ and $\Delta S$ have the same sign, then $\Delta G = \Delta H - T\Delta S$ represents the compromise between the tendency of a system to decrease its enthalpy and to increase its entropy in a constant $T$ and $P$ process. Because of the factor of $T$ multiplying $\Delta S$ in Equation 8.15, the $\Delta H$ term can dominate at low temperatures, whereas the $T\Delta S$ term can dominate at high temperatures. Of course if $\Delta H > 0$ and $\Delta S < 0$, then $\Delta G > 0$ at all temperatures and the process is never spontaneous.

An example of a reaction favored by its value of $\Delta_r H$ but disfavored by its value of $\Delta_r S$ is

$$NH_3(g) + HCl(g) \longrightarrow NH_4Cl(s)$$

The value of $\Delta_r H$ for this reaction at 298.15 K and one bar is $-176.2$ kJ, whereas the corresponding value of $\Delta_r S$ is $-0.285$ kJ·K$^{-1}$, giving $\Delta_r G = \Delta_r H - T\Delta_r S = -91.21$ kJ at 298.15 K. Therefore, this reaction proceeds spontaneously at 298.15 K and one bar.

A process for which the sign of $\Delta G$ changes with a small change in temperature is the vaporization of a liquid at its normal boiling point. We represent this process by

$$H_2O(l) \longrightarrow H_2O(g)$$

The expression for the molar Gibbs energy of vaporization, $\Delta_{vap}\overline{G}$, for this process is

$$\begin{aligned}\Delta_{vap}\overline{G} &= \overline{G}[H_2O(g)] - \overline{G}[H_2O(l)] \\ &= \Delta_{vap}\overline{H} - T\Delta_{vap}\overline{S}\end{aligned}$$

The molar enthalpy of vaporization of water at one atm near 100°C, $\Delta_{vap}\overline{H}$, is equal to 40.65 kJ·mol$^{-1}$ and $\Delta_{vap}\overline{S} = 108.9$ J·K$^{-1}$·mol$^{-1}$. Thus, we can write $\Delta_{vap}\overline{G}$ as

$$\Delta_{vap}\overline{G} = 40.65 \text{ kJ·mol}^{-1} - T(108.9 \text{ J·K}^{-1}\text{·mol}^{-1})$$

At $T = 373.15$ K,

$$\begin{aligned}\Delta_{vap}\overline{G} &= 40.65 \text{ kJ·mol}^{-1} - (373.15 \text{ K})(108.9 \text{ J·K}^{-1}\text{·mol}^{-1}) \\ &= 40.65 \text{ kJ·mol}^{-1} - 40.65 \text{ kJ·mol}^{-1} = 0\end{aligned}$$

The fact that $\Delta_{vap}\overline{G} = 0$ means that liquid and vapor water are in equilibrium with each other at one atm and 373.15 K. The molar Gibbs energy of liquid water is equal to the molar Gibbs energy of water vapor at 373.15 K and one atm. The transfer of one mole of liquid water to water vapor under these conditions is a reversible process, and so $\Delta_{\text{vap}}\overline{G} = 0$.

Now let's consider a temperature less than the normal boiling point, say 363.15 K. At this temperature, $\Delta_{\text{vap}}\overline{G} = +1.10\ \text{kJ}\cdot\text{mol}^{-1}$. The positive sign means that the formation of one mole of water vapor at one atm from one mole of liquid water at one atm and 363.15 K is not a spontaneous process. If the temperature is above the normal boiling point, however, say 383.15 K, then $\Delta_{\text{vap}}\overline{G} = -1.08\ \text{kJ}\cdot\text{mol}^{-1}$. The negative sign means that the formation of one mole of water vapor from one mole of liquid water at one atm and 383.15 K is a spontaneous process.

**EXAMPLE 8–1**

The molar enthalpy of fusion of ice at 273.15 K and one atm is $\Delta_{\text{fus}}\overline{H} = 6.01\ \text{kJ}\cdot\text{mol}^{-1}$, and the molar entropy of fusion under the same conditions is $\Delta_{\text{fus}}\overline{S} = 22.0\ \text{J}\cdot\text{K}^{-1}\cdot\text{mol}^{-1}$. Show that $\Delta_{\text{fus}}\overline{G} = 0$ at 273.15 K and one atm, that $\Delta_{\text{fus}}\overline{G} < 0$ when the temperature is greater than 273.15 K, and that $\Delta_{\text{fus}}\overline{G} > 0$ when the temperature is less than 273.15 K.

SOLUTION: Assuming that $\Delta_{\text{fus}}\overline{H}$ and $\Delta_{\text{fus}}\overline{S}$ do not vary appreciably around 273.15 K, we can write

$$\Delta_{\text{fus}}\overline{G} = 6010\ \text{J}\cdot\text{mol}^{-1} - T(22.0\ \text{J}\cdot\text{K}^{-1}\cdot\text{mol}^{-1})$$

If $T = 273.15$ K, then $\Delta_{\text{fus}}\overline{G} = 0$, indicating that ice and liquid water are in equilibrium with each other at 273.15 K and one atm. If $T < 273.15$ K, then $\Delta_{\text{fus}}\overline{G} > 0$, indicating that ice will not spontaneously melt under these conditions. If $T > 273.15$ K, then $\Delta_{\text{fus}}\overline{G} < 0$, indicating that ice will melt under these conditions.

The value of $\Delta G$ can be related to the maximum work that can be obtained from a spontaneous process carried out at constant $T$ and $P$. To show this, we start by differentiating $G = U - TS + PV$ to get

$$dG = dU - TdS - SdT + PdV + VdP$$

and substitute $dU = TdS + \delta w_{\text{rev}}$ for $dU$ to get

$$dG = -SdT + VdP + \delta w_{\text{rev}} + PdV$$

Because the reversible pressure-volume ($P$-$V$) work is $-PdV$, the quantity $\delta w_{\text{rev}} + PdV$ is the reversible work other than $P$-$V$ work (such as electrical work). Therefore, we can write $dG$ as

$$dG = -SdT + VdP + \delta w_{\text{nonPV}}$$

where $\delta w_{\text{nonPV}}$ represents the total work exclusive of $P$-$V$ work. For a reversible process taking place at constant $T$ and $P$, $dG = \delta w_{\text{nonPV}}$, or

$$\Delta G = w_{\text{nonPV}} \qquad \text{(reversible, constant } T \text{ and } P) \tag{8.16}$$

If $\Delta G < 0$, the process will occur spontaneously, and $w_{\text{nonPV}}$ is the work exclusive of $P$-$V$ work that can be done by the system if the change is carried out reversibly. This is the maximum work that can be obtained from the process. If any irreversibility occurs in the process, the quantity of work obtained will be less than the maximum. If $\Delta G > 0$, the process will not occur spontaneously and $w_{\text{nonPV}}$ is the minimum work, exclusive of $P$-$V$ work, that must be done on the system to make the process occur. For example, it is known experimentally that $\Delta G$ for the formation of one mole of $H_2O(l)$ at 298.15 K and one bar from $H_2(g)$ and $O_2(g)$ at 298.15 K and one bar is $-237.1\ \text{kJ}\cdot\text{mol}^{-1}$. Thus, a maximum of $237.1\ \text{kJ}\cdot\text{mol}^{-1}$ of useful work (that is, work exclusive of $P$-$V$ work) can be obtained from the spontaneous reaction

$$\text{H}_2(\text{g, 1 bar, 298.15 K}) + \tfrac{1}{2}\text{O}_2(\text{g, 1 bar, 298.15 K}) \longrightarrow \text{H}_2\text{O}(\text{l, 1 bar, 298.15 K})$$

Conversely, it would require at least $237.1\ \text{kJ}\cdot\text{mol}^{-1}$ of energy to drive the (nonspontaneous) reaction

$$\text{H}_2\text{O}(\text{l, 1 bar, 298.15 K}) \longrightarrow \text{H}_2(\text{g, 1 bar, 298.15 K}) + \tfrac{1}{2}\text{O}_2(\text{g, 1 bar, 298.15 K})$$

**EXAMPLE 8–2**

The value of $\Delta\overline{G}$ for the decomposition of one mole of $H_2O(l)$ to $H_2(g)$ and $O_2(g)$ at one bar and 298.15 K is $+237.1\ \text{kJ}\cdot\text{mol}^{-1}$. Calculate the minimum voltage required to decompose one mole of $H_2O(l)$ to $H_2(g)$ and $O_2(g)$ at one bar and 298.15 K by electrolysis.

SOLUTION: Electrolysis represents the non-$P$-$V$ work required to carry out the decomposition, so we write

$$\Delta\overline{G} = w_{\text{nonPV}} = +237.1\ \text{kJ}\cdot\text{mol}^{-1}$$

You might remember from physics that electrical work is given by charge × voltage. The charge involved in electrolyzing one mole of $H_2O(l)$ can be determined from the chemical equation of the reaction

$$\text{H}_2\text{O(l)} \longrightarrow \text{H}_2(\text{g}) + \tfrac{1}{2}\text{O}_2(\text{g})$$

The oxidation state of hydrogen goes from +1 to 0 and that of oxygen goes from −2 to 0. Thus two electrons are transferred per $H_2O(l)$ molecule, or two times the Avogadro constant of electrons per mole. The total charge of two moles of electrons is

$$\text{total charge} = (1.602 \times 10^{-19}\ \text{C}\cdot\text{e}^{-1})(12.044 \times 10^{23}\ \text{e}) = 1.929 \times 10^{5}\ \text{C}$$

The minimum voltage, $\mathcal{E}$, required to decompose one mole is given by

$$\mathcal{E} = \frac{\Delta\overline{G}}{1.929 \times 10^{5}\ \text{C}} = \frac{237.1 \times 10^{3}\ \text{J}\cdot\text{mol}^{-1}}{1.929 \times 10^{5}\ \text{C}} = 1.23\ \text{volts}$$

where we have used the fact that one joule is a coulomb times a volt ($1J = 1CV$).

## 8–3. Maxwell Relations Provide Several Useful Thermodynamic Formulas

A number of the thermodynamic functions we have defined cannot be measured directly. Consequently, we need to be able to express these quantities in terms of others that can be experimentally determined. To do so, we start with the definitions of $A$ and $G$, Equations 8.4 and 8.11. Differentiate Equation 8.4 to obtain

$$dA = dU - TdS - SdT$$

For a reversible process, $dU = TdS - PdV$, so

$$dA = -PdV - SdT \tag{8.17}$$

By comparing Equation 8.17 with the formal total derivative of $A = A(V, T)$,

$$dA = \left(\frac{\partial A}{\partial V}\right)_T dV + \left(\frac{\partial A}{\partial T}\right)_V dT$$

we see that

$$\left(\frac{\partial A}{\partial V}\right)_T = -P \quad \text{and} \quad \left(\frac{\partial A}{\partial T}\right)_V = -S \tag{8.18$a$, $b$}$$

Now if we use the fact that the cross derivatives of $A$ are equal (MathChapter D),

$$\left(\frac{\partial^2 A}{\partial T \partial V}\right) = \left(\frac{\partial^2 A}{\partial V \partial T}\right)$$

we find that

$$\left(\frac{\partial P}{\partial T}\right)_V = \left(\frac{\partial S}{\partial V}\right)_T \tag{8.19}$$

Equation 8.19, which is obtained by equating the second cross partial derivatives of $A$, is called a *Maxwell relation*. There are many useful Maxwell relations involving various thermodynamic quantities. Equation 8.19 is particularly useful because it allows us to determine how the entropy of a substance changes with volume if we know its equation of state. Integrating Equation 8.19 at constant $T$, we have

$$\Delta S = \int_{V_1}^{V_2} \left(\frac{\partial P}{\partial T}\right)_V dV \qquad \text{(constant } T\text{)} \tag{8.20}$$

We have applied the condition of constant $T$ to Equation 8.20 because we have integrated $(\partial S/\partial V)_T$; in other words, $T$ is held constant in the derivative, so $T$ must be held constant when we integrate.

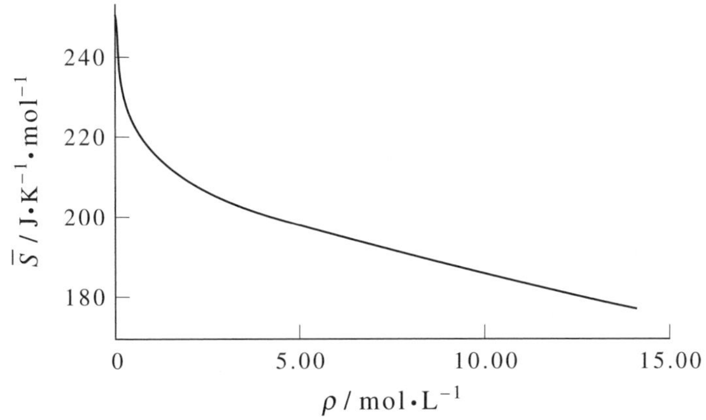


**FIGURE 8.2**
The molar entropy of ethane at 400 K plotted against density ($\rho = 1/\overline{V}$). The value of $\overline{S}^{\text{id}}$ at 400 K is 246.45 J·mol$^{-1}$·K$^{-1}$.

Equation 8.20 allows us to determine the entropy of a substance as a function of volume or density (recall that $\rho = 1/V$) from $P$-$V$-$T$ data. If we let $V_1$ in Equation 8.20 be very large, where the gas is sure to behave ideally, then Equation 8.20 becomes

$$S(T,V) - S^{\text{id}} = \int_{V^{\text{id}}}^{V} \left(\frac{\partial P}{\partial T}\right)_V dV'$$

Figure 8.2 plots the molar entropy of ethane at 400 K versus density. (Problem 8–3 involves calculating the molar entropy as a function of density using the van der Waals equation and the Redlich-Kwong equation.)

We can also use Equation 8.20 to derive an equation we derived earlier in Section 6–3 by another method. For an ideal gas, $(\partial P/\partial T)_V = nR/V$, so

$$\Delta S = nR\int_{V_1}^{V_2} \frac{dV}{V} = nR \ln \frac{V_2}{V_1} \qquad \text{(isothermal process)} \qquad (8.21)$$

**EXAMPLE 8–3**
Calculate $\Delta\overline{S}$ for an isothermal expansion from $\overline{V}_1$ to $\overline{V}_2$ for a gas that obeys the equation of state

$$P(\overline{V} - b) = RT$$

SOLUTION: We use Equation 8.20 to obtain

$$\Delta\overline{S} = \int_{\overline{V}_1}^{\overline{V}_2} \left(\frac{\partial P}{\partial T}\right)_{\overline{V}} d\overline{V} = R\int_{\overline{V}_1}^{\overline{V}_2} \frac{d\overline{V}}{\overline{V} - b} = R \ln \frac{\overline{V}_2 - b}{\overline{V}_1 - b}$$

Note that we derived this equation in Example 6–2, but we had to be told that $dU = 0$ in an isothermal process for a gas obeying the above equation of state. We did not need this information to derive our result here.

We have previously stated that the energy of an ideal gas depends only upon temperature. This statement is not generally true for real gases. Suppose we want to know how the energy of a gas changes with volume at constant temperature. Unfortunately, this quantity cannot be measured directly. We can use Equation 8.19, however, to derive a practical equation for $(\partial U/\partial V)_T$; in other words, we can derive an equation that tells us how the energy of a substance varies with its volume at constant temperature in terms of readily measurable quantities. We differentiate Equation 8.4 with respect to $V$ at constant temperature to obtain

$$\left(\frac{\partial A}{\partial V}\right)_T = \left(\frac{\partial U}{\partial V}\right)_T - T\left(\frac{\partial S}{\partial V}\right)_T$$

Substituting Equation 8.18*a* for $(\partial A/\partial V)_T$ and Equation 8.19 for $(\partial S/\partial V)_T$ gives

$$\left(\frac{\partial U}{\partial V}\right)_T = -P + T\left(\frac{\partial P}{\partial T}\right)_V \tag{8.22}$$

Equation 8.22 gives $(\partial U/\partial V)_T$ in terms of $P$-$V$-$T$ data. Equations like Equation 8.22 that relate thermodynamic functions to functions of $P$, $V$, and $T$ are sometimes called thermodynamic equations of state.

We can integrate Equation 8.22 with respect to $V$ to determine $U$ relative to the ideal gas value,

$$U(T, V) - U^{\text{id}} = \int_{V^{\text{id}}}^{V'} \left[T\left(\frac{\partial P}{\partial T}\right)_V - P\right] dV' \qquad \text{(constant } T\text{)}$$

where $V^{\text{id}}$ is a large volume, where the gas is sure to behave ideally. This equation along with the $P$-$V$-$T$ data gives us $U$ as a function of pressure. Figure 8.3 shows $\overline{U}$ plotted against pressure for ethane at 400 K. Problem 8–4 involves calculating $\overline{U}$ as a function of volume for the van der Waals equation and the Redlich-Kwong equation.

We can also use Equation 8.22 to show that the energy of an ideal gas is independent of the volume at constant temperature. For an ideal gas, $(\partial P/\partial T)_V = nR/V$, so

$$\left(\frac{\partial U}{\partial V}\right)_T = -P + T\frac{nR}{V} = -P + P = 0$$

which proves that the energy of an ideal gas depends only upon temperature.

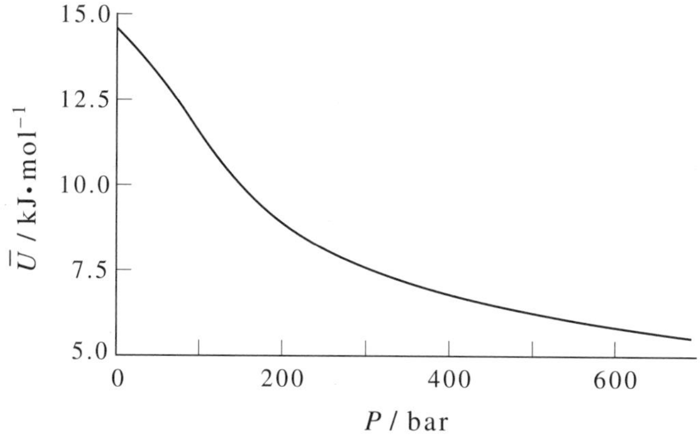


**FIGURE 8.3**
The molar energy $\overline{U}$ plotted against pressure for ethane at 400 K. The value of $\overline{U}^{\text{id}}$ is equal to 14.55 kJ·mol$^{-1}$.

**EXAMPLE 8–4**
In Example 6–2, we stated we would prove later that the energy of a gas that obeys the equation of state

$$P(\overline{V} - b) = RT$$

is independent of the volume. Use Equation 8.22 to prove this.

SOLUTION: For $P(\overline{V} - b) = RT$,

$$\left(\frac{\partial P}{\partial T}\right)_{\overline{V}} = \frac{R}{\overline{V} - b}$$

and so

$$\left(\frac{\partial U}{\partial \overline{V}}\right)_T = -P + \frac{RT}{\overline{V} - b} = -P + P = 0$$

**EXAMPLE 8–5**
Evaluate $(\partial U/\partial \overline{V})_T$ for one mole of a Redlich-Kwong gas.

SOLUTION: Recall that the Redlich-Kwong equation (Equation 2.7) is

$$P = \frac{RT}{\overline{V} - B} - \frac{A}{T^{1/2}\overline{V}(\overline{V} + B)}$$

Therefore,

$$\left(\frac{\partial P}{\partial T}\right)_{\overline{V}} = \frac{R}{\overline{V} - B} + \frac{A}{2T^{3/2}\overline{V}(\overline{V} + B)}$$

and so

$$\left(\frac{\partial U}{\partial \overline{V}}\right)_T = T\left(\frac{\partial P}{\partial T}\right)_{\overline{V}} - P = \frac{3A}{2T^{1/2}\overline{V}(\overline{V}+B)}$$

We derived the equation

$$C_P - C_V = \left[P + \left(\frac{\partial U}{\partial V}\right)_T\right]\left(\frac{\partial V}{\partial T}\right)_P$$

in Problem 3–27. Using Equation 8.22 for $(\partial U/\partial V)_T$, we obtain

$$C_P - C_V = T\left(\frac{\partial P}{\partial T}\right)_V\left(\frac{\partial V}{\partial T}\right)_P \tag{8.23}$$

For an ideal gas $(\partial P/\partial T)_V = nR/V$ and $(\partial V/\partial T)_P = nR/P$, and so $C_P - C_V = nR$, in agreement with Equation 3.43.

An alternative equation for $C_P - C_V$ that is more convenient than Equation 8.23 for solids and liquids is (Problem 8–11)

$$C_P - C_V = -T\left(\frac{\partial V}{\partial T}\right)_P^2\left(\frac{\partial P}{\partial V}\right)_T \tag{8.24}$$

Each of the partial derivatives here can be expressed in terms of familiar tabulated physical quantities. The isothermal compressibility of a substance is defined as

$$\kappa = -\frac{1}{V}\left(\frac{\partial V}{\partial P}\right)_T \tag{8.25}$$

and the coefficient of thermal expansion is defined as

$$\alpha = \frac{1}{V}\left(\frac{\partial V}{\partial T}\right)_P \tag{8.26}$$

Using these definitions, Equation 8.24 becomes

$$C_P - C_V = \frac{\alpha^2 TV}{\kappa} \tag{8.27}$$

**EXAMPLE 8–6**

The coefficient of thermal expansion, $\alpha$, of copper at 298 K is $5.00 \times 10^{-5}\ \text{K}^{-1}$, and its isothermal compressibility, $\kappa$, is $7.85 \times 10^{-7}\ \text{atm}^{-1}$. Given that the density of copper is $8.92\ \text{g}\cdot\text{cm}^{-3}$ at 298 K, calculate the value of $\overline{C}_P - \overline{C}_V$ for copper.

SOLUTION: For copper, the molar volume, $\overline{V}$, is given by

$$\overline{V} = \frac{63.54\ \text{g}\cdot\ \text{mol}^{-1}}{8.92\ \text{g}\cdot\text{cm}^{-3}}$$
$$= 7.12\ \text{cm}^3\cdot\text{mol}^{-1} = 7.12 \times 10^{-3}\ \text{L}\cdot\text{mol}^{-1}$$

and

$$\overline{C}_P - \overline{C}_V = \frac{(5.00 \times 10^{-5}\ \text{K}^{-1})^2(298\ \text{K})(7.12 \times 10^{-3}\ \text{L}\cdot\text{mol}^{-1})}{7.85 \times 10^{-7}\ \text{atm}^{-1}}$$
$$= 6.76 \times 10^{-3}\ \text{L}\cdot\text{atm}\cdot\text{K}^{-1}\cdot\text{mol}^{-1}$$
$$= 0.684\ \text{J}\cdot\text{K}^{-1}\cdot\text{mol}^{-1}$$

The experimental value of $\overline{C}_P$ is 24.43 $\text{J}\cdot\text{K}^{-1}\cdot\text{mol}^{-1}$. Note that $\overline{C}_P - \overline{C}_V$ is small compared with $\overline{C}_P$ (or $\overline{C}_V$) and is also much smaller for solids than for gases, as you might expect.

## 8–4. The Enthalpy of an Ideal Gas Is Independent of Pressure

Equation 8.18$a$ can be used directly to give the volume dependence of the Helmholtz energy. By integrating at constant temperature, we have

$$\Delta A = -\int_{V_1}^{V_2} P dV \qquad \text{(constant } T) \tag{8.28}$$

For the case of an ideal gas, we have

$$\Delta A = -nRT \int_{V_1}^{V_2} \frac{dV}{V} = -nRT \ln \frac{V_2}{V_1} \qquad \text{(constant } T) \tag{8.29}$$

Notice that this result is $-T$ times Equation 8.21 for $\Delta S$. This result must be so because $\Delta U = 0$ for an ideal gas at constant $T$, so $\Delta A = -T\Delta S$.

If we differentiate Equation 8.11, $G = U - TS + PV$, and substitute $dU = TdS - PdV$, we get

$$dG = -SdT + VdP \tag{8.30}$$

By comparing Equation 8.30 with

$$dG = \left(\frac{\partial G}{\partial T}\right)_P dT + \left(\frac{\partial G}{\partial P}\right)_T dP$$

we see that

$$\left(\frac{\partial G}{\partial T}\right)_P = -S \quad \text{and} \quad \left(\frac{\partial G}{\partial P}\right)_T = V \tag{8.31$a, b$}$$

Note that Equation 8.31$a$ says that $G$ decreases with increasing temperature (because $S \geq 0$) and that Equation 8.31$b$ says that $G$ increases with increasing pressure (because $V > 0$).

If we now take cross derivatives of $G$ as we did for $A$ in the previous section, we find that

$$-\left(\frac{\partial S}{\partial P}\right)_T = \left(\frac{\partial V}{\partial T}\right)_P \tag{8.32}$$

This Maxwell relation gives us an equation we can use to calculate the pressure dependence of $S$. We integrate Equation 8.32 with $T$ constant to get

$$\Delta S = -\int_{P_1}^{P_2} \left(\frac{\partial V}{\partial T}\right)_P dP \qquad \text{(constant } T\text{)} \tag{8.33}$$

Equation 8.33 can be used to obtain the molar entropy as a function of pressure by integrating $(\partial V/\partial T)_P$ data from some low pressure, where the gas is sure to behave ideally, to some arbitrary pressure. Figure 8.4 shows the molar entropy of ethane at 400 K obtained in this way plotted against pressure.

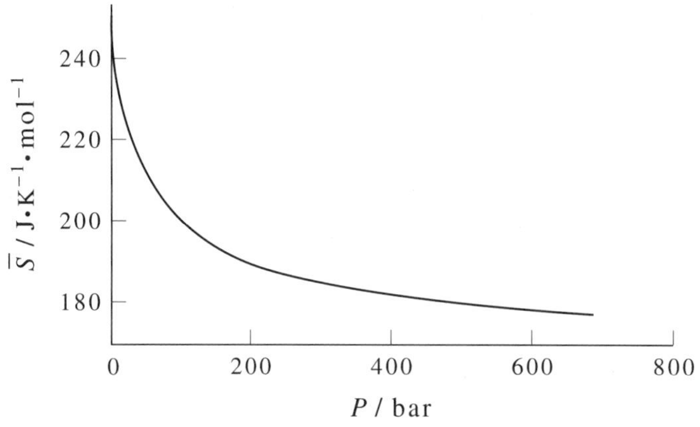


**FIGURE 8.4**
The molar entropy of ethane at 400 K plotted against pressure. The value of $\overline{S}^{\text{id}}$ at 400 K is 246.45 J·mol$^{-1}$·K$^{-1}$.

For an ideal gas, $(\partial V/\partial T)_P = nR/P$, so Equation 8.33 gives us

$$\Delta S = -nR\int_{P_1}^{P_2} \frac{dP}{P} = -nR\ln\frac{P_2}{P_1}$$

This results is not really a new one for us because if we let $P_2 = nRT/V_2$ and $P_1 = nRT/V_1$, we obtain Equation 8.21.

**EXAMPLE 8–7**

Use the virial expansion in the pressure

$$Z = 1 + B_{2P}P + B_{3P}P^2 + \cdots$$

to derive a virial expansion for $\Delta\overline{S}$ for a reversible isothermal change in pressure.

SOLUTION: Solve the above equation for $\overline{V}$:

$$\overline{V} = \frac{RT}{P} + RTB_{2P} + RTB_{3P}P + \cdots$$

and write

$$\left(\frac{\partial\overline{V}}{\partial T}\right)_P = \frac{R}{P} + R\left(B_{2P} + T\frac{dB_{2P}}{dT}\right) + R\left(B_{3P} + T\frac{dB_{3P}}{dT}\right)P + \cdots$$

Substitute this result into Equation 8.33 and integrate from $P_1$ to $P_2$ to obtain

$$\Delta\overline{S} = -\ln\frac{P_2}{P_1} - R\left(B_{2P} + T\frac{dB_{2P}}{dT}\right)P - \frac{R}{2}\left(B_{3P} + T\frac{dB_{3P}}{dT}\right)P^2 + \cdots$$

We can also use Equations 8.31 to show that the enthalpy of an ideal gas is independent of the pressure, just as its energy is independent of the volume. First, we differentiate Equation 8.13 with respect to $P$ at constant $T$ to obtain

$$\left(\frac{\partial G}{\partial P}\right)_T = \left(\frac{\partial H}{\partial P}\right)_T - T\left(\frac{\partial S}{\partial P}\right)_T$$

Now use Equation 8.31*b* for $(\partial G/\partial P)_T$ and Equation 8.32 for $(\partial S/\partial P)_T$ to obtain

$$\left(\frac{\partial H}{\partial P}\right)_T = V - T\left(\frac{\partial V}{\partial T}\right)_P \tag{8.34}$$

Note that Equation 8.34 is the analog of Equation 8.22. Equation 8.34 is also called a thermodynamic equation of state. It allows us to calculate the pressure dependence of $H$ from $P$-$V$-$T$ data (Such data for ethane at 400 K are shown in Figure 8.5). For an ideal gas, $(\partial V/\partial T)_P = nR/P$, so $(\partial H/\partial P)_T = 0$.

**EXAMPLE 8–8**

Evaluate $(\partial\overline{H}/\partial P)_T$ for a gas whose equation of state is

$$P\overline{V} = RT + B(T)P$$

SOLUTION: We have

$$\left(\frac{\partial\overline{V}}{\partial T}\right)_P = \frac{R}{P} + \frac{dB}{dT}$$

so Equation 8.34 gives us

$$\left(\frac{\partial \overline{H}}{\partial P}\right)_T = \overline{V} - T\left(\frac{\partial \overline{V}}{\partial T}\right)_P = \frac{RT}{P} + B(T) - \frac{RT}{P} - T\frac{dB}{dT}$$

or

$$\left(\frac{\partial \overline{H}}{\partial P}\right)_T = B(T) - T\frac{dB}{dT}$$

Note that $(\partial \overline{H}/\partial P)_T = 0$ when $B(T) = 0$.

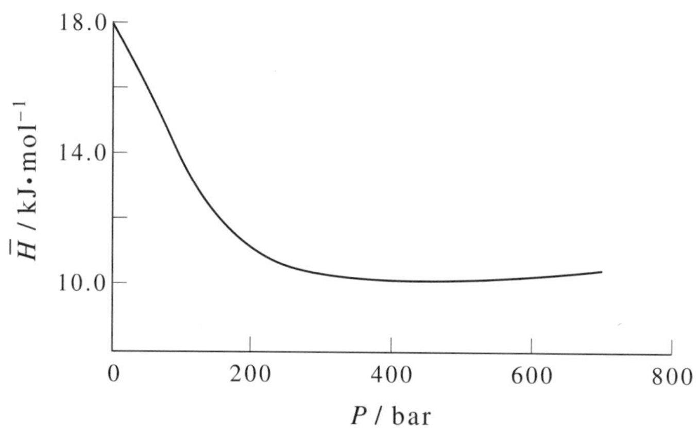


**FIGURE 8.5**
The molar enthalpy of ethane at 400 K plotted against pressure. The value of $\overline{H}^{\text{id}}$ at 400 K is 17.867 kJ·mol$^{-1}$.

## 8–5. The Various Thermodynamic Functions Have Natural Independent Variables

We may seem to be deriving a lot of equations in this chapter, but they can be organized neatly by recognizing that the energy, enthalpy, entropy, Helmholtz energy, and Gibbs energy depend upon natural sets of variables. For example, Equation 7.1 summarizes the First and Second Laws of Thermodynamics by

$$dU = TdS - PdV \tag{8.35}$$

Note that when $S$ and $V$ are considered to be the independent variables of $U$, then the total derivative of $U$,

$$dU = \left(\frac{\partial U}{\partial S}\right)_V dS + \left(\frac{\partial U}{\partial V}\right)_S dV \tag{8.36}$$

takes on a simple form, in the sense that the coefficients of $dS$ and $dV$ are simple thermodynamic functions. Consequently, we say that the natural variables of $U$ are $S$ and $V$, and we have

$$\left(\frac{\partial U}{\partial S}\right)_V = T \quad \text{and} \quad \left(\frac{\partial U}{\partial V}\right)_S = -P \tag{8.37}$$

This concept of natural variables is particularly clear if we consider $V$ and $T$ instead of $S$ and $V$ to be the independent variables of $U$, in which case we get (cf. Equation 8.22)

$$dU = \left[T\left(\frac{\partial P}{\partial T}\right)_V - P\right]dV + C_V dT \tag{8.38}$$

Certainly $U$ can be considered to be a function of $V$ and $T$, but its total derivative is not as simple as if it were considered to be a function of $S$ and $V$ (cf. Equation 8.36). Equation 8.35 also gives us that a criterion for a spontaneous process is that $dU < 0$ for a system at constant $S$ and $V$.

We can write Equation 8.35 in terms of $dS$ rather than $dU$ to get

$$dS = \frac{1}{T}dU + \frac{P}{T}dV \tag{8.39}$$

which suggests that the natural variables of $S$ are $U$ and $V$. Furthermore, the criterion for a spontaneous process is that $dS \geq 0$ at constant $U$ and $V$ (Equation 8.2 for an isolated system). Equation 8.39 gives us

$$\left(\frac{\partial S}{\partial U}\right)_V = \frac{1}{T} \quad \text{and} \quad \left(\frac{\partial S}{\partial V}\right)_U = \frac{P}{T} \tag{8.40}$$

The total derivative of the enthalpy is given by (Equation 7.6)

$$dH = TdS + VdP \tag{8.41}$$

which suggests that the natural variables of $H$ are $S$ and $P$. The criterion of spontaneity involving $H$ is that $dH < 0$ at constant $S$ and $P$.

The total derivative of the Helmholtz energy is

$$dA = -SdT - PdV \tag{8.42}$$

from which we obtain

$$\left(\frac{\partial A}{\partial T}\right)_V = -S \quad \text{and} \quad \left(\frac{\partial A}{\partial V}\right)_T = -P \tag{8.43}$$

Equation 8.42, plus the spontaneity criterion that $dA < 0$ at constant $T$ and $V$, suggest that $T$ and $V$ are the natural variables of $A$. The Maxwell relations obtained from Equation 8.43 are useful because the variables held constant are more experimentally controllable than are $S$ and $V$, as in Equations 8.37, or $U$ and $V$, as in Equations 8.40. The Maxwell relation from Equations 8.43 is

$$\left(\frac{\partial S}{\partial V}\right)_T = \left(\frac{\partial P}{\partial T}\right)_V \tag{8.44}$$

which allows us to calculate the volume dependence of $S$ in terms of $P$-$V$-$T$ data (see Figure 8.2).

Last, let's consider the Gibbs energy, whose total derivative is

$$dG = -SdT + VdP \tag{8.45}$$

Equation 8.45, plus the spontaneity criterion $dG < 0$ for a system at constant $T$ and $P$, tell us that the natural variables of $G$ are $T$ and $P$. Equation 8.45 gives us

$$\left(\frac{\partial G}{\partial T}\right)_P = -S \quad \text{and} \quad \left(\frac{\partial G}{\partial P}\right)_T = V \tag{8.46}$$

The Maxwell relation we obtain from Equations 8.46 is

$$\left(\frac{\partial S}{\partial P}\right)_T = -\left(\frac{\partial V}{\partial T}\right)_P \tag{8.47}$$

which we can use to calculate the pressure dependence of $S$ in terms of $P$-$V$-$T$ data (Figure 8.4).

This section is meant to provide both a summary of many of the equations we have derived so far and a way to bring some order to them. You do not need to memorize these equations because they can all be obtained from Equation 8.35:

$$dU = TdS - PdV \tag{8.48}$$

which is nothing more than the First and Second Laws of Thermodynamics expressed as one equation. If we add $d(PV)$ to both sides of this equation, we obtain

$$d(U + PV) = TdS - PdV + VdP + PdV$$

or

$$dH = TdS + VdP \tag{8.49}$$

If we subtract $d(TS)$ from both sides of Equation 8.48, we have

$$d(U - TS) = TdS - PdV - TdS - SdT$$

or

$$dA = -SdT - PdV \tag{8.50}$$

If we add $d(PV)$ and subtract $d(TS)$ from Equation 8.48, or subtract $d(TS)$ from Equation 8.49, or add $d(PV)$ to Equation 8.50, we get

$$dG = -SdT + VdP \tag{8.51}$$

The other equations of this section follow by comparing the total derivative of each function in terms of its natural variables to the above equations for $dU$, $dH$, $dA$, and $dG$. Table 8.1 summarizes some of the principal equations we have derived in this and previous chapters.

**TABLE 8.1**
The four principal thermodynamic energies, their differential expressions, and the corresponding Maxwell relations.

| Thermodynamic energy | Differential expression | Corresponding Maxwell relations |
|---|---|---|
| $U$ | $dU = TdS - PdV$ | $\left(\frac{\partial T}{\partial V}\right)_S = -\left(\frac{\partial P}{\partial S}\right)_V$ |
| $H$ | $dH = TdS + VdP$ | $\left(\frac{\partial T}{\partial P}\right)_S = \left(\frac{\partial V}{\partial S}\right)_P$ |
| $A$ | $dA = -SdT - PdV$ | $\left(\frac{\partial S}{\partial V}\right)_T = \left(\frac{\partial P}{\partial T}\right)_V$ |
| $G$ | $dG = -SdT + VdP$ | $\left(\frac{\partial S}{\partial P}\right)_T = -\left(\frac{\partial V}{\partial T}\right)_P$ |

## 8–6. The Standard State for a Gas at Any Temperature Is the Hypothetical Ideal Gas at One Bar

One of the most important applications of Equation 8.33 involves the correction for nonideality that we make to obtain the standard molar entropies of gases. The standard molar entropies of gases tabulated in the literature are expressed in terms of a hypothetical ideal gas at one bar and at the same temperature. This correction is usually small and is obtained in the following two-step procedure (Figure 8.6). We first take our real gas from its pressure of one bar to some very low pressure $P^{\text{id}}$, where it is sure to behave ideally. We use Equation 8.33 to do this, giving

$$\begin{aligned}\overline{S}(P^{\text{id}}) - \overline{S}(1\ \text{bar}) &= -\int_{1\ \text{bar}}^{P^{\text{id}}} \left(\frac{\partial \overline{V}}{\partial T}\right)_P dP \\ &= \int_{P^{\text{id}}}^{1\ \text{bar}} \left(\frac{\partial \overline{V}}{\partial T}\right)_P dP \qquad \text{(constant } T\text{)}\end{aligned} \tag{8.52}$$

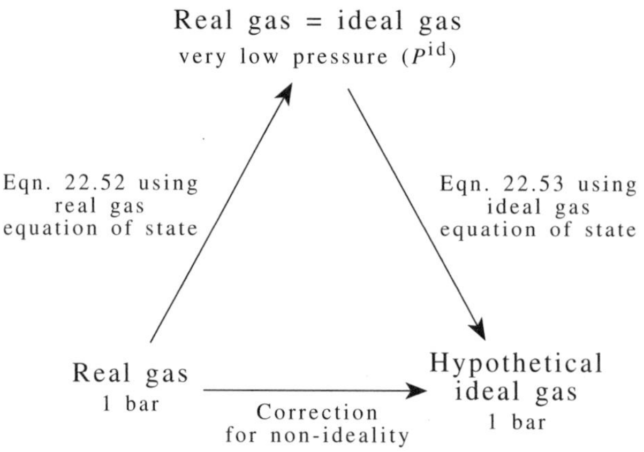


**FIGURE 8.6**
The scheme to bring the experimental entropies of gases to the standard state of a (hypothetical) ideal gas at the same temperature.

The superscript "id" on $P$ emphasizes that this value is for conditions for which the gas behaves ideally. The quantity $(\partial \overline{V}/\partial T)_P$ can be determined from the equation of state of the actual gas. We now calculate the change in entropy as we increase the pressure back to one bar, but *taking the gas to be ideal*. We use Equation 8.52 for this process, but with $(\partial \overline{V}/\partial T)_P = R/P$, giving

$$S^{\circ}(1 \text{ bar}) - \overline{S}(P^{\text{id}}) = -\int_{P^{\text{id}}}^{1 \text{ bar}} \frac{R}{P} dP \tag{8.53}$$

The superscript $\circ$ of $S^{\circ}(1 \text{ bar})$ emphasizes that this is the standard molar entropy of the gas. We add Equations 8.52 and 8.53 to get

$$S^{\circ}(\text{at 1 bar}) - \overline{S}(\text{at 1 bar}) = \int_{P^{\text{id}}}^{1 \text{ bar}} \left[ \left( \frac{\partial \overline{V}}{\partial T} \right)_P - \frac{R}{P} \right] dP \tag{8.54}$$

In Equation 8.54, $\overline{S}$ is the molar entropy we calculate from heat-capacity data and heats of transitions (Section 7–3), and $S^{\circ}$ is the molar entropy of the corresponding hypothetical ideal gas at one bar.

Equation 8.54 tells us that we can calculate the necessary correction to obtain the standard entropy if we know the equation of state. Because the pressures involved are around one bar, we can use the virial expansion using only the second virial coefficient. Using Equation 2.22,

$$\frac{P\overline{V}}{RT} = 1 + \frac{B_{2V}(T)}{RT} P + \cdots \tag{8.55}$$

we have

$$\left(\frac{\partial \overline{V}}{\partial T}\right)_P = \frac{R}{P} + \frac{dB_{2V}}{dT} + \cdots$$

Substituting this result into Equation 8.54 gives

$$S^{\circ}(\text{at 1 bar}) = \overline{S}(\text{at 1 bar}) + \frac{dB_{2V}}{dT} \times (1 \text{ bar}) + \cdots \tag{8.56}$$

where we have neglected $P^{\text{id}}$ with respect to one bar. The second term on the right side of Equation 8.56 represents the correction that we add to $\overline{S}$ to get $S^{\circ}$.

We can use Equation 8.56 to calculate the nonideality correction to the entropy of $N_2(g)$ at 298.15 K that we used in Table 7.1. The experimental value of $dB_{2V}/dT$ for $N_2(g)$ at 298.15 K and one bar is $0.192 \text{ cm}^3\cdot\text{mol}^{-1}\cdot\text{K}^{-1}$. Therefore, the correction for nonideality is given by

$$\begin{aligned}\text{correction for nonideality} &= (0.192 \text{ cm}^3\cdot\text{mol}^{-1}\cdot\text{K}^{-1})(1 \text{ bar}) \\ &= 0.192 \text{ cm}^3\cdot\text{bar}\cdot\text{mol}^{-1}\cdot\text{K}^{-1} \\ &= (0.192 \text{ cm}^3\cdot\text{bar}\cdot\text{mol}^{-1}\cdot\text{K}^{-1}) \\ &\quad \times \left(\frac{1 \text{ dm}^3}{10 \text{ cm}}\right)^3 \left(\frac{8.314 \text{ J}\cdot\text{mol}^{-1}\cdot\text{K}^{-1}}{0.08314 \text{ dm}^3\cdot\text{bar}\cdot\text{mol}^{-1}\cdot\text{K}^{-1}}\right) \\ &= 0.02 \text{ J}\cdot\text{K}^{-1}\cdot\text{mol}^{-1}\end{aligned}$$

which is what was used in Table 7.1. The correction in this case is rather small, but that is not always so. If second virial coefficent data are not available, then an approximate equation of state can be used (Problems 8–20 through 8–22).

## 8–7. The Gibbs–Helmholtz Equation Describes the Temperature Dependence of the Gibbs Energy

Both of Equations 8.31 are useful because they tell us how the Gibbs energy varies with pressure and with temperature. Let's look at Equation 8.31*b* first. We can use Equation 8.31*b* to calculate the pressure dependence of the Gibbs energy:

$$\Delta G = \int_{P_1}^{P_2} V dP \qquad (\text{constant } T) \tag{8.57}$$

For one mole of an ideal gas, we have

$$\Delta \overline{G} = RT \int_{P_1}^{P_2} \frac{dP}{P} = RT \ln \frac{P_2}{P_1} \tag{8.58}$$

We could have obtained the same result by using

$$\Delta\overline{G} = \Delta\overline{H} - T\Delta\overline{S} \qquad \text{(isothermal)}$$

For an isothermal change in an ideal gas, $\Delta\overline{H} = 0$ and $\Delta\overline{S}$ is given by Equation 8.21.

It is customary to let $P_1 = 1$ bar (exactly) in Equation 8.58 and to write it in the form

$$\overline{G}(T, P) = G^{\circ}(T) + RT \ln(P/1 \text{ bar}) \tag{8.59}$$

where $G^{\circ}(T)$ is called the standard molar Gibbs energy. The standard molar Gibbs energy in this case is the Gibbs energy of one mole of the ideal gas at a pressure of one bar. Note that $G^{\circ}(T)$ depends upon only the temperature. Equation 8.59 gives the Gibbs energy of an ideal gas relative to the standard Gibbs energy. According to Equation 8.59, $\overline{G}(T, P) - G^{\circ}(T)$ increases logarithmically with $P$, which we have seen is entirely an entropic effect for an ideal gas (because $H$ is independent of $P$ for an ideal gas). We will see in Chapter 12 that Equation 8.59 plays a central role in chemical equilibria involving gas-phase reactions.

**EXAMPLE 8–9**

Solids and liquids are fairly incompressible, so $V$ in Equation 8.57 may be taken to be constant to a good approximation in this case. Derive an expression for $\overline{G}(T, P)$ analogous to Equation 8.59 for a solid or a liquid.

SOLUTION: We integrate Equation 8.57 at constant $T$ to get

$$\overline{G}(P_2, T) - \overline{G}(P_1, T) = \overline{V}(P_2 - P_1)$$

We let $P_1 = 1$ bar and $\overline{G}(P_1 = 1 \text{ bar}, T) = G^{\circ}(T)$ to get

$$\overline{G}(T, P) = G^{\circ}(T) + \overline{V}(P - 1)$$

where $P$ must be expressed in bars. In this case, $\overline{G}(T, P)$ increases linearly with $P$, but because the volume of a condensed phase is much smaller than that of a gas, the slope of $\overline{G}(T, P)$ versus $P$, or $(\partial\overline{G}/\partial P)_T = \overline{V}$, is very small. Consequently, at ordinary pressures $\overline{G}(T, P)$ is almost independent of pressure and is approximately equal to $G^{\circ}(T)$ for a condensed phase.

Equation 8.31$a$ determines the temperature dependence of the Gibbs energy. We can derive a useful equation for the temperature dependence of $G$ by starting with Equation 8.31$a$ (Problem 8–24), but an easier way is to start with $G = H - TS$ and divide by $T$ to obtain

$$\frac{G}{T} = \frac{H}{T} - S$$

Now differentiate partially with respect to $T$ keeping $P$ fixed:

$$\left(\frac{\partial G/T}{\partial T}\right)_P = -\frac{H}{T^2} + \frac{1}{T}\left(\frac{\partial H}{\partial T}\right)_P - \left(\frac{\partial S}{\partial T}\right)_P$$

These last two terms cancel because of the relation $(\partial S/\partial T)_P = C_P(T)/T$ (Equation 7.7), so we have

$$\left(\frac{\partial G/T}{\partial T}\right)_P = -\frac{H}{T^2} \tag{8.60}$$

Equation 8.60 is called the *Gibbs-Helmholtz equation*. This equation can be directly applied to any process, in which case it becomes

$$\left(\frac{\partial \Delta G/T}{\partial T}\right)_P = -\frac{\Delta H}{T^2} \tag{8.61}$$

This equation is simply another form of the Gibbs-Helmholtz equation. We will use Equations 8.60 and 8.61 a number of times in the following chapters. For example, Equation 8.61 is used in Chapter 12 to derive an equation for the temperature dependence of an equilibrium constant.

We can determine the Gibbs energy as a function of temperature directly from equations we derived in Chapters 5 and 7. In Chapter 5, we learned how to calculate the enthalpy of a substance as a function of temperature in terms of its heat capacity and its various heats of transition. For example, if there exists only one solid phase, so that there are no solid-solid phase transitions between $T = 0$ K and its melting point, then (Equation 5.46)

$$\begin{aligned} H(T) - H(0) = &\int_0^{T_{\text{fus}}} C_P^s(T)dT + \Delta_{\text{fus}}H \\ &+ \int_{T_{\text{fus}}}^{T_{\text{vap}}} C_P^l(T)dT + \Delta_{\text{vap}}H \\ &+ \int_{T_{\text{vap}}}^{T} C_P^g(T')dT' \end{aligned} \tag{8.62}$$

for a temperature above the boiling point, where Figure 5.7 shows $\overline{H}(T) - \overline{H}(0)$ versus $T$ for benzene. We calculate $H(T)$ relative to $H(0)$ because it is not possible to calculate an absolute enthalpy; $H(0)$ is essentially our zero of energy.

In Chapter 7, we learned to calculate absolute entropies according to Equation 7.17),

$$S(T)=\int_0^{T_{\text{fus}}}\frac{C_P^s(T)}{T}dT+\frac{\Delta_{\text{fus}}H}{T_{\text{fus}}}$$
$$+\int_{T_{\text{fus}}}^{T_{\text{vap}}}\frac{C_P^l(T)}{T}dT+\frac{\Delta_{\text{vap}}H}{T_{\text{vap}}} \tag{8.63}$$
$$+\int_{T_{\text{vap}}}^{T}\frac{C_P^g(T')}{T'}dT'$$

Figure 7.2 shows $\overline{S}(T)$ versus $T$ for benzene. We can use Equations 8.62 and 8.63 to calculate $\overline{G}(T)-\overline{H}(0)$ because

$$\overline{G}(T)-\overline{H}(0)=\overline{H}(T)-\overline{H}(0)-T\overline{S}(T)$$

Figure 8.7 shows $\overline{G}(T)-\overline{H}(0)$ versus $T$ for benzene. There are several features of Figure 8.7 to appreciate. First note that $\overline{G}(T)-\overline{H}(0)$ decreases with increasing $T$. Furthermore, $\overline{G}(T)-\overline{H}(0)$ is a continuous function of temperature, even at a phase transition. To see that this is so, consider the equation (Equation 7.16)

$$\Delta_{\text{trs}}S=\frac{\Delta_{\text{trs}}H}{T_{\text{trs}}}$$

Because $\Delta_{\text{trs}}G=\Delta_{\text{trs}}H-T_{\text{trs}}\Delta_{\text{trs}}S$, we see that $\Delta_{\text{trs}}G=0$, indicating that the two phases are in equilibrium with each other. Two phases in equilibrium with each other have the same value of $G$, so $G(T)$ is continuous at a phase transition. Figure 8.7

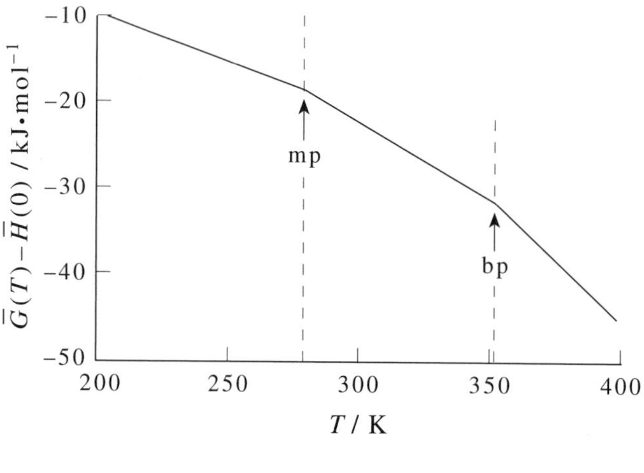


**FIGURE 8.7**
A plot of $\overline{G}(T)-\overline{H}(0)$ versus $T$ for benzene. Note that $\overline{G}(T)-\overline{H}(0)$ is continuous but its derivative (the slope of the curve) is discontinuous at the phase transitions.

also shows that there is a discontinuity in the slope at each phase transition. (Benzene melts at 278.7 K and boils at 353.2 K at one atm.) We can understand why there is a discontinuity in the slope of $G(T)$ versus $T$ at each phase transition by looking at Equation 8.31$a$

$$\left(\frac{\partial G}{\partial T}\right)_P = -S$$

Because entropy is an intrinsically positive quantity, the slope of $G(T)$ versus $T$ is negative. Furthermore, because $S(\text{gas}) > S(\text{liquid}) > S(\text{solid})$, the slopes within each single phase region increase in going from solid to liquid to gas, so the slope, $(\partial G/\partial T)_P$, is discontinuous in passing from one phase to another.

The values of $H^\circ(T) - H^\circ(0)$, $S^\circ(T)$, and $G^\circ(T) - H^\circ(0)$ are tabulated for a variety of substances. We will use these values to calculate equilibrium constants in Chapter 12.

## 8–8. Fugacity Is a Measure of the Nonideality of a Gas

In the previous section, we showed that the molar Gibbs energy of an ideal gas is given by

$$\overline{G}(T, P) = G^\circ(T) + RT \ln \frac{P}{P^\circ} \tag{8.64}$$

The pressure $P^\circ$ is equal to one bar and $G^\circ(T)$ is called the standard molar Gibbs energy. Recall that this equation is derived by starting with

$$\left(\frac{\partial \overline{G}}{\partial P}\right)_T = \overline{V} \tag{8.65}$$

and then integrating, using the ideal gas expression, $RT/P$, for $\overline{V}$. Let's now generalize Equation 8.64 to the case of a real gas.

We could start with the virial expansion,

$$\frac{P\overline{V}}{RT} = 1 + B_{2P}(T)P + B_{3P}(T)P^2 + \cdots$$

and substitute this into Equation 8.65 to obtain a virial expansion for the molar Gibbs energy,

$$\int_{P^{\text{id}}}^{P} d\overline{G} = RT \int_{P^{\text{id}}}^{P} \frac{dP'}{P'} + RT B_{2P}(T) \int_{P^{\text{id}}}^{P} dP' + RT B_{3P}(T) \int_{P^{\text{id}}}^{P} P' dP'$$

where we are integrating from some low pressure, say $P^{\text{id}}$, where the gas is sure to behave ideally, to some arbitrary pressure $P$. The result of the integration is

$$\overline{G}(T, P) = \overline{G}(T, P^{\text{id}}) + RT \ln \frac{P}{P^{\text{id}}} + RT B_{2P}(T)P + \frac{RT B_{3P}(T)P^2}{2} + \cdots \tag{8.66}$$

Now according to Equation 8.64, $\overline{G}(T, P^{\text{id}}) = G^\circ(T) + RT \ln P^{\text{id}}/P^\circ$, where $G^\circ(T)$ is the molar Gibbs energy of an ideal gas at a pressure of $P^\circ = 1$ bar. Therefore, Equation 8.66 can be written as

$$\overline{G}(T, P) = G^\circ(T) + RT \ln \frac{P}{P^\circ} + RT B_{2P}(T)P + \frac{RT B_{3P}(T)P^2}{2} + \cdots \tag{8.67}$$

Equation 8.67 is the generalization of Equation 8.64 to any real gas. Although Equation 8.67 is exact, it differs for each gas, depending upon the values of $B_{2P}(T)$, $B_{3P}(T)$, and so on. It turns out to be much more convenient, particularly for calculations involving chemical equilibria, as we will see in Chapter 12, to maintain the form of Equation 8.64 by defining a new thermodynamic function, $f(P, T)$, called *fugacity*, by the equation

$$\overline{G}(T, P) = G^\circ(T) + RT \ln \frac{f(P, T)}{f^\circ} \tag{8.68}$$

The nonideality is buried in $f(P, T)$. Because all gases behave ideally as $P \to 0$, fugacity must have the property that

$$f(P, T) \to P \qquad \text{as} \qquad P \to 0$$

so that Equation 8.68 reduces to Equation 8.64.

Equations 8.67 and 8.68 are equivalent if

$$\frac{f(P, T)}{f^\circ} = \frac{P}{P^\circ} \exp\left[B_{2P}(T)P + B_{3P}(T)P^2 + \cdots\right] \tag{8.69}$$

It might seem at this point that we are just going in circles, but by incorporating the nonideality of a gas through its fugacity, we can preserve the thermodynamic equations we have derived for ideal gases and write those corresponding to a real gas by simply replacing $P/P^\circ$ by $f/f^\circ$. All we need at this stage is a straightforward way to determine the fugacity of a gas at any pressure and temperature. Before looking into this, however, we must discuss the choice of the standard state in Equation 8.68. Being a type of energy, the Gibbs energy must always be taken relative to some chosen standard state.

Note that the standard molar Gibbs energy $G^\circ(T)$ is taken to be the same quantity in Equations 8.64 and 8.68. The standard state in Equation 8.64 is the ideal gas at one bar, so this must be the standard state in Equation 8.68 as well. Thus, the standard state of the real gas in Equation 8.68 is taken to be the corresponding ideal gas at one bar; in other words, the standard state of the real gas is one bar after it has been adjusted to ideal behavior. In an equation, we have that $f^\circ = P^\circ$. Note that this choice is also suggested by Equation 8.69, because otherwise $f(P, T)$ would not reduce to $P$ when $B_{2P}(T) = B_{3P}(T) = 0$.

This choice of standard state not only allows all gases to be brought to a single common state, but also leads to a procedure to calculate $f(P, T)$ at any pressure and temperature. To do so, consider the scheme in Figure 8.8, which depicts the difference in molar Gibbs energy between a real gas at $P$ and $T$ and an ideal gas at $P$ and $T$. We can calculate this difference by starting with the real gas at $P$ and $T$ and then calculating the change in Gibbs energy when the pressure is reduced to essentially zero (step 2), where the gas is certain to behave ideally. Then we calculate the change in Gibbs energy as we compress the gas back to pressure $P$, but taking the gas to behave ideally (step 3). The sum of steps 2 and 3, then, will be the difference in Gibbs energy of an ideal gas at $P$ and $T$ and the real gas at $P$ and $T$ (step 1). In an equation, we have

$$\Delta\overline{G}_1 = \overline{G}^{\text{id}}(T, P) - \overline{G}(T, P) \tag{8.70}$$

Substituting Equations 8.64 and 8.68 into Equation 8.70, we have

$$\Delta\overline{G}_1 = RT \ln \frac{P}{P^\circ} - RT \ln \frac{f}{f^\circ}$$

But the standard state of the real gas has been chosen such that $f^\circ = P^\circ = 1$ bar, so

$$\Delta\overline{G}_1 = RT \ln \frac{P}{f} \tag{8.71}$$

We now use Equation 8.65 to calculate the change in the Gibbs energy along steps 2 and 3:

$$\Delta\overline{G}_2 = \int_P^{P\to 0} \left(\frac{\partial G}{\partial P}\right)_T dG = \int_P^{P\to 0} \overline{V} dP'$$

$$\Delta\overline{G}_3 = \int_{P\to 0}^{P} \overline{V}^{\text{id}} dP' = \int_{P\to 0}^{P} \frac{RT}{P'} dP'$$

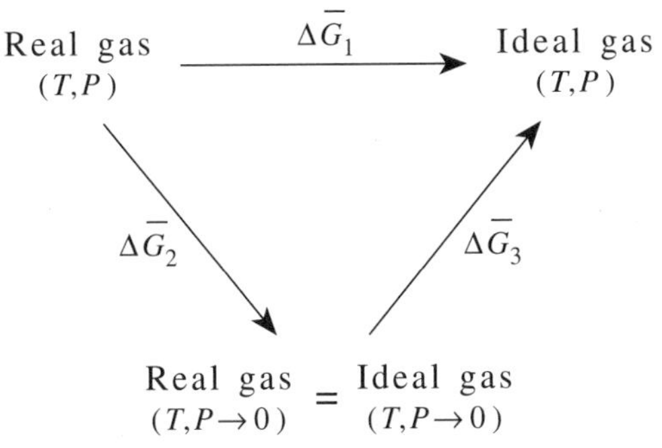


**FIGURE 8.8**
An illustration of the scheme used to relate the fugacity of a gas to its standard state, which is a (hypothetical) ideal gas at $P = 1$ bar and the temperature $T$ of interest.

The sum of $\Delta\overline{G}_2$ and $\Delta\overline{G}_3$ gives another expression for $\Delta\overline{G}_1$

$$\Delta\overline{G}_1 = \Delta\overline{G}_2 + \Delta\overline{G}_3 = \int_{P\to 0}^{P} \left(\frac{RT}{P'} - \overline{V}\right) dP'$$

Equating this expression for $\Delta\overline{G}_1$ to $\Delta\overline{G}_1$ in Equation 8.71, we have

$$\ln\frac{P}{f} = \int_0^P \left(\frac{1}{P'} - \frac{\overline{V}}{RT}\right) dP'$$

or

$$\ln\frac{f}{P} = \int_0^P \left(\frac{\overline{V}}{RT} - \frac{1}{P'}\right) dP' \tag{8.72}$$

Given either $P$-$V$-$T$ data or the equation of state of the real gas, Equation 8.72 allows us to calculate the ratio of the fugacity to the pressure of a gas at any pressure and temperature. Note that if the gas behaves ideally under the conditions of interest (in other words, if $\overline{V} = \overline{V}^{\text{id}}$ in Equation 8.72), then $\ln f/P = 0$, or $f = P$. Therefore, the extent of the deviation of $f/P$ from unity is a direct indication of the extent of the deviation of the gas from ideal behavior. The ratio $f/P$ is called the *fugacity coefficient*, $\gamma$,

$$\gamma = \frac{f}{P} \tag{8.73}$$

For an ideal gas, $\gamma = 1$.

By introducing the compressibility factor, $Z = P\overline{V}/RT$, Equation 8.72 can be written as

$$\ln\gamma = \int_0^P \frac{Z-1}{P'} dP' \tag{8.74}$$

Even though the lower limit here is $P = 0$, the integrand is finite (Problem 8–27). Furthermore, $(Z-1)/P = 0$ for an ideal gas (Problem 8–27), and hence $\ln\gamma = 0$ and $f = P$. Figure 8.9 shows $(Z-1)/P$ plotted against $P$ for CO(g) at 200 K. According to Equation 8.74, the area under this curve from 0 to $P$ is equal to $\ln\gamma$ at the pressure $P$. Figure 8.10 shows the resulting values of $\gamma = f/P$ plotted against the pressure for CO(g) at 200 K.

We can also calculate the fugacity if we know the equation of state of the gas.

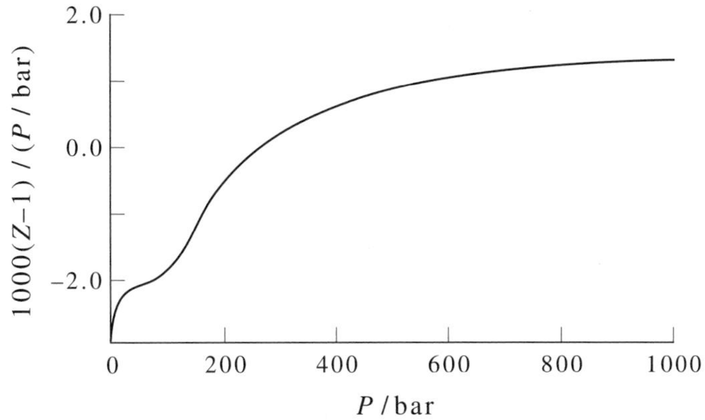


**FIGURE 8.9**
A plot of $(Z-1)/P$ versus $P$ for CO(g) at 200 K. The area under this curve from $P=0$ to $P$ gives $\ln\gamma$ at the pressure $P$.

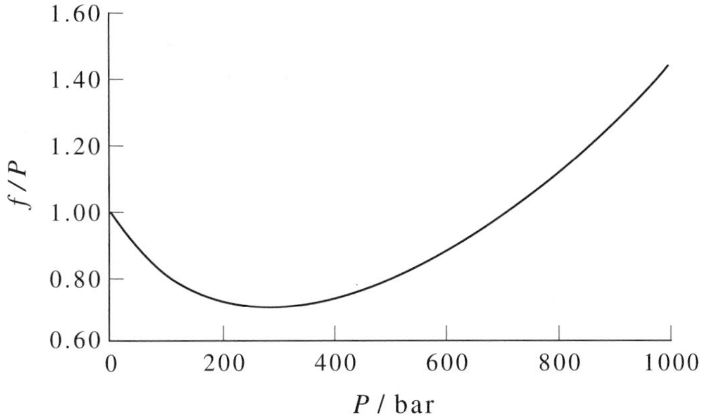


**FIGURE 8.10**
A plot of $\gamma = f/P$ against $P$ for CO(g) at 200 K. These values of $f/P$ were obtained from a numerical integration of $(Z-1)/P$ shown in Figure 8.9.

**EXAMPLE 8–10**

Derive an expression for the fugacity of a gas that obeys the equation of state

$$P(\overline{V}-b)=RT$$

where $b$ is a constant.

SOLUTION: We solve for $\overline{V}$ and substitute into Equation 8.72 to get

$$\ln \gamma = \int_0^P \frac{b}{RT} dP = \frac{bP}{RT}$$

or

$$\gamma = e^{bP/RT}$$

Problems 8–33 through 8–38 derive expressions for $\ln \gamma$ for the van der Waals equation and the Redlich-Kwong equation.

We can write Equation 8.74 in a form that shows that the fugacity coefficient is a function of the reduced pressure and the reduced temperature. If we change the integration variable to $P_R = P/P_c$, where $P_c$ is the critical pressure of the gas, then Equation 8.74 takes the form

$$\ln \gamma = \int_0^{P_R} \left( \frac{Z-1}{P_R'} \right) dP_R' \tag{8.75}$$

Now recall from Chapter 2 that, to a good approximation for most gases, the compressibility factor $Z$ is a universal function of $P_R$ and $T_R$ (see Figure 2.10). Therefore, the right side of Equation 8.75, and so $\ln \gamma$ itself, is also a universal function of $P_R$ and $T_R$. Figure 8.11 shows the experimental values of $\gamma$ for many gases as a family of curves of constant $T_R$ plotted against $P_R$.

**EXAMPLE 8–11**

Use Figure 8.11 and Table 2.5 to estimate the fugacity of nitrogen at 623 K and 1000 atm.

SOLUTION: We find from Table 2.5 that $T_c = 126.2$ K and $P_c = 33.6$ atm for $N_2(g)$. Therefore $T_R = 4.94$ at 623 K and $P_R = 29.8$ at 1000 atm. Reading from the curves in Figure 8.11, we find that $\gamma \approx 1.7$. At 1000 atm and 623 K, the fugacity of nitrogen is 1700 atm.

## Problems

**8-1.** The molar enthalpy of vaporization of benzene at its normal boiling point (80.09°C) is 30.72 kJ·mol$^{-1}$. Assuming that $\Delta_{vap}\overline{H}$ and $\Delta_{vap}\overline{S}$ stay constant at their values at 80.09°C, calculate the value of $\Delta_{vap}\overline{G}$ at 75.0°C, 80.09°C, and 85.0°C. Interpret these results physically.

**8-2.** Redo Problem 8–1 without assuming that $\Delta_{vap}\overline{H}$ and $\Delta_{vap}\overline{S}$ do not vary with temperature. Take the molar heat capacities of liquid and gaseous benzene to be 136.3 J·K$^{-1}$·mol$^{-1}$

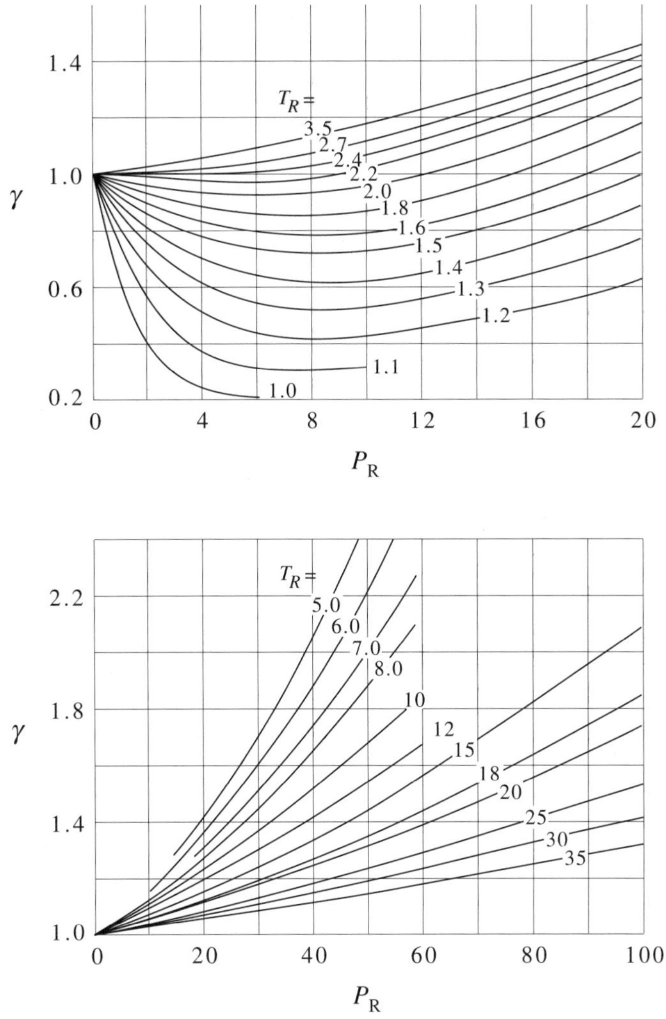


**FIGURE 8.11**
The fugacity coefficients of gases plotted against the reduced pressure, $P/P_c$, for various values of the reduced temperature, $T/T_c$.

and 82.4 $J \cdot K^{-1} \cdot mol^{-1}$, respectively. Compare your results with those you obtained in Problem 8–1. Are any of your physcial interpretations different?

**8-3.** Substitute $(\partial P/\partial T)_{\overline{V}}$ from the van der Waals equation into Equation 8.19 and integrate from $\overline{V}^{\text{id}}$ to $\overline{V}$ to obtain

$$\overline{S}(T, \overline{V}) - \overline{S}^{\text{id}}(T) = R \ln \frac{\overline{V} - b}{\overline{V}^{\text{id}} - b}$$

Now let $\overline{V}^{\mathrm{id}} = RT/P^{\mathrm{id}}$, $P^{\mathrm{id}} = P^{\circ} =$ one bar, and $\overline{V}^{\mathrm{id}} \gg b$ to obtain

$$\overline{S}(T,\overline{V}) - \overline{S}^{\mathrm{id}}(T) = -R \ln \frac{RT/P^{\circ}}{\overline{V} - b}$$

Given that $\overline{S}^{\mathrm{id}} = 246.35\ \mathrm{J\cdot mol^{-1}\cdot K^{-1}}$ for ethane at 400 K, show that

$$\overline{S}(\overline{V})/\mathrm{J\cdot mol^{-1}\cdot K^{-1}} = 246.35 - 8.3145 \ln \frac{33.258\ \mathrm{L\cdot mol^{-1}}}{\overline{V} - 0.065144\ \mathrm{L\cdot mol^{-1}}}$$

Calculate $\overline{S}$ as a function of $\rho = 1/\overline{V}$ for ethane at 400 K and compare your results with the experimental results shown in Figure 8.2.

Show that

$$\overline{S}(\overline{V})/\mathrm{J\cdot mol^{-1}\cdot K^{-1}} = 246.35 - 8.3145 \ln \frac{33.258\ \mathrm{L\cdot mol^{-1}}}{\overline{V} - 0.045153\ \mathrm{L\cdot mol^{-1}}} - 13.68 \ln \frac{\overline{V} + 0.045153\ \mathrm{L\cdot mol^{-1}}}{\overline{V}}$$

for the Redlich-Kwong equation for ethane at 400 K. Calculate $\overline{S}$ as a function of $\rho = 1/\overline{V}$ and compare your results with the experimental results shown in Figure 8.2.

**8-4.** Use the van der Waals equation to derive

$$\overline{U}(T,\overline{V}) - \overline{U}^{\mathrm{id}}(T) = -\frac{a}{\overline{V}}$$

Use this result along with the van der Waals equation to calculate the value of $\overline{U}$ as a function of $\overline{V}$ for ethane at 400 K, given that $\overline{U}^{\mathrm{id}} = 14.55\ \mathrm{kJ\cdot mol^{-1}}$. To do this, specify $\overline{V}$ (from $0.0700\ \mathrm{L\cdot mol^{-1}}$ to $7.00\ \mathrm{L\cdot mol^{-1}}$, see Figure 8.2), calculate both $\overline{U}(\overline{V})$ and $P(\overline{V})$, and plot $\overline{U}(\overline{V})$ versus $P(\overline{V})$. Compare your result with the experimental data in Figure 8.3.

Use the Redlich-Kwong equation to derive

$$\overline{U}(T,\overline{V}) - \overline{U}^{\mathrm{id}}(T) = -\frac{3A}{2BT^{1/2}} \ln \frac{\overline{V} + B}{\overline{V}}$$

Repeat the above calculation for ethane at 400 K.

**8-5.** Show that $(\partial U/\partial V)_T = 0$ for a gas that obeys an equation of state of the form $Pf(V) = RT$. Give two examples of such equations of state that appear in the text.

**8-6.** Show that

$$\left(\frac{\partial \overline{U}}{\partial \overline{V}}\right)_T = \frac{RT^2}{\overline{V}^2}\frac{dB_{2V}}{dT} + \frac{RT^2}{\overline{V}^3}\frac{dB_{3V}}{dT} + \cdots$$

**8-7.** Use the result of the previous problem to show that

$$\Delta\overline{U} = -T\frac{dB_{2V}}{dT}(P_2 - P_1) + \cdots$$

Use Equation 2.41 for the square-well potential to show that

$$\Delta\overline{U} = -\frac{2\pi\sigma^3 N_{\mathrm{A}}}{3}(\lambda^3 - 1)\frac{\varepsilon}{k_{\mathrm{B}}T}e^{\varepsilon/k_{\mathrm{B}}T}(P_2 - P_1) + \cdots$$

Given that $\sigma = 327.7$ pm, $\varepsilon/k_B = 95.2$ K, and $\lambda = 1.58$ for $N_2(g)$, calculate the value of $\Delta\overline{U}$ for a pressure increase from 1.00 bar to 10.0 bar at 300 K.

**8-8.** Determine $\overline{C}_P - \overline{C}_V$ for a gas that obeys the equation of state $P(\overline{V} - b) = RT$.

**8-9.** The coefficient of thermal expansion of water at 25°C is $2.572 \times 10^{-4}\ \mathrm{K}^{-1}$, and its isothermal compressibility is $4.525 \times 10^{-5}\ \mathrm{bar}^{-1}$. Calculate the value of $C_P - C_V$ for one mole of water at 25°C. The density of water at 25°C is $0.99705\ \mathrm{g \cdot mL^{-1}}$.

**8-10.** Use Equation 8.22 to show that

$$\left(\frac{\partial C_V}{\partial V}\right)_T = T\left(\frac{\partial^2 P}{\partial T^2}\right)_V$$

Show that $(\partial C_V/\partial V)_T = 0$ for an ideal gas and a van der Waals gas, and that

$$\left(\frac{\partial C_V}{\partial V}\right)_T = -\frac{3A}{4T^{3/2}\overline{V}(\overline{V} + B)}$$

for a Redlich-Kwong gas.

**8-11.** In this problem you will derive the equation (Equation 8.24)

$$C_P - C_V = -T\left(\frac{\partial V}{\partial T}\right)_P^2\left(\frac{\partial P}{\partial V}\right)_T$$

To start, consider $V$ to be a function of $T$ and $P$ and write out $dV$. Now divide through by $dT$ at constant volume ($dV = 0$) and then substitute the expression for $(\partial P/\partial T)_V$ that you obtain into Equation 8.23 to get the above expression.

**8-12.** The quantity $(\partial U/\partial V)_T$ has units of pressure and is called the *internal pressure*, which is a measure of the intermolecular forces within the body of a substance. It is equal to zero for an ideal gas, is nonzero but relatively small for dense gases, and is relatively large for liquids, particularly those whose molecular interactions are strong. Use the following data to calculate the internal pressure of ethane as a function of pressure at 280 K. Compare your values with the values you obtain from the van der Waals equation and the Redlich-Kwong equation.

| $P$/bar | $(\partial P/\partial T)_V/\mathrm{bar \cdot K^{-1}}$ | $\overline{V}/\mathrm{dm^3 \cdot mol^{-1}}$ | $P$/bar | $(\partial P/\partial T)_V/\mathrm{bar \cdot K^{-1}}$ | $\overline{V}/\mathrm{dm^3 \cdot mol^{-1}}$ |
|---|---|---|---|---|---|
| 4.458 | 0.01740 | 5.000 | 307.14 | 6.9933 | 0.06410 |
| 47.343 | 4.1673 | 0.07526 | 437.40 | 7.9029 | 0.06173 |
| 98.790 | 4.9840 | 0.07143 | 545.33 | 8.5653 | 0.06024 |
| 157.45 | 5.6736 | 0.06849 | 672.92 | 9.2770 | 0.05882 |

**8-13.** Show that

$$\left(\frac{\partial \overline{H}}{\partial P}\right)_T = -RT^2\left(\frac{dB_{2P}}{dT} + \frac{dB_{3P}}{dT}P + \cdots\right)$$

$$= B_{2V}(T) - T\frac{dB_{2V}}{dT} + O(P)$$

Use Equation 2.41 for the square-well potential to obtain

$$\left(\frac{\partial \overline{H}}{\partial P}\right)_T = \frac{2\pi\sigma^3 N_A}{3}\left[\lambda^3 - (\lambda^3 - 1)\left(1 + \frac{\varepsilon}{k_B T}\right)e^{\varepsilon/k_B T}\right]$$

Given that $\sigma = 327.7$ pm, $\varepsilon/k_B = 95.2$ K, and $\lambda = 1.58$ for $N_2$(g), calculate the value of $(\partial \overline{H}/\partial P)_T$ at 300 K. Evaluate $\Delta \overline{H} = \overline{H}(P = 10.0 \text{ bar}) - \overline{H}(P = 1.0 \text{ bar})$. Compare your result with 8.724 $\text{J}\cdot\text{mol}^{-1}$, the value of $\overline{H}(T) - \overline{H}(0)$ for nitrogen at 300 K.

**8-14.** Show that the enthalpy is a function of only the temperature for a gas that obeys the equation of state $P(\overline{V} - bT) = RT$, where $b$ is a constant.

**8-15.** Use your results for the van der Waals equation and the Redlich-Kwong equation in Problem 8–4 to calculate $\overline{H}(T, \overline{V})$ as a function of volume for ethane at 400 K. In each case, use the equation $\overline{H} = \overline{U} + P\overline{V}$. Compare your results with the experimental data shown in Figure 8.5.

**8-16.** Use Equation 8.34 to show that

$$\left(\frac{\partial C_P}{\partial P}\right)_T = -T\left(\frac{\partial^2 V}{\partial T^2}\right)_P$$

Use a virial expansion in $P$ to show that

$$\left(\frac{\partial \overline{C}_P}{\partial P}\right)_T = -T\frac{d^2 B_{2V}}{dT^2} + O(P)$$

Use the square-well second virial coefficient (Equation 2.41) and the parameters given in Problem 8–13 to calculate the value of $(\partial \overline{C}_P/\partial P)_T$ for $N_2$(g) at 0°C. Now calculate $\overline{C}_P$ at 100 atm and 0°C, using $\overline{C}_P^{\text{id}} = 5R/2$.

**8-17.** Show that the molar enthalpy of a substance at pressure $P$ relative to its value at one bar is given by

$$\overline{H}(T, P) = \overline{H}(T, P = 1 \text{ bar}) + \int_1^P \left[\overline{V} - T\left(\frac{\partial \overline{V}}{\partial T}\right)_P\right] dP'$$

Calculate the value of $\overline{H}(T, P) - \overline{H}(T, P = 1 \text{ bar})$ at 0°C and 100 bar for mercury given that the molar volume of mercury varies with temperature according to

$$\overline{V}(t) = (14.75 \text{ mL}\cdot\text{mol}^{-1})(1 + 0.182 \times 10^{-3}t + 2.95 \times 10^{-9}t^2 + 1.15 \times 10^{-10}t^3)$$

where $t$ is the Celsius temperature. Assume that $\overline{V}(0)$ does not vary with pressure over this range and express your answer in units of $\text{kJ}\cdot\text{mol}^{-1}$.

**8-18.** Show that

$$dH = \left[V - T\left(\frac{\partial V}{\partial T}\right)_P\right] dP + C_P dT$$

What does this equation tell you about the natural variables of $H$?

**8-19.** What are the natural variables of the entropy?

**8-20.** Experimentally determined entropies are commonly adjusted for nonideality by using an equation of state called the (modified) Berthelot equation:

$$\frac{P\overline{V}}{RT} = 1 + \frac{9}{128}\frac{PT_c}{P_c T}\left(1 - 6\frac{T_c^2}{T^2}\right)$$

Show that this equation leads to the correction

$$S^\circ(\text{at one bar}) = \overline{S}(\text{at one bar}) + \frac{27}{32}\frac{RT_c^3}{P_c T^3}(1\text{ bar})$$

This result needs only the critical data for the substance. Use this equation along with the critical data in Table 2.5 to calculate the nonideality correction for $N_2(g)$ at 298.15 K. Compare your result with the value used in Table 7.1.

**8-21.** Use the result of Problem 8–20 along with the critical data in Table 2.5 to determine the nonideality correction for CO(g) at its normal boiling point, 81.6 K. Compare your result with the value used in Problem 7–24.

**8-22.** Use the result of Problem 8–20 along with the critical data in Table 2.5 to determine the nonideality correction for $Cl_2(g)$ at its normal boiling point, 239 K. Compare your result with the value used in Problem 7–16.

**8-23.** Derive the equation

$$\left(\frac{\partial(A/T)}{\partial T}\right)_V = -\frac{U}{T^2}$$

which is a Gibbs-Helmholtz equation for $A$.

**8-24.** We can derive the Gibbs-Helmholtz equation directly from Equation 8.31*a* in the following way. Start with $(\partial G/\partial T)_P = -S$ and substitue for $S$ from $G = H - TS$ to obtain

$$\frac{1}{T}\left(\frac{\partial G}{\partial T}\right)_P - \frac{G}{T^2} = -\frac{H}{T^2}$$

Now show that the left side is equal to $(\partial[G/T]/\partial T)_P$ to get the Gibbs-Helmholtz equation.

**8-25.** Use the following data for benzene to plot $\overline{G}(T) - \overline{H}(0)$ versus $T$. [In this case we will ignore the (usually small) corrections due to nonideality of the gas phase.]

$$\overline{C}_P^{\text{s}}(T)/R = \frac{12\pi^4}{5}\left(\frac{T}{\Theta_D}\right)^3 \qquad \Theta_D = 130.5\text{ K} \qquad 0\text{ K} < T < 13\text{ K}$$

$$\overline{C}_P^{\text{s}}(T)/R = -0.6077 + (0.1088\text{ K}^{-1})T - (5.345\times10^{-4}\text{ K}^{-2})T^2 + (1.275\times10^{-6}\text{ K}^{-3})T^3$$
$$13\text{ K} < T < 278.6\text{ K}$$

$$\overline{C}_P^{\text{l}}(T)/R = 12.713 + (1.974\times10^{-3}\text{ K}^{-1})T - (4.766\times10^{-5}\text{ K}^{-2})T^2$$
$$278.6\text{ K} < T < 353.2\text{ K}$$

$$\overline{C}_P^{\rm g}(T)/R = -4.077 + (0.05676\ \text{K}^{-1})T - (3.588 \times 10^{-5}\ \text{K}^{-2})T^2 + (8.520 \times 10^{-9}\ \text{K}^{-3})T^3$$

$$353.2\ \text{K} < T < 1000\ \text{K}$$

$$T_{\rm fus} = 278.68\ \text{K} \qquad \Delta_{\rm fus}\overline{H} = 9.95\ \text{kJ}\cdot\text{mol}^{-1}$$

$$T_{\rm vap} = 353.24\ \text{K} \qquad \Delta_{\rm vap}\overline{H} = 30.72\ \text{kJ}\cdot\text{mol}^{-1}$$

**8-26.** Use the following data for propene to plot $\overline{G}(T) - \overline{H}(0)$ versus $T$. [In this case we will ignore the (usually small) corrections due to nonideality of the gas phase.]

$$\overline{C}_P^{\rm s}(T)/R = \frac{12\pi^4}{5}\left(\frac{T}{\Theta_{\rm D}}\right)^3 \qquad \Theta_{\rm D} = 100\ \text{K} \qquad 0\ \text{K} < T < 15\ \text{K}$$

$$\overline{C}_P^{\rm s}(T)/R = -1.616 + (0.08677\ \text{K}^{-1})T - (9.791 \times 10^{-4}\ \text{K}^{-2})T^2 + (2.611 \times 10^{-6}\ \text{K}^{-3})T^3$$

$$15\ \text{K} < T < 87.90\ \text{K}$$

$$\overline{C}_P^{\rm l}(T)/R = 15.935 - (0.08677\ \text{K}^{-1})T + (4.294 \times 10^{-4}\ \text{K}^{-2})T^2 - (6.276 \times 10^{-7}\ \text{K}^{-3})T^3$$

$$87.90\ \text{K} < T < 225.46\ \text{K}$$

$$\overline{C}_P^{\rm g}(T)/R = 1.4970 + (2.266 \times 10^{-2}\ \text{K}^{-1})T - (5.725 \times 10^{-6}\ \text{K}^{-2})T^2$$

$$225.46\ \text{K} < T < 1000\ \text{K}$$

$$T_{\rm fus} = 87.90\ \text{K} \qquad \Delta_{\rm fus}\overline{H} = 3.00\ \text{kJ}\cdot\text{mol}^{-1}$$

$$T_{\rm vap} = 225.46\ \text{K} \qquad \Delta_{\rm vap}\overline{H} = 18.42\ \text{kJ}\cdot\text{mol}^{-1}$$

**8-27.** Use a virial expansion for $Z$ to prove (a) that the integrand in Equation 8.74 is finite as $P \to 0$, and (b) that $(Z-1)/P = 0$ for an ideal gas.

**8-28.** Derive a virial expansion in the pressure for $\ln\gamma$.

**8-29.** The compressibility factor for ethane at 600 K can be fit to the expression

$$Z = 1.0000 - 0.000612(P/\text{bar}) + 2.661 \times 10^{-6}(P/\text{bar})^2 - 1.390 \times 10^{-9}(P/\text{bar})^3 - 1.077 \times 10^{-13}(P/\text{bar})^4$$

for $0 \le P/\text{bar} \le 600$. Use this expression to determine the fugacity coefficient of ethane as a function of pressure at 600 K.

**8-30.** Use Figure 8.11 and the data in Table 2.5 to estimate the fugacity of ethane at 360 K and 1000 atm.

**8-31.** Use the following data for ethane at 360 K to plot the fugacity coefficient against pressure.

| $\rho/\text{mol}\cdot\text{dm}^{-3}$ | $P/\text{bar}$ | $\rho/\text{mol}\cdot\text{dm}^{-3}$ | $P/\text{bar}$ | $\rho/\text{mol}\cdot\text{dm}^{-3}$ | $P/\text{bar}$ |
|---|---|---|---|---|---|
| 1.20 | 31.031 | 6.00 | 97.767 | 10.80 | 197.643 |
| 2.40 | 53.940 | 7.20 | 112.115 | 12.00 | 266.858 |
| 3.60 | 71.099 | 8.40 | 130.149 | 13.00 | 381.344 |
| 4.80 | 84.892 | 9.60 | 156.078 | 14.40 | 566.335 |

Compare your result with the result you obtained in Problem 8–30.

**8-32.** Use the following data for $N_2(g)$ at 0°C to plot the fugacity coefficient as a function of pressure.

| $P/\text{atm}$ | $Z = P\overline{V}/RT$ | $P/\text{atm}$ | $Z = P\overline{V}/RT$ | $P/\text{atm}$ | $Z = P\overline{V}/RT$ |
|---|---|---|---|---|---|
| 200 | 1.0390 | 1000 | 2.0700 | 1800 | 3.0861 |
| 400 | 1.2570 | 1200 | 2.3352 | 2000 | 3.3270 |
| 600 | 1.5260 | 1400 | 2.5942 | 2200 | 3.5640 |
| 800 | 1.8016 | 1600 | 2.8456 | 2400 | 3.8004 |

**8-33.** It might appear that we can't use Equation 8.72 to determine the fugacity of a van der Waals gas because the van der Waals equation is a cubic equation in $\overline{V}$, so we can't solve it analytically for $\overline{V}$ to carry out the integration in Equation 8.72. We can get around this problem, however, by integrating Equation 8.72 by parts. First show that

$$RT \ln \gamma = P\overline{V} - RT - \int_{\overline{V}^{\text{id}}}^{\overline{V}} P d\overline{V}' - RT \ln \frac{P}{P^{\text{id}}}$$

where $P^{\text{id}} \to 0$, $\overline{V}^{\text{id}} \to \infty$, and $P^{\text{id}}\overline{V}^{\text{id}} \to RT$. Substitute $P$ from the van der Waals equation into the first term and the integral on the right side of the above equation and integrate to obtain

$$RT \ln \gamma = \frac{RT\overline{V}}{\overline{V} - b} - \frac{a}{\overline{V}} - RT - RT \ln \frac{\overline{V} - b}{\overline{V}^{\text{id}} - b} - \frac{a}{\overline{V}} - RT \ln \frac{P}{P^{\text{id}}}$$

Now use the fact that $\overline{V}^{\text{id}} \to \infty$ and that $P^{\text{id}}\overline{V}^{\text{id}} = RT$ to show that

$$\ln \gamma = -\ln \left[1 - \frac{a(\overline{V} - b)}{RT\overline{V}^2}\right] + \frac{b}{\overline{V} - b} - \frac{2a}{RT\overline{V}}$$

This equation gives the fugacity coefficient of a van der Waals gas as a function of $\overline{V}$. You can use the van der Waals equation itself to calculate $P$ from $\overline{V}$, so the above equation, in conjunction with the van der Waals equation, gives $\ln \gamma$ as a function of pressure.

**8-34.** Use the final equation in Problem 8–33 along with the van der Waals equation to plot $\ln \gamma$ against pressure for CO(g) at 200 K. Compare your result with Figure 8.10.

**8-35.** Show that the expression for $\ln\gamma$ for the van der Waals equation (Problem 8–33) can be written in the reduced form

$$\ln\gamma = \frac{1}{3\overline{V}_R - 1} - \frac{9}{4\overline{V}_R T_R} - \ln\left[1 - \frac{3(3\overline{V}_R - 1)}{8T_R\overline{V}_R^2}\right]$$

Use this equation along with the van der Waals equation in reduced form (Equation 2.19) to plot $\gamma$ against $P_R$ for $T_R = 1.00$ and $2.00$ and compare your results with Figure 8.11.

**8-36.** Use the method outlined in Problem 8–33 to show that

$$\ln\gamma = \frac{B}{\overline{V} - B} - \frac{A}{RT^{3/2}(\overline{V} + B)} - \frac{A}{BRT^{3/2}}\ln\frac{\overline{V} + B}{\overline{V}} - \ln\left[1 - \frac{A(\overline{V} - B)}{RT^{3/2}\overline{V}(\overline{V} + B)}\right]$$

for the Redlich-Kwong equation. You need to use the standard integral

$$\int\frac{dx}{x(a + bx)} = -\frac{1}{a}\ln\frac{a + bx}{x}$$

**8-37.** Show that $\ln\gamma$ for the Redlich-Kwong equation (see Problem 8–36) can be written in the reduced form

$$\ln\gamma = \frac{0.25992}{\overline{V}_R - 0.25992} - \frac{1.2824}{T_R^{3/2}(\overline{V}_R + 0.25992)} - \frac{4.9340}{T_R^{3/2}}\ln\frac{\overline{V}_R + 0.25992}{\overline{V}_R} - \ln\left[1 - \frac{1.2824(\overline{V}_R - 0.25992)}{T_R^{3/2}\overline{V}_R(\overline{V}_R + 0.25992)}\right]$$

**8-38.** Use the expression for $\ln\gamma$ in reduced form given in Problem 8–37 along with the Redlich-Kwong equation in reduced form (Example 2–7) to plot $\ln\gamma$ versus $P_R$ for $T_R = 1.00$ and $2.00$ and compare your results with those you obtained in Problem 8–35 for the van der Waals equation.

**8-39.** Compare $\ln\gamma$ for the van der Waals equation (Problem 8–33) with the values of $\ln\gamma$ for ethane at 600 K (Problem 8–29).

**8-40.** Compare $\ln\gamma$ for the Redlich-Kwong equation (Problem 8–36) with the values of $\ln\gamma$ for ethane at 600 K (Problem 8–29).

**8-41.** We can use the equation $(\partial S/\partial U)_V = 1/T$ to illustrate the consequence of the fact that entropy always increases during an irreversible adiabatic process. Consider a two-compartment system enclosed by rigid adiabatic walls, and let the two compartments be separated by a rigid heat-conducting wall. We assume that each compartment is at equilibrium but that they are not in equilibrium with each other. Because no work can be done by this two-compartment system (rigid walls) and no energy as heat can be exchanged with the surroundings (adiabatic walls),

$$U = U_1 + U_2 = \text{constant}$$

Show that

$$dS = \left(\frac{\partial S_1}{\partial U_1}\right) dU_1 + \left(\frac{\partial S_2}{\partial U_2}\right) dU_2$$

because the entropy of each compartment can change only as a result of a change in energy. Now show that

$$dS = dU_1 \left(\frac{1}{T_1} - \frac{1}{T_2}\right) \geq 0$$

Use this result to discuss the direction of the flow of energy as heat from one temperature to another.

**8-42.** Modify the argument in Problem 8–41 to the case in which the two compartments are separated by a nonrigid, insulating wall. Derive the result

$$dS = \left(\frac{P_1}{T_1} - \frac{P_2}{T_2}\right) dV_1$$

Use this result to discuss the direction of a volume change under an isothermal pressure difference.

**8-43.** In this problem, we will derive virial expansions for $\overline{U}$, $\overline{H}$, $\overline{S}$, $\overline{A}$, and $\overline{G}$. Substitute

$$Z = 1 + B_{2P}P + B_{3P}P^2 + \cdots$$

into Equation 8.65 and integrate from a small pressure, $P^{\text{id}}$, to $P$ to obtain

$$\overline{G}(T, P) - \overline{G}(T, P^{\text{id}}) = RT \ln \frac{P}{P^{\text{id}}} + RTB_{2P}P + \frac{RTB_{3P}}{2}P^2 + \cdots$$

Now use Equation 8.64 (realize that $P = P^{\text{id}}$ in Equation 8.64) to get

$$\overline{G}(T, P) - G^\circ(T) = RT \ln P + RTB_{2P}P + \frac{RTB_{3P}}{2}P^2 + \cdots \qquad (1)$$

at $P^\circ = 1$ bar. Now use Equation 8.31$a$ to get

$$\overline{S}(T, P) - S^\circ(T) = -R \ln P - \frac{d(RTB_{2P})}{dT}P - \frac{1}{2}\frac{d(RTB_{3P})}{dT}P^2 + \cdots \qquad (2)$$

at $P^\circ = 1$ bar. Now use $\overline{G} = \overline{H} - T\overline{S}$ to get

$$\overline{H}(T, P) - H^\circ(T) = -RT^2\frac{dB_{2P}}{dT}P - \frac{RT^2}{2}\frac{dB_{3P}}{dT}P^2 + \cdots \qquad (3)$$

Now use the fact that $\overline{C}_P = (\partial \overline{H}/\partial T)_P$ to get

$$\overline{C}_P(T, P) - C_P^\circ(T) = -RT\left[2\frac{dB_{2P}}{dT} + T\frac{d^2B_{2P}}{dT^2}\right]P - \frac{RT}{2}\left[2\frac{dB_{3P}}{dT} + T\frac{d^2B_{3P}}{dT^2}\right]P^2 + \cdots \qquad (4)$$

We can obtain expansions for $\overline{U}$ and $\overline{A}$ by using the equation $\overline{H} = \overline{U} + P\overline{V} = \overline{U} + RTZ$ and $\overline{G} = \overline{A} + P\overline{V} = \overline{A} + RTZ$. Show that

$$\overline{U} - U^\circ = -RT\left(B_{2P} + T\frac{dB_{2P}}{dT}\right)P - RT\left(B_{3P} + \frac{T}{2}\frac{dB_{3P}}{dT}\right)P^2 + \cdots \qquad (5)$$

and

$$\overline{A} - A^\circ = RT \ln P - \frac{RTB_{3p}}{2} P^2 + \cdots \tag{6}$$

at $P^\circ = 1$ bar.

**8-44.** In this problem, we will derive the equation

$$\overline{H}(T, P) - H^\circ(T) = RT(Z - 1) + \int_{\overline{V}^{\text{id}}}^{\overline{V}} \left[ T \left( \frac{\partial P}{\partial T} \right)_V - P \right] d\overline{V}'$$

where $\overline{V}^{\text{id}}$ is a very large (molar) volume, where the gas is sure to behave ideally. Start with $dH = TdS + VdP$ to derive

$$\left( \frac{\partial H}{\partial V} \right)_T = T \left( \frac{\partial S}{\partial V} \right)_T + V \left( \frac{\partial P}{\partial V} \right)_T$$

and use one of the Maxwell relations for $(\partial S/\partial V)_T$ to obtain

$$\left( \frac{\partial H}{\partial V} \right)_T = T \left( \frac{\partial P}{\partial T} \right)_V + V \left( \frac{\partial P}{\partial V} \right)_T$$

Now integrate by parts from an ideal-gas limit to an arbitrary limit to obtain the desired equation.

**8-45.** Using the result of Problem 8–44, show that $H$ is independent of volume for an ideal gas. What about a gas whose equation of state is $P(\overline{V} - b) = RT$? Does $U$ depend upon volume for this equation of state? Account for any difference.

**8-46.** Using the result of Problem 8–44, show that

$$\overline{H} - H^\circ = \frac{RTb}{\overline{V} - b} - \frac{2a}{\overline{V}}$$

for the van der Waals equation.

**8-47.** Using the result of Problem 8–44, show that

$$\overline{H} - H^\circ = \frac{RTB}{\overline{V} - B} - \frac{A}{T^{1/2}(\overline{V} + B)} - \frac{3A}{2BT^{1/2}} \ln \frac{\overline{V} + B}{\overline{V}}$$

for the Redlich-Kwong equation.

**8-48.** We introduced the Joule-Thomson effect and the Joule-Thomson coefficient in Problems 5–52 through 5–54. The Joule-Thomson coefficient is defined by

$$\mu_{\text{JT}} = \left( \frac{\partial T}{\partial P} \right)_H = -\frac{1}{C_P} \left( \frac{\partial H}{\partial P} \right)_T$$

and is a direct measure of the expected temperature change when a gas is expanded through a throttle. We can use one of the equations derived in this chapter to obtain a convenient working equation for $\mu_{\mathrm{JT}}$. Show that

$$\mu_{\mathrm{JT}} = \frac{1}{C_P}\left[T\left(\frac{\partial V}{\partial T}\right)_P - V\right]$$

Use this result to show that $\mu_{\mathrm{JT}} = 0$ for an ideal gas.

**8-49.** Use the virial equation of state of the form

$$\frac{P\overline{V}}{RT} = 1 + \frac{B_{2V}(T)}{RT}P + \cdots$$

to show that

$$\mu_{\mathrm{JT}} = \frac{1}{C_P^{\mathrm{id}}}\left[T\frac{dB_{2V}}{dT} - B_{2V}\right] + O(P)$$

It so happens that $B_{2V}$ is negative and $dB_{2V}/dT$ is positive for $T^* < 3.5$ (see Figure 2.15) so that $\mu_{\mathrm{JT}}$ is positive for low temperatures. Therefore, the gas will cool upon expansion under these conditions. (See Problem 8–48.)

**8-50.** Show that

$$\mu_{\mathrm{JT}} = -\frac{b}{C_P}$$

for a gas that obeys the equation of state $P(\overline{V} - b) = RT$. (See Problem 8–48).

**8-51.** The second virial coefficient for a square-well potential is (Equation 2.41)

$$B_{2V}(T) = b_0[1 - (\lambda^3 - 1)(e^{\varepsilon/k_BT} - 1)]$$

Show that

$$\mu_{\mathrm{JT}} = \frac{b_0}{C_P}\left[(\lambda^3 - 1)\left(1 + \frac{\varepsilon}{k_BT}\right)e^{\varepsilon/k_BT} - \lambda^3\right]$$

where $b_0 = 2\pi\sigma^3 N_A/3$. Given the following square-well parameters, calculate $\mu_{\mathrm{JT}}$ at 0°C and compare your values with the given experimental values. Take $C_P = 5R/2$ for Ar and $7R/2$ for $N_2$ and $CO_2$.

| Gas | $b_0/\mathrm{cm^3\cdot mol^{-1}}$ | $\lambda$ | $\varepsilon/k_B$ | $\mu_{\mathrm{JT}}(\mathrm{exptl})/\mathrm{K\cdot atm^{-1}}$ |
|---|---|---|---|---|
| Ar | 39.87 | 1.85 | 69.4 | 0.43 |
| $N_2$ | 45.29 | 1.87 | 53.7 | 0.26 |
| $CO_2$ | 75.79 | 1.83 | 119 | 1.3 |

**8-52.** The temperature at which the Joule-Thomson coefficient changes sign is called the *Joule-Thomson inversion temperature*, $T_{\mathrm{i}}$. The low-pressure Joule-Thomson inversion temperature for the square-well potential is obtained by setting $\mu_{\mathrm{JT}} = 0$ in Problem 8–51. This procedure leads to an equation for $k_BT/\varepsilon$ in terms of $\lambda^3$ that cannot be solved analytically. Solve the equation numerically to calculate $T_{\mathrm{i}}$ for the three gases given in the previous problem. The experimental values are 794 K, 621 K, and 1500 K for Ar, $N_2$, and $CO_2$, respectively.

**8-53.** Use the data in Problem 8–51 to estimate the temperature drop when each of the gases undergoes an expansion for 100 atm to one atm.

**8-54.** When a rubber band is stretched, it exerts a restoring force, $f$, which is a function of its length $L$ and its temperature $T$. The work involved is given by

$$w = \int f(L, T)dL \tag{1}$$

Why is there no negative sign in front of the integral, as there is in Equation 5.2 for $P$-$V$ work? Given that the volume change upon stretching a rubber band is negligible, show that

$$dU = TdS + fdL \tag{2}$$

and that

$$\left(\frac{\partial U}{\partial L}\right)_T = T\left(\frac{\partial S}{\partial L}\right)_T + f \tag{3}$$

Using the definition $A = U - TS$, show that Equation 2 becomes

$$dA = -SdT + fdL \tag{4}$$

and derive the Maxwell relation

$$\left(\frac{\partial f}{\partial T}\right)_L = -\left(\frac{\partial S}{\partial L}\right)_T \tag{5}$$

Substitute Equation 5 into Equation 3 to obtain the analog of Equation 8.22

$$\left(\frac{\partial U}{\partial L}\right)_T = f - T\left(\frac{\partial f}{\partial T}\right)_L$$

For many elastic systems, the observed temperature-dependence of the force is linear. We define an *ideal rubber band* by

$$f = T\phi(L) \qquad \text{(ideal rubber band)} \tag{6}$$

Show that $(\partial U/\partial L)_T = 0$ for an ideal rubber band. Compare this result with $(\partial U/\partial V)_T = 0$ for an ideal gas.

Now let's consider what happens when we stretch a rubber band quickly (and, hence, adiabatically). In this case, $dU = dw = fdL$. Use the fact that $U$ depends upon only the temperature for an ideal rubber band to show that

$$dU = \left(\frac{\partial U}{\partial T}\right)_L dT = fdL \tag{7}$$

The quantity $(\partial U/\partial T)_L$ is a heat capacity, so Equation 7 becomes

$$C_L dT = fdL \tag{8}$$

Argue now that if a rubber band is suddenly stretched, then its temperature will rise. Verify this result by holding a rubber band against your upper lip and stretching it quickly.

**8-55.** Derive an expression for $\Delta S$ for the reversible, isothermal change of one mole of a gas that obeys van der Waals equation. Use your result to calculate $\Delta S$ for the isothermal compression of one mole of ethane from $10.0\,\text{dm}^3$ to $1.00\,\text{dm}^3$ at 400 K. Compare your result to what you would get using the ideal-gas equation.

**8-56.** Derive an expression for $\Delta S$ for the reversible, isothermal change of one mole of a gas that obeys the Redlich-Kwong equation (Equation 2.7). Use your result to calculate $\Delta S$ for the isothermal compression of one mole of ethane from $10.0\,\text{dm}^3$ to $1.00\,\text{dm}^3$ at 400 K. Compare your result with the result you would get using the ideal-gas equation.

*In the last eight problems, we explore the consequences of the fact that equilibrium states are stable; that is, if a system at equilibrium is perturbed in some way, it will spontaneously return to its equilibrium state. The stability of equilibrium states leads to a number of general inequalities called stability conditions. Examples of stability conditions that we derive in the following problems are* $(\partial P/\partial V)_T < 0$, $C_V > 0$, $C_P > C_V$, *and* $(\partial S/\partial T)_P > 0$.

**8-57.** Equation 8.2 says that $dU \leq 0$ for any spontaneous process that occurs at constant $S$ and $V$. Thus, the energy will always decrease as a result of a spontaneous process and will attain a minimum value at equilibrium. Consider now a two-compartment system like that shown in the figure below.

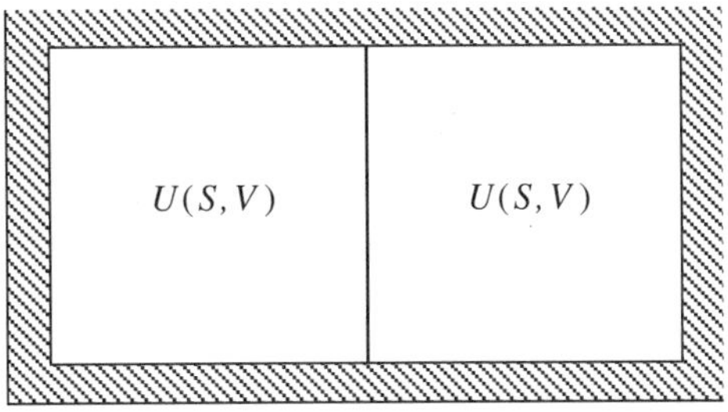


The entire system is enclosed by a rigid, adiabatic wall (so that it is isolated). The wall that separates the system into two compartments is totally restrictive (rigid, nonheat-conducting, and nonpermeable). The two identical compartments are at equilibrium with $U = U(S, V)$ at constant $N$.

Now let's partition the volume unequally between the two compartments (by moving the wall slightly), so that one compartment has a volume $V + \Delta V$ and the other has a volume $V - \Delta V$. Argue that

$$U(S, V + \Delta V) + U(S, V - \Delta V) > 2U(S, V) \tag{1}$$

Expand both $U(S, V + \Delta V)$ and $U(S, V - \Delta V)$ in a Taylor series (MathChapter C) in $\Delta V$ to obtain

$$\left(\frac{\partial^2 U}{\partial V^2}\right)_S > 0 \tag{2}$$

Equation 2 is one of several conditions for an equilibrium state to be stable with respect to small changes in its extensive variables.

Starting with $dU = TdS - PdV$, show that Equation 2 implies that the so-called adiabatic compressibility, $\kappa_S$, must obey the inequality

$$\kappa_S = -\frac{1}{V}\left(\frac{\partial V}{\partial P}\right)_S > 0$$

or that $\kappa_S$ must be a positive quantity.

**8-58.** In the previous problem, we showed that one of the stability conditions of an isolated system is that $(\partial^2 U/\partial V^2)_S > 0$. We did this by repartitioning the volume between the two compartments. Now consider the effect of repartioning the entropy between the two compartments keeping their volumes equal. Show that

$$\left(\frac{\partial^2 U}{\partial S^2}\right)_V > 0$$

Now use the equation $dU = TdS - PdV$ to show that this inequality implies that

$$\left(\frac{\partial T}{\partial S}\right)_V = \frac{T}{C_V} > 0$$

or that $C_V > 0$.

**8-59.** The enthalpy, $H$, (at constant $N$) has natural variables $S$ and $P$, one an extensive variable ($S$) and the other an intensive variable ($P$). To determine the sign of $(\partial^2 H/\partial S^2)_P$, we use the repartioning approach that we described in Problem 8–57. Show that

$$\left(\frac{\partial^2 H}{\partial S^2}\right)_P > 0$$

Now show that this inequality implies that

$$\left(\frac{\partial T}{\partial S}\right)_P = \frac{T}{C_P} > 0$$

or that $C_P > 0$.

The pressure is an intensive variable, so we are unable to repartion it. To determine the sign of $(\partial^2 H/\partial P^2)_S$, we first determine $(\partial H/\partial P)_S$. Show that

$$\left(\frac{\partial H}{\partial P}\right)_S = V$$

Now differentiate once again with respect to $P$ keeping $S$ constant.

$$\left(\frac{\partial V}{\partial P}\right)_S = \left(\frac{\partial^2 H}{\partial P^2}\right)_S$$

Now show that the reciprocal of $(\partial V/\partial P)_S$ is given by

$$\left(\frac{\partial P}{\partial V}\right)_S = -\left(\frac{\partial^2 U}{\partial V^2}\right)_S$$

Put this all together to show that

$$\left(\frac{\partial V}{\partial P}\right)_S = \left(\frac{\partial^2 H}{\partial P^2}\right)_S = -\frac{1}{\left(\frac{\partial^2 U}{\partial V^2}\right)_S} < 0$$

Thus, we see once again (see Problem 8–57) that

$$\kappa_S = -\frac{1}{V}\left(\frac{\partial V}{\partial P}\right)_S > 0$$

Notice that we determine the sign of the second derivative of $H$ with respect to the extensive variable, $S$, by using the repartioning that we used in Problem 8–57 and the sign of the second derivative with respect to the intensive variable, $P$, by first determining $(\partial H/\partial P)_S$, differentiating it, and then relating its reciprocal to $(\partial^2 U/\partial V^2)_S$, whose sign we already knew.

**8-60.** Of the two natural variables of the Helmholtz energy (at constant $N$), one is an extensive variable, $V$, and the other is an intensive variable, $T$. Using the repartioning approach shown in Problem 8–57, show that

$$\left(\frac{\partial^2 A}{\partial V^2}\right)_T > 0$$

and that

$$\left(\frac{\partial P}{\partial V}\right)_T < 0$$

Now show that the isothermal compressibility, $\kappa_T$, is always positive, or that

$$\kappa_T = -\frac{1}{V}\left(\frac{\partial P}{\partial V}\right)_T > 0$$

The Helmholtz energy is also a function of one intensive natural variable, $T$. Show that (see problem 8–59)

$$\left(\frac{\partial S}{\partial T}\right)_V = -\left(\frac{\partial^2 A}{\partial T^2}\right)_V = \frac{1}{\left(\frac{\partial T}{\partial S}\right)_V} = \frac{1}{\left(\frac{\partial^2 U}{\partial S^2}\right)_V} > 0$$

Thus, we see that the entropy increases monotonically with temperature at constant volume.

**8-61.** The two natural variables of the Gibbs energy (at constant $N$) are both intensive variables, $T$ and $P$. First show that

$$\left(\frac{\partial S}{\partial T}\right)_P = -\left(\frac{\partial^2 G}{\partial T^2}\right)_P = \frac{1}{\left(\frac{\partial T}{\partial S}\right)_P} = \frac{1}{\left(\frac{\partial^2 H}{\partial S^2}\right)_P} > 0$$

Thus, we see that the entropy increases monotonically with temperature at constant pressure.

Now show that

$$\left(\frac{\partial V}{\partial P}\right)_T = \left(\frac{\partial^2 G}{\partial P^2}\right)_T = \frac{1}{\left(\frac{\partial P}{\partial V}\right)_T} = -\frac{1}{\left(\frac{\partial^2 A}{\partial V^2}\right)_T} < 0$$

or that

$$\kappa_T = -\frac{1}{V}\left(\frac{\partial V}{\partial P}\right)_T > 0$$

**8-62.** Use the results of the previous problem and Equation 8.27 to prove that $C_P > C_V$.

**8-63.** In Problems 8–57 and 8–58, we repartioned the volume of each compartment without changing the entropy of each compartment, or we repartioned the entropy without changing the volume. Let's now look at the case in which both $V$ and $S$ change. Show that in this case,

$$U(S + \Delta S, V + \Delta V) + U(S - \Delta S, V - \Delta V) > 2U(S, V) \qquad (1)$$

The Taylor expansion of a function of two variables is

$$\begin{aligned} f(x + \Delta x, y + \Delta y) = f(x, y) &+ \left(\frac{\partial f}{\partial x}\right)_y \Delta x + \left(\frac{\partial f}{\partial y}\right)_x \Delta x \\ &+ \frac{1}{2}\left(\frac{\partial^2 f}{\partial x^2}\right)_y (\Delta x)^2 + \left(\frac{\partial^2 f}{\partial x \partial y}\right)(\Delta x)(\Delta y) \\ &+ \frac{1}{2}\left(\frac{\partial^2 f}{\partial y^2}\right)_x (\Delta y)^2 \\ &+ \text{cubic terms in } \Delta x, \Delta y, \text{ and their products} \end{aligned}$$

Using this result, show that Equation 1 becomes

$$\left(\frac{\partial^2 U}{\partial S^2}\right)_V (\Delta S)^2 + 2\left(\frac{\partial^2 U}{\partial S \partial V}\right)(\Delta S)(\Delta V) + \left(\frac{\partial^2 U}{\partial V^2}\right)_S (\Delta V)^2 > 0 \qquad (2)$$

Equation 2 is valid for any values of $\Delta S$ and $\Delta V$. Let $\Delta V = 0$ and $\Delta S = 0$ in turn to show that

$$\left(\frac{\partial^2 U}{\partial S^2}\right)_V > 0 \qquad \left(\frac{\partial^2 U}{\partial V^2}\right)_S > 0 \qquad (3)$$

Now let

$$\Delta S = \frac{(\partial^2 U/\partial S \partial V)}{\partial^2 U/\partial S^2)_V} \qquad \text{and} \qquad \Delta V = 1$$

to obtain

$$\left(\frac{\partial^2 U}{\partial S^2}\right)_V \left(\frac{\partial^2 U}{\partial V^2}\right)_S - \left(\frac{\partial^2 U}{\partial S \partial V}\right) > 0 \qquad (4)$$

It turns out that Equations 3 and 4 are the mathematical conditions that the extremum given by $dU = 0$ is actually a minimum rather than a maximum or a saddle point.

**8-64.** Equation 8.45 gives $\overline{S}$ for a monatomic ideal gas in terms of $T$ and $V$. Use the fact that $\overline{U} = 3RT/2$ to show that Equation 8.45 can be written as

$$\overline{U} = \overline{U}(\overline{S}, \overline{V}) = \frac{\alpha e^{2\overline{S}/3R}}{\overline{V}^{2/3}} \tag{5}$$

where $\alpha$ is a constant independent of $\overline{S}$ and $\overline{V}$. Show that $\overline{U}$ given by Equation 5 satisfies Equations 3 and 4 of Problem 8–63. Show also that

$$\left(\frac{\partial \overline{U}}{\partial \overline{S}}\right)_{\overline{V}} = T \qquad \left(\frac{\partial \overline{U}}{\partial \overline{V}}\right)_{\overline{S}} = -P$$

**Josiah Willard Gibbs** was born in New Haven, Connecticut, on February 11, 1839, and died there in 1903. He received his Ph.D. in engineering from Yale University in 1863, the second doctorate in science and the first in engineering awarded in the United States. He stayed on at Yale, for years without salary, and remained there for the rest of his life. In 1878, Gibbs published a long, original treatise on thermodynamics titled "On the Equilibrium of Heterogeneous Substances" in the *Transactions of the Connecticut Academy of Sciences*. In addition to introducing the concept of chemical potential, Gibbs introduced what is now called the Gibbs phase rule, which relates the number of components ($C$) and the number of phases ($P$) in a system to the number of degrees of freedom ($F$, the number of variables such as temperature and pressure that can be varied independently) by the equation $F = C + 2 - P$. Between its austere writing style and the obscurity of the journal in which it was published, however, this important work was not as widely appreciated as it deserved. Fortunately, Gibbs sent copies to a number of prominent European scientists. Maxwell and van der Waals immediately appreciated the signifance of the work and made it known in Europe. Eventually, Gibbs received the recognition that was his due, and Yale finally offered him a salaried position in 1880. Gibbs was an unassuming, modest person, living in New Haven in his family home his entire life.

CHAPTER 9

# Phase Equilibria

The relation between all the phases of a substance at various temperatures and pressures can be concisely represented by a phase diagram. In this chapter, we will study the information presented by phase diagrams and the thermodynamic consequences of this information. In particular, we will analyze the temperature and pressure dependence of a substance in terms of its Gibbs energy, particularly using the fact that a phase with the lower Gibbs energy will always be the more stable one.

Many thermodynamic systems of interest consist of two or more phases in equilibrium with each other. For example, both the solid and liquid phases of a substance are in equilibrium with each other at its melting point. Thus, an analysis of such a system as a function of temperature and pressure gives the pressure-dependence of the melting point. One of the many unusual properties of water is that the melting point of ice decreases with increasing pressure. We will see in this chapter that this property is a direct consequence of the fact that water expands upon freezing, or that the molar volume of liquid water is less than that of ice. We will also derive an expression that allows us to calculate the vapor pressure of a liquid as a function of temperature from a knowledge of its enthalpy of vaporization. These results can all be understood using a quantity called the chemical potential, which is one of the most useful functions of chemical thermodynamics. We will see that chemical potential is analogous to electric potential. Just as electric current flows from a region of high electric potential to a region of low electric potential, matter flows from a region of high chemical potential to a region of low chemical potential. In the last section of the chapter, we will derive a statistical thermodynamic expression for the chemical potential and show how to calculate it in terms of molecular or spectroscopic quantities.

## 9–1. A Phase Diagram Summarizes the Solid–Liquid–Gas Behavior of a Substance

You might recall from general chemistry that we can summarize the solid-liquid-gas behavior of a substance by means of a phase diagram, which indicates under what conditions of pressure and temperature the various states of matter of a substance exist in equilibrium. Figure 9.1 shows the phase diagram of benzene, a typical substance. Note that there are three principal regions in this phase diagram. Any point within one of these regions specifies a pressure and a temperature at which the single phase exists in equilibrium. For example, according to Figure 9.1, benzene exists as a solid at 60 torr and 260 K (point A), and as a gas at 60 torr and 300 K (point B).

The lines that separate the three regions indicate pressures and temperatures at which two phases can coexist at equilibrium. For example, at all points along the line that separates the solid and gas regions (line CF), benzene exists as a solid and a gas in equilibrium with each other. This line is called the solid-gas coexistence curve. As such, it specifies the vapor pressure of solid benzene as a function of temperature. Similarly, the line that separates the liquid and gas regions (line FD) gives the vapor pressure of liquid benzene as a function of temperature, and the line that separates the solid and liquid regions (line FE) gives the melting point of benzene as a function of pressure. Notice that the three lines in the phase diagram intersect at one point (point F), at which solid, liquid, and gaseous benzene coexist at equilibrium. This point is called the *triple point*, which occurs at 278.7 K (5.5°C) and 36.1 torr for benzene.

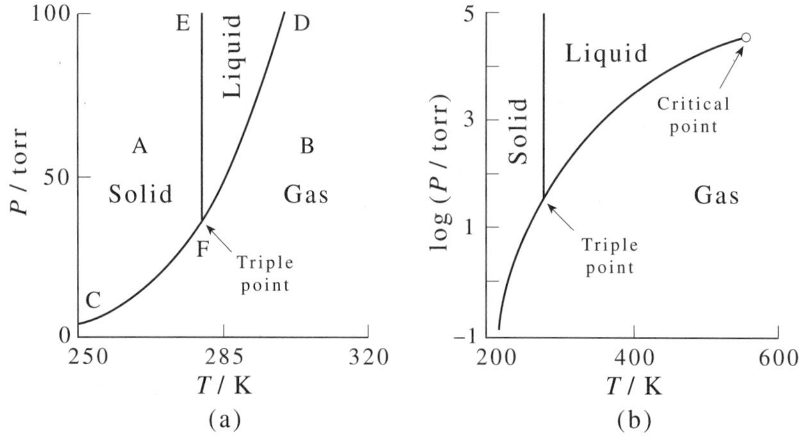


**FIGURE 9.1**
The phase diagram of benzene, (a) displayed as $P$ against $T$, and (b) log $P$ versus $T$. The log $P$ versus $T$ display condenses the vertical axis.

**EXAMPLE 9–1**

Experimentally, the vapor pressure of liquid benzene is given by

$$\ln(P/\text{torr}) = -\frac{4110\text{ K}}{T} + 18.33 \qquad 273\text{ K} < T < 300\text{ K}$$

and the vapor pressure of solid benzene is given by

$$\ln(P/\text{torr}) = -\frac{5319\text{ K}}{T} + 22.67 \qquad 250\text{ K} < T < 280\text{ K}$$

Calculate the pressure and the temperature at the triple point of benzene.

SOLUTION: Solid, liquid, and gaseous benzene coexist at the triple point. Therefore, at the triple point, these two equations for the vapor pressure must give the same value. Setting the two expressions above for $\ln P$ equal to each other gives

$$-\frac{4110\text{ K}}{T} + 18.33 = -\frac{5319\text{ K}}{T} + 22.67$$

or $T = 278.7$ K, or 5.5°C. The pressure at the triple point is given by $\ln(P/\text{torr}) = 3.58$ or $P = 36.1$ torr.

Within a single-phase region, both the pressure and the temperature must be specified, and we say that there are two degrees of freedom within a single-phase region of a pure substance. Along any of the coexistence curves, either the pressure or the temperature alone is sufficient to specify a point on the curve, so we say that there is one degree of freedom. The triple point is a fixed point, so there are no degrees of freedom there. If we think of $P$ and $T$ as degrees of freedom of the system, then the number of degrees of freedom, $f$, at any point in a phase diagram of a pure substance is given by $f = 3 - p$, where $p$ is the number of phases that coexist at equilibrium at that point.

If we start on the pressure axis at 760 torr (2.88 on the vertical axis in Figure 9.1*b*) and move horizontally to the right in the phase diagram of benzene, we can see how benzene behaves with increasing temperature at a constant pressure of 760 torr (one atmosphere). For temperatures below 278.7 K, benzene exists as a solid. At 278.7 K (5.5°C), we reach the solid-liquid coexistence curve, and benzene melts at this point. This point is called the *normal melting point*. (The melting point at a pressure of one bar is called the *standard melting point*.) Then for temperatures between 278.7 K and 353.2 K (80.1°C), benzene exists as a liquid. At the liquid-gas coexistence curve (353.2 K), benzene boils and then exists as a gas at temperatures higher than 353.2 K. Note that if we were to start at a pressure less than 760 torr (but above the triple point), the melting point is about the same as it is at 760 torr (because the solid-liquid coexistence curve is so steep), but the boiling point is lower than 353.2 K. Similarly, at a pressure greater than 760 torr, the melting point is about the same as it is at 760 torr, but the boiling point is greater than 353.2 K. Thus, the liquid-gas coexistence curve may also be interpreted as the boiling point of benzene as a function of pressure and the solid-liquid coexistence curve as the melting point as a function of pressure. Figure 9.2 shows the melting point of benzene plotted against pressure up to 10 000 atm. The slope of this curve is 0.0293 $°C \cdot atm^{-1}$ around 760 torr, which shows that the melting point is fairly insensitive to pressure. The melting point of benzene increases by about one degree in going from a pressure of 1 atm to 34 atm. By contrast, Figure 9.3 is a plot of the boiling point of benzene as a function of pressure; it shows that the boiling point depends strongly upon pressure. For example, the normal atmospheric pressure at an elevation of 10 000 feet (3100 meters) is 500 torr, so according to Figure 9.3, benzene boils at 67°C at this elevation. (Recall that the boiling point is defined to be that temperature at which the vapor pressure equals the atmospheric pressure.) The boiling point at exactly one atm is called the normal boiling point. The boiling point at exactly one bar is called the standard boiling point.

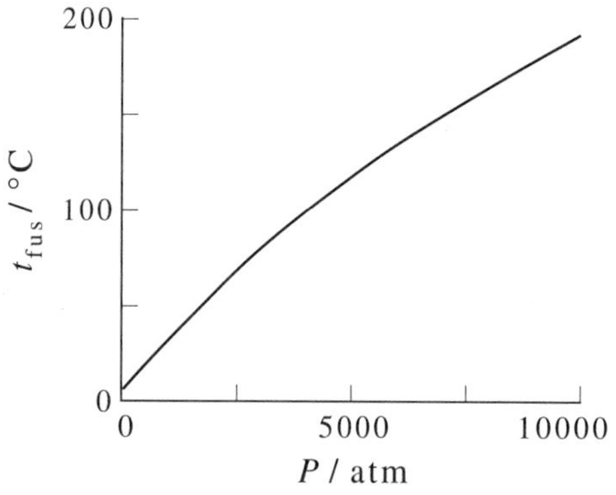


**FIGURE 9.2**
A plot of the melting point of benzene as a function of pressure. Notice that the melting point increases slowly with pressure. (Note that the scales of the horizontal axes in Figures 9.2 and 9.3 are very different.)

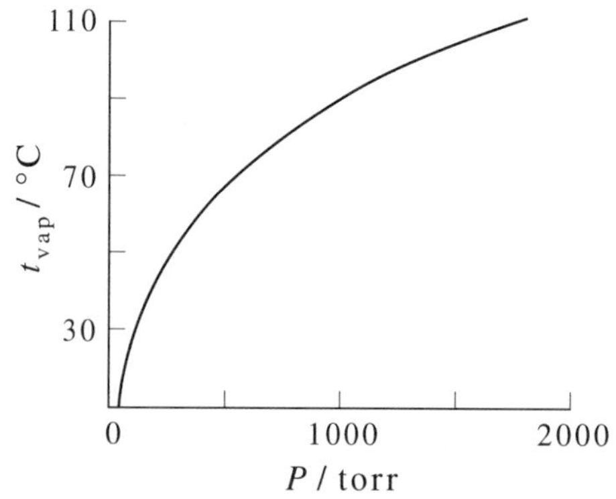


**FIGURE 9.3**
A plot of the boiling point of benzene as a function of pressure. Notice that the boiling point depends strongly on the pressure. (Note that the scales of the horizontal axes in Figures 9.2 and 9.3 are different.)

**EXAMPLE 9–2**

The vapor pressure of benzene can be expressed by the empirical formula

$$\ln(P/\text{torr}) = -\frac{3884\ \text{K}}{T} + 17.63$$

Use this formula to show that benzene boils at 67°C when the atmospheric pressure is 500 torr.

SOLUTION: Benzene boils when its vapor pressure is equal to the atmospheric pressure. Therefore $P = 500$ torr, so we have

$$\ln 500 = -\frac{3884\ \text{K}}{T} + 17.63$$

or $T = 340.2$ K, or 67.1°C.

Example 9–1 shows that the pressure at the triple point of benzene is 36.1 torr. Note from Figure 9.1 that if the pressure is less than 36.1 torr, benzene does not melt as we increase the temperature, but rather *sublimes*; that is, it passes directly from the solid phase to the gaseous phase. If the pressure at the triple point happens to be greater than one atm for a substance, it will sublime rather than melt at one atm. A noted substance with this property is carbon dioxide, whose solid phase is called dry ice because it doesn't liquefy at atmospheric pressure. Figure 9.4 shows the phase diagram for carbon dioxide. The triple point pressure of $CO_2$ is 5.11 atm, and so we see that $CO_2$ sublimes at one atm. The normal sublimation temperature of $CO_2$ is 195 K (−78°C).

Figure 9.5 shows the phase diagram for water. Water has the unusual property that its melting point decreases with increasing pressure (Figure 9.6). This behavior is reflected in the phase diagram of water by the slope of the solid-liquid coexistence curve. Although it is difficult to see in the phase diagram because the slope of the solid-liquid coexistence curve is so large, it does point upward to the left (has a negative slope). Numerically, the slope of the curve around one atm is $-130\ \text{atm}\cdot\text{K}^{-1}$. We will

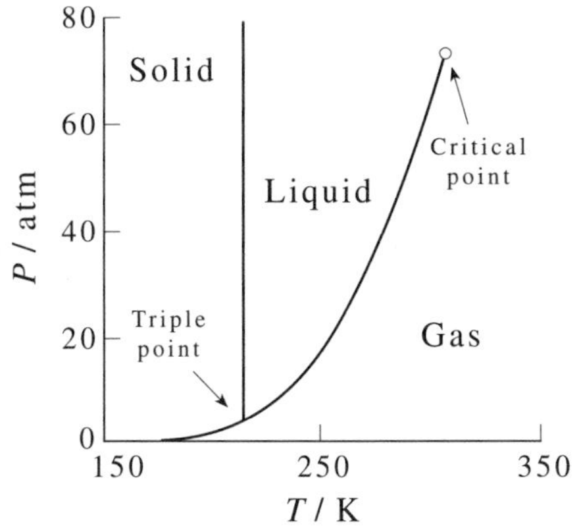


**FIGURE 9.4**
The phase diagram of carbon dioxide. Note that the triple point pressure of carbon dioxide is greater than one atm. Consequently, carbon dioxide sublimes at atmospheric pressure.

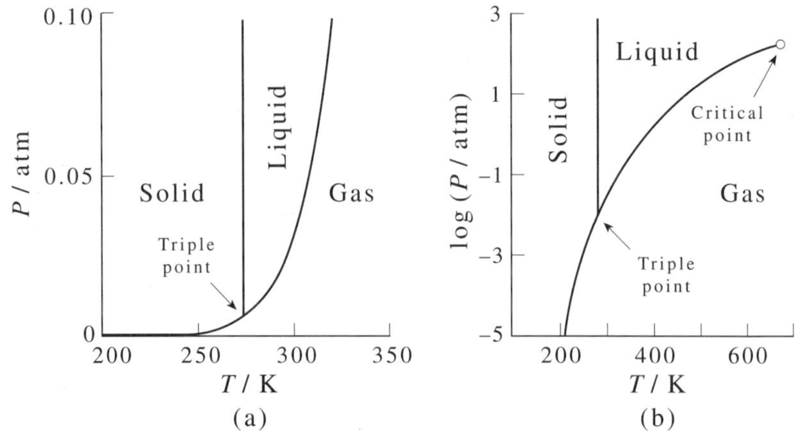


**FIGURE 9.5**
The phase diagram of water, (a) displayed as $P$ against $T$, and (b) $\log P$ versus $T$. The $\log P$ versus $T$ display condenses the horizontal axis. Although it is difficult to discern because of the scale of the figure, the melting point of water decreases with increasing pressure.

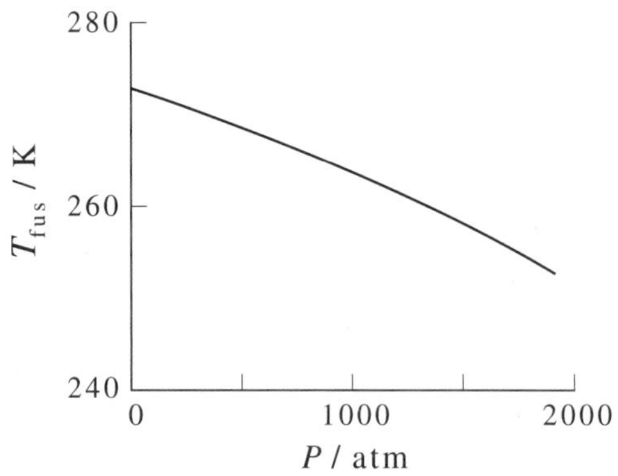


**FIGURE 9.6**
A plot of the melting point of water versus pressure. The melting point of water decreases with increasing pressure.

see in Section 9–3 that the reason the melting point of water decreases with increasing pressure is that the molar volume of ice is greater than that of water under the same conditions. Antimony and bismuth are two other such substances that expand upon freezing. Most substances, however, contract upon freezing.

In each of Figures 9.1 (benzene), 9.4 (carbon dioxide), and 9.5 (water), the liquid-gas coexistence curve ends abruptly at the critical point. (Recall that we discussed the critical behavior of gases in Section 2–3.) As the critical point is approached along the liquid-gas coexistence curve, the difference between the liquid phase and gaseous phase becomes increasingly less distinct until the difference disappears entirely at the critical point. For example, if we plot the densities of the liquid and vapor phases in equilibrium with each other along the liquid-vapor coexistence curve (such densities are called *orthobaric densities*), we see that these densities approach each other and become equal at the critical point (Figure 9.7). The liquid phase and vapor phase simply merge into a single fluid phase. Similarly, the molar enthalpy of vaporization decreases along this curve.

Figure 9.8 shows experimental values of the molar enthalpy of vaporization of benzene plotted against temperature. Notice that the value of $\Delta_{vap}\overline{H}$ decreases with increasing temperature and drops to zero at the critical temperature (289°C for benzene). The data in Figure 9.8 reflect the fact that the difference between a liquid and its vapor

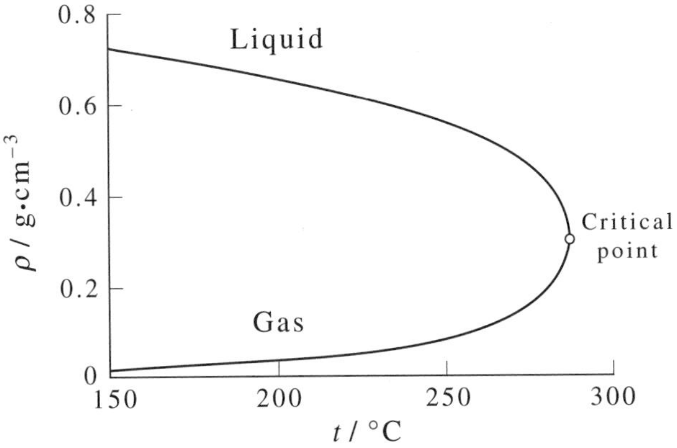


**FIGURE 9.7**
A plot of the orthobaric densities of the liquid and vapor phases of benzene in equilibrium along the liquid-vapor coexistence curve. Notice that the densities of the liquid and vapor phases approach one another and become equal at the critical point (289°C).

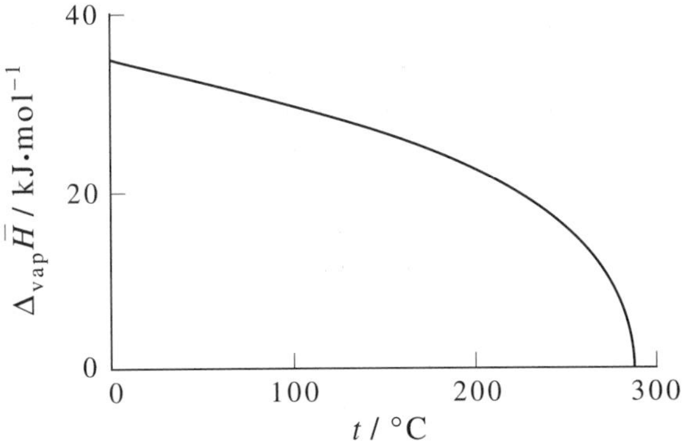


**FIGURE 9.8**
Experimental values of the molar enthalpy of vaporization of benzene plotted against temperature. The value of $\Delta_{vap}\overline{H}$ decreases with increasing temperature and drops to zero at the critical temperature, 289°C.

decreases as the critical point is approached. Because the two phases become less and less distinct as the critical point is approached and then merge into one phase at the critical point, $\Delta_{vap}S = S(\text{gas}) - S(\text{liquid})$ becomes zero at the critical point. Therefore $\Delta_{vap}H = T\Delta_{vap}S$ also becomes zero there. Above the critical point, there is no distinction between a liquid and a gas, and a gas cannot be liquefied no matter how great the pressure.

A nice lecture demonstration illustrates the idea of the critical temperature. First, fill a glass tube with a liquid such as sulfur hexafluoride. (The critical temperature of sulfur hexafluoride is 45.5°C, which is a convenient temperature to achieve.) After evacuating all the air so that the tube contains only pure sulfur hexafluoride, seal off the tube. Below 45.5°C, the tube will contain two layers, the liquid phase and the gas phase separated by a meniscus. Now, as the tube and its contents are warmed, the meniscus becomes less distinct and just as the critical temperature is reached, the meniscus disappears entirely and the tube becomes transparent [$SF_6$(g) is colorless]. When the tube and its contents are cooled, the liquid phase and the meniscus suddenly appear at the critical temperature.

A fluid very near its critical point constantly changes from a liquid to a vapor state, causing fluctuations in the density from one region to another. These fluctuations scatter light very strongly (somewhat like a finely dispersed fog) and the system appears milky. This effect is known as *critical opalescence*. These fluctuations are difficult to study experimentally because gravity causes the density fluctuations to be distorted. To overcome the effect of gravity, a team of scientists, engineers, and technicians designed an experiment to measure the laser light scattered by xenon at its critical point on board the Columbia space shuttle. After several preliminary experiments, they were able to measure the details of the fluctuations to within microkelvins of the critical temperature of xenon (289.72 K) on the March 1996 flight of Columbia. No other microgravity experiment has logged as many hours as this one, and the results will provide us with a detailed understanding of the liquid-vapor phase transition and the liquid-vapor interface.

Because of the existence of a critical point, a gas can be transformed into a liquid without ever passing through a two-phase state. Simply start in the gas region of the phase diagram and go into the liquid region by traveling out around the critical point. The gas passes gradually and continuously into the liquid state without a two-phase region appearing and without any apparent condensation.

You might wonder if the solid-liquid coexistence curve ends abruptly as the liquid-gas coexistence curve does. The very nature of a critical point requires that we pass from one phase to the other in a gradual, continual manner. Because the gas and liquid phases are both fluid phases, the difference between them is purely one of degree rather than actual structure. On the other hand, a liquid phase and a solid phase, or two different solid phases for that matter, are qualitatively different because they have intrinsically different structures. It is not possible to pass from one phase to the other in a gradual, continual manner. A critical point, therefore, cannot exist for such phases, and the coexistence curve separating these phases must continue indefinitely or intersect the coexistence curves of other phases. In fact, many substances exhibit a variety of solid phases at high pressures, and Figure 9.9 shows the high-pressure phase diagram of

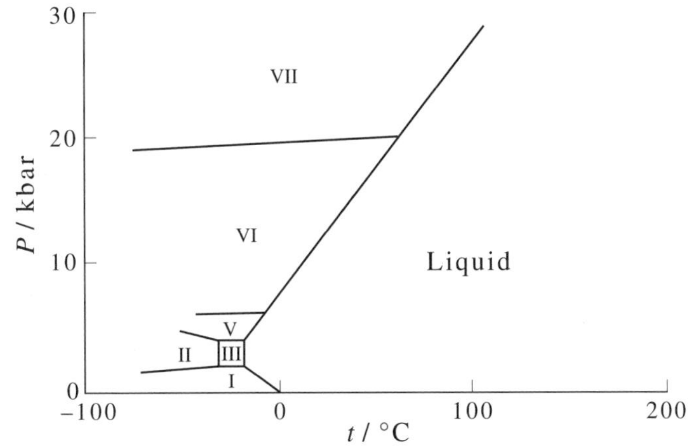


**FIGURE 9.9**
The phase diagram for water at high pressures showing six stable phases of ice.

water, showing various distinct solid phases. Ice (I) is the "normal" ice that occurs at one atm, and the other ices are different crystalline forms of solid $H_2O$ that are stable at very high pressures. Note, for example, that ice (VII) is stable at temperatures well above 0°C, and even above 100°C, but it is formed only at high pressures.

## 9–2. The Gibbs Energy of a Substance Has a Close Connection to Its Phase Diagram

Recall Figure 8.7 where the molar Gibbs energy of benzene is plotted against temperature. As the figure shows, the molar Gibbs energy is a continuous function of temperature, but there is a discontinuity in the slope of $\overline{G}(T)$ versus $T$ at each phase transition. Figure 9.10a is a magnification of a plot of $\overline{G}(T)$ versus $T$ in the region around the melting point of benzene (279 K). The dashed extensions represent the Gibbs energy of the supercooled liquid and the (hypothetical) superheated solid. Picture moving along the curve of $\overline{G}(T)$ versus $T$ in Figure 9.10a with increasing temperature. Along the solid-phase branch, $\overline{G}(T)$ decreases with a slope $(\partial\overline{G}/\partial T)_P = -\overline{S}^{\mathrm{s}}$. When we reach the melting point, we switch to the liquid branch because the Gibbs energy of the liquid phase is lower than that of the solid phase. The slope of the liquid branch is steeper than that of the solid branch because $(\partial\overline{G}/\partial T)_P = -\overline{S}^{\mathrm{l}}$ and $\overline{S}^{\mathrm{l}} > \overline{S}^{\mathrm{s}}$. Therefore, the molar Gibbs energy of the liquid phase must be lower than that of the solid phase at higher temperatures. The dashed extension of the solid branch represents the (hypothetical) superheated solid, and even if it were to occur, it would be unstable relative to the liquid and would convert to liquid. The dashed lines represent what are called metastable states. Figure 9.10b shows the transition from liquid to gas at the normal boiling point (353 K) of benzene. The boiling point occurs when the liquid and gas branches of the $\overline{G}(T)$ versus $T$ curves intersect. The slope of the gas branch is

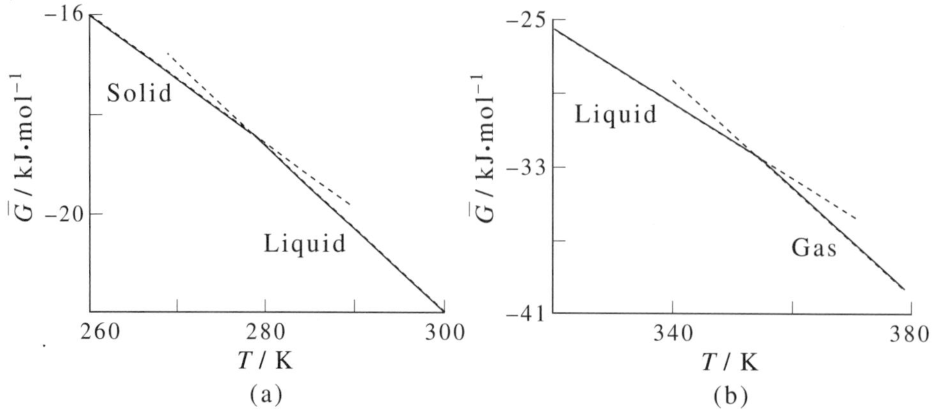


**FIGURE 9.10**
A plot of $\overline{G}(T)$ versus $T$ for benzene in the region around (a) its melting point (279 K) and (b) its boiling point (353 K).

steeper than that of the liquid branch because $\overline{S}^{\mathrm{g}} > \overline{S}^{\mathrm{l}}$, and so the molar Gibbs energy of the gas must be lower than that of the liquid at higher temperatures.

We can see from the equation $G = H - TS$ why the solid phase is favored at low temperatures whereas the gaseous phase is favored at high temperatures. At low temperatures, the $TS$ term is small compared with $H$; thus, a solid phase is favored at low temperatures because it has the lowest enthalpy of the three phases. At high temperatures, on the other hand, $H$ is small compared with the $TS$ term, so we see that the gas phase with its relatively large entropy is favored at high temperatures. The liquid phase, which is intermediate in both energy and disorder to the solid and gaseous phases, is favored at intermediate temperatures.

It is also instructive to look at the molar Gibbs energy as a function of pressure at a fixed temperature. Recall that $(\partial\overline{G}/\partial P)_T = \overline{V}$, so that the slope of $G$ versus $P$ is always positive. For most substances, $\overline{V}^{\mathrm{g}} \gg \overline{V}^{\mathrm{l}} > \overline{V}^{\mathrm{s}}$, so the slope of the gas branch is much greater than that of a liquid branch, which in turn is greater than that of a solid branch. Figure 9.11a sketches a plot of $\overline{G}(P)$ against $P$ showing the gas, liquid, and solid branches at a temperature just greater than the triple-point temperature. As we increase the pressure, we move along the gas branch of $\overline{G}(P)$ until we hit the liquid branch, at which point the gas condenses to a liquid. As we continue to increase the pressure, we reach the solid branch, which necessarily lies lower than that of the liquid branch. The path we have just followed in Figure 9.11a corresponds to moving up along a vertical line that lies just to the right of the triple point in the phase diagram of a "normal" substance like benzene. For a substance such as water, however, $\overline{V}^{\mathrm{s}} > \overline{V}^{\mathrm{l}}$ at least for moderate pressures, so a plot of $\overline{G}(P)$ against $P$ looks like that given in Figure 9.11b. Tracing along the curve for $\overline{G}(P)$ for increasing pressures in Figure 9.11b corresponds to moving up a vertical line just to the left of the triple point in the phase diagram of water.

Figure 9.12 shows the behavior of $\overline{G}(P)$ versus $P$ at a number of temperatures for a normal substance such as benzene. Part (a) shows $\overline{G}(P)$ versus $P$ for a temperature

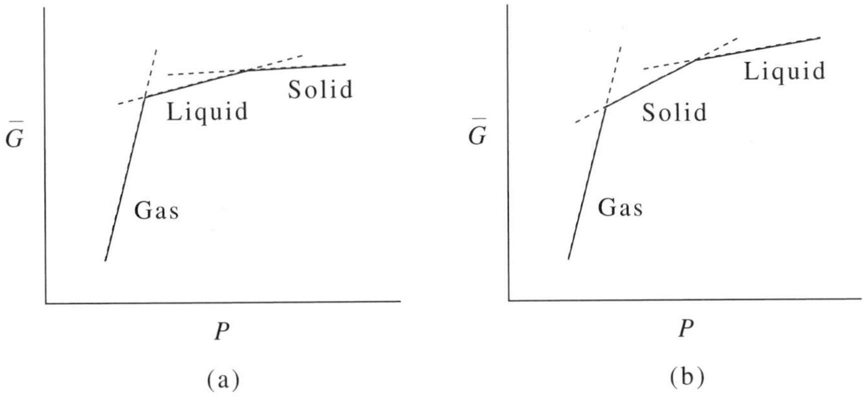


**FIGURE 9.11**
A plot of $\overline{G}(P)$ against $P$ showing the gas, liquid, and solid branches at a temperature near the triple point. (a) A "normal" substance ($\overline{V}^{\mathrm{s}} < \overline{V}^{\mathrm{l}}$) is depicted at a temperature above the triple-point temperature, where we see a gas-liquid-solid progression with increasing pressure. (b) A substance like water ($\overline{V}^{\mathrm{s}} > \overline{V}^{\mathrm{l}}$) is depicted at a temperature lower than the triple point temperature, where we see a gas-solid-liquid progression.

less than the triple-point temperature in Figure 9.1. In this case, we go directly from the gas phase to the solid phase as we increase the pressure. The molar Gibbs energy of the liquid phase at these temperatures lies higher than that of either the solid or gas phase and does not enter the picture. Part (b) shows the molar Gibbs energy situation at the triple-point temperature. At the triple point, the curves for the Gibbs energies of each of the three phases intersect, and for a "normal" substance like benzene, the Gibbs energy of the solid phase lies lower than that of the liquid phase for pressures above the triple-point pressure. Part (c) shows the Gibbs energies at a temperature slightly less than the critical temperature. Notice that the slopes of the gas and liquid branches are almost the same at the point of intersection. The reason for this similarity is that the slopes of the curves, $(\partial \overline{G}/\partial P)_T$, are equal to the molar volumes of the two phases, which are approaching each other as the critical point is approached. Part (d) shows the Gibbs energies at a temperature greater than the critical temperature. In this case, $\overline{G}(P)$ varies smoothly with pressure. There is no discontinuity in the slope in this case because only a single fluid phase is involved.

## 9–3. The Chemical Potentials of a Pure Substance in Two Phases in Equilibrium Are Equal

Consider a system consisting of two phases of a pure substance in equilibrium with each other. For example, we might have water vapor in equilibrium with liquid water. The Gibbs energy of this system is given by $G = G^{\mathrm{l}} + G^{\mathrm{g}}$, where $G^{\mathrm{l}}$ and $G^{\mathrm{g}}$ are the Gibbs energies of the liquid phase and the gas phase, respectively. Now, suppose that

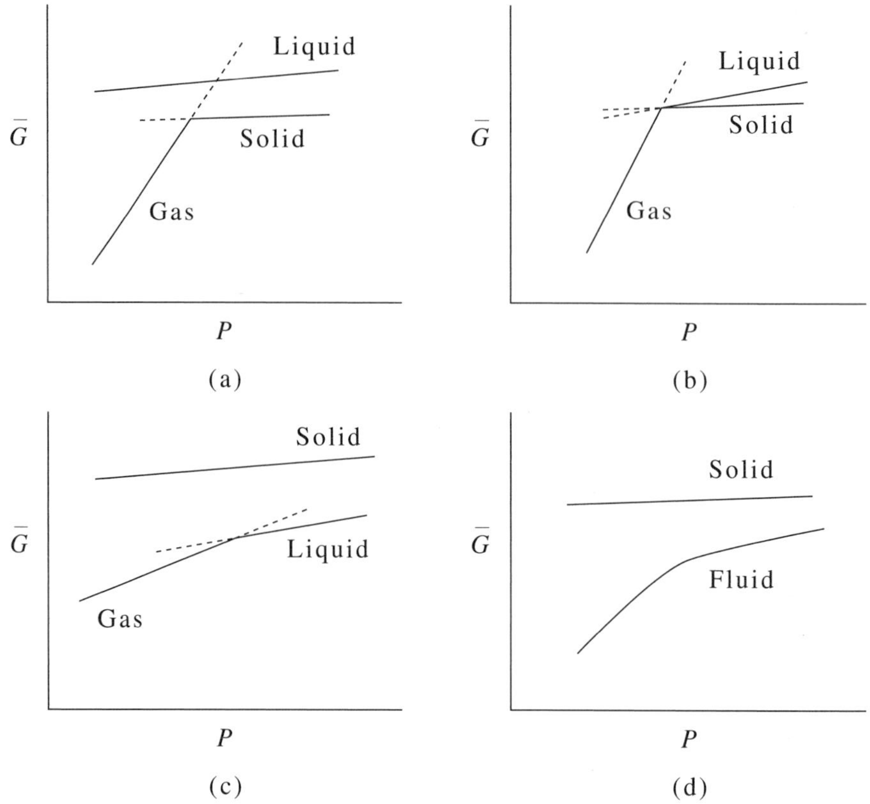


**FIGURE 9.12**
A plot of $\overline{G}(P)$ against $P$ at a number of temperatures for a "normal" substance like benzene. In (a) the temperature is less than the triple-point temperature; in (b) the temperature is equal to the triple-point temperature; in (c) the temperature is a little less than the critical temperature; and in (d) the temperature is greater than the critical temperature.

$dn$ moles are transferred from the liquid phase to the vapor phase, while $T$ and $P$ are kept constant. The infinitesimal change in Gibbs energy for this process is

$$dG = \left(\frac{\partial G^{\mathrm{g}}}{\partial n^{\mathrm{g}}}\right)_{P,T} dn^{\mathrm{g}} + \left(\frac{\partial G^{\mathrm{l}}}{\partial n^{\mathrm{l}}}\right)_{P,T} dn^{\mathrm{l}} \tag{9.1}$$

But $dn^{\mathrm{l}} = -dn^{\mathrm{g}}$ for the transfer of $dn$ moles from the liquid phase to the vapor phase, so Equation 9.1 becomes

$$dG = \left[\left(\frac{\partial G^{\mathrm{g}}}{\partial n^{\mathrm{g}}}\right)_{P,T} - \left(\frac{\partial G^{\mathrm{l}}}{\partial n^{\mathrm{l}}}\right)_{P,T}\right] dn^{\mathrm{g}} \tag{9.2}$$

The partial derivatives in Equation 9.2 are central quantities in the treatment of equilibria. They are called *chemical potentials* and are denoted by $\mu^{\mathrm{g}}$ and $\mu^{\mathrm{l}}$:

$$\mu^{\mathrm{g}} = \left(\frac{\partial G^{\mathrm{g}}}{\partial n^{\mathrm{g}}}\right)_{P,T} \quad \text{and} \quad \mu^{\mathrm{l}} = \left(\frac{\partial G^{\mathrm{l}}}{\partial n^{\mathrm{l}}}\right)_{P,T} \tag{9.3}$$

In terms of chemical potentials, then, Equation 9.2 reads

$$dG = (\mu^{\mathrm{g}} - \mu^{\mathrm{l}})dn^{\mathrm{g}} \qquad \text{(constant } T \text{ and } P) \tag{9.4}$$

If the two phases are in equilibrium with each other, then $dG = 0$, and because $dn^{\mathrm{g}} \neq 0$, we find that $\mu^{\mathrm{g}} = \mu^{\mathrm{l}}$. Thus, we find that if two phases of a single substance are in equilibrium with each other, then the chemical potentials of that substance in the two phases are equal.

If the two phases are not in equilibrium with each other, a spontaneous transfer of matter from one phase to the other will occur in the direction such that $dG < 0$. If $\mu^{\mathrm{g}} > \mu^{\mathrm{l}}$, the term in parentheses in Equation 9.4 is positive, so $dn^{\mathrm{g}}$ must be negative in order that $dG < 0$. In other words, matter will transfer from the vapor phase to the liquid phase, or from the phase with higher chemical potential to the phase with lower chemical potential. If, on the other hand, $\mu^{\mathrm{g}} < \mu^{\mathrm{l}}$, then $dn^{\mathrm{g}}$ will be positive, meaning that matter will transfer from the liquid phase to the vapor phase. Once again, the transfer occurs from the phase with higher chemical potential to the phase with lower chemical potential. Notice that chemical potential is analogous to electric potential. Just as electric current flows from a higher electric potential to a lower electric potential, matter "flows" from a higher chemical potential to a lower chemical potential (see Problem 9–19).

Although we have defined chemical potential quite generally in Equation 9.3, it takes on a simple, familiar form for a pure substance. Because $G$ , like $U$, $H$, and $S$, is an extensive thermodynamic function, $G$ is proportional to the size of a system, or $G \propto n$. We can express this proportionality by the equation $G = n\mu(T, P)$. Note that this equation is consistent with the definition of $\mu(T, P)$ because

$$\mu = \left(\frac{\partial G}{\partial n}\right)_{P,T} = \left(\frac{\partial n\mu(T, P)}{\partial n}\right)_{T,P} = \mu(T, P) \tag{9.5}$$

Therefore, for a single, pure substance, $\mu$ is the same quantity as the molar Gibbs energy and $\mu(T, P)$ is an intensive quantity like temperature and pressure.

We can use the fact that the chemical potentials of a single substance in two phases in equilibrium are equal to derive an expression for the variation of equilibrium pressure with temperature for any two phases of a given pure substance. Let the two phases be $\alpha$ and $\beta$, so that

$$\mu^{\alpha}(T, P) = \mu^{\beta}(T, P) \qquad \text{(equilibrium between phases)} \tag{9.6}$$

Now take the total derivative of both sides of Equation 9.6

$$\left(\frac{\partial \mu^{\alpha}}{\partial P}\right)_T dP + \left(\frac{\partial \mu^{\alpha}}{\partial T}\right)_P dT = \left(\frac{\partial \mu^{\beta}}{\partial P}\right)_T dP + \left(\frac{\partial \mu^{\beta}}{\partial T}\right)_P dT \tag{9.7}$$

Because $\mu$ is simply the molar Gibbs energy for a single substance, we have in analogy with Equations 8.31

$$\left(\frac{\partial G}{\partial P}\right)_T = V \quad \text{and} \quad \left(\frac{\partial G}{\partial T}\right)_P = -S$$

that

$$\left(\frac{\partial \mu}{\partial P}\right)_T = \left(\frac{\partial \overline{G}}{\partial P}\right)_T = \overline{V} \quad \text{and} \quad \left(\frac{\partial \mu}{\partial T}\right)_P = \left(\frac{\partial \overline{G}}{\partial T}\right)_P = -\overline{S} \tag{9.8}$$

where $\overline{V}$ and $\overline{S}$ are the molar volume and the molar entropy, respectively. We substitute this result into Equation 9.7 to obtain

$$\overline{V}^{\alpha} dP - \overline{S}^{\alpha} dT = \overline{V}^{\beta} dP - \overline{S}^{\beta} dT$$

Solving for $dP/dT$ gives

$$\frac{dP}{dT} = \frac{\overline{S}^{\beta} - \overline{S}^{\alpha}}{\overline{V}^{\beta} - \overline{V}^{\alpha}} = \frac{\Delta_{\text{trs}}\overline{S}}{\Delta_{\text{trs}}\overline{V}} \tag{9.9}$$

Equation 9.9 applies to two phases in equilibrium with each other, so we may use the fact that $\Delta_{\text{trs}}\overline{S} = \Delta_{\text{trs}}\overline{H}/T$ and write

$$\frac{dP}{dT} = \frac{\Delta_{\text{trs}}\overline{H}}{T\Delta_{\text{trs}}\overline{V}} \tag{9.10}$$

Equation 9.10 is called the *Clapeyron equation*, and relates the slope of the two-phase boundary line in a phase diagram with the values of $\Delta_{\text{trs}}\overline{H}$ and $\Delta_{\text{trs}}\overline{V}$ for a transition between these two phases.

Let's use Equation 9.10 to calculate the slope of the solid-liquid coexistence curve for benzene around one atm (Figure 9.1). The molar enthalpy of fusion of benzene at its normal melting point (278.7 K) is 9.95 $\text{kJ}\cdot\text{mol}^{-1}$, and $\Delta_{\text{fus}}\overline{V}$ under the same conditions is 10.3 $\text{cm}^3\cdot\text{mol}^{-1}$. Thus, $dP/dT$ at the normal melting point of benzene is

$$\begin{aligned}\frac{dP}{dT} &= \frac{9950\ \text{J}\cdot\text{mol}^{-1}}{(278.68\ \text{K})(10.3\ \text{cm}^3\cdot\text{mol}^{-1})}\left(\frac{10\ \text{cm}}{1\ \text{dm}}\right)^3\left(\frac{0.08206\ \text{dm}^3\cdot\text{atm}\cdot\text{mol}^{-1}\cdot\text{K}^{-1}}{8.314\ \text{J}\cdot\text{mol}^{-1}\cdot\text{K}^{-1}}\right)\\ &= 34.2\ \text{atm}\cdot\text{K}^{-1}\end{aligned}$$

We can take the reciprocal of this result to obtain

$$\frac{dT}{dP} = 0.0292 \text{ K}\cdot\text{atm}^{-1}$$

Thus, we see that the melting point of benzene increases by 0.0292 K per atmosphere of pressure around one atm. If $\Delta_{\text{fus}}\overline{H}$ and $\Delta_{\text{fus}}\overline{V}$ were independent of pressure, we could use this result to predict that the melting point of benzene at 1000 atm is 29.2 K higher than it is at one atm, or 307.9 K. The experimental value is 306.4 K, so our assumption of constant $\Delta_{\text{fus}}\overline{H}$ and $\Delta_{\text{fus}}\overline{V}$ is fairly satisfactory. Nevertheless, Equation 9.10 is strictly valid only over a limited range because $\Delta_{\text{fus}}\overline{H}$ and $\Delta_{\text{fus}}\overline{V}$ do vary with $T$ and $P$. Figure 9.2 shows the experimental melting point of benzene versus pressure up to 10 000 atm. You can see from the figure that the slope is not quite constant.

**EXAMPLE 9–3**

Determine the value of $dT/dP$ for ice at its normal melting point. The molar enthalpy of fusion of ice at 273.15 K and one atm is 6010 $\text{J}\cdot\text{mol}^{-1}$, and $\Delta_{\text{fus}}\overline{V}$ under the same conditions is $-1.63 \text{ cm}^3\cdot\text{mol}^{-1}$ (recall that unlike most substances water expands upon freezing, so that $\Delta_{\text{fus}}\overline{V} = \overline{V}^{\text{l}} - \overline{V}^{\text{s}} < 0$.) Estimate the melting point of ice at 1000 atm.

SOLUTION: We use the reciprocal of Equation 9.10:

$$\begin{aligned}\frac{dT}{dP} &= \frac{T\Delta_{\text{fus}}\overline{V}}{\Delta_{\text{fus}}\overline{H}} \\ &= \frac{(273.2\text{ K})(-1.63\text{ cm}^3\cdot\text{mol}^{-1})}{6010\text{ J}\cdot\text{mol}^{-1}}\left(\frac{10\text{ cm}}{1\text{ dm}}\right)^3 \\ &\quad\times\left(\frac{8.314\text{ J}\cdot\text{mol}^{-1}\cdot\text{K}^{-1}}{0.08206\text{ dm}^3\cdot\text{atm}\cdot\text{mol}^{-1}\cdot\text{K}^{-1}}\right) \\ &= -0.00751\text{ K}\cdot\text{atm}^{-1}\end{aligned}$$

Assuming that $dT/dP$ is constant up to 1000 atm, we find that $\Delta T = -7.51$ K, or that the melting point of ice at 1000 atm is 265.6 K. The experimental value is 263.7 K. The discrepancy arises from our assumption that the values of $\Delta_{\text{fus}}\overline{V}$ and $\Delta_{\text{fus}}\overline{H}$ are independent of pressure. Figure 9.6 shows the experimental melting point of ice versus pressure up to 2000 atm.

Notice that the melting point of ice decreases with increasing pressure, so that the slope of the solid-liquid equilibrium curve in the pressure-temperature phase diagram of water has a negative slope. Equation 9.10 shows that this slope is a direct result of the fact that $\Delta_{\text{fus}}\overline{V}$ is negative for this case.

Equation 9.10 can be used to estimate the molar volume of a liquid at its boiling point.

**EXAMPLE 9–4**

The vapor pressure of benzene is found to obey the empirical equation

$$\ln(P/\text{torr}) = 16.725 - \frac{3229.86\ \text{K}}{T} - \frac{118345\ \text{K}^2}{T^2}$$

from 298.2 K to its normal boiling point 353.24 K. Given that the molar enthalpy of vaporization at 353.24 K is $30.8\ \text{kJ}\cdot\text{mol}^{-1}$ and that the molar volume of liquid benzene at 353.24 K is $96.0\ \text{cm}^3\cdot\text{mol}^{-1}$, use the above equation to determine the molar volume of the vapor at its equilibrium pressure at 353.24 K and compare this value with the ideal-gas value.

SOLUTION: We start with Equation 9.10, which we solve for $\Delta_{\text{vap}}\overline{V}$

$$\Delta_{\text{vap}}\overline{V} = \frac{\Delta_{\text{vap}}\overline{H}}{T(dP/dT)}$$

Using the above empirical vapor pressure equation at $T = 353.24$ K,

$$\begin{aligned}\frac{dP}{dT} &= P\left(\frac{3229.86\ \text{K}}{T^2} + \frac{236690\ \text{K}^2}{T^3}\right)\\ &= (760\ \text{torr})(0.0312\ \text{K}^{-1}) = 23.75\ \text{torr}\cdot\text{K}^{-1} = 0.0312\ \text{atm}\cdot\text{K}^{-1}\end{aligned}$$

Therefore,

$$\begin{aligned}\Delta_{\text{vap}}\overline{V} &= \frac{30800\ \text{J}\cdot\text{mol}^{-1}}{(353.24\ \text{K})(0.0312\ \text{atm}\cdot\text{K}^{-1})}\\ &= (2790\ \text{J}\cdot\text{atm}^{-1}\cdot\text{mol}^{-1})\left(\frac{0.08206\ \text{L}\cdot\text{atm}}{8.314\ \text{J}}\right)\\ &= 27.6\ \text{L}\cdot\text{mol}^{-1}\end{aligned}$$

The molar volume of the vapor is

$$\begin{aligned}\overline{V}^{\text{g}} &= \Delta_{\text{vap}}\overline{V} + \overline{V}^{\text{l}} = 27.5\ \text{L}\cdot\text{mol}^{-1} + 0.0960\ \text{L}\cdot\text{mol}^{-1}\\ &= 27.7\ \text{L}\cdot\text{mol}^{-1}\end{aligned}$$

The corresponding value from the ideal gas equation is

$$\begin{aligned}\overline{V}^{\text{g}} &= \frac{RT}{P}\\ &= \frac{(0.08206\ \text{L}\cdot\text{atm}\cdot\text{K}^{-1}\cdot\text{mol}^{-1})(353.24\ \text{K})}{1\ \text{atm}}\\ &= 29.0\ \text{L}\cdot\text{mol}^{-1}\end{aligned}$$

which is slightly larger than the actual value.

## 9–4. The Clausius–Clapeyron Equation Gives the Vapor Pressure of a Substance As a Function of Temperature

When we used Equation 9.10 to calculate the variation of the melting points of ice (Example 9–3) and benzene, we assumed that $\Delta_{\text{trs}}\overline{H}$ and $\Delta_{\text{trs}}\overline{V}$ do not vary appreciably with pressure. Although this approximation is fairly satisfactory for solid-liquid and solid-solid transitions over a small $\Delta T$, it is not satisfactory for liquid-gas and solid-gas transitions because the molar volume of a gas varies strongly with pressure. If the temperature is not too near the critical point, however, Equation 9.10 can be cast into a very useful form for condensed phase-gas phase transitions.

Let's apply Equation 9.10 to a liquid-vapor equilibrium. In this case, we have

$$\frac{dP}{dT} = \frac{\Delta_{\text{vap}}\overline{H}}{T(\overline{V}^{\text{g}} - \overline{V}^{\text{l}})} \tag{9.11}$$

Equation 9.11 gives the slope of the liquid-vapor equilibrium line in the phase diagram of the substance. As long as we are not too near the critical point, $\overline{V}^{\text{g}} \gg \overline{V}^{\text{l}}$, so that we can neglect $\overline{V}^{\text{l}}$ compared with $\overline{V}^{\text{g}}$ in the denominator of Equation 9.11. Furthermore, if the vapor pressure is not too high (once again, if we are not too close to the critical point), we can assume the vapor is ideal and replace $\overline{V}^{\text{g}}$ by $RT/P$, so that Equation 9.11 becomes

$$\frac{1}{P}\frac{dP}{dT} = \frac{d\ln P}{dT} = \frac{\Delta_{\text{vap}}\overline{H}}{RT^2} \tag{9.12}$$

This equation, which was first derived by Clausius in 1850, is known as the *Clausius-Clapeyron equation*. Remember that we have neglected the molar volume of the liquid compared with the molar volume of the gas and that we assumed the vapor can be treated as an ideal gas. Nevertheless, Equation 9.12 has the advantage of being more convenient to use than Equation 9.10. As might be expected, however, Equation 9.10 is more accurate than Equation 9.12.

The real advantage of Equation 9.12 is that it can be readily integrated. If we assume $\Delta_{\text{vap}}\overline{H}$ does not vary with temperature over the integration limits of $T$, Equation 9.12 becomes

$$\ln\frac{P_2}{P_1} = -\frac{\Delta_{\text{vap}}\overline{H}}{R}\left(\frac{1}{T_2} - \frac{1}{T_1}\right) = \frac{\Delta_{\text{vap}}\overline{H}}{R}\left(\frac{T_2 - T_1}{T_1 T_2}\right) \tag{9.13}$$

Equation 9.13 can be used to calculate the vapor pressure at some temperature given the molar enthalpy of vaporization and the vapor pressure at some other temperature. For example, the normal boiling point of benzene is 353.2 K and $\Delta_{\text{vap}}\overline{H} = 30.8\ \text{kJ}\cdot\text{mol}^{-1}$. Assuming $\Delta_{\text{vap}}\overline{H}$ does not vary with temperature, let's calculate the vapor pressure of benzene at 373.2 K. We substitute $P_1 = 760$ torr, $T_1 = 353.2$ K, and $T_2 = 373.2$ K into Equation 9.13 to obtain

$$\ln \frac{P}{760} = \left( \frac{30800\ \text{J}\cdot\text{mol}^{-1}}{8.314\ \text{J}\cdot\text{K}^{-1}\cdot\text{mol}^{-1}} \right) \left( \frac{19.8\ \text{K}}{(353.2\ \text{K})(373.2\ \text{K})} \right)$$
$$= 0.556$$

or $P = 1330$ torr. The experimental value is 1360 torr.

**EXAMPLE 9–5**

The vapor pressure of water at 363.2 K is 529 torr. Use Equation 9.13 to determine the average value of $\Delta_{\text{vap}}\overline{H}$ of water between 363.3 K and 373.2 K.

SOLUTION: We use the fact that the normal boiling point of water is 373.2 K ($P = 760$ torr) and write

$$\ln \frac{760}{529} = \frac{\Delta_{\text{vap}}\overline{H}}{8.314\ \text{J}\cdot\text{K}^{-1}\cdot\text{mol}^{-1}} \frac{10.0\ \text{K}}{(363.2\ \text{K})(373.2\ \text{K})}$$

or

$$\Delta_{\text{vap}}\overline{H} = 40.8\ \text{kJ}\cdot\text{mol}^{-1}$$

The value of $\Delta_{\text{vap}}\overline{H}$ for water at its normal boiling point is 40.65 $\text{kJ}\cdot\text{mol}^{-1}$.

If we integrate Equation 9.12 indefinitely rather than between definite limits, we obtain (assuming $\Delta_{\text{vap}}\overline{H}$ is constant)

$$\ln P = -\frac{\Delta_{\text{vap}}\overline{H}}{RT} + \text{constant} \tag{9.14}$$

Equation 9.14 says that a plot of the logarithm of the vapor pressure against the reciprocal of the kelvin temperature should be a straight line with a slope of $-\Delta_{\text{vap}}\overline{H}/R$. Figure 9.13 shows such a plot for benzene over the temperature range 313 K to 353 K. The slope of the line gives $\Delta_{\text{vap}}\overline{H} = 32.3\ \text{kJ}\cdot\text{mol}^{-1}$. This value represents an average value of $\Delta_{\text{vap}}\overline{H}$ over the given temperature interval. The value of $\Delta_{\text{vap}}\overline{H}$ at the normal boiling point (353 K) is 30.8 $\text{kJ}\cdot\text{mol}^{-1}$.

We can recognize that $\Delta_{\text{vap}}\overline{H}$ varies with temperature by writing $\Delta_{\text{vap}}\overline{H}$ in the form

$$\Delta_{\text{vap}}\overline{H} = A + BT + CT^2 + \cdots$$

where $A$, $B$, $C$, ... are constants. If this equation is substituted into Equation 9.12, then integration gives

$$\ln P = -\frac{A}{RT} + \frac{B}{R} \ln T + \frac{C}{R} T + k + O(T^2) \tag{9.15}$$

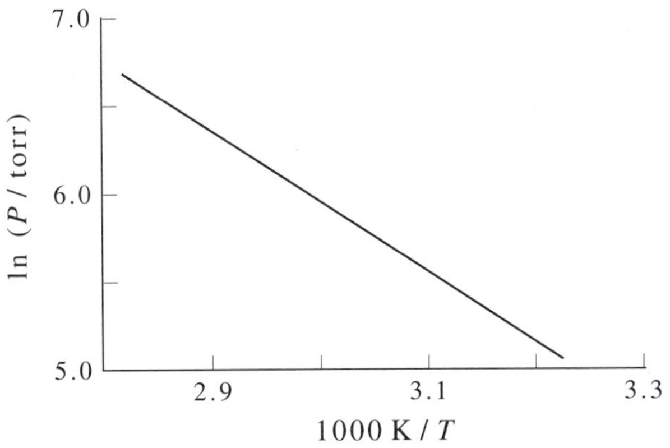


**FIGURE 9.13**
A plot of the logarithm of the vapor pressure of liquid benzene against the reciprocal kelvin temperature over a temperature range of 313 K to 353 K.

where $k$ is an integration constant. Equation 9.15 expresses the variation of vapor pressure over a larger temperature range than Equation 9.14. Thus, a plot of $\ln P$ against $1/T$ will not be exactly linear, in agreement with the experimental data for most liquids and solids over an extended temperature range. For example, the vapor pressure of solid ammonia in torr is found to obey the equation

$$\ln(P/\text{torr}) = -\frac{4124.4\ \text{K}}{T} - 1.81630 \ln(T/K) + 34.4834 \tag{9.16}$$

from 146 K to 195 K.

**EXAMPLE 9–6**

Use the Clausius-Clapeyron equation and Equation 9.16 to determine the molar enthalpy of sublimation of ammonia from 146 K to 195 K.

SOLUTION: According to Equation 9.12

$$\frac{d \ln P}{dT} = \frac{\Delta_{\text{sub}}\overline{H}}{RT^2}$$

Using Equation 9.16 for $\ln P$ gives us

$$\frac{\Delta_{\text{sub}}\overline{H}}{RT^2} = \frac{4124.4\text{K}^2}{T^2} - \frac{1.8163\text{K}}{T}$$

or

$$\begin{aligned}\Delta_{\text{sub}}\overline{H} &= (4124.4\ \text{K})R - (1.8163)RT \\ &= 34.29\ \text{kJ}\cdot\text{mol}^{-1} - (0.0151\ \text{kJ}\cdot\text{mol}^{-1}\cdot\text{K}^{-1})T \\ &\qquad 146\ \text{K} < T < 195\ \text{K}\end{aligned}$$

The Clausius-Clapeyron equation can be used to show that the slope of the solid-gas coexistence curve must be greater than the slope of the liquid-gas coexistence curve near the triple point, where these two curves meet. According to Equation 9.12, the slope of the solid-gas curve is given by

$$\frac{dP^{\mathrm{s}}}{dT} = P^{\mathrm{s}}\frac{\Delta_{\mathrm{sub}}\overline{H}}{RT^2} \tag{9.17}$$

and the slope of the liquid-gas curve is given by

$$\frac{dP^{\mathrm{l}}}{dT} = P^{\mathrm{l}}\frac{\Delta_{\mathrm{vap}}\overline{H}}{RT^2} \tag{9.18}$$

At the triple point, $P^{\mathrm{s}}$ and $P^{\mathrm{l}}$, the vapor pressures of the solid and liquid, respectively, are equal, so the ratio of the slopes from Equations 9.17 and 9.18 is

$$\frac{dP^{\mathrm{s}}/dT}{dP^{\mathrm{l}}/dT} = \frac{\Delta_{\mathrm{sub}}\overline{H}}{\Delta_{\mathrm{vap}}\overline{H}} \tag{9.19}$$

at the triple point. Because enthalpy is a state function, the enthalpy change in going directly from the solid phase to the gas phase is the same as first going from the solid phase to the liquid phase and then going from the liquid phase to the gas phase. In an equation, we have

$$\Delta_{\mathrm{sub}}\overline{H} = \Delta_{\mathrm{fus}}\overline{H} + \Delta_{\mathrm{vap}}\overline{H} \tag{9.20}$$

where the three $\Delta\overline{H}$s must all be evaluated at the same temperature. If we substitute Equation 9.20 into Equation 9.19, we see that

$$\frac{dP^{\mathrm{s}}/dT}{dP^{\mathrm{l}}/dT} = 1 + \frac{\Delta_{\mathrm{fus}}\overline{H}}{\Delta_{\mathrm{vap}}\overline{H}}$$

Thus, we see that the slope of the solid-gas curve is greater than that of the liquid-gas curve at the triple point.

**EXAMPLE 9–7**

The vapor pressures of solid and liquid ammonia near the triple point are given by

$$\log(P^{\mathrm{s}}/\mathrm{torr}) = 10.0 - \frac{1630\ \mathrm{K}}{T}$$

and

$$\log(P^{\mathrm{l}}/\mathrm{torr}) = 8.46 - \frac{1330\ \mathrm{K}}{T}$$

Calculate the ratio of the slopes of the solid-gas curve and the liquid-gas curve at the triple point.

SOLUTION: The derivatives of both expressions at the triple point are

$$\frac{dP^{\mathrm{s}}}{dT} = (2.303P_{\mathrm{tp}})\left(\frac{1630\ \mathrm{K}}{T_{\mathrm{tp}}^2}\right) = 4.31\ \mathrm{torr\cdot K^{-1}}$$

and

$$\frac{dP^{\mathrm{l}}}{dT} = (2.303P_{\mathrm{tp}})\left(\frac{1330\ \mathrm{K}}{T_{\mathrm{tp}}^2}\right) = 3.52\ \mathrm{torr\cdot K^{-1}}$$

so the ratio of the slopes is $4.31/3.52 = 1.22$.

## 9–5. Chemical Potential Can Be Evaluated From a Partition Function

In this section, we will derive a convenient formula for the chemical potential in terms of a partition function. Recall that the corresponding formulas for the energy and entropy are (see Equations 3.21 and 6.43)

$$U = k_{\mathrm{B}}T^2\left(\frac{\partial \ln Q}{\partial T}\right)_{N,V} \tag{9.21}$$

and

$$S = k_{\mathrm{B}}T\left(\frac{\partial \ln Q}{\partial T}\right)_{N,V} + k_{\mathrm{B}}\ln Q \tag{9.22}$$

Using the fact that the Helmholtz energy $A$ is equal to $U - TS$, Equations 9.21 and 9.22 give

$$A = -k_{\mathrm{B}}T\ln Q \tag{9.23}$$

Let's now include $N$ in our discussion of natural variables, and write

$$\begin{aligned} dA &= \left(\frac{\partial A}{\partial T}\right)_{N,V} dT + \left(\frac{\partial A}{\partial V}\right)_{N,T} dV + \left(\frac{\partial A}{\partial N}\right)_{T,V} dN \\ &= -SdT - PdV + \left(\frac{\partial A}{\partial N}\right)_{T,V} dN \end{aligned} \tag{9.24}$$

The last term in Equation 9.24 is expressed in terms of $N$, the number of molecules in the system. It is more conventional to express this quantity in terms of $n$, the number of moles in the system. We can do this by noting that

$$\left(\frac{\partial A}{\partial N}\right)_{T,V} dN = \left(\frac{\partial A}{\partial n}\right)_{T,V} dn$$

because $n$ and $N$ differ by a constant factor of the Avogadro constant. Therefore, we may write Equation 9.24 in the form

$$dA = -SdT - PdV + \left(\frac{\partial A}{\partial n}\right)_{T,V} dn \tag{9.25}$$

We'll now show that $(\partial A/\partial n)_{T,V}$ is just another way of writing the chemical potential, $\mu$. If we add $d(PV)$ to both sides of Equation 9.25 and use the equation $G = A + PV$, we get

$$dG = dA + d(PV) = -SdT + VdP + \left(\frac{\partial A}{\partial n}\right)_{T,V} dn$$

But if we compare this result to the total derivative of $G = G(T, P, n)$,

$$\begin{aligned} dG &= \left(\frac{\partial G}{\partial T}\right)_{P,N} dT + \left(\frac{\partial G}{\partial P}\right)_{T,N} dP + \left(\frac{\partial G}{\partial n}\right)_{T,P} dn \\ &= -SdT + VdP + \mu dn \end{aligned}$$

we see that

$$\mu = \left(\frac{\partial G}{\partial n}\right)_{T,P} = \left(\frac{\partial A}{\partial n}\right)_{T,V} \tag{9.26}$$

Thus, we can use either $G$ or $A$ to determine $\mu$ as long as we keep the natural variables of each one fixed when we take the partial derivative with respect to $n$.

We can now substitute Equation 9.23 into Equation 9.26 to obtain

$$\mu = -k_B T \left(\frac{\partial \ln Q}{\partial n}\right)_{V,T} = -RT \left(\frac{\partial \ln Q}{\partial N}\right)_{V,T} \tag{9.27}$$

We have gone from the second term to the third term by multipying $k_B$ and $n$ by the Avogadro constant. Equation 9.27 takes on a fairly simple form for an ideal gas. If we substitute the ideal-gas expression

$$Q(N, V, T) = \frac{[q(V, T)]^N}{N!}$$

into $\ln Q$, we can write

$$\ln Q = N \ln q - N \ln N + N$$

where we have used Stirling's approximation for $\ln N!$. If we substitute this result into Equation 9.27, we obtain

$$\begin{aligned} \mu &= -RT(\ln q - \ln N - 1 + 1) \\ &= -RT \ln \frac{q(V, T)}{N} \qquad \text{(ideal gas)} \end{aligned} \tag{9.28}$$

Recall now that $q(V, T) \propto V$ for an ideal gas, and so we can write Equation 9.28 as

$$\mu = -RT \ln\left[\left(\frac{q}{V}\right)\frac{V}{N}\right] \tag{9.29}$$

where $q(V, T)/V$ is a function of temperature only. Equation 9.29 also gives us an equation for $G$ because $G = n\mu$. We can make Equation 9.29 look exactly like Equation 8.59 if we substitute $k_B T/P$ for $V/N$:

$$\begin{aligned}\mu &= -RT \ln\left[\left(\frac{q}{V}\right)\frac{k_B T}{P}\right]\\ &= -RT \ln\left[\left(\frac{q}{V}\right)k_B T\right] + RT \ln P\end{aligned} \tag{9.30}$$

If we compare this equation with

$$\mu(T, P) = \mu^\circ(T) + RT \ln P \tag{9.31}$$

we see that

$$\mu^\circ(T) = -RT \ln\left[\left(\frac{q}{V}\right)k_B T\right] \tag{9.32}$$

Once again, recall that $q/V$ is a function of $T$ only for an ideal gas.

To calculate $\mu^\circ(T)$, we must remember that $P$ is expressed relative to the standard state pressure $P^\circ$, which is equal to one bar or $10^5$ Pa. We emphasize this convention by writing Equation 9.31 as

$$\mu(T, P) = \mu^\circ(T) + RT \ln \frac{P}{P^\circ} \tag{9.33}$$

If we compare Equation 9.33 with Equation 9.30, we see that

$$\begin{aligned}\mu^\circ(T) &= -RT \ln\left[\left(\frac{q}{V}\right)k_B T\right] + RT \ln P^\circ\\ &= -RT \ln\left[\left(\frac{q}{V}\right)\frac{k_B T}{P^\circ}\right]\end{aligned} \tag{9.34}$$

The argument of the logarithm in Equation 9.34 is unitless, as it must be. Equation 9.34 gives us a molecular formula to calculate $\mu^\circ(T)$, or $G^\circ(T)$. For example, for Ar(g) at 298.15 K:

$$\begin{aligned}\frac{q(V, T)}{V} &= \left(\frac{2\pi m k_B T}{h^2}\right)^{3/2}\\ &= \left[\frac{(2\pi)(0.03995\ \text{kg}\cdot\text{mol}^{-1})(1.3806\times10^{-23}\ \text{J}\cdot\text{K}^{-1})(298.15\ \text{K})}{(6.022\times10^{23}\ \text{mol}^{-1})(6.626\times10^{-34}\ \text{J}\cdot\text{s})^2}\right]^{3/2}\\ &= 2.444\times10^{32}\ \text{m}^{-3}\end{aligned}$$

$$\frac{k_B T}{P^\circ} = \frac{RT}{N_A P^\circ} = \frac{(8.314\ \text{J}\cdot\text{mol}^{-1}\cdot\text{K}^{-1})(298.15\ \text{K})}{(6.022 \times 10^{23}\ \text{mol}^{-1})(1.00 \times 10^5\ \text{Pa})}$$
$$= 4.116 \times 10^{-26}\ \text{m}^3$$

and

$$RT = (8.314\ \text{J}\cdot\text{K}^{-1}\cdot\text{mol}^{-1})(298.15\ \text{K}) = 2479\ \text{J}\cdot\text{mol}^{-1}$$

and so

$$\begin{aligned}\mu^\circ(298.15\ \text{K}) &= -(2479\ \text{J}\cdot\text{mol}^{-1}) \ln\left[(2.444 \times 10^{32}\ \text{m}^{-3})(4.116 \times 10^{-26}\ \text{m}^3)\right] \\ &= -(2479\ \text{J}\cdot\text{mol}^{-1}) \ln[1.006 \times 10^7] \\ &= -3.997 \times 10^4\ \text{J}\cdot\text{mol}^{-1} = -39.97\ \text{kJ}\cdot\text{mol}^{-1}\end{aligned}$$

This result is in excellent agreement with the experimental value of $-39.97\ \text{kJ}\cdot\text{mol}^{-1}$.

Being essentially an energy, the value of the chemical potential must be based upon some choice of a zero of energy. The chemical potential we have just calculated is based upon the ground state of the atom being zero. For diatomic molecules, we have chosen the ground-state energy (vibrational and electronic) to be $-D_0$, as illustrated in Figure 4.2. In tabulating values of $\mu^\circ(T)$, it is customary to take the ground-state energy of the molecule rather than the separated atoms as in Figure 4.2 to be the zero of energy. To see how this definition of the zero of energy changes the form of the partition function, write

$$\begin{aligned}q(V, T) &= \sum_j e^{-\varepsilon_j/k_B T} \\ &= e^{-\varepsilon_0/k_B T} + e^{-\varepsilon_1/k_B T} + \cdots\end{aligned}$$

If we factor out $e^{-\varepsilon_0/k_B T}$, we have

$$\begin{aligned}q(V, T) &= e^{-\varepsilon_0/k_B T}[1 + e^{-(\varepsilon_1-\varepsilon_0)/k_B T} + e^{-(\varepsilon_2-\varepsilon_0)/k_B T} + \cdots] \\ &= e^{-\varepsilon_0/k_B T} q^0(V, T)\end{aligned} \tag{9.35}$$

where we have written $q^0(V, T)$ to emphasize that the ground-state energy of the molecule is taken to be zero. Substituting this result into Equation 9.34 gives

$$\begin{aligned}\mu^\circ(T) - E_0 &= -RT \ln\left[\left(\frac{q^0}{V}\right)\frac{k_B T}{P^\circ}\right] \\ &= -RT \ln\left[\left(\frac{q^0}{V}\right)\frac{RT}{N_A P^\circ}\right]\end{aligned} \tag{9.36}$$

where $E_0 = N_A \varepsilon_0$ and $P^\circ = 1\ \text{bar} = 10^5\ \text{Pa}$.

The partition function $q^0(V, T)$ for a diatomic molecule is

$$q^0(V,T) = \left(\frac{2\pi m k_B T}{h^2}\right)^{3/2} V \cdot \frac{T}{\sigma \Theta_{rot}} \cdot \frac{1}{1 - e^{-\Theta_{vib}/T}} \cdot g_{e1} \tag{9.37}$$

Notice that this expression is the same as Equation 4.39 except for the factor of $e^{-h\nu/k_B T} e^{D_e/k_B T} = e^{D_0/RT}$ in Equation 4.39, which accounts for the ground-state energy being taken to be $-D_0$. The ground-state energy associated with $q^0(V, T)$ given by Equation 9.37 is zero. Let's use Equation 9.36 along with Equation 9.37 to calculate $\mu^\circ - E_0$ for HI(g) at 298.15 K in the harmonic oscillator-rigid rotator approximation, with $\Theta_{rot} = 9.25$ K and $\Theta_{vib} = 3266$ K (Table 4.2). Therefore,

$$\begin{aligned}\frac{q^0(V,T)}{V} &= \left[\frac{(2\pi)(0.1279\ \text{kg}\cdot\text{mol}^{-1})(1.3806 \times 10^{-23}\ \text{J}\cdot\text{K}^{-1})(298.15\ \text{K})}{(6.022 \times 10^{23}\ \text{mol}^{-1})(6.626 \times 10^{-34}\ \text{J}\cdot\text{s})}\right]^{3/2} \\ &\quad \times \left(\frac{298.15\ \text{K}}{9.25\ \text{K}}\right) \frac{1}{1 - e^{-3266\ \text{K}/298.15\ \text{K}}} \\ &= 4.51 \times 10^{34}\ \text{m}^{-3}\end{aligned}$$

$$\begin{aligned}\frac{RT}{N_A P^\circ} &= \frac{(8.314\ \text{J}\cdot\text{mol}^{-1}\cdot\text{K}^{-1})(298.15\ \text{K})}{(6.022 \times 10^{23}\ \text{mol}^{-1})(10^5\ \text{Pa})} \\ &= 4.116 \times 10^{-26}\ \text{m}^3\end{aligned}$$

and

$$\begin{aligned}\mu^\circ(298.15\ \text{K}) - E_0 &= -(8.314\ \text{J}\cdot\text{mol}^{-1}\cdot\text{K}^{-1})(298.15\ \text{K}) \ln(1.86 \times 10^9) \\ &= -52.90\ \text{kJ}\cdot\text{mol}^{-1}\end{aligned}$$

The literature value, which includes anharmonic and nonrigid rotator effects, is $-52.94$ kJ·mol$^{-1}$. We will use values of $\mu^\circ(T) - E_0$ when we discuss chemical equilibria in Chapter 12.

# Problems

**9-1.** Sketch the phase diagram for oxygen using the following data: triple point, 54.3 K and 1.14 torr; critical point, 154.6 K and 37 828 torr; normal melting point, −218.4°C; and normal boiling point, −182.9°C. Does oxygen melt under an applied pressure as water does?

**9-2.** Sketch the phase diagram for $I_2$ given the following data: triple point, 113°C and 0.12 atm; critical point, 512°C and 116 atm; normal melting point, 114°C; normal boiling point, 184°C; and density of liquid > density of solid.

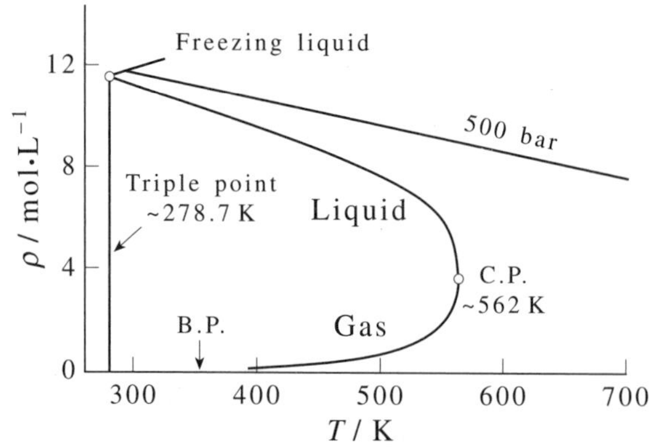


**FIGURE 9.14**
A density-temperature phase diagram of benzene.

**9-3.** Figure 9.14 shows a density-temperature phase diagram for benzene. Using the following data for the triple point and the critical point, interpret this phase diagram. Why is the triple point indicated by a line in this type of phase diagram?

| | $T$/K | $P$/bar | $\rho$/mol·L$^{-1}$ Vapor | Liquid |
|---|---|---|---|---|
| Triple point | 278.680 | 0.04785 | 0.002074 | 11.4766 |
| Critical point | 561.75 | 48.7575 | 3.90 | 3.90 |
| Normal freezing point | 278.68 | 1.01325 | | |
| Normal boiling point | 353.240 | 1.01325 | 0.035687 | 10.4075 |

**9-4.** The vapor pressures of solid and liquid chlorine are given by

$$\ln(P^{s}/\text{torr}) = 24.320 - \frac{3777 \text{ K}}{T}$$

$$\ln(P^{l}/\text{torr}) = 17.892 - \frac{2669 \text{ K}}{T}$$

where $T$ is the absolute temperature. Calculate the temperature and pressure at the triple point of chlorine.

**9-5.** The pressure along the melting curve from the triple-point temperature to an arbitrary temperature can be fit empirically by the Simon equation, which is

$$(P - P_{\text{tp}})/\text{bar} = a\left[\left(\frac{T}{T_{\text{tp}}}\right)^{\alpha} - 1\right]$$

where $a$ and $\alpha$ are constants whose values depend upon the substance. Given that $P_{\text{tp}} =$ 0.04785 bar, $T_{\text{tp}} = 278.68$ K, $a = 4237$, and $\alpha = 2.3$ for benzene, plot $P$ against $T$ and compare your result with that given in Figure 9.2.

**9-6.** The slope of the melting curve of methane is given by

$$\frac{dP}{dT} = (0.08446\ \text{bar}\cdot\text{K}^{-1.85})T^{0.85}$$

from the triple point to arbitrary temperatures. Using the fact that the temperature and pressure at the triple point are 90.68 K and 0.1174 bar, calculate the melting pressure of methane at 300 K.

**9-7.** The vapor pressure of methanol along the entire liquid-vapor coexistence curve can be expressed very accurately by the empirical equation

$$\ln(P/\text{bar}) = -\frac{10.752849}{x} + 16.758207 - 3.603425x + 4.373232x^2 - 2.381377x^3 + 4.572199(1-x)^{1.70}$$

where $x = T/T_c$, and $T_c = 512.60$ K. Use this formula to show that the normal boiling point of methanol is 337.67 K.

**9-8.** The standard boiling point of a liquid is the temperature at which the vapor pressure is exactly one bar. Use the empirical formula given in the previous problem to show that the standard boiling point of methanol is 337.33 K.

**9-9.** The vapor pressure of benzene along the liquid-vapor coexistence curve can be accurately expressed by the empirical expression

$$\ln(P/\text{bar}) = -\frac{10.655375}{x} + 23.941912 - 22.388714x + 20.2085593x^2 - 7.219556x^3 + 4.84728(1-x)^{1.70}$$

where $x = T/T_c$, and $T_c = 561.75$ K. Use this formula to show that the normal boiling point of benzene is 353.24 K. Use the above expression to calculate the standard boiling point of benzene.

**9-10.** Plot the following data for the densities of liquid and gaseous ethane in equilibrium with each other as a function of temperature, and determine the critical temperature of ethane.

| $T$/K | $\rho^l$/mol·dm$^{-3}$ | $\rho^g$/mol·dm$^{-3}$ | $T$/K | $\rho^l$/mol·dm$^{-3}$ | $\rho^g$/mol·dm$^{-3}$ |
|---|---|---|---|---|---|
| 100.00 | 21.341 | $1.336 \times 10^{-3}$ | 283.15 | 12.458 | 2.067 |
| 140.00 | 19.857 | 0.03303 | 293.15 | 11.297 | 2.880 |
| 180.00 | 18.279 | 0.05413 | 298.15 | 10.499 | 3.502 |
| 220.00 | 16.499 | 0.2999 | 302.15 | 9.544 | 4.307 |
| 240.00 | 15.464 | 0.5799 | 304.15 | 8.737 | 5.030 |
| 260.00 | 14.261 | 1.051 | 304.65 | 8.387 | 5.328 |
| 270.00 | 13.549 | 1.401 | 305.15 | 7.830 | 5.866 |

**9-11.** Use the data in the preceding problem to plot $(\rho^l + \rho^g)/2$ against $T_c - T$, with $T_c = 305.4$ K. The resulting straight line is an empirical law called the *law of rectilinear diameters*. If this curve is plotted on the same figure as in the preceding problem, the intersection of the two curves gives the critical density, $\rho_c$.

**9-12.** Use the data in Problem 9–10 to plot $(\rho^{\rm l} - \rho^{\rm g})$ against $(T_{\rm c} - T)^{1/3}$ with $T_{\rm c} = 305.4$ K. What does this plot tell you?

**9-13.** The densities of the coexisting liquid and vapor phases of methanol from the triple point to the critical point are accurately given by the empirical expressions

$$\frac{\rho^{\rm l}}{\rho_{\rm c}} - 1 = 2.51709(1-x)^{0.350} + 2.466694(1-x) - 3.066818(1-x^2) + 1.325077(1-x^3)$$

and

$$\ln\frac{\rho^{\rm g}}{\rho_{\rm c}} = -10.619689\frac{1-x}{x} - 2.556682(1-x)^{0.350} + 3.881454(1-x) + 4.795568(1-x)^2$$

where $\rho_{\rm c} = 8.40$ mol·L$^{-1}$ and $x = T/T_{\rm c}$, where $T_{\rm c} = 512.60$ K. Use these expressions to plot $\rho^{\rm l}$ and $\rho^{\rm g}$ against temperature, as in Figure 9.7. Now plot $(\rho^{\rm l} + \rho^{\rm g})/2$ against $T$. Show that this line intersects the $\rho^{\rm l}$ and $\rho^{\rm g}$ curves at $T = T_{\rm c}$.

**9-14.** Use the expressions given in the previous problem to plot $(\rho^{\rm l} - \rho^{\rm g})/2$ against $(T_{\rm c} - T)^{1/3}$. Do you get a reasonably straight line? If not, determine the value of the exponent of $(T_{\rm c} - T)$ that gives the best straight line.

**9-15.** The molar enthalpy of vaporization of ethane can be expressed as

$$\Delta_{\rm vap}\overline{H}(T)/{\rm kJ\cdot mol^{-1}} = \sum_{j=1}^{6} A_j x^j$$

where $A_1 = 12.857$, $A_2 = 5.409$, $A_3 = 33.835$, $A_4 = -97.520$, $A_5 = 100.849$, $A_6 = -37.933$, and $x = (T_{\rm c} - T)^{1/3}/(T_{\rm c} - T_{\rm tp})^{1/3}$ where the critical temperature $T_{\rm c} = 305.4$ K and the triple point temperature $T_{\rm tp} = 90.35$ K. Plot $\Delta_{\rm vap}\overline{H}(T)$ versus $T$ and show that the curve is similar to that of Figure 9.8.

**9-16.** Fit the following data for argon to a fifth degree in $T$. Use your result to determine the critical temperature. Repeat using a fifth-degree polynomial.

| $T$/K | $\Delta_{\rm vap}\overline{H}$/J·mol$^{-1}$ | $T$/K | $\Delta_{\rm vap}\overline{H}$/J·mol$^{-1}$ |
|---|---|---|---|
| 83.80 | 6573.8 | 122.0 | 4928.7 |
| 86.0 | 6508.4 | 126.0 | 4665.0 |
| 90.0 | 6381.8 | 130.0 | 4367.7 |
| 94.0 | 6245.2 | 134.0 | 4024.7 |
| 98.0 | 6097.7 | 138.0 | 3618.8 |
| 102.0 | 5938.8 | 142.0 | 3118.2 |
| 106.0 | 5767.6 | 146.0 | 2436.3 |
| 110.0 | 5583.0 | 148.0 | 1944.5 |
| 114.0 | 5383.5 | 149.0 | 1610.2 |
| 118.0 | 5166.5 | 150.0 | 1131.5 |

**9-17.** Use the following data for methanol at one atm to plot $\overline{G}$ versus $T$ around the normal boiling point (337.668 K). What is the value of $\Delta_{\text{vap}}\overline{H}$?

| $T/\text{K}$ | $\overline{H}/\text{kJ}\cdot\text{mol}^{-1}$ | $\overline{S}/\text{J}\cdot\text{mol}^{-1}\cdot\text{K}^{-1}$ |
|---|---|---|
| 240 | 4.7183 | 112.259 |
| 280 | 7.7071 | 123.870 |
| 300 | 9.3082 | 129.375 |
| 320 | 10.9933 | 134.756 |
| 330 | 11.8671 | 137.412 |
| 337.668 | 12.5509 | 139.437 |
| 337.668 | 47.8100 | 243.856 |
| 350 | 48.5113 | 245.937 |
| 360 | 49.0631 | 247.492 |
| 380 | 50.1458 | 250.419 |
| 400 | 51.2257 | 253.189 |

**9-18.** In this problem, we will sketch $\overline{G}$ versus $P$ for the solid, liquid, and gaseous phases for a generic ideal substance as in Figure 9.11. Let $\overline{V}^{\text{s}} = 0.600$, $\overline{V}^{\text{l}} = 0.850$, and $RT = 2.5$, in arbitrary units. Now show that

$$\overline{G}^{\text{s}} = 0.600(P - P_0) + \overline{G}_0^{\text{s}}$$
$$\overline{G}^{\text{l}} = 0.850(P - P_0) + \overline{G}_0^{\text{l}}$$

and

$$\overline{G}^{\text{g}} = 2.5 \ln(P/P_0) + \overline{G}_0^{\text{g}}$$

where $P_0 = 1$ and $\overline{G}_0^{\text{s}}$, $\overline{G}_0^{\text{l}}$, and $\overline{G}_0^{\text{g}}$ are the respective zeros of energy. Show that if we (arbitrarily) choose the solid and liquid phases to be in equilibrium at $P = 2.00$ and the liquid and gaseous phases to be in equilibrium at $P = 1.00$, then we obtain

$$\overline{G}_0^{\text{s}} - \overline{G}_0^{\text{l}} = 0.250$$

and

$$\overline{G}_0^{\text{l}} = \overline{G}_0^{\text{g}}$$

from which we obtain (by adding these two results)

$$\overline{G}_0^{\text{s}} - \overline{G}_0^{\text{g}} = 0.250$$

Now we can express $\overline{G}^{\text{s}}$, $\overline{G}^{\text{l}}$, and $\overline{G}^{\text{g}}$ in terms of a common zero of energy, $\overline{G}_0^{\text{g}}$, which we must do to compare them with each other and to plot them on the same graph. Show that

$$\overline{G}^{\text{s}} - \overline{G}_0^{\text{g}} = 0.600(P - 1) + 0.250$$
$$\overline{G}^{\text{l}} - \overline{G}_0^{\text{g}} = 0.850(P - 1)$$
$$\overline{G}^{\text{g}} - \overline{G}_0^{\text{g}} = 2.5 \ln P$$

Plot these on the same graph from $P = 0.100$ to $3.00$ and compare your result with Figure 9.11.

**9-19.** In this problem, we will demonstrate that entropy always increases when there is a material flow from a region of higher concentration to one of lower concentration. (Compare with Problems 8–41 and 8–42.) Consider a two-compartment system enclosed by rigid, impermeable, adiabatic walls, and let the two compartments be separated by a rigid, insulating, but permeable wall. We assume that the two compartments are in equilibrium but that they are not in equilibirum with each other. Show that

$$U_1 = \text{constant}, \qquad U_2 = \text{constant}, \qquad V_1 = \text{constant}, \qquad V_2 = \text{constant},$$

and

$$n_1 + n_2 = \text{constant}$$

for this system. Now show that

$$dS = \frac{dU}{T} + \frac{P}{T}dV - \frac{\mu}{T}dn$$

in general, and that

$$\begin{aligned} dS &= \left(\frac{\partial S_1}{\partial n_1}\right) dn_1 + \left(\frac{\partial S_2}{\partial n_2}\right) dn_2 \\ &= dn_1 \left(\frac{\mu_2}{T} - \frac{\mu_1}{T}\right) \geq 0 \end{aligned}$$

for this system. Use this result to discuss the direction of a (isothermal) material flow under a chemical potential difference.

**9-20.** Determine the value of $dT/dP$ for water at its normal boiling point of 373.15 K given that the molar enthalpy of vaporization is 40.65 $\text{kJ}\cdot\text{mol}^{-1}$, and the densities of the liquid and vapor are 0.9584 $\text{g}\cdot\text{mL}^{-1}$ and 0.6010 $\text{g}\cdot\text{L}^{-1}$, respectively. Estimate the boiling point of water at 2 atm.

**9-21.** The orthobaric densities of liquid and gaseous ethyl acetate are 0.826 $\text{g}\cdot\text{mL}^{-1}$ and 0.00319 $\text{g}\cdot\text{mL}^{-1}$, respectively, at its normal boiling point (77.11°C). The rate of change of vapor pressure with temperature is 9.0 $\text{torr}\cdot\text{K}^{-1}$ at the normal boiling point. Estimate the molar enthalpy of vaporization of ethyl acetate at its normal boiling point.

**9-22.** The vapor pressure of mercury from 400°C to 1300°C can be expressed by

$$\ln(P/\text{torr}) = -\frac{7060.7\ \text{K}}{T} + 17.85$$

The density of the vapor at its normal boiling point is 3.82 $\text{g}\cdot\text{L}^{-1}$ and that of the liquid is 12.7 $\text{g}\cdot\text{mL}^{-1}$. Estimate the molar enthalpy of vaporization of mercury at its normal boiling point.

**9-23.** The pressures at the solid-liquid coexistence boundary of propane are given by the empirical equation

$$P = -718 + 2.38565T^{1.283}$$

where $P$ is in bars and $T$ is in kelvins. Given that $T_{\text{fus}} = 85.46$ K and $\Delta_{\text{fus}}\overline{H} = 3.53\ \text{kJ}\cdot\text{mol}^{-1}$, calculate $\Delta_{\text{fus}}\overline{V}$ at 85.46 K.

**9-24.** Use the vapor pressure data given in Problem 9–7 and the density data given in Problem 9–13 to calculate $\Delta_{\text{vap}}\overline{H}$ for methanol from the triple point (175.6 K) to the critical point (512.6 K). Plot your result.

**9-25.** Use the result of the previous problem to plot $\Delta_{\text{vap}}\overline{S}$ of methanol from the triple point to the critical point.

**9-26.** Use the vapor pressure data for methanol given in Problem 9–7 to plot $\ln P$ against $1/T$. Using your calculations from Problem 9–24, over what temperature range do you think the Clausius-Clapeyron equation will be valid?

**9-27.** The molar enthalpy of vaporization of water is $40.65\ \text{kJ}\cdot\text{mol}^{-1}$ at its normal boiling point. Use the Clausius-Clapeyron equation to calculate the vapor pressure of water at 110°C. The experimental value is 1075 torr.

**9-28.** The vapor pressure of benzaldehyde is 400 torr at 154°C and its normal boiling point is 179°C. Estimate its molar enthalpy of vaporization. The experimental value is $42.50\ \text{kJ}\cdot\text{mol}^{-1}$.

**9-29.** Use the following data to estimate the normal boiling point and the molar enthalpy of vaporization of lead.

| $T$/K | 1500 | 1600 | 1700 | 1800 | 1900 |
|---|---|---|---|---|---|
| $P$/torr | 19.72 | 48.48 | 107.2 | 217.7 | 408.2 |

**9-30.** The vapor pressure of solid iodine is given by

$$\ln(P/\text{atm}) = -\frac{8090.0\ \text{K}}{T} - 2.013\ln(T/\text{K}) + 32.908$$

Use this equation to calculate the normal sublimation temperature and the molar enthalpy of sublimation of $I_2(s)$ at 25°C. The experimental value of $\Delta_{\text{sub}}\overline{H}$ is $62.23\ \text{kJ}\cdot\text{mol}^{-1}$.

**9-31.** Fit the following vapor pressure data of ice to an equation of the form

$$\ln P = -\frac{a}{T} + b\ln T + cT + d$$

where $T$ is temperature in kelvins. Use your result to determine the molar enthalpy of sublimation of ice at 0°C.

| $t/°C$ | $P/\text{torr}$ | $t/°C$ | $P/\text{torr}$ |
|---|---|---|---|
| −10.0 | 1.950 | −4.8 | 3.065 |
| − 9.6 | 2.021 | −4.4 | 3.171 |
| − 9.2 | 2.093 | −4.0 | 3.280 |
| − 8.8 | 2.168 | −3.6 | 3.393 |
| − 8.4 | 2.246 | −3.2 | 3.509 |
| − 8.0 | 2.326 | −2.8 | 3.360 |
| − 7.6 | 2.408 | −2.4 | 3.753 |
| − 7.2 | 2.493 | −2.0 | 3.880 |
| − 6.8 | 2.581 | −1.6 | 4.012 |
| − 6.4 | 2.672 | −1.2 | 4.147 |
| − 6.0 | 2.765 | −0.8 | 4.287 |
| − 5.6 | 2.862 | −0.4 | 4.431 |
| − 5.2 | 2.962 | 0.0 | 4.579 |

**9-32.** The following table gives the vapor pressure data for liquid palladium as a function of temperature:

| $T/\text{K}$ | $P/\text{bar}$ |
|---|---|
| 1587 | $1.002 \times 10^{-9}$ |
| 1624 | $2.152 \times 10^{-9}$ |
| 1841 | $7.499 \times 10^{-8}$ |

Estimate the molar enthalpy of vaporization of palladium.

**9-33.** The sublimation pressure of $CO_2$ at 138.85 K and 158.75 K is $1.33 \times 10^{-3}$ bar and $2.66 \times 10^{-2}$ bar, respectively. Estimate the molar enthalpy of sublimation of $CO_2$.

**9-34.** The vapor pressures of solid and liquid hydrogen iodide can be expressed empirically as

$$\ln(P^{\text{s}}/\text{torr}) = -\frac{2906.2\text{ K}}{T} + 19.020$$

and

$$\ln(P^{\text{l}}/\text{torr}) = -\frac{2595.7\text{ K}}{T} + 17.572$$

Calculate the ratio of the slopes of the solid-gas curve and the liquid-gas curve at the triple point.

**9-35.** Given that the normal melting point, the critical temperature, and the critical pressure of hydrogen iodide are 222 K, 424 K and 82.0 atm, respectively, use the data in the previous problem to sketch the phase diagram of hydrogen iodide.

**9-36.** Consider the phase change

$$\text{C(graphite)} \rightleftharpoons \text{C(diamond)}$$

Given that $\Delta_r G^\circ/\text{J}\cdot\text{mol}^{-1} = 1895 + 3.363T$, calculate $\Delta_r H^\circ$ and $\Delta_r S^\circ$. Calculate the pressure at which diamond and graphite are in equilibrium with each other at 25°C. Take the density of diamond and graphite to be 3.51 g·cm$^{-3}$ and 2.25 g·cm$^{-3}$, respectively. Assume that both diamond and graphite are incompressible.

**9-37.** Use Equation 9.36 to calculate $\mu^\circ - E_0$ for Kr(g) at 298.15 K. The literature value is $-42.72$ kJ·mol$^{-1}$.

**9-38.** Show that Equations 9.30 and 9.32 for $\mu(T, P)$ for a monatomic ideal gas are equivalent to using the relation $\overline{G} = \overline{H} - T\overline{S}$ with $\overline{H} = 5RT/2$ and $S$ given by Equation 6.45.

**9-39.** Use Equation 9.37 and the molecular parameters in Table 4.2 to calculate $\mu^\circ - E_0$ for $N_2$(g) at 298.15 K. The literature value is $-48.46$ kJ·mol$^{-1}$.

**9-40.** Use Equation 9.37 and the molecular parameters in Table 4.2 to calculate $\mu^\circ - E_0$ for CO(g) at 298.15 K. The literature value is $-50.26$ kJ·mol$^{-1}$.

**9-41.** Use Equation 4.60 [without the factor of $\exp(D_e/k_B T)$] and the molecular parameters in Table 4.4 to calculate $\mu^\circ - E_0$ for $CH_4$(g) at 298.15 K. The literature value is $-45.51$ kJ·mol$^{-1}$.

**9-42.** When we refer to the equilibrium vapor pressure of a liquid, we tacitly assume that some of the liquid has evaporated into a vacuum and that equilibrium is then achieved. Suppose, however, that we are able by some means to exert an additional pressure on the surface of the liquid. One way to do this is to introduce an insoluble, inert gas into the space above the liquid. In this problem, we will investigate how the equilibrium vapor pressure of a liquid depends upon the total pressure exerted on it.

Consider a liquid and a vapor in equilibrium with each other, so that $\mu^\text{l} = \mu^\text{g}$. Show that

$$\overline{V}^\text{l} dP^\text{l} = \overline{V}^\text{g} dP^\text{g}$$

because the two phases are at the same temperature. Assuming that the vapor may be treated as an ideal gas and that $\overline{V}^\text{l}$ does not vary appreciably with pressure, show that

$$\ln \frac{P^\text{g}(\text{at } P^\text{l} = P)}{P^\text{g}(\text{at } P^\text{l} = 0)} = \frac{\overline{V}^\text{l} P^\text{l}}{RT}$$

Use this equation to calculate the vapor pressure of water at a total pressure of 10.0 atm at 25°C. Take $P^\text{g}$ (at $P^\text{l} = 0$) = 0.0313 atm.

**9-43.** Using the fact that the vapor pressure of a liquid does not vary appreciably with the total pressure, show that the final result of the previous problem can be written as

$$\frac{\Delta P^\text{g}}{P^\text{g}} = \frac{\overline{V}^\text{l} P^\text{l}}{RT}$$

*Hint*: Let $P^\text{g}(\text{at } P = P^\text{l}) = P^\text{g}(\text{at } P = 0) + \Delta P$ and use the fact that $\Delta P$ is small. Calculate $\Delta P$ for water at a total pressure of 10.0 atm at 25°C. Compare your answer with the one you obtained in the previous problem.

**9-44.** In this problem, we will show that the vapor pressure of a droplet is not the same as the vapor pressure of a relatively large body of liquid. Consider a spherical droplet of liquid of radius $r$ in equilibrium with a vapor at a pressure $P$, and a flat surface of the same liquid in equilibrium with a vapor at a pressure $P_0$. Show that the change in Gibbs energy for the isothermal transfer of $dn$ moles of the liquid from the flat surface to the droplet is

$$dG = dnRT \ln \frac{P}{P_0}$$

This change in Gibbs energy is due to the change in surface energy of the droplet (the change in surface energy of the large, flat surface is negligible). Show that

$$dnRT \ln \frac{P}{P_0} = \gamma dA$$

where $\gamma$ is the surface tension of the liquid and $dA$ is the change in the surface area of a droplet. Assuming the droplet is spherical, show that

$$dn = \frac{4\pi r^2 dr}{\overline{V}^{\,1}}$$

$$dA = 8\pi r dr$$

and finally that

$$\ln \frac{P}{P_0} = \frac{2\gamma \overline{V}^{\,1}}{rRT} \qquad (1)$$

Because the right side is positive, we see that the vapor pressure of a droplet is greater than that of a planar surface. What if $r \to \infty$?

**9-45.** Use Equation 1 of Problem 9–44 to calculate the vapor pressure at 25°C of droplets of water of radius $1.0 \times 10^{-5}$ cm. Take the surface tension of water to be $7.20 \times 10^{-4}$ J·m$^{-2}$.

**9-46.** Figure 9.15 shows reduced pressure, $P_R$, plotted against reduced volume, $\overline{V}_R$, for the van der Waals equation at a reduced temperature, $T_R$, of 0.85. The so-called van der Waals loop apparent in the figure will occur for any reduced temperature less than unity and is a consequence of the simplified form of the van der Waals equation. It turns out that any analytic equation of state (one that can be written as a Maclaurin expansion in the reduced density, $1/\overline{V}_R$) will give loops for subcritical temperatures ($T_R < 1$). The correct behavior as the pressure is increased is given by the path abdfg in Figure 9.15. The horizontal region bdf, not given by the van der Waals equation, represents the condensation of the gas to a liquid at a fixed pressure. We can draw the horizontal line (called a *tie line*) at the correct position by recognizing that the chemical potentials of the liquid and the vapor must be equal at the points b and f. Using this requirement, Maxwell showed that the horizontal line representing condensation should be drawn such that the areas of the loops above and below the line must be equal. To prove *Maxwell's equal-area construction rule*, integrate $(\partial \mu / \partial P)_T = \overline{V}$ by parts along the path bcdef and use the fact that $\mu^1$ (the value of $\mu$ at point $f$) $= \mu^g$ (the value of $\mu$ at point $b$) to obtain

$$\mu^1 - \mu^g = P_0(\overline{V}^{\,1} - \overline{V}^{\,g}) - \int_{\text{bcdef}} P d\overline{V}$$

$$= \int_{\text{bcdef}} (P_0 - P) d\overline{V}$$

where $P_0$ is the pressure corresponding to the tie line. Interpret this result.

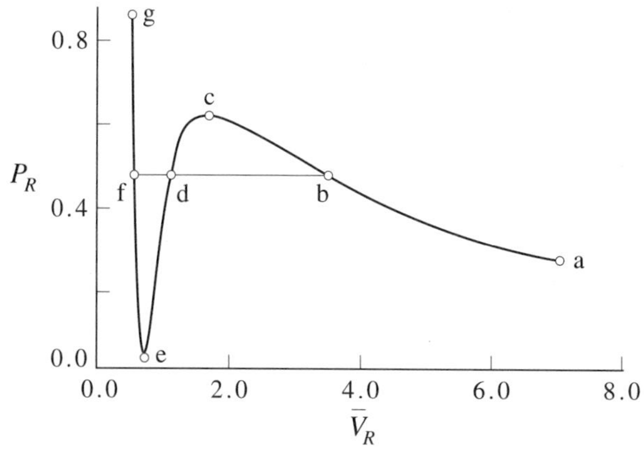


**FIGURE 9.15**
A plot of reduced pressure, $P_R$, versus reduced volume, $\overline{V}_R$, for the van der Waals equation at a reduced temperature, $T_R$, of 0.85.

**9-47.** The isothermal compressibility, $\kappa_T$, is defined by

$$\kappa_T = -\frac{1}{V}\left(\frac{\partial V}{\partial P}\right)_T$$

Because $(\partial P/\partial V)_T = 0$ at the critical point, $\kappa_T$ diverges there. A question that has generated a great deal of experimental and theoretical research is the question of the manner in which $\kappa_T$ diverges as $T$ approaches $T_c$. Does it diverge as $\ln(T - T_c)$ or perhaps as $(T - T_c)^{-\gamma}$ where $\gamma$ is some *critical exponent*? An early theory of the behavior of thermodynamic functions such as $\kappa_T$ very near the critical point was proposed by van der Waals, who predicted that $\kappa_T$ diverges as $(T - T_c)^{-1}$. To see how van der Waals arrived at this prediction, we consider the (double) Taylor expansion of the pressure $P(\overline{V}, T)$ about $T_c$ and $V_c$:

$$P(\overline{V}, T) = P(\overline{V}_c, T_c) + (T - T_c)\left(\frac{\partial P}{\partial T}\right)_c + \frac{1}{2}(T - T_c)^2\left(\frac{\partial^2 P}{\partial T^2}\right)_c$$
$$+ (T - T_c)(\overline{V} - \overline{V}_c)\left(\frac{\partial^2 P}{\partial \overline{V}\partial T}\right)_c + \frac{1}{6}(\overline{V} - \overline{V}_c)^3\left(\frac{\partial^3 P}{\partial \overline{V}^3}\right)_c + \cdots$$

Why are there no terms in $(\overline{V} - \overline{V}_c)$ or $(\overline{V} - \overline{V}_c)^2$? Write this Taylor series as

$$P = P_c + a(T - T_c) + b(T - T_c)^2 + c(T - T_c)(\overline{V} - \overline{V}_c) + d(\overline{V} - \overline{V}_c)^3 + \cdots$$

Now show that

$$\left(\frac{\partial P}{\partial \overline{V}}\right)_T = c(T - T_c) + 3d(\overline{V} - \overline{V}_c)^2 + \cdots \qquad \begin{pmatrix} T \to T_c \\ \overline{V} \to \overline{V}_c \end{pmatrix}$$

and that

$$\kappa_T = \frac{-1/\overline{V}}{c(T - T_c) + 3d(\overline{V} - \overline{V}_c)^2 + \cdots}$$

Now let $\overline{V} = \overline{V}_c$ to obtain

$$\kappa_T \propto \frac{1}{T - T_c} \qquad T \to (T_c)$$

Accurate experimental measurements of $\kappa_T$ as $T \to T_c$ suggest that $\kappa_T$ diverges a little more strongly than $(T - T_c)^{-1}$. In particular, it is found that $\kappa_T \to (T - T_c)^{-\gamma}$ where $\gamma = 1.24$. Thus, the theory of van der Waals, although qualitatively correct, is not quantitatively correct.

**9-48.** We can use the ideas of the previous problem to predict how the difference in the densities ($\rho^{\text{l}}$ and $\rho^{\text{g}}$) of the coexisting liquid and vapor states (*orthobaric densities*) behave as $T \to T_c$. Substitute

$$P = P_c + a(T - T_c) + b(T - T_c)^2 + c(T - T_c)(\overline{V} - \overline{V}_c) + d(\overline{V} - \overline{V}_c)^3 + \cdots \qquad (1)$$

into the Maxwell equal-area construction (Problem 9–46) to get

$$P_0 = P_c + a(T - T_c) + b(T - T_c)^2 + \frac{c}{2}(T - T_c)(\overline{V}^{\text{l}} + \overline{V}^{\text{g}} - 2\overline{V}_c)$$
$$+ \frac{d}{4}[(\overline{V}^{\text{g}} - \overline{V}_c)^2 + (\overline{V}^{\text{l}} - \overline{V}_c)^2](\overline{V}^{\text{l}} + \overline{V}^{\text{g}} - 2\overline{V}_c) + \cdots \qquad (2)$$

For $P < P_c$, Equation 1 gives loops and so has three roots, $\overline{V}^{\text{l}}$, $\overline{V}_c$, and $\overline{V}^{\text{g}}$ for $P = P_0$. We can obtain a first approximation to these roots by assuming that $\overline{V}_C \approx \frac{1}{2}(\overline{V}^{\text{l}} + \overline{V}^{\text{g}})$ in Equation 2 and writing

$$P_0 = P_c + a(T - T_c) + b(T - T_c)^2$$

To this approximation, the three roots to Equation 1 are obtained from

$$d(\overline{V} - \overline{V}_c)^3 + c(T - T_c)(\overline{V} - \overline{V}_c) = 0$$

Show that the three roots are

$$\overline{V}_1 = \overline{V}^{\text{l}} = \overline{V}_c - \left(\frac{c}{d}\right)^{1/2}(T_c - T)^{1/2}$$
$$\overline{V}_2 = \overline{V}_c$$
$$\overline{V}_3 = \overline{V}^{\text{g}} = \overline{V}_c + \left(\frac{c}{d}\right)^{1/2}(T_c - T)^{1/2}$$

Now show that

$$\overline{V}^{\text{g}} - \overline{V}^{\text{l}} = 2\left(\frac{c}{d}\right)^{1/2}(T_c - T)^{1/2} \qquad \begin{pmatrix} T < T_c \\ T \to T_c \end{pmatrix}$$

and that this equation is equivalent to

$$\rho^{\text{l}} - \rho^{\text{g}} \longrightarrow A(T_c - T)^{1/2} \qquad \begin{pmatrix} T < T_c \\ T \to T_c \end{pmatrix}$$

Thus, the van der Waals theory predicts that the critical exponent in this case is 1/2. It has been shown experimentally that

$$\rho^{\mathrm{l}} - \rho^{\mathrm{g}} \longrightarrow A(T_{\mathrm{c}} - T)^{\beta}$$

where $\beta = 0.324$. Thus, as in the previous problem, although qualitatively correct, the van der Waals theory is not quantitatively correct.

**9-49.** The following data give the temperature, the vapor pressure, and the density of the coexisting vapor phase of butane. Use the van der Waals equation and the Redlich-Kwong equation to calculate the vapor pressure and compare your result with the experimental values given below.

| $T/\mathrm{K}$ | $P/\mathrm{bar}$ | $\rho^{\mathrm{g}}/\mathrm{mol\cdot L^{-1}}$ |
|---|---|---|
| 200 | 0.0195 | 0.00117 |
| 210 | 0.0405 | 0.00233 |
| 220 | 0.0781 | 0.00430 |
| 230 | 0.1410 | 0.00746 |
| 240 | 0.2408 | 0.01225 |
| 250 | 0.3915 | 0.01924 |
| 260 | 0.6099 | 0.02905 |
| 270 | 0.9155 | 0.04239 |
| 280 | 1.330 | 0.06008 |

**9-50.** The following data give the temperature, the vapor pressure, and the density of the coexisting vapor phase of benzene. Use the van der Waals equation and the Redlich-Kwong equation to calculate the vapor pressure and compare your result with the experimental values given below. Use Equations 2.17 and 2.18 with $T_{\mathrm{c}} = 561.75$ K and $P_{\mathrm{c}} = 48.7575$ bar to calculate the van der Waals parameters and the Redlich-Kwong parameters.

| $T/\mathrm{K}$ | $P/\mathrm{bar}$ | $\rho^{\mathrm{g}}/\mathrm{mol\cdot L^{-1}}$ |
|---|---|---|
| 290.0 | 0.0860 | 0.00359 |
| 300.0 | 0.1381 | 0.00558 |
| 310.0 | 0.2139 | 0.00839 |
| 320.0 | 0.3205 | 0.01223 |
| 330.0 | 0.4666 | 0.01734 |
| 340.0 | 0.6615 | 0.02399 |
| 350.0 | 0.9161 | 0.03248 |

**9-51.** In Problems 8–57 to 8–63, we considered only cases with constant $n$. Use the approach of Problem 8–57 to repartition $n$ between two initially identical compartments to show that $(\partial^2 A/\partial n^2)_{V,T} > 0$ and that $(\partial \mu/\partial n)_{V,T} > 0$.

**Joel Hildebrand** was born in Camden, NJ, on November 16, 1881, and died in 1983. He received his Ph.D. in chemistry from the University of Pennsylvania in 1906. After spending a year at the University of Berlin with Nernst, he returned to the University of Pennsylvania as an instructor. In 1913, he joined the Department of Chemistry at the University of California at Berkeley, where he stayed for the remainder of his life. Although he officially retired in 1952, he remained professionally active until his death, publishing his last paper in 1981. Hildebrand made significant contributions to the fields of liquids and nonelectrolyte solutions. He retained a long interest in deviations from ideal solutions (Raoult's law) and the theory of regular solutions. His books, *The Solubility of Nonelectrolytes* and *Regular Solutions*, published with Robert Scott, were standard references in the field. Hildebrand was a famed, excellent teacher of general chemistry at Berkeley. His general chemistry text, *Principles of Chemistry*, influenced other schools to place greater emphasis on principles and less on the memorization of specific material in the teaching of general chemistry. Hildebrand was a great lover of the outdoors and especially enjoyed skiing and camping. He managed the U.S. Olympic Ski Team in 1936, was President of the Sierra Club from 1937 to 1940, and wrote a book on camping with his daughter Louise, *Camp Catering* or, *How to rustle grub for hikers, campers, mountaineers, canoeists, hunters, skiers, and fishermen.*

# CHAPTER 10

# Solutions I: Liquid-Liquid Solutions

In this and the next chapter, we will apply our thermodynamic principles to solutions. This chapter focuses on solutions that consist of two volatile liquids, such as alcohol–water solutions. We will first discuss partial molar quantities, which provide the most convenient set of thermodynamic variables to describe solutions. This discussion will lead to the Gibbs–Duhem equation, which gives us a relation between the change in the properties of one component of a solution in terms of the change in the properties of the other component. The simplest model of a solution is an ideal solution, in which both components obey Raoult's law over the entire composition range. Although a few solutions behave almost ideally, most solutions are not ideal. Just as nonideal gases can be described in terms of fugacity, nonideal solutions can be described in terms of a quantity called activity. Activity must be calculated with respect to a specific standard state, and in Section 10–8 we introduce two commonly-used standard states: a solvent, or Raoult's law standard state, and a solute, or Henry's law standard state.

## 10–1. Partial Molar Quantities Are Important Thermodynamic Properties of Solutions

Up to this point, we have discussed the thermodynamics of only one-component systems. We will now discuss the thermodynamics of multicomponent systems, although, for simplicity, we will discuss only systems of two components. Most of the concepts and results we will develop are applicable to multicomponent systems. Let's consider a solution consisting of $n_1$ moles of component 1 and $n_2$ moles of component 2. The Gibbs energy of this solution is a function of $T$ and $P$ and the two mole numbers $n_1$ and $n_2$. We emphasize this dependence of $G$ on these variables by writing $G = G(T, P, n_1, n_2)$. The total derivative of $G$ is given by

$$dG = \left(\frac{\partial G}{\partial T}\right)_{P,n_1,n_2} dT + \left(\frac{\partial G}{\partial P}\right)_{T,n_1,n_2} dP + \left(\frac{\partial G}{\partial n_1}\right)_{P,T,n_2} dn_1 + \left(\frac{\partial G}{\partial n_2}\right)_{P,T,n_1} dn_2 \tag{10.1}$$

If the composition of the solution is fixed, so that $dn_1 = dn_2 = 0$, then Equation 10.1 is the same as Equation 8.30, and we have

$$\left(\frac{\partial G}{\partial T}\right)_{P,n_1,n_2} = -S(P, T, n_1, n_2)$$

and

$$\left(\frac{\partial G}{\partial P}\right)_{T,n_1,n_2} = V(P, T, n_1, n_2)$$

As in the previous chapter, the partial derivatives of $G$ with respect to mole numbers are called chemical potentials, or partial molar Gibbs energies. The standard notation for chemical potential is $\mu$, so we can write Equation 10.1 as

$$dG = -SdT + VdP + \mu_1 dn_1 + \mu_2 dn_2 \tag{10.2}$$

where

$$\mu_j = \mu_j(T, P, n_1, n_2) = \left(\frac{\partial G}{\partial n_j}\right)_{T,P,n_{i\neq j}} = \overline{G}_j \tag{10.3}$$

We will see that the chemical potential of each component in the solution plays a central role in determining the thermodynamic properties of the solution.

Other extensive thermodynamic variables have associated partial molar values, although only the partial molar Gibbs energy is given a special symbol and name. For example, $(\partial S/\partial n_j)_{T,P,n_{i\neq j}}$ is called the partial molar entropy and is denoted by $\overline{S}_j$, and $(\partial V/\partial n_j)_{T,P,n_{i\neq j}}$ is called the partial molar volume and is denoted by $\overline{V}_j$. Generally, if $Y = Y(T, P, n_1, n_2)$ is some extensive thermodynamic property, then its associated partial molar quantity, denoted by $\overline{Y}_j$, is by definition

$$\overline{Y}_j = \overline{Y}_j(T, P, n_1, n_2) = \left(\frac{\partial Y}{\partial n_j}\right)_{T,P,n_{i\neq j}} \tag{10.4}$$

Physically, the partial molar quantity $\overline{Y}_j$ is a measure of how $Y$ changes when $n_j$ is changed while keeping $T$, $P$, and the other mole numbers fixed.

Partial molar quantities are intensive thermodynamic quantities. In fact, for a pure system, the chemical potential is just the Gibbs energy per mole. We can use the intensive property of partial molar quantities to derive one of the most important relations for solutions. As a concrete example, we will consider a *binary solution*, that is, one composed of two different liquids. The Gibbs energy of a binary solution (Equation 10.2) is

$$dG = -SdT + VdP + \mu_1 dn_1 + \mu_2 dn_2$$

At constant $T$ and $P$, we have

$$dG = \mu_1 dn_1 + \mu_2 dn_2 \tag{10.5}$$

Now, imagine that we increase the size of the system uniformly by means of a scale parameter $\lambda$ such that $dn_1 = n_1 d\lambda$ and $dn_2 = n_2 d\lambda$. Note that as we vary $\lambda$ from 0 to 1, the number of moles of components 1 and 2 varies from 0 to $n_1$ and 0 to $n_2$, respectively. Because $G$ depends extensively on $n_1$ and $n_2$, we must have that $dG = G d\lambda$. Therefore, the total Gibbs energy varies from 0 to some final value $G$ as $\lambda$ is varied. Introducing $d\lambda$ into Equation 10.5 gives

$$\int_0^1 G d\lambda = \int_0^1 n_1 \mu_1 d\lambda + \int_0^1 n_2 \mu_2 d\lambda$$

Because $G$, $n_1$, and $n_2$ are final values (and so do not depend upon $\lambda$) and $\mu_1$ and $\mu_2$ are intensive variables (and so do not depend upon the size parameter $\lambda$), we can write the above equation as

$$G \int_0^1 d\lambda = n_1 \mu_1 \int_0^1 d\lambda + n_2 \mu_2 \int_0^1 d\lambda$$

or, upon integration,

$$G(T, P, n_1, n_2) = \mu_1 n_1 + \mu_2 n_2 \tag{10.6}$$

Note that $G = \mu n$ for a one-component system, which shows once again that $\mu$ is the Gibbs energy per mole for a pure system, or more generally, that the partial molar quantity of any extensive thermodynamic quantity of a pure substance is its molar value.

Partial molar quantities have a particularly nice physical interpretation in terms of volume, for which the equivalent equation to Equation 10.6 would be

$$V(T, P, n_1, n_2) = \overline{V}_1 n_1 + \overline{V}_2 n_2 \tag{10.7}$$

Now, when 1-propanol and water are mixed, the final volume of the solution is not equal to the sum of the volumes of pure 1-propanol and water. We can use Equation 10.7 to calculate the final volume of a solution of any composition if we know the partial molar volumes of 1-propanol and water at that composition. Figure 10.1 shows the partial molar volumes of 1-propanol and water as a function of the mole fraction of 1-propanol in 1-propanol/water solutions at 20°C. We can use this figure to estimate the final volume of solution when 100 mL of 1-propanol is mixed with 100 mL of water at 20°C. The densities of 1-propanol and water at 20°C are 0.803 $\text{g}\cdot\text{mL}^{-1}$ and 0.998 $\text{g}\cdot\text{mL}^{-1}$, respectively. Using these densities, we see that 100 mL each of 1-propanol and water corresponds to a mole fraction of 1-propanol of 0.194. Referring

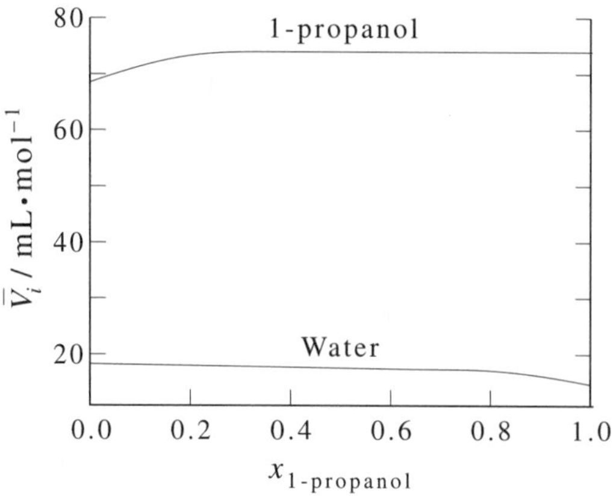


**FIGURE 10.1**
The partial molar volumes of 1-propanol and water in a 1-propanol/water solution at 20°C plotted against the mole fraction of 1-propanol in the solution.

to Figure 10.1, we see that this corresponds to roughly $\overline{V}_{\text{1-propanol}} = 72\text{ mL}\cdot\text{mol}^{-1}$ and $\overline{V}_{\text{water}} = 18\text{ mL}\cdot\text{mol}^{-1}$. Thus, the final volume of the solution is

$$\begin{aligned} V &= n_1\overline{V}_{\text{1-propanol}} + n_2\overline{V}_{\text{water}} \\ &= \left(\frac{80.3\text{ g}}{60.09\text{ g}\cdot\text{mol}^{-1}}\right)(72\text{ mL}\cdot\text{mol}^{-1}) + \left(\frac{99.8\text{ g}}{18.02\text{ g}\cdot\text{mol}^{-1}}\right)(18\text{ mL}\cdot\text{mol}^{-1}) \\ &= 196\text{ mL} \end{aligned}$$

compared with a total volume of 200 mL before mixing. Problems 10–8 through 10–12 involve the determination of partial molar volumes from solution data.

## 10–2. The Gibbs–Duhem Equation Relates the Change in the Chemical Potential of One Component of a Solution to the Change in the Chemical Potential of the Other

Most of our thermodynamic formulas for single-component systems (pure substances) have analogous formulas in terms of partial molar quantities. For example, if we start with $G = H - TS$ and differentiate with respect to $n_j$ keeping $T$, $P$, and $n_{i\neq j}$ fixed, we obtain

$$\left(\frac{\partial G}{\partial n_j}\right)_{T,P,n_{i\neq j}} = \left(\frac{\partial H}{\partial n_j}\right)_{T,P,n_{i\neq j}} - T\left(\frac{\partial S}{\partial n_j}\right)_{T,P,n_{i\neq j}}$$

or

$$\mu_j = \overline{G}_j = \overline{H}_j - T\overline{S}_j \qquad (10.8)$$

Furthermore, by using the fact that cross second partial derivatives are equal, we get

$$\overline{S}_j = \left(\frac{\partial S}{\partial n_j}\right)_{T,P,n_{i\neq j}} = \frac{\partial}{\partial n_j}\left(-\frac{\partial G}{\partial T}\right)_{P,n_i} = -\frac{\partial}{\partial T}\left(\frac{\partial G}{\partial n_j}\right)_{T,P,n_{i\neq j}} = -\left(\frac{\partial \mu_j}{\partial T}\right)_{P,n_i}$$

and

$$\overline{V}_j = \left(\frac{\partial V}{\partial n_j}\right)_{T,P,n_{i\neq j}} = \frac{\partial}{\partial n_j}\left(\frac{\partial G}{\partial P}\right)_{T,n_i} = \frac{\partial}{\partial P}\left(\frac{\partial G}{\partial n_j}\right)_{T,P,n_{i\neq j}} = \left(\frac{\partial \mu_j}{\partial P}\right)_{T,n_i}$$

If we substitute these two results into

$$d\mu_j = \left(\frac{\partial \mu_j}{\partial T}\right)_{P,n_i} dT + \left(\frac{\partial \mu_j}{\partial P}\right)_{T,n_i} dP$$

we obtain

$$d\mu_j = -\overline{S}_j dT + \overline{V}_j dP \tag{10.9}$$

which is an extension of Equation 8.30 to multicomponent systems.

---

**EXAMPLE 10–1**

Derive an equation for the temperature dependence of $\mu_j(T, P)$ in analogy with the Gibbs–Helmholtz equation (Equation 8.60).

SOLUTION: The Gibbs–Helmholtz equation is (Equation 8.60)

$$\left(\frac{\partial G/T}{\partial T}\right)_{P,n_i} = -\frac{H}{T^2}$$

Now differentiate with respect to $n_j$ and interchange the order of differentiation on the left side to get

$$\left(\frac{\partial \mu_j/T}{\partial T}\right)_P = -\frac{\overline{H}_j}{T^2}$$

where $\overline{H}_j$ is the partial molar enthalpy of component $j$.

---

We will now derive one of the most useful equations involving partial molar quantities. First we differentiate Equation 10.6

$$dG = \mu_1 dn_1 + \mu_2 dn_2 + n_1 d\mu_1 + n_2 d\mu_2$$

and subtract Equation 10.5 to get

$$n_1 d\mu_1 + n_2 d\mu_2 = 0 \qquad \text{(constant } T \text{ and } P\text{)} \tag{10.10}$$

If we divide both sides by $n_1 + n_2$, we have

$$x_1 d\mu_1 + x_2 d\mu_2 = 0 \qquad \text{(constant } T \text{ and } P\text{)} \qquad (10.11)$$

where $x_1$ and $x_2$ are mole fractions. Either of Equations 10.10 or 10.11 is called the *Gibbs–Duhem equation*. The Gibbs–Duhem equation tells us that if we know the chemical potential of one component as a function of composition, we can determine the other. For example, suppose we were to know that

$$\mu_2 = \mu_2^* + RT \ln x_2 \qquad 0 \le x_2 \le 1$$

over the whole range of $x_2$ (0 to 1). A superscript * is the IUPAC notation for a property of a pure substance, so in this equation, $\mu_2^* = \mu_2(x_2 = 1)$ is the chemical potential of pure component 2. We can differentiate $\mu_2$ with respect to $x_2$ and substitute into Equation 10.11 to get

$$\begin{aligned} d\mu_1 &= -\frac{x_2}{x_1} d\mu_2 = -RT\frac{x_2}{x_1} d\ln x_2 \\ &= -RT\frac{x_2}{x_1}\frac{dx_2}{x_2} = -RT\frac{dx_2}{x_1} \qquad (0 \le x_2 \le 1) \end{aligned}$$

But $dx_2 = -dx_1$ (because $x_1 + x_2 = 1$), so

$$d\mu_1 = RT\frac{dx_1}{x_1} \qquad (0 \le x_1 \le 1)$$

where $0 \le x_1 \le 1$ because $0 \le x_2 \le 1$. Now integrate both sides from $x_1 = 1$ (pure component 1) to arbitrary $x_1$ to get

$$\mu_1 = \mu_1^* + RT \ln x_1 \qquad (0 \le x_1 \le 1)$$

where $\mu_1^* = \mu_1(x_1 = 1)$. We will see later in this chapter that this result says that if one component of a binary solution obeys Raoult's law over the complete concentration range, the other component does also.

**EXAMPLE 10–2**

Derive a Gibbs–Duhem type of equation for the volume of a binary solution.

SOLUTION: We start with Equation 10.7, which is the analog of Equation 10.6

$$V(T, P, n_1, n_2) = n_1\overline{V}_1 + n_2\overline{V}_2$$

and differentiate (at constant $T$ and $P$) to obtain

$$dV = n_1 d\overline{V}_1 + \overline{V}_1 dn_1 + n_2 d\overline{V}_2 + \overline{V}_2 dn_2$$

Subtract the analog of Equation 10.5

$$dV = \overline{V}_1 dn_1 + \overline{V}_2 dn_2 \qquad \text{(constant } T \text{ and } P\text{)}$$

to obtain

$$n_1 d\overline{V}_1 + n_2 d\overline{V}_2 = 0 \qquad \text{(constant } T \text{ and } P\text{)}$$

This equation says that if we know the change in the partial molar volume of one component of a binary system over a range of composition, we can determine the change in the partial molar volume of the other component over the same range.

## 10–3. At Equilibrium, the Chemical Potential of Each Component Has the Same Value in Each Phase in Which the Component Appears

Consider a binary solution of two liquids that is in equilibrium with its vapor phase, which contains both components. Examples are a solution of 1-propanol and water or a solution of benzene and toluene, each in equilibrium with its vapor. We wish to generalize our treatment in the previous chapter, in which we treated a pure liquid in equilibrium with its vapor phase, and develop the criterion for equilibrium in a binary solution. The Gibbs energy of the solution and its vapor is

$$G = G^{\text{sln}} + G^{\text{vap}}$$

Let $n_1^{\text{sln}}$, $n_2^{\text{sln}}$ and $n_1^{\text{vap}}$, $n_2^{\text{vap}}$ be the mole numbers of each component in each phase. For generality, let $j$ denote either component 1 or 2, so $n_j$ denotes the number of moles of component $j$. Now suppose that $dn_j$ moles of component $j$ are transferred from the solution to the vapor at constant $T$ and $P$, so that $dn_j^{\text{vap}} = +dn_j$ and $dn_j^{\text{sln}} = -dn_j$. The accompanying change in the Gibbs energy is

$$\begin{aligned} dG &= dG^{\text{sln}} + dG^{\text{vap}} \\ &= \left(\frac{\partial G^{\text{sln}}}{\partial n_j^{\text{sln}}}\right)_{T,P,n_{i\neq j}} dn_j^{\text{sln}} + \left(\frac{\partial G^{\text{vap}}}{\partial n_j^{\text{vap}}}\right)_{T,P,n_{i\neq j}} dn_j^{\text{vap}} \\ &= \mu_j^{\text{sln}} dn_j^{\text{sln}} + \mu_j^{\text{vap}} dn_j^{\text{vap}} = (\mu_j^{\text{vap}} - \mu_j^{\text{sln}}) dn_j^{\text{vap}} \end{aligned}$$

If the transfer from the solution to the vapor occurs spontaneously, then $dG < 0$. Furthermore, $dn_j^{\text{vap}} > 0$, so $\mu_j^{\text{vap}}$ must be less than $\mu_j^{\text{sln}}$ in order that $dG < 0$. Therefore, molecules of component $j$ move spontaneously from the phase of higher chemical potential (solution) to that of lower chemical potential (vapor). Similarly, if $\mu_j^{\text{vap}} > \mu_j^{\text{sln}}$, then molecules of component $j$ move spontaneously from the vapor phase to the solution phase ($dn_j^{\text{vap}} < 0$). At equilibrium, where $dG = 0$, we have that

$$\mu_j^{\text{vap}} = \mu_j^{\text{sln}} \tag{10.12}$$

Equation 10.12 holds for each component. Although we have discussed a solution in equilibrium with its vapor phase, our choice of phases was arbitrary, so Equation 10.12 is valid for the equilibrium between any two phases in which component $j$ occurs.

The important result here is that Equation 10.12 says that the chemical potential of each component in the liquid solution phase can be measured by the chemical potential of that component in the vapor phase. If the pressure of the vapor phase is low enough that we can consider it to be ideal, then Equation 10.12 becomes

$$\mu_j^{\text{sln}} = \mu_j^{\text{vap}} = \mu_j^{\circ}(T) + RT \ln P_j \tag{10.13}$$

where the standard state is taken to be $P_j^{\circ} = 1$ bar. For *pure* component $j$, Equation 10.13 becomes

$$\mu_j^*(l) = \mu_j^*(\text{vap}) = \mu_j^{\circ}(T) + RT \ln P_j^* \tag{10.14}$$

where the superscript $^*$ represents pure (liquid) component $j$. Thus, for example, $\mu_j^*(l)$ is the chemical potential and $P_j^*$ is the vapor pressure of pure $j$. If we subtract Equation 10.14 from Equation 10.13, we obtain

$$\mu_j^{\text{sln}} = \mu_j^*(l) + RT \ln \frac{P_j}{P_j^*} \tag{10.15}$$

Equation 10.15 is a central equation in the study of binary solutions. Note that $\mu_j^{\text{sln}} \to \mu_j^*$ as $P_j \to P_j^*$. Strictly speaking, we should use fugacities (Section 8–8) instead of pressures in Equation 10.15, but usually the magnitudes of vapor pressures are such that pressures are quite adequate. For example, the vapor pressure of water at 293.15 K is 17.4 torr, or 0.0232 bar.

## 10–4. The Components of an Ideal Solution Obey Raoult's Law for All Concentrations

A few solutions have the property that the partial vapor pressure of each component is given by the simple equation

$$P_j = x_j P_j^* \tag{10.16}$$

Equation 10.16 is called *Raoult's law*, and a solution that obeys Raoult's law over the entire composition range is said to be an *ideal solution.*

The molecular picture behind an ideal binary solution is that the two types of molecules are randomly distributed throughout the solution. Such a distribution will occur if (1) the molecules are roughly the same size and shape, and (2) the intermolecular forces in the pure liquids 1 and 2 and in a mixture of 1 and 2 are all similar. We expect ideal-solution behavior only when the molecules of the two components are similar. For example, benzene and toluene, *o*-xylene and *p*-xylene, hexane and heptane, and bromoethane and iodoethane form essentially ideal solutions. Figure 10.2 depicts an ideal solution, in which the two types of molecules are randomly distributed. The mole fraction $x_j$ reflects the fraction of the solution surface that is occupied by

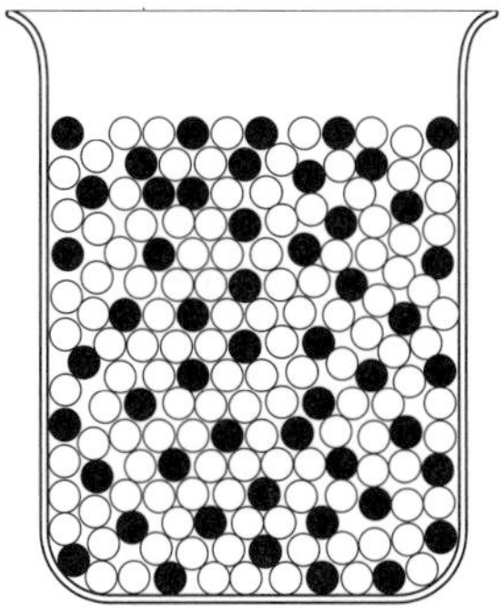

**FIGURE 10.2**
A molecular depiction of an ideal solution. The two types of molecules are distributed throughout the solution in a random manner.

$j$ molecules. Because the $j$ molecules on the surface are the molecules that can escape into the vapor phase, the partial pressure $P_j$ is just $x_j P_j^*$.

According to Raoult's law (Equation 10.16) and Equation 10.15, the chemical potential of component $j$ in the solution is given by

$$\mu_j^{\text{sln}} = \mu_j^*(l) + RT \ln x_j \tag{10.17}$$

Equation 10.17 also serves to define an ideal solution if it is valid for all values of $x_j$ ($0 \leq x_j \leq 1$). Furthermore, we showed in Section 10–2 that if one component obeys Equation 10.17 from $x_j = 0$ to $x_j = 1$, then so does the other.

The total vapor pressure over an ideal solution is given by

$$\begin{aligned} P_{\text{total}} &= P_1 + P_2 = x_1 P_1^* + x_2 P_2^* = (1 - x_2) P_1^* + x_2 P_2^* \\ &= P_1^* + x_2 (P_2^* - P_1^*) \end{aligned} \tag{10.18}$$

Therefore, a plot of $P_{\text{total}}$ against $x_2$ (or $x_1$) will be a straight line as shown in Figure 10.3.

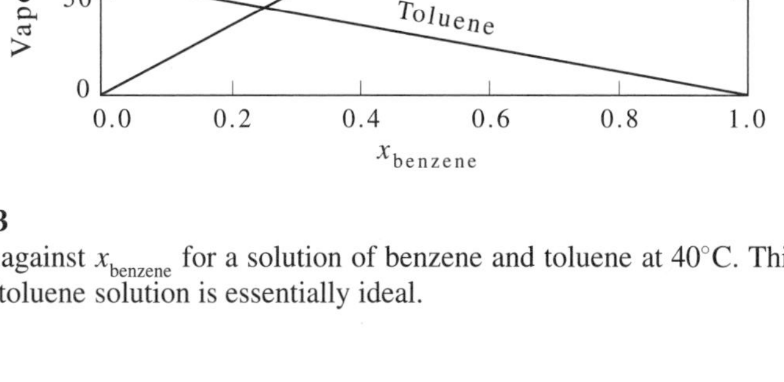

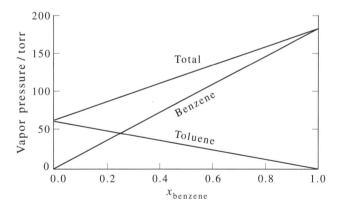

**FIGURE 10.3**
A plot of $P_{\text{total}}$ against $x_{\text{benzene}}$ for a solution of benzene and toluene at 40°C. This plot shows that a benzene/toluene solution is essentially ideal.

**EXAMPLE 10–3**

1-propanol and 2-propanol form essentially an ideal solution at all concentrations at 25°C. Letting the subscripts 1 and 2 denote 1-propanol and 2-propanol, respectively, and given that $P_1^* = 20.9$ torr and $P_2^* = 45.2$ torr at 25°C, calculate the total vapor pressure and the composition of the vapor phase at $x_2 = 0.75$.

SOLUTION: We use Equation 10.18:

$$\begin{aligned} P_{\text{total}}(x_2 = 0.75) &= x_1 P_1^* + x_2 P_2^* \\ &= (0.25)(20.9 \text{ torr}) + (0.75)(45.2 \text{ torr}) \\ &= 39.1 \text{ torr} \end{aligned}$$

Let $y_j$ denote the mole fraction of each component in the vapor phase. Then, by Dalton's law of partial pressures,

$$y_1 = \frac{P_1}{P_{\text{total}}} = \frac{x_1 P_1^*}{P_{\text{total}}} = \frac{(0.25)(20.9 \text{ torr})}{39.1 \text{ torr}} = 0.13$$

Similarly,

$$y_2 = \frac{P_2}{P_{\text{total}}} = \frac{x_2 P_2^*}{P_{\text{total}}} = \frac{(0.75)(45.2 \text{ torr})}{39.1 \text{ torr}} = 0.87$$

Note that $y_1 + y_2 = 1$. Also note that the vapor is richer than the solution in the more volatile component.

Problem 10–15 has you expand Example 10–3 by calculating $P_{\text{total}}$ as a function of $x_2$ (the mole fraction of 2-propanol in the liquid phase) and as a function of $y_2$ (the mole fraction of 2-propanol in the vapor phase), and then plotting $P_{\text{total}}$ against $x_2$ and $y_2$. The resulting plot, which is shown in Figure 10.4, is called a *pressure-composition diagram*. The upper curve shows the total vapor pressure as a function of the composition of the liquid phase (the liquid curve), and the lower curve shows the total vapor pressure as a function of the composition of the vapor phase (the vapor curve). Now let's see what happens when you start at the point $P_a$, $x_a$ in Figure 10.4 and lower the pressure. At the point $P_a$, $x_a$, the pressure exceeds the vapor pressure of the solution, so the region above the liquid curve consists of one (liquid) phase. As the pressure is lowered, we reach the point A, where liquid starts to vaporize. Along the line AB, the system consists of liquid and vapor in equilibrium with each other. At the point B, all the liquid has vaporized, and the region below the vapor curve consists of one (vapor) phase.

Let's consider the point C in the liquid-vapor region. Point C lies on a line connecting the composition of liquid ($x_2 = 0.75$) and vapor ($y_2 = 0.87$) phases that we calculated in Example 10–3. Such a line is called a *tie line*. The overall composition of the two-phase (liquid-vapor) system is $x_a$. We can determine the relative amounts

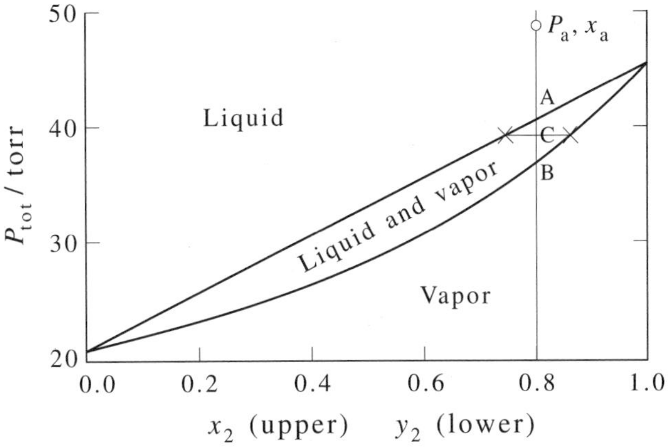


**FIGURE 10.4**
A pressure-composition diagram for a 1-propanol/2-propanol solution, which forms an essentially ideal solution at 25°C. This figure can be calculated using the approach in Example 10–3. The upper curve (called the liquid curve) represents $P_{\text{total}}$ versus $x_2$, the mole fraction of 2-propanol in the liquid phase, and the lower curve (called the vapor curve) represents $P_{\text{total}}$ versus $y_2$, the mole fraction of 2-propanol in the vapor phase. The two points marked by × represent the values of $x_2$ and $y_2$ from Example 10–3.

of liquid and vapor phase in the following way. The mole fractions in the liquid and vapor phases are

$$x_2 = \frac{n_2^{\text{l}}}{n_1^{\text{l}} + n_2^{\text{l}}} = \frac{n_2^{\text{l}}}{n^{\text{l}}} \quad \text{and} \quad y_2 = \frac{n_2^{\text{vap}}}{n_1^{\text{vap}} + n_2^{\text{vap}}} = \frac{n_2^{\text{vap}}}{n^{\text{vap}}}$$

where $n^{\text{vap}}$ and $n^{\text{l}}$ are the total number of moles in the vapor and liquid phases, respectively. The overall mole fraction at $x_a$ is given by the total number of moles of component 2 divided by the total number of moles

$$x_a = \frac{n_2^{\text{l}} + n_2^{\text{vap}}}{n^{\text{l}} + n^{\text{vap}}}$$

Using a material balance of the number of moles of component 2, we have

$$x_a(n^{\text{l}} + n^{\text{vap}}) = x_2 n^{\text{l}} + y_2 n^{\text{vap}}$$

or

$$\frac{n^{\text{l}}}{n^{\text{vap}}} = \frac{y_2 - x_a}{x_a - x_2} \tag{10.19}$$

This equation represents what is called the *lever rule* because $n^{\text{vap}}(y_2 - x_a) = n^{\text{l}}(x_a - x_2)$ can be interpreted as a balance of each value of "$n$" times the distance from each curve to the point C in Figure 10.4. Note that $n^{\text{l}} = 0$ when $x_a = y_2$ (vapor curve) and that $n^{\text{vap}} = 0$ when $x_a = x_2$ (liquid curve).

**EXAMPLE 10–4**

Calculate the relative amounts of liquid and vapor phases at an overall composition of 0.80 for the values in Example 10–3.

SOLUTION: In this case, $x_a = 0.80$, $x_2 = 0.75$, and $y_2 = 0.87$ (see Example 10–3), so

$$\frac{n^{\text{l}}}{n^{\text{vap}}} = \frac{0.87 - 0.80}{0.80 - 0.75} = 1.6$$

According to Example 10–3, the mole fraction of 2-propanol in the vapor phase in equilibrium with a 1-propanol/2-propanol solution is greater than the mole fraction of 2-propanol in the solution. We can display the composition of the solution and vapor phases at various temperatures by a diagram called a *temperature-composition diagram*. To construct such a diagram, we choose some total ambient pressure such as 760 torr and write

$$\begin{aligned} 760 \text{ torr} &= x_1 P_1^* + x_2 P_2^* = x_1 P_1^* + (1 - x_1) P_2^* \\ &= P_2^* - x_1 (P_2^* - P_1^*) \end{aligned}$$

or

$$x_1 = \frac{P_2^* - 760 \text{ torr}}{P_2^* - P_1^*}$$

We then choose some temperature between the boiling points of the two components and solve the above equation for $x_1$, the compositon of the solution that will give a total pressure of 760 torr. A plot of temperature against $x_1$ shows the boiling temperature (at $P_{\text{total}} = 760$ torr) of a solution as a function of its composition ($x_1$). Such a curve, labeled the solution curve, is shown in Figure 10.5. For example, at $t = 90°\text{C}$, $P_1^*$ (the vapor pressure of 1-propanol) $= 575$ torr and $P_2^*$ (the vapor pressure of 2-propanol) $= 1027$ torr. Therefore,

$$x_1 = \frac{P_2^* - 760 \text{ torr}}{P_2^* - P_1^*} = \frac{1027 \text{ torr} - 760 \text{ torr}}{1027 \text{ torr} - 575 \text{ torr}} = 0.59$$

The point corresponding to $t = 90°\text{C}$ and $x_1 = 0.59$ is labeled by point $a$ in Figure 10.5. We can also calculate the corresponding composition of the vapor phase as a function of temperature. The mole fraction of component 1 in the vapor phase is given by Dalton's law

$$y_1 = \frac{P_1}{760 \text{ torr}} = \frac{x_1 P_1^*}{760 \text{ torr}}$$

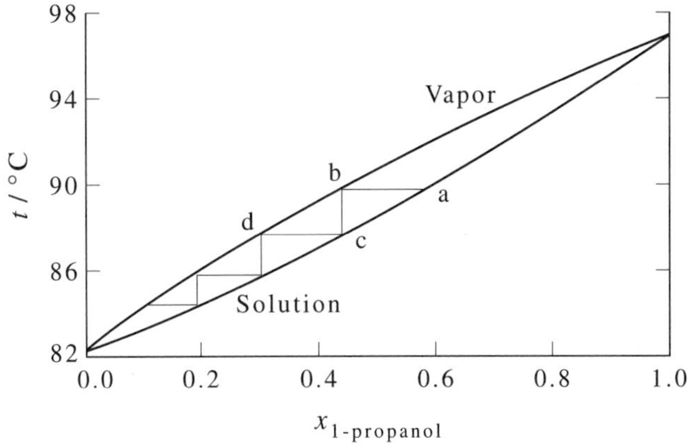


**FIGURE 10.5**
A temperature-composition diagram of a 1-propanol/2-propanol solution, which is essentially an ideal solution. The boiling point of 1-propanol is 97.2°C and that of 2-propanol is 82.3°C.

because the total pressure is taken (arbitrarily) to be 760 torr. We saw above that $x_1 = 0.59$ at 90°C, so we have that

$$y_1 = (0.59)(575 \text{ torr})/(760 \text{ torr}) = 0.45$$

which is labelled by point *b* in Figure 10.5.

---

**EXAMPLE 10–5**
The vapor pressures (in torr) of 1-propanol and 2-propanol as a function of the Celsius temperature, $t$, are given by the empirical formulas

$$\ln P_1^* = 18.0699 - \frac{3452.06}{t + 204.64}$$

and

$$\ln P_2^* = 18.6919 - \frac{3640.25}{t + 219.61}$$

Use these formulas to calculate $x_1$ and $y_1$ at 93.0°C, and compare your results with the values given in Figure 10.5.

SOLUTION: At 93.0°C,

$$\ln P_1^* = 18.0699 - \frac{3452.06}{93.0 + 204.64} = 6.472$$

or $P_1^* = 647$ torr. Similarly, $P_2^* = 1150$ torr. Therefore,

$$x_1 = \frac{P_2^* - 760 \text{ torr}}{P_2^* - P_1^*} = \frac{1150 \text{ torr} - 760 \text{ torr}}{1150 \text{ torr} - 647 \text{ torr}} = 0.77$$

and

$$y_1 = \frac{x_1 P_1^*}{760 \text{ torr}} = \frac{(0.77)(647 \text{ torr})}{760 \text{ torr}} = 0.65$$

in agreement with the values shown in Figure 10.5.

The temperature-composition diagram can be used to illustrate the process of fractional distillation, in which a vapor is condensed and then re-evaporated many times (Figure 10.6). If we were to start with a 1-propanol/2-propanol solution that has a mole fraction of 0.59 in 1-propanol (point a in Figure 10.5), the mole fraction of 1-propanol in the vapor will be 0.45 (point *b*). If this vapor is condensed (point *c*) and then re-evaporated, then the mole fraction of 1-propanol in the vapor phase will be about 0.30 (point *d*). As this process is continued, the vapor becomes increasingly richer in 2-propanol, eventually resulting in pure 2-propanol. A fractional distillation column differs from an ordinary distillation column in that the former is packed with glass beads, which provide a large surface area for the repeated condensation-evaporation process.

We can calculate the change in thermodynamic properties upon forming an ideal solution from its pure components. Let's take the Gibbs energy as an example. We define the Gibbs energy of mixing by

$$\Delta_{\text{mix}} G = G^{\text{sln}}(T, P, n_1, n_2) - G_1^*(T, P, n_1) - G_2^*(T, P, n_2) \tag{10.20}$$

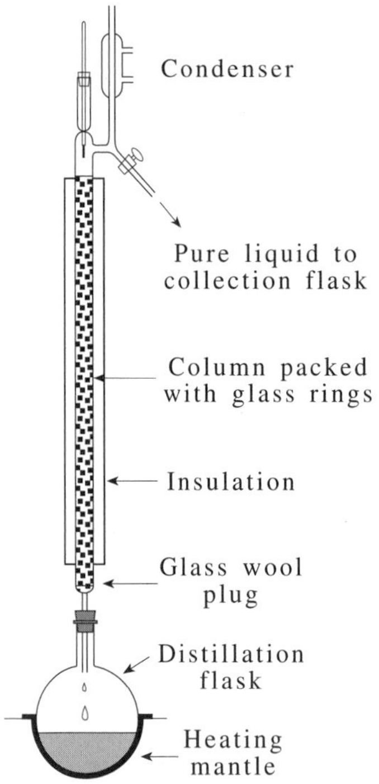


**FIGURE 10.6**
A simple fractional distillation column. Because repeated condensation and re-evaporation occur along the entire column, the vapor becomes progresively richer in the more volatile component as it moves up the column.

where $G_1^*$ and $G_2^*$ are the Gibbs energies of the pure components. Using Equation 10.17 for an ideal solution gives

$$\begin{aligned}\Delta_{\text{mix}} G^{\text{id}} &= n_1\mu_1^{\text{sln}} + n_2\mu_2^{\text{sln}} - n_1\mu_1^* - n_2\mu_2^* \\ &= RT(n_1 \ln x_1 + n_2 \ln x_2)\end{aligned} \tag{10.21}$$

This quantity is always negative because $x_1$ and $x_2$ are less than one. In other words, an ideal solution will always form spontaneously from its separate components. The entropy of mixing of an ideal solution is given by

$$\Delta_{\text{mix}} S^{\text{id}} = -\left(\frac{\partial \Delta_{\text{mix}} G^{\text{id}}}{\partial T}\right)_{P,n_1,n_2} = -R(n_1 \ln x_1 + n_2 \ln x_2) \tag{10.22}$$

Note that this result for an ideal solution is the same as Equation 6.26 for the mixing of ideal gases. This similarity is due to the fact that in both cases the molecules in the final solution are randomly mixed. Nevertheless, you should realize that an ideal solution and a mixture of ideal gases differ markedly in the interactions involved. Although the molecules do not interact in a mixture of ideal gases, they interact strongly in an ideal solution. In an ideal solution, the interactions in the mixture and those in the pure liquids are essentially identical.

The volume change upon mixing of an ideal solution is given by

$$\Delta_{\text{mix}} V^{\text{id}} = \left(\frac{\partial \Delta_{\text{mix}} G^{\text{id}}}{\partial P}\right)_{T,n_1,n_2} = 0 \tag{10.23}$$

and the enthalpy of mixing is (see Equations 10.21 and 10.22)

$$\Delta_{\text{mix}} H^{\text{id}} = \Delta_{\text{mix}} G^{\text{id}} + T\Delta_{\text{mix}} S^{\text{id}} = 0 \tag{10.24}$$

Therefore, there is no volume change upon mixing, nor is there any energy as heat absorbed or evolved when an ideal solution is formed from its pure components. Both Equations 10.23 and 10.24 result from the facts that the molecules are roughly the same size and shape (hence $\Delta_{\text{mix}} V^{\text{id}} = 0$) and that the various interaction energies are the same (hence $\Delta_{\text{mix}} H^{\text{id}} = 0$). Equations 10.23 and 10.24 are indeed observed to be true experimentally for ideal solutions. For most solutions, however, $\Delta_{\text{mix}} H$ and $\Delta_{\text{mix}} V$ do not equal zero.

## 10–5. Most Solutions Are Not Ideal

Ideal solutions are not very common. Figures 10.7 and 10.8 show vapor pressure diagrams for carbon disulfide/dimethoxymethane [$(CH_3O)_2CH_2$] solutions and trichloromethane/acetone solutions, respectively. The behavior in Figure 10.7 shows so-called positive deviations from Raoult's law because the partial vapor pressures

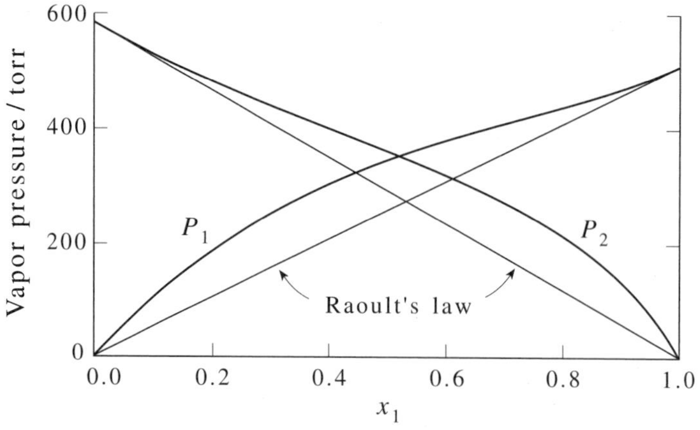


**FIGURE 10.7**
The vapor pressure diagram of a carbon disulfide/dimethoxymethane solution at 25°C. This system shows positive deviations from ideal, or Raoult's law, behavior.

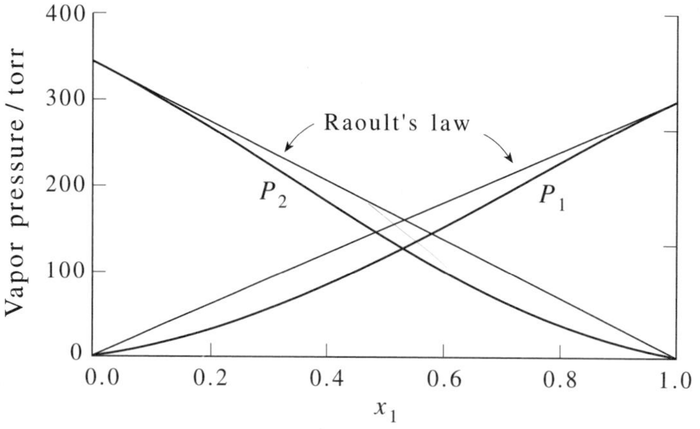


**FIGURE 10.8**
The vapor pressure diagram of a trichloromethane/acetone solution at 25°C. This system shows negative deviations from ideal, or Raoult's law, behavior.

of carbon disulfide and dimethoxymethane are greater than predicted on the basis of Raoult's law. Physically, positive deviations occur because carbon disulfide–dimethoxymethane interactions are more repulsive than either carbon disulfide–carbon disulfide or dimethoxymethane–dimethoxymethane interactions. Negative deviations, on the other hand, like those shown in Figure 10.8 for a trichloromethane/acetone solution, are due to stronger unlike-molecule interactions than like-molecule interactions. Problem 10–36 asks you to show that if one component of a binary solution exhibits positive deviations from ideal behavior, then the other component must do likewise.

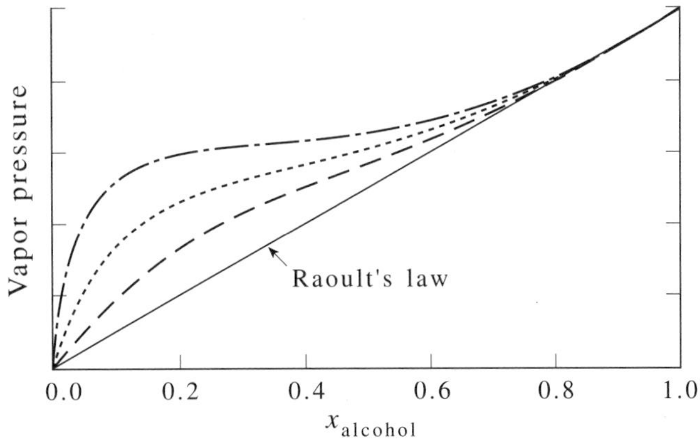


**FIGURE 10.9**
The vapor pressure diagram of alcohol/water solutions as a function of the number of carbon atoms in the alcohols, showing increasing deviation from ideal behavior. The dashed line corresponds to methanol, the dotted line to ethanol, and the dashed-dotted line to 1-propanol.

Figure 10.9 shows plots of methanol, ethanol, and 1-propanol vapor pressures in alcohol/water solutions. Note that the positive deviation from ideal behavior increases with the size of the hydrocarbon part of the alcohol. This behavior occurs because the water–hydrocarbon (repulsive) interactions become increasingly prevalent as the size of the hydrocarbon chain increases.

There are some important features to notice in Figures 10.7 and 10.8. Let's focus on component 1. The vapor pressure of component 1 approaches its Raoult's law value as $x_1$ approaches 1. In an equation, we have that

$$P_1 \longrightarrow x_1 P_1^* \quad \text{as} \quad x_1 \longrightarrow 1 \tag{10.25}$$

Although we deduced Equation 10.25 from Figures 10.7 and 10.8, it is generally true. Physically, this behavior may be attributed to the fact that there are so few component-2 molecules that most component-1 molecules see only other component-1 molecules, so that the solution behaves ideally. Raoult's law behavior is *not* observed for component 1 as $x_1 \to 0$ in Figures 10.7 and 10.8, however. Although not easily seen in Figures 10.7 and 10.8, the vapor pressure of component 1 as $x_1 \to 0$ is linear in $x_1$, but the slope is not equal to $P_1^*$ as in Equation 10.25. We emphasize this behavior by writing

$$P_1 \longrightarrow k_{\mathrm{H},1} x_1 \quad \text{as} \quad x_1 \longrightarrow 0 \tag{10.26}$$

In the special case of an ideal solution, $k_{\mathrm{H},1} = P_1^*$, but ordinarily $k_{\mathrm{H},1} \neq P_1^*$. Equation 10.26 is called *Henry's law*, and $k_{\mathrm{H},1}$ is called the *Henry's law constant* of component 1. As $x_1 \to 0$, the component-1 molecules are completely surrounded by component-2 molecules, and the value of $k_{\mathrm{H},1}$ reflects the intermolecular interactions between the two components. As $x_1 \to 1$, on the other hand, the component-1 molecules are completely surrounded by component-1 molecules, and $P_1^*$ is what reflects the intermolecular interactions in the pure liquid. Although we have focussed our discussion on component 1 in Figures 10.7 and 10.8, the same situtation holds for component 2. Equations 10.25 and 10.26 can be written as

$$\begin{aligned} P_j &\longrightarrow x_j P_j^* \quad \text{as } x_j \longrightarrow 1 \\ P_j &\longrightarrow x_j k_{\mathrm{H},j} \quad \text{as } x_j \longrightarrow 0 \end{aligned} \qquad (10.27)$$

Thus, in a vapor pressure diagram of a solution of two volatile liquids, the vapor pressure of each component approaches Raoult's law as the mole fraction of that component approaches one and Henry's law as the mole fraction approaches zero.

**EXAMPLE 10–6**

The vapor pressure (in torr) of component 1 over a binary solution is given by

$$P_1 = 180x_1 e^{x_2^2 + \frac{1}{2}x_2^3} \qquad 0 \le x_1 \le 1$$

Determine the vapor pressure ($P_1^*$) and the Henry's law constant ($k_{\mathrm{H},1}$) of pure component 1.

SOLUTION: In the limit that $x_1 \to 1$, the exponential factor $\to 1$ because $x_2 \to 0$ as $x_1 \to 1$. Therefore,

$$P_1 \longrightarrow 180x_1 \quad \text{as} \quad x_1 \longrightarrow 1$$

so $P_1^* = 180$ torr. As $x_1 \to 0$, on the other hand, the exponential factor approaches $e^{3/2}$ because $x_2 \to 1$ as $x_1 \to 0$. Thus, we have

$$P_1 \longrightarrow 180e^{3/2}x_1 = 807x_1 \quad \text{as} \quad x_1 \longrightarrow 0$$

and $k_{\mathrm{H},1} = 807$ torr.

We will now show that the Henry's law behavior of component 2 as $x_2 \to 0$ is a thermodynamic consequence of the Raoult's law behavior of component 1 as $x_1 \to 1$. To prove this connection, we will start with the Gibbs–Duhem equation (Equation 10.11)

$$x_1 d\mu_1 + x_2 d\mu_2 = 0 \qquad \text{(constant } T \text{ and } P\text{)}$$

Now, assuming that the vapor phase may be treated as an ideal gas, both chemical potentials can be expressed as

$$\mu_j(T, P) = \mu_j^\circ(T) + RT \ln P_j$$

(Recall that the argument of the logarithm is actually $P_j/P^\circ$, where $P^\circ$ is one bar.) Now this form of $\mu_j(T, P)$ allows us to write

$$d\mu_1 = RT\left(\frac{\partial \ln P_1}{\partial x_1}\right)_{T,P} dx_1$$

and

$$d\mu_2 = RT\left(\frac{\partial \ln P_2}{\partial x_2}\right)_{T,P} dx_2$$

Substitute these two expressions into the Gibbs–Duhem equation to get

$$x_1\left(\frac{\partial \ln P_1}{\partial x_1}\right)_{T,P} dx_1 + x_2\left(\frac{\partial \ln P_2}{\partial x_2}\right)_{T,P} dx_2 = 0 \tag{10.28}$$

But $dx_1 = -dx_2$ (because $x_1 + x_2 = 1$), so Equation 10.28 becomes

$$x_1\left(\frac{\partial \ln P_1}{\partial x_1}\right)_{T,P} = x_2\left(\frac{\partial \ln P_2}{\partial x_2}\right)_{T,P} \tag{10.29}$$

which is another form of the Gibbs–Duhem equation. If component 1 obeys Raoult's law as $x_1 \rightarrow 1$, then $P_1 \rightarrow x_1 P_1^*$ and $(\partial \ln P_1/\partial x_1)_{T,P} = 1/x_1$, so the left side of Equation 10.29 becomes unity. Thus, we have the condition

$$x_2\left(\frac{\partial \ln P_2}{\partial x_2}\right)_{T,P} = 1 \quad \text{as} \quad x_1 \rightarrow 1 \text{ or } x_2 \rightarrow 0$$

We now integrate this expression indefinitely to get

$$\ln P_2 = \ln x_2 + \text{constant} \quad \text{as} \quad x_1 \rightarrow 1 \text{ or } x_2 \rightarrow 0$$

or

$$P_2 = k_{\text{H},2} x_2 \quad \text{as} \quad x_2 \rightarrow 0$$

Thus, we see that if component 1 obeys Raoult's law as $x_1 \rightarrow 1$, then component 2 must obey Henry's law as $x_2 \rightarrow 0$. Problem 10–32 has you prove the converse: if component 2 obeys Henry's law as $x_2 \rightarrow 0$, then component 1 must obey Raoult's law as $x_1 \rightarrow 1$.

## **10–6.** The Gibbs–Duhem Equation Relates the Vapor Pressures of the Two Components of a Volatile Binary Solution

The following example shows that if we know the vapor pressure curve of one of the components over the entire composition range, we can calculate the vapor pressure of the other component.

**EXAMPLE 10–7**

The vapor pressure curve of one of the components (say component 1) of a nonideal binary solution can often be represented empirically by (see Figure 10.10)

$$P_1 = x_1 P_1^* e^{\alpha x_2^2 + \beta x_2^3} \qquad 0 \le x_1 \le 1$$

where $\alpha$ and $\beta$ are parameters that are used to fit the data. Show that the vapor pressure of component 2 is necessarily given by

$$P_2 = x_2 P_2^* e^{\gamma x_1^2 + \delta x_1^3} \qquad 0 \le x_2 \le 1$$

where $\gamma = \alpha + 3\beta/2$ and $\delta = -\beta$. Notice that the parameters $\alpha$ and $\beta$ must in some manner reflect the extent of the nonideality of the solution because both $P_1$ and $P_2$ reduce to the ideal solution expressions when $\alpha = \beta = 0$. Furthermore, note that $P_1 \to x_1 P_1^* e^{\alpha+\beta}$ as $x_1 \to 0$ ($x_2 \to 1$), so the Henry's law constant of component 1 is $k_{\mathrm{H},1} = P_1^* e^{\alpha+\beta}$. Similarly, we find that $k_{\mathrm{H},2} = P_2^* e^{\alpha+\beta/2}$.

SOLUTION: We use the Gibbs–Duhem equation

$$d\mu_2 = -\frac{x_1}{x_2} d\mu_1$$

along with (Equation 10.13)

$$\begin{aligned} \mu_1 &= \mu_1^\circ + RT \ln P_1 \\ &= \mu_1^\circ + RT \ln P_1^* + RT \ln x_1 \\ &\qquad + \alpha RT(1-x_1)^2 + \beta RT(1-x_1)^3 \end{aligned}$$

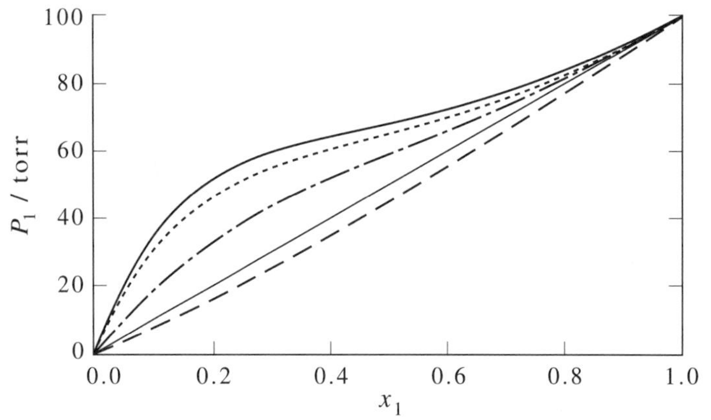


**FIGURE 10.10**
A plot of $P_1 = x_1 P_1^* e^{\alpha x_2^2 + \beta x_2^3}$ for $P_1^* = 100$ torr and various values of $\alpha$ and $\beta$. The values of $\alpha$ and $\beta$ for the five curves, top to bottom, are 1.0, 0.60; 0.80, 0.60; 0.60, 0.20; 0,0 (ideal solution); and −0.80, 0.60.

Differentiate this equation with respect to $x_1$ and substitute the result into the above Gibbs–Duhem equation to obtain

$$
\begin{aligned}
d\mu_2 &= -\frac{x_1}{x_2}RT\left[\frac{dx_1}{x_1} - 2\alpha(1-x_1)dx_1 - 3\beta(1-x_1)^2dx_1\right] \\
&= RT\left[-\frac{dx_1}{x_2} + 2\alpha x_1 dx_1 + 3\beta x_1(1-x_1)dx_1\right]
\end{aligned}
$$

Now change variables from $x_1$ to $x_2$

$$
d\mu_2 = RT\left[\frac{dx_2}{x_2} - 2\alpha(1-x_2)dx_2 - 3\beta x_2(1-x_2)dx_2\right]
$$

and integrate from $x_2 = 1$ to arbitrary $x_2$ and use the fact that $\mu_2 = \mu_2^*$ when $x_2 = 1$ to get

$$
\begin{aligned}
\mu_2 - \mu_2^* &= RT\left[\ln x_2 + \alpha(1-x_2)^2 - \frac{3\beta}{2}(x_2^2 - 1) + \beta(x_2^3 - 1)\right] \\
&= RT\left[\ln x_2 + \alpha x_1^2 + \frac{3\beta}{2}x_1^2 - \beta x_1^3\right]
\end{aligned}
$$

Using the fact that $\mu_2 = \mu_2^\circ + RT\ln P_2$ and that $\mu_2^* = \mu_2^\circ + RT\ln P_2^*$, we see that

$$
\ln P_2 = \ln P_2^* + \ln x_2 + \alpha x_1^2 + \frac{3\beta}{2}x_1^2 - \beta x_1^3
$$

or

$$
P_2 = x_2 P_2^* e^{(\alpha+3\beta/2)x_1^2 - \beta x_1^3}
$$

We could also have used Equation 10.29 to do this problem (Problem 10–33).

Figure 10.11 shows the boiling-point diagram of a benzene/ethanol system, in which the boiling points of benzene/ethanol solutions (at one atm) are plotted against the mole fraction of ethanol. Figure 10.11 shows that if you were to start with a solution with an ethanol mole fraction of 0.2, for example, then repeated evaporation–condensation would lead to a mixture consisting of a mole fraction of about 0.4 that cannot be separated by further fractional distillation.

Such a mixture, for which there is no change in composition upon boiling, is called an *azeotrope*. Thus, it is not possible to achieve a separation of a benzene/ethanol solution by distillation into pure benzene and pure ethanol. If we start out at an ethanol mole fraction of 0.2, we would obtain a separation of pure benzene and the azeotrope. Similarly, if we started out with an ethanol mole fraction of 0.8, we would achieve a separation of pure ethanol and the benzene/ethanol azeotrope.

As our final topic in this section on nonideal solutions, let's consider the case in which the positive deviations from ideal behavior become increasingly large, as often occurs as the temperature is lowered. Figure 10.12 illustrates typical vapor pressure behavior for a series of temperatures, where $T_3 > T_c > T_2 > T_1$. The vertical axis is $P_2/P_2^*$, so each curve is "normalized" by the vapor pressure of pure component 2 at each

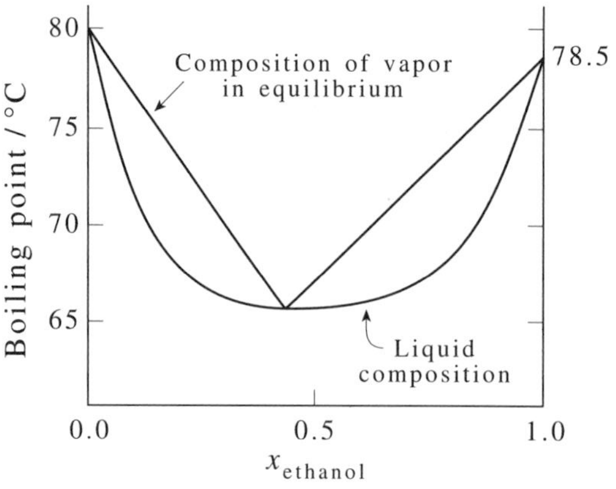


**FIGURE 10.11**
The boiling-point diagram of a benzene/ethanol solution, showing the occurrence of an azeotropic solution at an ethanol mole fraction of about 0.4. The quantity $x_1$ is the mole fraction of ethanol.

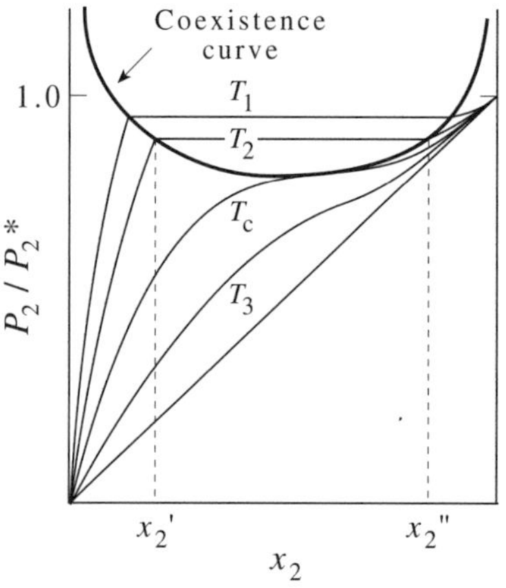


**FIGURE 10.12**
An illustration of the critical behavior of a binary solution as a function of temperature ($T_3 > T_c > T_2 > T_1$).

temperature. Therefore, all the curves meet at $P_2/P_2^* = 1$ at $x_2 = 1$. For temperature $T_3$, which is greater than $T_c$, the slope of the $P_2$ versus $x_2$ curve is everywhere positive. At $T_c$, the curve has an inflection point, where $\partial P_2/\partial x_2 = 0$ and $\partial^2 P_2/\partial x_2^2 = 0$. For the temperatures $T_1$ and $T_2$, which are less than $T_c$, the curves have a horizontal portion that becomes wider as the temperature is lowered. The temperature $T_c$ is called the *critical temperature* or *consulate temperature*, and as we will now discuss, the consulate temperature is the temperature below which the two liquids are not miscible in all proportions.

Let's follow the $T_2$ curve in Figure 10.12 as we start with pure component 1 ($x_2 = 0$) and add component 2. Up to the point $x_2'$, the added component 2 simply dissolves in component 1 to form a single solution phase. Above the concentration $x_2'$, however, two separate or immiscible solution phases form, one of composition $x_2'$ and one of composition $x_2''$. As $x_2$ is increased from $x_2'$ to $x_2''$, the two phases must maintain a constant mole fraction of component 2 ($x_2'$ and $x_2''$) and therefore, the relative proportions of the two phases change, with the volume of the phase of composition $x_2''$ increasing and the volume of the phase of composition $x_2'$ decreasing. The overall composition of the two phases together is given by the value of $x_2$. When $x_2 > x_2''$, we obtain a single solution phase.

We can derive a lever rule to calculate the relative amounts of the two phases in the following way. Consider some overall composition $x_2$, which lies between $x_2'$ and $x_2''$. Let $n_1'$, $n_2'$ and $n_1''$, $n_2''$ be the number of moles of the two components in the phases of composition $x_2'$ and $x_2''$, respectively. Then, the mole fraction of component 2 in each phase is

$$x_2' = \frac{n_2'}{n_1' + n_2'} \qquad \text{and} \qquad x_2'' = \frac{n_2''}{n_1'' + n_2''}$$

and the overall mole fraction of component-2 is

$$x_2 = \frac{n_2' + n_2''}{n_1' + n_1'' + n_2' + n_2''}$$

Using material balance of the number of moles of component 2 allows us to write

$$x_2(n_1' + n_1'' + n_2' + n_2'') = x_2'(n_1' + n_2') + x_2''(n_1'' + n_2'')$$

We can rearrange this material balance equation to give

$$\frac{n'}{n''} = \frac{n_1' + n_2'}{n_1'' + n_2''} = \frac{x_2'' - x_2}{x_2 - x_2'} \tag{10.30}$$

Equation 10.30 gives the relative total number of moles in each phase. Note that if $x_2 = x_2''$, then $n' = 0$, and if $x_2 = x_2'$, then $n'' = 0$. As Equation 10.30 shows, when $x_2$ reaches $x_2''$, the phase of composition $x_2'$ disappears, and there is a single solution phase of composition $x_2 = x_2''$. For $x_2 \geq x_2''$, there is a single solution phase of composition $x_2$. Thus, at a temperature $T_2$, the two liquids are immiscible when $x_2$ is between $x_2'$ and $x_2''$ but are miscible for $x_2 < x_2'$ and $x_2 > x_2''$. Similar behavior occurs at other temperatures less than $T_c$, and Figure 10.12 summarizes this behavior. The heavy curve in Figure 10.12 is called a *coexistence curve*. Points inside the coexistence curve represent two solution phases, whereas points below the coexistence curve represent one solution phase. Problem 10–43 has you determine the coexistence curve for a simple model system.

We can display the results illustrated by Figure 10.12 in a temperature-composition diagram (Figure 10.13*a*). The curve separating the one-phase region from the two-phase region is the coexistence curve. The temperature $T_c$, the temperature above which the two liquids are totally miscible, is the consulate temperature. The coexistence curve in

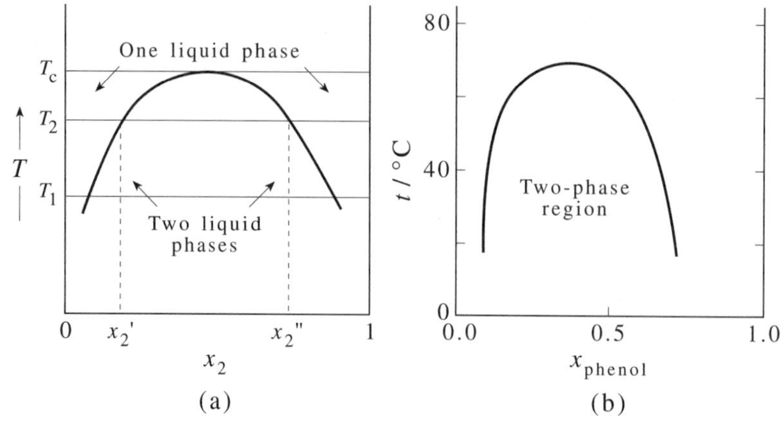


**FIGURE 10.13**
(a) A temperature-composition diagram for the system illustrated in Figure 10.12.
(b) A temperature-composition diagram for a water/phenol system.

Figure 10.13*a* looks "upside down" compared with the one in Figure 10.12, but note that the temperature decreases as you go up in Figure 10.12, whereas they decrease as you go down in Figure 10.13. Figure 10.13*b* shows a coexistence curve for a water/phenol system.

## 10–7. The Central Thermodynamic Quantity for Nonideal Solutions Is the Activity

The chemical potential of component $j$ in a liquid solution is given by (Equation 10.15)

$$\mu_j^{\text{sln}} = \mu_j^* + RT \ln \frac{P_j}{P_j^*} \tag{10.31}$$

if we assume, as usual, that the vapor pressures involved are low enough that the vapors can be considered to behave ideally (otherwise, we replace the partial pressures by partial fugacities). An ideal solution is one in which $P_j = x_j P_j^*$ for all concentrations, so that Equation 10.31 becomes

$$\mu_j^{\text{sln}} = \mu_j^* + RT \ln x_j \qquad \text{(ideal solution)} \tag{10.32}$$

Equation 10.31 is still valid for a nonideal solution, but the relation between $P_j/P_j^*$ and composition is more complicated than simply $P_j = x_j P_j^*$. For example, we saw in Example 10–7 that partial vapor pressure data are often fit by an expression like

$$P_1 = x_1 P_1^* \exp(\alpha x_2^2 + \beta x_2^3 + \cdots) \tag{10.33}$$

The exponential factor here accounts for the nonideality of the system. The chemical potential of component 1 in this case is given by

$$\mu_1 = \mu_1^* + RT \ln x_1 + \alpha RT x_2^2 + \beta RT x_2^3 + \cdots \tag{10.34}$$

In Section 8–8, we introduced the idea of fugacity to preserve the form of the thermodynamic equations we had derived for ideal gases. We will follow a similar procedure for solutions, using an ideal solution as our standard.

To carry over the form of Equation 10.32 to nonideal solutions, we define a quantity called the *activity* by the equation

$$\mu_j^{\text{sln}} = \mu_j^* + RT \ln a_j \tag{10.35}$$

where $\mu_j^*$ is the chemical potential, or the molar Gibbs energy, of the pure liquid. Equation 10.35 is the generalization of Equation 10.32 to nonideal solutions. The first of Equations 10.27 says that $P_j = x_j P_j^*$, as $x_j \to 1$. If we substitute this result into Equation 10.31, we obtain

$$\mu_j^{\text{sln}} = \mu_j^* + RT \ln x_j \qquad (\text{as } x_j \to 1)$$

If we compare this equation with Equation 10.35, which is valid at all concentrations, we can define the activity of component $j$ by

$$a_j = \frac{P_j}{P_j^*} \qquad (\text{ideal vapor}) \tag{10.36}$$

such that $a_j \to x_j$ as $x_j \to 1$. In other words, the activity of a pure liquid is unity (at a total pressure of one bar and at the temperature of interest). For an ideal solution, $P_j = x_j P_j^*$ for all concentrations, and so the activity of component $i$ in an ideal solution is given by $a_j = x_j$. In a nonideal solution, $a_j$ still is equal to $P_j/P_j^*$, but this ratio is no longer equal to $x_j$, although $a_j \to x_j$ as $x_j \to 1$.

According to Equations 10.33 and 10.36, the activity of component 1 can be represented empirically by

$$a_1 = x_1 e^{\alpha x_2^2 + \beta x_2^3 + \cdots}$$

Note that $a_1 \to 1$ as $x_1 \to 1$ ($x_2 \to 0$). The ratio $a_j/x_j$ can be used as a measure of the deviation of the solution from ideality. This ratio is called the *activity coefficient* of component $j$ and is denoted by $\gamma_j$:

$$\gamma_j = \frac{a_j}{x_j} \tag{10.37}$$

If $\gamma_j = 1$ for all concentrations, the solution is ideal. If $\gamma_j \neq 1$, the solution is not ideal. For example, the partial vapor pressures of chlorobenzene in equilibrium with a chlorobenzene/1-nitropropane solution at 75°C are listed below:

| | | | | | |
|---|---|---|---|---|---|
| $x_1$ | 0.119 | 0.289 | 0.460 | 0.691 | 1.00 |
| $P_1$/torr | 19.0 | 41.9 | 62.4 | 86.4 | 119 |

According to these data, the vapor pressure of pure chlorobenzene at 75°C is 119 torr, so the activities and activity coefficients are as follows:

| | | | | | |
|---|---|---|---|---|---|
| $x_1$ | 0.119 | 0.289 | 0.460 | 0.691 | 1.00 |
| $a_1(= P_1/P_1^*)$ | 0.160 | 0.352 | 0.524 | 0.726 | 1.00 |
| $\gamma_1(= a_1/x_1)$ | 1.34 | 1.22 | 1.14 | 1.05 | 1.00 |

Figure 10.14 shows the activity coefficient of chlorobenzene in 1-nitropropane at 75°C plotted against the mole fraction of chlorobenzene.

Activity is really just another way of expressing chemical potential because the two quantities are directly related to each other through $\mu_j = \mu_j^* + RT \ln a_j$. Therefore, just as the chemical potential of one component of a binary solution is related to the chemical potential of the other component by way of the Gibbs-Duhem equation, the activities are related to each other by

$$x_1 d \ln a_1 + x_2 d \ln a_2 = 0 \tag{10.38}$$

For example, if $a_1 = x_1$ over the entire composition range, meaning that component 1 obeys Raoult's law over the entire composition range, then

$$d \ln a_2 = -\frac{x_1}{x_2}\frac{dx_1}{x_1} = -\frac{dx_1}{x_2} = \frac{dx_2}{x_2}$$

Integrate from $x_2 = 1$ to arbitrary $x_2$ and use the fact that $a_2 \to 1$ as $x_2 \to 1$ to get

$$\ln a_2 = \ln x_2$$

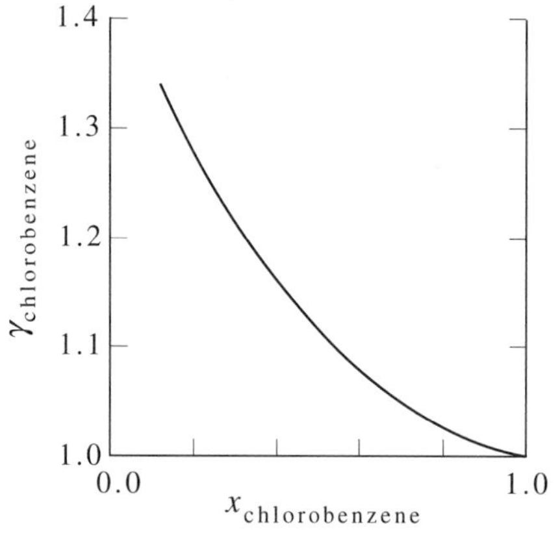


**FIGURE 10.14**
The activity coefficient of chlorobenzene in 1-nitropropane at 75°C plotted against the mole fraction of chlorobenzene.

or $a_2 = x_2$. Thus, we see once again that if one component obeys Raoult's law over the entire composition range, the other component will also.

**EXAMPLE 10–8**

Show that if

$$a_1 = x_1 e^{\alpha x_2^2}$$

then

$$a_2 = x_2 e^{\alpha x_1^2}$$

SOLUTION: We first differentiate $\ln a_1$ with respect to $x_1$:

$$d \ln a_1 = \frac{dx_1}{x_1} - 2\alpha(1 - x_1)dx_1$$

and substitute into Equation 10.38 to obtain

$$\begin{aligned} d \ln a_2 &= -\frac{x_1}{x_2}\left(\frac{dx_1}{x_1} - 2\alpha x_2 dx_1\right) \\ &= -\frac{dx_1}{x_2} + 2\alpha x_1 dx_1 \end{aligned}$$

Now change the integration variable from $x_1$ to $x_2$:

$$d \ln a_2 = \frac{dx_2}{x_2} - 2\alpha(1 - x_2)dx_2$$

and integrate from $x_2 = 1$ (where $a_2 = 1$) to arbitrary $x_2$:

$$\ln a_2 = \ln x_2 + \alpha(1 - x_2)^2$$

or

$$a_2 = x_2 e^{\alpha x_1^2}$$

## 10–8. Activities Must Be Calculated with Respect to Standard States

In one sense, there are two types of binary solutions, those in which the two components are miscible in all proportions and those in which they are not. Only in the latter case are the designations "solvent" and "solute" unambiguous. As we will see in this section, the different nature of these two types of solutions leads us to define different standard states.

Although we have not said so explicitly, we have tacitly assumed both components of the solutions we have considered thus far exist as pure liquids at the temperatures of the solutions. We have defined the activity of each component by (Equation 10.36)

$$a_j = \frac{P_j}{P_j^*} \qquad \text{(ideal vapor)} \tag{10.39}$$

so that $a_j \to x_j$ as $x_j \to 1$ and $a_j = 1$ when $P_j = P_j^*$. An activity defined by Equation 10.39 is said to be based upon a solvent, or Raoult's law standard state. Because of the relation (Equation 10.35) $\mu_j = \mu_j^* + RT \ln a_j$, the chemical potential of component $j$ is also based upon a solvent, or Raoult's law, standard state. You need to realize that activities or chemical potentials are meaningless unless it is clear just what has been used as the standard state. If the two liquids are miscible in all proportions, there is no distinction between solvent and solute and a solvent standard state is normally used. If, on the other hand, one component is sparingly soluble in the other, then picking a standard state based upon Henry's law instead of Raoult's law is more convenient. To see how we define the activity in this case, we start with Equation 10.31

$$\mu_j^{\text{sln}} = \mu_j^* + RT \ln \frac{P_j}{P_j^*} \tag{10.40}$$

Because component $j$ is sparingly soluble, we use the second of Equations 10.27, which says that $P_j \to x_j k_{\text{H},j}$ as $x_j \to 0$, where $k_{\text{H},j}$ is the Henry's law constant of component $j$. If we substitute the limiting value of $x_j k_{\text{H},j}$ into Equation 10.40 for $P_j$, we obtain

$$\begin{aligned} \mu_j^{\text{sln}} &= \mu_j^* + RT \ln \frac{x_j k_{\text{H},j}}{P_j^*} \qquad (x_j \to 0) \\ &= \mu_j^* + RT \ln \frac{k_{\text{H},j}}{P_j^*} + RT x_j \qquad (x_j \to 0) \end{aligned} \tag{10.41}$$

We define the activity of component $j$ by

$$\mu_j^{\text{sln}} = \mu_j^* + RT \ln \frac{k_{\text{H},j}}{P_j^*} + RT \ln a_j \tag{10.42}$$

so that $a_j \to x_j$ as $x_j \to 0$, as can be seen by comparing Equations 10.41 and 10.42. Equation 10.42 becomes equivalent to Equation 10.35 if we define $a_j$ by

$$a_j = \frac{P_j}{k_{\text{H},j}} \qquad \text{(ideal vapor)} \tag{10.43}$$

and choose the standard state such that

$$\mu_j^* = \mu_j^* + RT \ln \frac{k_{\text{H},j}}{P_j^*}$$

or $k_{\mathrm{H},j} = P_j^*$. The standard state in this case requires that $k_{\mathrm{H},j} = P_j^*$. This standard state may not exist in practice, so it is called a hypothetical standard state. Nevertheless, the definition of activity involving Henry's law for dilute components given by Equation 10.43 is natural and useful.

The numerical value of an activity or an activity coefficient depends upon the choice of standard state. Table 10.1 lists vapor pressure data for carbon disulfide/dimethoxymethane solutions at 35.2°C, and these data are plotted in Figure 10.15. Notice that both curves approach Raoult's law as their corresponding mole fractions approach unity. The dashed lines in the figure represent the linear regions as the corresponding mole fractions approach zero. The slopes of these lines give the Henry's law constant for each component. The values come out to be $k_{\mathrm{H,CS_2}} = 1130$ torr and $k_{\mathrm{H,dimeth}} = 1500$ torr. We can use these values and the values of the vapor pressures of the pure components to calculate activities and activity coefficients based upon each standard state. For example, Table 10.1 gives $P_{\mathrm{CS_2}} = 407.0$ torr and $P_{\mathrm{dimeth}} = 277.8$ torr at $x_{\mathrm{CS_2}} = 0.6827$. Therefore,

$$a_{\mathrm{CS_2}}^{(\mathrm{R})} = \frac{P_{\mathrm{CS_2}}}{P_{\mathrm{CS_2}}^*} = \frac{407.0 \text{ torr}}{514.5 \text{ torr}} = 0.7911$$

**TABLE 10.1**

Vapor pressure data of carbon disulfide/dimethoxymethane solutions at 35.2°C.

| $x_{\mathrm{CS_2}}$ | $P_{\mathrm{CS_2}}$/torr | $P_{\mathrm{dimeth}}$/torr |
|---|---|---|
| 0.0000 | 0.000 | 587.7 |
| 0.0489 | 54.5 | 558.3 |
| 0.1030 | 109.3 | 529.1 |
| 0.1640 | 159.5 | 500.4 |
| 0.2710 | 234.8 | 451.2 |
| 0.3470 | 277.6 | 412.7 |
| 0.4536 | 324.8 | 378.0 |
| 0.4946 | 340.2 | 360.8 |
| 0.5393 | 357.2 | 342.2 |
| 0.6071 | 381.9 | 313.3 |
| 0.6827 | 407.0 | 277.8 |
| 0.7377 | 424.3 | 250.1 |
| 0.7950 | 442.3 | 217.4 |
| 0.8445 | 458.1 | 184.9 |
| 0.9108 | 481.8 | 124.2 |
| 0.9554 | 501.0 | 65.1 |
| 1.0000 | 514.5 | 0.000 |

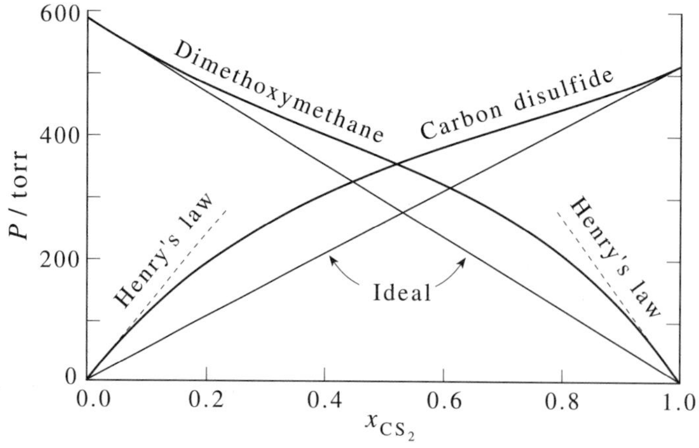


**FIGURE 10.15**
Vapor pressures of carbon disulfide and dimethoxymethane over their solutions at 35.2°C. The solid straight lines represent ideal behavior, and the dashed lines represent the Henry's law behavior for each component as the corresponding mole fractions approach zero.

and

$$a_{\text{dimeth}}^{(\text{R})} = \frac{P_{\text{dimeth}}}{P_{\text{dimeth}}^*} = \frac{277.8 \text{ torr}}{587.7 \text{ torr}} = 0.4727$$

with

$$\gamma_{\text{CS}_2}^{(\text{R})} = \frac{a_{\text{CS}_2}^{(R)}}{x_{\text{CS}_2}} = \frac{0.7911}{0.6827} = 1.159$$

and

$$\gamma_{\text{dimeth}}^{(\text{R})} = \frac{a_{\text{dimeth}}^{(R)}}{x_{\text{dimeth}}} = \frac{0.4727}{0.3173} = 1.490$$

where the superscript (R) simply emphasizes that these values are based upon a Raoult's law, or solvent, standard state.

Similarly,

$$a_{\text{CS}_2}^{(\text{H})} = \frac{P_{\text{CS}_2}}{k_{\text{H,CS}_2}} = \frac{407.0 \text{ torr}}{1130 \text{ torr}} = 0.360$$

$$a_{\text{dimeth}}^{(\text{H})} = \frac{P_{\text{dimeth}}}{k_{\text{H,dimeth}}} = \frac{277.8 \text{ torr}}{1500 \text{ torr}} = 0.185$$

$$\gamma_{\text{CS}_2}^{(\text{H})} = \frac{a_{\text{CS}_2}^{(\text{H})}}{x_{\text{CS}_2}} = \frac{0.360}{0.6827} = 0.527$$

and

$$\gamma^{(H)}_{\text{dimeth}} = \frac{a^{(H)}_{\text{dimeth}}}{x_{\text{dimeth}}} = \frac{0.185}{0.3173} = 0.583$$

where the superscript (H) simply emphasizes that these values are based upon a Henry's law, or solute, standard state. Figure 10.16a shows the Raoult's law, or solvent-based, activities, and Figure 10.16b shows the Henry's law, or solute-based, activities plotted against the mole fraction of carbon disulfide. We will see in the next chapter that a solute, or Henry's law, standard state is particularly appropriate for a substance that does not exist as a liquid at one bar and at the temperature of the solution under study.

The activity coefficients based upon the Raoult's law standard state (which is the usual standard state for miscible liquids) are plotted in Figure 10.17. Notice that $\gamma_{CS_2} \to 1$ as $x_{CS_2} \to 1$ and that it goes to 2.2 as $x_{CS_2} \to 0$. Both of these limiting values may be deduced from the definition of $\gamma$ (Equation 10.37)

$$\gamma_j = \frac{a_j}{x_j} = \frac{P_j}{x_j P_j^*}$$

Now $P_j \to P_j^*$ as $x_j \to 1$, and so $\gamma_j \to 1$ as $x_j \to 1$. At the other limit, however, $P_j \to x_j k_{H,j}$ as $x_j \to 0$, so we see that $\gamma_j \to k_{H,j}/P_j^*$ as $x_j \to 0$. The value of $k_H$ for $CS_2(l)$ is 1130 torr, so $\gamma_{CS_2} \to k_{H,CS_2}/P^*_{CS_2} = (1130\text{ torr}/514.5\text{ torr}) = 2.2$, in agreement with Figure 10.17.

The activity coefficient of dimethoxymethane approaches 2.5 as $x_{\text{dimeth}} \to 0$ ($x_{CS_2} \to 1$), in agreement with $\gamma_{\text{dimeth}} \to k_{H,\text{dimeth}}/P^*_{\text{dimeth}} = (1500\text{ torr}/587.7\text{ torr}) = 2.5$.

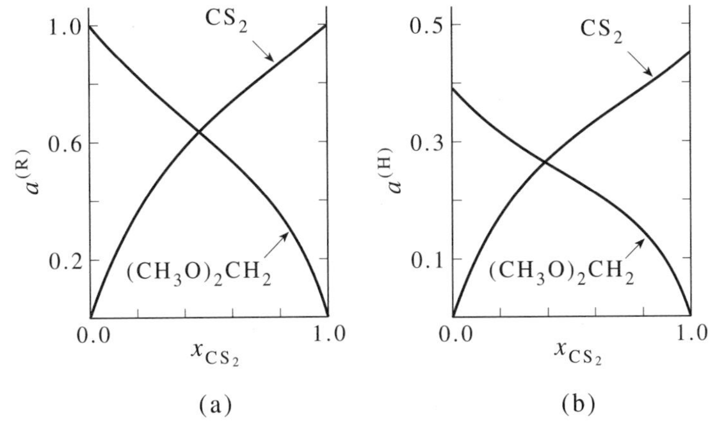


**FIGURE 10.16**
(a) The Raoult's law activities of carbon disulfide and dimethoxymethane in carbon disulfide/dimethoxymethane solutions at 35.2°C plotted against the mole fraction of carbon disulfide. (b) The Henry's law activities for the same system.

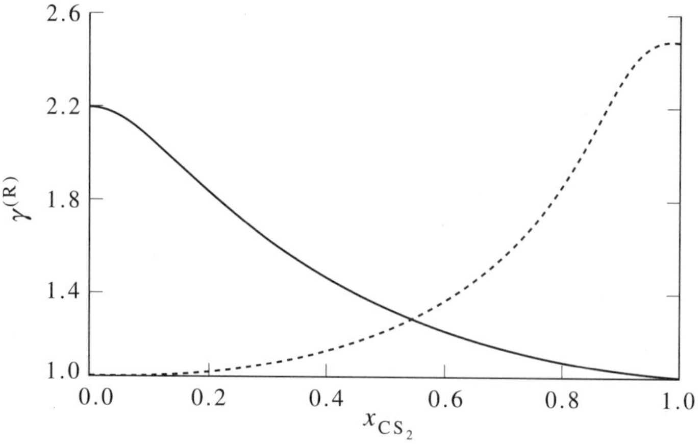


**FIGURE 10.17**
The Raoult's law activity coefficients of carbon disulfide (solid line) and dimethoxymethane (dashed line) plotted against $x_{CS_2}$ for carbon disulfide/dimethoxymethane solutions at 35.2°C.

We can calculate the Gibbs energy of mixing of binary solutions in terms of the activity coefficients. Recall from Equation 10.20 that

$$\Delta_{\text{mix}} G = n_1 \mu_1^{\text{sln}} + n_2 \mu_2^{\text{sln}} - n_1 \mu_1^* - n_2 \mu_2^*$$

But, according to Equations 10.35 and 10.37,

$$\mu_j^{\text{sln}} = \mu_j^* + RT \ln a_j = \mu_j^* + RT \ln x_j + RT \ln \gamma_j$$

so

$$\Delta_{\text{mix}} G/RT = n_1 \ln x_1 + n_2 \ln x_2 + n_1 \ln \gamma_1 + n_2 \ln \gamma_2 \tag{10.44}$$

If we divide $\Delta_{\text{mix}} G/RT$ by the total number of moles, $n_1 + n_2$, we obtain the *molar Gibbs energy of mixing*, $\Delta_{\text{mix}} \overline{G}$.

$$\frac{\Delta_{\text{mix}} \overline{G}}{RT} = x_1 \ln x_1 + x_2 \ln x_2 + x_1 \ln \gamma_1 + x_1 \ln \gamma_2 \tag{10.45}$$

The first two terms in Equation 10.44 represent the Gibbs energy of mixing of an ideal solution. To focus on the effect of nonideality, we define an *excess Gibbs energy of mixing*, $G^{\text{E}}$:

$$G^{\text{E}} = \Delta_{\text{mix}} G - \Delta_{\text{mix}} G^{\text{id}} \tag{10.46}$$

We see from Equation 10.44 that

$$G^{\text{E}}/RT = n_1 \ln \gamma_1 + n_2 \ln \gamma_2$$

If we divide $G^{\text{E}}$ by the total number of moles $n_1 + n_2$, we obtain the *molar excess Gibbs energy of mixing*, $\overline{G}^{\text{E}}$:

$$\overline{G}^{\text{E}}/RT = x_1 \ln \gamma_1 + x_2 \ln \gamma_2 \tag{10.47}$$

**EXAMPLE 10–9**

Derive a formula for $\overline{G}^{\text{E}}$ for a binary solution in which the vapor pressures can be expressed by

$$P_1 = x_1 P_1^* e^{\alpha x_2^2} \quad \text{and} \quad P_2 = x_2 P_2^* e^{\alpha x_1^2}$$

and show that $\overline{G}^{\text{E}}$ is symmetric about the line $x_1 = x_2 = 1/2$.

SOLUTION: According to the above expression for $P_1$ and $P_2$,

$$\gamma_1 = \frac{P_1}{x_1 P_1^*} = e^{\alpha x_2^2} \quad \text{and} \quad \gamma_2 = \frac{P_2}{x_2 P_2^*} = e^{\alpha x_1^2}$$

Substitute these expressions into Equation 10.47 to obtain

$$\overline{G}^{\text{E}}/RT = \alpha x_1 x_2^2 + \alpha x_2 x_1^2$$

But

$$x_1 x_2^2 + x_2 x_1^2 = x_1 x_2 (x_1 + x_2) = x_1 x_2$$

so

$$\overline{G}^{\text{E}}/RT = \alpha x_1 x_2$$

which is symmetric in $x_1$ and $x_2$, and therefore, about the line $x_1 = x_2 = 1/2$.

Many solutions can be described by the equations in Example 10–9, and such solutions are called *regular solutions*. Problems 10–37 through 10–45 involve regular solutions.

We can use $\gamma_{\text{CS}_2}$ and $\gamma_{\text{dimeth}}$ that we calculated for Figure 10.17 to calculate $\overline{G}^{\text{E}}$, which is shown in Figure 10.18. Note that the plot of $\overline{G}^{\text{E}}$ versus $x_{\text{CS}_2}$ is not symmetric about $x_{\text{CS}_2} = 1/2$. This asymmetry implies that $\beta \neq 0$ in the empirical vapor pressure formula used in Example 10–7, and that carbon disulfide and dimethoxymethane do not form a regular solution at 35.2°C.

## 10–9. We Can Construct a Molecular Model of Non-ideal Solutions

In this section, we shall introduce a simple molecular model for a non-ideal solution that qualitatively displays many of the experimental properties of non-ideal solutions. We shall assume that the molecules of components 1 and 2 are distributed randomly

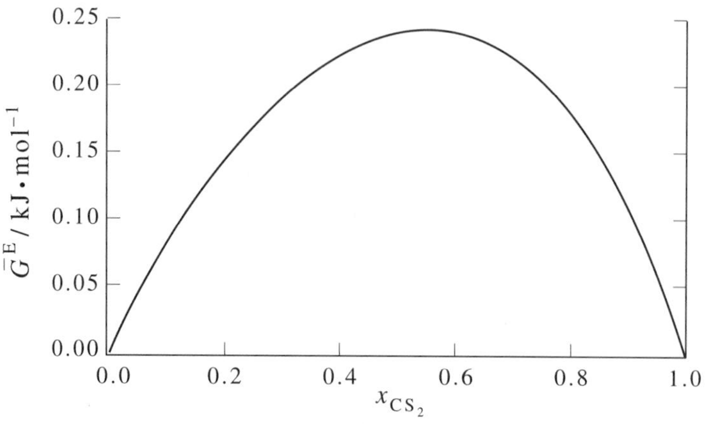


**FIGURE 10.18**
The molar excess Gibbs energy of mixing of carbon disulfide/dimethoxymethane solutions at 35.2°C plotted against the mole fraction of carbon disulfide.

throughout the solution, so that the entropy will be the same as that of an ideal solution. Therefore, the difference between the Gibbs energy of our model solution and that of an ideal solution will be due to an energy term (actually a potential energy term). We express the potential energy of the solution in the form

$$U = N_{11}\epsilon_{11} + N_{12}\epsilon_{12} + N_{22}\epsilon_{22}$$

where $N_{ij}$ is the number of neighboring pairs of molecules of type $i$ and $j$ and where $\epsilon_{ij}$ is the interaction energy of an $i$ and $j$ molecule when they are next to each other. Because we have assumed that the molecules are randomly distributed, we can derive simple expressions for the $N_{ij}$. Let's consider $N_{11}$, the number of neighboring 1–1 pairs, first. We can interpret the mole fraction $x_1$ as the probability that any neighbor of a molecule is a component 1 molecule. On the average, the total number of type 1 neighbors of any given molecule, then, is given by $zx_1$, where $z$ is the coordination number of molecules around a central molecule. Typically $z$ is around 6 to 10, but we shall not require a precise value. There are $N_1$ component 1 molecules in the solution, so therefore the number of 1–1 neighboring pairs is $(N_1)(zx_1)/2$, where the factor of 2 is inserted to avoid counting each 1–1 pair twice. Similarly, we have $N_{22} = zx_2N_2/2$. We use the same value of $z$ because we assume that molecular sizes are about the same. The number of 1–2 neighboring pairs is given by $zx_2N_1$, or $zx_1N_2$, the two expressions being equivalent. The total interaction energy in the solution, then, is given by

$$U = \frac{N_1 z x_1}{2}\epsilon_{11} + \frac{N_2 z x_1}{2}\epsilon_{22} + N_1 z x_2 \epsilon_{12}$$

Using the fact that $x_1 = N_1/(N_1 + N_2)$ and $x_2 = N_2/(N_1 + N_2)$ gives

$$U = \frac{zN_1^2}{2(N_1 + N_2)}\epsilon_{11} + \frac{zN_2^2}{2(N_1 + N_2)}\epsilon_{22} + \frac{zN_1N_2}{N_1 + N_2}\epsilon_{12} \tag{10.48}$$

We can focus on the non-ideality of the solution by introducing the quantity

$$w = 2\epsilon_{12} - \epsilon_{11} - \epsilon_{22} \tag{10.49}$$

For an ideal solution, $\epsilon_{12} = (\epsilon_{11} + \epsilon_{22})/2$, and so $w = 0$. The magnitude of $w$, then, is a measure of the deviation of the solution from ideal behavior. Equation 10.48 takes on a fairly simple form if we eliminate $\epsilon_{12}$ in favor of $w$ using Equation 10.49

$$U = \frac{z\epsilon_{11}N_1}{2} + \frac{z\epsilon_{22}N_2}{2} + \frac{zwN_1N_2}{2(N_1 + N_2)} \tag{10.50}$$

The last term here is equal to zero when $w = 0$ and so represents the non-ideal behavior of the solution. Therefore, we can express the Gibbs energy of the solution in the form

$$G_{\text{soln}} = G_{\text{ideal}} + \frac{zwN_1N_2}{2(N_1 + N_2)} \tag{10.51}$$

We can express Equation 10.51 in terms of numbers of moles rather than numbers of molecules by dividing $N_1$ and $N_2$ by the Avogadro constant ($N_A$) to obtain

$$G_{\text{soln}} = G_{\text{ideal}} + \frac{zwN_An_1n_2}{2(n_1 + n_2)} \tag{10.52}$$

The chemical potential of component 1 is given by

$$\mu_1 = \left(\frac{\partial G}{\partial n_1}\right)_{T,P,n_2} = \left(\frac{\partial G_{\text{ideal}}}{\partial n_1}\right)_{T,P,n_2} + \frac{zwN_A}{2}\left(\frac{\partial n_1n_2/(n_1 + n_2)}{\partial n_1}\right)_{n_2} \tag{10.53}$$

The term $(\partial G_{\text{ideal}}/\partial n_1)_{T,P,n_2}$ is the chemical potential of an ideal solution (Equation 10.17)

$$\mu_1 = \mu_1^* + RT \ln x_1$$

and the derivative in the second term in Equation 10.53 is equal to

$$\begin{aligned}\left(\frac{\partial n_1n_2/(n_1 + n_2)}{\partial n_1}\right)_{n_2} &= \frac{n_2}{n_1 + n_2} - \frac{n_1n_2}{(n_1 + n_2)^2} \\ &= x_2 - x_1x_2 = x_2(1 - x_1) = x_2^2\end{aligned}$$

If we substitute the above two equations into Equation 10.53, then we obtain

$$\begin{aligned}\mu_1 &= \mu_1^* + RT \ln x_1 + \frac{zwN_Ax_2^2}{2} \\ &= \mu_1^* + RT \ln(x_1 e^{ux_2^2/RT})\end{aligned} \tag{10.54}$$

where $u = zN_A w/2$. If we compare Equation 10.54 to Equations 10.35 and 10.36, then we see that

$$a_1 = \frac{P_1}{P_1^*} = x_1 e^{ux_2^2/RT} \tag{10.55}$$

If we had differentiated Equation 10.52 with respect to $n_2$ instead of $n_1$, then we would have found that

$$a_2 = \frac{P_2}{P_2^*} = x_2 e^{ux_1^2/RT} \tag{10.56}$$

Equations 10.55 and 10.56 are the empirical expressions for $P_1$ and $P_2$ that we have used several times in earlier sections. For example, Example 10–8 shows that Equations 10.55 and 10.56 are related by the Gibbs-Duhem equation.

Equation 10.55 is plotted in Figure 10.19 for $u/RT = +1,\ 0$, and $-1$. The value of $u = 0$ gives ideal behavior. For positive values of $u$, we have positive deviations from ideal behavior. From Equation 10.49, positive values of $u$ mean that 1–1 and 2–2 interactions are more favorable than 1–2 interactions. The increasing number of 1–2 interactions that result when the solution is formed from its components causes molecules to escape into the vapor phase, thus giving a positive deviation from ideal behavior. For negative values of $u$, the 1–2 interactions are more favorable than the 1–1 and 2–2 interactions. In this case, the molecules mix well, thus producing negative deviations from ideal (Raoult's law) behavior.

The following Example shows that we can use Equations 10.55 and 10.56 to derive an expression for the molar Gibbs energy of mixing.

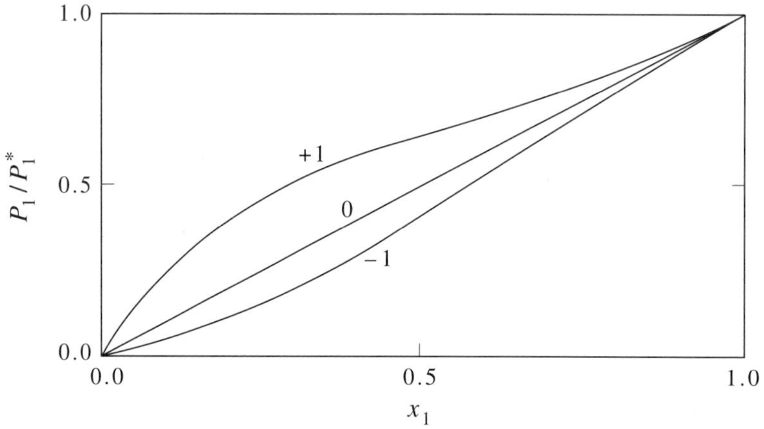


**FIGURE 10.19**
A plot of $P_1/P_1^*$ versus $x_1$ given by Equation 10.55 for $u/RT = +1,\ 0$, and $-1$. The value $u = 0$ yields an ideal solution. Positive values of $u/RT$ yield positive deviations from ideal (Raoult's law) behavior, and negative values of $u/RT$ yield negative deviations from ideal behavior.

**EXAMPLE 10–10**

Use Equations 10.55 and 10.56 to derive formulas for $\Delta_{mix}\overline{G}$ and $\overline{G}^E$ for a binary solution.

SOLUTION: According Equations 10.55 and 10.56,

$$\gamma_1 = \frac{P_1}{x_1 P_1^*} = e^{ux_2^2/RT} \quad \text{and} \quad \gamma_2 = \frac{P_2}{x_2 P_2^*} = e^{ux_1^2/RT}$$

Substitute these expressions into Equation 10.45 to obtain

$$\Delta_{mix}\overline{G} = RT(x_1 \ln x_1 + x_2 \ln x_2) + ux_1x_2^2 + ux_2x_1^2$$

But

$$ux_1x_2^2 + ux_2x_1^2 = ux_1x_2(x_1 + x_2) = ux_1x_2$$

so

$$\Delta_{mix}\overline{G} = RT(x_1 \ln x_1 + x_2 \ln x_2) + ux_1x_2 \tag{10.57}$$

Similarly, Equation 10.47 gives

$$\overline{G}^E = ux_1x_2 \tag{10.58}$$

Problem 10–61 has you show that

$$\frac{\Delta_{mix}\overline{S}}{R} = -x_1 \ln x_1 - x_2 \ln x_2 \tag{10.59}$$

$$\overline{S}^E = 0 \tag{10.60}$$

and

$$\overline{H}^E = \Delta_{mix}\overline{H} = ux_1x_2 \tag{10.61}$$

The entropy of mixing is the same as for an ideal solution because we have assumed that the molecules are distributed randomly throughout the solution. Unlike an ideal solution, $\Delta_{mix}H \neq 0$ for a non-ideal solution.

Equation 10.57 can be written as

$$\frac{\Delta_{mix}\overline{G}}{u} = \frac{RT}{u}(x_1 \ln x_1 + x_2 \ln x_2) + x_1x_2 \tag{10.62}$$

Figure 10.20 shows plots of $\Delta_{mix}\overline{G}/u$ for several values of $RT/u$. Note that the slopes of all the curves equal zero at the midpoint, $x_1 = x_2 = 1/2$. The curve for $RT/u = 0.50$ is special in the sense that curves for values of $RT/u$ greater than 0.50 are concave upwards for all values of $x_1$, whereas curves for values of $RT/u$ less than 0.50 are concave downward at $x_1 = 1/2$. In mathematical terms, $\partial^2(\Delta_{mix}\overline{G}/u)/\partial x_1^2$ is

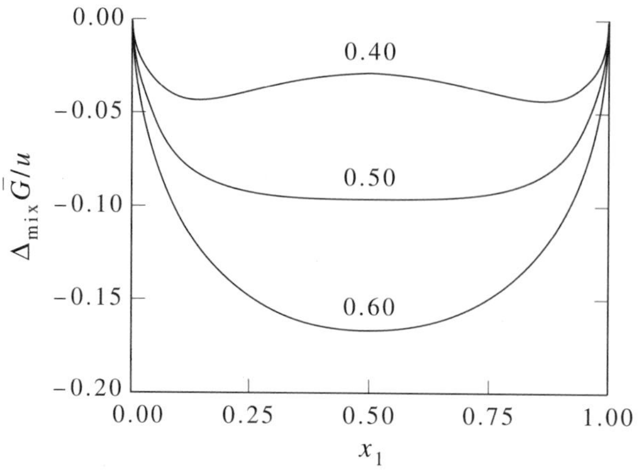


**FIGURE 10.20**
Plots of $\Delta_{\text{mix}}\overline{G}/u$ for $RT/u = 0.60$ (bottom curve), $RT/u = 0.50$ (middle curve), and $RT/u = 0.40$ (top curve).

positive (a minimum) at $x_1 = x_2 = 1/2$ for the curves that lie below the curve with $RT/u = 0.50$, whereas $\partial^2(\Delta_{\text{mix}}\overline{G}/u)/\partial x_1^2$ is negative (a maximum) at $x_1 = x_2 = 1/2$ for curves that lie above it. The region where $\partial^2(\Delta_{\text{mix}}\overline{G}/u)/\partial x_1^2$ is negative represents an unstable region (Problem 10–62) and is similar to the loops of the van der Waals equation or the Redlich-Kwong equation when $T < T_c$ (Figure 2.8), and in this case correspond to regions in which the two liquids are not miscible. The critical value $RT/u = 0.50$ corresponds to a solution critical temperature, $T_c$, where the two liquids are miscible in all proportions at temperatures above $T_c = 0.50u/R$ and immiscible at temperatures below $T_c = u/R$.

Let's consider the curve with $RT/u = 0.40$ in Figure 10.20. The two minima represent two immiscible solutions in equilibrium with each other. The compositions of these two solutions are given by the values of $x_1$ at each minimum. Using Equation 10.62, we have

$$\frac{\partial(\Delta_{\text{mix}}\overline{G}/u)}{\partial x_1} = \frac{RT}{u}[\ln x_1 - \ln(1 - x_1)] + (1 - 2x_1) = 0 \tag{10.63}$$

as the condition for the extrema of $\Delta_{\text{mix}}\overline{G}/u$. First note that $x_1 = 1/2$ solves Equation 10.63 for any value of $RT/u$, which accounts for the fact that all the curves in Figure 10.20 have either a maximum or a minimum at $x_1 = 1/2$. By plotting $(RT/u)[\ln x_1 - \ln(1 - x_1)] + (1 - 2x_1)$ against $x_1$ for various values of $RT/u$, you can see that only $x_1 = 1/2$ satisfies Equation 10.63 for $RT/u \geq 0.50$, whereas two other roots occur for $RT/u < 0.50$. The two roots give the composition of the two miscible solutions in equilibrium with each other. For the case in which $RT/u = 0.40$, the two values of $x_1$ are 0.145 and 0.855. Figure 10.21 shows the mole fraction of component 1 in each of the two immiscible solutions as a function of temperature $(RT/u)$. Note that Figure 10.21 is similar to Figure 10.13.

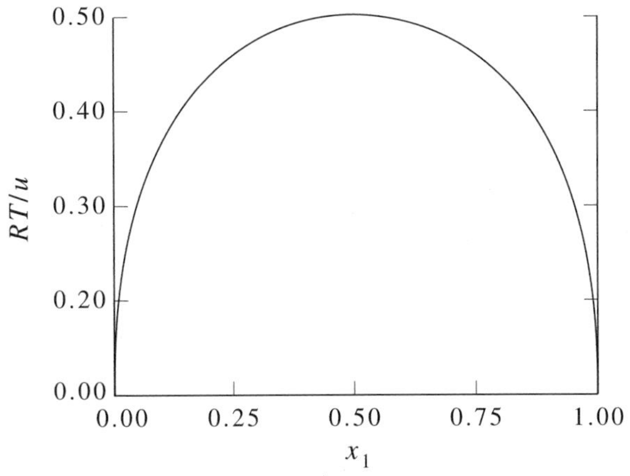


**FIGURE 10.21**
A temperature-composition diagram for a binary system for which $\Delta_{\text{mix}}\overline{G}/u = (RT/u)(x_1 \ln x_1 + x_2 \ln x_2) + x_1 x_2$ (Equation 10.62). The curve gives the compositions of the two immiscible solutions as a function of temperature. There is only one homogeneous phase in the region above the curve and there are two immiscible solutions in equilibrium with each other in the region below the curve.

**EXAMPLE 10–11**

Use Equation 10.63 to calculate the composition of the two immiscible solutions in equilibrium with each other at a temperature given by $RT/u = 0.40$.

SOLUTION: We use the Newton-Raphson method that we introduced in MathChapter A. The function $f(x)$ of Equation A.1 is

$$f(x) = \frac{RT}{u}[\ln x - \ln(1-x)] + 1 - 2x$$

Equation A.1 becomes

$$x_{n+1} = x_n - \frac{\dfrac{RT}{u}[\ln x_n - \ln(1-x_n)] + 1 - 2x_n}{\dfrac{RT}{u}\left[\dfrac{1}{x_n(1-x_n)}\right] - 2}$$

with $RT/u = 0.40$. For one of the solutions, we start with $x_0 = 0.100$ and get

| $n$ | $x_n$ | $f(x_n)$ | $f'(x_n)$ |
|---|---|---|---|
| 0 | 0.100 | −0.07889 | 2.4444 |
| 1 | 0.132 | −0.01695 | 1.4851 |
| 2 | 0.144 | −0.001370 | 1.2509 |
| 3 | 0.145 | −0.000017 | 1.2305 |
| 4 | 0.145 | | |

For the other solution, we start with $x_0 = 0.900$ and get

| $n$ | $x_n$ | $f(x_n)$ | $f'(x_n)$ |
|---|---|---|---|
| 0 | 0.900 | 0.07889 | 2.4444 |
| 1 | 0.868 | 0.01695 | 1.4851 |
| 2 | 0.856 | 0.00137 | 1.2509 |
| 3 | 0.855 | 0.000017 | 1.2305 |
| 4 | 0.855 | | |

We must emphasize once again that the results of this section are a result of the simple randomly-distributed model that we used. Although the model gives a number of qualitative results, it must be borne in mind that it is just an approximate model that we introduced to give some molecular or physical insight into the nature of non-ideal liquid binary solutions.

We will continue our discussion of solutions in the next chapter, where we focus on solutions in which the two components are not soluble in all proportions. In particular, we will discuss solutions of solids in liquids, where the terms solute and solvent are meaningful.

# Problems

**10-1.** In the text, we went from Equation 10.5 to 10.6 using a physical argument involving varying the size of the system while keeping $T$ and $P$ fixed. We could also have used a mathematical process called *Euler's theorem*. Before we can learn about Euler's theorem, we must first define a *homogeneous function*. A function $f(z_1, z_2, \ldots, z_N)$ is said to be homogeneous if

$$f(\lambda z_1, \lambda z_2, \ldots, \lambda z_N) = \lambda f(z_1, z_2, \ldots, z_N)$$

Argue that extensive thermodynamic quantities are homogeneous functions of their extensive variables.

**10-2.** Euler's theorem says that if $f(z_1, z_2, \ldots, z_N)$ is homogeneous, then

$$f(z_1, z_2, \ldots, z_N) = z_1 \frac{\partial f}{\partial z_1} + z_2 \frac{\partial f}{\partial z_2} + \cdots + z_N \frac{\partial f}{\partial z_N}$$

Prove Euler's theorem by differentiating the equation in Problem 10–1 with respect to $\lambda$ and then setting $\lambda = 1$.

Apply Euler's theorem to $G = G(n_1, n_2, T, P)$ to derive Equation 10.6. (*Hint*: Because $T$ and $P$ are intensive variables, they are simply irrevelant variables in this case.)

**10-3.** Use Euler's theorem (Problem 10–2) to prove that

$$Y(n_1, n_2, \ldots, T, P) = \sum n_j \overline{Y}_j$$

for any extensive quantity $Y$.

**10-4.** Apply Euler's theorem to $U = U(S, V, n)$. Do you recognize the resulting equation?

**10-5.** Apply Euler's theorem to $A = A(T, V, n)$. Do you recognize the resulting equation?

**10-6.** Apply Euler's theorem to $V = V(T, P, n_1, n_2)$ to derive Equation 10.7.

**10-7.** The properties of many solutions are given as a function of the mass percent of the components. If we let the mass percent of component-2 be $A_2$, then derive a relation between $A_2$ and the mole fractions, $x_1$ and $x_2$.

**10-8.** The *CRC Handbook of Chemistry and Physics* gives the densities of many aqueous solutions as a function of the mass percentage of solute. If we denote the density by $\rho$ and the mass percentage of component-2 by $A_2$, the *Handbook* gives $\rho = \rho(A_2)$ (in g·mL$^{-1}$). Show that the quantity $V = (n_1 M_1 + n_2 M_2)/\rho(A_2)$ is the volume of the solution containing $n_1$ moles of component 1 and $n_2$ moles of component-2. Now show that

$$\overline{V}_1 = \frac{M_1}{\rho(A_2)}\left[1 + \frac{A_2}{\rho(A_2)}\frac{d\rho(A_2)}{dA_2}\right]$$

and

$$\overline{V}_2 = \frac{M_2}{\rho(A_2)}\left[1 + \frac{(A_2 - 100)}{\rho(A_2)}\frac{d\rho(A_2)}{dA_2}\right]$$

Show that

$$V = n_1\overline{V}_1 + n_2\overline{V}_2$$

in agreement with Equation 10.7.

**10-9.** The density (in g·mol$^{-1}$) of a 1-propanol-water solution at 20°C as a function of $A_2$, the mass percentage of 1-propanol, can be expressed as

$$\rho(A_2) = \sum_{j=0}^{7} \alpha_j A_2^j$$

where

$$\alpha_0 = 0.99823 \qquad \alpha_4 = 1.5312 \times 10^{-7}$$

$$\alpha_1 = -0.0020577 \qquad \alpha_5 = -2.0365 \times 10^{-9}$$

$$\alpha_2 = 1.0021 \times 10^{-4} \qquad \alpha_6 = 1.3741 \times 10^{-11}$$

$$\alpha_3 = -5.9518 \times 10^{-6} \qquad \alpha_7 = -3.7278 \times 10^{-14}$$

Use this expression to plot $\overline{V}_{H_2O}$ and $\overline{V}_{\text{1-propanol}}$ versus $A_2$ and compare your values with those in Figure 10.1.

**10-10.** Given the density of a binary solution as a function of the mole fraction of component 2 $[\rho = \rho(x_2)]$, show that the volume of the solution containing $n_1$ moles of component 1 and

$n_2$ moles of component 2 is given by $V = (n_1 M_1 + n_2 M_2)/\rho(x_2)$, where $M_j$ is the molar mass of component $j$. Now show that

$$\overline{V}_1 = \frac{M_1}{\rho(x_2)}\left[1 + \left(\frac{x_2(M_2 - M_1) + M_1}{M_1}\right)\frac{x_2}{\rho(x_2)}\frac{d\rho(x_2)}{dx_2}\right]$$

and

$$\overline{V}_2 = \frac{M_2}{\rho(x_2)}\left[1 - \left(\frac{x_2(M_2 - M_1) + M_1}{M_2}\right)\frac{1 - x_2}{\rho(x_2)}\frac{d\rho(x_2)}{dx_2}\right]$$

Show that

$$V = n_1\overline{V}_1 + n_2\overline{V}_2$$

in agreement with Equation 10.7.

**10-11.** The density (in $\text{g}\cdot\text{mol}^{-1}$) of a 1-propanol/water solution at 20°C as a function of $x_2$, the mole fraction of 1-propanol, can be expressed as

$$\rho(x_2) = \sum_{j=0}^{4} \alpha_j x_2^j$$

where

$$\alpha_0 = 0.99823 \qquad \alpha_3 = -0.17163$$

$$\alpha_1 = -0.48503 \qquad \alpha_4 = -0.01387$$

$$\alpha_2 = 0.47518$$

Use this expression to calculate the values of $\overline{V}_{\text{H}_2\text{O}}$ and $\overline{V}_{\text{1-propanol}}$ as a function of $x_2$ according to the equation in Problem 10–10.

**10-12.** Use the data in the *CRC Handbook of Chemistry and Physics* to curve fit the density of a water/glycerol solution to a fifth-order polynomial in the mole fraction of glycerol, and then determine the partial molar volumes of water and glycerol as a function of mole fraction. Plot your result.

**10-13.** Just before Example 10–2, we showed that if one component of a binary solution obeys Raoult's law over the entire composition range, the other component does also. Now show that if $\mu_2 = \mu_2' + RT\ln x_2$ for $x_{2,\text{min}} \le x_2 \le 1$, then $\mu_1 = \mu_1' + RT\ln x_1$ for $0 < x_1 < 1 - x_{2,\text{min}}$). Notice that for the range over which $\mu_2$ obeys the simple form given, $\mu_1$ obeys a similarly simple form. If we let $x_{2,\text{min}} = 0$, we obtain $\mu_1 = \mu_1^* + RT\ln x_1$ $(0 \le x_1 \le 1)$.

**10-14.** Continue the calculations in Example 10–3 to obtain $y_2$ as a function of $x_2$ by varying $x_2$ from 0 to 1. Plot your result.

**10-15.** Use your results from Problem 10–14 to construct the pressure-composition diagram in Figure 10.4.

**10-16.** Calculate the relative amounts of liquid and vapor phases at an overall composition of 0.50 for one of the pair of values, $x_2 = 0.38$ and $y_2 = 0.57$, that you obtained in Problem 10–14.

**10-17.** In this problem, we will derive analytic expressions for the pressure-composition curves in Figure 10.4. The liquid (upper) curve is just

$$P_{\text{total}} = x_1 P_1^* + x_2 P_2^* = (1 - x_2)P_1^* + x_2 P_2^* = P_1^* + x_2(P_2^* - P_1^*) \tag{1}$$

which is a straight line, as seen in Figure 10.4. Solve the equation

$$y_2 = \frac{x_2 P_2^*}{P_{\text{total}}} = \frac{x_2 P_2^*}{P_1^* + x_2(P_2^* - P_1^*)} \tag{2}$$

for $x_2$ in terms of $y_2$ and substitute into Equation (1) to obtain

$$P_{\text{total}} = \frac{P_1^* P_2^*}{P_2^* - y_2(P_2^* - P_1^*)}$$

Plot this result versus $y_2$ and show that it gives the vapor (lower) curve in Figure 10.4.

**10-18.** Prove that $y_2 > x_2$ if $P_2^* > P_1^*$ and that $y_2 < x_2$ if $P_2^* < P_1^*$. Interpret this result physically.

**10-19.** Tetrachloromethane and trichloroethylene form an essentially ideal solution at 40°C at all concentrations. Given that the vapor pressure of tetrachloromethane and trichloroethylene at 40°C are 214 torr and 138 torr, respectively, plot the pressure-composition diagram for this system (see Problem 10–17).

**10-20.** The vapor pressures of tetrachloromethane (1) and trichloroethylene (2) between 76.8°C and 87.2°C can be expressed empirically by the formulas

$$\ln(P_1^*/\text{torr}) = 15.8401 - \frac{2790.78}{t + 226.4}$$

and

$$\ln(P_2^*/\text{torr}) = 15.0124 - \frac{2345.4}{t + 192.7}$$

where $t$ is the Celsius temperature. Assuming that tetrachloromethane and trichloroethylene form an ideal solution at all compositions between 76.8°C and 87.2°C, calculate the values of $x_1$ and $y_1$ at 82.0°C (at an ambient pressure of 760 torr).

**10-21.** Use the data in Problem 10–20 to construct the entire temperature-composition diagram of a tetrachloromethane/trichlororethylene solution.

**10-22.** The vapor pressures of benzene and toluene between 80°C and 110°C as a function of the Kelvin temperature are given by the empirical formulas

$$\ln(P_{\text{benz}}^*/\text{torr}) = -\frac{3856.6\text{ K}}{T} + 17.551$$

and

$$\ln(P^*_{\rm tol}/\text{torr}) = -\frac{4514.6\ \text{K}}{T} + 18.397$$

Assuming that benzene and toluene form an ideal solution, use these formulas to construct a temperature-composition diagram of this system at an ambient pressure of 760 torr.

**10-23.** Construct the temperature-composition diagram for 1-propanol and 2-propanol in Figure 10.5 by varying $t$ from 82.3°C (the boiling point of 2-propanol) to 97.2°C (the boiling point of 1-propanol), calculating (1) $P^*_1$ and $P^*_2$ at each temperature (see Example 10–5), (2) $x_1$ according to $x_1 = (P^*_2 - 760)/(P^*_2 - P^*_1)$, and (3) $y_1$ according to $y_1 = x_1 P^*_1/760$. Now plot $t$ versus $x_1$ and $y_1$ on the same graph to obtain the temperature-composition diagram.

**10-24.** Prove that $\overline{V}_j = \overline{V}^*_j$ for an ideal solution, where $\overline{V}^*_j$ is the molar volume of pure component $j$.

**10-25.** The volume of mixing of miscible liquids is defined as the volume of the solution minus the volume of the individual pure components. Show that

$$\Delta_{\rm mix}\overline{V} = \sum x_1(\overline{V}_i - \overline{V}^*_i)$$

at constant $P$ and $T$, where $\overline{V}^*_i$ is the molar volume of pure component $i$. Show that $\Delta_{\rm mix}\overline{V} = 0$ for an ideal solution (see Problem 10–24).

**10-26.** Suppose the vapor pressures of the two components of a binary solution are given by

$$P_1 = x_1 P^*_1 e^{x_2^2/2}$$

and

$$P_2 = x_2 P^*_2 e^{x_1^2/2}$$

Given that $P^*_1 = 75.0$ torr and $P^*_2 = 160$ torr, calculate the total vapor pressure and the composition of the vapor phase at $x_1 = 0.40$.

**10-27.** Plot $y_1$ versus $x_1$ for the system described in the previous problem. Why does the curve lie below the straight line connecting the origin with the point $x_1 = 1$, $y_1 = 1$? Describe a system for which the curve would lie above the diagonal line.

**10-28.** Use the expressions for $P_1$ and $P_2$ given in Problem 10–26 to construct a pressure-composition diagram.

**10-29.** The vapor pressure (in torr) of the two components in a binary solution are given by

$$P_1 = 120x_1 e^{0.20x_2^2+0.10x_2^3}$$

and

$$P_2 = 140x_2 e^{0.35x_1^2-0.10x_1^3}$$

Determine the values of $P^*_1$, $P^*_2$, $k_{\rm H,1}$, and $k_{\rm H,2}$.

**10-30.** Suppose the vapor pressure of the two components of a binary solution are given by

$$P_1 = x_1 P_1^* e^{\alpha x_2^2 + \beta x_2^3}$$

and

$$P_2 = x_2 P_2^* e^{(\alpha + 3\beta/2)x_1^2 - \beta x_1^3}$$

Show that $k_{H,1} = P_1^* e^{\alpha+\beta}$ and $k_{H,2} = P_2^* e^{\alpha+\beta/2}$.

**10-31.** The empirical expression for the vapor pressure that we used in Examples 10–6 and 10–7, for example,

$$P_1 = x_1 P_1^* e^{\alpha x_2^2 + \beta x_2^3 + \cdots}$$

is sometimes called the *Margules equation*. Use Equation 10.29 to prove that there can be no linear term in the exponential factor in $P_1$, for otherwise $P_2$ will not satisfy Henry's law as $x_2 \to 0$.

**10-32.** In the text, we showed that the Henry's law behavior of component-2 as $x_2 \to 0$ is a direct consequence of the Raoult's law behavior of component 1 as $x_1 \to 1$. In this problem, we will prove the converse: the Raoult's law behavior of component 1 as $x_1 \to 1$ is a direct consequence of the Henry's law behavior of component-2 as $x_2 \to 0$. Show that the chemical potential of component-2 as $x_2 \to 0$ is

$$\mu_2(T, P) = \mu_2^\circ(T) + RT \ln k_{H,2} + RT \ln x_2 \qquad x_2 \to 0$$

Differentiate $\mu_2$ with respect to $x_2$ and substitute the result into the Gibbs-Duhem equation to obtain

$$d\mu_1 = RT \frac{dx_1}{x_1} \qquad x_2 \longrightarrow 0$$

Integrate this expression from $x_1 = 1$ to $x_1 \approx 1$ and use the fact that $\mu_1(x_1 = 1) = \mu_1^*$ to obtain

$$\mu_1(T, P) = \mu_1^*(T) + RT \ln x_1 \qquad x_1 \to 1$$

which is the Raoult's law expression for chemical potential.

**10-33.** In Example 10–7, we saw that if

$$P_1 = x_1 P_1^* e^{\alpha x_2^2 + \beta x_2^3}$$

then

$$P_2 = x_2 P_2^* e^{(\alpha + 3\beta/2)x_1^2 - \beta x_1^3}$$

Show that this result follows directly from Equation 10.29.

**10-34.** Suppose we express the vapor pressures of the components of a binary solution by

$$P_1 = x_1 P_1^* e^{\alpha x_2^2}$$

and

$$P_2 = x_2 P_2^* e^{\beta x_1^2}$$

Use the Gibbs-Duhem equation or Equation 10.29 to prove that $\alpha$ must equal $\beta$.

**10-35.** Use Equation 10.29 to show that if one component of a binary solution obeys Raoult's law for all concentrations, then the other component also obeys Raoult's law for all concentrations.

**10-36.** Use Equation 10.29 to show that if one component of a binary solution has positive deviations from Raoult's law, then the other component must also.

*The following nine problems develop the idea of a regular solution.*

**10-37.** If the vapor pressures of the two components in a binary solution are given by

$$P_1 = x_1 P_1^* e^{ux_2^2/RT} \quad \text{and} \quad P_2 = x_2 P_2^* e^{ux_1^2/RT}$$

show that

$$\Delta_{\text{mix}}\overline{G}/u = \Delta_{\text{mix}}G/(n_1 + n_2)u = \frac{RT}{u}(x_1 \ln x_1 + x_2 \ln x_2) + x_1 x_2$$

$$\Delta_{\text{mix}}\overline{S}/R = \Delta_{\text{mix}}S/(n_1 + n_2)R = -(x_1 \ln x_1 + x_2 \ln x_2)$$

and

$$\Delta_{\text{mix}}\overline{H}/u = \Delta_{\text{mix}}H/(n_1 + n_2)u = x_1 x_2$$

A solution that satisfies these equations is called a *regular solution*. The statistical thermodynamic model of binary solutions presented in Section 10–9 shows that $u$ is proportional to $2\varepsilon_{12} - \varepsilon_{11} - \varepsilon_{22}$, where $\varepsilon_{ij}$ is the interaction energy between molecules of components $i$ and $j$. Note that $u = 0$ if $\varepsilon_{12} = (\varepsilon_{11} + \varepsilon_{22})/2$, which means that energetically, molecules of components 1 and 2 "like" the opposite molecules as well as their own.

**10-38.** Prove that $\Delta_{\text{mix}}\overline{G}$, $\Delta_{\text{mix}}\overline{S}$, and $\Delta_{\text{mix}}\overline{H}$ in the previous problem are symmetric about the point $x_1 = x_2 = 1/2$.

**10-39.** Plot $P_1/P_1^* = x_1 e^{ux_2^2/RT}$ versus $x_1$ for $RT/u = 0.60, 0.50, 0.45, 0.40$, and $0.35$. Note that some of the curves have regions where the slope is negative. The following problem has you show that this behavior occurs when $RT/u < 0.50$. These regions are similar to the loops of the van der Waals equation or the Redlich-Kwong equation when $T < T_c$ (Figure 2.8), and in this case correspond to regions in which the two liquids are not miscible. The critical value $RT/u = 0.50$ corresponds to a solution critical temperature, $0.50u/R$.

**10-40.** Differentiate $P_1 = x_1 P_1^* e^{u(1-x_1)^2/RT}$ with respect to $x_1$ to prove that $P_1$ has a maximum or a minimum at the points $x_1 = \frac{1}{2} \pm \frac{1}{2}(1 - \frac{2RT}{u})^{1/2}$. Show that $RT/u \leq 0.50$ for either a maximum or a minimum to occur. Do the positions of these extremes when $RT/u = 0.35$ correspond to the plot you obtained in the previous problem?

**10-41.** Plot $\Delta_{\text{mix}}\overline{G}/u$ in Problem 10–37 versus $x_1$ for $RT/u = 0.60, 0.50, 0.45, 0.40$, and $0.35$. Note that some of the curves have regions where $\partial^2\Delta_{\text{mix}}\overline{G}/\partial x_1^2 < 0$. These regions correspond to regions in which the two liquids are not miscible. Show that $RT/u = 0.50$ is a critical value, in the sense that unstable regions occur only when $RT/u < 0.50$. (See the previous problem.)

**10-42.** Plot both $P_1/P_1^* = x_1 e^{ux_2^2/RT}$ and $P_2/P_2^* = x_2 e^{ux_1^2/RT}$ for $RT/u = 0.60, 0.50, 0.45, 0.40$, and $0.35$. Prove that the loops occur for values of $RT/u < 0.50$.

**10-43.** Plot both $P_1/P_1^* = x_1 e^{ux_2^2/RT}$ and $P_2/P_2^* = x_2 e^{ux_1^2/RT}$ for $RT/u = 0.40$. The loops indicate regions in which the two liquids are not miscible, as explained in Problem 10–39. Draw a horizontal line connecting the left-side and the right-side intersections of the two curves. This line, which connects states in which the vapor pressure (or chemical potential) of each component is the same in the two solutions of different composition, corresponds to one of the horizontal lines in Figure 10.12. Now set $P_1/P_1^* = x_1 e^{ux_2^2/RT}$ equal to $P_2/P_2^* = x_2 e^{ux_1^2/RT}$ and solve for $RT/u$ in terms of $x_1$. Plot $RT/u$ against $x_1$ and obtain a coexistence curve like the one in Figure 10.12.

**10-44.** The molar enthalpies of mixing of solutions of tetrachloromethane (1) and cyclohexane (2) at 25°C are listed below.

| $x_1$ | $\Delta_{\text{mix}}\overline{H}/\text{J}\cdot\text{mol}^{-1}$ |
|---|---|
| 0.0657 | 37.8 |
| 0.2335 | 107.9 |
| 0.3495 | 134.9 |
| 0.4745 | 146.7 |
| 0.5955 | 141.6 |
| 0.7213 | 118.6 |
| 0.8529 | 73.6 |

Plot $\Delta_{\text{mix}}\overline{H}$ against $x_1x_2$ according to Problem 10–37. Do tetrachloromethane and cyclohexane form a regular solution?

**10-45.** The molar enthalpies of mixing of solutions of tetrahydrofuran and trichloromethane at 25°C are listed below.

| $x_{\text{THF}}$ | $\Delta_{\text{mix}}\overline{H}/\text{J}\cdot\text{mol}^{-1}$ |
|---|---|
| 0.0568 | −0.469 |
| 0.1802 | −1.374 |
| 0.3301 | −2.118 |
| 0.4508 | −2.398 |
| 0.5702 | −2.383 |
| 0.7432 | −1.888 |
| 0.8231 | −1.465 |
| 0.9162 | −0.802 |

Do tetrahydrofuran and trichloromethane form a regular solution?

**10-46.** Derive the equation

$$x_1 d \ln \gamma_1 + x_2 d \ln \gamma_2 = 0$$

by starting with Equation 10.11. Use this equation to obtain the same result as in Example 10–8.

**10-47.** The vapor pressure data for carbon disulfide in Table 10.1 can be curve fit by

$$P_1 = x_1(514.5 \text{ torr})e^{1.4967x_2^2 - 0.68175x_2^3}$$

Using the results of Example 10–7, show that the vapor pressure of dimethoxymethane is given by

$$P_2 = x_2(587.7 \text{ torr})e^{0.4741x_1^2 + 0.68175x_1^3}$$

Now plot $P_2$ versus $x_2$ and compare the result with the data in Table 10.1. Do carbon disulfide and dimethoxymethane form a regular solution at 35.2°C? Plot $\overline{G}^{\text{E}}$ against $x_1$. Is the plot symmetric about a vertical line at $x_1 = 1/2$?

**10-48.** A mixture of trichloromethane and acetone with $x_{\text{acet}} = 0.713$ has a total vapor pressure of 220.5 torr at 28.2°C, and the mole fraction of acetone in the vapor is $y_{\text{acet}} = 0.818$. Given that the vapor pressure of pure trichloromethane at 28.2°C is 221.8 torr, calculate the activity and the activity coefficient (based upon a Raoult's law standard state) of trichloromethane in the mixture. Assume the vapor behaves ideally.

**10-49.** Consider a binary solution for which the vapor pressure (in torr) of one of the components (say component 1) is given empirically by

$$P_1 = 78.8x_1e^{0.65x_2^2 + 0.18x_2^3}$$

Calculate the activity and the activity coefficient of component 1 when $x_1 = 0.25$ based on a solvent and a solute standard state.

**10-50.** Some vapor pressure data for ethanol/water solutions at 25°C are listed below.

| $x_{\text{ethanol}}$ | $P_{\text{ethanol}}/\text{torr}$ | $P_{\text{water}}/\text{torr}$ |
|---|---|---|
| 0.00 | 0.00 | 23.78 |
| 0.02 | 4.28 | 23.31 |
| 0.05 | 9.96 | 22.67 |
| 0.08 | 14.84 | 22.07 |
| 0.10 | 17.65 | 21.70 |
| 0.20 | 27.02 | 20.25 |
| 0.30 | 31.23 | 19.34 |
| 0.40 | 33.93 | 18.50 |
| 0.50 | 36.86 | 17.29 |
| 0.60 | 40.23 | 15.53 |
| 0.70 | 43.94 | 13.16 |
| 0.80 | 48.24 | 9.89 |
| 0.90 | 53.45 | 5.38 |
| 0.93 | 55.14 | 3.83 |
| 0.96 | 56.87 | 2.23 |
| 0.98 | 58.02 | 1.13 |
| 1.00 | 59.20 | 0.00 |

Plot these data to determine the Henry's law constant for ethanol in water and for water in ethanol at 25°C.

**10-51.** Using the data in Problem 10–50, plot the activity coefficients (based upon Raoult's law) of both ethanol and water against the mole fraction of ethanol.

**10-52.** Using the data in Problem 10–50, plot $\overline{G}^{\text{E}}/RT$ against $x_{\text{H}_2\text{O}}$. Is a water/ethanol solution at 25°C a regular solution?

**10-53.** Some vapor pressure data for a 2-propanol/benzene solution at 25°C are

| $x_{\text{2-propanol}}$ | $P_{\text{2-propanol}}/\text{torr}$ | $P_{\text{total}}/\text{torr}$ |
|---|---|---|
| 0.000 | 0.0 | 94.4 |
| 0.059 | 12.9 | 104.5 |
| 0.146 | 22.4 | 109.0 |
| 0.362 | 27.6 | 108.4 |
| 0.521 | 30.4 | 105.8 |
| 0.700 | 36.4 | 99.8 |
| 0.836 | 39.5 | 84.0 |
| 0.924 | 42.2 | 66.4 |
| 1.000 | 44.0 | 44.0 |

Plot the activities and the activity coefficients of 2-propanol and benzene relative to a Raoult's law standard state versus the mole fraction of 2-propanol.

**10-54.** Using the data in Problem 10–53, plot $\overline{G}^{\text{E}}/RT$ versus $x_{\text{2-propanol}}$.

**10-55.** *Excess thermodynamic quantities* are defined relative to the values the quantities would have if the pure components formed an ideal solution at the same given temperature and pressure. For example, we saw that (Equation 10.47)

$$\frac{\overline{G}^{\text{E}}}{RT} = \frac{G^{\text{E}}}{(n_1 + n_2)RT} = x_1 \ln \gamma_1 + x_2 \ln \gamma_2$$

Show that

$$\frac{\overline{S}^{\text{E}}}{R} = \frac{S^{\text{E}}}{(n_1 + n_2)R} = -(x_1 \ln \gamma_1 + x_2 \ln \gamma_2) - T\left(x_1 \frac{\partial \ln \gamma_1}{\partial T} + x_2 \frac{\partial \ln \gamma_2}{\partial T}\right)$$

**10-56.** Show that

$$\overline{G}^{\text{E}} = \frac{G^{\text{E}}}{n_1 + n_2} = u x_1 x_2$$

$$\overline{S}^{\text{E}} = \frac{S^{\text{E}}}{n_1 + n_2} = 0$$

and

$$\overline{H}^{\text{E}} = \frac{H^{\text{E}}}{n_1 + n_2} = u x_1 x_2$$

for a regular solution (see Problem 10–37).

**10-57.** Example 10–7 expresses the vapor pressures of the two components of a binary solution as

$$P_1 = x_1 P_1^* e^{\alpha x_2^2 + \beta x_2^3}$$

and

$$P_2 = x_2 P_2^* e^{(\alpha + 3\beta/2)x_1^2 - \beta x_1^3}$$

Show that these expressions are equivalent to

$$\gamma_1 = e^{\alpha x_2^2 + \beta x_2^3} \quad \text{and} \quad \gamma_2 = e^{(\alpha + 3\beta/2)x_1^2 - \beta x_1^3}$$

Using these expressions for the activity coefficients, derive an expression for $\overline{G}^{\text{E}}$ in terms of $\alpha$ and $\beta$. Show that your expression reduces to that for $\overline{G}^{\text{E}}$ for a regular solution.

**10-58.** Prove that the maxima or minima of $\Delta_{\text{mix}}\overline{G}$ defined in Problem 10–37 occur at $x_1 = x_2 = 1/2$ for any value of $RT/u$. Now prove that

$$\frac{\partial^2 \Delta_{\text{mix}}\overline{G}}{\partial x_1^2} \begin{cases} > 0 & \text{for } RT/u > 0.50 \\ = 0 & \text{for } RT/u = 0.50 \\ < 0 & \text{for } RT/u < 0.50 \end{cases}$$

at $x_1 = x_2 = 1/2$. Is this result consistent with the graphs you obtained in Problem 10–41?

**10-59.** Use the data in Table 10.1 to plot Figures 10.15 through 10.18.

**10-60.** Use Equation 10.62 to show that the slopes of all the curves in Figure 10.20 are equal to zero when $x_1 = x_2 = 1/2$.

**10-61.** Derive Equations 10.59 through 10.61.

**10-62.** In this problem, we will prove that $(\partial^2 \Delta_{\text{mix}} \overline{G}/\partial x_1^2)$ must be greater than zero in a stable region of a binary solution. First choose some point $x_1^0$ in Figure 10.20 and draw a straight line at $x_1^0$ tangent to the curve of $\Delta_{\text{mix}} \overline{G}$ against $x_1$. Now argue that $\Delta_{\text{mix}} \overline{G}$ must lie above the tangent line for the region around $x_1^0$ to be stable, or that

$$\Delta_{\text{mix}} \overline{G} \text{ at } x_1 \text{ around } x_1^0 > \text{tangent line at } x_1^0$$

Show that the equation for the tangent line is

$$y = (x_1 - x_1^0) \left( \frac{\partial \Delta_{\text{mix}} \overline{G}}{\partial x_1} \right)_{x_1 = x_1^0} + \Delta_{\text{mix}} \overline{G}(x_1^0)$$

Now expand $\Delta_{\text{mix}} \overline{G}$ in a Taylor series (MathChapter C) about the point $x_1^0$ and show that

$$\left( \frac{\partial^2 \Delta_{\text{mix}} \overline{G}}{\partial x_1^2} \right)_{x_1 = x_1^0} > 0$$

for the binary solution to be stable.

**Peter Debye** (left) was born in Maastricht, the Netherlands, on March 24, 1884 and died in 1966. Debye was originally trained as an electrical engineer but turned his attention to physics, receiving his Ph.D. from the University of Munich in 1908. After holding positions in Switzerland, the Netherlands, and Germany, he moved to the University of Berlin in the early 1930s. Although he had been assured that he would be able retain his Dutch citizenship, Debye found that he would be unable to continue his work in Berlin unless he became a German citizen. He refused and left Germany in 1939 for Cornell University, where he remained for the rest of his life, becoming an American citizen in 1946. Debye was awarded the Nobel Prize for chemistry in 1936 "for his contributions to our knowledge of molecular structure through his investigations on dipole moments and on the diffraction of X rays and electrons in gases." **Erich Hückel** (right) was born in Göttingen, Germany, on August 19, 1896 and died in 1980. He received his Ph.D. in physics from the University of Göttingen in 1921. He later worked with Peter Debye in Zurich, and together they developed a theory for the thermodynamic properties of solutions of strong electrolytes that is now known as the Debye–Hückel theory. Hückel also developed Hückel molecular orbital theory, which we learned in Chapter 10 applies to conjugated and aromatic molecules. Hückel was appointed professor of theoretical physics at the University of Marburg in 1937, where he remained until his retirement.

CHAPTER 11

# Solutions II: Solid–Liquid Solutions

In the previous chapter, we studied binary solutions, such as ethanol/water solutions, in which the two components were miscible in all proportions. In such solutions, either component can be treated as a solvent. In this chapter, we will study solutions in which one of the components is present at much smaller concentrations than the other, so that the terms "solute" and "solvent" are meaningful. We will introduce a solute standard state based upon Henry's law such that the activity of the solute becomes equal to its concentration as its concentration goes to zero. In the first few sections, we will study solutions of nonelectrolytes, and then solutions of electrolytes. Unlike for solutions of nonelectrolytes, we will be able to present exact expressions for the activities and activity coefficients in dilute solutions of electrolytes. In Sections 11–3 and 11–4, we will discuss the colligative properties of solutions, such as osmotic pressure, as well as the depression of the freezing point and elevation of the boiling point of a solvent by the addition of solute.

## 11–1. We Use a Raoult's Law Standard State for the Solvent and a Henry's Law Standard State for the Solute for Solutions of Solids Dissolved in Liquids

In Section 10–8, we considered solutions in which one of the components is only sparingly soluble in the other. In cases such as these, we use the terms *solute* for the sparingly soluble component and *solvent* for the component in excess. We customarily denote solvent quantities by a subscript 1 and solute quantities by a subscript 2. The activities we defined for the solvent and solute are such that $a_1 \to x_1$ as $x_1 \to 1$ and $a_2 \to x_2$ as $x_2 \to 0$. Recall that $a_1$ is defined with respect to a Raoult's law standard state (Equation 10.39)

$$a_1 = \frac{P_1}{P_1^*} \qquad \text{(Raoult's law standard state)} \qquad (11.1)$$

and that $a_2$ is defined with respect to a Henry's law standard state (Equation 10.43)

$$a_{2x} = \frac{P_2}{k_{\mathrm{H},x}} \qquad \text{(Henry's law standard state)} \tag{11.2}$$

where the subscript $x$ emphasizes that $a_{2x}$ and $k_{\mathrm{H},x}$ are based on a mole fraction scale ($P_2 = k_{\mathrm{H},x}x_2$). Even if the solute does not have a measurable vapor pressure, defining the activity by Equation 11.2 is nevertheless convenient because the ratio is still meaningful; even though $P_2$ and $k_{\mathrm{H},2}$ may be exceedingly small, the ratio $P_2/k_{\mathrm{H},2}$ is finite.

Although we have defined the activities of the solvent and solute in terms of mole fractions, the use of mole fractions to express the concentration of a solute in a dilute solution is not numerically convenient. A more convenient unit is *molality* ($m$), which is defined as the number of moles of solute per 1000 grams of solvent. In an equation, we have

$$m = \frac{n_2}{1000 \text{ g solvent}} \tag{11.3}$$

where $n_2$ is the number of moles of solute (subscript 2). Note that the units of molality are $\mathrm{mol \cdot kg^{-1}}$. We say that a solution containing 2.00 moles of NaCl in 1.00 kg of water is 2.00 molal, or that it is a 2.00 $\mathrm{mol \cdot kg^{-1}}$ NaCl(aq) solution. The relation between the mole fraction of solute ($x_2$) and molality ($m$) is

$$x_2 = \frac{n_2}{n_1 + n_2} = \frac{m}{\dfrac{1000\ \mathrm{g \cdot kg^{-1}}}{M_1} + m} \tag{11.4}$$

where $M_1$ is the molar mass ($\mathrm{g \cdot mol^{-1}}$) of the solvent. The term $1000\ \mathrm{g \cdot kg^{-1}}/M_1$ is the number of moles of solvent ($n_1$) in 1000 g of solvent and $m$, by definition, is the number of moles of solute in 1000 g of solvent. In the case of water, $1000\ \mathrm{mol \cdot kg^{-1}}/M_1$ is equal to 55.506 $\mathrm{mol \cdot kg^{-1}}$, so Equation 11.4 becomes

$$x_2 = \frac{m}{55.506\ \mathrm{mol \cdot kg^{-1}} + m} \tag{11.5}$$

Note that $x_2$ and $m$ are directly proportional to each other if $m \ll 55.506\ \mathrm{mol \cdot kg^{-1}}$, which is the case for dilute solutions.

**EXAMPLE 11–1**

Calculate the mole fraction of a 0.200 $\mathrm{mol \cdot kg^{-1}}$ $C_{12}H_{22}O_{11}$(aq) solution.

SOLUTION: The solution contains 0.200 moles of sucrose per 1000.0 kg of water. The mole fraction of sucrose is

$$x_2 = \frac{n_2}{n_1 + n_2} = \frac{0.200 \text{ mol}}{\dfrac{1000.0 \text{ g}}{18.02\ \mathrm{g \cdot mol^{-1}}} + 0.200 \text{ mol}} = 0.000359$$

We define the solute activity in terms of molality by requiring that

$$a_{2m} \longrightarrow m \quad \text{as} \quad m \longrightarrow 0 \tag{11.6}$$

where the subscript $m$ emphasizes that $a_{2m}$ is based on a molality scale. We can express Henry's law in terms of the molality rather than the mole fraction by $P_2 = k_{\mathrm{H},m} m$, where once again the subscript $m$ emphasizes that $k_{\mathrm{H},m}$ is based on a molality scale. In terms of $k_{\mathrm{H},m}$, the activity of the solute is defined by

$$a_{2m} = \frac{P_2}{k_{\mathrm{H},m}} \tag{11.7}$$

Another common concentration unit is *molarity* ($c$), which is the number of moles of solute per 1000 mL of solution. In an equation,

$$c = \frac{n_2}{1000 \text{ mL solution}} \tag{11.8}$$

Note that molarity has units of mol·L$^{-1}$. We say that a solution containing 2.00 moles of NaCl in 1.00 liter of solution is a 2.00-molar solution, or that it is a 2.00 mol·L$^{-1}$ NaCl(aq) solution.

We define the solute activity in terms of molarity by requiring that

$$a_{2c} \longrightarrow c \quad \text{as} \quad c \longrightarrow 0 \tag{11.9}$$

where the subscript $c$ emphasizes that $a_{2c}$ is based on a molarity scale. We can express Henry's law in terms of the molarity rather than the mole fraction of solute by $P_2 = k_{\mathrm{H},c} c$, where once again the subscript $c$ emphasizes that $k_{\mathrm{H},c}$ is based on a molarity scale. In terms of $k_{\mathrm{H},c}$, the activity of the solute is defined by

$$a_{2c} = \frac{P_2}{k_{\mathrm{H},c}} \tag{11.10}$$

Converting from molarity to molality is easy if we know the density of the solution, which is available for many solutions in handbooks. For example, the density of a 2.450 mol·L$^{-1}$ aqueous sucrose solution at 20°C is 1.3103 g·mL$^{-1}$. Thus, there are 838.6 g of sucrose in 1000 mL of solution, which has a total mass of 1310.3 g. Of these 1310.3 g, 838.6 g are due to sucrose, so 1310.3 g − 838.6 g = 471.7 g are due to water. The molality then is given by

$$m = \frac{2.450 \text{ mol sucrose}}{471.7 \text{ g H}_2\text{O}} \times \frac{1000 \text{ g H}_2\text{O}}{\text{kg H}_2\text{O}} = 5.194 \text{ mol·kg}^{-1}$$

**EXAMPLE 11–2**

The density (in g·mL$^{-1}$) of an aqueous sucrose solution can be expressed as

$$\rho/\text{g·mL}^{-1} = 0.9982 + (0.1160 \text{ kg·mol}^{-1})m - (0.0156 \text{ kg}^2\text{·mol}^{-2})m^2 + (0.0011 \text{ kg}^3\text{·mol}^{-3})m^3 \qquad 0 \le m \le 6 \text{ mol·kg}^{-1}$$

Calculate the molarity of a 2.00-molal aqueous sucrose solution.

SOLUTION: A 2.00-molal aqueous sucrose solution contains 2.00 moles (684.6 g) of sucrose per 1000 g of $H_2O$, or 2.00 moles of sucrose in 1684.6 g of solution. The density of the solution is given by

$$\begin{aligned}\rho/\text{g}\cdot\text{mL}^{-1} &= 0.9982 + (0.1160\ \text{kg}\cdot\text{mol}^{-1})(2.00\ \text{mol}\cdot\text{kg}^{-1}) \\ &\quad - (0.0156\ \text{kg}^2\cdot\text{mol}^{-2})(4.00\ \text{mol}^2\cdot\text{kg}^{-2}) \\ &\quad + (0.0011\ \text{kg}^3\cdot\text{mol}^{-3})(8.00\ \text{mol}^3\cdot\text{kg}^{-3}) \\ &= 1.177\end{aligned}$$

so the volume of the solution is

$$V = \frac{\text{mass}}{\text{density}} = \frac{1684.6\ \text{g}}{1.177\ \text{g}\cdot\text{mL}^{-1}} = 1432\ \text{mL}$$

Therefore, the molarity of the solution is

$$c = \frac{2.00\ \text{mol sucrose}}{1.432\ \text{L}} = 1.40\ \text{mol}\cdot\text{L}^{-1}$$

Problem 11–5 asks you to derive a general relation between $c$ and $m$.

**EXAMPLE 11–3**

Given the density ($\rho$) of the solution in $\text{g}\cdot\text{mL}^{-1}$, derive a general relation between $x_2$ and $c$.

SOLUTION: Consider exactly a one liter sample of the solution. In this case, $c = n_2$, the number of moles of solute in the one-liter sample. The mass of the solution is given by

$$\text{mass of the solution per liter} = (1000\ \text{mL}\cdot\text{L}^{-1})\rho$$

so the mass of the solvent is

$$\begin{aligned}\text{mass of the solvent per liter} &= \text{mass of the solution} - \text{mass of the solute} \\ &= (1000\ \text{mL}\cdot\text{L}^{-1})\rho - cM_2\end{aligned}$$

where $M_2$ is the molar mass ($\text{g}\cdot\text{mol}^{-1}$) of the solute. Therefore, $n_1$, the number of moles of solvent, is

$$n_1 = \frac{(1000\ \text{mL}\cdot\text{L}^{-1})\rho - cM_2}{M_1}$$

so

$$x_2 = \frac{n_2}{n_1 + n_2} = \frac{c}{\dfrac{(1000\ \text{mL}\cdot\text{L}^{-1})\rho - cM_2}{M_1} + c}$$

$$= \frac{cM_1}{(1000\ \text{mL}\cdot\text{L}^{-1})\rho + c(M_1 - M_2)} \qquad (11.11)$$

Table 11.1 summarizes the equations for the activities we have defined for the various concentration scales. In each case, the activity coefficient $\gamma$ is defined by dividing the activity by the appropriate concentration. Thus, for example, $\gamma_m = a_{2m}/m$. Problem 11–12 asks you to derive a relation between the various solute activity coefficients in Table 11.1.

## 11–2. The Activity of a Nonvolatile Solute Can Be Obtained from the Vapor Pressure of the Solvent

The equations for the solute activities in Table 11.1 are applicable to nonvolatile as well as volatile solutes. The vapor pressure of a nonvolatile solute is so low, however, that these equations are not practical to use. Fortunately, the Gibbs–Duhem equation provides us with a way to determine the activity of a nonvolatile solute from a measurement of the activity of the solvent. We will illustrate this procedure using an

**TABLE 11.1**
A summary of the equations for the activities used for the various concentration scales for dilute solutions.

| | | |
|---|---|---|
| **Solvent—Raoult's law standard state** | | |
| $a_1 = \dfrac{P_1}{P_1^*}$ | $a_1 \to x_1$ as $x_1 \to 1$ | |
| $\gamma_1 = \dfrac{a_1}{x_1}$ | $P_1 \to P_1^* x_1$ as $x_1 \to 1$ | (Raoult's law) |
| **Solute—Henry's law standard state** | | |
| *Mole fraction scale* | | |
| $a_{2x} = \dfrac{P_2}{k_{\mathrm{H},x}}$ | $a_{2x} \to x_2$ as $x_2 \to 0$ | |
| $\gamma_{2x} = \dfrac{a_{2x}}{x_2}$ | $P_2 \to k_{\mathrm{H},x} x_2$ as $x_2 \to 0$ | (Henry's law) |
| *Molality scale* | | |
| $a_{2m} = \dfrac{P_2}{k_{\mathrm{H},m}}$ | $a_{2m} \to m$ as $m \to 0$ | |
| $\gamma_{2m} = \dfrac{a_{2m}}{m}$ | $P_2 \to k_{\mathrm{H},m} m$ as $m \to 0$ | (Henry's law) |
| *Molarity scale* | | |
| $a_{2c} = \dfrac{P_2}{k_{\mathrm{H},c}}$ | $a_{2c} \to c$ as $c \to 0$ | |
| $\gamma_{2c} = \dfrac{a_{2c}}{c}$ | $P_2 \to k_{\mathrm{H},c} c$ as $c \to 0$ | (Henry's law) |

aqueous solution of sucrose. According to a Raoult's law standard state, the activity of the water is given by $P_1/P_1^*$. Now let's consider a dilute solution, in which case $a_1 = x_1$. We now want to relate $a_1$ to the molality of the solute, $m$. For a dilute solution, $m \ll 55.506\ \text{mol}\cdot\text{kg}^{-1}$, so we can neglect $m$ compared with $55.506\ \text{mol}\cdot\text{kg}^{-1}$ in the denominator of Equation 11.5 and write

$$x_2 \approx \frac{m}{55.506\ \text{mol}\cdot\text{kg}^{-1}}$$

Therefore, for small concentrations,

$$\ln a_1 = \ln x_1 = \ln(1 - x_2) \approx -x_2 \approx -\frac{m}{55.506\ \text{mol}\cdot\text{kg}^{-1}} \tag{11.12}$$

where we have used the fact that $\ln(1 - x_2) \approx -x_2$ for small values of $x_2$.

Table 11.2 and Figure 11.1 give experimental data for the vapor pressure of water in equilibrium with an aqueous sucrose solution at 25°C as a function of molality and mole fraction, respectively. The equilibrium vapor pressure of pure water at 25°C is 23.756 torr, so $a_1 = P_1/P_1^* = P_1/23.756$ is given in the third column of Table 11.2.

Equation 11.12 relates $a_1$ to the molality $m$ for only a dilute solution. For example, Table 11.2 shows that $a_1 = 0.93276$ at 3.00 molal, whereas Equation 11.12 gives $\ln a_1 = -0.054048$, or $a_1 = 0.9474$. To account for this discrepancy, we now define a quantity $\phi$, called the *osmotic coefficient*, by

$$\ln a_1 = -\frac{m\phi}{55.506\ \text{mol}\cdot\text{kg}^{-1}} \tag{11.13}$$

Note that $\phi = 1$ if the solution behaves as an ideal dilute solution. Thus, the deviation of $\phi$ from unity is a measure of the nonideality of the solution.

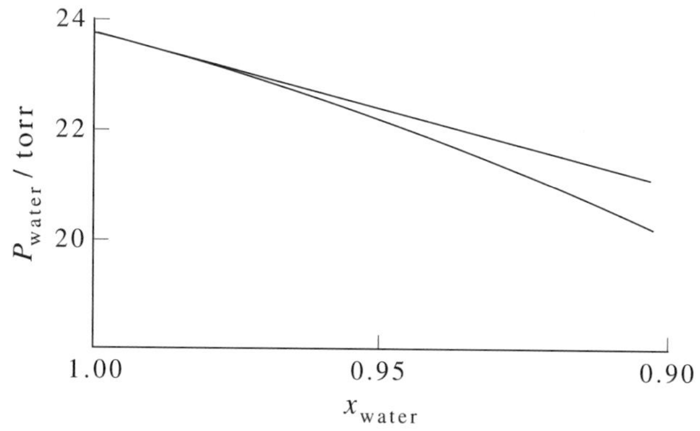


**FIGURE 11.1**
The vapor pressure of water in equilibrium with an aqueous sucrose solution at 25°C plotted against the mole fraction of water. Note that Raoult's law (the straight line in the figure) holds from $x_{water} = 1.00$ to about 0.97, but that deviations occur at lower values of $x_{water}$.

**TABLE 11.2**

The vapor pressure of water ($P_1$) in equilibrium with an aqueous sucrose solution at 25°C as a function of molality ($m$). Additional data are the activity of the water ($a_1$), the osmotic coefficient ($\phi$), and the activity coefficient ($\gamma_{2m}$) of the sucrose.

| $m/\text{mol}\cdot\text{kg}^{-1}$ | $P_1$/torr | $a_1$ | $\phi$ | $\gamma_{2m}$ | $\ln \gamma_{2m}$ |
|---|---|---|---|---|---|
| 0.00 | 23.756 | 1.00000 | 1.0000 | 1.000 | 0.0000 |
| 0.10 | 23.713 | 0.99819 | 1.0056 | 1.017 | 0.0169 |
| 0.20 | 23.669 | 0.99634 | 1.0176 | 1.034 | 0.0334 |
| 0.30 | 23.625 | 0.99448 | 1.0241 | 1.051 | 0.0497 |
| 0.40 | 23.580 | 0.99258 | 1.0335 | 1.068 | 0.0658 |
| 0.50 | 23.534 | 0.99067 | 1.0406 | 1.085 | 0.0816 |
| 0.60 | 23.488 | 0.98872 | 1.0494 | 1.105 | 0.0998 |
| 0.70 | 23.441 | 0.98672 | 1.0601 | 1.125 | 0.1178 |
| 0.80 | 23.393 | 0.98472 | 1.0683 | 1.144 | 0.1345 |
| 0.90 | 23.344 | 0.98267 | 1.0782 | 1.165 | 0.1527 |
| 1.00 | 23.295 | 0.98059 | 1.0880 | 1.188 | 0.1723 |
| 1.20 | 23.194 | 0.97634 | 1.1075 | 1.233 | 0.2095 |
| 1.40 | 23.089 | 0.97193 | 1.1288 | 1.283 | 0.2492 |
| 1.60 | 22.982 | 0.96740 | 1.1498 | 1.335 | 0.2889 |
| 1.80 | 22.872 | 0.96280 | 1.1690 | 1.387 | 0.3271 |
| 2.00 | 22.760 | 0.95807 | 1.1888 | 1.442 | 0.3660 |
| 2.50 | 22.466 | 0.94569 | 1.2398 | 1.590 | 0.4637 |
| 3.00 | 22.159 | 0.93276 | 1.2879 | 1.751 | 0.5602 |
| 3.50 | 21.840 | 0.91933 | 1.3339 | 1.924 | 0.6544 |
| 4.00 | 21.515 | 0.90567 | 1.3749 | 2.101 | 0.7424 |
| 4.50 | 21.183 | 0.89170 | 1.4139 | 2.310 | 0.8372 |
| 5.00 | 20.848 | 0.87760 | 1.4494 | 2.481 | 0.9087 |
| 5.50 | 20.511 | 0.86340 | 1.4823 | 2.680 | 0.9858 |
| 6.00 | 20.176 | 0.84930 | 1.5111 | 3.878 | 1.3553 |

**EXAMPLE 11–4**

Using the data in Table 11.2, calculate the value of $\phi$ at 1.00 mol·kg$^{-1}$.

SOLUTION: We simply use Equation 11.13 and find that

$$\phi = -\frac{(55.506\ \text{mol}\cdot\text{kg}^{-1})\ln(0.98059)}{1.00\ \text{mol}\cdot\text{kg}^{-1}} = 1.0880$$

in agreement with the entry in Table 11.2.

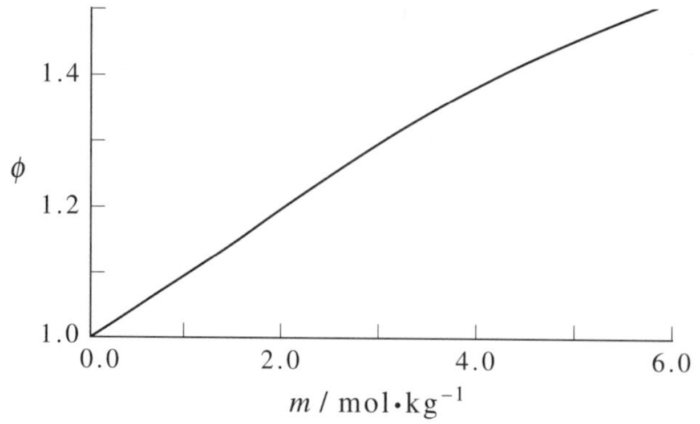


**FIGURE 11.2**
The osmotic coefficient ($\phi$) of an aqueous sucrose solution at 25°C plotted against the molality ($m$). The magnitutde of the deviation of the value of $\phi$ from unity is a measure of the nonideality of the solution.

Figure 11.2 shows $\phi$ for an aqueous sucrose solution at 25°C plotted against $m$. Note that the solution becomes increasingly nonideal as $m$ increases.

The fifth column in Table 11.2 gives the activity coefficient of the sucrose calculated from the activity of the water, or from the osmotic coefficient, by means of the Gibbs–Duhem equation,

$$n_1 d \ln a_1 + n_2 d \ln a_2 = 0$$

In terms of molality, $m$, $n_1 = 55.506$ mol and $n_2 = m$, so the Gibbs–Duhem equation becomes

$$(55.506 \text{ mol·kg}^{-1}) d \ln a_1 + m d \ln a_2 = 0 \tag{11.14}$$

Using Equation 11.13, we see that $(55.506 \text{ mol·kg}^{-1}) d \ln a_1 = -d(m\phi)$. If we substitute this result and $a_{2m} = \gamma_{2m} m$ (Table 11.1) into Equation 11.14, we obtain

$$d(m\phi) = m d \ln(\gamma_{2m} m)$$

or

$$m d\phi + \phi dm = m(d \ln \gamma_{2m} + d \ln m)$$

We can rewrite this equation as

$$d \ln \gamma_{2m} = d\phi + \frac{\phi - 1}{m} dm$$

We now integrate from $m = 0$ (where $\gamma_{2m} = \phi = 1$) to arbitrary $m$ to get

$$\ln \gamma_{2m} = \phi - 1 + \int_0^m \left( \frac{\phi - 1}{m'} \right) dm' \tag{11.15}$$

Equation 11.15 allows us to calculate the activity coefficient of the solute from the data on the vapor pressure of the solvent. The vapor pressure of the solvent gives us the activity of the solvent from Equation 11.1; then the osmotic coefficient $\phi$ is calculated from Equation 11.13, and $\ln \gamma_{2m}$ is determined from Equation 11.15.

The data for $\phi$ in Table 11.2 can be fit with a polynomial in the molality. If we choose (arbitrarily) a 5th-degree polynomial, we find that (Problem 11–18)

$$\begin{aligned}\phi = 1.00000 &+ (0.07349\ \text{kg}\cdot\text{mol}^{-1})m + (0.019783\ \text{kg}^2\cdot\text{mol}^{-2})m^2 \\ &- (0.005688\ \text{kg}^3\cdot\text{mol}^{-3})m^3 + (6.036\times 10^{-4}\ \text{kg}^4\cdot\text{mol}^{-4})m^4 \\ &- (2.517\times 10^{-5}\ \text{kg}^5\cdot\text{mol}^{-5})m^5 \qquad 0 \le m \le 6\ \text{mol}\cdot\text{kg}^{-1}\end{aligned}$$

We can substitute this expression into Equation 11.15 to obtain $\ln \gamma_{2m}$.

**EXAMPLE 11–5**

Use the above polynomial fit for $\phi$ and Equation 11.15 to calculate the value of $\gamma_{2m}$ for a 1.00-molal aqueous sucrose solution.

SOLUTION: First, we need to evaluate the integral in Equation 11.15 (neglecting to write the units in the coefficients of the powers of $m$):

$$\begin{aligned}\int_0^1 \left(\frac{\phi - 1}{m}\right) dm &= \int_0^1 [0.07349 + 0.019783m - 0.005688m^2 \\ &\qquad + 6.036\times 10^{-4}m^3 - 2.517\times 10^{-5}m^4]dm \\ &= 0.07349 + \frac{0.019783}{2} - \frac{0.005688}{3} \\ &\qquad + \frac{6.036\times 10^{-4}}{4} - \frac{2.517\times 10^{-5}}{5} \\ &= 0.08163\end{aligned}$$

so

$$\begin{aligned}\ln \gamma_{2m} &= \phi - 1 + \int_0^1 \left(\frac{\phi - 1}{m}\right) dm \\ &= 0.08816 + 0.08163 = 0.1698\end{aligned}$$

or $\gamma_{2m} = 1.185$, in agreement with the entry in Table 11.2.

The values of $\ln \gamma_{2m}$ and $\gamma_{2m}$ given in Table 11.2 have been calculated using the procedure in Example 11–5. Figure 11.3 shows $\ln \gamma_{2m}$ plotted against $m$ for an aqueous sucrose solution at 25°C.

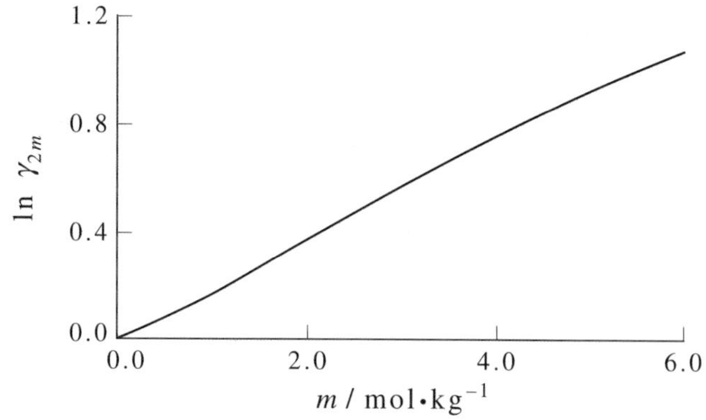


**FIGURE 11.3**
The logarithm of the activity coefficient ($\ln \gamma_{2m}$) of sucrose in an aqueous sucrose solution at 25°C plotted against the molality ($m$).

## 11–3. Colligative Properties Are Solution Properties That Depend Only Upon the Number Density of Solute Particles

A number of solution properties, called *colligative properties*, depend, at least in dilute solution, upon only the number of solute particles, and not upon their kind. Colligative properties include the lowering of the vapor pressure of a solvent by the addition of a solute, the elevation of the boiling point of a solution by a nonvolatile solute, the depression of the freezing point of a solution by a solute, and osmotic pressure. We will discuss only freezing-point depression and osmotic pressure.

At the freezing point of a solution, solid solvent is in equilibrium with the solvent in solution. The thermodynamic condition of this equilibrium is that

$$\mu_1^{\text{s}}(T_{\text{fus}}) = \mu_1^{\text{sln}}(T_{\text{fus}})$$

where as usual the subscript 1 denotes solvent and $T_{\text{fus}}$ is the freezing point of the solution. We use Equation 10.35 for $\mu_1$ to obtain

$$\mu_1^{\text{s}} = \mu_1^* + RT \ln a_1 = \mu_1^{\text{l}} + RT \ln a_1$$

We have written $\mu_1^{\text{l}}$ for $\mu_1^*$ simply to compare it with $\mu_1^{\text{s}}$. Solving for $\ln a_1$, we get

$$\ln a_1 = \frac{\mu_1^{\text{s}} - \mu_1^{\text{l}}}{RT} \tag{11.16}$$

Now differentiate with respect to temperature and use the Gibbs–Helmholtz equation (Example 10–1),

$$\left[\frac{\partial(\mu_1/T)}{\partial T}\right]_{P,x_1} = -\frac{\overline{H}_1}{T^2}$$

to obtain

$$\left(\frac{\partial \ln a_1}{\partial T}\right)_{P,x_1} = \frac{\overline{H}_1^{\mathrm{l}} - \overline{H}_1^{\mathrm{s}}}{RT^2} = \frac{\Delta_{\mathrm{fus}}\overline{H}}{RT^2} \tag{11.17}$$

where we have used the fact that $\overline{H}_1^{\mathrm{l}} - \overline{H}_1^{\mathrm{s}} = \Delta_{\mathrm{fus}}\overline{H}$ for the pure solvent. If we integrate Equation 11.17 from pure solvent, where $a_1 = 1$, $T = T_{\mathrm{fus}}^*$, to a solution with arbitrary values of $a_1$ and $T_{\mathrm{fus}}$, we obtain

$$\ln a_1 = \int_{T_{\mathrm{fus}}^*}^{T_{\mathrm{fus}}} \frac{\Delta_{\mathrm{fus}}\overline{H}}{RT^2}\,dT \tag{11.18}$$

Equation 11.18 can be used to determine the activity of the solvent in a solution (Problem 11–20).

You may have calculated freezing-point depressions in general chemistry using the formula

$$\Delta T_{\mathrm{fus}} = K_{\mathrm{f}} m \tag{11.19}$$

where $K_{\mathrm{f}}$ is a constant, called the *freezing-point depression constant*, whose value depends upon the solvent. We can derive Equation 11.19 from Equation 11.18 by making a few approximations appropriate to dilute solutions. If the solution is sufficiently dilute, then $\ln a_1 = \ln x_1 = \ln(1 - x_2) \approx -x_2$, and if we assume that $\Delta_{\mathrm{fus}}\overline{H}$ is independent of temperature over the temperature range $(T_{\mathrm{fus}}, T_{\mathrm{fus}}^*)$, we obtain

$$\begin{aligned} -x_2 &= \frac{\Delta_{\mathrm{fus}}\overline{H}}{R}\int_{T_{\mathrm{fus}}^*}^{T_{\mathrm{fus}}} \frac{dT}{T^2} = \frac{\Delta_{\mathrm{fus}}\overline{H}}{R}\left(\frac{1}{T_{\mathrm{fus}}^*} - \frac{1}{T_{\mathrm{fus}}}\right) \\ &= \frac{\Delta_{\mathrm{fus}}\overline{H}}{R}\left(\frac{T_{\mathrm{fus}} - T_{\mathrm{fus}}^*}{T_{\mathrm{fus}}T_{\mathrm{fus}}^*}\right) \end{aligned} \tag{11.20}$$

Because $x_2$ and $\Delta_{\mathrm{fus}}\overline{H}$ are positive quantities, we see immediately that $T_{\mathrm{fus}} - T_{\mathrm{fus}}^* < 0$, or that $T_{\mathrm{fus}} < T_{\mathrm{fus}}^*$. Thus, we find that the addition of a solute will lower the freezing point of a solution. We can express $x_2$ in terms of molality by using Equation 11.4,

$$x_2 = \frac{m}{\dfrac{1000\ \mathrm{g\cdot kg^{-1}}}{M_1} + m} \approx \frac{M_1 m}{1000\ \mathrm{g\cdot kg^{-1}}} \tag{11.21}$$

for small values of $m$ (dilute solution). Furthermore, because $T_{\mathrm{fus}}^* - T_{\mathrm{fus}}$ is usually only a few degrees (dilute solution once again), we can replace $T_{\mathrm{fus}}$ in the denominator of Equation 11.20 by $T_{\mathrm{fus}}^*$ to a good approximation to get finally (Problem 11–23)

$$\Delta T_{\mathrm{fus}} = T_{\mathrm{fus}}^* - T_{\mathrm{fus}} = K_{\mathrm{f}} m \tag{11.22}$$

where

$$K_f = \frac{M_1}{1000\ \text{g}\cdot\text{kg}^{-1}} \frac{R(T^*_{\text{fus}})^2}{\Delta_{\text{fus}}\overline{H}} \tag{11.23}$$

We can calculate the value of $K_f$ for water.

$$K_f = \left(\frac{18.02\ \text{g}\cdot\text{mol}^{-1}}{1000\ \text{g}\cdot\text{kg}^{-1}}\right) \frac{(8.314\ \text{J}\cdot\text{K}^{-1}\cdot\text{mol}^{-1})(273.2\ \text{K})^2}{6.01\ \text{kJ}\cdot\text{mol}^{-1}}$$
$$= 1.86\ \text{K}\cdot\text{kg}\cdot\text{mol}^{-1}$$

Equation 11.22 tells us that the freezing point of a 0.20-molal solution of sucrose in water is $-(1.86\ \text{K}\cdot\text{kg}\cdot\text{mol}^{-1})(0.20\ \text{mol}\cdot\text{kg}^{-1}) = -0.37$ K.

**EXAMPLE 11–6**

The freezing-point depression constant of water is $1.86\ \text{K}\cdot\text{kg}\cdot\text{mol}^{-1}$. Calculate the value of $K_f$ for cyclohexane, whose freezing point is 279.6 K and molar enthalpy of fusion is $2.68\ \text{kJ}\cdot\text{mol}^{-1}$.

SOLUTION: We use Equation 11-23 with $M_1 = 84.16\ \text{g}\cdot\text{mol}^{-1}$ and the above values of $T^*_{\text{fus}}$ and $\Delta_{\text{fus}}\overline{H}$.

$$K_f = \left(\frac{84.16\ \text{g}\cdot\text{mol}^{-1}}{1000\ \text{g}\cdot\text{kg}^{-1}}\right) \frac{(8.314\ \text{J}\cdot\text{K}^{-1}\cdot\text{mol}^{-1})(279.6\ \text{K})^2}{2680\ \text{J}\cdot\text{mol}^{-1}}$$
$$= 20.4\ \text{K}\cdot\text{kg}\cdot\text{mol}^{-1}$$

Thus, the freezing point of a 0.20-molal solution of hexane in cyclohexane is 4.1 K lower than the freezing point of pure cyclohexane, or $T_{\text{fus}} = 275.5$ K.

We can derive an expression for the boiling-point elevation of a solution containing a nonvolatile solute. The analog of Equation 11.22 is (Problem 11–25)

$$\Delta T_{\text{vap}} = T_{\text{vap}} - T^*_{\text{vap}} = K_b m \tag{11.24}$$

where the *boiling-point elevation constant* is given by

$$K_b = \frac{M_1}{1000\ \text{g}\cdot\text{kg}^{-1}} \frac{R(T^*_{\text{vap}})^2}{\Delta_{\text{vap}}\overline{H}} \tag{11.25}$$

The value of $K_b$ for water is only $0.512\ \text{K}\cdot\text{kg}\cdot\text{mol}^{-1}$, so the boiling point elevation is a rather small effect for aqueous solutions.

## 11–4. Osmotic Pressure Can Be Used to Determine the Molecular Masses of Polymers

Figure 11.4 illustrates the development of osmotic pressure. In the initial state, we have pure water on the left and an aqueous sucrose solution on the right. The two liquids are separated by a membrane containing pores that allow water molecules but not solute molecules to pass through. Such a membrane is called a *semipermeable membrane*. (Many biological cells are surrounded by membranes semipermeable to water.) The levels of the two liquids in Figure 11.4 are initially the same, but water will pass through the semipermeable membrane until the chemical potentials of the water on the two sides of the membrane are equal. This process results in the situation shown in the equilibrium state, where the two liquid levels are no longer equal. The hydrostatic pressure head that is built up is called *osmotic pressure*.

Because the water is free to pass through the semipermeable membrane, the chemical potential of the water must be the same on the two sides of the membrane at equilibrium. In other words, the chemical potential of the pure water at a pressure $P$ must equal the chemical potential of the water in the solution at a pressure $P + \Pi$ and an activity $a_1$. In an equation,

$$\begin{aligned}\mu_1^*(T, P) &= \mu_1^{\text{sln}}(T, P + \Pi, a_1) \\ &= \mu_1^*(T, P + \Pi) + RT \ln a_1 \end{aligned} \qquad (11.26)$$

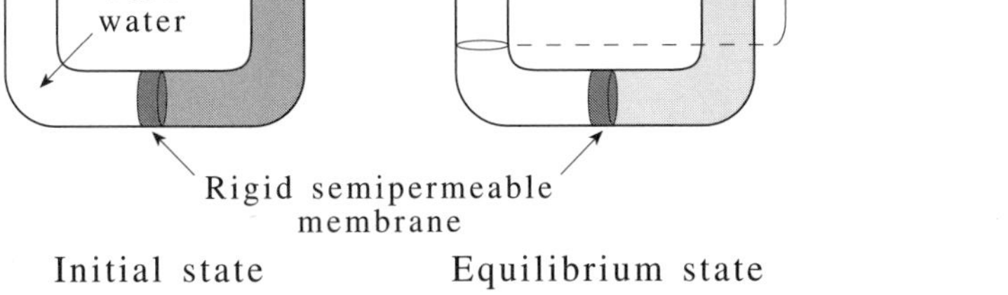

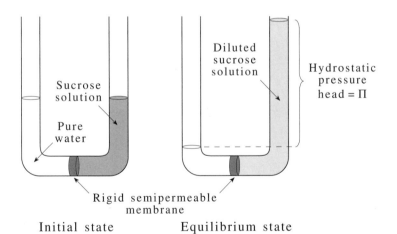

**FIGURE 11.4**
Passage of water through a rigid, semipermeable membrane separating pure water from an aqueous sucrose solution. The water passes through the membrane until the chemical potential of the water in the aqueous sucrose solution equals that of the pure water. The chemical potential of water in the sucrose solution increases as the hydrostatic pressure above the solution increases.

where $a_1 = P_1/P_1^*$. We can rewrite Equation 11.26. as

$$\mu_1^*(T, P+\Pi) - \mu_1^*(T, P) + RT \ln a_1 = 0 \tag{11.27}$$

The first two terms in Equation 11.27 are the difference in the chemical potential of the pure solvent at two different pressures. Equation 9.8

$$\left(\frac{\partial \mu_1^*}{\partial P}\right)_T = \overline{V}_1^* \tag{9.8}$$

where $\overline{V}_1^*$ is the molar volume of the pure solvent, tells us how the chemical potential varies with pressure. We can use Equation 9.8 to evaluate $\mu_1^*(T, P+\Pi) - \mu_1^*(T, P)$ by integrating both sides from $P$ to $P+\Pi$ to get

$$\mu_1^*(T, P+\Pi) - \mu_1^*(T, P) = \int_P^{P+\Pi} \left(\frac{\partial \mu_1^*}{\partial P'}\right)_T dP' = \int_P^{P+\Pi} \overline{V}_1^* dP' \tag{11.28}$$

If we substitute Equation 11.28 into Equation 11.27, we obtain

$$\int_P^{P+\Pi} \overline{V}^* dP' + RT \ln a_1 = 0 \tag{11.29}$$

Assuming $\overline{V}_1^*$ does not vary with applied pressure, we can write Equation 11.29 as

$$\Pi \overline{V}_1^* + RT \ln a_1 = 0 \tag{11.30}$$

Furthermore, if the solution is dilute, then $a_1 \approx x_1 = 1 - x_2$, with $x_2$ small. Therefore, we can write $\ln a_1$ as $\ln(1-x_2) \approx -x_2$, so that Equation 11.30 becomes

$$\Pi \overline{V}_1^* = RT x_2$$

Furthermore, because $x_2$ is small, $n_2 \ll n_1$ and

$$x_2 = \frac{n_2}{n_1 + n_2} \approx \frac{n_2}{n_1}$$

Substitute this into the above equation to get

$$\Pi = \frac{n_2 RT}{n_1 \overline{V}_1^*} \approx \frac{n_2 RT}{V}$$

where we have replaced $n_1 \overline{V}_1^*$ by the total volume of the solution, $V$ (dilute solution). The above equation is usually written as

$$\Pi = cRT \tag{11.31}$$

where $c$ is the molarity, $n_2/V$, of the solution. Equation 11.31 is called the van't Hoff equation for osmotic pressure. Using this equation, we calculate the osmotic pressure of a 0.100-molar aqueous solution of sucrose at 20°C to be

$$\begin{aligned}\Pi &= (0.100\ \text{mol}\cdot\text{L}^{-1})(0.08206\ \text{L}\cdot\text{atm}\cdot\text{K}^{-1}\cdot\text{mol}^{-1})(293.2\ \text{K})\\ &= 2.40\ \text{atm}\end{aligned}$$

Thus, we see that osmotic pressure is a large effect. Because of this, osmotic pressure can be used to determine molecular masses of solutes, particularly solutes with large molecular masses such as polymers and proteins.

**EXAMPLE 11–7**

It is found that 2.20 g of a certain polymer dissolved in enough water to make 300 mL of solution has an osmotic pressure of 7.45 torr at 20°C. Determine the molecular mass of the polymer.

SOLUTION: The molarity of the solution is given by

$$\begin{aligned}c = \frac{\Pi}{RT} &= \frac{7.45\ \text{torr}/760\ \text{torr}\cdot\text{atm}^{-1}}{(0.08206\ \text{L}\cdot\text{atm}\cdot\text{K}^{-1}\cdot\text{mol}^{-1})(293.2\ \text{K})}\\ &= 4.07\times10^{-4}\ \text{mol}\cdot\text{L}^{-1}\end{aligned}$$

Therefore, there are $4.07\times10^{-4}$ moles of polymer per liter of solution, or $(0.300)(4.07\times10^{-4}) = 1.22\times10^{-4}$ moles per 300 mL of solution. Thus, we find that $1.22\times10^{-4}$ moles corresponds to 2.20 g, or that the molecular mass is 18,000.

If a pressure in excess of 26 atm is applied to seawater at 15°C, the chemical potential of the water in the seawater will exceed that of pure water. Consequently, pure water can be obtained from seawater by using a rigid semipermeable membrane and an applied pressure in excess of the osmotic pressure of 26 atm. This process is known as *reverse osmosis*. Reverse osmosis units are commercially available and are used to obtain fresh water from salt water using a variety of semipermeable membranes, the most common of which is cellulose acetate.

## 11–5. Solutions of Electrolytes Are Nonideal at Relatively Low Concentrations

When sodium chloride dissolves in water, the solution contains sodium ions and chloride ions and essentially no undissociated sodium chloride. The ions interact with each other through a coulombic potential, which varies as $1/r$. We should compare this interaction with the one between neutral solute molecules (nonelectrolytes) such as sucrose, where the interaction varies as something like $1/r^6$. Thus, the interaction between ions in solution is effective over a much greater distance than the interaction between neutral solute particles, so solutions of electrolytes deviate from ideal behavior more strongly and at lower concentrations than do solutions of nonelectrolytes. Figure 11.5 shows $\ln \gamma_{2m}$ for sucrose, sodium chloride, and calcium chloride plotted versus molality. Note that $CaCl_2$(aq) appears to behave more nonideally than NaCl(aq), which in turn behaves more nonideally than sucrose. The charge of +2 on the calcium ion leads to a stronger coulombic interaction and hence a stronger deviation from ideality than for NaCl. At 0.100 $mol \cdot kg^{-1}$, the activity coefficient of sucrose is 0.998, whereas that of $CaCl_2$(aq) is 0.518 and that of NaCl(aq) is 0.778.

Before we discuss the determination of activity coefficients for electrolytes, we must first introduce notation needed to describe the thermodynamic properties of solutions of electrolytes. Consider the general salt $C_{\nu_+}A_{\nu_-}$, which dissociates into $\nu_+$ cations and $\nu_-$ anions per formula unit as in

$$C_{\nu_+}A_{\nu_-}(s) \xrightarrow{H_2O(l)} \nu_+C^{z_+}(aq) + \nu_-A^{z_-}(aq)$$

where $\nu_+z_+ + \nu_-z_- = 0$ by electronegativity. For example, $\nu_+ = 1$ and $\nu_- = 2$ for $CaCl_2$ and $\nu_+ = 2$ and $\nu_- = 1$ for $Na_2SO_4$. Therefore, $CaCl_2$ is called a 1–2 electrolyte and $Na_2SO_4$ is called a 2–1 electrolyte. We write the chemical potential of the salt in terms of the chemical potentials of its constituent ions according to

$$\mu_2 = \nu_+\mu_+ + \nu_-\mu_- \tag{11.32}$$

where

$$\mu_2 = \mu_2^\circ + RT \ln a_2 \tag{11.33}$$

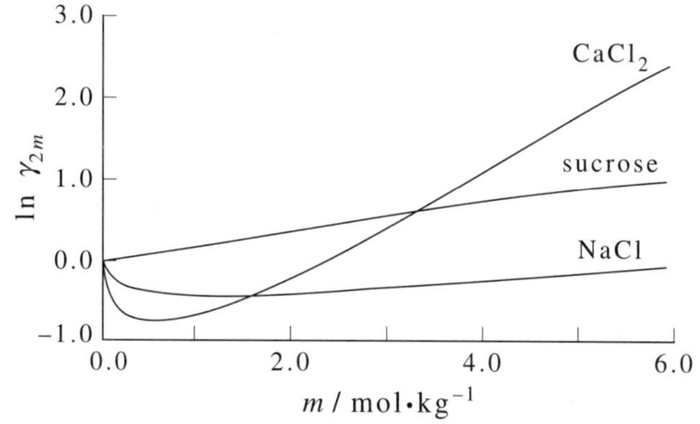


**FIGURE 11.5**
The logarithm of the activity coefficient ($\ln \gamma_{2m}$) of aqueous solutions of sucrose, sodium chloride, and calcium chloride plotted against molality ($m$) at 25°C. Note that the electrolyte solutions deviate from ideality ($\ln \gamma_{2m} = 0$) much more strongly than does sucrose at small concentrations.

and

$$\mu_+ = \mu_+^\circ + RT \ln a_+$$
$$\mu_- = \mu_-^\circ + RT \ln a_- \tag{11.34}$$

The superscript zeros here represent the chosen standard state, which we can leave unspecified at this point but is usually taken to be the solute or Henry's law standard state. If we substitute Equations 11.34 into Equation 11.32 and equate the result to Equation 11.33, we obtain

$$\nu_+ \ln a_+ + \nu_- \ln a_- = \ln a_2$$

where we have used the relation $\mu_2^\circ = \nu_+\mu_+^\circ + \nu_-\mu_-^\circ$ in analogy with Equation 11.32. We can rewrite the above equation as

$$a_2 = a_+^{\nu_+} a_-^{\nu_-} \tag{11.35}$$

For many of the formulas that occur in the thermodynamics of solutions of electrolytes, it is convenient to define a quantity $a_\pm$, called the *mean ionic activity*, by

$$a_2 = a_\pm^\nu = a_+^{\nu_+} a_-^{\nu_-} \tag{11.36}$$

where $\nu = \nu_+ + \nu_-$. Note that $a_\pm$ is raised to the same power as the sum of the exponents in the last term in Equation 11.36, For example, we write

$$a_{\text{NaCl}} = a_\pm^2 = a_+ a_-$$

and

$$a_{\text{CaCl}_2} = a_\pm^3 = a_+ a_-^2$$

Even though we cannot determine activities of single ions, we can still *define* single-ion activity coefficients by

$$a_+ = m_+\gamma_+ \qquad \text{and} \qquad a_- = m_-\gamma_-$$

where $m_+$ and $m_-$ are the molalities of the individual ions, which are given by $m_+ = \nu_+ m$ and $m_- = \nu_- m$. If we substitute these expressions for $a_+$ and $a_-$ into Equation 11.36, we get

$$a_2 = a_\pm^\nu = (m_+^{\nu_+} m_-^{\nu_-})(\gamma_+^{\nu_+} \gamma_-^{\nu_-}) \tag{11.37}$$

In analogy with the definition of the *mean ionic activity* $a_\pm$ in Equation 11.36, we define a mean ionic molality $m_\pm$ by

$$m_\pm^\nu = m_+^{\nu_+} m_-^{\nu_-} \tag{11.38}$$

and a *mean ionic activity coefficient* $\gamma_{\pm}$ by

$$\gamma_{\pm}^{\nu} = \gamma_{+}^{\nu_{+}}\gamma_{-}^{\nu_{-}} \tag{11.39}$$

Again, notice that the sum of the exponents on both sides of Equations 11.38 and 11.39 are the same. Given these definitions, we can now write Equation 11.37 as

$$a_2 = a_{\pm}^{\nu} = m_{\pm}^{\nu}\gamma_{\pm}^{\nu} \tag{11.40}$$

**EXAMPLE 11–8**

Write out Equation 11.40 explicitly for $CaCl_2$.

SOLUTION: In this case, $\nu_{+} = 1$ and $\nu_{-} = 2$. Furthermore, according to the equation

$$CaCl_2(s) \xrightarrow{H_2O(l)} Ca^{2+}(aq) + 2\,Cl^{-}(aq)$$

we see that $m_{+} = m$ and $m_{-} = 2m$. Thus,

$$a_2 = a_{\pm}^3 = (m)(2m)^2\gamma_{\pm}^3 = 4m^3\gamma_{\pm}^3$$

The relations between $a_2$, $m$, and $\gamma_{\pm}$ for other types of electrolytes are given in Table 11.3.

**TABLE 11.3**

The relations between the activity of a strong electrolyte, its molality, and its mean ionic activity coefficient for various types of strong electrolytes.

| Type | |
|---|---|
| 1–1 | |
| KCl(aq) | $a_2 = a_{+}a_{-} = a_{\pm}^2 = m_{\pm}^2\gamma_{\pm}^2 = (m_{+})(m_{-})\gamma_{\pm}^2 = m^2\gamma_{\pm}^2$ |
| 1–2 | |
| $CaCl_2$(aq) | $a_2 = a_{+}a_{-}^2 = a_{\pm}^3 = m_{\pm}^3\gamma_{\pm}^3 = (m_{+})(m_{-})^2\gamma_{\pm}^3 = (m)(2m)^2\gamma_{\pm}^3 = 4m^3\gamma_{\pm}^3$ |
| 1–3 | |
| $LaCl_3$(aq) | $a_2 = a_{+}a_{-}^3 = a_{\pm}^4 = m_{\pm}^4\gamma_{\pm}^4 = (m_{+})(m_{-})^4\gamma_{\pm}^4 = (m)(3m)^3\gamma_{\pm}^4 = 27m^4\gamma_{\pm}^4$ |
| 2–1 | |
| $Na_2SO_4$(aq) | $a_2 = a_{+}^2a_{-} = a_{\pm}^3 = (m_{+})^2(m_{-})\gamma_{\pm}^3 = (2m)^2(m)\gamma_{\pm}^3 = 4m^3\gamma_{\pm}^3$ |
| 2–2 | |
| $ZnSO_4$(aq) | $a_2 = a_{+}a_{-} = a_{\pm}^2 = m_{\pm}^2\gamma_{\pm}^2 = (m_{+})(m_{-})\gamma_{\pm}^2 = m^2\gamma_{\pm}^2$ |
| 3–1 | |
| $Na_3Fe(CN)_6$(aq) | $a_2 = a_{+}^3a_{-} = a_{\pm}^4 = m_{\pm}^4\gamma_{\pm}^4 = (m_{+})^3(m_{-})\gamma_{\pm}^4 = (3m)^3(m)\gamma_{\pm}^4 = 27m^4\gamma_{\pm}^4$ |

Mean ionic activity coefficients can be determined experimentally by the same methods used for the activity coefficients of nonelectrolytes. We will illustrate their determination from the measurement of the vapor pressure of the solvent as we did for an aqueous sucrose solution in Section 11–2. In analogy with Equation 11.13, we define an osmotic coefficient for aqueous electrolyte solutions by

$$\ln a_1 = -\frac{\nu m \phi}{55.506\ \text{mol}\cdot\text{kg}^{-1}} \tag{11.41}$$

Notice that this equation differs from Equation 11.13 by the inclusion of a factor of $\nu$ here. Equation 11.41 reduces to Equation 11.13 for nonelectrolyte solutions because $\nu = 1$ in that case. Problem 11–34 asks you to show that with this factor of $\nu$, $\phi \to 1$ as $m \to 0$ for solutions of electrolytes or nonelectrolytes. Starting with Equation 11.41 and the Gibbs–Duhem equation, you can derive the analog of Equation 11.15 straightforwardly:

$$\ln \gamma_{\pm} = \phi - 1 + \int_0^m \left(\frac{\phi - 1}{m'}\right) dm' \tag{11.42}$$

Table 11.4 gives the vapor pressure of an aqueous solution of NaCl as a function of molality. Also included in the table are activities of the water (calculated from

**TABLE 11.4**
The vapor pressure ($P_{H_2O}$), activity of the water ($a_w$), osmotic coefficient ($\phi$), and logarithm of the mean ionic activity coefficient ($\ln \gamma_{\pm}$) of the NaCl in an aqueous solution of NaCl at 25°C as a function of molality ($m$).

| $m/\text{mol}\cdot\text{kg}^{-1}$ | $P_{H_2O}/\text{torr}$ | $a_w$ | $\phi$ | $\ln \gamma_{\pm}$ |
|---|---|---|---|---|
| 0.000 | 23.76 | 1.0000 | 1.0000 | 0.0000 |
| 0.200 | 23.60 | 0.9934 | 0.9245 | −0.3079 |
| 0.400 | 23.44 | 0.9868 | 0.9205 | −0.3685 |
| 0.600 | 23.29 | 0.9802 | 0.9227 | −0.3977 |
| 0.800 | 23.13 | 0.9736 | 0.9285 | −0.4143 |
| 1.000 | 22.97 | 0.9669 | 0.9353 | −0.4234 |
| 1.400 | 22.64 | 0.9532 | 0.9502 | −0.4267 |
| 1.800 | 22.30 | 0.9389 | 0.9721 | −0.4166 |
| 2.200 | 21.96 | 0.9242 | 0.9944 | −0.3972 |
| 2.600 | 21.59 | 0.9089 | 1.0196 | −0.3709 |
| 3.000 | 21.22 | 0.8932 | 1.0449 | −0.3396 |
| 3.400 | 20.83 | 0.8769 | 1.0723 | −0.3046 |
| 3.800 | 20.43 | 0.8600 | 1.1015 | −0.2666 |
| 4.400 | 19.81 | 0.8339 | 1.1457 | −0.2053 |
| 5.000 | 19.17 | 0.8068 | 1.1916 | −0.1389 |

$a_1 = P_1/P_1^*$), osmotic coefficients (calculated from Equation 11.41), and mean ionic activity coefficients (calculated from Equation 11.42).

For sucrose in Section 11–2, we curve fit $\phi$ to a polynomial in $m$ and then used that polynomial to calculate the value of $\gamma_{2m}$. As we will see in Section 11–6, the osmotic coefficient of electrolytes is better described by an expression of the form (a polynomial in $m^{1/2}$)

$$\phi = 1 + am^{1/2} + bm + cm^{3/2} + \cdots$$

The osmotic coefficient data for sodium chloride given in Table 11.4 can be fit by

$$\begin{aligned} \phi = 1 &- (0.3920\ \text{kg}^{1/2}\cdot\text{mol}^{-1/2})m^{1/2} + (0.7780\ \text{kg}\cdot\text{mol}^{-1})m \\ &- (0.8374\ \text{kg}^{3/2}\cdot\text{mol}^{-3/2})m^{3/2} + (0.5326\ \text{kg}^{2}\cdot\text{mol}^{-2})m^{2} \\ &- (0.1673\ \text{kg}^{5/2}\cdot\text{mol}^{-5/2})m^{5/2} + (0.0206\ \text{kg}^{3}\cdot\text{mol}^{-3})m^{3} \\ &\qquad 0 \le m \le 5.0\ \text{mol}\cdot\text{kg}^{-1} \end{aligned} \tag{11.43}$$

This expression for $\phi$ along with Equation 11.42 were used to calculate the values of $\ln \gamma_\pm$ given in Table 11.4.

**EXAMPLE 11–9**

Verify the entry for $\ln \gamma_\pm$ at 1.00 molal in Table 11.4.

SOLUTION: We first write (neglecting the units in the coefficients of the powers of $m$ in Equation 11.43)

$$\begin{aligned} \int_0^m \left(\frac{\phi - 1}{m'}\right) dm' &= -(0.3920)(2m^{1/2}) + 0.7780m - (0.8374)\frac{2m^{3/2}}{3} \\ &+ (0.5326)\frac{m^2}{2} - (0.1673)\frac{2m^{5/2}}{5} + (0.0206)\frac{m^3}{3} \end{aligned}$$

and add this result to $\phi - 1$ to obtain

$$\begin{aligned} \ln \gamma_\pm &= -(0.3920)(3m^{1/2}) + (0.7780)(2m) - (0.8374)\frac{5m^{3/2}}{3} \\ &+ (0.5326)\frac{3m^2}{2} - (0.1673)\frac{7m^{5/2}}{5} + (0.0206)\frac{4m^3}{3} \end{aligned}$$

Thus, at 1.00 molal, $\ln \gamma_\pm = -0.4234$, or $\gamma_\pm = 0.655$.

The formulas we derived in Section 11–3 for the colligative properties of solutions of nonelectrolytes take on a slightly different form for solutions of electrolytes. The difference lies in Equation 11.21 for $x_2$. For a strong electrolyte that dissociates into $\nu_+$ cations and $\nu_-$ anions per formula unit, the mole fraction of solute particles is given by

$$x_2 = \frac{\nu m}{\dfrac{1000\ \mathrm{g\cdot kg^{-1}}}{M_1} + \nu m} \approx \frac{\nu m M_1}{1000\ \mathrm{g\cdot kg^{-1}}} \tag{11.44}$$

Note that the right side here contains a factor of $\nu$. If this expression for $x_2$ is carried through in derivations of the formulas for the colligative effects, we obtain

$$\Delta T_{\mathrm{fus}} = \nu K_{\mathrm{f}} m \tag{11.45}$$

$$\Delta T_{\mathrm{vap}} = \nu K_{\mathrm{b}} m \tag{11.46}$$

and

$$\Pi = \nu M R T \tag{11.47}$$

**EXAMPLE 11–10**

A 0.050-molal aqueous solution of $K_3Fe(CN)_6$ has a freezing point of $-0.36^\circ$C. How many ions are formed per formula unit of $K_3Fe(CN)_6$?

SOLUTION: We can solve Equation 11.45 for $\nu$ to obtain

$$\nu = \frac{\Delta T_{\mathrm{fus}}}{K_{\mathrm{f}} m} = \frac{0.36^\circ\mathrm{C}}{(1.86\ ^\circ\mathrm{C\cdot kg\cdot mol^{-1}})(0.050\ \mathrm{mol\cdot kg^{-1}})} = 3.9$$

Thus, the dissolution process of $K_3Fe(CN)_6$ can be written as

$$\mathrm{K_3Fe(CN)_6} \xrightarrow{\mathrm{H_2O(l)}} 3\ \mathrm{K^+(aq)} + \mathrm{Fe(CN)_6^{3-}(aq)}$$

## 11–6. The Debye–Hückel Theory Gives an Exact Expression for $\ln \gamma_\pm$ for Very Dilute Solutions

In the previous section, we expressed the osmotic coefficient for solutions of electrolytes in the form $\phi = 1 + am^{1/2} + bm + \cdots$ rather than as a simple polynomial in $m$ as we did for sucrose in Section 11–2. The reason we did so is that in 1925, Peter Debye and Erich Hückel showed theoretically that at low concentrations, the logarithm of the activity coefficient of ion $j$ is given by

$$\ln \gamma_j = -\frac{\kappa q_j^2}{8\pi \varepsilon_0 \varepsilon_r k_{\mathrm{B}} T} \tag{11.48}$$

and that the logarithm of the mean ionic activity coefficient is given by (see Problems 11–50 through 11–58)

$$\ln \gamma_{\pm} = -|q_+ q_-| \frac{\kappa}{8\pi \varepsilon_0 \varepsilon_r k_B T} \tag{11.49}$$

where $q_+ = z_+ e$ and $q_- = z_- e$ are the charges on the cations and anions, $\varepsilon_r$ is the (unitless) relative permittivity of the solvent, and $\kappa$ is given by

$$\kappa^2 = \sum_{j=1}^{s} \frac{q_j^2}{\varepsilon_0 \varepsilon_r k_B T} \left( \frac{N_j}{V} \right) \tag{11.50}$$

where $s$ is the number of ionic species and $N_j/V$ is the number density of species $j$. If we convert $N_j/V$ to molarity, Equation 11.50 becomes

$$\kappa^2 = N_A (1000\ \text{L}\cdot\text{m}^{-3}) \sum_{j=1}^{s} \frac{q_j^2 c_j}{\varepsilon_0 \varepsilon_r k_B T} \tag{11.51}$$

It is customary to define a quantity $I_c$, called the *ionic strength*, by

$$I_c = \frac{1}{2} \sum_{j=1}^{s} z_j^2 c_j \tag{11.52}$$

where $c_j$ is the molarity of the $j$th ionic species, in which case (Problem 11–46)

$$\kappa^2 = \frac{2e^2 N_A (1000\ \text{L}\cdot\text{m}^{-3})}{\varepsilon_0 \varepsilon_r k_B T} (I_c/\text{mol}\cdot\text{L}^{-1}) \tag{11.53}$$

**EXAMPLE 11–11**

First show that $\kappa$ has units of $\text{m}^{-1}$ and then show that $\ln \gamma_{\pm}$ in Equation 11.49 is unitless, as it must be.

SOLUTION: We start with Equation 11.50. The units of $q_j$ are C, $\varepsilon_0$ are $\text{C}^2\cdot\text{s}^2\cdot\text{kg}^{-1}\cdot\text{m}^{-3}$, $k_B$ are $\text{J}\cdot\text{K}^{-1} = \text{kg}\cdot\text{m}^2\cdot\text{s}^{-2}\cdot\text{K}^{-1}$, $T$ are K, and $N_j/V$ are $\text{m}^{-3}$. Therefore, the units of $\kappa^2$ are

$$\kappa^2 \sim \frac{(\text{C}^2)(\text{m}^{-3})}{(\text{C}^2\cdot\text{s}^2\cdot\text{kg}^{-1}\cdot\text{m}^{-3})(\text{kg}\cdot\text{m}^2\cdot\text{s}^{-2}\cdot\text{K}^{-1})(\text{K})} = m^{-2}$$

or

$$\kappa \sim \text{m}^{-1}$$

Using Equation 11.49 for $\ln \gamma_{\pm}$,

$$\ln \gamma_{\pm} \sim \frac{(\text{C}^2)(\text{m}^{-1})}{(\text{C}^2\cdot\text{s}^2\cdot\text{kg}^{-1}\cdot\text{m}^{-3})(\text{kg}\cdot\text{m}^2\cdot\text{s}^{-2}\cdot\text{K}^{-1})(\text{K})} = \text{unitless}$$

Equation 11.49 is called the Debye–Hückel limiting law because it is the exact form that $\ln \gamma_{\pm}$ takes on for all electrolyte solutions for sufficiently low concentrations. Just what is meant by "sufficiently low concentrations" depends upon the system. Note that $\ln \gamma_{\pm}$ goes as $\kappa$ in Equation 11.49, that $\kappa$ goes as $I_c^{1/2}$ in Equation 11.53, and that $I_c^{1/2}$ goes as $c^{1/2}$ in Equation 11.52. Consequently, $\ln \gamma_{\pm}$ varies as $c^{1/2}$. This $c^{1/2}$ dependence is typical for electrolyte solutions, so when we curve fit $\phi$ in Section 11–5, we fit it to a polynomial in $c^{1/2}$ (or $m^{1/2}$) instead of $c$ (or $m$).

Most of the experimental data for $\ln \gamma_{\pm}$ are given in terms of molality rather than molarity. In Figure 11.6, we plot $\ln \gamma_{\pm}$ versus $m^{1/2}$ for a number of 1–1 electrolytes. Note that all the curves merge into a single straight line at small concentrations, in accord with the limiting law nature of Equation 11.49. At small concentrations where the limiting law is valid, the molality and molarity scales differ by only a multiplicative constant, so a linear plot in $c^{1/2}$ is also linear in $m^{1/2}$ (Problem 11–5).

The quantity $\kappa$ in Equation 11.50 is a central quantity in the Debye–Hückel theory and has the following physical interpretation. Consider an ion with charge $q_i$ situated at the origin of a spherical coordinate system. According to Debye and Hückel (see also Problem 11–51), the net charge in a spherical shell of radius $r$ and thickness $dr$ surrounding this central ion is

$$p_i(r)dr = -q_i\kappa^2 r e^{-\kappa r} dr \tag{11.54}$$

If we integrate this expression from 0 to $\infty$, we obtain

$$\int_0^\infty p_i(r)dr = -q_i\kappa^2 \int_0^\infty r e^{-\kappa r} dr = -q_i$$

This result simply says that the total charge surrounding an ion of charge $q_i$ is equal and of the opposite sign to $q_i$. In other words, it expresses the electroneutrality of the

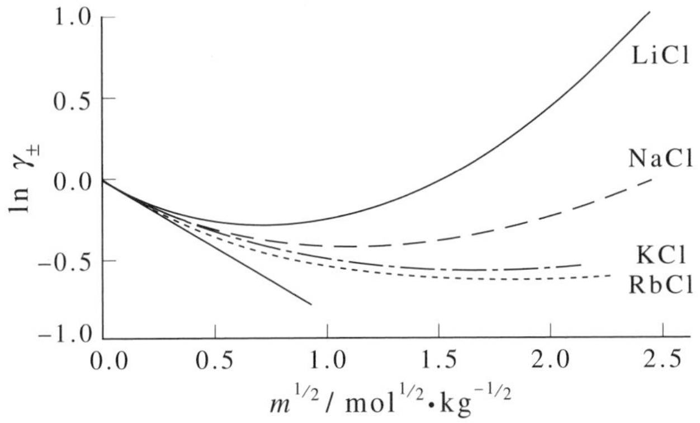


**FIGURE 11.6**
Values of $\ln \gamma_{\pm}$ versus $m^{1/2}$ for aqueous alkali halide solutions at 25°C. Note that even though the four curves are different, they all merge into one, the Debye–Hückel limiting law (Equation 11.49) at small concentrations.

solution. Equation 11.54, which is plotted in Figure 11.7, shows that there is a diffuse shell of net charge of opposite sign surrounding any given ion in solution. We say that Equation 11.54 describes an *ionic atmosphere* about the central ion. Furthermore, the maximum in the curve in Figure 11.7 occurs at $r = \kappa^{-1}$, so we say that $\kappa^{-1}$, which Example 11–11 shows has units of m, is a measure of the thickness of the ionic atmosphere.

For a 1–1 electrolyte in aqueous solution at 25°C, a handy formula for $\kappa$ is (Problem 11–53)

$$\frac{1}{\kappa} = \frac{304 \text{ pm}}{(c/\text{mol}\cdot\text{L}^{-1})^{1/2}} \tag{11.55}$$

where $c$ is the molarity of the solution. The thickness of the ionic atmosphere in a 0.010 molar solution is approximately 3000 pm, or about 10 times the size of a typical ion.

For an aqueous solution at 25°C, Equation 11.49 becomes (Problem 11–59)

$$\ln \gamma_{\pm} = -1.173|z_+ z_-|(I_c/\text{mol}\cdot\text{L}^{-1})^{1/2} \tag{11.56}$$

According to Equation 11.52, $I_c$ is related to the concentration, but the relation itself depends upon the type of electrolyte. For example, for a 1–1 electrolyte, $z_+ = 1$, $z_- = -1$, $c_+ = c$, and $c_- = c$, so $I = c$. For a 1–2 electrolyte such as $CaCl_2$, $z_+ = 2$, $z_- = -1$, $c_+ = c$, and $c_- = 2c$, so $I_c = \frac{1}{2}(4c + 2c) = 3c$. Generally, $I_c$ is equal to some numerical factor times $c$, where the value of the numerical factor depends upon the type of salt. Therefore, Equation 11.56 says that a plot of $\ln \gamma_{\pm}$ versus $c^{1/2}$ should be a straight line and that the slope of the line should depend upon the type of electrolyte. The slope will be $-1.173$ for a 1–1 electrolyte and $-(1.173)(2)(3^{1/2}) = -4.06$ for

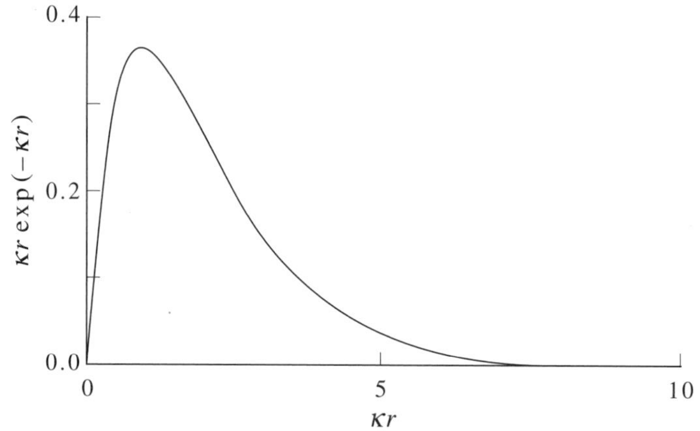


**FIGURE 11.7**
A plot of the net charge in a spherical shell of radius $r$ and thickness $dr$ surrounding a central ion of charge $q_i$. This plot illustrates the ionic atmosphere that surrounds each ion in solution. The maximum here corresponds to $r = \kappa^{-1}$.

a 1–2 electrolyte. Figure 11.8 shows a plot of $\ln \gamma_{\pm}$ versus $c^{1/2}$ for NaCl(aq) and $CaCl_2$(aq). Notice that the plots are indeed linear for small concentrations and that deviations from linear behavior occur at higher concentrations [$c^{1/2} \approx 0.05\ \text{mol}\cdot\text{L}^{-1}$ or $c = 0.003\ \text{mol}\cdot\text{L}^{-1}$ for $CaCl_2$(aq) and $c^{1/2} \approx 0.15\ \text{mol}\cdot\text{L}^{-1}$ or $c = 0.02\ \text{mol}\cdot\text{L}^{-1}$ for NaCl(aq)]. The slopes of the two linear portions are in the ratio of 4.06 to 1.17.

## 11–7. The Mean Spherical Approximation Is an Extension of the Debye–Hückel Theory to Higher Concentrations

The Debye–Hückel theory assumes that the ions are simply point ions (zero radii) and that they interact with a purely coulombic potential $[U(r) = z_+ z_- e^2/4\pi\varepsilon_0\varepsilon_r r]$. In addition, the solvent is considered a continuous medium with a uniform relative permittivity $\varepsilon_r$ (78.54 for water at 25°C). Although the assumptions of point ions and a continuum solvent may seem crude, they are quite satisfactory when the ions are far apart from each other on the average, as they are in very dilute solutions. Consequently, the Debye–Hückel expression for $\ln \gamma_{\pm}$ given by Equation 11.49 is exact in the limit of small concentrations. There is no corresponding theory for solutions of nonelectrolytes because, being neutral species, nonelectrolyte molecules do not interact with each other to any significant extent until they approach each other relatively closely, where the solvent can hardly be assumed to be a continuous medium.

Figure 11.8 emphasizes that the Debye–Hückel theory is a limiting law. It should not be considered a quantitative theory with which to calculate activity coefficients except at very low concentrations. Nevertheless, the Debye–Hückel theory has played an invaluable role as a strict limiting law that all electrolyte solutions obey. In addition, any theory that attempts to describe solutions at higher concentrations must reduce to Equation 11.49 for small concentrations. Many attempts have been made to construct

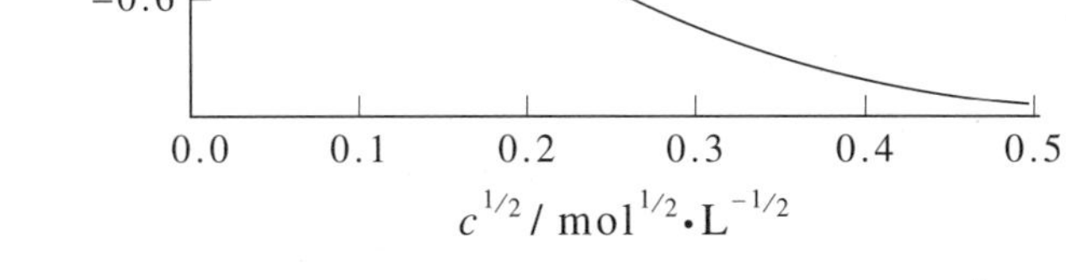

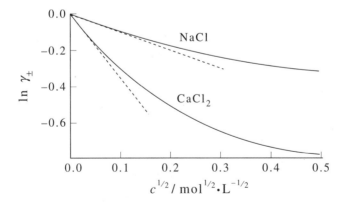

**FIGURE 11.8**
A plot of the logarithm of the mean ionic activity coefficient ($\ln \gamma_{\pm}$) for NaCl(aq) and $CaCl_2$(aq) at 25°C versus $c^{1/2}$. Note that both curves approach the Debye–Hückel limiting law (the straight lines) as the molarity goes to zero.

theories for more concentrated electrolyte solutions, but most have met with only limited success. One early attempt is called the Extended Debye–Hückel theory, in which Equation 11.49 is modified to be

$$\ln \gamma_{\pm} = -\frac{1.173|z_+ z_-|(I_c/\text{mol}\cdot\text{L}^{-1})^{1/2}}{1 + (I_c/\text{mol}\cdot\text{L}^{-1})^{1/2}} \tag{11.57}$$

This expression becomes Equation 11.49 in the limit of small concentrations because $I_c^{1/2}$ becomes negligible compared with unity in the denominator of Equation 11.57 in this limit.

**EXAMPLE 11–12**

Use Equation 11.57 to calculate $\ln \gamma_{\pm}$ for 0.050 molar LiCl(aq), and compare the result with that obtained from Equation 11.49. The accepted experimental value is −0.191.

SOLUTION: For a 1–1 salt such as LiCl, $I_c = c$, so

$$\ln \gamma_{\pm} = -1.173(0.050)^{1/2} = -0.262$$

and

$$\ln \gamma_{\pm} = -\frac{1.173(0.050)^{1/2}}{1 + (0.050)^{1/2}} = -0.214$$

Although Equation 11.57 provides some improvement over the Debye–Hückel limiting law, it is not very accurate even at 0.050 molar. At 0.200 molar, Equation 11.57 gives −0.362 for $\ln \gamma_{\pm}$ versus the experimental value of −0.274.

Another semiempirical expression for $\ln \gamma_{\pm}$ that has been widely used to fit experimental data is

$$\ln \gamma_{\pm} = -\frac{1.173|z_+ z_-|(I_c/\text{ mol}\cdot\text{L}^{-1})^{1/2}}{1 + (I_c/\text{ mol}\cdot\text{L}^{-1})^{1/2}} + Cm \tag{11.58}$$

where $C$ is a parameter whose value depends upon the electrolyte. Although Equation 11.58 can be used to fit experimental $\ln \gamma_{\pm}$ data up to one molar or so, $C$ is still strictly an adjustable parameter.

In the 1970s, significant advances were made in the theory of electrolyte solutions. Most of the work on these theories is based on a model called the *primitive model*, in which the ions are considered hard spheres with charges at their centers and the solvent is considered a continuous medium with a uniform relative permittivity. In spite of the obvious deficiencies of this model, it addresses the long-range coulombic interactions between the ions and their short-range repulsion. These turn out to be major considerations, and as we will see, the primitive model can give quite satisfactory agreement with experimental data over a fairly large concentration range.

Most of these theories that have been developed require numerical solutions to fairly complicated equations, but one is notable in that it provides analytic expressions for the various thermodynamic properties of electrolyte solutions. The name of this theory, the mean spherical approximation (MSA), derives from its original formulation, and the theory can be viewed as a Debye–Hückel theory in which the finite (nonzero) size of the ions is accounted for in a fairly rigorous manner. A central result of the mean spherical approximation is that

$$\ln \gamma_{\pm} = \ln \gamma_{\pm}^{\text{el}} + \ln \gamma^{\text{HS}} \tag{11.59}$$

where $\ln \gamma_{\pm}^{\text{el}}$ is an electrostatic (coulombic) contribution to $\ln \gamma_{\pm}$ and $\ln \gamma^{\text{HS}}$ is a hard-sphere (finite-size) contribution. For solutions of 1–1 electrolytes, $\ln \gamma_{\pm}^{\text{el}}$ is given by

$$\ln \gamma_{\pm}^{\text{el}} = \frac{x(1+2x)^{1/2} - x - x^2}{4\pi \rho d^3} \tag{11.60}$$

where $\rho$ is the number density of charged particles, $d$ is the sum of the radius of a cation and an anion, and $x = \kappa d$, where $\kappa$ is given by Equation 11.53. Although it is not obvious by casual inspection, Equation 11.59 reduces to the Debye–Hückel limiting law, Equation 11.49, in the limit of small concentrations (Problem 11–60). The hard sphere contribution to $\ln \gamma_{\pm}$ is given by

$$\ln \gamma^{\text{HS}} = \frac{4y - \frac{9}{4}y^2 + \frac{3}{8}y^3}{\left(1 - \frac{y}{2}\right)^3} \tag{11.61}$$

where $y = \pi \rho d^3/6$.

In spite of the fact that Equations 11.60 and 11.61 are somewhat lengthy, they are easy to use because once $d$ has been chosen, they give $\ln \gamma_{\pm}$ in terms of the molarity $c$. Figure 11.9 shows experimental values of $\ln \gamma_{\pm}$ for NaCl(aq) at 25°C and $\ln \gamma_{\pm}$ as calculated from Equation 11.59 with $d = 320$ pm.

Given essentially one adjustable parameter (the sum of the ionic radii), the agreement is seen to be quite good. We also show the results for the more commonly seen Equation 11.57 in Figure 11.9.

# Problems

**11-1.** The density of a glycerol/water solution that is 40.0% glycerol by mass is $1.101\ \text{g}\cdot\text{mL}^{-1}$ at 20°C. Calculate the molality and the molarity of glycerol in the solution at 20°C. Calculate the molality at 0°C.

**11-2.** Concentrated sulfuric acid is sold as a solution that is 98.0% sulfuric acid and 2.0% water by mass. Given that the density is $1.84\ \text{g}\cdot\text{mL}^{-1}$, calculate the molarity of concentrated sulfuric acid.

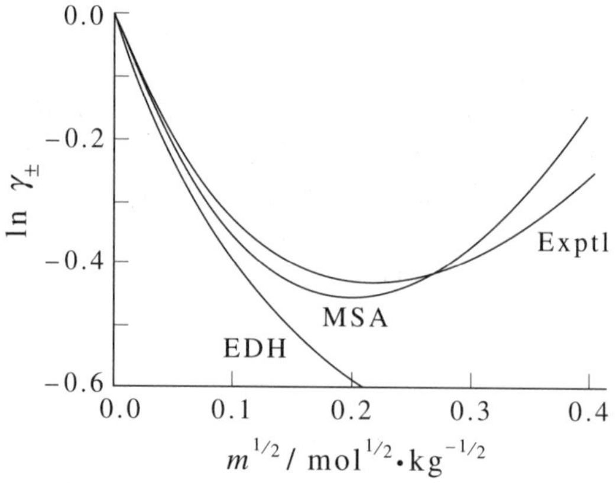


**FIGURE 11.9**
A comparison of $\ln \gamma_{\pm}$ from the mean spherical approximation (Equation 11.59) with experimental data for NaCl(aq) at 25°C. The line labelled EDH is the extended Debye–Hückel theory result, Equation 11.57. The value of $d$, the sum of the radii of the cation and anion, is taken to be 320 pm.

**11-3.** Concentrated phosphoric acid is sold as a solution that is 85% phosphoric acid and 15% water by mass. Given that the molarity is 15 $\text{mol·L}^{-1}$, calculate the density of concentrated phosphoric acid.

**11-4.** Calculate the mole fraction of glucose in an aqueous solution that is 0.500 molal in glucose.

**11-5.** Show that the relation between molarity and molality for a solution with a single solute is

$$c = \frac{(1000\ \text{mL·L}^{-1})\rho m}{1000\ \text{g·kg}^{-1} + mM_2}$$

where $c$ is the molarity, $m$ is the molality, $\rho$ is the density of the solution in $\text{g·mL}^{-1}$, and $M_2$ is the molar mass ($\text{g·mol}^{-1}$) of the solute.

**11-6.** The *CRC Handbook of Chemistry and Physics* has tables of "concentrative properties of aqueous solutions" for many solutions. Some entries for CsCl(s) are

| $A/\%$ | $\rho/\text{g·mL}^{-1}$ | $c/\text{mol·L}^{-1}$ |
|---|---|---|
| 1.00 | 1.0058 | 0.060 |
| 5.00 | 1.0374 | 0.308 |
| 10.00 | 1.0798 | 0.641 |
| 20.00 | 1.1756 | 1.396 |
| 40.00 | 1.4226 | 3.380 |

where $A$ is the mass percent of the solute, $\rho$ is the density of the solution, and $c$ is the molarity. Using these data, calculate the molality at each concentration.

**11-7.** Derive a relation between the mass percentage ($A$) of a solute in a solution and its molality ($m$). Calculate the molality of an aqueous sucrose solution that is 18% sucrose by mass.

**11-8.** Derive a relation between the mole fraction of the solvent and the molality of a solution.

**11-9.** The volume of an aqueous sodium chloride solution at 25°C can be expressed as

$$V/\text{mL} = 1001.70 + (17.298\ \text{kg}\cdot\text{mol}^{-1})m + (0.9777\ \text{kg}^2\cdot\text{mol}^{-2})m^2 - (0.0569\ \text{kg}^3\cdot\text{mol}^{-3})m^3$$

$$0 \le m \le 6\ \text{mol}\cdot\text{kg}^{-1}$$

where $m$ is the molality. Calculate the molarity of a solution that is 3.00 molar in sodium chloride.

**11-10.** If $x_2^\infty$, $m^\infty$, and $c^\infty$ are the mole fraction, molality, and molarity, respectively, of a solute at infinite dilution, show that

$$x_2^\infty = \frac{m^\infty M_1}{1000\ \text{g}\cdot\text{kg}^{-1}} = \frac{c^\infty M_1}{(1000\ \text{mL}\cdot\text{L}^{-1})\rho_1}$$

where $M_1$ is the molar mass ($\text{g}\cdot\text{mol}^{-1}$) and $\rho_1$ is the density ($\text{g}\cdot\text{mL}^{-1}$) of the solvent. Note that mole fraction, molality, and molarity are all directly proportional to each other at low concentrations.

**11-11.** Consider two solutions whose solute activities are $a_2'$ and $a_2''$, referred to the same standard state. Show that the difference in the chemical potentials of these two solutions is independent of the standard state and depends only upon the ratio $a_2'/a_2''$. Now choose one of these solutions to be at an arbitrary concentration and the other at a very dilute concentration (essentially infinitely dilute) and argue that

$$\frac{a_2'}{a_2''} = \frac{\gamma_{2x}x_2}{x_2^\infty} = \frac{\gamma_{2m}m}{m^\infty} = \frac{\gamma_{2c}c}{c^\infty}$$

**11-12.** Use Equations 11.4, 11.11, and the results of the previous two problems to show that

$$\gamma_{2x} = \gamma_{2m}\left(1 + \frac{mM_1}{1000\ \text{g}\cdot\text{kg}^{-1}}\right) = \gamma_{2c}\left(\frac{\rho}{\rho_1} + \frac{c[M_1 - M_2]}{\rho_1[1000\ \text{mL}\cdot\text{L}^{-1}]}\right)$$

where $\rho$ is the density of the solution. Thus, we see that the three different activity coefficients are related to one another.

**11-13.** Use Equations 11.4, 11.11, and the results of Problem 11–12 to derive

$$\gamma_{2m} = \gamma_{2c}\left(\frac{\rho}{\rho_1} - \frac{cM_2}{\rho_1[1000\ \text{mL}\cdot\text{L}^{-1}]}\right)$$

Given that the density of an aqueous citric acid ($M_2 = 192.12\ \text{g}\cdot\text{mol}^{-1}$) solution at 20°C is given by

$$\rho/\text{g}\cdot\text{mL}^{-1} = 0.99823 + (0.077102\ \text{L}\cdot\text{mol}^{-1})c$$

$$0 \le c < 1.772\ \text{mol}\cdot\text{L}^{-1}$$

plot $\gamma_{2m}/\gamma_{2c}$ versus $c$. Up to what concentration do $\gamma_{2m}$ and $\gamma_{2c}$ differ by 2%?

**11-14.** The *CRC Handbook of Chemistry and Physics* gives a table of mass percent of sucrose in an aqueous solution and its corresponding molarity at 25°C. Use these data to plot molality versus molarity for an aqueous sucrose solution.

**11-15.** Using the data in Table 11.2, calculate the activity coefficient of water (on a mole fraction basis) at a sucrose concentration of 3.00 molal.

**11-16.** Using the data in Table 11.2, plot the activity coefficient of water against the mole fraction of water.

**11-17.** Using the data in Table 11.2, calculate the value of $\phi$ at each value of $m$ and reproduce Figure 11.2.

**11-18.** Fit the data for the osmotic coefficient of sucrose in Table 11.2 to a 4th-degree polynomial and calculate the value of $\gamma_{2m}$ for a 1.00-molal solution. Compare your result with the one obtained in Example 11–5.

**11-19.** Using the data for sucrose given in Table 11.2, determine the value of $\ln \gamma_{2m}$ at 3.00 molal by plotting $(\phi - 1)/m$ versus $m$ and determining the area under the curve by numerical integration (MathChapter A) rather than by curve fitting $\phi$ first. Compare your result with the value given in Table 11.2.

**11-20.** Equation 11.18 can be used to determine the activity of the solvent at its freezing point. Assuming that $\Delta C_P^*$ is independent of temperature, show that

$$\Delta_{\text{fus}}\overline{H}(T) = \Delta_{\text{fus}}\overline{H}(T_{\text{fus}}^*) + \Delta\overline{C}_P^*(T - T_{\text{fus}}^*)$$

where $\Delta_{\text{fus}}\overline{H}(T_{\text{fus}}^*)$ is the molar enthalpy of fusion at the freezing point of the pure solvent $(T_{\text{fus}}^*)$ and $\Delta\overline{C}_P^*$ is the difference in the molar heat capacities of liquid and solid solvent. Using Equation 11.18, show that

$$-\ln a_1 = \frac{\Delta_{\text{fus}}\overline{H}(T_{\text{fus}}^*)}{R(T_{\text{fus}}^*)^2}\theta + \frac{1}{R(T_{\text{fus}}^*)^2}\left(\frac{\Delta_{\text{fus}}\overline{H}(T_{\text{fus}}^*)}{T_{\text{fus}}^*} - \frac{\Delta\overline{C}_P^*}{2}\right)\theta^2 + \cdots$$

where $\theta = T_{\text{fus}}^* - T_{\text{fus}}$.

**11-21.** Take $\Delta_{\text{fus}}\overline{H}(T_{\text{fus}}^*) = 6.01\ \text{kJ}\cdot\text{mol}^{-1}$, $\overline{C}_P^{\,l} = 75.2\ \text{J}\cdot\text{K}^{-1}\cdot\text{mol}^{-1}$, and $\overline{C}_P^{\,s} = 37.6\ \text{J}\cdot\text{K}^{-1}\cdot\text{mol}^{-1}$ to show that the equation for $-\ln a_1$ in the previous problem becomes

$$-\ln a_1 = (0.00968\ \text{K}^{-1})\theta + (5.2 \times 10^{-6}\ \text{K}^{-2})\theta^2 + \cdots$$

for an aqueous solution. The freezing point depression of a 1.95-molal aqueous sucrose solution is 4.45°C. Calculate the value of $a_1$ at this concentration. Compare your result with the value in Table 11.2. The value you calculated in this problem is for 0°C, whereas the value in Table 11.2 is for 25°C, but the difference is fairly small because $a_1$ does not vary greatly with temperature (Problem 11–61).

**11-22.** The freezing point of a 5.0-molal aqueous glycerol (1,2,3-propanetriol) solution is −10.6°C. Calculate the activity of water at 0°C in this solution. (See Problems 11–20 and 11–21.)

**11-23.** Show that replacing $T_{\text{fus}}$ by $T_{\text{fus}}^*$ in the denominator of $(T_{\text{fus}} - T_{\text{fus}}^*)/T_{\text{fus}}^* T_{\text{fus}}$ (see Equation 11.20) gives $-\theta/(T_{\text{fus}}^*)^2 - \theta^2/(T_{\text{fus}}^*)^3 + \cdots$ where $\theta = T_{\text{fus}}^* - T_{\text{fus}}$.

**11-24.** Calculate the value of the freezing point depression constant for nitrobenzene, whose freezing point is 5.7°C and whose molar enthalpy of fusion is $11.59 \text{ kJ}\cdot\text{mol}^{-1}$.

**11-25.** Use an argument similar to the one we used to derive Equations 11.22 and 11.23 to derive Equations 11.24 and 11.25.

**11-26.** Calculate the boiling point elevation constant for cyclohexane given that $T_{\text{vap}} = 354 \text{ K}$ and $\Delta_{\text{vap}}\overline{H} = 29.97 \text{ kJ}\cdot\text{mol}^{-1}$.

**11-27.** A solution containing 1.470 g of dichlorobenzene in 50.00 g of benzene boils at 80.60°C at a pressure of 1.00 bar. The boiling point of pure benzene is 80.09°C, and the molar enthalpy of vaporization of pure benzene is $32.0 \text{ kJ}\cdot\text{mol}^{-1}$. Determine the molecular mass of dichlorobenzene from these data.

**11-28.** Consider the following phase diagram for a typical pure substance. Label the region corresponding to each phase. Illustrate how this diagram changes for a dilute solution of a nonvolatile solute.

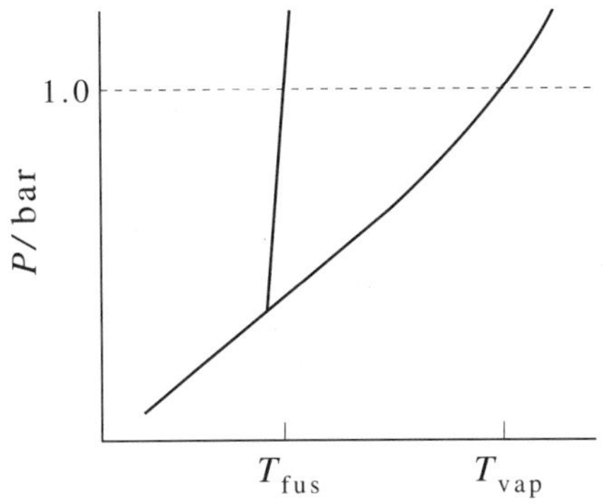


Now demonstrate that the boiling point increases and the freezing point decreases as a result of the dissolution of the solute.

**11-29.** A solution containing 0.80 g of a protein in 100 mL of a solution has an osmotic pressure of 2.06 torr at 25°C. What is the molecular mass of the protein?

**11-30.** Show that the osmotic pressure of an aqueous solution can be written as

$$\Pi = \frac{RT}{\overline{V}^*}\left(\frac{m}{55.506 \text{ mol}\cdot\text{kg}^{-1}}\right)\phi$$

**11-31.** According to Table 11.2, the activity of the water in a 2.00-molal sucrose solution is 0.95807. What external pressure must be applied to the solution at 25.0°C to make the activity of the water in the solution the same as that in pure water at 25.0°C and 1 atm? Take the density of water to be $0.99707 \text{ g}\cdot\text{mL}^{-1}$.

**11-32.** Show that $a_2 = a_\pm^2 = m^2\gamma_\pm^2$ for a 2–2 salt such as $CuSO_4$ and that $a_2 = a_\pm^4 = 27m^4\gamma_\pm^4$ for a 1–3 salt such as $LaCl_3$.

**11-33.** Verify the following table:

| Type of salt | Example | $I_m$ |
|---|---|---|
| 1–1 | KCl | $m$ |
| 1–2 | $CaCl_2$ | $3m$ |
| 2–1 | $K_2SO_4$ | $3m$ |
| 2–2 | $MgSO_4$ | $4m$ |
| 1–3 | $LaCl_3$ | $6m$ |
| 3–1 | $Na_3PO_4$ | $6m$ |

Show that the general result for $I_m$ is $|z_+z_-|(\nu_+ + \nu_-)m/2$.

**11-34.** Show that the inclusion of the factor $\nu$ in Equation 11.41 allows $\phi \to 1$ as $m \to 0$ for solutions of electrolytes as well as nonelectrolytes. [*Hint*: Realize that $x_2$ involves the total number of moles of solute particles (see Equation 11.44)].

**11-35.** Use Equation 11.41 and the Gibbs–Duhem equation to derive Equation 11.42.

**11-36.** The osmotic coefficient of $CaCl_2$(aq) solutions can be expressed as

$$\begin{aligned}\phi = 1.0000 &- (1.2083\ \text{kg}^{1/2}\cdot\text{mol}^{-1/2})m^{1/2} + (3.2215\ \text{kg}\cdot\text{mol}^{-1})m \\ &- (3.6991\ \text{kg}^{3/2}\cdot\text{mol}^{-3/2})m^{3/2} + (2.3355\ \text{kg}^2\cdot\text{mol}^{-2})m^2 \\ &- (0.67218\ \text{kg}^{5/2}\cdot\text{mol}^{-5/2})m^{5/2} + (0.069749\ \text{kg}^3\cdot\text{mol}^{-3})m^3 \\ &\qquad 0 \le m \le 5.00\ \text{mol}\cdot\text{kg}^{-1}\end{aligned}$$

Use this expression to calculate and plot $\ln\gamma_\pm$ as a function of $m^{1/2}$.

**11-37.** Use Equation 11.43 to calculate $\ln\gamma_\pm$ for NaCl(aq) at 25°C as a function of molality and plot it versus $m^{1/2}$. Compare your results with those in Table 11.4.

**11-38.** In Problem 11–19, you determined $\ln\gamma_{2m}$ for sucrose by calculating the area under the curve of $\phi - 1$ versus $m$. When dealing with solutions of electrolytes, it is better numerically to plot $(\phi - 1)/m^{1/2}$ versus $m^{1/2}$ because of the natural dependence of $\phi$ on $m^{1/2}$. Show that

$$\ln\gamma_\pm = \phi - 1 + 2\int_0^{m^{1/2}} \frac{\phi - 1}{m^{1/2}}\, dm^{1/2}$$

**11-39.** Use the data in Table 11.4 to calculate $\ln\gamma_\pm$ for NaCl(aq) at 25°C by plotting $(\phi - 1)/m^{1/2}$ against $m^{1/2}$ and determine the area under the curve by numerical integration (MathChapter A). Compare your values of $\ln\gamma_\pm$ with those you obtained in Problem 11-37 where you calculated $\ln\gamma_\pm$ from a curve-fit expression of $\phi$ as a polynomial in $m^{1/2}$.

**11-40.** Don Juan Pond in the Wright Valley of Antarctica freezes at $-57°C$. The major solute in the pond is $CaCl_2$. Estimate the concentration of $CaCl_2$ in the pond water.

**11-41.** A solution of mercury(II) chloride is a poor conductor of electricity. A 40.7-g sample of $HgCl_2$ is dissolved in 100.0 g of water, and the freezing point of the solution is found to be $-2.83°C$. Explain why $HgCl_2$ in solution is a poor conductor of electricity.

**11-42.** The freezing point of a 0.25-molal aqueous solution of Mayer's reagent, $K_2HgI_4$, is found to be $-1.41°C$. Suggest a possible dissociation reaction that takes place when $K_2HgI_4$ is dissolved in water.

**11-43.** Given the following freezing-point depression data, determine the number of ions produced per formula unit when the indicated substance is dissolved in water to produce a 1.00-molal solution.

| Formula | $\Delta T/\text{K}$ |
|---|---|
| $PtCl_2 \cdot 4NH_3$ | 5.58 |
| $PtCl_2 \cdot 3NH_3$ | 3.72 |
| $PtCl_2 \cdot 2NH_3$ | 1.86 |
| $KPtCl_3 \cdot NH_3$ | 3.72 |
| $K_2PtCl_4$ | 5.58 |

Interpret your results.

**11-44.** An aqueous solution of NaCl has an ionic strength of $0.315\ \text{mol}\cdot\text{L}^{-1}$. At what concentration will an aqueous solution of $K_2SO_4$ have the same ionic strength?

**11-45.** Derive the "practical" formula for $\kappa^2$ given by Equation 11.53.

**11-46.** Some authors define ionic strength in terms of molality rather than molarity, in which case

$$I_m = \frac{1}{2}\sum_{j=1}^{s} z_j^2 m_j$$

Show that this definition modifies Equation 11.53 for dilute solutions to be

$$\kappa^2 = \frac{2e^2 N_A (1000\ \text{L}\cdot\text{m}^{-3})\rho}{\varepsilon_0 \varepsilon_r kT}(I_m/\text{mol}\cdot\text{kg}^{-1})$$

for an aqueous solution at 25°C, where $\rho$ is the density of the solvent (in $\text{g}\cdot\text{mL}^{-1}$).

**11-47.** Show that

$$\ln \gamma_\pm = -1.171|z_+ z_-|(I_m/\text{mol}\cdot\text{kg}^{-1})^{1/2}$$

for an aqueous solution at 25°C, where $I_m$ is the ionic strength expressed in terms of molality. Take $\varepsilon_r$ to be 78.54 and the density at water to be $0.99707\ \text{g}\cdot\text{mL}^{-1}$.

**11-48.** Use the Debye-Hückel theory to calculate the value of $\ln \gamma_\pm$ for a 0.010-molar NaCl(aq) solution at 25°C. The experimental value of $\gamma_\pm$ is 0.902. Take $\varepsilon_r = 78.54$ for $H_2O(l)$ at 25.0°C.

**11-49.** Derive the general equation

$$\phi = 1 + \frac{1}{m}\int_0^m m' d \ln \gamma_\pm$$

(*Hint*: See the derivation in Problem 11–35.) Use this result to show that

$$\phi = 1 + \frac{\ln \gamma_\pm}{3}$$

for the Debye–Hückel theory.

*In the next nine problems we will develop the Debye–Hückel theory of ionic solutions and derive Equations 11.48 and 11.49.*

**11-50.** In the Debye–Hückel theory, the ions are modeled as point ions, and the solvent is modeled as a continuous medium (no structure) with a relative permittivity $\varepsilon_r$. Consider an ion of type $i$ ($i$ = a cation or an anion) situated at the origin of a spherical coordinate system. The presence of this ion at the origin will attract ions of opposite charge and repel ions of the same charge. Let $N_{ij}(r)$ be the number of ions of type $j$ ($j$ = a cation or an anion) situated at a distance $r$ from the central ion of type $i$ (a cation or an anion). We can use a Boltzmann factor to say that

$$N_{ij}(r) = N_j e^{-w_{ij}(r)/k_\mathrm{B}T}$$

where $N_j/V$ is the bulk number density of $j$ ions and $w_{ij}(r)$ is the interaction energy of an $i$ ion with a $j$ ion. This interaction energy will be electrostatic in origin, so let $w_{ij}(r) = q_j\psi_i(r)$, where $q_j$ is the charge on the ion of type $j$ and $\psi_i(r)$ is the electrostatic potential due to the central ion of type $i$.

A fundamental equation from physics that relates a spherically symmetric electrostatic potential $\psi_i(r)$ to a spherically symmetric charge density $\rho_i(r)$ is Poisson's equation

$$\frac{1}{r^2}\frac{d}{dr}\left(r^2\frac{d\psi_i}{dr}\right) = -\frac{\rho_i(r)}{\varepsilon_0\varepsilon_r} \tag{1}$$

where $\varepsilon_r$ is the relative permittivity of the solvent. In our case, $\rho_i(r)$ is the charge density around the central ion. First, show that

$$\rho_i(r) = \frac{1}{V}\sum_j q_j N_{ij}(r) = \sum_j q_j C_j e^{-q_j\psi_i(r)/k_\mathrm{B}T}$$

where $C_j$ is the bulk number density of species $j$ ($C_j = N_j/V$). Linearize the exponential term and use the condition of electroneutrality to show that

$$\rho_i(r) = -\psi_i(r)\sum_j \frac{q_j^2 C_j}{k_\mathrm{B}T} \tag{2}$$

Now substitute $\rho_i(r)$ into Poisson's equation to get

$$\frac{1}{r^2}\frac{d}{dr}\left(r^2\frac{d\psi_i}{dr}\right) = \kappa^2\psi_i(r) \tag{3}$$

where

$$\kappa^2 = \sum_j \frac{q_j^2 C_j}{\varepsilon_0 \varepsilon_r k_B T} = \sum_j \frac{q_j^2}{\varepsilon_0 \varepsilon_r k_B T} \left( \frac{N_j}{V} \right) \tag{4}$$

Show that Equation 3 can be written as

$$\frac{d^2}{dr^2}[r\psi_i(r)] = \kappa^2 [r\psi_i(r)]$$

Now show that the only solution for $\psi_i(r)$ that is finite for large values of $r$ is

$$\psi_i(r) = \frac{Ae^{-\kappa r}}{r} \tag{5}$$

where $A$ is a constant. Use the fact that if the concentration is very small, then $\psi_i(r)$ is just Coulomb's law and so $A = q_i/4\pi\varepsilon_0\varepsilon_r$ and

$$\psi_i(r) = \frac{q_i e^{-\kappa r}}{4\pi\varepsilon_0\varepsilon_r r} \tag{6}$$

Equation 6 is a central result of the Debye–Hückel theory. The factor of $e^{-\kappa r}$ modulates the resulting Coulombic potential, so Equation 6 is called a *screened Coulombic potential.*

**11-51.** Use Equations 2 and 6 of the previous problem to show that the net charge in a spherical shell of radius $r$ surrounding a central ion of type $i$ is

$$p_i(r)dr = \rho_i(r)4\pi r^2 dr = -q_i\kappa^2 r e^{-\kappa r} dr$$

as in Equation 11.54. Why is

$$\int_0^\infty p_i(r)dr = -q_i$$

**11-52.** Use the result of the previous problem to show that the most probable value of $r$ is $1/\kappa$.

**11-53.** Show that

$$r_{\mathrm{mp}} = \frac{1}{\kappa} = \frac{304\ \mathrm{pm}}{(c/\mathrm{mol\cdot L^{-1}})^{1/2}}$$

where $c$ is the molarity of an aqueous solution of a 1–1 electrolyte at 25°C. Take $\varepsilon_r = 78.54$ for $H_2O(l)$ at 25°C.

**11-54.** Show that

$$r_{\mathrm{mp}} = \frac{1}{\kappa} = 430\ \mathrm{pm}$$

for a 0.50-molar aqueous solution of a 1–1 electrolyte at 25°C. Take $\varepsilon_r = 78.54$ for $H_2O(l)$ at 25°C.

**11-55.** How does the thickness of the ionic atmosphere compare for a 1–1 electrolyte and a 2–2 electrolyte?

**11-56.** In this problem, we will calculate the total electrostatic energy of an electrolyte solution in the Debye–Hückel theory. Use the equations in Problem 11–50 to show that the number of ions of type $j$ in a spherical shell of radii $r$ and $r + dr$ about a central ion of type $i$ is

$$\left(\frac{N_{ij}(r)}{V}\right)4\pi r^2 dr = C_j e^{-q_j\psi_i(r)/k_\mathrm{B}T}4\pi r^2 dr \approx C_j\left(1-\frac{q_j\psi_i(r)}{k_\mathrm{B}T}\right)4\pi r^2 dr \qquad (1)$$

The total Coulombic interaction between the central ion of type $i$ and the ions of type $j$ in the spherical shell is $N_{ij}(r)u_{ij}(r)4\pi r^2 dr/V$ where $u_{ij}(r) = q_i q_j/4\pi\varepsilon_0\varepsilon_r r$. To determine the electrostatic interaction energy of all the ions in the solution with the central ion (of type $i$), $U_i^{\mathrm{el}}$, sum $N_{ij}(r)u_{ij}(r)/V$ over all types of ions in a spherical shell and then integrate over all spherical shells to get

$$U_i^{\mathrm{el}} = \int_0^\infty \left(\sum_j \frac{N_{ij}(r)u_{ij}(r)}{V}\right)4\pi r^2 dr$$

$$= \sum_j \frac{C_j q_i q_j}{\varepsilon_0\varepsilon_r}\int_0^\infty \left(1-\frac{q_j\psi_i(r)}{k_\mathrm{B}T}\right) r dr$$

Use electroneutrality to show that

$$U_i^{\mathrm{el}} = -q_i\kappa^2\int_0^\infty \psi_i(r) r dr$$

Now, using Equation 6 of Problem 11–50, show that the interaction of all ions with the central ion (of type $i$) is given by

$$U_i^{\mathrm{el}} = -\frac{q_i^2\kappa^2}{4\pi\varepsilon_0\varepsilon_r}\int_0^\infty e^{-\kappa r}dr = -\frac{q_i^2\kappa}{4\pi\varepsilon_0\varepsilon_r}$$

Now argue that the total electrostatic energy is

$$U^{\mathrm{el}} = \frac{1}{2}\sum_i N_i U_i^{\mathrm{el}} = -\frac{Vk_\mathrm{B}T\kappa^3}{8\pi}$$

Why is there a factor of 1/2 in this equation? Wouldn't you be overcounting the energy otherwise?

**11-57.** We derived an expression for $U^{\mathrm{el}}$ in the previous problem. Use the Gibbs–Helmholtz equation for $A$ (Problem 8–23) to show that

$$A^{\mathrm{el}} = -\frac{Vk_\mathrm{B}T\kappa^3}{12\pi}$$

**11-58.** If we assume that the electrostatic interactions are the sole cause of the nonideality of an electrolyte solution, then we can say that

$$\mu_j^{\mathrm{el}} = \left(\frac{\partial A^{\mathrm{el}}}{\partial n_j}\right)_{T,V} = RT\ln\gamma_j^{\mathrm{el}}$$

or that

$$\mu_j^{\mathrm{el}} = \left(\frac{\partial A^{\mathrm{el}}}{\partial N_j}\right)_{T,V} = k_\mathrm{B}T\ln\gamma_j^{\mathrm{el}}$$

Use the result you got for $A^{\text{el}}$ in the previous problem to show that

$$k_{\text{B}}T \ln \gamma_j^{\text{el}} = -\frac{\kappa q_j^2}{8\pi\varepsilon_0\varepsilon_r}$$

Use the formula

$$\gamma_{\pm}^{\nu} = \gamma_{+}^{\nu_+}\gamma_{-}^{\nu_-}$$

to show that

$$\ln \gamma_{\pm} = -\left(\frac{\nu_+ q_+^2 + \nu_- q_-^2}{\nu_+ + \nu_-}\right)\frac{\kappa}{8\pi\varepsilon_0\varepsilon_r k_{\text{B}}T}$$

Use the electroneutrality condition $\nu_+ q_+ + \nu_- q_- = 0$ to rewrite $\ln \gamma_{\pm}$ as

$$\ln \gamma_{\pm} = -|q_+ q_-|\frac{\kappa}{8\pi\varepsilon_0\varepsilon_r k_{\text{B}}T}$$

in agreement with Equation 11.49.

**11-59.** Derive Equation 11.56 from Equation 11.49.

**11-60.** Show that Equation 11.60 reduces to Equation 11.49 for small concentrations.

**11-61.** In this problem, we will investigate the temperature dependence of activities. Starting with the equation $\mu_1 = \mu_1^* + RT \ln a_1$, show that

$$\left(\frac{\partial \ln a_1}{\partial T}\right)_{P,x_1} = \frac{\overline{H}_1^* - \overline{H}_1}{RT^2}$$

where $\overline{H}_1^*$ is the molar enthalpy of the pure solvent (at one bar) and $\overline{H}_1$ is its partial molar enthalpy in the solution. The difference between $\overline{H}_1^*$ and $\overline{H}_1$ is small for dilute solutions, so $a_1$ is fairly independent of temperature.

**11-62.** Henry's law says that the pressure of a gas in equilibrium with a nonelectrolyte solution of the gas in a liquid is proportional to the molality of the gas in the solution for sufficiently dilute solutions. What form do you think Henry's law takes on for a gas such as HCl(g) dissolved in water? Use the following data for HCl(g) at 25°C to test your prediction.

| $P_{\text{HCl}}/10^{-11}$ bar | $m_{\text{HCl}}/10^{-3}$ mol·kg$^{-1}$ |
|---|---|
| 0.147 | 1.81 |
| 0.238 | 2.32 |
| 0.443 | 3.19 |
| 0.663 | 3.93 |
| 0.851 | 4.47 |
| 1.08 | 5.06 |
| 1.62 | 6.25 |
| 1.93 | 6.84 |
| 2.08 | 7.12 |

**Jacobus Henricus van't Hoff** was born in Rotterdam, the Netherlands, on August 30, 1852, and died in 1911. He had hoped to pursue a career in chemistry, but his parents persuaded him to study engineering because they believed that the prospects for pure research in chemistry were poor. Nonetheless, he later studied chemistry at several universities and finally completed his doctorate at the University of Utrecht in 1874. In the same year, he published a paper on the tetrahedral model of the carbon atom, which he did not present as his thesis for his doctorate because he feared that it was too controversial to be accepted. His work was strongly attacked by some prominent organic chemists, but eventually the model of the tetrahedral carbon atom became the foundation of modern stereochemistry. van't Hoff then turned his attention to investigating the rates of chemical reactions and chemical equilibrium. He determined the order of many chemical reactions and used the concept of reaction rates to describe the dynamics of chemical equilibrium. In 1887, Wilhelm Ostwald and van't Hoff founded the first journal devoted to physical chemistry, *Zeitschrift für Physikalische Chemie*. In 1901, van't Hoff was awarded the first-ever Nobel Prize for chemistry "for his discovery of the laws of chemical dynamics and osmotic pressure in solutions." Throughout his life, he was interested in philosophy and poetry, especially that of Lord Byron.

# CHAPTER 12

# Chemical Equilibrium

One of the most important applications of thermodynamics is to chemical reactions at equilibrium. Thermodynamics enables us to predict with confidence the equilibrium pressures or concentrations of reaction mixtures. In this chapter we shall derive a relation between the standard Gibbs energy change and the equilibrium constant for a chemical reaction. We shall also learn how to predict the direction in which a chemical reaction will proceed if we start with arbitrary concentrations of reactants and products. We have developed all the necessary thermodynamic concepts in previous chapters. The underlying fundamental idea is that $\Delta G = 0$ for a system in equilibrium at constant temperature and pressure, and that the sign of $\Delta G$ determines whether or not a given process or chemical reaction will occur spontaneously at constant $T$ and $P$.

## 12–1. Chemical Equilibrium Results When the Gibbs Energy Is a Minimum with Respect to the Extent of Reaction

For simplicity, we shall discuss gas-phase reactions first. Consider the general gas-phase reaction, which is described by the balanced equation

$$\nu_A A(g) + \nu_B B(g) \rightleftharpoons \nu_Y Y(g) + \nu_Z Z(g)$$

We define a quantity $\xi$, called the *extent of reaction*, such that the numbers of moles of the reactants and products are given by

$$\underbrace{\begin{aligned} n_A &= \nu_{A0} - \nu_A\xi \\ n_B &= \nu_{B0} - \nu_B\xi \end{aligned}}_{\text{reactants}} \qquad \underbrace{\begin{aligned} n_Y &= \nu_{Y0} + \nu_Y\xi \\ n_Z &= \nu_{Z0} + \nu_Z\xi \end{aligned}}_{\text{products}} \tag{12.1}$$

where $n_{j0}$ is the initial number of moles for each species. Recall from Chapter 5 that stoichiometric coefficients do not have units. Consequently, Equations 12.1 indicate that $\xi$ has units of moles. As the reaction proceeds from reactants to products,

$\xi$ varies from 0 to some maximum value dictated by the stoichiometry of the reaction. For example, if $n_{A0}$ and $n_{B0}$ in Equations 12.1 are equal to $\nu_A$ moles and $\nu_B$ moles, respectively, then $\xi$ will vary from 0 to one mole. Differentiation of Equations 12.1 gives

$$\underbrace{\begin{aligned} dn_A &= -\nu_A d\xi \\ dn_B &= -\nu_B d\xi \end{aligned}}_{\text{reactants}} \qquad \underbrace{\begin{aligned} dn_Y &= \nu_Y d\xi \\ dn_Z &= \nu_Z d\xi \end{aligned}}_{\text{products}} \tag{12.2}$$

The negative signs indicate that the reactants are disappearing and the positive signs indicate that the products are being formed as the reaction progresses from reactants to products.

Now let's consider a system containing reactants and products at constant $T$ and $P$. The Gibbs energy for this multicomponent system is a function of $T$, $P$, $n_A$, $n_B$, $n_Y$, and $n_Z$, which we can express mathematically as $G = G(T, P, n_A, n_B, n_Y, n_Z)$. The total derivative of $G$ is given by

$$\begin{aligned} dG = {} & \left(\frac{\partial G}{\partial T}\right)_{P,n_j} dT + \left(\frac{\partial G}{\partial P}\right)_{T,n_j} dP + \left(\frac{\partial G}{\partial n_A}\right)_{T,P,n_{j\neq A}} dn_A \\ & + \left(\frac{\partial G}{\partial n_B}\right)_{T,P,n_{j\neq B}} dn_B + \left(\frac{\partial G}{\partial n_Y}\right)_{T,P,n_{j\neq Y}} dn_Y + \left(\frac{\partial G}{\partial n_Z}\right)_{T,P,n_{j\neq Z}} dn_Z \end{aligned}$$

where the subscript $n_j$ in the first two partial derivatives stands for $n_A$, $n_B$, $n_Y$, and $n_Z$. Using Equations 8.31 for $(\partial G/\partial T)_{P,n_j}$ and $(\partial G/\partial P)_{T,n_j}$, $dG$ becomes

$$dG = -SdT + VdP + \mu_A dn_A + \mu_B dn_B + \mu_Y dn_Y + \mu_Z dn_Z$$

where

$$\mu_A = \left(\frac{\partial G}{\partial n_A}\right)_{T,P,n_B,n_Y,n_Z}$$

with similar expressions for $\mu_B$, $\mu_Y$, and $\mu_Z$. For a reaction that takes place at constant $T$ and $P$, $dG$ becomes

$$dG = \sum_j \mu_j dn_j = \mu_A dn_A + \mu_B dn_B + \mu_Y dn_Y + \mu_Z dn_Z \quad \text{(constant } T \text{ and } P\text{)} \tag{12.3}$$

Substitute Equations 12.2 into Equation 12.3 to obtain

$$\begin{aligned} dG &= -\nu_A\mu_A d\xi - \nu_B\mu_B d\xi + \nu_Y\mu_Y d\xi + \nu_Z\mu_Z d\xi \\ &= (\nu_Y\mu_Y + \nu_Z\mu_Z - \nu_A\mu_A - \nu_B\mu_B)d\xi \quad \text{(constant } T \text{ and } P\text{)} \end{aligned} \tag{12.4}$$

or

$$\left(\frac{\partial G}{\partial \xi}\right)_{T,P} = \nu_Y\mu_Y + \nu_Z\mu_Z - \nu_A\mu_A - \nu_B\mu_B \tag{12.5}$$

We shall denote the right side of Equation 12.5 by $\Delta_r G$, so that

$$\left(\frac{\partial G}{\partial \xi}\right)_{T,P} = \Delta_r G = \nu_Y \mu_Y + \nu_Z \mu_Z - \nu_A \mu_A - \nu_B \mu_B \tag{12.6}$$

The quantity $\Delta_r G$ is defined as the change in Gibbs energy when the extent of reaction changes by one mole. The units of $\Delta_r G$ are then $J \cdot mol^{-1}$. The quantity $\Delta_r G$ has meaning only if the balanced chemical equation is specified.

If we assume that all the partial pressures are low enough that we can consider each species to behave ideally, then we can use Equation 9.33 $[\mu_j(T, P) = \mu_j^\circ(T) + RT \ln(P_j/P^\circ)]$ for the $\mu_j(T, P)$, in which case Equation 12.6 becomes

$$\Delta_r G = \nu_Y \mu_Y^\circ(T) + \nu_Z \mu_Z^\circ(T) - \nu_A \mu_A^\circ(T) - \nu_B \mu_B^\circ(T) \\ + RT\left(\nu_Y \ln\frac{P_Y}{P^\circ} + \nu_Z \ln\frac{P_Z}{P^\circ} - \nu_A \ln\frac{P_A}{P^\circ} - \nu_B \ln\frac{P_B}{P^\circ}\right)$$

or

$$\Delta_r G = \Delta_r G^\circ + RT \ln Q \tag{12.7}$$

where

$$\Delta_r G^\circ(T) = \nu_Y \mu_Y^\circ(T) + \nu_Z \mu_Z^\circ(T) - \nu_A \mu_A^\circ(T) - \nu_B \mu_B^\circ(T) \tag{12.8}$$

and

$$Q = \frac{(P_Y/P^\circ)^{\nu_Y}(P_Z/P^\circ)^{\nu_Z}}{(P_A/P^\circ)^{\nu_A}(P_B/P^\circ)^{\nu_B}} \tag{12.9}$$

The quantity $\Delta_r G^\circ(T)$ is the change in standard Gibbs energy for the reaction between unmixed reactants in their standard states at temperature $T$ and a pressure of one bar to form unmixed products in their standard states at the same temperature $T$ and a pressure of one bar. Because the standard pressure $P^\circ$ in Equation 12.9 is taken to be one bar, the $P^\circ$'s are usually not displayed. It must be remembered, however, that all the pressures are referred to one bar, and that $Q$ consequently is unitless.

When the reaction system is in equilibrium, the Gibbs energy must be a minimum with respect to any displacement of the reaction from its equilibrium position, and so Equation 12.5 becomes

$$\left(\frac{\partial G}{\partial \xi}\right)_{T,P} = \Delta_r G = 0 \quad \text{(equilibrium)} \tag{12.10}$$

Setting $\Delta_r G = 0$ in Equation 12.7 gives

$$\Delta_r G^\circ(T) = -RT \ln\left(\frac{P_Y^{\nu_Y} P_Z^{\nu_Z}}{P_A^{\nu_A} P_B^{\nu_B}}\right)_{eq} = -RT \ln K_P(T) \tag{12.11}$$

where

$$K_P(T) = \left( \frac{P_{\text{Y}}^{\nu_{\text{Y}}} P_{\text{Z}}^{\nu_{\text{Z}}}}{P_{\text{A}}^{\nu_{\text{A}}} P_{\text{B}}^{\nu_{\text{B}}}} \right)_{\text{eq}} \tag{12.12}$$

and where the subscript eq emphasizes that the pressures in Equations 12.11 and 12.12 are the pressures at *equilibrium*. The quantity $K_P(T)$ is called the *equilibrium constant* of the reaction. Although we have used an eq subscript for emphasis, this notation is not normally used and $K_P(T)$ is written without the subscript. Equilibrium-constant expressions imply that the pressures are their equilibrium values. The value of $K_P$ cannot be evaluated unless the balanced chemical reaction to which it refers and the standard states of each of the reactants and products are given.

### EXAMPLE 12–1

Write out the equilibrium-constant expression for the reaction that is represented by the equation

$$3\,\text{H}_2(\text{g}) + \text{N}_2(\text{g}) \rightleftharpoons 2\,\text{NH}_3(\text{g})$$

SOLUTION: According to Equation 12.12,

$$K_P(T) = \frac{P_{\text{NH}_3}^2}{P_{\text{H}_2}^3 P_{\text{N}_2}}$$

where all the pressures are referred to the standard pressure of one bar. Note that if we had written the equation for the reaction as

$$\tfrac{3}{2}\,\text{H}_2(\text{g}) + \tfrac{1}{2}\,\text{N}_2(\text{g}) \rightleftharpoons \text{NH}_3(\text{g})$$

then we would have obtained

$$K_P(T) = \frac{P_{\text{NH}_3}}{P_{\text{H}_2}^{3/2} P_{\text{N}_2}^{1/2}}$$

which is the square root of our previous expression. Thus, we see that the form of $K_P(T)$ and its subsequent numerical value depend upon how we write the chemical equation that describes the reaction.

## 12–2. An Equilibrium Constant Is a Function of Temperature Only

Equation 12.11 says that regardless of the initial pressures of the reactants and products, at equilibrium the ratio of their partial pressures raised to their respective stoichiometric coefficients will be a fixed value at a given temperature. Consider the reaction described by

$$PCl_5(g) \rightleftharpoons PCl_3(g) + Cl_2(g) \tag{12.13}$$

The equilibrium-constant expression for this reaction is

$$K_P(T) = \frac{P_{PCl_3} P_{Cl_2}}{P_{PCl_5}} \tag{12.14}$$

Suppose that initially we have one mole of $PCl_5(g)$ and no $PCl_3(g)$ or $Cl_2(g)$. When the reaction occurs to an extent $\xi$, there will be $(1 - \xi)$ moles of $PCl_5(g)$, $\xi$ moles of $PCl_3(g)$, and $\xi$ moles of $Cl_2(g)$ in the reaction mixture and the total number of moles will be $(1 + \xi)$. If we let $\xi_{eq}$ be the extent of reaction at equilibrium, then the partial pressures of each species will be

$$P_{PCl_3} = P_{Cl_2} = \frac{\xi_{eq} P}{1 + \xi_{eq}}$$

$$P_{PCl_5} = \frac{(1 - \xi_{eq})P}{1 + \xi_{eq}}$$

where $P$ is the total pressure. The equilibrium-constant expression is

$$K_P(T) = \frac{\xi_{eq}^2}{1 - \xi_{eq}^2} P \tag{12.15}$$

It might appear from this result that $K_P(T)$ depends upon the total pressure, but this is not so. As Equation 12.11 shows, $K_P(T)$ is a function of only the temperature, and so is a constant value at a fixed temperature. Therefore, if $P$ changes, then $\xi_{eq}$ must change so that $K_P(T)$ in Equation 12.15 remains constant. Figure 12.1 shows $\xi_{eq}$ plotted against $P$ at 200°C, where $K_P = 5.4$. Note that $\xi_{eq}$ decreases uniformly with increasing $P$, indicating that the equilibrium is shifted from the product side to the reactant side of Equation 12.13 or that less $PCl_5$ is dissociated. This effect of pressure on the position of equilibrium is an example of *Le Châtelier's principle*, which you learned in general chemistry. Le Châtelier's principle can be stated as follows: If a chemical reaction at equilibrium is subjected to a change in conditions that displaces it from equilibrium, then the reaction adjusts toward a new equilibrium state. The reaction proceeds in the direction that — at least partially — offsets the change in conditions. Thus, an increase in pressure shifts the equilibrium in Equation 12.13 such that the total number of moles decreases.

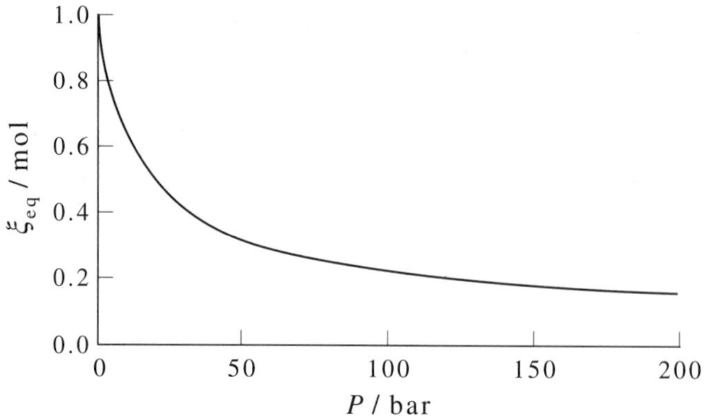


**FIGURE 12.1**
A plot of the fraction of $PCl_5$(g) that is dissociated at equilibrium, $\xi_{eq}$, against total pressure $P$ for the reaction given by Equation 12.13 at 200°C.

---

**EXAMPLE 12–2**
Consider the association of potassium atoms in the vapor phase to form dimers.

$$2\,K(g) \rightleftharpoons K_2(g)$$

Suppose we start with 2 moles of K(g) and no dimers. Derive an expression for $K_P(T)$ in terms of $\xi_{eq}$, the extent of reaction at equilibrium, and the pressure $P$.

SOLUTION: At equilibrium, there will be $2(1-\xi_{eq})$ moles of K(g) and $\xi_{eq}$ moles of $K_2$(g). The total number of moles will be $(2-\xi_{eq})$. The partial pressure of each species will be

$$P_K = \frac{2(1-\xi_{eq})P}{2-\xi_{eq}}$$

$$P_{K_2} = \frac{\xi_{eq}P}{2-\xi_{eq}}$$

and

$$K_P(T) = \frac{P_{K_2}}{P_K^2} = \frac{\xi_{eq}(2-\xi_{eq})}{4(1-\xi_{eq})^2 P}$$

If $P$ decreases, then $\xi_{eq}(2-\xi_{eq})/4(1-\xi_{eq})^2$ must decrease, which occurs by $\xi_{eq}$ decreasing. If $P$ increases, then $\xi_{eq}(2-\xi_{eq})/4(1-\xi_{eq})^2$ must increase, which occurs by $\xi_{eq}$ increasing $[(1-\xi_{eq})$ becoming smaller].

---

We subscripted the equilibrium constant defined by Equation 12.12 with a $P$ to emphasize that it is expressed in terms of equilibrium pressures. We can also express the equilibrium constant in terms of densities or concentrations by using the ideal-gas relation $P = cRT$ where $c$ is the concentration, $n/V$. Thus, we can rewrite $K_P$ as

$$K_P = \frac{c_Y^{\nu_Y} c_Z^{\nu_Z}}{c_A^{\nu_A} c_B^{\nu_B}} \left(\frac{RT}{P^\circ}\right)^{\nu_Y+\nu_Z-\nu_A-\nu_B} \tag{12.16}$$

Just as we relate the pressures in the expression for $K_P$ to some standard pressure $P^\circ$, we must relate the concentrations in Equation 12.16 to some standard concentration $c^\circ$, often taken to be 1 mol·L$^{-1}$. If we multiply and divide each concentration in Equation 12.16 by $c^\circ$, we can write

$$K_P = K_c \left(\frac{c^\circ RT}{P^\circ}\right)^{\nu_Y+\nu_Z-\nu_A-\nu_B} \tag{12.17}$$

where

$$K_c = \frac{(c_Y/c^\circ)^{\nu_Y}(c_Z/c^\circ)^{\nu_Z}}{(c_A/c^\circ)^{\nu_A}(c_B/c^\circ)^{\nu_B}} \tag{12.18}$$

Both $K_P$ and $K_c$ in Equation 12.17 are unitless, as is the factor $(c^\circ RT/P^\circ)^{\nu_Y+\nu_Z-\nu_A-\nu_B}$. The actual choices of $P^\circ$ and $c^\circ$ determine the units of $R$ to use in Equation 12.17. If $P^\circ$ is taken to be one bar and $c^\circ$ to be one mol·L$^{-1}$ (as is often the case), then the factor $c^\circ RT/P^\circ = RT/\text{L·bar·mol}^{-1}$ and $R$ must be expressed as 0.083145 L·bar·mol$^{-1}$.

Equation 12.17 provides a relation between $K_P$ and $K_c$ for ideal gases. Just as we don't display the $P^\circ$'s in Equation 12.9 because most often $P^\circ$ = one bar, we don't display the $P^\circ$s and $c^\circ$s in Equation 12.18 because most often $c^\circ$ = one mol·L$^{-1}$. You must always be aware, however, of which reference states are being used in $K_P$ and $K_c$ when converting the numerical value of one to the other.

**EXAMPLE 12–3**

The value of $K_P(T)$ (based upon a standard state of one bar) for the reaction described by

$$NH_3(g) \rightleftharpoons \tfrac{3}{2}H_2(g) + \tfrac{1}{2}N_2(g)$$

is $1.36 \times 10^{-3}$ at 298.15 K. Determine the corresponding value of $K_c(T)$ (based upon a standard state of one mol·L$^{-1}$).

SOLUTION: In this case, $\nu_A = 1$, $\nu_Y = 3/2$, and $\nu_Z = 1/2$, so Equation 12.17 gives

$$K_P(T) = K_c(T)\left(\frac{c^\circ RT}{P^\circ}\right)^1$$

The conversion factor at 298.15 K is

$$\frac{c^\circ RT}{P^\circ} = \frac{(1\ \text{mol}\cdot\text{L}^{-1})(0.083145\ \text{L}\cdot\text{bar}\cdot\text{mol}^{-1}\cdot\text{K}^{-1})(298.15\ \text{K})}{1\ \text{bar}}$$
$$= 24.79$$

and so $K_c = K_P/24.79 = 5.49 \times 10^{-5}$.

## 12–3. Standard Gibbs Energies of Formation Can Be Used to Calculate Equilibrium Constants

Notice that combining Equations 12.8 and 12.11 gives a relation between $\mu_j^\circ(T)$, the standard chemical potentials of the reactants and products, and the equilibrium constant, $K_P$. In particular, $K_P$ is related to the difference between the standard chemical potentials of the products and reactants. Because a chemical potential is an energy (it is the molar Gibbs energy of a pure substance), its value must be referred to some (arbitrary) zero of energy. A convenient choice of a zero of energy is based on the procedure that we used to set up a table of standard molar enthalpies of formation (Table 5.2) in Section 5–11. Recall that we defined the standard molar enthalpy of formation of a substance as the energy as heat involved when one mole of the substance is formed directly from its constituent elements in their most stable form at one bar and the temperature of interest. For example, the value of $\Delta_r H$ for

$$H_2(g) + \tfrac{1}{2}\, O_2(g) \rightleftharpoons H_2O(l)$$

is $-285.8$ kJ when all the species are at 298.15 K and one bar, and so we write $\Delta_f H^\circ[H_2O(l)] = -285.8\ \text{kJ}\cdot\text{mol}^{-1}$ at 298.15 K. By convention, we also have that $\Delta_f H^\circ[H_2(g)] = \Delta_f H^\circ[O_2(g)] = 0$ for $H_2(g)$ and $O_2(g)$ at 298.15 K and one bar. We also set up a table of practical absolute entropies of substances (Table 7.2) in Section 7–9, and so because

$$\Delta_r G^\circ = \Delta_r H^\circ - T\Delta_r S^\circ$$

we can also set up a table of values of $\Delta_f G^\circ$. Then for a reaction such as

$$\nu_A A + \nu_B B \longrightarrow \nu_Y Y + \nu_Z Z$$

we have

$$\Delta_r G^\circ = \nu_Y \Delta_f G^\circ[Y] + \nu_Z \Delta_f G^\circ[Z] - \nu_A \Delta_f G^\circ[A] - \nu_B \Delta_f G^\circ[B] \qquad (12.19)$$

Table 12.1 lists values of $\Delta_f G^\circ$ at 298.15 K and one bar for a variety of substances, and much more extensive tables are available (see Section 12.9).

**EXAMPLE 12–4**

Using the data in Table 12.1, calculate $\Delta_r G^\circ(T)$ and $K_P$ at 298.15 K for

$$NH_3(g) \rightleftharpoons \tfrac{3}{2}\, H_2(g) + \tfrac{1}{2}\, N_2(g)$$

SOLUTION: From Equation 12.19,

$$\begin{aligned}\Delta_r G^\circ &= \left(\tfrac{3}{2}\right)\Delta_f G^\circ[H_2(g)] + \left(\tfrac{1}{2}\right)\Delta_f G^\circ[N_2(g)] - (1)\Delta_f G^\circ[NH_3(g)]\\ &= \left(\tfrac{3}{2}\right)(0) + \left(\tfrac{1}{2}\right)(0) - (1)(-16.367\ \text{kJ}\cdot\text{mol}^{-1})\\ &= 16.367\ \text{kJ}\cdot\text{mol}^{-1}\end{aligned}$$

**TABLE 12.1**
Standard molar Gibbs energies of formation, $\Delta_f G^\circ$, for various substances at 298.15 K and one bar.

| Substance | Formula | $\Delta_f G^\circ/\text{kJ}\cdot\text{mol}^{-1}$ |
|---|---|---|
| acetylene | $C_2H_2(g)$ | 209.20 |
| ammonia | $NH_3(g)$ | −16.367 |
| benzene | $C_6H_6(l)$ | 124.35 |
| bromine | $Br_2(g)$ | 3.126 |
| butane | $C_4H_{10}(g)$ | −17.15 |
| carbon(diamond) | $C(s)$ | 2.900 |
| carbon(graphite) | $C(s)$ | 0 |
| carbon dioxide | $CO_2(g)$ | −394.389 |
| carbon monoxide | $CO(g)$ | −137.163 |
| ethane | $C_2H_6(g)$ | −32.82 |
| ethanol | $C_2H_5OH(l)$ | −174.78 |
| ethene | $C_2H_4(g)$ | 68.421 |
| glucose | $C_6H_{12}O_6(s)$ | −910.52 |
| hydrogen bromide | $HBr(g)$ | −53.513 |
| hydrogen chloride | $HCl(g)$ | −95.300 |
| hydrogen fluoride | $HF(g)$ | −274.646 |
| hydrogen iodide | $HI(g)$ | 1.560 |
| hydrogen peroxide | $H_2O_2(l)$ | −105.445 |
| iodine | $I_2(g)$ | 19.325 |
| methane | $CH_4(g)$ | −50.768 |
| methanol | $CH_3OH(l)$ | −166.27 |
| | $CH_3OH(g)$ | −161.96 |
| nitrogen oxide | $NO(g)$ | 86.600 |
| nitrogen dioxide | $NO_2(g)$ | 51.258 |
| dinitrogen tetraoxide | $N_2O_4(g)$ | 97.787 |
| | $N_2O_4(l)$ | 97.521 |
| propane | $C_3H_8(g)$ | −23.47 |
| sucrose | $C_{12}H_{22}O_{11}(s)$ | −1544.65 |
| sulfur dioxide | $SO_2(g)$ | −300.125 |
| sulfur trioxide | $SO_3(g)$ | −371.016 |
| tetrachloromethane | $CCl_4(l)$ | −65.21 |
| | $CCl_4(g)$ | −53.617 |
| water | $H_2O(l)$ | −237.141 |
| | $H_2O(g)$ | −228.582 |

and from Equation 12.11

$$\ln K_p(T) = -\frac{\Delta_r G^\circ}{RT} = -\frac{16.367 \times 10^3 \text{ J}\cdot\text{mol}^{-1}}{(8.3145 \text{ J}\cdot\text{K}^{-1}\cdot\text{mol}^{-1})(298.15 \text{ K})}$$
$$= -6.602$$

or $K_p = 1.36 \times 10^{-3}$ at 298.15 K.

## 12–4. A Plot of the Gibbs Energy of a Reaction Mixture Against the Extent of Reaction Is a Minimum at Equilibrium

In this section we shall treat a concrete example of the Gibbs energy of a reaction mixture as a function of the extent of reaction. Consider the thermal decomposition of $N_2O_4(g)$ to $NO_2(g)$ at 298.15 K, which we represent by the equation

$$N_2O_4(g) \rightleftharpoons 2\,NO_2(g)$$

Suppose we start with one mole of $N_2O_4(g)$ and no $NO_2(g)$. Then as the reaction proceeds, $n_{N_2O_4}$, the number of moles of $N_2O_4(g)$, will be given by $1 - \xi$ and $n_{NO_2}$ will be given by $2\xi$. Note that $n_{N_2O_4} = 1$ mol and $n_{NO_2} = 0$ when $\xi = 0$ and that $n_{N_2O_4} = 0$ and $n_{NO_2} = 2$ mol when $\xi = 1$ mol. The Gibbs energy of the reaction mixture is given by

$$\begin{aligned} G(\xi) &= (1-\xi)\overline{G}_{N_2O_4} + 2\xi\overline{G}_{NO_2} \\ &= (1-\xi)G^\circ_{N_2O_4} + 2\xi G^\circ_{NO_2} + (1-\xi)RT\ln P_{N_2O_4} + 2\xi RT\ln P_{NO_2} \end{aligned} \tag{12.20}$$

If the reaction is carried out at a constant total pressure of one bar, then

$$P_{N_2O_4} = x_{N_2O_4}P_{\text{total}} = x_{N_2O_4} \quad \text{and} \quad P_{NO_2} = x_{NO_2}$$

The total number of moles in the reaction mixture is $(1-\xi) + 2\xi = 1 + \xi$, and so we have

$$P_{N_2O_4} = x_{N_2O_4} = \frac{1-\xi}{1+\xi} \quad \text{and} \quad P_{NO_2} = x_{NO_2} = \frac{2\xi}{1+\xi}$$

Thus, Equation 12.20 becomes

$$G(\xi) = (1-\xi)G^\circ_{N_2O_4} + 2\xi G^\circ_{NO_2} + (1-\xi)RT\ln\frac{1-\xi}{1+\xi} + 2\xi RT\ln\frac{2\xi}{1+\xi}$$

According to Section 12–3, we can choose our standard states such that $G^\circ_{N_2O_4} = \Delta_f G^\circ_{N_2O_4}$ and $G^\circ_{NO_2} = \Delta_f G^\circ_{NO_2}$, so $G(\xi)$ becomes

$$G(\xi) = (1-\xi)\Delta_f G^\circ_{N_2O_4} + 2\xi\Delta_f G^\circ_{NO_2} + (1-\xi)RT\ln\frac{1-\xi}{1+\xi} + 2\xi RT\ln\frac{2\xi}{1+\xi} \tag{12.21}$$

Equation 12.21 gives the Gibbs energy of the reaction mixture, $G$, as a function of the extent of the reaction, $\xi$. Using the values of $\Delta_f G^\circ_{N_2O_4}$ and $\Delta_f G^\circ_{NO_2}$ given in Table 12.1, Equation 12.21 becomes

$$G(\xi) = (1-\xi)(97.787\ \text{kJ}\cdot\text{mol}^{-1}) + 2\xi(51.258\ \text{kJ}\cdot\text{mol}^{-1}) + (1-\xi)RT\ln\frac{1-\xi}{1+\xi} + 2\xi RT\ln\frac{2\xi}{1+\xi} \tag{12.22}$$

where $RT = 2.4790\ \text{kJ}\cdot\text{mol}^{-1}$. Figure 12.2 shows $G(\xi)$ plotted against $\xi$. The minimum in the plot, or the equilibrium state, occurs at $\xi_{eq} = 0.1892$ mol. Thus, the reaction will proceed from $\xi = 0$ to $\xi = \xi_{eq} = 0.1892$ mol, where equilibrium is established.

The equilibrium constant is given by

$$K_P = \frac{P^2_{NO_2}}{P_{N_2O_4}} = \frac{[2\xi_{eq}/(1+\xi_{eq})]^2}{(1-\xi_{eq})/(1+\xi_{eq})} = \frac{4\xi^2_{eq}}{1-\xi^2_{eq}} = 0.148$$

We can compare this result to the one that we obtain from $\Delta_r G^\circ = -RT\ln K_P$, or

$$\begin{aligned}\ln K_P &= -\frac{\Delta_r G^\circ}{RT} \\ &= \frac{(2)(\Delta_f G^\circ[NO_2(g)]) - (1)(\Delta_f G^\circ[N_2O_4(g)])}{(8.3145\ \text{J}\cdot\text{K}^{-1}\cdot\text{mol}^{-1})(298.15\ \text{K})} \\ &= -\frac{4.729\times 10^3\ \text{J}\cdot\text{mol}^{-1}}{(8.3145\ \text{J}\cdot\text{K}^{-1}\cdot\text{mol}^{-1})(298.15\ \text{K})} = -1.9076\end{aligned}$$

or $K_P = 0.148$.

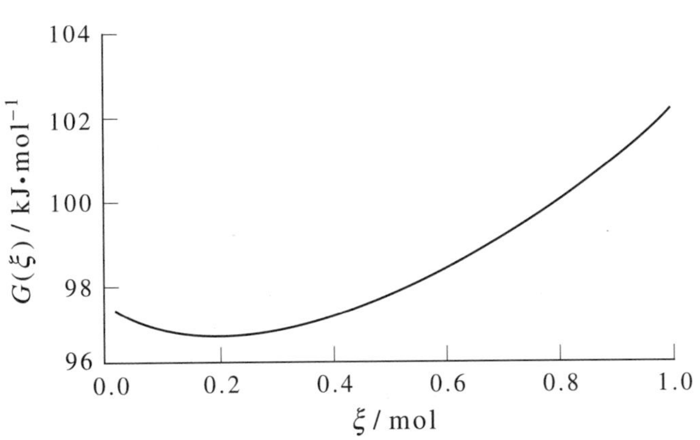


**FIGURE 12.2**
A plot of the Gibbs energy of the reaction mixture versus the extent of reaction for $N_2O_4(g) \rightleftharpoons 2\,NO_2(g)$ at 298.15 K and one bar.

We can also differentiate Equation 12.22 with respect to $\xi$ explicitly to obtain

$$\left(\frac{\partial G}{\partial \xi}\right)_{T,P} = (2)(51.258\ \text{kJ}\cdot\text{mol}^{-1}) - 97.787\ \text{kJ}\cdot\text{mol}^{-1} - RT\ln\frac{1-\xi}{1+\xi}$$
$$+2RT\ln\frac{2\xi}{1+\xi} + (1-\xi)RT\left(\frac{1+\xi}{1-\xi}\right)\left[-\frac{1}{1+\xi} - \frac{1-\xi}{(1+\xi)^2}\right]$$
$$+2\xi RT\left(\frac{1+\xi}{2\xi}\right)\left[\frac{2}{1+\xi} - \frac{2\xi}{(1+\xi)^2}\right] \qquad (12.23)$$

We can replace $(1-\xi)/(1+\xi)$ in the first logarithm term by $P_{N_2O_4}$ and $2\xi/(1+\xi)$ in the second logarithm term by $P_{NO_2}$. Furthermore, a little algebra shows that the last two terms add up to zero, and so Equation 12.23 becomes

$$\left(\frac{\partial G}{\partial \xi}\right)_{T,P} = \Delta_r G^\circ + RT\ln\frac{P_{NO_2}^2}{P_{N_2O_4}}$$

At equilibrium, $\partial G/\partial\xi = 0$ and we get Equation 12.11.

We can also evaluate $\xi_{eq}$ explicitly by setting Equation 12.23 equal to zero. Using the fact that the last two terms in Equation 12.23 add up to zero, we have

$$\frac{(2)(51.258\ \text{kJ}\cdot\text{mol}^{-1}) - 97.787\ \text{kJ}\cdot\text{mol}^{-1}}{(8.3145\ \text{J}\cdot\text{mol}^{-1}\cdot\text{K}^{-1})(298.15\ \text{K})} = \ln\left(\frac{1-\xi_{eq}}{1+\xi_{eq}}\right) - \ln\frac{4\xi_{eq}^2}{(1+\xi_{eq})^2}$$

or

$$1.9076 = \ln\left(\frac{1-\xi_{eq}^2}{4\xi_{eq}^2}\right)$$

or

$$\frac{1-\xi_{eq}^2}{4\xi_{eq}^2} = e^{1.9076} = 6.7371$$

or $\xi_{eq} = 0.1892$, in agreement with Figure 12.2. Problems 12–18 through 12–21 ask you to carry out a similar analysis for two other gas-phase reactions.

## 12–5. The Ratio of the Reaction Quotient to the Equilibrium Constant Determines the Direction in Which a Reaction Will Proceed

Consider the general reaction described by the equation

$$\nu_A A(g) + \nu_A B(g) \rightleftharpoons \nu_Y Y(g) + \nu_Z Z(g)$$

Equation 12.7 for this reaction scheme is

$$\Delta_r G(T) = \Delta_r G^\circ(T) + RT \ln \frac{P_Y^{\nu_Y} P_Z^{\nu_Z}}{P_A^{\nu_A} P_B^{\nu_B}} \tag{12.24}$$

Realize that the pressures in this equation are not necessarily equilibrium pressures, but are arbitrary. Equation 12.24 gives the value of $\Delta_r G$ when $\nu_A$ moles of A(g) at pressure $P_A$ react with $\nu_B$ moles of B(g) at pressure $P_B$ to produce $\nu_Y$ moles of Y(g) at pressure $P_Y$ and $\nu_Z$ moles of Z(g) at pressure $P_Z$. If all the pressures happen to be one bar, then the logarithm term in Equation 12.24 will be zero and $\Delta_r G$ will be equal to $\Delta_r G^\circ$; in other words, the Gibbs energy change will be equal to the standard Gibbs energy change. If, on the other hand, the pressures are the equilibrium pressures, then $\Delta_r G$ will equal zero and we obtain Equation 12.11.

We can write Equation 12.24 in more concise form by introducing a quantity called the *reaction quotient* $Q_P$ (see Equation 12.9)

$$Q_P = \frac{P_Y^{\nu_Y} P_Z^{\nu_Z}}{P_A^{\nu_A} P_B^{\nu_B}} \tag{12.25}$$

and using Equation 12.11 for $\Delta_r G^\circ$:

$$\begin{aligned} \Delta_r G &= -RT \ln K_P + RT \ln Q_P \\ &= RT \ln(Q_P/K_P) \end{aligned} \tag{12.26}$$

Realize that even though $Q_P$ has the *form* of an equilibrium constant, the pressures are arbitrary.

At equilibrium, $\Delta_r G = 0$ and $Q_P = K_P$. If $Q_P < K_P$, then $Q_P$ must increase as the system proceeds toward equilibrium, which means that the partial pressures of the products must increase and those of the reactants must decrease. In other words, the reaction proceeds from left to right as written. In terms of $\Delta_r G$, if $Q_P < K_P$, then $\Delta_r G < 0$, indicating that the reaction is spontaneous from left to right as written. Conversely, if $Q_P > K_P$, then $Q_P$ must decrease as the reaction proceeds to equilibrium and so the pressures of the products must decrease and those of the reactants must increase. In terms of $\Delta_r G$, if $Q_P > K_P$, then $\Delta_r G > 0$, indicating that the reaction is spontaneous from right to left as written.

**EXAMPLE 12–5**

The equilibrium constant for the reaction described by

$$2\,SO_2(g) + O_2(g) \rightleftharpoons 2\,SO_3(g)$$

is $K_P = 10$ at 960 K. Calculate $\Delta_r G$ and indicate in which direction the reaction will proceed spontaneously for

$$2\,SO_2(1.0 \times 10^{-3}\text{ bar}) + O_2(0.20\text{ bar}) \rightleftharpoons 2\,SO_3(1.0 \times 10^{-4}\text{ bar})$$

SOLUTION: We first calculate the reaction quotient under these conditions. According to Equation 12.25,

$$Q_P = \frac{P^2_{SO_3}}{P^2_{SO_2} P_{O_2}} = \frac{(1.0 \times 10^{-4})^2}{(1.0 \times 10^{-3})^2(0.20)} = 5.0 \times 10^{-2}$$

Note that these quantities are unitless because the pressures are taken relative to one bar. Using Equation 12.26, we have

$$\begin{aligned}\Delta_r G &= RT \ln \frac{Q_P}{K_P} \\ &= (8.314\ \text{J}\cdot\text{K}^{-1}\cdot\text{mol}^{-1})(900\ \text{K}) \ln \frac{5.0 \times 10^{-2}}{10} \\ &= -13\ \text{kJ}\cdot\text{mol}^{-1}\end{aligned}$$

The fact that $\Delta_r G < 0$ implies that the reaction will proceed from left to right as written. This may also be seen from the fact that $Q_P < K_P$.

## 12–6. The Sign of $\Delta_r G$ And Not That of $\Delta_r G^\circ$ Determines the Direction of Reaction Spontaneity

It is important to appreciate the difference between $\Delta_r G$ and $\Delta_r G^\circ$. The superscript $^\circ$ on $\Delta_r G^\circ$ emphasizes that this is the value of $\Delta_r G$ when all the reactants and products are unmixed at partial pressures equal to one bar; $\Delta_r G^\circ$ is the *standard* Gibbs energy change. If $\Delta_r G^\circ < 0$, then $K_P > 1$, meaning that the reaction will proceed from reactants to products if all the species are mixed at one bar partial pressures. If $\Delta_r G^\circ > 0$, then $K_P < 1$, meaning that the reaction will proceed from products to reactants if all the species are mixed at one bar partial pressures. The fact that $\Delta_r G^\circ > 0$ does *not* mean that the reaction will not proceed from reactants to products if the species are mixed under all conditions. For example, consider the reaction described by

$$N_2O_4(g) \rightleftharpoons 2\ NO_2(g)$$

for which $\Delta_r G^\circ = 4.729\ \text{kJ}\cdot\text{mol}^{-1}$ at 298.15 K. The corresponding value of $K_P(T)$ is 0.148. The fact that $\Delta_r G^\circ = +4.729\ \text{kJ}\cdot\text{mol}^{-1}$ does *not* mean that no $N_2O_4(g)$ will dissociate when we place some of it in a reaction vessel at 298.15 K. The value of $\Delta_r G$ for the dissociation of $N_2O_4(g)$ is given by

$$\begin{aligned}\Delta_r G &= \Delta_r G^\circ + RT \ln Q_P \\ &= 4.729\ \text{kJ}\cdot\text{mol}^{-1} + (2.479\ \text{kJ}\cdot\text{mol}^{-1}) \ln \frac{P^2_{NO_2}}{P_{N_2O_4}} \qquad (12.27)\end{aligned}$$

Let's say that we fill a container with $N_2O_4(g)$ and no $NO_2(g)$. Initially then, the logarithm term and $\Delta_r G$ in Equation 12.27 will be essentially negative infinity. Therefore, the dissociation of $N_2O_4(g)$ takes place spontaneously. The partial pressure of $N_2O_4(g)$ decreases and that of $NO_2(g)$ increases until equilibrium is reached. The equilibrium state is determined by the condition $\Delta_r G = 0$, at which point $Q_P = K_P$. Thus, initially $\Delta_r G$ has a large negative value and increases to zero as the reaction goes to equilibrium.

We should point out here that even though $\Delta_r G < 0$, the reaction may not occur at a detectable rate. For example, consider the reaction given by

$$2\,H_2(g) + O_2(g) \rightleftharpoons 2\,H_2O(l)$$

The value of $\Delta_r G^\circ$ at 25°C for this reaction is −237 kJ per mole of $H_2O(l)$ formed. Consequently, $H_2O(l)$ at one bar and 25°C is much more stable than a mixture of $H_2(g)$ and $O_2(g)$ under those conditions. Yet, a mixture of $H_2(g)$ and $O_2(g)$ can be kept indefinitely. If a spark or a catalyst is introduced into this mixture, however, then the reaction occurs explosively. This observation serves to illustrate an important point: The "no" of thermodynamics is emphatic. If thermodynamics says that a certain process will not occur spontaneously, then it will not occur. The "yes" of thermodynamics, on the other hand, is actually a "maybe". The fact that a process will occur spontaneously does not imply that it will necessarily occur at a detectable rate.

## 12–7. The Variation of an Equilibrium Constant with Temperature Is Given by the Van't Hoff Equation

We can use the Gibbs-Helmoltz equation (Equation 8.61)

$$\left(\frac{\partial \Delta G^\circ/T}{\partial T}\right)_P = -\frac{\Delta H^\circ}{T^2} \tag{12.28}$$

to derive an equation for the temperature dependence of $K_P(T)$. Substitute $\Delta G^\circ(T) = -RT \ln K_P(T)$ into Equation 12.28 to obtain

$$\left(\frac{\partial \ln K_P(T)}{\partial T}\right)_P = \frac{d \ln K_P(T)}{dT} = \frac{\Delta_r H^\circ}{RT^2} \tag{12.29}$$

Note that if $\Delta_r H^\circ > 0$ (endothermic reaction), then $K_P(T)$ increases with temperature, and if $\Delta_r H^\circ < 0$ (exothermic reaction), then $K_P(T)$ decreases with increasing temperature. This is another example of Le Châtelier's principle.

Equation 12.29 can be integrated to give

$$\ln \frac{K_P(T_2)}{K_P(T_1)} = \int_{T_1}^{T_2} \frac{\Delta_r H^\circ(T)dT}{RT^2} \tag{12.30}$$

If the temperature range is small enough that we can consider $\Delta_r H^\circ$ to be a constant, then we can write

$$\ln \frac{K_P(T_2)}{K_P(T_1)} = -\frac{\Delta_r H^\circ}{R}\left(\frac{1}{T_2} - \frac{1}{T_1}\right) \tag{12.31}$$

Equation 12.31 suggests that a plot of $\ln K_P(T)$ versus $1/T$ should be a straight line with a slope of $-\Delta_r H^\circ/R$ over a sufficiently small temperature range. Figure 12.3 shows such a plot for the reaction $H_2(g) + CO_2(g) \rightleftharpoons CO(g) + H_2O(g)$ over the temperature range 600°C to 900°C.

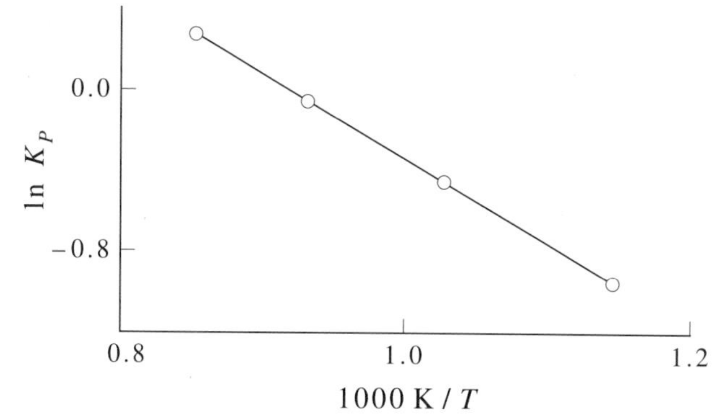


**FIGURE 12.3**
A plot of $\ln K_P(T)$ versus $1/T$ for the reaction $H_2(g) + CO_2(g) \rightleftharpoons CO(g) + H_2O(g)$ over the temperature range 600°C to 900°C. The circles represent experimental data

**EXAMPLE 12–6**

Given that $\Delta_r H^\circ$ has an average value of $-69.8\ \text{kJ}\cdot\text{mol}^{-1}$ over the temperature range 500 K to 700 K for the reaction described by

$$PCl_3(g) + Cl_2(g) \rightleftharpoons PCl_5(g)$$

estimate $K_P$ at 700 K given that $K_P = 0.0408$ at 500 K.

SOLUTION: We use Equation 12.31 with the above values

$$\ln \frac{K_P}{0.0408} = -\frac{-69.8 \times 10^3\ \text{J}\cdot\text{mol}^{-1}}{8.3145\ \text{J}\cdot\text{K}^{-1}\cdot\text{mol}^{-1}}\left(\frac{1}{700\ \text{K}} - \frac{1}{500\ \text{K}}\right)$$
$$= -4.80$$

or

$$K_P(T) = (0.0408)e^{-4.80} = 3.36 \times 10^{-4}$$

Note that the reaction is exothermic and so $K_P(T = 700\ \text{K})$ is less than $K_P(T = 500\ \text{K})$.

In Section 5–12 we discussed the temperature variation of $\Delta_r H^\circ$. In particular, we derived the equation

$$\Delta_r H^\circ(T_2) = \Delta_r H^\circ(T_1) + \int_{T_1}^{T_2} \Delta C_P^\circ(T)dT \tag{12.32}$$

where $\Delta C_P^\circ$ is the difference between the heat capacities of the products and reactants. Experimental heat capacity data over temperature ranges are often presented as polynomials in the temperature, and if this is the case, then $\Delta_r H^\circ(T)$ can be expressed in the form (see Example 5–13)

$$\Delta_r H^\circ(T) = \alpha + \beta T + \gamma T^2 + \delta T^3 + \cdots \tag{12.33}$$

If this form for $\Delta_r H^\circ(T)$ is substituted into Equation 12.29, and both sides integrated indefinitely, then we find that

$$\ln K_P(T) = -\frac{\alpha}{RT} + \frac{\beta}{R}\ln T + \frac{\gamma}{R}T + \frac{\delta T^2}{2R} + A \tag{12.34}$$

The constants $\alpha$ through $\delta$ are known from Equation 12.33 and $A$ is an integration constant that can be evaluated from a knowledge of $K_P(T)$ at some particular temperature. We could also have integrated Equation 12.29 from some temperature $T_1$ at which the value of $K_P(T)$ is known to an arbitrary temperature $T$ to obtain

$$\ln K_P(T) = \ln K_P(T_1) + \int_{T_1}^{T} \frac{\Delta_r H^\circ(T')dT'}{RT'^2} \tag{12.35}$$

Equations 12.34 and 12.35 are generalizations of Equation 12.31 to the case where the temperature dependence of $\Delta_r H^\circ$ is not ignored. Equation 12.34 shows that if $\ln K_P(T)$ is plotted against $1/T$, then the slope is not constant, but has a slight curvature. Figure 12.4 shows $\ln K_P(T)$ plotted versus $1/T$ for the ammonia synthesis reaction. Note that $\ln K_P(T)$ does not vary linearly with $1/T$, showing that $\Delta_r H^\circ$ is temperature dependent.

**EXAMPLE 12–7**

Consider the reaction described by

$$\tfrac{1}{2}N_2(g) + \tfrac{3}{2}H_2(g) \rightleftharpoons NH_3(g)$$

The molar heat capacities of $N_2(g)$, $H_2(g)$, and $NH_3(g)$ can be expressed in the form

$$C_P^\circ[N_2(g)]/J\cdot K^{-1}\cdot mol^{-1} = 24.98 + 5.912 \times 10^{-3}T - 0.3376 \times 10^{-6}T^2$$

$$C_P^\circ[H_2(g)]/J\cdot K^{-1}\cdot mol^{-1} = 29.07 - 0.8368 \times 10^{-3}T + 2.012 \times 10^{-6}T^2$$

$$C_P^\circ[NH_3(g)]/J\cdot K^{-1}\cdot mol^{-1} = 25.93 + 32.58 \times 10^{-3}T - 3.046 \times 10^{-6}T^2$$

over the temperature range 300 K to 1500 K. Given that $\Delta_f H^\circ[NH_3(g)] = -46.11\ kJ\cdot mol^{-1}$ at 300 K and that $K_P = 6.55 \times 10^{-3}$ at 725 K, derive a general expression for the variation of $K_P(T)$ with temperature in the form of Equation 12.34.

SOLUTION: We first use Equation 12.32

$$\Delta_r H^\circ(T_2) = \Delta_r H^\circ(T_1) + \int_{T_1}^{T_2} \Delta C_P^\circ(T)dT$$

with $T_1 = 300$ K and $\Delta_r H^\circ(T_1 = 300\ K) = -46.11\ kJ\cdot mol^{-1}$ and

$$\Delta C_P^\circ = C_P^\circ[NH_3(g)] - \frac{1}{2}C_P^\circ[N_2(g)] - \frac{3}{2}C_P^\circ[H_2(g)]$$

Integration gives

$$\begin{aligned}\Delta_r H^\circ(T)/J\cdot mol^{-1} &= -46.11 \times 10^3 + \int_{300\ K}^{T} \Delta C_P^\circ(T)dT \\ &= -46.11 \times 10^3 - 31.17(T - 300) \\ &\quad + \frac{30.88 \times 10^{-3}}{2}(T^2 - (300)^2) - \frac{5.895 \times 10^{-6}}{3}(T^3 - (300)^3)\end{aligned}$$

or

$$\Delta_r H^\circ(T)/J\cdot mol^{-1} = -38.10 \times 10^3 - 31.17T + 15.44 \times 10^{-3}T^2 - 1.965 \times 10^{-6}T^3$$

Now we use Equation 12.35 with $T_1 = 725$ K and $K_P(T = 725\ K) = 6.55 \times 10^{-3}$.

$$\begin{aligned}\ln K_P(T) &= \ln K_P(T = 725\ K) + \int_{725}^{T} \frac{\Delta_r H^\circ(T')}{RT'^2}dT' \\ &= -5.028 + \frac{1}{R}\left[+38.10\left(\frac{1}{T} - \frac{1}{725}\right) - 31.17(\ln T - \ln 725)\right. \\ &\quad \left. +15.44 \times 10^{-3}(T - 725) - \frac{1.965 \times 10^{-6}}{2}(T^2 - (725)^2)\right] \\ &= 12.06 + \frac{4583}{T} - 3.749 \ln T + 1.857 \times 10^{-3}T - 0.118 \times 10^{-6}T^2\end{aligned}$$

This equation was used to generate Figure 12.4. At 600 K, $\ln K_P = -3.21$, or $K_P = 0.040$, in excellent agreement with the experimental value of 0.041.

It is interesting to compare the results of this section to those of Section 9–4, where we derived the Clausius-Clapeyron equation, Equation 9.13. Note that Equations 12.31 and 9.13 are essentially the same because the vaporization of a liquid can be represented by the "chemical equation"

$$X(l) \rightleftharpoons X(g)$$

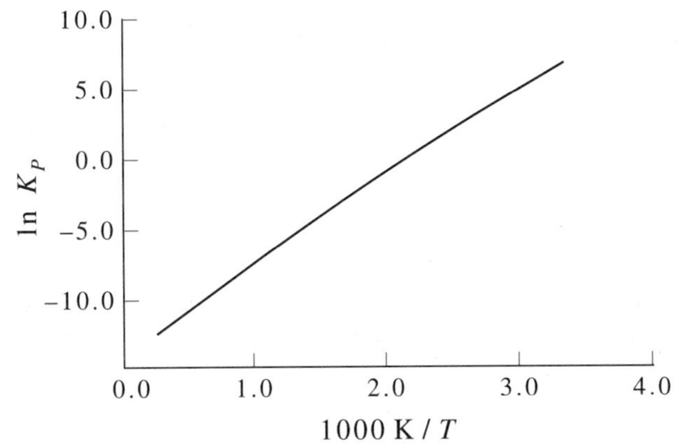


**FIGURE 12.4**
A plot of $\ln K_P(T)$ versus $1/T$ for the ammonia synthesis reaction, $\frac{3}{2}H_2(g) + \frac{1}{2}N_2(g) \rightleftharpoons NH_3(g)$.

## 12–8. We Can Calculate Equilibrium Constants in Terms of Partition Functions

An important chemical application of statistical thermodynamics is the calculation of equilibrium constants in terms of molecular parameters. Consider the general homogeneous gas-phase chemical reaction

$$\nu_A A(g) + \nu_B B(g) \rightleftharpoons \nu_Y Y(g) + \nu_Z Z(g)$$

in a reaction vessel at fixed volume and temperature. In this case we have (cf. Equation 9.26)

$$dA = \mu_A dn_A + \mu_B dn_B + \mu_Y dn_Y + \mu_Z dn_Z \qquad \text{(constant } T \text{ and } V)$$

instead of Equation 12.3. Introducing the extent of reaction through Equations 12.2, however, leads to the same condition for chemical equilbrium as in Section 12–1,

$$\nu_Y \mu_Y + \nu_Z \mu_Z - \nu_A \mu_A - \nu_B \mu_B = 0 \tag{12.36}$$

We now introduce statistical thermodynamics through the relation between the chemical potential and a partition function. In a mixture of ideal gases, the species are independent, and so the partition function of the mixture is a product of the partition functions of the individual components. Thus

$$Q(N_A, N_B, N_Y, N_Z, V, T) = Q(N_A, V, T)Q(N_B, V, T)Q(N_Y, V, T)Q(N_Z, V, T)$$

$$= \frac{q_A(V,T)^{N_A}}{N_A!}\frac{q_B(V,T)^{N_B}}{N_B!}\frac{q_Y(V,T)^{N_Y}}{N_Y!}\frac{q_Z(V,T)^{N_Z}}{N_Z!}$$

The chemical potential of each species is given by an equation such as (Problem 12–33)

$$\mu_A = -RT \left( \frac{\partial \ln Q}{\partial N_A} \right)_{N_j, V, T} = -RT \ln \frac{q_A(V,T)}{N_A} \tag{12.37}$$

where Stirling's approximation has been used for $N_A!$. The $N_j$ subscript on the partial derivative indicates that the numbers of particles of the other species are held fixed. Equation 12.37 simply says that the chemical potential of one species of an ideal gas mixture is calculated as if the other species were not present. This, of course, is the case for an ideal gas mixture.

If we substitute Equation 12.37 into Equation 12.36, then we get

$$\frac{N_Y^{\nu_Y} N_Z^{\nu_Z}}{N_A^{\nu_A} N_B^{\nu_B}} = \frac{q_Y^{\nu_Y} q_Z^{\nu_Z}}{q_A^{\nu_A} q_B^{\nu_B}} \tag{12.38}$$

For an ideal gas, the molecular partition function is of the form $f(T)V$ (Section 4–6) so that $q/V$ is a function of temperature only. If we divide each factor on both sides of Equation 12.38 by $V^{\nu_j}$ and denote the number density $N_j/V$ by $\rho_j$, then we have

$$K_c(T) = \frac{\rho_Y^{\nu_Y} \rho_Z^{\nu_Z}}{\rho_A^{\nu_A} \rho_B^{\nu_B}} = \frac{(q_Y/V)^{\nu_Y} (q_Z/V)^{\nu_Z}}{(q_A/V)^{\nu_A} (q_B/V)^{\nu_B}} \tag{12.39}$$

Note that $K_c$ is a function of temperature only. Recall that $K_P(T)$ and $K_c(T)$ are related by (Equation 12.17)

$$K_P(T) = \frac{P_Y^{\nu_Y} P_Z^{\nu_Z}}{P_A^{\nu_A} P_B^{\nu_B}} = K_c(T) \left( \frac{c^\circ RT}{P^\circ} \right)^{\nu_Y + \nu_Z - \nu_A - \nu_B}$$

By means of Equation 12.17 and Equation 12.39, along with the results of Chapter 4, we can calculate equilibrium constants in terms of molecular parameters. This is best illustrated by means of examples.

## A. A Chemical Reaction Involving Diatomic Molecules

We shall calculate the equilibrium constant for the reaction

$$H_2(g) + I_2(g) \rightleftharpoons 2\,HI(g)$$

from 500 K to 1000 K. The equilibrium constant is given by

$$K(T) = \frac{(q_{HI}/V)^2}{(q_{H_2}/V)(q_{I_2}/V)} = \frac{q_{HI}^2}{q_{H_2} q_{I_2}} \tag{12.40}$$

Using Equation 4.39 for the molecular partition functions gives

$$K(T) = \left(\frac{m_{\mathrm{HI}}^2}{m_{\mathrm{H}_2} m_{\mathrm{I}_2}}\right)^{3/2} \left(\frac{4\Theta_{\mathrm{rot}}^{\mathrm{H}_2}\Theta_{\mathrm{rot}}^{\mathrm{I}_2}}{(\Theta_{\mathrm{rot}}^{\mathrm{HI}})^2}\right) \frac{(1-e^{-\Theta_{\mathrm{vib}}^{\mathrm{H}_2}/T})(1-e^{-\Theta_{\mathrm{vib}}^{\mathrm{I}_2}/T})}{(1-e^{-\Theta_{\mathrm{vib}}^{\mathrm{HI}}/T})^2} \times \exp\frac{2D_0^{\mathrm{HI}} - D_0^{\mathrm{H}_2} - D_0^{\mathrm{I}_2}}{RT} \qquad (12.41)$$

where we have replaced $D_e$ in Equation 4.39 by $D_0 + h\nu/2$ (Figure 4.2). All the necessary parameters are given in Table 4.2. Table 12.2 gives the numerical values of $K_P(T)$ and Figure 12.5 shows $\ln K$ plotted versus $1/T$. From the slope of the line in Figure 12.5 we get $\Delta_r\overline{H} = -12.9\ \mathrm{kJ\cdot mol^{-1}}$ compared to the experimental value of $-13.4\ \mathrm{kJ\cdot mol^{-1}}$. The discrepancy is due to the inadequacy of the rigid rotator-harmonic oscillator approximation at these temperatures.

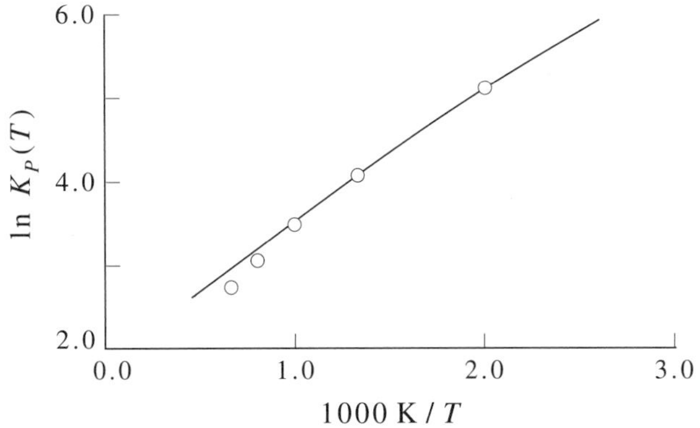


**FIGURE 12.5**
The logarithm of the equilibrium constant versus $1/T$ for the reaction $H_2(g) + I_2(g) \rightleftharpoons 2\,HI(g)$. The line is calculated from Equation 12.41 and the circles are the experimental values.

**TABLE 12.2**
The values of $K_P(T)$ for the reaction described by $H_2(g) + I_2(g) \rightleftharpoons 2\,HI(g)$ calculated according to Equation 12.41.

| $T$/K | $K_P(T)$ | $\ln K_P(T)$ |
|---|---|---|
| 500 | 138 | 4.92 |
| 750 | 51.1 | 3.93 |
| 1000 | 28.5 | 3.35 |
| 1250 | 19.1 | 2.95 |
| 1500 | 14.2 | 2.65 |

## B. A Reaction Involving Polyatomic Molecules

As an example of a reaction involving a polyatomic molecule, consider the reaction

$$H_2(g) + \tfrac{1}{2}\, O_2(g) \rightleftharpoons H_2O(g)$$

whose equilibrium constant is given by

$$K_c(T) = \frac{(q_{H_2O}/V)}{(q_{H_2}/V)(q_{O_2}/V)^{1/2}} \tag{12.42}$$

It is almost as convenient to calculate each partition function separately as to substitute them into $K_c$ first. The necessary parameters are given in Tables 4.2 and 4.4. At 1500 K, the three partition functions are (Equations 4.39 and 4.60)

$$\frac{q_{H_2}(T,V)}{V} = \left(\frac{2\pi m_{H_2} k_B T}{h^2}\right)^{3/2} \left(\frac{T}{2\Theta_{\rm rot}^{H_2}}\right) (1 - e^{-\Theta_{\rm vib}^{H_2}/T})^{-1} e^{D_0^{H_2}/RT}$$
$$= 2.80 \times 10^{32} e^{D_0^{H_2}/RT} \ {\rm m}^{-3} \tag{12.43}$$

$$\frac{q_{O_2}(T,V)}{V} = \left(\frac{2\pi m_{O_2} k_B T}{h^2}\right)^{3/2} \left(\frac{T}{2\Theta_{\rm rot}^{O_2}}\right) (1 - e^{-\Theta_{\rm vib}^{O_2}/T})^{-1} 3 e^{D_0^{O_2}/RT}$$
$$= 2.79 \times 10^{36} e^{D_0^{O_2}/RT} \ {\rm m}^{-3} \tag{12.44}$$

and

$$\frac{q_{H_2O}(T,V)}{V} = \left(\frac{2\pi m_{H_2O} k_B T}{h^2}\right)^{3/2} \frac{\pi^{1/2}}{\sigma} \left(\frac{T^3}{\Theta_{\rm rot,A}^{H_2O}\Theta_{\rm rot,B}^{H_2O}\Theta_{\rm rot,C}^{H_2O}}\right)^{1/2} \prod_{j=1}^{3} (1 - e^{-\Theta_{{\rm vib},j}^{H_2O}/T})^{-1} e^{D_0^{H_2O}/RT}$$
$$= 5.33 \times 10^{35} e^{D_0^{H_2O}/RT} \ {\rm m}^{-3} \tag{12.45}$$

The factor of 3 occurs in $q_{O_2}/V$ because the degeneracy of the ground electronic state of $O_2$ is 3. (Table 1.4).

Notice that each of the above $q(T,V)/V$ has units of $\mathrm{m}^{-3}$. This tells us that the reference state in this (molecular) case is a concentrations of one molecule per cubic meter, or that $c^\circ$ = one molecule·$\mathrm{m}^{-3}$. Using the values of $D_0$ from Table 4.2 and 4.4, the value of $K_c$ at 1500 K is $K_c = 2.34 \times 10^{-7}$. To convert to $K_P$, we divide $K_c$ by

$$\left(\frac{c^\circ RT}{N_A P^\circ}\right)^{1/2} = \left[\frac{(1\ \mathrm{m}^{-3})(8.3145\ \mathrm{J\cdot mol^{-1}\cdot K^{-1}})(1500\ \mathrm{K})}{(6.022 \times 10^{23}\ \mathrm{mol}^{-1})(10^5\ \mathrm{Pa})}\right]^{1/2}$$
$$= 4.55 \times 10^{-13}$$

to obtain $K_P = 5.14 \times 10^5$, based upon a one bar standard state.

**TABLE 12.3**
The logarithm of the equilibrium constant for the reaction $H_2(g) + \frac{1}{2} O_2(g) \rightleftharpoons H_2O\,(g)$

| $T$/K | $\ln K_P$(calc) | $\ln K_P$(exp) |
|---|---|---|
| 1000 | 23.5 | 23.3 |
| 1500 | 13.1 | 13.2 |
| 2000 | 8.52 | 8.15 |

Table 12.3 compares the calculated values of $\ln K_P$ with experimental data. Although the agreement is fairly good, the agreement can be considerably improved by using more sophisticated spectroscopic models. At high temperatures, the rotational energies of the molecules are high enough to warrant centrifugal distortion effects and other extensions of the simple rigid rotator-harmonic oscillator approximation.

## 12–9. Molecular Partition Functions and Related Thermodynamic Data Are Extensively Tabulated

In the previous section we have seen that the rigid rotator-harmonic oscillator approximation can be used to calculate equilibrium constants in reasonably good agreement with experiment, and because of the simplicity of the model, the calculations involved are not extensive. If greater accuracy is desired, however, one must include corrections to the rigid rotator-harmonic oscillator model, and the calculations become increasingly more laborious. It is natural, then, that a number of numerical tables of partition functions has evolved, and in this section we shall discuss the use of such tables. These tables are actually much more extensive than a compilation of partition functions. They include many experimentally determined values of thermodynamic properties, often complemented by theoretical calculations. The thermodynamic tables that we are about to discuss in this section, then, represent a collection of the thermodynamic and/or statistical thermodynamic properties of many substances.

One of the most extensive tabulations of the thermochemical properties of substances is an American Chemical Society publication, *Journal of Physical Chemical Reference Data*, volume 14, supplement 1, 1985, usually referred to as the JANAF (*j*oint, *a*rmy, *n*avy, *a*ir *f*orce) tables. Each species listed has about a full page of thermodynamic/spectroscopic data, and Table 12.4 is a replica of the entry for ammonia. Note that the fourth and fifth columns of thermodynamic data are headed by $-\{G^\circ - H^\circ(T_r)\}/T$ and $H^\circ - H^\circ(T_r)$. Recall that the value of an energy must be referred to some fixed reference point (such as a zero of energy). The reference point used in the JANAF tables is the standard molar enthalpy at 298.15 K. Consequently, $G^\circ(T)$ and $H^\circ(T)$ are expressed relative to that value, as expressed

**TABLE 12.4**
A replica of the page of $NH_3(g)$ data in the JANAF tables

**JANAF THERMOCHEMICAL TABLES** **1293**

AMMONIA ($NH_3$) IDEAL GAS $M_r$ = 17.03052

$\Delta_f H°$(0 K) = -38.907 ± 0.4 kJ $mol^{-1}$

S°(298.15 K) = 192.774 ± 0.025 J $K^{-1}$ $mol^{-1}$ $\Delta_f H°$(298.15 K) = -45.898 ± 0.4 kJ $mol^{-1}$

Vibrational Frequencies and Degeneracies

| ν, $cm^{-1}$ | ν, $cm^{-1}$ | ν, $cm^{-1}$ | ν, $cm^{-1}$ |
|---|---|---|---|
| 3506(1) | 1022(1) | 3577(2) | 1691(2) |

Ground State Quantum Weight: 1 σ(internal) = 2
Point Group: $C_{3v}$ σ(external) = 3
Bond Length: N-H = 1.0124 Å
Bond Angle: 106.67°
Product of the Moments of Inertia: $I_A I_B I_C$ = 0.0348 x $10^{-117}$ $g^3$ $cm^6$

Enthalpy of Formation

2nd and 3rd law analyses of equilibrium data for the reaction $1/2N_2(g)+3/2H_2(g) = NH_3(g)$ cited in the previous JANAF evaluation (1) plus more recent work of Schulz and Schaefer (6) were made using the revised thermal functions for $NH_3(g)$. All of the previously cited work in reaction calorimetry plus the early work of Berthelot (7, 8) and Thomsen (9) were reevaluated. No significant differences in the 3rd law calculations of the equilibrium data or in the corrections to the flow calorimetry data of Haber and Tamaru (12) and Wittig and Schmatz (13) were found. Thus, the 0.1 kcal discrepancy between the results of the equilibrium and reaction calorimetry measurements remains unresolved. The previous JANAF selection (1) for $\Delta_f H°$(298.15 K) of $NH_3(g)$ was adopted. A recent evaluation (14) which includes new indirect calorimetry (unpublished) further confirms this selection.

| Source | Method | $\Delta_f H°$(298.15 K) kcal $mol^{-1}$ | $\Delta_f H°$(298.15 K) kcal $mol^{-1}$* | ΔS°(obs.-calc. 298.15 K)* cal $K^{-1}$ $mol^{-1}$ |
|---|---|---|---|---|
| Larson, Dodge (2) | $K_p°(4)$ from $K_p°$(10-1,000 atm, 600-800 K) | -10.88 | -10.70±0.11 | +0.24±0.15[a] |
| Haber et al. (3) | $K_p°(4)$ from $K_p°$(30 atm, 800-1200 K) | -10.86 | -10.88±0.15 | -0.02±0.15[b] |
| Haber, Maschke (5) | $K_p°$(1 atm, 900-1400 K) | -10.85 | -10.62±0.22 | 0.20±0.19[c] |
| Schulz, Schaefer (6) | $K_p°$(1 atm, 567-673 K) | -10.87 | -10.78±0.20 | 0.14±0.3[d] |
| Berthelot (7) | Indirect; Reaction of $Br_2(aq)$ and $NH_3(aq)$ | -11.4 | | |
| Berthelot (8) | Indirect; Reaction of $O_2(g)$ with $NH_3(g)$ | -12.1 | | |
| Thomsen (9) | Indirect; Reaction of $O_2(g)$ with $NH_3(g)$ | -11.9 | | |
| Becker, Roth (10) | Indirect; Heat of combustion oxalates | -11.00±0.15 | | |
| Haber et al. (11) | Flow calorimetry at 298 K | -11.10±0.05 | | |
| Haber, Tamaru (12) | Flow calorimetry (739-932 K) | -10.97±0.008 | | |
| Wittig, Schmatz (13) | Flow calorimetry at 832 K | -10.99±0.05 | | |

*2nd law analysis assuming $\Delta C_p°$(cal $K^{-1}$ $mol^{-1}$) equals (a) -2.672+0.00591(T-700), (b) -1.236+0.00404(T-1000), (c) -0.855+0.00305(T-1100), (d) 3.287+0.00651(T-600).

Heat Capacity and Entropy

The thermodynamic functions differ from those of the 1965 JANAF table (1) in being taken directly form the later and more complete work of Haar (15). Haar treated in detail the contribution of the highly anharmonic out-of-plane vibrational mode, including its large coupling with rotation and its coupling with the other vibrational modes. Haar's values of $C_p°$ pass through a shallow maximum between 4000 and 5000 K; they were extrapolated from 5000 to 6000 K by assuming a constant value (19.300 cal $K^{-1}$ $mol^{-1}$). A summary of Haar's estimated uncertainties and of the differences of the 1965 table from the present table (in cal $K^{-1}$ $mol^{-1}$) is as follows:

| | Uncertainties (Haar, 15) | | 1965 Table minus This Table | |
|---|---|---|---|---|
| T, K | $C_p°$ | S° | $C_p°$ | S° |
| 1000 | 0.006 | 0.006 | -0.034 | -0.033 |
| 3000 | 0.10 | 0.06 | +0.142 | -0.122 |
| 5000 | 0.6 | 0.04 | +1.775 | +0.265 |

The National Bureau of Standards prepared this table (16) by critical analysis of data existing in 1972. Using the results of Haar (15) and $\Delta_f H°$ selected by NBS (16), we recalculate the table in terms of R=1.987192 cal $K^{-1}$ $mol^{-1}$ (17) and current JANAF reference states for the elements.

References


1. JANAF Thermochemical Tables, 2nd ed., NSRDS-NBS 37, 1971.
2. A. T. Larson and R. L. Dodge, J. Amer. Chem. Soc. 45, 2918 (1923).
3. F. Haber, S. Tamaru, and Ch. Ponnaz, Z. Elektrochem. 21, 89 (1914).
4. C. C. Stephenson and H. O. McMahan, J. Amer. Chem. Soc. 61, 437 (1939).
5. F. Haber and A. Maschke, Z. Electrochem. 21, 128 (1915).
6. G. Schulz and H. Schaefer, Ber. Bunsenges. Physik. Chem. 70, 21 (1966).
7. M. Berthelot, Compt. Rend. 89, 877 (1879); Ann. Chim. Phys. [5] 20, 247 (1880).
8. M. Berthelot, Ann. Chim. Phys. [5] 20, 244 (1880).
9. J. Thomsen, "Thermochemical Investigations," Vol. II, p. 68, Johann A. Barth, Leipzig, 1883.
10. G. Becker and W. A. Rogh, Z. Elektrochem. 40, 836 (1934).
11. F. Haber, S. Tamaru, and L. W. Oeholm. Z. Elektrochem. 21, 206 (1915).
12. F. Haber and S. Tamaru, Z. Elektrochem. 21, 191 (1915).
13. E. Wittig and W. Schmatz, Z. Elektrochem. 63, 475 (1959).
14. W. H. Evans, Personal Communication, U. S. Nat. Bur. Stand, July 1972; evaluation of ICSU-CODATA Task Group on Key Values for Thermodynamics, June (1972).
15. L. Haar, J. Res. Nat. Bur. Stand. 72A, 207 (1968).
16. S. Abramowitz et al., U. S. Nat. Bur. Stand., Rept. 10904, 239, July, 1972.
17. CODATA Task Group on Fundamental Constants, CODATA Bulletin 11, December, 1973.


Ammonia ($NH_3$) $H_3N_1(g)$

Enthalpy Reference Temperature = $T_r$ = 298.15 K Standard State Pressure = p° = 0.1 MPa

| T/K | $C_p°$ (J $K^{-1}mol^{-1}$) | S° (J $K^{-1}mol^{-1}$) | $-[G°-H°(T_r)]/T$ (J $K^{-1}mol^{-1}$) | $H°-H°(T_r)$ (kJ $mol^{-1}$) | $\Delta_f H°$ (kJ $mol^{-1}$) | $\Delta_f G°$ (kJ $mol^{-1}$) | Log $K_f$ |
|---|---|---|---|---|---|---|---|
| 0 | 0. | 0. | INFINITE | -10.045 | -38.907 | -38.907 | INFINITE |
| 100 | 33.284 | 155.840 | 223.211 | -6.737 | -41.550 | -34.034 | 17.777 |
| 200 | 33.757 | 178.990 | 195.962 | -3.394 | -43.703 | -25.679 | 6.707 |
| 298.15 | 35.652 | 192.774 | 192.774 | 0. | -45.898 | -16.367 | 2.867 |
| 300 | 35.701 | 192.995 | 192.775 | 0.066 | -45.939 | -16.183 | 2.818 |
| 400 | 38.716 | 203.663 | 194.209 | 3.781 | -48.041 | -5.941 | 0.776 |
| 500 | 42.048 | 212.659 | 197.021 | 7.819 | -49.857 | 4.800 | -0.501 |
| 600 | 45.293 | 220.615 | 200.302 | 12.188 | -51.374 | 15.879 | -1.382 |
| 700 | 48.354 | 227.829 | 203.727 | 16.872 | -52.618 | 27.190 | -2.029 |
| 800 | 51.235 | 234.476 | 207.160 | 21.853 | -53.621 | 38.662 | -2.524 |
| 900 | 53.948 | 240.669 | 210.543 | 27.113 | -54.411 | 50.247 | -2.916 |
| 1000 | 56.491 | 246.486 | 213.849 | 32.637 | -55.013 | 61.910 | -3.234 |
| 1100 | 58.859 | 251.983 | 217.069 | 38.406 | -55.451 | 73.625 | -3.496 |
| 1200 | 61.048 | 257.199 | 220.197 | 44.402 | -55.746 | 85.373 | -3.716 |
| 1300 | 63.057 | 262.166 | 223.236 | 50.609 | -55.917 | 97.141 | -3.903 |
| 1400 | 64.893 | 266.907 | 226.187 | 57.008 | -55.982 | 108.918 | -4.064 |
| 1500 | 66.564 | 271.442 | 229.054 | 63.582 | -55.954 | 120.696 | -4.203 |
| 1600 | 68.079 | 275.788 | 231.840 | 70.315 | -55.847 | 132.469 | -4.325 |
| 1700 | 69.452 | 279.957 | 234.549 | 77.193 | -55.672 | 144.234 | -4.432 |
| 1800 | 70.695 | 283.962 | 237.184 | 84.201 | -55.439 | 155.986 | -4.527 |
| 1900 | 71.818 | 287.815 | 239.748 | 91.328 | -55.157 | 167.725 | -4.611 |
| 2000 | 72.833 | 291.525 | 242.244 | 98.561 | -54.833 | 179.447 | -4.687 |
| 2100 | 73.751 | 295.101 | 244.677 | 105.891 | -54.473 | 191.152 | -4.755 |
| 2200 | 74.581 | 298.552 | 247.048 | 113.309 | -54.084 | 202.840 | -4.816 |
| 2300 | 75.330 | 301.884 | 249.360 | 120.805 | -53.671 | 214.509 | -4.872 |
| 2400 | 76.009 | 305.104 | 251.616 | 128.372 | -53.238 | 226.160 | -4.922 |
| 2500 | 76.626 | 308.220 | 253.818 | 136.005 | -52.789 | 237.792 | -4.968 |
| 2600 | 77.174 | 311.236 | 255.969 | 143.695 | -52.329 | 249.406 | -5.011 |
| 2700 | 77.672 | 314.158 | 258.070 | 151.438 | -51.860 | 261.003 | -5.049 |
| 2800 | 78.132 | 316.991 | 260.124 | 159.228 | -51.386 | 272.581 | -5.085 |
| 2900 | 78.529 | 319.740 | 262.132 | 167.062 | -50.909 | 284.143 | -5.118 |
| 3000 | 78.902 | 322.409 | 264.097 | 174.933 | -50.433 | 295.689 | -5.148 |
| 3100 | 79.228 | 325.001 | 266.020 | 182.840 | -49.959 | 307.218 | -5.177 |
| 3200 | 79.521 | 327.521 | 267.903 | 190.778 | -49.491 | 318.733 | -5.203 |
| 3300 | 79.785 | 329.972 | 269.747 | 198.744 | -49.030 | 330.233 | -5.227 |
| 3400 | 80.011 | 332.358 | 271.554 | 206.734 | -48.578 | 341.719 | -5.250 |
| 3500 | 80.216 | 334.680 | 273.324 | 214.745 | -48.139 | 353.191 | -5.271 |
| 3600 | 80.400 | 336.942 | 275.060 | 222.776 | -47.713 | 364.652 | -5.291 |
| 3700 | 80.550 | 339.147 | 276.763 | 230.824 | -47.302 | 376.101 | -5.310 |
| 3800 | 80.684 | 341.297 | 278.433 | 238.886 | -46.908 | 387.539 | -5.327 |
| 3900 | 80.793 | 343.395 | 280.072 | 246.960 | -46.534 | 398.967 | -5.344 |
| 4000 | 80.881 | 345.441 | 281.680 | 255.043 | -46.180 | 410.385 | -5.359 |
| 4100 | 80.956 | 347.439 | 283.260 | 263.136 | -45.847 | 421.795 | -5.374 |
| 4200 | 81.006 | 349.391 | 284.811 | 271.234 | -45.539 | 433.198 | -5.388 |
| 4300 | 81.048 | 351.297 | 286.335 | 279.337 | -45.254 | 444.593 | -5.401 |
| 4400 | 81.065 | 353.161 | 287.833 | 287.442 | -44.996 | 455.981 | -5.413 |
| 4500 | 81.073 | 354.983 | 289.305 | 295.550 | -44.764 | 467.364 | -5.425 |
| 4600 | 81.057 | 356.765 | 290.752 | 303.656 | -44.561 | 478.743 | -5.436 |
| 4700 | 81.032 | 358.508 | 292.175 | 311.761 | -44.387 | 490.117 | -5.447 |
| 4800 | 80.990 | 360.213 | 293.575 | 319.862 | -44.242 | 501.488 | -5.457 |
| 4900 | 80.931 | 361.882 | 294.952 | 327.958 | -44.129 | 512.856 | -5.467 |
| 5000 | 80.856 | 363.517 | 296.307 | 336.048 | -44.047 | 524.223 | -5.477 |
| 5100 | 80.751 | 365.117 | 297.641 | 344.127 | -43.999 | 535.587 | -5.486 |
| 5200 | 80.751 | 366.685 | 298.954 | 352.202 | -43.979 | 546.951 | -5.494 |
| 5300 | 80.751 | 368.223 | 300.246 | 360.277 | -43.982 | 558.315 | -5.503 |
| 5400 | 80.751 | 369.732 | 301.519 | 368.352 | -44.006 | 569.680 | -5.511 |
| 5500 | 80.751 | 371.214 | 302.773 | 376.428 | -44.049 | 581.044 | -5.518 |
| 5600 | 80.751 | 372.669 | 304.008 | 384.503 | -44.112 | 592.410 | -5.526 |
| 5700 | 80.751 | 374.098 | 305.225 | 392.578 | -44.193 | 603.778 | -5.533 |
| 5800 | 80.751 | 375.503 | 306.425 | 400.653 | -44.291 | 615.147 | -5.540 |
| 5900 | 80.751 | 376.883 | 307.607 | 408.728 | -44.404 | 626.516 | -5.547 |
| 6000 | 80.751 | 378.240 | 308.773 | 416.803 | -44.531 | 637.889 | -5.553 |

PREVIOUS: June 1977 (1 atm) CURRENT: June 1977 (1 bar)

Ammonia ($NH_3$) $H_3N_1(g)$

by the headings $-\{G^\circ(T) - H^\circ(298.15\text{ K})\}/T$ and $H^\circ(T) - H^\circ(298.15\text{ K})$. Table 12.4 gives $-\{G^\circ(T) - H^\circ(298.15\text{ K})\}/T$ for ammonia at a number of temperatures. The ratio $\{G^\circ(T) - H^\circ(298.15\text{ K})\}/T$ rather than $\{G^\circ(T) - H^\circ(298.15\text{ K})\}$ is given because $\{G^\circ(T) - H^\circ(298.15\text{ K})\}/T$ varies more slowly with temperature, and hence the tables are easier to interpolate. It is not necessary to specify a reference point for the heat capacity or the entropy, as indicated by the headings to the second and third columns. The sixth and seventh columns give values of $\Delta_f H^\circ$ and $\Delta_f G^\circ$ at various temperatures. We learned in Section 12–3 that these data can be used to calculate values of $\Delta_r H^\circ$, $\Delta_r G^\circ$, and equilibrium constants of reactions.

Because $G^\circ(T)$ and $H^\circ(T)$ are expressed relative to $H^\circ(298.15\text{ K})$ in Table 12.4, we must express the molecular partition function $q(V, T)$ relative to a zero of energy. Recall that in Section 12–5 we wrote $q(V, T)$ as

$$
\begin{aligned}
q(V, T) &= \sum_j e^{-\varepsilon_j/k_\text{B}T} = e^{-\varepsilon_0/k_\text{B}T} + e^{-\varepsilon_1/k_\text{B}T} + \cdots \\
&= e^{-\varepsilon_0/k_\text{B}T}(1 + e^{-(\varepsilon_1-\varepsilon_0)/k_\text{B}T} + \cdots) \\
&= e^{-\varepsilon_0/k_\text{B}T} q^0(V, T)
\end{aligned}
\tag{12.46}
$$

where $q^0(V, T)$ is a molecular partition function in which the ground state energy is taken to be zero. If we substitute Equation 12.46 into Equation 3.41, then we obtain

$$
\begin{aligned}
U = \langle E \rangle &= N k_\text{B} T^2 \left(\frac{\partial \ln q}{\partial T}\right)_V \\
&= N\varepsilon_0 + N k_\text{B} T^2 \left(\frac{\partial \ln q^0}{\partial T}\right)_V
\end{aligned}
\tag{12.47}
$$

For one mole of an ideal gas, $\overline{H} = H^\circ(T) = \overline{U} + P\overline{V} = \overline{U} + RT$, and so Equation 12.47 becomes

$$
H^\circ(T) = H_0^\circ + RT^2 \left(\frac{\partial \ln q^0}{\partial T}\right)_V + RT \tag{12.48}
$$

where $H_0^\circ = N_\text{A}\varepsilon_0$. Because $q^0(V, T)$ is the molecular partition function in which the ground state energy is taken to be zero, $q^0(V, T)$ is given by either Equation 4.57 or 4.60, without the factors of $e^{-\Theta_{\text{vib},j}/2T}$ and $e^{D_e/k_\text{B}T}$, which represent the ground state of the molecule. Using either Equation 4.57 or 4.60, Equation 12.48 becomes

$$
\begin{aligned}
H^\circ(T) - H_0^\circ &= \frac{3}{2}RT + \frac{2}{2}RT + \sum_j \frac{R\Theta_{\text{vib},j}}{e^{\Theta_{\text{vib},j}/T} - 1} + RT \\
&= \frac{7}{2}RT + \sum_j \frac{R\Theta_{\text{vib},j}}{e^{\Theta_{\text{vib},j}/T} - 1} \quad \text{(linear molecule)}
\end{aligned}
\tag{12.49a}
$$

or

$$H^\circ(T) - H_0^\circ = \frac{3}{2}RT + \frac{3}{2}RT + \sum_j \frac{R\Theta_{\text{vib},j}}{e^{\Theta_{\text{vib},j}/T} - 1} + RT$$

$$= 4RT + \sum_j \frac{R\Theta_{\text{vib},j}}{e^{\Theta_{\text{vib},j}/T} - 1} \quad \text{(nonlinear molecule)} \qquad (12.49b)$$

Note that there are no terms involving $\Theta_{\text{vib},j}/2T$ or $D_e/k_BT$ in Equations 12.49 as there are in Equations 4.58 and 4.61 because we have taken the energy of the ground vibrational state to be zero.

We can use Equation 12.49*b* and the parameters in Table 4.4 to calculate $H^\circ(298.15\text{ K}) - H_0^\circ$ for ammonia

$$\begin{aligned} H^\circ(298.15\text{ K}) - H_0^\circ &= 4(8.3145\text{ J}\cdot\text{mol}^{-1}\cdot\text{K}^{-1})(298.15\text{ K}) \\ &\quad +(8.3145\text{ J}\cdot\text{mol}^{-1}\cdot\text{K}^{-1})\left[\frac{4800\text{ K}}{e^{4800/298.15} - 1}\right. \\ &\quad \left. + \frac{1360\text{ K}}{e^{1360/298.15} - 1} + \frac{(2)(4880\text{ K})}{e^{4880/298.15} - 1} + \frac{(2)(2330\text{ K})}{e^{2330/298.15} - 1}\right] \\ &= 10.05\text{ kJ}\cdot\text{mol}^{-1} \end{aligned}$$

The very first entry in the fifth column in Table 12.4 is $-10.045\text{ kJ}\cdot\text{mol}^{-1}$. This value represents $H^\circ(0\text{ K}) - H^\circ(298.15\text{ K})$, which is the negative of $H^\circ(298.15\text{ K}) - H^\circ(0\text{ K})$ that we just calculated because $H_0^\circ = H^\circ\ (0\text{ K})$. Thus, the value given by Equation 12.49*b* and the value given in Table 12.4 are in excellent agreement.

**EXAMPLE 12–8**

Use Equation 12.49*b* and the parameters in Table 4.4 to calculate $H^\circ(T) - H_0^\circ$ for $NH_3(g)$ at 1000 K and one bar. Compare your result to Table 12.4.

SOLUTION: Equation 12.49b gives

$$H^\circ(1000\text{ K}) - H_0^\circ = 42.290\text{ kJ}\cdot\text{mol}^{-1}$$

Table 12.4 gives

$$H_0^\circ - H^\circ(298.15\text{ K}) = H^\circ(0\text{ K}) - H^\circ(298.15\text{ K}) = -10.045\text{ kJ}\cdot\text{mol}^{-1} \qquad (1)$$

and

$$H^\circ(1000\text{ K}) - H^\circ(298.15\text{ K}) = 32.637\text{ kJ}\cdot\text{mol}^{-1} \qquad (2)$$

If we subtract Equation 1 from Equation 2, then we obtain

$$H^\circ(1000\text{ K}) - H_0^\circ = 42.682\text{ kJ}\cdot\text{mol}^{-1}$$

The value obtained from Table 12.4 is more accurate than the value calculated from Equation 12.49b. At 1000 K, the ammonia molecule is excited enough that the rigid rotator-harmonic oscillator approximation begins to become unsatisfactory.

We can also use the data in Table 12.4 to calculate the value of $q^0(V, T)$ for ammonia. Recall from Section 9–5 that we derived the equation (Equation 9.36)

$$\mu^\circ(T) - E_0^\circ = -RT \ln \left\{ \left( \frac{q^0}{V} \right) \frac{RT}{N_A P^\circ} \right\} \tag{12.50}$$

where $E_0^\circ = N_A \varepsilon_0 = H_0^\circ$ and $P^\circ = 1$ bar $= 10^5$ Pa. Equation 12.50 is valid only for an ideal gas, and recall that $q(V, T)/V$, or $q^0(V, T)/V$, is a function of temperature only for an ideal gas. Equation 12.50 clearly displays the fact that the chemical potential is calculated relative to some zero of energy.

Because $G^\circ = \mu^\circ$ for a pure substance, we can write Equation 12.50 as

$$G^\circ - H_0^\circ = -RT \ln \left\{ \left( \frac{q^0}{V} \right) \frac{RT}{N_A P^\circ} \right\} \tag{12.51}$$

It is easy to show that $G^\circ \to H_0^\circ$ as $T \to 0$ (because $T \ln T \to 0$ as $T \to 0$), and so $H_0^\circ$ is also the standard Gibbs energy at 0 K.

According to Equation 12.51

$$\frac{q^0}{V} \frac{RT}{N_A P^\circ} = e^{-(G^\circ - H_0^\circ)/RT}$$

or

$$\frac{q^0(V, T)}{V} = \frac{N_A P^\circ}{RT} e^{-(G^\circ - H_0^\circ)/RT} \tag{12.52a}$$

where $P^\circ = 10^5$ Pa. The fourth column in Table 12.4 gives $-\{G^\circ - H^\circ(298.15 \text{ K})\}/T$ instead of $-(G^\circ - H_0^\circ)/T$, but the first entry of the fifth column gives $H_0^\circ - H^\circ(298.15 \text{ K})$. Therefore, the exponential in Equation 12.52a can be obtained from

$$\underbrace{-\frac{(G^\circ - H_0^\circ)}{T}}_{\substack{\text{exponent in} \\ \text{Equation 12.52a}}} = \underbrace{-\frac{(G^\circ - H^\circ(298.15 \text{ K}))}{T}}_{\substack{\text{fourth column in} \\ \text{Table 12.4}}} + \underbrace{\frac{(H_0^\circ - H^\circ(298.15 \text{ K}))}{T}}_{\substack{\text{first entry of fifth} \\ \text{column in Table 12.4} \\ \text{divided by } T}} \tag{12.52b}$$

Let's use Equations 12.52 to calculate $q^0(V, T)$ for ammonia at 500 K. Substituting the data in Table 12.4 into Equation 12.52$b$ gives

$$-\frac{(G^\circ - H_0^\circ)}{500\text{ K}} = 197.021\text{ J}\cdot\text{K}^{-1}\cdot\text{mol}^{-1} + \frac{-10.045\text{ kJ}\cdot\text{mol}^{-1}}{500\text{ K}}$$
$$= 176.931\text{ J}\cdot\text{K}^{-1}\cdot\text{mol}^{-1}$$

If we substitute this value into Equation 12.52$a$, then we obtain

$$\frac{q^0(V, T)}{V} = \frac{(6.022 \times 10^{23}\text{ mol}^{-1})(10^5\text{ Pa})}{(8.314\text{ J}\cdot\text{mol}^{-1}\cdot\text{K}^{-1})(500\text{ K})} e^{(176.931\text{ J}\cdot\text{K}^{-1}\cdot\text{mol}^{-1})/8.314\text{ J}\cdot\text{mol}^{-1}\cdot\text{K}^{-1}}$$
$$= 2.53 \times 10^{34}\text{ m}^{-3}$$

Equation 4.60 gives (Problem 12–48)

$$\frac{q^0(V, T)}{V} = 2.59 \times 10^{34}\text{ m}^{-3}$$

The value given by Equations 12.52 is the more accurate because Equation 4.60 is based on the rigid rotator-harmonic oscillator approximation.

**EXAMPLE 12–9**

The JANAF tables give $-[G^\circ - H^\circ(298.15\text{ K})]/T = 231.002\text{ J}\cdot\text{mol}^{-1}\cdot\text{K}^{-1}$ and $H_0^\circ - H^\circ(298.15\text{ K}) = -8.683\text{ kJ}\cdot\text{mol}^{-1}$ for $O_2(g)$ at 1500 K. Use these data and Equations 12.52 to calculate $q^0(V, T)/V$ for $O_2(g)$ at 1500 K.

SOLUTION: Equation 12.52$b$ gives

$$-\frac{G^\circ - H_0^\circ}{T} = 231.002\text{ J}\cdot\text{mol}^{-1}\cdot\text{K}^{-1} + \frac{-8.683\text{ kJ}\cdot\text{mol}^{-1}}{1500\text{ K}}$$
$$= 225.093\text{ J}\cdot\text{mol}^{-1}\cdot\text{K}^{-1}$$

and Equation 12.52$a$ gives

$$\frac{q^0(V, T)}{V} = \frac{(6.022 \times 10^{23}\text{ mol}^{-1})(10^5\text{ Pa})}{(8.314\text{ J}\cdot\text{mol}^{-1}\cdot\text{K}^{-1})(1500\text{ K})} e^{(225.093\text{ J}\cdot\text{K}^{-1}\cdot\text{mol}^{-1})/8.314\text{ J}\cdot\text{mol}^{-1}\cdot\text{K}^{-1}}$$
$$= 2.76 \times 10^{36}\text{ m}^{-3}$$

The value calculated in the previous section is $2.79 \times 10^{36}\text{ m}^{-3}$.

Lastly, the thermodynamic data in the JANAF tables can also be used to calculate values of $D_0$ for molecules. Table 12.4 gives $\Delta_f H^\circ(0\text{ K}) = -38.907\text{ kJ}\cdot\text{mol}^{-1}$ for $NH_3(g)$. The chemical equation that represents this process is

$$\tfrac{3}{2}\,H_2(g) + \tfrac{1}{2}\,N_2(g) \rightleftharpoons NH_3(g) \quad \Delta_f H^\circ(0\text{ K}) = -38.907\text{ kJ}\cdot\text{mol}^{-1} \qquad (1)$$

The entries in the JANAF tables for H(g) and N(g) give $\Delta_f H^\circ(0\text{ K}) = 216.035\text{ kJ}\cdot\text{mol}^{-1}$ and $470.82\text{ kJ}\cdot\text{mol}^{-1}$, respectively. These values correspond to the equations

$$\tfrac{1}{2}\text{H}_2(\text{g}) \rightleftharpoons \text{H(g)} \qquad \Delta_f H^\circ(0\text{ K}) = 216.035\text{ kJ}\cdot\text{mol}^{-1} \tag{2}$$

and

$$\tfrac{1}{2}\text{N}_2(\text{g}) \rightleftharpoons \text{N(g)} \qquad \Delta_f H^\circ(0\text{ K}) = 470.82\text{ kJ}\cdot\text{mol}^{-1} \tag{3}$$

If we subtract Equation 1 from the sum of Equation 3 and three times Equation 2, then we obtain

$$\text{NH}_3(\text{g}) \rightleftharpoons \text{N(g)} + 3\,\text{H(g)}$$

$$\begin{aligned}\Delta_f H^\circ(0\text{ K}) &= 38.907\text{ kJ}\cdot\text{mol}^{-1} + (3)(216.035\text{ kJ}\cdot\text{mol}^{-1}) + 470.82\text{ kJ}\cdot\text{mol}^{-1}\\ &= 1157.83\text{ kJ}\cdot\text{mol}^{-1}\end{aligned}$$

The value given in Table 4.4 is $1158\text{ kJ}\cdot\text{mol}^{-1}$.

**EXAMPLE 12–10**

The JANAF tables give $\Delta_f H^\circ(0\text{ K})$ for HI(g), H(g), and I(g) to be $28.535\text{ kJ}\cdot\text{mol}^{-1}$, $216.035\text{ kJ}\cdot\text{mol}^{-1}$, and $107.16\text{ kJ}\cdot\text{mol}^{-1}$, respectively. Calculate the value of $D_0$ for HI(g).

SOLUTION: The above data can be presented as

$$\tfrac{1}{2}\text{H}_2(\text{g}) + \tfrac{1}{2}\text{I}_2(\text{s}) \rightleftharpoons \text{HI(g)} \qquad \Delta_f H^\circ(0\text{ K}) = 28.535\text{ kJ}\cdot\text{mol}^{-1} \tag{1}$$

$$\tfrac{1}{2}\text{H}_2(\text{g}) \rightleftharpoons \text{H(g)} \qquad \Delta_f H^\circ(0\text{ K}) = 216.035\text{ kJ}\cdot\text{mol}^{-1} \tag{2}$$

$$\tfrac{1}{2}\text{I}_2(\text{s}) \rightleftharpoons \text{I(g)} \qquad \Delta_f H^\circ(0\text{ K}) = 107.16\text{ kJ}\cdot\text{mol}^{-1} \tag{3}$$

If we subtract Equation 1 from the sum of Equations 2 and 3, then we obtain

$$\text{HI(g)} \rightleftharpoons \text{H(g)} + \text{I(g)} \quad \Delta_f H^\circ(0\text{ K}) = 294.66\text{ kJ}\cdot\text{mol}^{-1}$$

The value given in Table 4.2 is $294.7\text{ kJ}\cdot\text{mol}^{-1}$.

The thermodynamic tables contain a great deal of thermodynamic and/or statistical thermodynamic data. Their use requires a little practice, but it is well worth the effort. Problems 12–45 through 12–58 are meant to supply this practice.

## 12–10. Equilibrium Constants for Real Gases Are Expressed in Terms of Partial Fugacities

Up to this point in this chapter, we have discussed equilibria in systems of ideal gases only. In this section, we shall discuss equilibria in systems of nonideal gases. In Section 8–8 we introduced the idea of fugacity through the equation

$$\mu(T, P) = \mu^{\circ}(T) + RT \ln \frac{f}{f^{\circ}} \tag{12.53}$$

where $\mu^{\circ}(T)$ is the chemical potential of the corresponding ideal gas at one bar. Once again to simplify the notation we shall not display the $f^{\circ}$ in the rest of this chapter. Therefore, Equation 12.53 can be written in the form

$$\mu(T, P) = \mu^{\circ}(T) + RT \ln f \tag{12.54}$$

Consequently, we must keep in mind that $f$ is taken relative to its standard state. In a mixture of gases, we would have

$$\mu_j(T, P) = \mu_j^{\circ}(T) + RT \ln f_j \tag{12.55}$$

Because the molecules in a mixture of gases in which the gases do not behave ideally are not independent of one another, the partial fugacity of each gas generally depends upon the concentrations of all the other gases in the mixture.

Now let's consider the general gas-phase reaction

$$\nu_{\mathrm{A}}\mathrm{A(g)} + \nu_{\mathrm{B}}\mathrm{B(g)} \rightleftharpoons \nu_{\mathrm{Y}}\mathrm{Y(g)} + \nu_{\mathrm{Z}}\mathrm{Z(g)}$$

The change in Gibbs energy upon converting the reactants at arbitrary partial pressures to products at arbitrary partial pressures is

$$\Delta_{\mathrm{r}}G = \nu_{\mathrm{Y}}\mu_{\mathrm{Y}} + \nu_{\mathrm{Z}}\mu_{\mathrm{Z}} - \nu_{\mathrm{A}}\mu_{\mathrm{A}} - \nu_{\mathrm{B}}\mu_{\mathrm{B}}$$

If we substitute Equation 12.55 into this equation, then we get

$$\Delta_{\mathrm{r}}G = \Delta_{\mathrm{r}}G^{\circ} + RT \ln \frac{f_{\mathrm{Y}}^{\nu_{\mathrm{Y}}} f_{\mathrm{Z}}^{\nu_{\mathrm{Z}}}}{f_{\mathrm{A}}^{\nu_{\mathrm{A}}} f_{\mathrm{B}}^{\nu_{\mathrm{B}}}} \tag{12.56}$$

where

$$\Delta_{\mathrm{r}}G^{\circ} = \nu_{\mathrm{Y}}\mu_{\mathrm{Y}}^{\circ} + \nu_{\mathrm{Z}}\mu_{\mathrm{Z}}^{\circ} - \nu_{\mathrm{A}}\mu_{\mathrm{A}}^{\circ} - \nu_{\mathrm{B}}\mu_{\mathrm{B}}^{\circ}$$

Note that Equation 12.56 is the generalization of Equation 12.24 to a system of nonideal gases. Realize that the values of the fugacities at this point are arbitrary, and not necessarily equilibrium values. If the reaction system is in equilibrium, then $\Delta_{\mathrm{r}}G = 0$ and all the fugacities take on their equilibrium values. Equation 12.56 becomes

$$\Delta_{\mathrm{r}}G^{\circ}(T) = -RT \ln K_f \tag{12.57}$$

where the equilibrium constant $K_f$ is given by

$$K_f(T) = \left( \frac{f_Y^{\nu_Y} f_Z^{\nu_Z}}{f_A^{\nu_A} f_B^{\nu_B}} \right)_{\text{eq}} \tag{12.58}$$

Once again notice that the equilibrium constant is a function of temperature only, as dictated by Equation 12.57.

The equilibrium constant defined by Equation 12.57 is called a *thermodynamic equilibrium constant*. Equation 12.57, which relates $K_f$ to $\Delta_r G^\circ$ is exact, being valid for real gases as well as ideal gases. At low pressures we can replace the partial fugacities by partial pressures to obtain $K_P$, but we should expect this approximation to fail at high pressures. The formulas to calculate partial fugacities from equation-of-state data are extensions of the formulas in Section 8–8 where we calculated fugacities for pure gases. In order to obtain the partial fugacities to use in Equation 12.58 we need rather extensive pressure-volume data for the mixture of reacting gases. These data are available for the important industrial reaction

$$\tfrac{1}{2}\,N_2(g) + \tfrac{3}{2}\,H_2(g) \rightleftharpoons NH_3(g)$$

Table 12.5 shows both $K_P$ and $K_f$ as a function of the total pressure of the reaction mixture. Note that $K_P$ is not a constant, but that $K_f$ is fairly constant with increasing total pressure. The results shown in Table 12.5 emphasize that we must use fugacities and not pressures when dealing with systems at high pressures.

**TABLE 12.5**
Values of $K_P$ and $K_f$ as a function of total pressure for the ammonia synthesis equilibrium at 450°C.

| total pressure/bar | $K_P/10^{-3}$ | $K_f/10^{-3}$ |
|---|---|---|
| 10 | 6.59 | 6.55 |
| 30 | 6.76 | 6.59 |
| 50 | 6.90 | 6.50 |
| 100 | 7.25 | 6.36 |
| 300 | 8.84 | 6.08 |
| 600 | 12.94 | 6.42 |

**EXAMPLE 12–11**

The equilibrium constants $K_P$ and $K_f$ can be related by a quantity $K_\gamma$, such that $K_f = K_\gamma K_P$ and $K_\gamma$ has the form of an equilibrium constant, but involving activity coefficients, $\gamma_j$. First derive an expression for $K_\gamma$ and then evaluate it at the various pressures given in Table 12.5.

SOLUTION: The relation between pressure and fugacity is given by

$$f_j = \gamma_j P_j$$

If we substitute this expression into Equation 12.58, then we obtain

$$K_f = \frac{(\gamma_Y^{\nu_Y} P_Y^{\nu_Y})(\gamma_Z^{\nu_Z} P_Z^{\nu_Z})}{(\gamma_A^{\nu_A} P_A^{\nu_A})(\gamma_B^{\nu_B} P_B^{\nu_B})}$$

$$= \left(\frac{\gamma_Y^{\nu_Y}\gamma_Z^{\nu_Z}}{\gamma_A^{\nu_A}\gamma_B^{\nu_B}}\right) \cdot \left(\frac{P_Y^{\nu_Y}P_Z^{\nu_Z}}{P_A^{\nu_A}P_B^{\nu_B}}\right) = K_\gamma \cdot K_P$$

where we have used the standard state $f^\circ = P^\circ = 1$ bar. Using the data in Table 12.5, we see that

| $P$/bar | 10 | 30 | 50 | 100 | 300 | 600 |
|---|---|---|---|---|---|---|
| $K_\gamma$ | 0.994 | 0.975 | 0.942 | 0.877 | 0.688 | 0.496 |

The deviation of $K_\gamma$ from unity is a measure of the nonideality of the system.

## 12–11. Thermodynamic Equilibrium Constants Are Expressed in Terms of Activities

In the previous section we discussed the condition of equilibrium for a reaction system consisting of real gases. The central result was the introduction of $K_f$, in which the equilibrium constant is expressed in terms of partial fugacites. In this section we shall derive a similar expression for general equilibrium systems, consisting of gases, solids, liquids, and/or solutions. The starting point is Equation 12.35, which we write as

$$\mu_j = \mu_j^\circ(T) + RT \ln a_j \tag{12.59}$$

where $a_j$ is the activity of species $j$ and $\mu_j^\circ$ is the chemical potential of the standard state. This equation essentially defines the activity, $a_j$. Recall that we discussed two different standard states in Chapters 10 and 11: a Raoult's law standard state, in which $a_j \to x_j$ as $x_j \to 1$, in which case $\mu_j^\circ = \mu_j^*$, the chemical potential of pure component $j$; and a Henry's law standard state, in which $a_j \to m_j$ or $a_j \to c_j$ as $m_j \to 0$ or $c_j \to 0$, in which case $\mu_j^\circ$ is the chemical potential of the (hypothetical) corresponding ideal solution at unit molality or unit molarity. Although Equation 12.53 is restricted to gases, Equation 12.59 is general. In fact, we can include Equation 12.53 as a special case of Equation 12.59 by defining the activity of a gas by the relation $a_j = f_j/f_j^\circ$. In this case, $\mu_j^\circ(T)$ in Equation 12.59 is the corresponding (hypothetical) ideal gas at one bar and at the temperature of interest. Agreeing to set $a_j = f_j/f_j^\circ$ simply allows us to treat gases, liquids, solids, (and solutions) in the same notation.

Now let's consider the general reaction

$$\nu_A A + \nu_B B \rightleftharpoons \nu_Y Y + \nu_Z Z$$

The change in Gibbs energy for converting A and B in arbitrary states to Y and Z in arbitrary states is given by

$$\Delta_r G = \nu_Y \mu_Y + \nu_Z \mu_Z - \nu_A \mu_A - \nu_B \mu_B$$

If we substitute Equation 12.59 into this equation, then we obtain

$$\Delta_r G = \Delta_r G^\circ + RT \ln \frac{a_Y^{\nu_Y} a_Z^{\nu_Z}}{a_A^{\nu_A} a_B^{\nu_B}} \tag{12.60}$$

where

$$\Delta_r G^\circ = \nu_Y \mu_Y^\circ + \nu_Z \mu_Z^\circ - \nu_A \mu_A^\circ - \nu_B \mu_B^\circ$$

Equation 12.60 is called the *Lewis equation*, after the great thermodynamicist G. N. Lewis, who first introduced the concept of activity and pioneered the rigorous thermodynamic analysis of chemical equilibria. Note that Equation 12.60 is a generalization of Equation 12.56 to a non-ideal system, which may consist of condensed phases and solutions as well as gases. Realize that the activities at this point are arbitrary, and not necessarily the equilibrium activites. Just as we did in Section 12–5 for the case of a reaction system of ideal gases, we introduce a reaction quotient, or an *activity quotient*, in this case, by

$$Q_a = \frac{a_Y^{\nu_Y} a_Z^{\nu_Z}}{a_A^{\nu_A} a_B^{\nu_B}} \tag{12.61}$$

Using this notation, we can write Equation 12.60 as

$$\Delta_r G = \Delta_r G^\circ + RT \ln Q_a \tag{12.62}$$

According to Equation 12.59, $a_j = 1$ when a substance is in its standard state. Therefore, if all the reactants and products in a reaction mixture are in their standard states, then all the $a_j = 1$ in Equation 12.61 and so $Q_a = 1$, giving $\Delta_r G = \Delta_r G^\circ$. If the reaction system is at equilibrium at fixed $T$ and $P$, then $\Delta_r G = 0$, and we have

$$\Delta_r G^\circ = -RT \ln Q_{a,\text{eq}} \tag{12.63}$$

where $Q_{a,\text{eq}}$ denotes $Q_a$ in which all the activities have their equilibrium values. In analogy with Section 12–5, we denote $Q_{a,\text{eq}}$ by $K_a$

$$K_a(T) = \left( \frac{a_Y^{\nu_Y} a_Z^{\nu_Z}}{a_A^{\nu_A} a_B^{\nu_B}} \right)_{\text{eq}} \tag{12.64}$$

which we call a *thermodynamic equilibrium constant*. Equation 12.57 becomes

$$\Delta_r G^\circ = -RT \ln K_a \tag{12.65}$$

Equation 12.65 is completely general and rigorous, and applies to any system in equilibrium. Note that for a reaction involving only gases, $a_i = f_i$, and $K_a(T) = K_f(T)$, Equation 12.58, and Equation 12.65 is equivalent to Equation 12.57. Equations 12.64 and 12.65 are more general than Equations 12.57 and 12.58 because the reactants can be in any phase. The application of this equation is best done by example.

Let's consider a heterogeneous system such as the water-gas reaction

$$\text{C(s)} + \text{H}_2\text{O(g)} \overset{1000^\circ\text{C}}{\rightleftharpoons} \text{CO(g)} + \text{H}_2\text{(g)}$$

which is used in the industrial production of hydrogen. The (thermodynamic) equilibrium constant for this equation is

$$K_a = \frac{a_{\text{CO(g)}} a_{\text{H}_2\text{(g)}}}{a_{\text{C(s)}} a_{\text{H}_2\text{O(g)}}} = \frac{f_{\text{CO(g)}} f_{\text{H}_2\text{(g)}}}{a_{\text{C(s)}} f_{\text{H}_2\text{O(g)}}}$$

Although we have dealt with fugacities of gases earlier, we have not dealt with activities of pure solids and liquids. We must first choose a standard state for a pure condensed phase, which we choose to be the pure substance in its normal state at one bar and at the temperature of interest. To calculate the activity, we start with

$$\left(\frac{\partial \mu}{\partial P}\right)_T = \overline{V} \tag{12.66}$$

and the constant-temperature derivative of Equation 12.59

$$d\mu = RT\, d\ln a \qquad \text{(constant } T\text{)} \tag{12.67}$$

If we write Equation 12.66 as

$$d\mu = \overline{V} dP \qquad \text{(constant } T\text{)}$$

and introduce Equation 12.67, then we have

$$d\ln a = \frac{\overline{V}}{RT} dP \qquad \text{(constant } T\text{)}$$

We now integrate from the chosen standard state ($a = 1$, $P = 1$ bar) to an arbitrary state to obtain

$$\int_{a=1}^{a} d\ln a' = \int_1^P \frac{\overline{V}}{RT} dP' \qquad \text{(constant } T\text{)}$$

or

$$\ln a = \frac{1}{RT}\int_1^P \overline{V}\,dP' \qquad \text{(constant } T\text{)} \tag{12.68}$$

For a condensed phase, $\overline{V}$ is essentially a constant over a moderate pressure range, and so Equation 12.68 becomes

$$\ln a = \frac{\overline{V}}{RT}(P-1) \tag{12.69}$$

**EXAMPLE 12–12**

Calculate the activity of C($s$) in the form of coke at 100 bar and 1000°C.

SOLUTION: The density of coke at 1000°C is about 1.5 g·cm$^{-3}$, and so its molar volume, $\overline{V}$, is 8.0 cm$^3$·mol$^{-1}$. From Equation 12.69

$$\ln a = \frac{(8.0\ \text{cm}^3\cdot\text{mol}^{-1})(1\ \text{dm}^3/1000\ \text{cm}^3)(99\ \text{bar})}{(0.08206\ \text{dm}^3\cdot\text{bar}\cdot\text{K}^{-1}\cdot\text{mol}^{-1})(1273\ \text{K})} = 0.0076$$

or $a = 1.01$. Note that the activity is essentially unity even at 100 bar.

According to Example 12–12, the activity of a pure condensed phase is unity at moderate pressures. Consequently, the activities of pure solids and liquids are normally not included in equilibrium constant expressions (as you may recall from general chemistry). For example, for the reaction

$$\text{C(s)} + \text{H}_2\text{O(g)} \rightleftharpoons \text{CO(g)} + \text{H}_2\text{(g)}$$

the equilibrium constant is given by

$$K = \frac{f_{\text{CO(g)}} f_{\text{H}_2\text{(g)}}}{f_{\text{H}_2\text{O(g)}}} \approx \frac{P_{\text{CO(g)}} P_{\text{H}_2\text{(g)}}}{P_{\text{H}_2\text{O(g)}}}$$

if the pressures are low enough. However, there are cases where the activities cannot be set to unity, as the following Example shows.

**EXAMPLE 12–13**

The change in the standard molar Gibbs energy for the conversion of graphite into diamond is 2.900 kJ·mol$^{-1}$ at 298.15 K. The density of graphite is 2.27 g·cm$^{-3}$ and that of diamond is 3.52 g·cm$^{-3}$ at 298.15 K. At what pressure will these two forms of carbon be at equilibrium at 298.15 K?

SOLUTION: We can represent the process by the chemical equation

$$\text{C(graphite)} \rightleftharpoons \text{C(diamond)}$$

for which

$$\Delta_r G^\circ = -RT \ln K_a = -RT \ln \frac{a_{\text{diamond}}}{a_{\text{graphite}}}$$

Using Equation 12.69, we have

$$\Delta_r G^\circ = -RT \left[ \frac{\Delta \overline{V}}{RT}(P-1) \right]$$

or

$$\frac{2900 \text{ J}\cdot\text{mol}^{-1}}{(8.3145 \text{ J}\cdot\text{mol}^{-1}\cdot\text{K}^{-1})(298.15 \text{ K})} =$$

$$-\frac{(3.41 \text{ cm}^3\cdot\text{mol}^{-1} - 5.29 \text{ cm}^3\cdot\text{mol}^{-1})(1 \text{ dm}^3/1000 \text{ cm}^3)(P-1) \text{ bar}}{(0.083145 \text{ dm}^3\cdot\text{bar}\cdot\text{mol}^{-1}\cdot\text{K}^{-1})(298.15 \text{ K})}$$

Solving the expression for $P$ gives

$$P = 1.54 \times 10^4 \text{ bar} \approx 15\,000 \text{ bar}$$

## 12–12. The Use of Activities Makes a Significant Difference in Solubility Calculations Involving Ionic Species

Equation 12.65 can also be applied to reactions that take place in solution. For example, let's consider the dissociation of an aqueous solution that is 0.100 molar in acetic acid, $CH_3COOH(aq)$, for which $K = 1.74 \times 10^{-5}$ on a molarity scale. The equation for the reaction is

$$CH_3COOH(aq) + H_2O(l) \rightleftharpoons H_3O^+(aq) + CH_3COO^-(aq)$$

and the equilibirum-constant expression is

$$K_a = \frac{a_{H_3O^+} a_{CH_3COO^-}}{a_{CH_3COOH} a_{H_2O}} = \frac{a_{H_3O^+} a_{CH_3COO^-}}{a_{CH_3COOH}} = 1.74 \times 10^{-5} \tag{12.70}$$

Being a neutral species at a concentration of around 0.100 molar, the undissociated acetic acid has an activity coefficient of essentially unity and so $a_{HAc} = c_{HAc}$. For the ions, we use the fact that (Table 11.3)

$$a_{H^+} a_{CH_3COO^-} = c_{H^+} c_{Ac^-} \gamma_\pm^2$$

and so Equation 12.70 becomes

$$\frac{c_{H_3O^+} c_{Ac^-}}{c_{HAc}} = \frac{1.74 \times 10^{-5}}{\gamma_\pm^2} \tag{12.71}$$

As a first approximation, we shall set all the activity coefficients equal to unity and write

$$K_c = \frac{c_{H_3O^+}c_{Ac^-}}{c_{HAc}} = 1.74 \times 10^{-5}\ \text{mol}\cdot\text{L}^{-1}$$

From the following set-up

| | $CH_3COOH(aq)$ | + $H_2O(l)$ | ⇌ $H_3O(aq)$ | + $CH_3COO^-(aq)$ |
|---|---|---|---|---|
| initial | $0.100\ \text{mol}\cdot\text{L}^{-1}$ | — | $\approx 0$ | 0 |
| equilibrium | $0.100\ \text{mol}\cdot\text{L}^{-1} - x$ | — | $x$ | $x$ |

we get

$$\frac{x^2}{0.100\ \text{mol}\cdot\text{L}^{-1} - x} = 1.74 \times 10^{-5}\ \text{mol}\cdot\text{L}^{-1}$$

or $x = 1.31 \times 10^{-3}\ \text{mol}\cdot\text{L}^{-1}$, for a pH of 2.88. This is the type of calculation that is done in general chemistry.

Now let's not set $\gamma_\pm$ equal to unity. For $\gamma_\pm$, we shall use Equation 11.57

$$\ln \gamma_\pm = -\frac{1.173|z_+z_-|(I_c/\ \text{mol}\cdot\text{L}^{-1})^{1/2}}{1 + (I_c/\ \text{mol}\cdot\text{L}^{-1})^{1/2}}$$

where the ionic strength $I_c$ is given by

$$I_c = \frac{1}{2}(c_{H^+} + c_{Ac^-}) = c_{H^+} = c_{Ac^-}$$

In order to calculate $I_c$ we must know $c_{H^+}$ or $c_{Ac^-}$, but we cannot determine either of these from Equation 12.71 because it contains $\gamma_\pm^2$. We can solve this problem by iteration, however. We first calculate $\gamma_\pm$ using the values of $c_{H^+}$ and $c_{Ac^-}$ that we obtained above by letting $\gamma_\pm = 1$:

$$\ln \gamma_\pm = -\frac{1.173(1.31 \times 10^{-3})^{1/2}}{1 + (1.31 \times 10^{-3})^{1/2}} = -0.0410$$

or $\gamma_\pm^2 = 0.921$. We now use this value in the right-hand side of Equation 12.71, and write

$$\frac{x^2}{0.100\ \text{mol}\cdot\text{L}^{-1} - x} = \frac{1.74 \times 10^{-5}\ \text{mol}\cdot\text{L}^{-1}}{0.921}$$

Solving for $x$, we find that $x = 1.365 \times 10^{-3}\ \text{mol}\cdot\text{L}^{-1}$. We now use this value to calculate a new value of $\gamma_\pm^2$ (= 0.920), and use this value in Equation 12.71 to calculate a new value of $x$ (= $1.366 \times 10^{-3}\ \text{mol}\cdot\text{L}^{-1}$). Cycling through once more gives $\gamma_\pm^2 =$ 0.920 and $x = 1.366 \times 10^{-3}\ \text{mol}\cdot\text{L}^{-1}$, and so we find that $x = 1.37 \times 10^{-3}\ \text{mol}\cdot\text{L}^{-1}$ (to three significant figures) and pH = 2.86. Thus we see that we calculate a pH of

2.86 using activities and a pH of 2.88 ignoring activities, not a significant difference. Fortunately the myriad of pH calculations that you did in general chemistry were sufficiently accurate. This is not necessarily the case for solubility calculations, as we shall now see.

The solubility product, $K_{sp}$, of $BaF_2(s)$ in water at 25°C is $1.7 \times 10^{-6}$, and the associated chemical equation is

$$BaF_2(s) \rightleftharpoons Ba^{2+}(aq) + 2\,F^-(aq)$$

The equilibrium constant expression is

$$a_{Ba^{2+}} a_{F^-}^2 = K_{sp} = 1.7 \times 10^{-6}$$

Using the formula (Table 11.3)

$$a_{Ba^{2+}} a_{F^-}^2 = c_{Ba^{2+}} c_{F^-}^2 \gamma_\pm^3$$

we have

$$c_{Ba^{2+}} c_{F^-}^2 = \frac{1.7 \times 10^{-6}}{\gamma_\pm^3} \tag{12.72}$$

If we set $\gamma_\pm = 1$, and let $s$ be the solubility of $BaF_2(s)$, then $c_{Ba^{2+}} = s$ and $c_{F^-} = 2s$, and we have

$$(s)(2s)^2 = 1.7 \times 10^{-6}\ \text{mol}^3\cdot\text{L}^{-3}$$

or $s = (1.7 \times 10^{-6}\ \text{mol}^3\cdot\text{L}^{-3}/4)^{1/3} = 7.52 \times 10^{-3}\ \text{mol}\cdot\text{L}^{-1}$. We now calculate the ionic strength using this value of $s$ to obtain

$$I_c = \frac{1}{2}(4s + 2s) = 3s = 0.0226\ \text{mol}\cdot\text{L}^{-1}$$

Using this value of $I_c$ in Equation 11.57 gives $\gamma_\pm = 0.736$. Substitute this value into Equation 12.55 to get

$$4s^3 = \frac{1.7 \times 10^{-6}\ \text{mol}^3\cdot\text{L}^{-3}}{0.399}$$

and so $s = 0.0102\ \text{mol}\cdot\text{L}^{-1}$. Cycling through again gives $\gamma_\pm = 0.705$ and $s = 0.0107\ \text{mol}\cdot\text{L}^{-1}$. Once more gives $\gamma_\pm = 0.700$ and $s = 0.0107\ \text{mol}\cdot\text{L}^{-1}$ and one last iteration gives $\gamma_\pm = 0.700$ and $s = 0.011\ \text{mol}\cdot\text{L}^{-1}$ to two significant figures. Notice that in this case there is over a 30% difference between calculating $s$ with and without the inclusion of activity coefficients.

**EXAMPLE 12–14**

Calculate the solubility of $TlBrO_3(s)$ in pure water and in an aqueous solution that is $0.500\ \text{mol}\cdot\text{L}^{-1}$ in $KNO_3(aq)$. $K_{sp} = 1.72 \times 10^{-4}$ for $TlBrO_3(s)$.

SOLUTION: The equation for the dissolution of $TlBrO_3(s)$ is

$$TlBrO_3(s) \rightleftharpoons Tl^+(aq) + BrO_3^-(aq)$$

with

$$a_{Tl^+}a_{BrO_3^-} = c_{Tl^+}c_{BrO_3^-}\gamma_\pm^2 = s^2\gamma_\pm^2 = 1.72 \times 10^{-4}$$

Letting $\gamma_\pm = 1$ at first, we find that $s = 0.0131\ \text{mol·L}^{-1}$. Using this value of $s$, we get $I_c = s$ and $\gamma_\pm = 0.887$ for $TlBrO_3(s)$ in pure water. Using this value of $\gamma_\pm$ in the $K_{sp}$ expression gives $s = 0.0148\ \text{mol·L}^{-1}$. Subsequent iterations give $s = 0.0149\ \text{mol·L}^{-1}$.

For the case with $0.500\ \text{mol·L}^{-1}$ $KNO_3(aq)$, we write

$$I_c = \frac{1}{2}(s + s + 0.500\ \text{mol·L}^{-1} + 0.500\ \text{mol·L}^{-1}) = s + 0.500\ \text{mol·L}^{-1}$$

Because $s$ is much less than $0.500\ \text{mol·L}^{-1}$, we intially let $I_c = 0.500\ \text{mol·L}^{-1}$, which gives $\gamma_\pm = 0.616$. Using this value in the solubility product expression gives $s = 0.0213\ \text{mol·L}^{-1}$. Now $I_c = 0.5213\ \text{mol·L}^{-1}$ and $\gamma_\pm = 0.612$ and $s = 0.0214\ \text{mol·L}^{-1}$. Subsequent iterations give $s = 0.0214\ \text{mol·L}^{-1}$. Notice that the solubility of $TlBrO_3(s)$ is significantly enhanced in the 0.500 molar $KNO_3(aq)$ even though the $KNO_3(aq)$ does not participate in the dissolution reaction. If we had not included the activity coefficients, we would have gotten no effect at all.

# Problems

**12-1.** Express the concentrations of each species in the following chemical equations in terms of the extent of reaction, $\xi$. The initial conditions are given under each equation.

**a.**

| | $SO_2Cl_2(g)$ | $\rightleftharpoons$ | $SO_2(g)$ | + | $Cl_2(g)$ |
|---|---|---|---|---|---|
| (1) | $n_0$ | | 0 | | 0 |
| (2) | $n_0$ | | $n_1$ | | 0 |

**b.**

| | $2\,SO_3(g)$ | $\rightleftharpoons$ | $2\,SO_2(g)$ | + | $O_2(g)$ |
|---|---|---|---|---|---|
| (1) | $n_0$ | | 0 | | 0 |
| (2) | $n_0$ | | 0 | | $n_1$ |

**c.**

| | $N_2(g)$ | + | $2\,O_2(g)$ | $\rightleftharpoons$ | $N_2O_4(g)$ |
|---|---|---|---|---|---|
| (1) | $n_0$ | | $2n_0$ | | 0 |
| (2) | $n_0$ | | $n_0$ | | 0 |

**12-2.** Write out the equilibrium-constant expression for the reaction that is described by the equation

$$2\,SO_2(g) + O_2(g) \rightleftharpoons 2\,SO_3(g)$$

Compare your result to what you get if the reaction is represented by

$$SO_2(g) + \tfrac{1}{2}\,O_2(g) \rightleftharpoons SO_3(g)$$

**12-3.** Consider the dissociation of $N_2O_4(g)$ into $NO_2(g)$ described by

$$N_2O_4(g) \rightleftharpoons 2\,NO_2(g)$$

Assuming that we start with $n_0$ moles of $N_2O_4(g)$ and no $NO_2(g)$, show that the extent of reaction, $\xi_{eq}$, at equilibrium is given by

$$\frac{\xi_{eq}}{n_0} = \left(\frac{K_P}{K_P + 4P}\right)^{1/2}$$

Plot $\xi_{eq}/n_0$ against $P$ given that $K_P = 6.1$ at 100°C. Is your result in accord with Le Châtelier's principle?

**12-4.** In Problem 12–3 you plotted the extent of reaction at equilibrium against the total pressure for the dissociation of $N_2O_4(g)$ to $NO_2(g)$. You found that $\xi_{eq}$ decreases as $P$ increases, in accord with Le Châtelier's principle. Now let's introduce $n_{inert}$ moles of an inert gas into the system. Assuming that we start with $n_0$ moles of $N_2O_4(g)$ and no $NO_2(g)$, derive an expression for $\xi_{eq}/n_0$ in terms of $P$ and the ratio $r = n_{inert}/n_0$. As in Problem 12–3, let $K_P = 6.1$ and plot $\xi_{eq}/n_0$ versus $P$ for $r = 0$, $r = 0.50$, $r = 1.0$, and $r = 2.0$. Show that introducing an inert gas into the reaction mixture at constant pressure has the same effect as lowering the pressure. What is the effect of introducing an inert gas into a reaction system at constant volume?

**12-5.** Re-do Problem 12–3 with $n_0$ moles of $N_2O_4(g)$ and $n_1$ moles of $NO_2(g)$ initially. Let $n_1/n_0 = 0.50$ and 2.0.

**12-6.** Consider the ammonia-synthesis reaction, which can be described by

$$N_2(g) + 3\,H_2(g) \rightleftharpoons 2\,NH_3(g)$$

Suppose initially there are $n_0$ moles of $N_2(g)$ and $3n_0$ moles of $H_2(g)$ and no $NH_3(g)$. Derive an expression for $K_P(T)$ in terms of the equilibrium value of the extent of reaction, $\xi_{eq}$, and the pressure, $P$. Use this expression to discuss how $\xi_{eq}/n_0$ varies with $P$ and relate your conclusions to Le Châtelier's principle.

**12-7.** Nitrosyl chloride, NOCl, decomposes according to

$$2\,NOCl(g) \rightleftharpoons 2\,NO(g) + Cl_2(g)$$

Assuming that we start with $n_0$ moles of NOCl(g) and no NO(g) or $Cl_2(g)$, derive an expression for $K_P$ in terms of the equilibirum value of the extent of reaction, $\xi_{eq}$, and the pressure, $P$. Given that $K_P = 2.00 \times 10^{-4}$, calculate $\xi_{eq}/n_0$ when $P = 0.080$ bar. What is the new value of $\xi_{eq}/n_0$ at equilibrium when $P = 0.160$ bar? Is this result in accord with Le Châtelier's principle?

**12-8.** The value of $K_P$ at 1000°C for the decomposition of carbonyl dichloride (phosgene) according to

$$COCl_2(g) \rightleftharpoons CO(g) + Cl_2(g)$$

is 34.8 if the standard state is taken to be one bar. What would the value of $K_P$ be if for some reason the standard state were taken to be 0.500 bar? What does this result say about the numerical values of equilibrium constants?

**12-9.** Most gas-phase equilibrium constants in the recent chemical literature were calculated assuming a standard state pressure of one atmosphere. Show that the corresponding equilibrium constant for a standard state pressure of one bar is given by

$$K_P(\text{bar}) = K_P(\text{atm})(1.01325)^{\Delta\nu}$$

where $\Delta\nu$ is the sum of the stoichiometric coefficients of the products minus that of the reactants.

**12-10.** Using the data in Table 12.1, calculate $\Delta_r G^\circ(T)$ and $K_P(T)$ at 25°C for

$$\begin{array}{ll} \text{(a)} & N_2O_4(g) \rightleftharpoons 2\,NO_2(g) \\ \text{(b)} & H_2(g)+I_2(g) \rightleftharpoons 2\,HI(g) \\ \text{(c)} & 3\,H_2(g) + N_2(g) \rightleftharpoons 2\,NH_3(g) \end{array}$$

**12-11.** Calculate the value of $K_c(T)$ based upon a one $\text{mol}\cdot\text{L}^{-1}$ standard state for each of the equations in Problem 12–10.

**12-12.** Derive a relation between $K_P$ and $K_c$ for the following:

$$\begin{array}{ll} \text{(a)} & CO(g) + Cl_2(g) \rightleftharpoons COCl_2(g) \\ \text{(b)} & CO(g) + 3\,H_2(g) \rightleftharpoons CH_4(g) + H_2O(g) \\ \text{(c)} & 2\,BrCl(g) \rightleftharpoons Br_2(g) + Cl_2(g) \end{array}$$

**12-13.** Consider the dissociation reaction of $I_2(g)$ described by

$$I_2(g) \rightleftharpoons 2\,I(g)$$

The total pressure and the partial pressure of $I_2(g)$ at 1400°C have been measured to be 36.0 torr and 28.1 torr, respectively. Use these data to calculate $K_P$ (one bar standard state) and $K_c$ (one $\text{mol}\cdot\text{L}^{-1}$ standard state) at 1400°C.

**12-14.** Show that

$$\frac{d \ln K_c}{dT} = \frac{\Delta_r U^\circ}{RT^2}$$

for a reaction involving ideal gases.

**12-15.** Consider the gas-phase reaction for the synthesis of methanol from CO(g) and $H_2(g)$

$$CO(g) + 2\,H_2(g) \rightleftharpoons CH_3OH(g)$$

The value of the equilibrium constant $K_P$ at 500 K is $6.23 \times 10^{-3}$. Initially equimolar amounts of CO(g) and $H_2(g)$ are introduced into the reaction vessel. Determine the value of $\xi_{eq}/n_0$ at equilibrium at 500 K and 30 bar.

**12-16.** Consider the two equations

(1) $CO(g) + H_2O(g) \rightleftharpoons CO_2(g) + H_2(g) \quad K_1$

(2) $CH_4(g) + H_2O(g) \rightleftharpoons CO(g) + 3\,H_2(g) \quad K_2$

Show that $K_3 = K_1K_2$ for the sum of these two equations

(3) $CH_4(g) + 2\,H_2O(g) \rightleftharpoons CO_2(g) + 4\,H_2(g) \quad K_3$

How do you explain the fact that you would add the values of $\Delta_r G^\circ$ but multiply the equilibrium constants when adding Equations 1 and 2 to get Equation 3.

**12-17.** Given:

$$2\,BrCl(g) \rightleftharpoons Cl_2(g) + Br_2(g) \qquad K_P = 0.169$$
$$2\,IBr(g) \rightleftharpoons Br_2(g) + I_2(g) \qquad K_P = 0.0149$$

Determine $K_P$ for the reaction

$$BrCl(g) + \tfrac{1}{2}\,I_2(g) \rightleftharpoons IBr(g) + \tfrac{1}{2}\,Cl_2(g)$$

**12-18.** Consider the reaction described by

$$Cl_2(g) + Br_2(g) \rightleftharpoons 2\,BrCl(g)$$

at 500 K and a total pressure of one bar. Suppose that we start with one mole each of $Cl_2(g)$ and $Br_2(g)$ and no BrCl(g). Show that

$$G(\xi) = (1-\xi)G^\circ_{Cl_2} + (1-\xi)G^\circ_{Br_2} + 2\xi G^\circ_{BrCl} + 2(1-\xi)RT\ln\frac{1-\xi}{2} + 2\xi RT\ln\xi$$

where $\xi$ is the extent of reaction. Given that $G^\circ_{BrCl} = -3.694\ \text{kJ}\cdot\text{mol}^{-1}$ at 500 K, plot $G(\xi)$ versus $\xi$. Differentiate $G(\xi)$ with respect to $\xi$ and show that the minimum value of $G(\xi)$ occurs at $\xi_{eq} = 0.549$. Also show that

$$\left(\frac{\partial G}{\partial \xi}\right)_{T,P} = \Delta_r G^\circ + RT\ln\frac{P^2_{BrCl}}{P_{Cl_2}P_{Br_2}}$$

and that $K_P = 4\xi^2_{eq}/(1-\xi_{eq})^2 = 5.9$.

**12-19.** Consider the reaction described by

$$2\,H_2O(g) \rightleftharpoons 2\,H_2(g) + O_2(g)$$

at 4000 K and a total pressure of one bar. Suppose that we start with two moles of $H_2O(g)$ and no $H_2(g)$ or $O_2(g)$. Show that

$$G(\xi) = 2(1-\xi)G^\circ_{H_2O} + 2\xi G^\circ_{H_2} + \xi G^\circ_{O_2} + 2(1-\xi)RT\ln\frac{2(1-\xi)}{2+\xi}$$
$$+ 2\xi RT\ln\frac{2\xi}{2+\xi} + \xi RT\ln\frac{\xi}{2+\xi}$$

where $\xi$ is the extent of reaction. Given that $\Delta_f G^\circ[H_2O(g)] = -18.334\ kJ\cdot mol^{-1}$ at 4000 K, plot $G(\xi)$ against $\xi$. Differentiate $G(\xi)$ with respect to $\xi$ and show that the minimum value of $G(\xi)$ occurs at $\xi_{eq} = 0.553$. Also show that

$$\left(\frac{\partial G}{\partial \xi}\right)_{T,P} = \Delta_r G^\circ + RT \ln \frac{P_{H_2}^2 P_{O_2}}{P_{H_2O}^2}$$

and that $K_P = \xi_{eq}^3/(2+\xi_{eq})(1-\xi_{eq})^2 = 0.333$.

**12-20.** Consider the reaction described by

$$3\,H_2(g) + N_2(g) \rightleftharpoons 2\,NH_3(g)$$

at 500 K and a total pressure of one bar. Suppose that we start with three moles of $H_2(g)$, one mole of $N_2(g)$, and no $NH_3(g)$. Show that

$$\begin{aligned} G(\xi) = {} & (3-3\xi)G_{H_2}^\circ + (1-\xi)G_{N_2}^\circ + 2\xi G_{NH_3}^\circ \\ & +(3-3\xi)RT\ln\frac{3-3\xi}{4-2\xi} + (1-\xi)RT\ln\frac{1-\xi}{4-2\xi} + 2\xi RT\ln\frac{2\xi}{4-2\xi} \end{aligned}$$

where $\xi$ is the extent of reaction. Given that $G_{NH_3}^\circ = 4.800\ kJ\cdot mol^{-1}$ at 500 K (see Table 12.4), plot $G(\xi)$ versus $\xi$. Differentiate $G(\xi)$ with respect to $\xi$ and show that the minimum value of $G(\xi)$ occurs at $\xi_{eq} = 0.158$. Also show that

$$\left(\frac{\partial G}{\partial \xi}\right)_{T,P} = \Delta_r G^\circ + RT \ln \frac{P_{NH_3}^2}{P_{H_2}^3 P_{N_2}}$$

and that $K_P = 16\xi_{eq}^2(2-\xi_{eq})^2/27(1-\xi_{eq})^4 = 0.10$.

**12-21.** Suppose that we have a mixture of the gases $H_2(g)$, $CO_2(g)$, $CO(g)$, and $H_2O(g)$ at 1260 K, with $P_{H_2} = 0.55$ bar, $P_{CO_2} = 0.20$ bar, $P_{CO} = 1.25$ bar, and $P_{H_2O} = 0.10$ bar. Is the reaction described by the equation

$$H_2(g) + CO_2(g) \rightleftharpoons CO(g) + H_2O(g) \qquad K_P = 1.59$$

at equilibrium under these conditions? If not, in what direction will the reaction proceed to attain equilibrium?

**12-22.** Given that $K_P = 2.21 \times 10^4$ at 25°C for the equation

$$2\,H_2(g) + CO(g) \rightleftharpoons CH_3OH(g)$$

predict the direction in which a reaction mixture for which $P_{CH_3OH} = 10.0$ bar, $P_{H_2} = 0.10$ bar, and $P_{CO} = 0.0050$ bar proceeds to attain equilibrium.

**12-23.** The value of $K_P$ at 500 K for a gas-phase reaction doubles when the temperature is increased from 300 K to 400 K at a fixed pressure. What is the value of $\Delta_r H^\circ$ for this reaction?

**12-24.** The value of $\Delta_r H^\circ$ is $34.78\ kJ\cdot mol^{-1}$ at 1000 K for the reaction described by

$$H_2(g) + CO_2(g) \rightleftharpoons CO(g) + H_2O(g)$$

Given that the value of $K_P$ is 0.236 at 800 K, estimate the value of $K_P$ at 1200 K, assuming that $\Delta_r H^\circ$ is independent of temperature.

**12-25.** The value of $\Delta_r H^\circ$ is $-12.93\ \text{kJ}\cdot\text{mol}^{-1}$ at 800 K for

$$H_2(g) + I_2(g) \rightleftharpoons 2\,HI(g)$$

Assuming that $\Delta_r H^\circ$ is independent of temperature, calculate $K_P$ at 700 K given that $K_P = 29.1$ at 1000 K.

**12-26.** The equilibrium constant for the reaction described by

$$2\,HBr(g) \rightleftharpoons H_2(g) + Br_2(g)$$

can be expressed by the empirical formula

$$\ln K = -6.375 + 0.6415\ln(T/\text{K}) - \frac{11790\ \text{K}}{T}$$

Use this formula to determine $\Delta_r H^\circ$ as a function of temperature. Calculate $\Delta_r H^\circ$ at 25°C and compare your result to the one you obtain from Table 5.2.

**12-27.** Use the following data for the reaction described by

$$2\,HI(g) \rightleftharpoons H_2(g) + I_2(g)$$

to obtain $\Delta_r H^\circ$ at 400°C.

| $T/\text{K}$ | 500 | 600 | 700 | 800 |
|---|---|---|---|---|
| $K_P/10^{-2}$ | 0.78 | 1.24 | 1.76 | 2.31 |

**12-28.** Consider the reaction described by

$$CO_2(g) + H_2(g) \rightleftharpoons CO(g) + H_2O(g)$$

The molar heat capacitites of $CO_2(g)$, $H_2(g)$, $CO(g)$, and $H_2O(g)$ can be expressed by

$$\overline{C}_P[CO_2(g)]/R = 3.127 + (5.231\times10^{-3}\ \text{K}^{-1})T - (1.784\times10^{-6}\ \text{K}^{-2})T^2$$

$$\overline{C}_P[H_2(g)]/R = 3.496 - (1.006\times10^{-4}\ \text{K}^{-1})T + (2.419\times10^{-7}\ \text{K}^{-2})T^2$$

$$\overline{C}_P[CO(g)]/R = 3.191 + (9.239\times10^{-4}\ \text{K}^{-1})T - (1.41\times10^{-7}\ \text{K}^{-2})T^2$$

$$\overline{C}_P[H_2O(g)]/R = 3.651 + (1.156\times10^{-3}\ \text{K}^{-1})T + (1.424\times10^{-7}\ \text{K}^{-2})T^2$$

over the temperature range 300 K to 1500 K. Given that

| substance | $CO_2(g)$ | $H_2(g)$ | $CO(g)$ | $H_2O(g)$ |
|---|---|---|---|---|
| $\Delta_f H^\circ/\text{kJ}\cdot\text{mol}^{-1}$ | $-393.523$ | 0 | $-110.516$ | $-241.844$ |

at 300 K and that $K_P = 0.695$ at 1000 K, derive a general expression for the variation of $K_P(T)$ with temperature in the form of Equation 12.34.

**12-29.** The temperature dependence of the equilibrium constant $K_P$ for the reaction described by

$$2\,C_3H_6(g) \rightleftharpoons C_2H_4(g) + C_4H_8(g)$$

is given by the equation

$$\ln K_P(T) = -2.395 - \frac{2505\text{ K}}{T} + \frac{3.477 \times 10^6\text{ K}^2}{T^2} \qquad 300\text{ K} < T < 600\text{ K}$$

Calculate the values of $\Delta_r G^\circ$, $\Delta_r H^\circ$, and $\Delta_r S^\circ$ for this reaction at 525 K.

**12-30.** At 2000 K and one bar, water vapor is 0.53% dissociated. At 2100 K and one bar, it is 0.88% dissociated. Calculate the value of $\Delta_r H^\circ$ for the dissociation of water at one bar, assuming that the enthalpy of reaction is constant over the range from 2000 K to 2100 K.

**12-31.** The following table gives the standard molar Gibbs energy of formation of Cl(g) at three different temperatures.

| $T$/K | 1000 | 2000 | 3000 |
|---|---|---|---|
| $\Delta_f G^\circ$/kJ·mol$^{-1}$ | 65.288 | 5.081 | −56.297 |

Use these data to determine the value of $K_P$ at each temperature for the reaction described by

$$\tfrac{1}{2}\,Cl_2(g) \rightleftharpoons Cl(g)$$

Assuming that $\Delta_r H^\circ$ is temperature independent, determine the value of $\Delta_r H^\circ$ from these data. Combine your results to determine $\Delta_r S^\circ$ at each temperature. Interpret your results.

**12-32.** The following experimental data were determined for the reaction described by

$$SO_3(g) \rightleftharpoons SO_2(g) + \tfrac{1}{2}\,O_2(g)$$

| $T$/K | 800 | 825 | 900 | 953 | 1000 |
|---|---|---|---|---|---|
| $\ln K_P$ | −3.263 | −3.007 | −1.899 | −1.173 | −0.591 |

Calculate $\Delta_r G^\circ$, $\Delta_r H^\circ$, and $\Delta_r S^\circ$ for this reaction at 900 K. State any assumptions that you make.

**12-33.** Show that

$$\mu = -RT \ln \frac{q(V,T)}{N}$$

if

$$Q(N,V,T) = \frac{[q(V,T)]^N}{N!}$$

**12-34.** Use Equation 12.40 to calculate $K(T)$ at 750 K for the reaction described by $H_2(g) + I_2(g) \rightleftharpoons 2\,HI(g)$. Use the molecular parameters given in Table 4.2. Compare your value to the one given in Table 12.2 and the experimental value shown in Figure 12.5.

**12-35.** Use the statistical thermodynamic formulas of Section 12–8 to calculate $K_P(T)$ at 900 K, 1000 K, 1100 K, and 1200 K for the association of Na(g) to form dimers, $Na_2(g)$ according to the equation

$$2\,Na(g) \rightleftharpoons Na_2(g)$$

Use your result at 1000 K to calculate the fraction of sodium atoms that form dimers at a total pressure of one bar. The experimental values of $K_P(T)$ are

| $T/K$ | 900 | 1000 | 1100 | 1200 |
|---|---|---|---|---|
| $K_P$ | 1.32 | 0.47 | 0.21 | 0.10 |

Plot $\ln K_P$ against $1/T$ to determine the value of $\Delta_r H^\circ$.

**12-36.** Using the data in Table 4.2, calculate $K_P$ at 2000 K for the reaction described by the equation

$$CO_2(g) \rightleftharpoons CO(g) + \tfrac{1}{2}\,O_2(g)$$

The experimental value is $1.3 \times 10^{-3}$.

**12-37.** Using the data in Tables 4.2 and 4.4, calculate the equilibrium constant for the water gas reaction

$$CO_2(g) + H_2(g) \rightleftharpoons CO(g) + H_2O(g)$$

at 900 K and 1200 K. The experimental values at these two temperatures are 0.43 and 1.37, respectively.

**12-38.** Using the data in Tables 4.2 and 4.4, calculate the equilibrium constant for the reaction

$$3\,H_2(g) + N_2(g) \rightleftharpoons 2\,NH_3(g)$$

at 700 K. The accepted value is $8.75 \times 10^{-5}$ (see Table 12.4).

**12-39.** Calculate the equilibrium constant $K_P$ for the reaction

$$I_2(g) \rightleftharpoons 2\,I(g)$$

using the data in Table 4.2 and the fact that the degeneracy of the ground electronic state of an iodine atom is 4 and that the degeneracy of its first excited electronic state is 2 and that its energy is 7580 $cm^{-1}$. The experimental values of $K_P$ are

| $T/K$ | 800 | 900 | 1000 | 1100 | 1200 |
|---|---|---|---|---|---|
| $K_P$ | $3.05 \times 10^{-5}$ | $3.94 \times 10^{-4}$ | $3.08 \times 10^{-3}$ | $1.66 \times 10^{-2}$ | $6.79 \times 10^{-2}$ |

Plot $\ln K_p$ against $1/T$ to determine the value of $\Delta_r H^\circ$. The experimental value is 153.8 $kJ \cdot mol^{-1}$.

**12-40.** Consider the reaction given by

$$H_2(g) + D_2(g) \rightleftharpoons 2\,HD(g)$$

Using the Born-Oppenheimer approximation and the molecular parameters in Table 4.2, show that

$$K(T) = 4.24e^{-77.7\ \mathrm{K}/T}$$

Compare your predictions using this equation to the data in the JANAF tables.

**12-41.** Using the harmonic oscillator-rigid rotator approximation, show that

$$K(T) = \left(\frac{m_{\mathrm{H_2}} m_{\mathrm{Br_2}}}{m_{\mathrm{HBr}}^2}\right)^{3/2} \left(\frac{\sigma_{\mathrm{HBr}}^2}{\sigma_{\mathrm{H_2}}\sigma_{\mathrm{Br_2}}}\right) \left(\frac{(\Theta_{\mathrm{rot}}^{\mathrm{HBr}})^2}{\Theta_{\mathrm{rot}}^{\mathrm{H_2}}\Theta_{\mathrm{rot}}^{\mathrm{Br_2}}}\right)$$

$$\times \frac{(1 - e^{-\Theta_{\mathrm{vib}}^{\mathrm{HBr}}/T})^2}{(1 - e^{-\Theta_{\mathrm{vib}}^{\mathrm{H_2}}/T})(1 - e^{-\Theta_{\mathrm{vib}}^{\mathrm{Br_2}}/T})} e^{(D_0^{\mathrm{H_2}} + D_0^{\mathrm{Br_2}} - 2D_0^{\mathrm{HBr}})/RT}$$

for the reaction described by

$$2\,\mathrm{HBr(g)} \rightleftharpoons \mathrm{H_2(g)} + \mathrm{Br_2(g)}$$

Using the values of $\Theta_{\mathrm{rot}}$, $\Theta_{\mathrm{vib}}$, and $D_0$ given in Table 4.2, calculate $K$ at 500 K, 1000 K, 1500 K, and 2000 K. Plot $\ln K$ against $1/T$ and determine the value of $\Delta_r H^\circ$.

**12-42.** Use Equation 12.49*b* to calculate $H^\circ(T) - H_0^\circ$ for $NH_3$(g) from 300 K to 6000 K and compare your values to those given in Table 12.4 by plotting them on the same graph.

**12-43.** Use the JANAF tables to calculate $K_P$ at 1000 K for the reaction described by

$$\mathrm{H_2(g)} + \mathrm{I_2(g)} \rightleftharpoons 2\,\mathrm{HI(g)}$$

Compare your results to the value given in Table 12.2.

**12-44.** Use the JANAF tables to plot $\ln K_P$ versus $1/T$ from 900 K to 1200 K for the reaction described by

$$2\,\mathrm{Na(g)} \rightleftharpoons \mathrm{Na_2(g)}$$

and compare your results to those obtained in Problem 12–35.

**12-45.** In Problem 12–36 we calculated $K_P$ for the decomposition of $CO_2$(g) to CO(g) and $O_2$(g) at 2000 K. Use the JANAF tables to calculate $K_P$ and compare your result to the one that you obtained in Problem 12–36.

**12-46.** You calculated $K_P$ at 700 K for the ammonia synthesis reaction in Problem 12–38. Use the data in Table 12.4 to calculate $K_P$ and compare your result to the one that you obtained in Problem 12–38.

**12-47.** The JANAF tables give the following data for I(g) at one bar:

| $T/\mathrm{K}$ | 800 | 900 | 1000 | 1100 | 1200 |
|---|---|---|---|---|---|
| $\Delta_f G^\circ/\mathrm{kJ\cdot mol^{-1}}$ | 34.580 | 29.039 | 24.039 | 18.741 | 13.428 |

Calculate $K_P$ for the reaction described by

$$I_2(g) \rightleftharpoons 2\,I(g)$$

and compare your results to the values given in Problem 12–39.

**12-48.** Use Equation 18.60 to calculate the value of $q^0(V, T)/V$ given in the text (page 504) for $NH_3(g)$ at 500 K.

**12-49.** The JANAF tables give the following data for Ar(g) at 298.15 K and one bar:

$$-\frac{G^\circ - H^\circ(298.15\text{ K})}{T} = 154.845\ \text{J}\cdot\text{mol}^{-1}\cdot\text{K}^{-1}$$

and

$$H^\circ(0\text{ K}) - H^\circ(298.15\text{ K}) = -6.197\ \text{kJ}\cdot\text{mol}^{-1}$$

Use these data to calculate $q^0(V, T)/V$ and compare your result to what you obtain using Equation 4.13.

**12-50.** Use the JANAF tables to calculate $q^0(V, T)/V$ for $CO_2(g)$ at 500 K and one bar and compare your result to what you obtain using Equation 4.57 (with the ground state energy taken to be zero).

**12-51.** Use the JANAF tables to calculate $q^0(V, T)/V$ for $CH_4(g)$ at 1000 K and one bar and compare your result to what you obtain using Equation 4.60 (with the ground state energy taken to be zero).

**12-52.** Use the JANAF tables to calculate $q^0(V, T)/V$ for $H_2O(g)$ at 1500 K and one bar and compare your result to what you obtain using Equation 12.45. Why do you think there is some discrepancy?

**12-53.** The JANAF tables give the following data:

| | H(g) | Cl(g) | HCl(g) |
|---|---|---|---|
| $\Delta_f H^\circ(0\text{ K})/\text{kJ}\cdot\text{mol}^{-1}$ | 216.035 | 119.621 | −92.127 |

Use these data to calculate $D_0$ for HCl(g) and compare your value to the one in Table 4.2.

**12-54.** The JANAF tables give the following data:

| | C(g) | H(g) | $CH_4$(g) |
|---|---|---|---|
| $\Delta_f H^\circ(0\text{ K})/\text{kJ}\cdot\text{mol}^{-1}$ | 711.19 | 216.035 | −66.911 |

Use these data to calculate $D_0$ for $CH_4(g)$ and compare your value to the one in Table 4.4.

**12-55.** Use the JANAF tables to calculate $D_0$ for $CO_2(g)$ and compare your result to the one given in Table 4.4.

**12-56.** A determination of $K_\gamma$ (see Example 12–11) requires a knowledge of the fugacity of each gas in the equilibrium mixture. These data are not usually available, but a useful approximation is to take the fugacity coefficient of a gaseous constituent of a mixture to be equal to the value for the pure gas at the *total pressure of the mixture*. Using this approximation, we can use Figure 8.11 to determine $\gamma$ for each gas and then calculate $K_\gamma$. In this problem we shall apply this approximation to the data in Table 12.5. First use Figure 8.11 to estimate that $\gamma_{H_2} = 1.05$, $\gamma_{N_2} = 1.05$, and that $\gamma_{NH_3} = 0.95$ at a total pressure of 100 bar and a temperature of 450°C. In this case $K_\gamma = 0.86$, in fairly good agreement with the value given in Example 12–11. Now calculate $K_\gamma$ at 600 bar and compare your result with the value given in Example 12–11.

**12-57.** Recall from general chemistry that Le Châtelier's principle says that pressure has no effect on a gaseous equilibrium system such as

$$CO(g) + H_2O(g) \rightleftharpoons H_2(g) + CO_2(g)$$

in which the total number of moles of reactants is equal to the total number of moles of product in the chemical equation. The thermodynamic equilibrium constant in this case is

$$K_f = \frac{f_{CO_2} f_{H_2}}{f_{CO} f_{H_2O}} = \frac{\gamma_{CO_2}\gamma_{H_2}}{\gamma_{CO}\gamma_{H_2O}} \frac{P_{CO_2} P_{H_2}}{P_{CO} P_{H_2O}} = K_\gamma K_P$$

If the four gases behaved ideally, then pressure would have no effect on the position of equilibrium. However, because of deviations from ideal behavior, a shift in the equilibrium composition will occur when the pressure is changed. To see this, use the approximation introduced in Problem 12–56 to estimate $K_\gamma$ at 900 K and 500 bar. Note that $K_\gamma$ under these conditions is greater than $K_\gamma$ at one bar, where $K_\gamma \approx 1$ (ideal behavior). Consequently, argue that an increase in pressure causes the equilibrium to shift to the left in this case.

**12-58.** Calculate the activity of $H_2O(l)$ as a function of pressure from one bar to 100 bar at 20.0°C. Take the density of $H_2O(l)$ to be $0.9982\ g\cdot mL^{-1}$ and assume that it is incompressible.

**12-59.** Consider the dissociation of HgO(s,red) to Hg(g) and $O_2(g)$ according to

$$HgO(s, red) \rightleftharpoons Hg(g) + \tfrac{1}{2} O_2(g)$$

If we start with only HgO(s,red), then assuming ideal behavior, show that

$$K_P = \frac{2}{3^{3/2}} P^{3/2}$$

where $P$ is the total pressure. Given the following "dissociation pressure" of HgO(s,red) at various temperatures, plot $\ln K_P$ versus $1/T$.

| $t/°C$ | $P/atm$ | $t/°C$ | $P/atm$ |
|---|---|---|---|
| 360 | 0.1185 | 430 | 0.6550 |
| 370 | 0.1422 | 440 | 0.8450 |
| 380 | 0.1858 | 450 | 1.067 |
| 390 | 0.2370 | 460 | 1.339 |
| 400 | 0.3040 | 470 | 1.674 |
| 410 | 0.3990 | 480 | 2.081 |
| 420 | 0.5095 | | |

An excellent curve fit to the plot of $\ln K_P$ against $1/T$ is given by

$$\ln K_P = -172.94 + \frac{4.0222 \times 10^5 \text{ K}}{T} - \frac{2.9839 \times 10^8 \text{ K}^2}{T^2} + \frac{7.0527 \times 10^{10} \text{ K}^3}{T^3}$$

$$630 \text{ K} < T < 750 \text{ K}$$

Use this expression to determine $\Delta_r H^\circ$ as a function of temperature in the interval 630 K < $T$ < 750 K. Given that

$$C_P^\circ[\text{O}_2(\text{g})]/R = 4.8919 - \frac{829.931 \text{ K}}{T} - \frac{127962 \text{ K}^2}{T^2}$$

$$C_P^\circ[\text{Hg(g)}]/R = 2.500$$

$$C_P^\circ[\text{HgO(s, red)}]/R = 5.2995$$

in the interval 298 K < $T$ < 750 K, calculate $\Delta_r H^\circ$, $\Delta_r S^\circ$, and $\Delta_r G^\circ$ at 298 K.

**12-60.** Consider the dissociation of $Ag_2O(s)$ to Ag(s) and $O_2(g)$ according to

$$\text{Ag}_2\text{O(s)} \rightleftharpoons 2\,\text{Ag(s)} + \tfrac{1}{2}\,\text{O}_2(\text{g})$$

Given the following "dissociation pressure" data:

| $t/°\text{C}$ | 173 | 178 | 183 | 188 |
|---|---|---|---|---|
| $P/\text{torr}$ | 422 | 509 | 605 | 717 |

Express $K_P$ in terms of $P$ (in torr) and plot $\ln K_P$ versus $1/T$. An excelllent curve fit to these data is given by

$$\ln K_P = 0.9692 + \frac{5612.7 \text{ K}}{T} - \frac{2.0953 \times 10^6 \text{ K}^2}{T^2}$$

Use this expression to derive an equation for $\Delta_r H^\circ$ from 445 K < $T$ < 460 K. Now use the following heat capacity data:

$$C_P^\circ[\text{O}_2(\text{g})]/R = 3.27 + (5.03 \times 10^{-4} \text{ K}^{-1})T$$
$$C_P^\circ[\text{Ag(s)}]/R = 2.82 + (7.55 \times 10^{-4} \text{ K}^{-1})T$$
$$C_P^\circ[\text{Ag}_2\text{O(s)}]/R = 6.98 + (4.48 \times 10^{-3} \text{ K}^{-1})T$$

to calculate $\Delta_r H^\circ$, $\Delta_r S^\circ$, and $\Delta_r G^\circ$ at 298 K.

**12-61.** Calcium carbonate occurs as two crystalline forms, calcite and aragonite. The value of $\Delta_r G^\circ$ for the transition

$$\text{CaCO}_3(\text{calcite}) \rightleftharpoons \text{CaCO}_3(\text{aragonite})$$

is $+1.04 \text{ kJ}\cdot\text{mol}^{-1}$ at 25°C. The density of calcite at 25°C is $2.710 \text{ g}\cdot\text{cm}^{-3}$ and that of aragonite is $2.930 \text{ g}\cdot\text{cm}^{-3}$. At what pressure will these two forms of $CaCO_3$ be at equilbrium at 25°C.

**12-62.** The decomposition of ammonium carbamate, $NH_2COONH_4$ takes place according to

$$\text{NH}_2\text{COONH}_4(\text{s}) \rightleftharpoons 2\,\text{NH}_3(\text{g}) + \text{CO}_2(\text{g})$$

Show that if all the $NH_3$(g) and $CO_2$(g) result from the decomposition of ammonium carbamate, then $K_P = (4/27)P^3$, where $P$ is the total pressure at equilibrium.

**12-63.** Calculate the solubility of LiF(s) in water at 25°C. Compare your result to the one you obtain by using concentrations instead of activities. Take $K_{sp} = 1.7 \times 10^{-3}$.

**12-64.** Calculate the solubility of $CaF_2$(s) in a solution that is 0.0150 molar in $MgSO_4$(aq). Take $K_{sp} = 3.9 \times 10^{-11}$ for $CaF_2$(s).

**12-65.** Calculate the solubility of $CaF_2$(s) in a solution that is 0.050-molar in NaF(aq). Compare your result to the one you obtain by using concentrations instead of activities. Take $K_{sp} = 3.9 \times 10^{-11}$ for $CaF_2$(s).

**Gilbert Newton Lewis** was born in West Newton, Massachusetts, on October 25, 1875, and died in 1946. In 1899, he received his Ph.D. from Harvard University, and after spending a year studying in Germany, he returned to Harvard as an instructor. Lewis left Harvard in 1904 to become Superintendent of Weights and Measures in the Philippines, and a year later he moved to The Massachusetts Institute of Technology. In 1912, he accepted the position of Dean of the College of Chemistry at the University of California at Berkeley, which he developed into one of the finest teaching and research departments in the world. He remained at Berkeley for the rest of his life, suffering a fatal heart attack in his laboratory. Lewis was one of America's outstanding chemists, certainly the finest not to receive a Nobel Prize. Lewis made many important contributions in chemistry. In the 1920s, he introduced Lewis formulas and described a covalent bond (which he named) as a shared pair of electrons. His work or the application of thermodynamics to physical chemistry culminated in his outstanding 1923 text, coauthored with Merle Randall, *Thermodynamics and the Free Energy of Chemical Substances*, from which a generation of chemists learned thermodynamics. Lewis was a dynamic individual who was responsible for development of many outstanding chemists, several of whom became members of the Berkeley faculty. His department produced a remarkable number of Nobel Prize winners.

CHAPTER 13

# Thermodynamics of Electrochemical Cells

Batteries produce electricity because the oxidizing agent and the reducing agent in an oxidation-reduction reaction are arranged such that the electrons that are transferred from one reactant to the other are forced to travel through an external circuit. The basic experimental setup that realizes such an electron transfer is called an electrochemical cell. In this chapter, we will see that electrochemical cells have a number of useful thermodynamic applications. Under readily attainable experimental conditions, the voltage that a cell produces can be directly related to the Gibbs energy change of the underlying oxidation-reduction reaction. This fundamental relation provides us with one of the most convenient and accurate methods for determining activity coefficients of electrolyte solutions. Using the relation between the equilibrium constant and the standard Gibbs energy change of a reaction, we can also use electrochemical cell measurements to determine solubility products and acid-dissociation constants. We conclude the chapter with a discussion of batteries and fuel cells.

## 13–1. An Electrochemical Cell Produces an Electric Current as the Result of a Chemical Reaction

In this section, we will show how a spontaneous chemical reaction can produce an electric current. Consider the reaction

$$\mathrm{Zn(s)} + \mathrm{Cu^{2+}(aq)} \longrightarrow \mathrm{Cu(s)} + \mathrm{Zn^{2+}(aq)} \tag{13.1}$$

If we place a zinc rod into a solution of $CuSO_4$(aq), for example, electrons will be transferred from the zinc atoms to the copper(II) ions, producing copper atoms and zinc ions in solution. No electric current is produced by this system because the reactants are in direct contact; the system is short-circuited. If we can somehow keep the reactants separated, however, we can make electrons from the zinc atoms travel through a wire to reach the copper(II) ions. A setup that allows this to be done is called an *electrochemical*

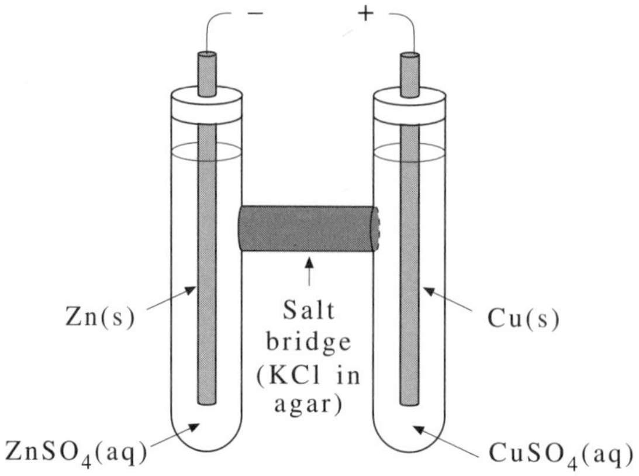


**FIGURE 13.1**
An illustration of a zinc-copper electrochemical cell. (This cell is called a *Daniel cell.*) The equation for the reaction is $Zn(s) + Cu^{2+}(aq) \rightarrow Zn^{2+}(aq) + Cu(s)$. The reaction is carried out such that the electrons are transferred from the zinc to the copper through an external circuit.

*cell.* An electrochemical cell for Equation 13.1 is shown in Figure 13.1. The cell consists of rods of zinc and copper (the *electrodes*), each immersed in an aqueous solution containing their respective ions, $Zn^{2+}$(aq) and $Cu^{2+}$(aq). When the reaction described by Equation 13.1 proceeds spontaneously, a zinc atom in the zinc electrode gives up two electrons to the external circuit and enters the solution as $Zn^{2+}$(aq). The spontaneity of the reaction drives the electrons through the external circuit to the copper electrode, where they are picked up by a $Cu^{2+}$(aq), which deposits on the electrode as a copper atom. Unless the two solutions are connected electrically in some manner, the reaction quickly "fizzles out" because of a separation of uncompensated positive charge resulting from the $Zn^{2+}$(aq) being produced in the $ZnSO_4$(aq) solution and uncompensated negative charge resulting from the $Cu^{2+}$(aq) leaving the $CuSO_4$(aq) solution. The two solutions can be connected electrically by a *salt bridge*, which consists of a saturated KCl(aq) solution mixed with agar, a substance that forms a gel similar to gelatin. The purpose of the gel is to prevent the two solutions from mixing while still permitting the passage of an electric current carried by the $K^+$(aq) and $Cl^-$(aq) ions. The salt bridge therefore provides an ionic current pathway between the $ZnSO_4$(aq) and $CuSO_4$(aq) solutions. As a $Zn^{2+}$(aq) ion enters the $ZnSO_4$(aq) solution from the Zn(s) electrode, two chloride ions pass from the salt bridge into the solution. Meanwhile, as a $Cu^{2+}$(aq) ion leaves the $CuSO_4$(aq) solution and deposits on the Cu(s) electrode, two $K^+$(aq) ions enter the solution.

In general, an electrode is any solid on whose surface oxidation-reduction reactions take place. The electrode at which reduction occurs [$Cu^{2+}(aq) + 2\,e^- \rightarrow Cu(s)$] is called the *cathode*; the electrode at which oxidation takes place [$Zn(s) \rightarrow Zn^{2+}(aq) +$

$2\,e^-$] is called the *anode*. These definitions are conveniently remembered with the aid of the mnemonic device

$$\text{consonants}\begin{cases}\textit{c}\text{athode}\\ \textit{r}\text{eduction}\end{cases}\qquad\text{vowels}\begin{cases}\textit{a}\text{node}\\ \textit{o}\text{xidation}\end{cases}$$

## 13–2. Half Cells Can Be Classified into Various Types

All electrochemical cells consist of two half cells. One of the simplest types of half cells consists of a metal electrode in contact with a solution of its own ions. Examples of this type of electrode are the $Zn^{+2}(aq)/Zn(s)$ electrode and the $Cu^{+2}(aq)/Cu(s)$ electrode that make up the Daniel cell (Figure 13.1). The equations of the corresponding half-cell reduction reactions are

$$Zn^{2+}(aq) + 2\,e^- \rightleftharpoons Zn(s)$$

and

$$Cu^{2+}(aq) + 2\,e^- \rightleftharpoons Cu(s)$$

The reactions may be made to go in either direction with a small change in voltage about its equilibrium value. Electrodes with this property are said to be *reversible*. In order to relate observed cell voltages to thermodynamic quantities, we will require that the cells that we use consist of reversible electrodes.

Another type of reversible electrode consists of a metal [such as Ag(s)] and a sparingly soluble salt of the metal [such as AgI(s)] in contact with a solution of a soluble salt consisting of the anion of the sparingly soluble salt [such as KI(aq)]. Examples of this type of electrode are the Ag(s)|AgCl(s)|HCl(aq) electrode and the $Pb(s)|PbSO_4(s)|H_2SO_4(aq)$ electrode. The (oxidation) reaction that occurs at the Ag(s)|AgCl(s)|HCl(aq) electrode may be thought of as taking place in two steps. The first step is

$$Ag(s) \rightleftharpoons Ag^+(aq) + e^-$$

which is followed by

$$Ag^+(aq) + Cl^-(aq) \rightleftharpoons AgCl(s)$$

to form the sparingly soluble salt. The net equation for the electrode reaction is

$$Ag(s) + Cl^-(aq) \rightleftharpoons AgCl(s) + e^-$$

Figure 13.2 is a schematic illustration of a silver-silver chloride electrode.

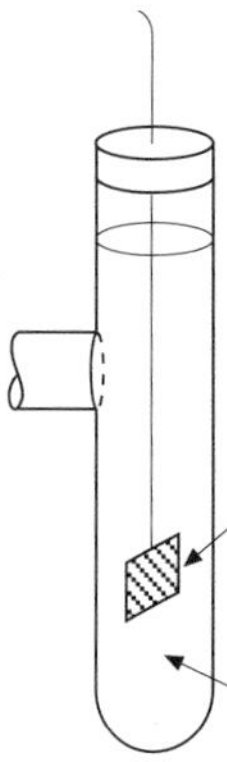


**FIGURE 13.2**
A schematic illustration of a silver-silver chloride electrode. The electrode consists of metallic silver in contact with silver chloride immersed in a solution of hydrochloric acid.

**EXAMPLE 13–1**
Write the equation for the electrode (oxidation) reaction that occurs in the $Pb(s)|PbSO_4(s)|H_2SO_4(aq)$ electrode.

SOLUTION: We may picture the reaction as occurring in two steps, as we did for the reaction that occurs at the Ag(s)|AgCl(s)|HCl(aq) electrode. The first step is

$$Pb(s) \rightleftharpoons Pb^{2+}(aq) + 2\,e^-$$

followed by the formation of the sparingly soluble $PbSO_4(s)$

$$Pb^{2+}(aq) + SO_4^{2-}(aq) \rightleftharpoons PbSO_4(s)$$

The net equation for the electrode reaction is

$$Pb(s) + SO_4^{2-}(aq) \rightleftharpoons PbSO_4(s) + 2\,e^-$$

A third type of reversible electrode is a gas electrode. In a gas electrode, a wire or a strip of a nonreactive metal such as platinum is immersed in a solution through which a gas is bubbled. The solute of the solution and the gas have a common atom. The classic example of a gas electrode is the hydrogen electrode, $Pt(s)|H_2(g)|H^+(aq)$ (Figure 13.3). The equation for the overall (oxidation) reaction that occurs at this electrode is

$$\tfrac{1}{2}\,H_2(g) \rightleftharpoons H^+(aq) + e^-$$

A chlorine electrode works in a similar manner.

$$\tfrac{1}{2}\,Cl_2(g) + e^- \rightleftharpoons Cl^-(aq)$$

One other type of reversible electrode is an oxidation-reduction electrode. Strictly speaking, all electrodes are oxidation-reduction electrodes, but the term is reserved for electrodes that consist of an inert metal immersed in a solution containing both

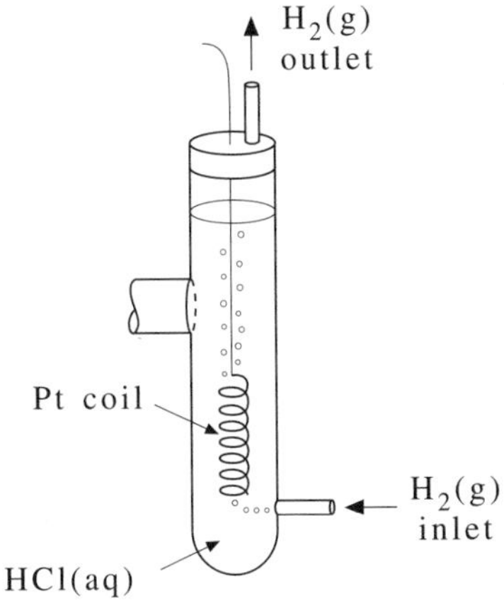


**FIGURE 13.3**
A schematic illustration of a hydrogen gas electrode. The electrode consists of an inert metal such as platinum immersed in an acid solution such as hydrochloric acid. Hydrogen gas is bubbled into the solution over the electrode.

the oxidized and reduced states of an oxidation-reduction couple. Examples of such electrodes involve the reactions

$$\mathrm{Tl^{+}(aq)} \rightleftharpoons \mathrm{Tl^{3+}(aq)} + 2\,e^{-}$$

and

$$\mathrm{Fe^{2+}(aq)} \rightleftharpoons \mathrm{Fe^{3+}(aq)} + e^{-}$$

The four types of electrodes discussed here do not include all possible electrodes, but most electrodes do fall into one of the above categories.

## **13–3.** A Cell Diagram Is Used to Represent an Electrochemical Cell

Electrochemical cells are often described by means of a *cell diagram*. For example, the cell diagram for the electrochemical cell shown in Figure 13.1 is

$$\mathrm{Zn(s)|ZnSO_4(aq)||CuSO_4(aq)|Cu(s)} \tag{13.2}$$

The single vertical bars indicate boundaries of phases that are in contact, and the double vertical bars indicate a salt bridge. Thus, in the cell represented by the above cell diagram, Zn(s) and $ZnSO_4$(aq) are separate phases in physical contact, as are Cu(s) and $CuSO_4$(aq), and a salt bridge connects the $ZnSO_4$(aq) and $CuSO_4$(aq) solutions.

The convention that allows us to deduce the chemical equation corresponding to the cell reaction from the cell diagram is that the reaction occurring at the left-hand electrode in the cell diagram is written as an oxidation reaction and that occurring at the right-hand electrode is written as a reduction reaction. This convention enables us to write the equation for the cell reaction unambigously. For the above cell diagram, then, we have

$$\mathrm{Zn(s)} \rightleftharpoons \mathrm{Zn^{2+}(aq)} + 2\,e^- \quad \text{(oxidation at the left-hand electrode)}$$

$$\mathrm{Cu^{2+}(aq)} + 2\,e^- \rightleftharpoons \mathrm{Cu(s)} \quad (\textit{r}\text{eduction at the } \textit{r}\text{ight-hand electrode})$$

This convention is easy to remember because *r*eduction and *r*ight both begin with *r*. The chemical equation corresponding to the cell reaction is the sum of the two electrode reactions

$$\mathrm{Zn(s)} + \mathrm{Cu^{2+}(aq)} \rightleftharpoons \mathrm{Zn^{2+}(aq)} + \mathrm{Cu(s)}$$

which is Equation 13.1.

**EXAMPLE 13–2**

Write the equations for the electrode reactions and the net cell reaction for the electrochemical cell

$$\mathrm{Zn(s)|Zn^{2+}(aq)||H^+(aq)|H_2(g)|Pt(s)}$$

(An illustration of this cell is shown in Figure 13.4.)

SOLUTION: Oxidation takes place at the left electrode. Thus, the equation for the reaction at the left electrode is

$$\mathrm{Zn(s)} \longrightarrow \mathrm{Zn^{2+}(aq)} + 2\,e^-$$

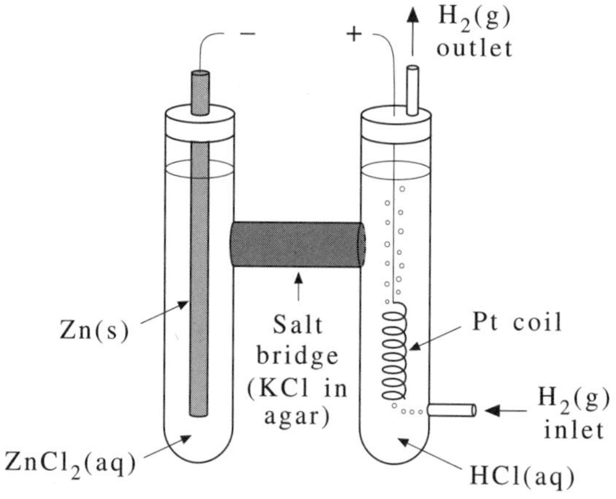


**FIGURE 13.4**
A zinc-hydrogen gas electrochemical cell. Note the H-type geometry of the cell, which holds the salt bridge and separates the two electrolytic solutions.

Reduction takes place at the right electrode, and the equation for the reaction is

$$2\,\text{H}^+(\text{aq}) + 2\,e^- \longrightarrow \text{H}_2(\text{g})$$

The equation for the net cell reaction is

$$\text{Zn(s)} + 2\,\text{H}^+(\text{aq}) \longrightarrow \text{Zn}^{2+}(\text{aq}) + \text{H}_2(\text{g})$$

**EXAMPLE 13–3**

Write the equations for the electrode reactions and the net cell reaction for the electrochemical cell

$$\text{Cd(s)}|\text{CdSO}_4(\text{aq})|\text{Hg}_2\text{SO}_4(\text{s})|\text{Hg(l)}$$

Note that this cell does not have a salt bridge because it has only one electrolytic solution.

SOLUTION: Oxidation takes place at the left electrode. Oxidation of Cd(s) yields $\text{Cd}^{2+}(\text{aq})$, so we write

$$\text{Cd(s)} \longrightarrow \text{Cd}^{2+}(\text{aq}) + 2\,e^-$$

The (reduction) reaction at the right electrode is

$$\text{Hg}_2\text{SO}_4(\text{s}) + 2\,e^- \longrightarrow 2\,\text{Hg(l)} + \text{SO}_4^{2-}(\text{aq})$$

The net cell reaction is

$$\text{Cd(s)} + \text{Hg}_2\text{SO}_4(\text{s}) \longrightarrow 2\,\text{Hg(l)} + \text{Cd}^{2+}(\text{aq}) + \text{SO}_4^{2-}(\text{aq})$$

The cell described here is very similar to the *Weston standard cell* (see Figure 13.5), which has been used as a source of accurately known voltage. The voltage of the cell is 1.018 V and is essentially independent of temperature, a very convenient feature for a voltage reference.

With the electrode convention we have adopted, electrons flow through the external circuit from left to right in the cell diagram if the reaction occurs spontaneously. The measured cell voltage is equal to the electrical potential of the right electrode minus that of the left electrode, or

$$\Delta V = V_\text{R} - V_\text{L} \tag{13.3}$$

Realize when we write $V_\text{R} - V_\text{L}$ here, we are referring to the cell diagram and not to the spatial arrangement of the cell. Because the reaction described by Equation 13.1 occurs spontaneously (unless the ratio of the concentrations of zinc ions to copper ions is extremely small), the left electrode (as written in the cell diagram) will take on

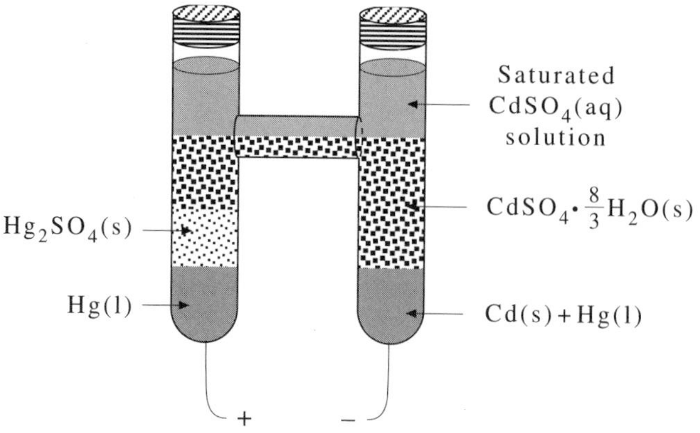


**FIGURE 13.5**
A Weston standard cell. The cell diagram is $Cd(s)|CdSO_4(aq)|Hg_2SO_4(s)|Hg(l)$ independent of the orientation of the cell in the figure.

electrons and pass them through the external circuit to the copper electrode, which will then pass them on to the copper ions and reduce them to metallic copper. We say, then, that the potential of the left electrode is negative with respect to the right electrode. Note that in this case, $V_R > V_L$, so the cell voltage, $\Delta V$, is positive.

A key property of a cell that allows us to apply thermodynamics to electrochemistry is the *electromotive force* (emf), $E$, which is defined by

$$E = (\Delta V)_{I=0} \tag{13.4}$$

Equation 13.4 says that the emf of a cell is the potential difference of the cell measured under the condition of no flow of electric current.

## 13–4. The Gibbs Energy Change of a Cell Reaction Is Directly Related to the Electromotive Force of the Cell

So far, we have not introduced thermodynamics into our discussion of electrochemical cells. Such an introduction must be done with some thought because the equations of thermodynamics, as we have developed them, apply only to systems in equilibrium or to reversible processes. This brings us to an important concept of electrochemistry, namely, that of a *reversible cell.*

The emf of a cell can be measured with a potentiometer, an instrument that can be used to measure an electric current under conditions in which no current is allowed to pass through the cell (see Equation 13.4). The potentiometer can be adjusted continuously such that a current can pass through the cell in either direction. If the emf (or strictly speaking, the potential difference, because $I$ is no longer exactly zero) changes only slightly when the current changes from one direction to the other, the cell is said to be reversible. The condition of reversibility can be expressed succinctly by the mathematical expression,

$$\Delta V(I) \longrightarrow E \quad \text{as} \quad I \longrightarrow 0 \pm \epsilon \quad \epsilon \longrightarrow 0 \tag{13.5}$$

If a cell is reversible, the chemical reaction occurring within the cell can proceed in either direction, depending upon the flow of current, and at the null point, the driving force of the reaction is exactly balanced by the emf of the potentiometer.

Because $E$ is associated with a balance point of the reaction, where it will go one direction or the other with small changes about $E$, one might suspect that $E$ is related to the change in Gibbs energy for the reaction. Furthermore, if $E > 0$, the cell reaction proceeds spontaneously and if $E < 0$, the reverse reaction does. This concept follows from Equations 13.3 and 13.4. For the cell diagram given by Equation 13.2, $E > 0$. Clearly, then, the relation between $E$ and $\Delta G$ must involve a negative sign because $\Delta G < 0$ for the reaction associated with the cell diagram.

Let's see just what the relation between $E$ and $\Delta G$ is. Consider a cell in which $n$ moles of electrons are transferred from the left electrode to the right electrode. The magnitude of the charge of one mole of electrons is given by the product of the magnitude of the charge on an electron ($1.6022 \times 10^{-19}$ C) and the Avogadro constant ($6.0221 \times 10^{23}\ \text{mol}^{-1}$), and is equal to 96 485 $\text{C}\cdot\text{mol}^{-1}$. This quantity is called the *Faraday constant* and is denoted by $F$. Thus, if $n$ moles of electrons pass through the external circuit, then the total electric charge passing through the external circuit is $nF$. If $\Delta V > 0$, then the electrons will flow from left to right (by our convention), and an amount of work equal to $nF\Delta V$ will be done by the cell. If this work is done reversibly, $\Delta V$ is replaced by $E$, and the electrical work done by the cell will be $nFE$. Note that if $E > 0$, the cell does work or is able to do work. According to Equation 8.16

$$\Delta G = -w_{\text{non}PV}$$

which gives us that

$$\Delta G = -nFE \tag{13.6}$$

which is the desired relation between $\Delta G$ and $E$, and is the central equation of the thermodynamics of electrochemical cells. As we will see, electrochemical measurements can be used to determine thermodynamic data of ionic reactions that occur in cells.

The relation $\Delta G = -nFE$ tells us that the emf of a cell depends upon the concentrations, or more precisely, upon the activities, of the reactants and products in the cell reaction. For the general chemical reaction described by

$$a\,\text{A} + b\,\text{B} \rightleftharpoons y\,\text{Y} + z\,\text{Z}$$

$\Delta G$ is given by (Equation 12.60)

$$\Delta G = \Delta G^\circ(T, P) + RT \ln \frac{a_\text{Y}^{\nu_\text{Y}} a_\text{Z}^{\nu_\text{Z}}}{a_\text{A}^{\nu_\text{A}} a_\text{B}^{\nu_\text{B}}}$$

Here $\Delta G^\circ = \nu_Y\mu_Y^\circ + \nu_Z\mu_Z^\circ - \nu_A\mu_A^\circ - \nu_B\mu_B^\circ$, where the $\mu_j^\circ$ are the chemical potentials in some appropriate and convenient standard state. Now because $\Delta G = -nFE$,

$$E = E^\circ - \frac{RT}{nF}\ln\frac{a_Y^{\nu_Y}a_Z^{\nu_Z}}{a_A^{\nu_A}a_B^{\nu_B}} \tag{13.7}$$

Equation 13.7 is called the *Nernst equation* and shows how the emf of a cell depends upon the activities, or the concentrations, of the species participating in the cell reaction. The quantity, $E^\circ$, called the *standard emf* of the cell, is the emf of the cell when the activities of the products and reactants are equal to unity.

## 13–5. The Standard Emf of an Electrochemical Cell Can Be Found by Extrapolation

Because we can calculate the value of $E$ for arbitrary activities once we know $E^\circ$ for a cell, it is important for us to be able to determine the value of $E^\circ$. The following calculation illustrates a standard procedure for doing this. Consider the cell pictured in Figure 13.6. The cell diagram for this cell is

$$\mathrm{Pt(s)|H_2(g)|HCl(aq)|AgCl(s)|Ag(s)} \tag{13.8}$$

The left electrode is a hydrogen electrode, and the right electrode is a silver-silver chloride electrode. According to our convention that oxidation occurs at the left electrode in the cell diagram, the two electrode reactions are

$$\tfrac{1}{2}\,\mathrm{H_2(g)} \rightleftharpoons \mathrm{H^+(aq)} + e^-$$
$$\mathrm{AgCl(s)} + e^- \rightleftharpoons \mathrm{Ag(s)} + \mathrm{Cl^-(aq)}$$

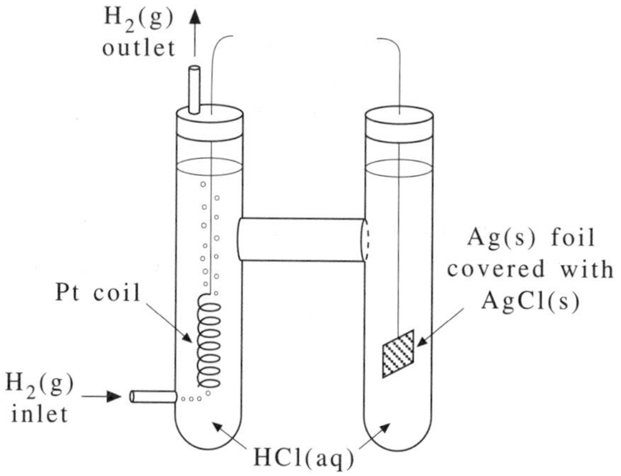


**FIGURE 13.6**
An electrochemical cell consisting of a hydrogen electrode and a silver-silver chloride electrode. The cell diagram for this cell is $\mathrm{Pt(s)|H_2(g)|HCl(aq)|AgCl(s)|Ag(s)}$.

and the overall reaction associated with the above cell diagram is

$$\mathrm{AgCl(s)} + \tfrac{1}{2}\,\mathrm{H_2(g)} \rightleftharpoons \mathrm{H^+(aq)} + \mathrm{Cl^-(aq)} + \mathrm{Ag(s)} \tag{13.9}$$

The Nernst equation for the emf of the cell is

$$E = E^\circ - \frac{RT}{nF}\ln\frac{a_{\mathrm{Ag}}a_{\mathrm{H^+}}a_{\mathrm{Cl^-}}}{a_{\mathrm{AgCl}}a_{\mathrm{H_2}}^{1/2}} \tag{13.10}$$

Because Ag(s) and AgCl(s) are solids, we set $a_{\mathrm{Ag}}$ and $a_{\mathrm{AgCl}}$ equal to unity in Equation 13.10. In addition, we regulate the pressure of the hydrogen gas over the platinum electrode such that $a_{\mathrm{H_2}} = 1$, which amounts to setting its fugacity equal to unity. For a gas such as hydrogen at ordinary pressures, the fugacity is essentially equal to the pressure, so we can set the pressure equal to unity (one bar). The number of electrons transferred in the overall reaction as written above is one, so $n = 1$ in Equation 13.10.

Finally, then, Equation 13.10 becomes

$$E = E^\circ - \frac{RT}{F}\ln a_{\mathrm{H^+}}a_{\mathrm{Cl^-}} \tag{13.11}$$

From Table 11.3, $a_+a_- = a_\pm^2$ and $a_\pm = \gamma_\pm m$, so Equation 13.11 becomes

$$E = E^\circ - \frac{2RT}{F}\ln \gamma_\pm m$$

or

$$E + \frac{2RT}{F}\ln m = E^\circ - \frac{2RT}{F}\ln \gamma_\pm \tag{13.12}$$

In our discussion of the Debye-Hückel theory, however, we saw that (Problem 11–47)

$$\ln \gamma_\pm = -1.171 m^{1/2}$$

for an aqueous solution of a 1–1 electrolyte such as HCl(aq) at 298.15 K, so we can write Equation 13.12 as

$$E + \frac{2RT}{F}\ln m = E^\circ + \frac{2.342RT}{F}m^{1/2} \tag{13.13}$$

The replacement of $\ln \gamma_\pm$ by $Am^{1/2}$ is valid only in the limit of infinite dilution, so the procedure is to plot the left side of Equation 13.13 versus $m^{1/2}$ and extrapolate to $m = 0$. The intercept, then, is $E^\circ$. Table 13.1 gives data for $E$ versus $m$ at 298.15 K for the cell pictured in Figure 13.6. The data in Table 13.1 are used to plot the left side of Equation 13.13 against $m^{1/2}$ in Figure 13.7. The linear portion at small values of $m$ extrapolates to $E^\circ = 0.222$ V.

**TABLE 13.1**
The emf versus $m$ at 298.15 K for the cell whose cell diagram is $Pt(s)|H_2(1\ bar)|HCl(m)|AgCl(s)|Ag(s)$.

| $m/mol \cdot kg^{-1}$ | 0.1238 | 0.05391 | 0.02563 | 0.013407 | 0.009138 | 0.005619 | 0.003215 |
|---|---|---|---|---|---|---|---|
| $E/V$ | 0.34199 | 0.38222 | 0.41824 | 0.44974 | 0.46860 | 0.49257 | 0.52053 |

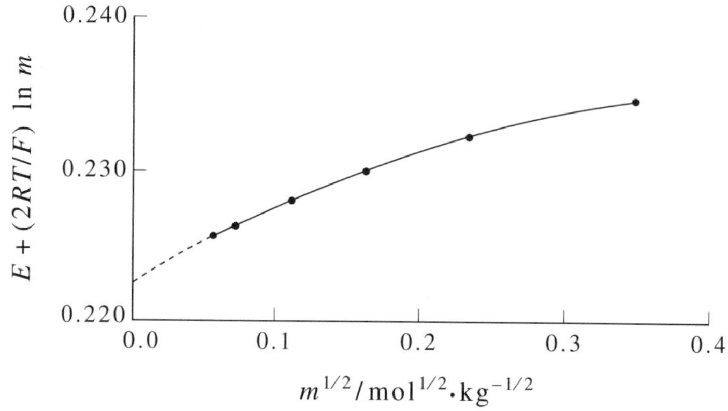


**FIGURE 13.7**
A plot of $E + (2RT/F)\ln m$ against $m^{1/2}$ at 298.15 K for the cell illustrated in Figure 13.6 and whose cell diagram is given by Equation 13.8. The extrapolation of this plot to $m \to 0$ yields the value of $E^\circ$, the standard emf of the cell.

Although the plot in Figure 13.7 can be extrapolated to $m = 0$ without too much difficulty, we can obtain a more reliable method by using an extension of the Debye-Hückel expression for $\ln \gamma_\pm$ to higher concentrations. For example, we can use the semiempirical expression given by Equation 11.58

$$\ln \gamma_\pm = -\frac{1.171|z_+z_-|(I_m/mol \cdot kg^{-1})^{1/2}}{1 + (I_m/mol \cdot kg^{-1})^{1/2}} + Cm \tag{13.14}$$

where $C$ is an adjustable parameter. Equation 13.14 becomes

$$\ln \gamma_\pm = -\frac{1.171 m^{1/2}}{1 + m^{1/2}} + Cm \tag{13.15}$$

for a 1–1 electrolyte such as HCl(aq). If we substitute Equation 13.15 into Equation 13.12 at 298.15 K, we obtain

$$E + 0.05139 \ln m - 0.06018 \frac{m^{1/2}}{1 + m^{1/2}} = E^\circ - 0.05139 Cm \tag{13.16}$$

Now if we plot the left side of Equation 13.16 against $m$, we can obtain the value of $E^\circ$ by extrapolating to $m = 0$. Figure 13.8 shows such a plot for the data given in

Table 13.1. Note that the plot is less curved than the one in Figure 13.7, making the required extrapolation much easier. The plot shown in Figure 13.8 is called a *Hitchcock plot* and should be used for an accurate determination of the value of $E^\circ$.

**EXAMPLE 13–4**

Use the following data to determine the value of $E^\circ$ for the cell whose cell diagram is

$$\text{Pt(s)}|\text{H}_2\text{(1 bar)}|\text{HBr}(m)|\text{AgBr(s)}|\text{Ag(s)}$$

| $m/\text{mol}\cdot\text{kg}^{-1}$ | 0.00100 | 0.00200 | 0.00500 | 0.0100 | 0.0200 | 0.0500 |
|---|---|---|---|---|---|---|
| $E/\text{V}$ | 0.4297 | 0.3949 | 0.3490 | 0.3147 | 0.2807 | 0.2360 |

SOLUTION: Using our convention that oxidation takes place at the left electrode in a cell diagram, we write the two electrode reactions as

$$\tfrac{1}{2}\,\text{H}_2\text{(g)} \longrightarrow \text{H}^+\text{(aq)} + e^-$$

and

$$\text{AgBr(s)} + e^- \longrightarrow \text{Ag(s)} + \text{Br}^-\text{(aq)}$$

The net reaction is

$$\text{AgBr(s)} + \tfrac{1}{2}\,\text{H}_2\text{(g)} \longrightarrow \text{Ag(s)} + \text{HBr(aq)}$$

This equation is analogous to Equation 13.9, so we can use Equation 13.13, or better yet, Equation 13.16. The resulting plots using Equation 13.13 (plotting the left side against $m^{1/2}$) or Equation 13.16 (plotting the left side against $m$) are shown in Figure 13.9. Notice that the Hitchcock plot is much easier to extrapolate to $m = 0$, yielding a value of $E^\circ = 0.0730$ V.

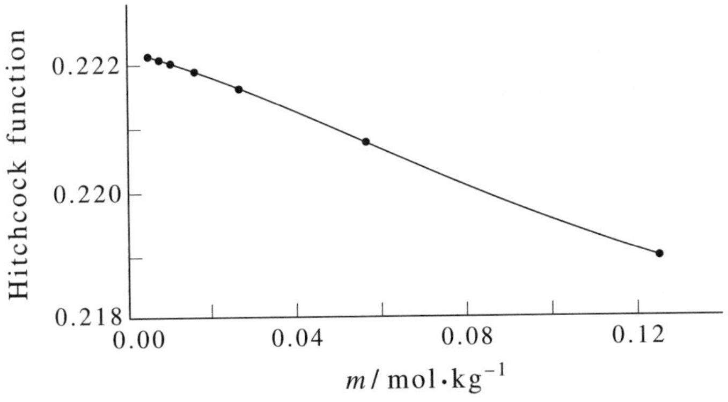


**FIGURE 13.8**
The determination of $E^\circ$ by means of a Hitchcock plot for the data given in Table 13.1.

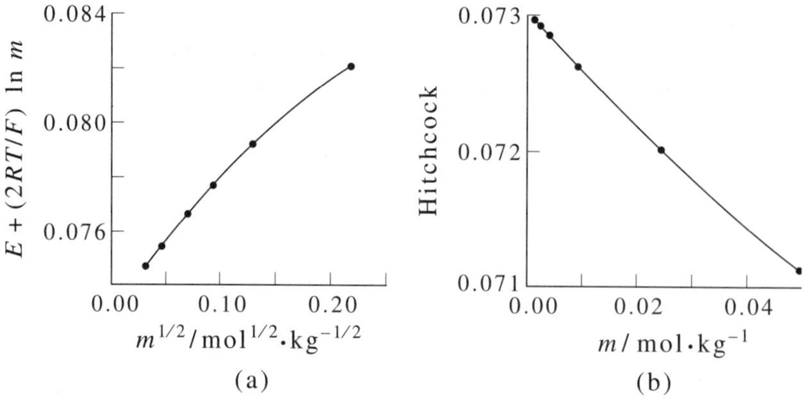


**FIGURE 13.9**
Two plots illustrating the determination of the value of $E^\circ$ at 298.15 K for the cell whose cell diagram is $\text{Pt(s)}|\text{H}_2\text{(1 bar)}|\text{HBr}(m)|\text{AgBr(s)}|\text{Ag(s)}$ from the data given in Example 13–4. (a) The left side of Equation 13.13 plotted against $m^{1/2}$. (b) The left side of Equation 13.16 plotted against $m$. Both plots can be extrapolated to 0.0730 V, but the extrapolation is more readily seen in (b).

## 13–6. We Can Assign Values of $E^\circ$ to Single Electrodes

Consider the cell whose cell diagram is

$$\text{Pt(s)}|\text{H}_2\text{(g)}|\text{H}^+\text{(aq)}||\text{Zn}^{2+}\text{(aq)}|\text{Zn(s)} \tag{13.17}$$

The value of $E^\circ$ for the corresponding cell is −0.723 V at 298.15 K. The equations for the electrode reactions are

$$\text{H}_2\text{(g)} \longrightarrow 2\,\text{H}^+\text{(aq)} + 2\,e^- \qquad \text{(oxidation at the left electrode)} \tag{13.18}$$

$$\text{Zn}^{2+}\text{(aq)} + 2\,e^- \longrightarrow \text{Zn(s)} \qquad \text{(reduction at the right electrode)} \tag{13.19}$$

and the equation for the overall reaction is

$$\text{Zn}^{2+}\text{(aq)} + \text{H}_2\text{(g)} \longrightarrow 2\,\text{H}^+\text{(aq)} + \text{Zn(s)} \tag{13.20}$$

If we let $\Delta_{\text{ox}}G^\circ$ and $\Delta_{\text{red}}G^\circ$ be the standard Gibbs energy change for Equations 13.18 and 13.19, then $\Delta_{\text{r}}G^\circ$ for Equation 13.20 is

$$\Delta_{\text{r}}G^\circ = \Delta_{\text{ox}}G^\circ + \Delta_{\text{red}}G^\circ \tag{13.21}$$

Because $\Delta G^\circ$ and $E^\circ$ are directly related by Equation 13.7, we can use Equation 13.21 to write

$$E^\circ_{\text{cell}} = E^\circ_{\text{ox}} + E^\circ_{\text{red}} \tag{13.22}$$

for a reaction in which the electrons on each side cancel (see Problem 13–57). If we apply Equation 13.22 to the cell diagram given by Equation 13.17, we have

$$\begin{aligned} E^\circ_{\text{cell}} &= E^\circ[\text{Zn}^{2+}(\text{aq})/\text{Zn(s)}] + E^\circ[\text{H}_2(\text{g})/\text{H}^+(\text{aq})] \\ &= -0.723\ \text{V} \end{aligned} \tag{13.23}$$

Note that we have indicated that reduction takes place at the zinc electrode by writing $E^\circ[Zn^{2+}(aq)/Zn(s)]$ and that oxidation takes place at the hydrogen electrode by writing $E^\circ[H_2(g)/H^+(aq)]$. By coupling other electrodes with a hydrogen electrode, we will obtain results similar to Equation 13.23. For example, if we use a AgCl(s)–Ag(s) electrode instead of a zinc electrode, we would obtain

$$\begin{aligned} E^\circ_{\text{cell}} &= E^\circ[\text{AgCl(s)}/\text{Ag(s)}] + E^\circ[\text{H}_2(\text{g})/\text{H}^+(\text{aq})] \\ &= 0.222\ \text{V} \end{aligned} \tag{13.24}$$

It is not possible to measure the voltage of a single electrode; only the difference in voltage between two electrodes can be measured. If, however, we agree to choose a numerical value for the standard voltage of some particular electrode, then we can assign standard voltages to single electrodes. If we take $E^\circ$ for the hydrogen electrode to be zero, we can write

$$E^\circ_{\text{red}}[\text{Zn}^{2+}(\text{aq})/\text{Zn(s)}] = -0.723\ \text{V} \tag{13.25}$$

and

$$E^\circ_{\text{red}}[\text{AgCl(s)}/\text{Ag(s)}] = 0.222\ \text{V} \tag{13.26}$$

and continue the process by coupling other electrodes with a hydrogen electrode. Thus, we will formally write

$$E^\circ_{\text{ox}}[\text{H}_2(\text{g})/\text{H}^+(\text{aq})] = 0 \qquad \text{(by convention)} \tag{13.27}$$

Because $E^\circ$ implies that all the reactants involved are in their standard states, we could write Equation 13.27 more explicitly as

$$E^\circ_{\text{ox}}[\text{H}_2(f = 1\ \text{bar})/\text{H}^+(a_\pm = 1)] = 0 \qquad \text{(by convention)} \tag{13.28}$$

to emphasize that $H_2(g)$ must be at unit fugacity (which is essentially equivalent to $P = 1$ bar) and that the strong acid corresponding to $H^+(aq)$ [for example, HCl(aq)] must be at unit activity.

Equations 13.25 and 13.26 are examples of the standard reduction potentials that we can assign to electrode reactions (Table 13.2). We can use the entries in Table 13.2 to calculate the value of $E^\circ$ for a cell whose cell reaction is the sum of the two electrode reactions. According to Equation 13.22

$$E^\circ_{\text{cell}} = E^\circ_{\text{red}} + E^\circ_{\text{ox}} \tag{13.29}$$

where reduction takes place at the right electrode and oxidation takes place at the left electrode. We can emphasize this right-left convention by writing Equation 13.29 as

$$E^\circ_{\text{cell}} = E^\circ_{\text{red,R}} + E^\circ_{\text{ox,L}} \tag{13.30}$$

But, for a particular electrode-reaction equation, oxidation is the reverse of reduction, so

$$E^\circ_{\text{ox,L}} = -E^\circ_{\text{red,L}}$$

Using the fact that $E^\circ_{\text{ox}} = -E^\circ_{\text{red}}$, Equation 13.30 can be written as

$$E^\circ_{\text{cell}} = E^\circ_{\text{red,R}} - E^\circ_{\text{red,L}} \tag{13.31}$$

Thus the emf of the cell is the standard reduction potential of the right electrode in the cell diagram minus the standard reduction potential of the left electrode.

Consider the cell whose cell diagram is

$$\text{Ag(s)}|\text{AgNO}_3(0.100\text{ m})|\text{Cu(NO}_3)_2(1.00\text{ m})|\text{Cu(s)}$$

The overall cell reaction equation is

$$2\text{ Ag(s)} + \text{Cu}^{2+}(1.00\text{ m}) \rightleftharpoons 2\text{ Ag}^+(0.100\text{ m}) + \text{Cu(s)} \tag{13.32}$$

and the two electrode reactions are

$$2\text{ Ag(s)} \rightleftharpoons 2\text{ Ag}^+(\text{aq}) + 2\,e^-$$

and

$$\text{Cu}^{2+}(\text{aq}) + 2\,e^- \rightleftharpoons \text{Cu(s)}$$

We see from Table 13.2 that

$$\text{Ag}^+(\text{aq}) + e^- \rightleftharpoons \text{Ag(s)} \qquad E^\circ = +0.799\text{ V}$$
$$\text{Cu}^{2+}(\text{aq}) + 2\,e^- \rightleftharpoons \text{Cu(s)} \qquad E^\circ = +0.337\text{ V}$$

so Equation 13.31 gives us

$$E^\circ = 0.337\text{ V} - 0.799\text{ V} = -0.462\text{ V}$$

Note that the $\text{Ag(s)}|\text{Ag}^+(\text{aq})$ half-cell emf is *not* multiplied by 2. The emf of an electrochemical cell is independent of the size of the cell; emf is an *intensive* property. (See also Problems 13–54 through 13–58.)

The emf of the cell whose cell reaction is described by Equation 13.32 is given by the Nernst equation (Equation 13.7) with $n = 2$:

$$E = E^\circ - \frac{RT}{2F} \ln \frac{a^2_{\text{AgNO}_3} a_{\text{Cu}}}{a^2_{\text{Ag}} a_{\text{Cu(NO}_3)_2}}$$

**TABLE 13.2**
Standard reduction potentials in water at 25°C.

| Electrode reaction | $E^\circ$/V |
|---|---|
| $K^+(aq) + e^- \longrightarrow K(s)$ | −2.925 |
| $Ca^{2+}(aq) + 2\,e^- \longrightarrow Ca(s)$ | −2.866 |
| $Na^+(aq) + e^- \longrightarrow Na(s)$ | −2.714 |
| $Al^{3+}(aq) + 3\,e^- \longrightarrow Al(s)$ | −1.66 |
| $2\,H_2O(l) + 2\,e^- \longrightarrow H_2(g) + 2\,OH^-(aq)$ | −0.828 |
| $Zn^{2+}(aq) + 2\,e^- \longrightarrow Zn(s)$ | −0.763 |
| $Ag_2S(s) + 2\,e^- \longrightarrow 2\,Ag(s) + S^{2-}(aq)$ | −0.7051 |
| $Fe^{2+}(aq) + 2\,e^- \longrightarrow Fe(s)$ | −0.440 |
| $Cr^{3+}(aq) + e^- \longrightarrow Cr^{2+}(aq)$ | −0.408 |
| $Cd^{2+}(aq) + 2\,e^- \longrightarrow Cd(s)$ | −0.403 |
| $PbI_2(s) + 2\,e^- \longrightarrow 2\,I^-(aq) + Pb(s)$ | −0.364 |
| $PbSO_4(s) + 2\,e^- \longrightarrow Pb(s) + SO_4^{2-}(aq)$ | −03583 |
| $PbBr_2(s) + 2\,e^- \longrightarrow 2\,Br^-(aq) + Pb(s)$ | −0.274 |
| $PbCl_2(s) + 2\,e^- \longrightarrow 2\,Cl^-(aq) + Pb(s)$ | −0.266 |
| $V^{3+}(aq) + e^- \longrightarrow V^{2+}(aq)$ | −0.255 |
| $Ni^{2+}(aq) + 2\,e^- \longrightarrow Ni(s)$ | −0.250 |
| $AgI(s) + e^- \longrightarrow Ag(s) + I^-(aq)$ | −0.151 |
| $Sn^{2+}(aq) + 2\,e^- \longrightarrow Sn(s)$ | −0.136 |
| $Pb^{2+}(aq) + 2\,e^- \longrightarrow Pb(s)$ | −0.126 |
| $Hg_2I_2(s) + 2\,e^- \longrightarrow 2\,Hg(l) + 2\,I^-(aq)$ | −0.0405 |
| $Fe^{3+}(aq) + 3\,e^- \longrightarrow Fe(s)$ | −0.036 |
| $2\,H^+(aq) + 2\,e^- \longrightarrow H_2(g)$ | 0.000 |
| $AgBr(s) + e^- \longrightarrow Ag(s) + Br^-(aq)$ | +0.0732 |
| $Hg_2Br_2(s) + 2\,e^- \longrightarrow 2\,Hg(l) + 2\,Br^-(aq)$ | +0.1396 |
| $Sn^{4+}(aq) + 2\,e^- \longrightarrow Sn^{2+}(aq)$ | +0.15 |
| $AgCl(s) + e^- \longrightarrow Ag(s) + Cl^-(aq)$ | +0.2224 |
| $Hg_2Cl_2(s) + 2\,e^- \longrightarrow 2\,Hg(l) + 2\,Cl^-(aq)$ | +0.268 |
| $Cu^{2+}(aq) + 2\,e^- \longrightarrow Cu(s)$ | +0.337 |
| $I_2(s) + 2\,e^- \longrightarrow 2\,I^-(aq)$ | +0.5345 |
| $Hg_2SO_4(s) + 2\,e^- \longrightarrow 2\,Hg(l) + SO_4^{2-}(aq)$ | +0.6155 |
| $Ag_2SO_4(s) + 2\,e^- \longrightarrow 2\,Ag(s) + SO_4^{2-}(aq)$ | +0.653 |
| $Fe^{3+}(aq) + e^- \longrightarrow Fe^{2+}(aq)$ | +0.771 |
| $Hg_2^{2}(aq) + 2\,e^- \longrightarrow 2\,Hg(l)$ | +0.796 |
| $Ag^+(aq) + e^- \longrightarrow Ag(s)$ | +0.799 |
| $Pd^{2+}(aq) + 2\,e^- \longrightarrow Pd(s)$ | +0.987 |
| $O_2(g) + 4\,H^+(aq) + 4\,e^- \longrightarrow 2\,H_2O(l)$ | +1.229 |
| $Cl_2(g) + 2\,e^- \longrightarrow 2\,Cl^-(aq)$ | +1.361 |
| $Co^{3+}(aq) + e^- \longrightarrow Co^{2+}(aq)$ | +1.84 |

Setting the activities of Ag(s) and Cu(s) equal to unity and using the relations $a_{AgNO_3} = a_\pm^2 = \gamma_\pm^2 m_\pm^2 = \gamma_\pm^2 m^2$ and $a_{Cu(NO_3)_2} = a_\pm^3 = \gamma_\pm^3 m_\pm^3 = \gamma_\pm^3 4m^3$ (Table 11.3), we get

$$E = E^\circ - \frac{RT}{2F} \ln \frac{\gamma_{\pm,AgNO_3}^4 (0.100)^4}{4\gamma_{\pm Cu(NO_3)_2}^3 (1.00)^3}$$

Given that $\gamma_{\pm,AgNO_3} = 0.721$ at 0.100 m and that $\gamma_{\pm Cu(NO_3)_2} = 0.456$ at 1.00 m, we have

$$\begin{aligned} E &= -0.462 \text{ V} - (0.01285 \text{ V}) \ln \frac{(0.721)^4 (0.100)^4}{4(0.456)^3 (1.00)^3} \\ &= -0.462 \text{ V} + 0.123 \text{ V} = -0.339 \text{ V} \end{aligned}$$

The negative sign indicates that Equation 13.32 will be spontaneous from right to left under the given conditions.

---

**EXAMPLE 13–5**

Use the data in Table 13.2 to calculate the value of the emf of a cell whose cell diagram is

$$\text{Zn(s)}|\text{ZnCl}_2(0.500 \text{ m})||\text{CuCl}_2(1.00 \text{ m})|\text{Cu(s)}$$

Take $\gamma_\pm = 0.396$ for 0.500 m $ZnCl_2$(aq) and $\gamma_\pm = 0.419$ for 1.00 m $CuCl_2$(aq).

SOLUTION: The two electrode reactions are

$$\text{Zn(s)} \longrightarrow \text{Zn}^{2+}(0.500 \text{ m}) + 2\,e^- \qquad \text{(left)}$$

and

$$\text{Cu}^{2+}(1.00 \text{ m}) + 2\,e^- \longrightarrow \text{Cu(s)} \qquad \text{(right)}$$

and the cell reaction is given by

$$\text{Zn(s)} + \text{CuCl}_2(1.00 \text{ m}) \longrightarrow \text{Cu(s)} + \text{ZnCl}_2(0.500 \text{ m})$$

We first calculate the value of $E^\circ_{cell}$ from Equation 13.31,

$$\begin{aligned} E^\circ_{cell} &= E^\circ_{red,R} + E^\circ_{ox,L} = E^\circ_{red,R} - E^\circ_{red,L} \\ &= E^\circ[\text{Cu}^{2+}(\text{aq})/\text{Cu(s)}] - E^\circ[\text{Zn}^{2+}(\text{aq})/\text{Zn(s)}] \\ &= 0.337 \text{ V} - (-0.763 \text{ V}) = 1.100 \text{ V} \end{aligned}$$

and then use the Nernst equation with $n = 2$ (Equation 13.7)

$$E = E^\circ - \frac{RT}{2F} \ln \frac{a_{Cu} a_{ZnCl_2}}{a_{Zn} a_{CuCl_2}}$$

We set the activity of the Cu(s) and Zn(s) equal to unity and write

$$E = 1.100\ \mathrm{V} - \frac{RT}{2F}\ln\frac{a_{\mathrm{ZnCl_2}}}{a_{\mathrm{CuCl_2}}}$$

According to Table 11.3, $a_{\mathrm{ZnCl_2}} = \gamma_\pm^3 m_\pm^3 = 4\gamma_\pm^3 m^3$ and $a_{\mathrm{CuCl_2}} = \gamma_\pm^3 m_\pm^3 = 4\gamma_\pm^3 m^3$. Using the fact that $\gamma_\pm = 0.396$ for $\mathrm{ZnCl_2}$(0.500 m) and $\gamma_\pm = 0.419$ for $\mathrm{CuCl_2}$(1.00 m), the Nernst equation gives

$$\begin{aligned} E &= 1.100\ \mathrm{V} - \frac{(8.3145\ \mathrm{J\cdot mol^{-1}\cdot K^{-1}})(298.15\ \mathrm{K})}{(2)(96\,485\ \mathrm{C\cdot mol^{-1}})}\ln\frac{(4)(0.396)^3(0.500)^3}{(4)(0.419)^3(1.00)^3} \\ &= 1.100\ \mathrm{V} + 0.0289\ \mathrm{V} = 1.129\ \mathrm{V} \end{aligned}$$

We should point out a limitation of Equation 13.31 before we go on. It turns out that there is always a difference in electrical potential across the interface of two solutions unless they have an identical composition. This potential is called a *liquid junction potential*. For example, consider the salt bridge connecting the two solutions in Figure 13.1. There is a liquid junction potential at each salt bridge-solution interface, so Equation 13.31 in this case should be written as

$$E^\circ = E^\circ_{\mathrm{red,R}} - E^\circ_{\mathrm{red,L}} + E^\circ_{\mathrm{LJ}}$$

where $E^\circ_{\mathrm{LJ}}$ is the sum of the two liquid junction potentials. The magnitude of the liquid junction potential caused by a salt bridge depends upon the relative mobilities of the cation and anion that constitute the salt bridge. Potassium chloride is used because a potassium ion and a chloride ion are about equally mobile in aqueous solution, so the total liquid junction potential due to a potassium chloride salt bridge is quite small (around a millivolt or less) and can be neglected. Generally, in designing electrochemical cells with a liquid junction, we must take care to minimize the liquid junction potential. We will discuss liquid junction potentials in Chapter 14.

## 13–7. Electrochemical Cells Can Be Used to Determine Activity Coefficients

In Chapter 11, we learned that activity coefficients can be determined from vapor pressure measurements and from freezing point measurements. One of the most convenient and accurate methods for determining the activity coefficients of solutions of electrolytes involves electrochemical cells. Let's go back to Equation 13.13

$$E + \frac{2RT}{F}\ln m = E^\circ + \frac{2.342RT}{F}m^{1/2} \qquad (13.13)$$

which we derived for a cell whose cell diagram is
Pt(s)|$\mathrm{H_2}$(1 atm)|HCl(aq)|AgCl(s)|Ag(s). We determined the value of $E^\circ$ to be 0.222 V in Section 13–3 by plotting the left side of Equation 13.13 against $m^{1/2}$ and extrapolating

to $m = 0$. Now that we know the value of $E^\circ$, either by extrapolation of $E$ versus $m$ data or, more conveniently, from a table of standard reduction potentials, we can use the same $E$ versus $m$ data (Table 13.1) and Equation 13.12 to determine $\gamma_\pm$ versus $m$ for HCl(aq). Solving Equation 13.12 for $\ln \gamma_\pm$ and using the first set of data in Table 13.1, we get

$$\begin{aligned}\ln \gamma_\pm &= \frac{F}{2RT}(E^\circ - E) - \ln m \\ &= \frac{(96\,485\ \text{C}\cdot\text{mol}^{-1})(0.2224\ \text{V} - 0.34199\ \text{V})}{2(8.3145\ \text{J}\cdot\text{mol}^{-1}\cdot\text{K}^{-1})(298.15\ \text{K})} - \ln 0.1238 \\ &= -0.238\ \text{V}\end{aligned}$$

or $\gamma_\pm = 0.788$ at $m = 0.1238\ \text{mol}\cdot\text{kg}^{-1}$. The other values obtained from the data in Table 13.1 are plotted in Figure 13.10.

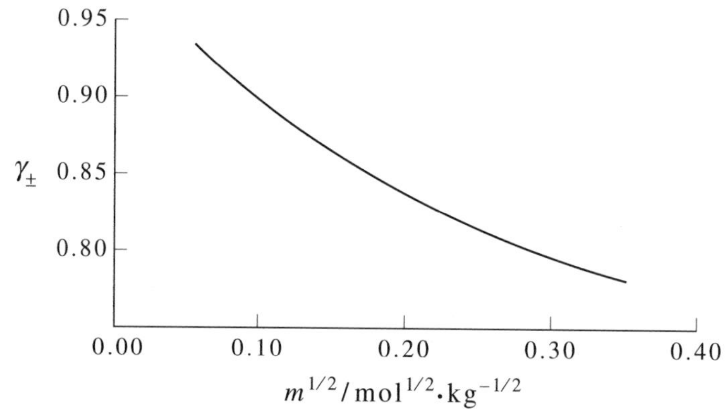


**FIGURE 13.10**
The mean ionic activity coefficient of HCl(aq) obtained from Equation 13.12 and the data in Table 13.1 plotted against $m^{1/2}$.

**EXAMPLE 13–6**

Show that the Nernst equation for a cell whose cell diagram is $\text{Pt(s)}|\text{H}_2\text{(1 bar)}|\text{HBr(aq)}|\text{AgBr(s)}|\text{Ag(s)}$ can be written as

$$E = E^\circ - \frac{2RT}{F} \ln \gamma_\pm m$$

where $\gamma_\pm m$ is the activity of HBr(aq). Use the result of Example 13–3 and the following cell data to calculate and plot $\gamma_\pm$ versus $m^{1/2}$.

| $m/\text{mol}\cdot\text{kg}^{-1}$ | 0.0010 | 0.0050 | 0.0100 | 0.0200 | 0.0500 | 0.100 | 0.200 | 0.500 |
|---|---|---|---|---|---|---|---|---|
| $E/\text{V}$ | 0.4297 | 0.3490 | 0.3147 | 0.2807 | 0.2360 | 0.2004 | 0.16662 | 0.1188 |

SOLUTION: The reaction at the left electrode (oxidation) is

$$\tfrac{1}{2}\,H_2(1\text{ bar}) \longrightarrow H^+(aq) + e^-$$

and that at the right electrode (reduction) is

$$AgBr(s) + e^- \longrightarrow Br^-(aq) + Ag(s)$$

The equation for the net cell reaction is

$$\tfrac{1}{2}\,H_2(1\text{ bar}) + AgBr(s) \longrightarrow H^+(aq) + Br^-(aq) + Ag(s)$$

The corresponding Nernst equation is

$$E = E^\circ - \frac{RT}{F}\ln\frac{a_{Ag}a_{H^+}a_{Br^-}}{a_{AgBr}a_{H_2}^{1/2}}$$

If we set $a_{Ag}$, $a_{AgBr}$, and $a_{H_2}$ equal to unity and use the relations $a_{H^+}a_{Br^-} = a_\pm^2 = \gamma_\pm^2 m^2$ (Table 11.3), we obtain

$$E = E^\circ - \frac{2RT}{F}\ln\gamma_\pm m$$

Solving this equation for $\ln\gamma_\pm$ gives

$$\ln\gamma_\pm = \frac{F}{2RT}(E^\circ - E) - \ln m$$

For the first set of data in the above table,

$$\begin{aligned}\ln\gamma_\pm &= \frac{(96\,485\ \text{C}\cdot\text{mol}^{-1})(0.0732\ \text{V} - 0.4297\ \text{V})}{2(8.3145\ \text{J}\cdot\text{mol}^{-1}\cdot\text{K}^{-1})(298.15\ \text{K})} - \ln 0.0010\\ &= -0.0300\ \text{V}\end{aligned}$$

or $\gamma_\pm = 0.970$. The other data are plotted in Figure 13.11.

Both calculations done in this section have involved 1–1 electrolytes. Let's now do an example involving $ZnCl_2(aq)$, a 2–1 electrolyte. Consider a cell whose cell diagram is

$$Zn(s)|ZnCl_2(aq)|AgCl(s)|Ag(s)$$

The reaction at the left electrode (oxidation) is

$$Zn(s) \longrightarrow Zn^{2+}(aq) + 2\,e^-$$

and that at the right electrode (reduction) is

$$AgCl(s) + e^- \longrightarrow Cl^-(aq) + Ag(s)$$

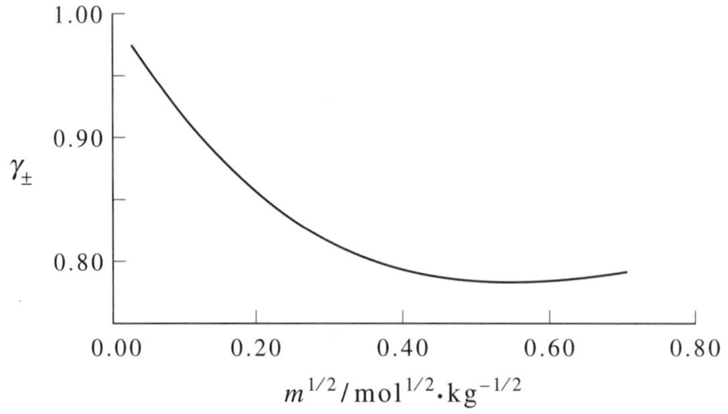


**FIGURE 13.11**
The mean ionic activity coefficient of HBr(aq) plotted against $m^{1/2}$. The values of $\gamma_{\pm}$ are obtained according to Example 13–5.

The equation for the overall cell reaction can be written as

$$\text{Zn(s)} + 2\ \text{AgCl(s)} \longrightarrow \text{Zn}^{2+}\text{(aq)} + 2\ \text{Cl}^-\text{(aq)} + 2\ \text{Ag(s)}$$

Because the activities of the three solid phases can be set equal to unity, the emf of the cell is given by

$$E = E^\circ - \frac{RT}{2F} \ln a_{\text{Zn}^{2+}} a^2_{\text{Cl}^-} \tag{13.33}$$

The 2 in the denominator of the $\ln a_{\text{Zn}^{2+}} a^2_{\text{Cl}^-}$ term results because two electrons are transferred in the cell reaction, as we have written it. Using Table 11.3, we write $a_{\text{Zn}^{2+}} a^2_{\text{Cl}^-} = 4m^3\gamma^3_{\pm}$, so Equation 13.33 becomes

$$E = E^\circ - \frac{RT}{2F} \ln 4m^3 - \frac{3RT}{2F} \ln \gamma_{\pm} \tag{13.34}$$

Given $E$ versus $m$ data at sufficiently low concentrations, we could plot $E + (RT/2F) \ln 4m^3$ versus $m^{1/2}$ and then extrapolate to $m \to 0$ to determine $E^\circ$, as we did in Section 13–5 for the cell whose cell diagram is given by Equation 13.8. Alternatively, we can use the data in Table 13.2 to obtain

$$\begin{aligned} E^\circ_{\text{cell}} &= E^\circ[\text{AgCl(s)/Ag(s)}] - E^\circ[\text{Zn}^{2+}\text{(aq)/Zn(s)}] \\ &= 0.222\ \text{V} - (-0.763\ \text{V}) = 0.985\ \text{V} \end{aligned}$$

Solving Equation 13.34 for $\ln \gamma_{\pm}$ gives

$$\ln \gamma_{\pm} = \frac{2F}{3RT}(E^\circ - E) - \frac{1}{3} \ln(4m^3)$$

The following values of $E$ at 298.15 K versus $m$ can be used to calculate the value of $\gamma_\pm$ as a function of $m$ (Problem 13–14):

| $m/\text{mol}\cdot\text{kg}^{-1}$ | 0.100 | 0.500 | 1.00 | 2.00 | 3.00 | 4.00 | 5.00 | 6.00 |
|---|---|---|---|---|---|---|---|---|
| $E/\text{V}$ | 1.0813 | 1.0296 | 1.0087 | 0.98805 | 0.97269 | 0.95902 | 0.94496 | 0.93156 |

These results are plotted in Figure 13.12.

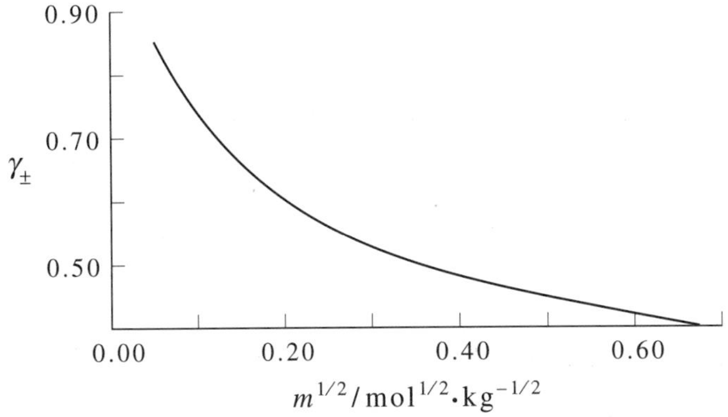


**FIGURE 13.12**
The mean ionic activity coefficient of $ZnCl_2$(aq) plotted against $m^{1/2}$.

## 13–8. Electrochemical Measurements Can Be Used to Determine Values of $\Delta_r H$ and $\Delta_r S$ of Cell Reactions

Equation 13.6 serves as the bridge between thermodynamics and electrochemistry. We can readily express $\Delta H$ and $\Delta S$ in terms of $E$ by using Equation 8.31a

$$\Delta S = -\left(\frac{\partial \Delta G}{\partial T}\right)_P = nF\left(\frac{\partial E}{\partial T}\right)_P \tag{13.35}$$

and then the relation $\Delta G = \Delta H - T\Delta S$ to obtain

$$\Delta H = -nFE + nFT\left(\frac{\partial E}{\partial T}\right)_P \tag{13.36}$$

We can use Equations 13.35 and 13.36 to calculate the values of $\Delta_r H$ and $\Delta_r S$ for any cell reaction. For example, the temperature dependence of $E^\circ$ for the cell

$$\text{Pt(s)}|\text{H}_2\text{(g)}|\text{HCl(aq)}|\text{AgCl(s)}|\text{Ag(s)}$$

is given by

$$E^\circ = 0.22239\ \text{V} - (643.29 \times 10^{-6}\ \text{V}\cdot\text{K}^{-1})(T - 298.15\ \text{K}) - (2.948 \times 10^{-6}\ \text{V}\cdot\text{K}^{-2})(T - 298.15\ \text{K})^2$$

from 0°C to 50°C. The equation for the cell reaction is

$$H_2(g) + 2\ AgCl(s) \longrightarrow 2\ HCl(aq) + 2\ Ag(s)$$

Using Equation 13.35, we have

$$\begin{aligned}\Delta_r S^\circ &= nF\left(\frac{\partial E^\circ}{\partial T}\right)_P \\ &= (2)(96\,485\ \mathrm{C\cdot mol^{-1}})[643.29\times10^{-6}\ \mathrm{V\cdot K^{-1}} \\ &\qquad - (5.896\times10^{-6}\ \mathrm{V\cdot K^{-2}})(T - 298.15\ \mathrm{K})] \\ &= -124.1\ \mathrm{J\cdot K^{-1}\cdot mol^{-1}} - 1.13\ \mathrm{J\cdot K^{-2}\cdot mol^{-1}}(T - 298.15\ \mathrm{K}) \\ &= -124.1\ \mathrm{J\cdot K^{-1}\cdot mol^{-1}}\end{aligned}$$

at 25°C, and using Equation 13.36, we have

$$\begin{aligned}\Delta_r H^\circ &= -nFE^\circ + T\Delta_r S^\circ \\ &= -(2)(96\,485\ \mathrm{C\cdot mol^{-1}})(0.22239\ \mathrm{V}) + (298.15\ \mathrm{K})(-0.1241\ \mathrm{kJ\cdot K^{-1}\cdot mol^{-1}}) \\ &= -79.92\ \mathrm{kJ\cdot mol^{-1}}\end{aligned}$$

at 25°C.

**EXAMPLE 13–7**

The cell whose cell diagram is

$$Pb(s)|PbCl_2(s)|KCl(aq)|Hg_2Cl_2(s)|Hg(l)$$

has a standard emf of 0.5359 V and a temperature coeffcient of $1.45\times10^{-4}\ \mathrm{V\cdot K^{-1}}$ at 298.15 K. Determine the cell reaction and the values of $\Delta_r G^\circ$, $\Delta_r H^\circ$, and $\Delta_r S^\circ$.

SOLUTION: The reaction at the left electrode (oxidation) is

$$Pb(s) + 2\ Cl^-(aq) \longrightarrow PbCl_2(s) + 2\ e^-$$

and that at the right electrode (reduction) is

$$Hg_2Cl_2(s) + 2\ e^- \longrightarrow 2\ Hg(l) + 2\ Cl^-(aq)$$

and the cell reaction is

$$Pb(s) + Hg_2Cl_2(s) \longrightarrow PbCl_2(s) + 2\ Hg(l)$$

The value of $\Delta_r G^\circ$ is

$$\begin{aligned}\Delta_r G^\circ = -nFE^\circ &= -(2)(96\,485\ \mathrm{C\cdot mol^{-1}})(0.5359\ \mathrm{V}) \\ &= -103.4\ \mathrm{kJ\cdot mol^{-1}}\end{aligned}$$

the value of $\Delta_r S^\circ$ is

$$\begin{aligned}\Delta_r S^\circ &= nF\left(\frac{\partial E^\circ}{\partial T}\right)_P \\ &= (2)(96\,485\ \mathrm{C{\cdot}mol^{-1}})(1.45\times10^{-4}\ \mathrm{V{\cdot}K^{-1}}) \\ &= 28.0\ \mathrm{J{\cdot}K^{-1}{\cdot}mol^{-1}}\end{aligned}$$

and the value of $\Delta_r H^\circ$ is

$$\Delta_r H^\circ = \Delta_r G^\circ + T\Delta_r S^\circ = -95.1\ \mathrm{kJ{\cdot}mol^{-1}}$$

## 13–9. Solubility Products Can Be Determined with Electrochemical Cells

Because $\Delta G = -nFE = 0$ at equilibrium, we can use Equation 13.7 to derive a relation between the standard emf and the thermodynamic equilibrium constant of the equation associated with the cell reaction. Setting $E = 0$ in Equation 13.7 gives

$$E^\circ = \frac{RT}{nF}\ln K_a \tag{13.37}$$

An important application of this equation is to the determination of the solubility products of sparingly soluble salts.

As a concrete example, let's consider the determination of the solubility product of AgCl(s). The equation for the dissolution of AgCl(s) is

$$\mathrm{AgCl(s)} \rightleftharpoons \mathrm{Ag^+(aq)} + \mathrm{Cl^-(aq)} \tag{13.38}$$

and the corresponding solubility-product expression is

$$K_{sp} = a_{\mathrm{Ag^+}}a_{\mathrm{Cl^-}} \tag{13.39}$$

We can construct an electrochemical cell whose net cell equation is Equation 13.38 by using a Ag(s)|AgCl(s)|$Cl^-$(aq) electrode and a Ag(s)|$Ag^+$(aq) electrode. The reduction electrode-reaction equations corresponding to these two electrodes are

$$\mathrm{AgCl(s)} + e^- \longrightarrow \mathrm{Ag(s)} + \mathrm{Cl^-(aq)} \tag{13.40}$$

and

$$\mathrm{Ag^+(aq)} + e^- \longrightarrow \mathrm{Ag(s)} \tag{13.41}$$

To obtain the equation for the dissolution of AgCl(s), we subtract Equation 13.41 from Equation 13.40. Therefore, the left electrode (the electrode at which oxidation takes place) in the cell diagram is the $Ag(s)|Ag^+(aq)$ electrode, and the cell diagram is

$$\mathrm{Ag(s)|AgCl(aq)|AgCl(s)|Ag(s)}$$

Notice that this cell has no liquid junction and hence has no liquid-junction potential. The value of $E^\circ$ at 298.15 K for this cell is given by

$$E^\circ = E^\circ_\mathrm{R} - E^\circ_\mathrm{L} = 0.222\ \mathrm{V} - 0.799\ \mathrm{V} = -0.577\ \mathrm{V}$$

According to Equation 13.37,

$$\begin{aligned} K_a = K_\mathrm{sp} = e^{nFE^\circ/RT} &= \exp\left[\frac{(1)(96\,485\ \mathrm{C\cdot mol^{-1}})(-0.577\ \mathrm{V})}{(8.3145\ \mathrm{J\cdot mol^{-1}\cdot K^{-1}})(298.15\ \mathrm{K})}\right] \\ &= 1.77 \times 10^{-10} \end{aligned}$$

at 298.15 K. The standard state here is a one-molal solution.

**EXAMPLE 13–8**

Use the data in Table 13.2 to calculate the value of the solubility product of $PbI_2(s)$.

SOLUTION: Using the above calculation as a guide, we use a $Pb(s)|PbI_2(s)|I^-(aq)$ electrode and a $Pb(s)|Pb^{2+}(aq)$ electrode, whose corresponding reduction electrode-reaction equations are

$$\mathrm{PbI_2(s)} + 2\,e^- \longrightarrow \mathrm{Pb(s)} + 2\,\mathrm{I^-(aq)} \qquad (1)$$

and

$$\mathrm{Pb^{2+}(aq)} + 2\,e^- \longrightarrow \mathrm{Pb(s)} \qquad (2)$$

To obtain the equation for the dissolution of $PbI_2(s)$, we subtract Equation 2 from Equation 1. Therefore, the left electrode (the one at which oxidation takes place) is the $Pb(s)|Pb^{2+}(aq)$ electrode, and the cell diagram is

$$\mathrm{Pb(s)|PbI_2(aq)|PbI_2(s)|Pb(s)}$$

The value of $E^\circ$ at 298.15 K for this cell is

$$E^\circ = E^\circ_\mathrm{R} - E^\circ_\mathrm{L} = -0.364\ \mathrm{V} - (-0.126\ \mathrm{V}) = -0.238\ \mathrm{V}$$

Therefore, according to Equation 13.37

$$\begin{aligned} K_\mathrm{sp} = e^{2FE^\circ/RT} &= \exp\left[\frac{(2)(96\,485\ \mathrm{C\cdot mol^{-1}})(-0.238\ \mathrm{V})}{(8.3145\ \mathrm{J\cdot mol^{-1}\cdot K^{-1}})(298.15\ \mathrm{K})}\right] \\ &= 8.99 \times 10^{-9} \end{aligned}$$

The preceding two calculations of the solubility product of a sparingly soluble salt illustrate the general method. Let the sparingly soluble salt be $M_{\nu_+}A_{\nu_-}(s)$, so that the equation for its dissolution is

$$M_{\nu_+}A_{\nu_+}(s) \rightleftharpoons \nu_+ M^{z_+}(aq) + \nu_- A^{z_-}(aq) \tag{13.42}$$

and its solubility-product expression is

$$K_{sp} = a_{M^{z_+}}^{\nu_+} a_{A^{z_-}}^{\nu_-} \tag{13.43}$$

We can construct an electrochemical cell whose net cell equation is Equation 13.42 by using an $M(s)|M_{\nu_+}A_{\nu_-}(s)|A^{z_-}(aq)$ electrode and an $M(s)|M^{z_+}(aq)$ electrode. The net chemical equation of the reaction occurring in a cell with the cell diagram

$$M(s)|M_{\nu_+}A_{\nu_-}(aq)|M_{\nu_+}A_{\nu_-}(s)|M(s)$$

is Equation 13.42, so the value of $E^\circ$ of this cell will give the solubility product of $M_{\nu_+}A_{\nu_-}(s)$

**EXAMPLE 13–9**

Devise an electrochemical cell (without a liquid junction) that can be used to determine the solubility product of $Hg_2Cl_2(s)$. Use the data in Table 13.2 to calculate the value of the solubility product of $Hg_2Cl_2(s)$ at 298.15 K.

SOLUTION: The equation for the dissolution of $Hg_2Cl_2(s)$ is

$$Hg_2Cl_2(s) \rightleftharpoons Hg_2^{2+}(aq) + 2\,Cl^-(aq)$$

Two electrode reactions that give this net equation are

$$Hg_2Cl_2(s) + 2\,e^- \longrightarrow 2\,Hg(l) + 2\,Cl^-(aq)$$

and

$$2\,Hg(l) \longrightarrow Hg_2^{2+}(aq) + 2\,e^-$$

We place the oxidation electrode reaction on the left and the reduction electrode reaction on the right, and so we have the cell diagram

$$Hg(l)|Hg_2Cl_2(aq)|Hg_2Cl_2(s)|Hg(l)$$

According to Table 13.2,

$$\begin{aligned} E^\circ_{cell} &= E^\circ[Hg_2Cl_2(s)/Hg(l)] - E^\circ[Hg_2^{2+}(aq)/Hg(l)] \\ &= 0.268\text{ V} - 0.796\text{ V} = -0.528\text{ V} \end{aligned}$$

The solubility product is given by Equation 13.34 with $n = 2$

$$K_{sp} = e^{nFE^\circ/RT} = \exp\left[\frac{(2)(96\,485\ \text{C}\cdot\text{mol}^{-1})(-0.528\ \text{V})}{(8.3145\ \text{J}\cdot\text{mol}^{-1}\cdot\text{K}^{-1})(298.15\ \text{K})}\right]$$
$$= 1.41 \times 10^{-18}$$

## 13–10. The Dissociation Constants of Weak Acids Can Be Determined with Electrochemical Cells

Consider the electrochemical cell

$$\text{Pt(s)}|\text{H}_2\text{(1 bar)}|\text{HA}(m_1), \text{NaA}(m_2), \text{NaCl}(m_3)|\text{AgCl(s)}|\text{Ag(s)} \tag{13.44}$$

where HA is a weak acid and NaA is its sodium salt. The equations for the electrode reactions are

$$\tfrac{1}{2}\,\text{H}_2\text{(g)} \longrightarrow \text{H}^+\text{(aq)} + e^- \qquad \text{(left)}$$

and

$$\text{AgCl(s)} + e^- \longrightarrow \text{Ag(s)} + \text{Cl}^-\text{(aq)} \qquad \text{(right)}$$

and the equation for the net cell reaction is

$$\tfrac{1}{2}\,\text{H}_2\text{(g)} + \text{AgCl(s)} \longrightarrow \text{H}^+\text{(aq)} + \text{Cl}^-\text{(aq)} + \text{Ag(s)}$$

Given that the fugacity (essentially the pressure) of the $H_2$(g) is one bar, the emf of this cell is given by

$$E = E^\circ - \frac{RT}{F}\ln a_{\text{H}^+}a_{\text{Cl}^-}$$

where $E^\circ = 0.2224$ V is the standard reduction potential of the silver-silver chloride electrode. Using the relation $a_{\text{H}^+}a_{\text{Cl}^-} = \gamma_{\text{H}^+}m_{\text{H}^+}\gamma_{\text{Cl}^-}m_{\text{Cl}^-}$ (Table 11.3), we have

$$E = E^\circ - \frac{RT}{F}\ln m_{\text{H}^+}m_{\text{Cl}^-} - \frac{RT}{F}\ln \gamma_{\text{H}^+}\gamma_{\text{Cl}^-} \tag{13.45}$$

The dissociation-constant expression of the weak acid, HA(aq), is

$$K_a = \frac{a_{\text{H}^+}a_{\text{A}^-}}{a_{\text{HA}}} = \frac{\gamma_{\text{H}^+}m_{\text{H}^+}\gamma_{\text{A}^-}m_{\text{A}^-}}{\gamma_{\text{HA}}m_{\text{HA}}} \tag{13.46}$$

If we solve Equation 13.46 for $a_{H^+} = m_{H^+}\gamma_{H^+}$ and substitute this result into Equation 13.45, we obtain

$$E = E^\circ - \frac{RT}{F}\ln K_a - \frac{RT}{F}\ln\frac{m_{HA}m_{Cl^-}}{m_{A^-}} - \frac{RT}{F}\ln\frac{\gamma_{HA}\gamma_{Cl^-}}{\gamma_{A^-}}$$

We can rearrange this equation into the form

$$\frac{F}{RT}(E - E^\circ) + \ln\frac{m_{HA}m_{Cl^-}}{m_{A^-}} = -\ln K_a' \qquad (13.47)$$

where

$$\ln K_a' = \ln K_a + \ln\frac{\gamma_{HA}\gamma_{Cl^-}}{\gamma_{A^-}} \qquad (13.48)$$

The procedure is to vary $m_{HA}$, $m_{Cl^-}$, and $m_{A^-}$ and to evaluate the left side of Equation 13.47 as a function of ionic strength

$$I_m = \frac{1}{2}\sum_j z_j^2 m_j$$

This procedure gives $\ln K_a'$ as a function of ionic strength. We now plot $(F/RT)(E - E^\circ) + \ln(m_{HA}m_{Cl^-}/m_{A^-})$ against ionic strength, and the extrapolation to zero ionic strength gives $\ln K_a$ because $\gamma_{HA}\gamma_{Cl^-}/\gamma_{A^-} \to 1$ as $I_m \to 0$. Figure 13.13 shows such a plot using the following data for acetic acid [HAc(aq)] at 298.15 K:

| $m_{HAc}$/mol·kg$^{-1}$ | $m_{NaAc}$/mol·kg$^{-1}$ | $m_{NaCl}$/mol·kg$^{-1}$ | $I_m$/mol·kg$^{-1}$ | $E$/V |
|---|---|---|---|---|
| 0.004779 | 0.004599 | 0.004896 | 0.00950 | 0.63959 |
| 0.012035 | 0.011582 | 0.012426 | 0.02401 | 0.61583 |
| 0.021006 | 0.020216 | 0.021516 | 0.04173 | 0.60154 |
| 0.04922 | 0.04737 | 0.05042 | 0.09779 | 0.57977 |
| 0.08101 | 0.07796 | 0.08297 | 0.16095 | 0.56712 |
| 0.09056 | 0.08716 | 0.09276 | 0.17992 | 0.56423 |

Because acetic acid is a fairly weak acid, we have neglected the small contribution that $H^+$(aq) and $Ac^-$(aq) make to $I_m$ in Figure 13.13. The extrapolation of the best linear fit to the data plotted in Figure 13.13 gives $-\ln K_a = 10.958$, or $K_a = 1.74 \times 10^{-5}$. If the $H^+$(aq) and $Ac^-$(aq) from the dissociation of HAc(aq) are taken into account, you obtain $K_a = 1.75 \times 10^{-5}$, in agreement with the "accepted" value of $K_a$ on a molality scale. Problem 13–41 treats the acid dissociation of formic acid, which is about ten times stronger than acetic acid. In this case the $H^+$(aq) and $Ac^-$(aq) from the dissociation of the formic acid cannot be neglected.

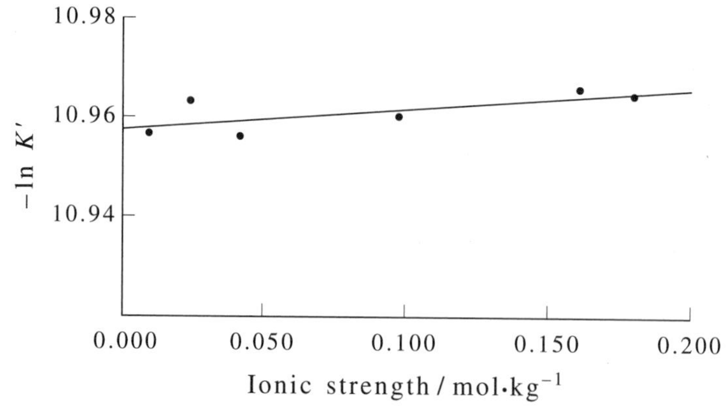


**FIGURE 13.13**
A plot of the left side of Equation 13.47 with $E^\circ = 0.2224$ V against ionic strength using the supplied data for the cell $\mathrm{Pt(s)|H_2(1\ bar)|HA}(m_1), \mathrm{NaA}(m_2), \mathrm{NaCl}(m_3)\mathrm{|AgCl(s)|Ag(s)}$. The extrapolation of the best linear fit to these data gives $-\ln K_a = 10.958$, or $K_a = 1.74 \times 10^{-5}$ for acetic acid.

**EXAMPLE 13–10**
The value of $K_a$ for propanoic acid can be determined using the cell whose cell diagram is

$$\mathrm{Pt(s)|H_2(g)|HP(aq),\ NaP(aq),\ NaCl(aq)|AgCl(s)|Ag(s)}$$

Use the following data to determine the value of $K_a$ at 25°C.

| $m_{\mathrm{HP}} = m_{\mathrm{NaP}} = m_{\mathrm{NaCl}}$ | $E/\mathrm{V}$ |
|---|---|
| 0.004899 | 0.64758 |
| 0.005918 | 0.64249 |
| 0.01281 | 0.62286 |
| 0.01605 | 0.61692 |
| 0.01643 | 0.61651 |
| 0.01715 | 0.61538 |
| 0.01811 | 0.61401 |
| 0.02220 | 0.60901 |
| 0.02555 | 0.60522 |
| 0.02648 | 0.60423 |
| 0.03113 | 0.60001 |
| 0.03179 | 0.59958 |
| 0.03569 | 0.59664 |

SOLUTION: We plot the left side of Equation 13.47 against ionic strength and extrapolate to zero ionic strength. Neglecting the small concentrations of $H^+$(aq) and $P^-$(aq) due to the dissociation of HP(aq), the expression for $-\ln K_a'$ is

$$\ln K_a' = \frac{F}{RT}(E - E^\circ) + \ln m$$

where $T = 298.15$ K, $E^\circ = 0.2224$ V, and $m = m_{\text{HP}} = m_{\text{NaP}} = m_{\text{NaCl}}$. The ionic strength is equal to $I_m = m_{\text{NaP}} + m_{\text{NaCl}} = 2m$. Figure 13.14 is a plot of $-\ln K_a'$ against $I_m$. Extrapolation of a linear curve fit gives an intercept of 11.227, or $K_a = 1.33 \times 10^{-5}$.

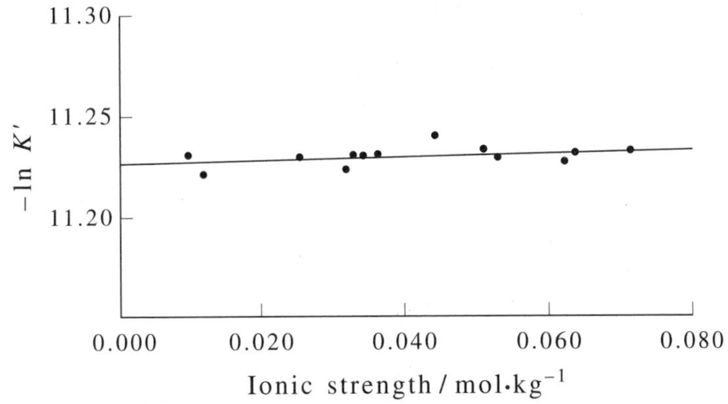


**FIGURE 13.14**
A plot of $-\ln K_a'$, given by the expression in Example 13–10 for propanoic acid, against ionic strength for the cell Pt(s)|$H_2$(g)|HP(aq), NaP(aq), NaCl(aq)|AgCl(s)|Ag(s). A linear fit gives an intercept of 11.227, or $K_a = 1.33 \times 10^{-5}$.

## 13–11. We Can Assign Thermodynamic Values to Individual Ions in Solution

The thermodynamic properties of single ionic species are not measureable quantities, but we can still set up a table of single-ion thermodynamic properties in the following manner. Consider the reaction

$$\text{Zn(s)} + 2\,\text{H}^+\text{(aq)} \longrightarrow \text{Zn}^{2+}\text{(aq)} + \text{H}_2\text{(g)} \tag{13.49}$$

The standard emf for the associated electrochemical cell is 0.763 V at 298.15 K. If all the species are in their standard states

$$\begin{aligned}\Delta_r G^\circ &= -nFE = -147\ \text{kJ·mol}^{-1}\\ &= \Delta_f G^\circ[\text{Zn}^{2+}\text{(aq)}] + \Delta_f G^\circ[\text{H}_2\text{(g)}] - \Delta_f G^\circ[\text{Zn(s)}] - 2\Delta_f G^\circ[\text{H}^+\text{(aq)}]\end{aligned}$$

The standard Gibbs energies of formation of $H_2$(g) and Zn(s) are equal to zero by convention, and so we can write

$$\Delta_f G^\circ[\text{Zn}^{2+}\text{(aq)}] - 2\Delta_f G^\circ[\text{H}^+\text{(aq)}] = -147\ \text{kJ·mol}^{-1} \tag{13.50}$$

We can continue this process and determine values of $\Delta_f G^\circ$ for various ions relative to $\Delta_f G^\circ[\text{H}^+(\text{aq})]$. For example, consider the reaction

$$\tfrac{1}{2}\,\text{H}_2(\text{g}) + \tfrac{1}{2}\,\text{Cl}_2(\text{g}) \longrightarrow \text{H}^+(\text{aq}) + \text{Cl}^-(\text{aq}) \tag{13.51}$$

The standard emf for the associated electrochemical cell is 1.360 V at 298.15 K. Thus, we can write

$$\begin{aligned}\Delta_r G^\circ &= -nFE = -131.2\ \text{kJ}\cdot\text{mol}^{-1}\\ &= \Delta_f G^\circ[\text{H}^+(\text{aq})] + \Delta_f G^\circ[\text{Cl}^-(\text{aq})] - \frac{1}{2}\Delta_f G^\circ[\text{H}_2(\text{g})] - \frac{1}{2}\Delta_f G^\circ[\text{Cl}_2(\text{g})]\end{aligned}$$

or

$$\Delta_f G^\circ[\text{Cl}^-(\text{aq})] + \Delta_f G^\circ[\text{H}^+(\text{aq})] = -131.2\ \text{kJ}\cdot\text{mol}^{-1} \tag{13.52}$$

Notice that the combination of $\text{Zn}^{2+}(\text{aq})$ and $\text{H}^+(\text{aq})$ is electrically neutral ($\text{Zn}^{2+} - 2\text{H}^+$) in Equation 13.50 and that the combination of $\text{Cl}^-(\text{aq})$ and $\text{H}^+(\text{aq})$ is electrically neutral ($\text{Cl}^- + \text{H}^+$) in Equation 13.52. This result along with many other similar results shows that $\text{H}^+(\text{aq})$ will cancel for any neutral salts. Consequently, we are at liberty to set the value of $\Delta_f G^\circ[\text{H}^+(\text{aq})]$ arbitrarily and we choose to set $\Delta_f G^\circ[\text{H}^+(\text{aq})] = 0$ at 298.15 K. Accepting this convention, we can write

$$\Delta_f G^\circ[\text{Zn}^{2+}(\text{aq})] = -147\ \text{kJ}\cdot\text{mol}^{-1}$$

from Equation 13.50 and

$$\Delta_f G^\circ[\text{Cl}^-(\text{aq})] = -131.2\ \text{kJ}\cdot\text{mol}^{-1}$$

from Equation 13.52.

We can determine $\Delta_f H^\circ$ and $\Delta_f S^\circ$ from the temperature variation of $E^\circ$ for the appropriate cell. For the cell associated with Equation 13.49, $(\partial E^\circ/\partial T)_P = -0.119\ \text{mV}\cdot\text{K}^{-1}$, so we can write

$$\begin{aligned}\Delta_r S^\circ &= nF\left(\frac{\partial E^\circ}{\partial T}\right)_P = -22.96\text{J}\cdot\text{K}^{-1}\cdot\text{mol}^{-1}\\ &= S^\circ[\text{Zn}^{2+}(\text{aq})] + S^\circ[\text{H}_2(\text{g})] - S^\circ[\text{Zn(s)}] - 2S^\circ[\text{H}^+(\text{aq})]\end{aligned} \tag{13.53}$$

The standard molar entropies of the neutral species $\text{H}_2(\text{g})$ and Zn(s) are $130.68\ \text{J}\cdot\text{K}^{-1}\cdot\text{mol}^{-1}$ and $41.63\ \text{J}\cdot\text{K}^{-1}\cdot\text{mol}^{-1}$, respectively, at 298.15 K, so Equation 13.53 becomes

$$S^\circ[\text{Zn}^{2+}(\text{aq})] - 2S^\circ[\text{H}^+(\text{aq})] = -112.0\ \text{J}\cdot\text{K}^{-1}\cdot\text{mol}^{-1} \tag{13.54}$$

The corresponding result for Equation 13.51 turns out to be

$$S^\circ[\text{Cl}^-(\text{aq})] + S^\circ[\text{H}^+(\text{aq})] = 56.5\ \text{J}\cdot\text{K}^{-1}\cdot\text{mol}^{-1} \tag{13.55}$$

These results suggest that we choose to set $S^\circ[\mathrm{H}^+(\mathrm{aq})] = 0$ at 298.15 K, and we write

$$S^\circ[\mathrm{Zn}^{2+}(\mathrm{aq})] = -112.0\ \mathrm{J\cdot K^{-1}\cdot mol^{-1}}$$

from Equation 13.54 and

$$S^\circ[\mathrm{Cl}^-(\mathrm{aq})] = 56.5\ \mathrm{J\cdot K^{-1}\cdot mol^{-1}}$$

from Equation 13.55.

We can calculate $\Delta_f H^\circ$ in each case from the equation

$$\Delta_f H^\circ = nF\left[T\left(\frac{\partial E^\circ}{\partial T}\right)_P - E^\circ\right]$$

For Equation 13.49, we have

$$\begin{aligned}\Delta_r H^\circ &= 2(96\,485\ \mathrm{C\cdot mol^{-1}})[(298.15\ \mathrm{K})(-0.119\ \mathrm{mV\cdot K^{-1}}) - 0.763\ \mathrm{V}] \\ &= -154\ \mathrm{kJ\cdot mol^{-1}}\end{aligned}$$

from which we obtain

$$\Delta_f H^\circ[\mathrm{Zn}^{2+}(\mathrm{aq})] - 2\Delta_f H^\circ[\mathrm{H}^+(\mathrm{aq})] = -154\ \mathrm{kJ\cdot mol^{-1}}$$

and for Equation 13.51, we have $\Delta_r H^\circ = -167\ \mathrm{kJ\cdot mol^{-1}}$ or

$$\Delta_f H^\circ[\mathrm{Cl}^-(\mathrm{aq})] + \Delta_f H^\circ[\mathrm{H}^+(\mathrm{aq})] = -167\ \mathrm{kJ\cdot mol^{-1}}$$

We set $\Delta_f H^\circ[\mathrm{H}^+(\mathrm{aq})] = 0$ at 298.15 K, and so we have

$$\Delta_f H^\circ[\mathrm{Zn}^{2+}(\mathrm{aq})] = -154\ \mathrm{kJ\cdot mol^{-1}}$$

and

$$\Delta_f H^\circ[\mathrm{Cl}^-(\mathrm{aq})] = -167\ \mathrm{kJ\cdot mol^{-1}}$$

Values of the thermodynamic properties of single ions are tabulated in a number of places, but one particularly valuable source is a joint American Chemical Society and American Institute of Physics publication [*Journal of Physical and Chemical Reference Data*, vol. II, suppl.2 (1982)], called the *NBS Tables of Chemical Thermodynamic Properties, Selected Values for Inorganic and $C_1$ and $C_2$ Organic Substances in SI Units.* Figure 13.15 shows a typical page of these tables. Unlike the JANAF tables (Section 12–9), which give data over a wide range of temperatures for about 1500 substances, the above tables give data only at 298.15 K, but for several thousand substances. We can use the data in Figure 13.15 along with $\Delta_f G^\circ[\mathrm{Cl}^-(\mathrm{aq})] = -131.228\ \mathrm{kJ\cdot mol^{-1}}$ (from page 2–47 of the tables) to calculate the value of the solubility product of TlCl(s) at 298.15 K. The equation for the dissolution of TlCl(s) is

$$\mathrm{TlCl(s)} \longrightarrow \mathrm{Tl^+(aq)} + \mathrm{Cl^-(aq)}$$

Table 32:Tl — **THALLIUM** (Prepared 1965) — Table 32:Tl

| Substance Formula | Description | State | Molar mass g mol$^{-1}$ | 0 K $\Delta_fH^\circ_0$ kJ mol$^{-1}$ | 298.15 K (25°C) and 0.1 MPa (1 bar) $\Delta_fH^\circ$ | $\Delta_fG^\circ$ kJ mol$^{-1}$ | $H^\circ-H^\circ_0$ | $S^\circ$ J mol$^{-1}$ K$^{-1}$ | $C_p$ |
|---|---|---|---|---|---|---|---|---|---|
| Tl | | cr | 204.3700 | 0 | 0 | 0 | 6.828 | 64.18 | 26.32 |
| | | g | 204.3700 | 182.845 | 182.21 | 147.41 | 6.197 | 180.963 | 20.786 |
| | in Hg:u, saturated | | 204.3700 | — | 0.318 | -0.259 | — | 66.11 | — |
| $Tl^+$ | | g | 204.3700 | 772.199 | 777.764 | — | 6.197 | — | — |
| $Tl^{2+}$ | | g | 204.3700 | 2743.210 | 2754.971 | — | 6.197 | — | — |
| $Tl^{3+}$ | | g | 204.3700 | 5621.2 | 5639.2 | — | 6.197 | — | — |
| $Tl^+$ | | ao | 204.3700 | — | 5.36 | -32.40 | — | 125.5 | — |
| $Tl^{3+}$ | | ao | 204.3700 | — | 196.6 | 214.6 | — | -192. | — |
| $Tl_2O$ | | cr | 424.7394 | — | -178.7 | -147.3 | — | 126. | — |
| $Tl_2O_3$ | | cr | 456.7382 | — | — | -311.7 | — | — | — |
| $Tl_2O_4$ | | cr | 472.7376 | — | — | -347.2 | — | — | — |
| TlOH | | cr | 221.3774 | — | -238.9 | -195.8 | — | 88. | — |
| | | ai | 221.3774 | — | -224.64 | -189.63 | — | 114.6 | — |
| | | ao | 221.3774 | — | — | -194.1 | — | — | — |
| | in 350 $H_2O$ | | 221.3774 | — | -226.44 | — | — | — | — |
| TlOH | in 500 $H_2O$ | | 221.3774 | — | -226.69 | — | — | — | — |
| | in 750 $H_2O$ | | 221.3774 | — | -227.02 | — | — | — | — |
| | in 1 000 $H_2O$ | | 221.3774 | — | -227.32 | — | — | — | — |
| | in 1 500 $H_2O$ | | 221.3774 | — | -227.69 | — | — | — | — |
| | in 2 000 $H_2O$ | | 221.3774 | — | -227.94 | — | — | — | — |
| TlOH | in ∞ $H_2O$ | | 221.3774 | — | -224.64 | — | — | — | — |
| $TlOH^{2+}$ | | ao | 221.3774 | — | — | -15.9 | — | — | — |
| $Tl(OH)_2^+$ | | ao | 238.3848 | — | — | -244.7 | — | — | — |
| $Tl(OH)_3$ | | cr | 255.3922 | — | — | -507.0 | — | — | — |
| TlF | | cr | 223.3684 | — | -324.7 | — | — | — | — |
| TlF | | g | 223.3684 | — | -182.4 | — | — | — | — |
| | | ai | 223.3684 | — | -327.27 | -311.19 | — | 111.7 | — |
| | | ao | 223.3684 | — | — | -311.7 | — | — | — |
| | in 800 $H_2O$ | | 223.3684 | — | -328.4 | — | — | — | — |
| $TlHF_2$ | | cr | 243.3748 | — | — | — | 18.025 | 146.11 | 88.91 |
| $TlHF_2$ | in 800 $H_2O$ | | 243.3748 | — | -645.6 | — | — | — | — |
| TlCl | | cr | 239.8230 | -205.397 | -204.14 | -184.92 | 12.678 | 111.25 | 50.92 |
| | | g | 239.8230 | — | -67.8 | — | — | — | — |
| $TlCl^+$ | | g | 239.8230 | — | 874.5 | — | — | — | — |
| | | ai | 239.8230 | — | -161.80 | -163.62 | — | 182.0 | — |
| TlCl | | ao | 239.8230 | — | -167.78 | -166.97 | — | 172.8 | — |
| $TlCl^{2+}$ | | ao | 239.8230 | — | 4.2 | 40.6 | — | -79. | — |
| $TlCl_2^+$ | | ao | 275.2760 | — | -179.9 | -123.8 | — | 29. | — |
| $TlCl_2^-$ | | ao | 275.2760 | — | — | -295.8 | — | — | — |
| $TlCl_3$ | | cr | 310.7290 | — | -315.1 | — | — | — | — |
| $TlCl_3$ | | ai | 310.7290 | — | -305.0 | -179.0 | — | -23.0 | — |
| | | ao | 310.7290 | — | -351.5 | -274.4 | — | 134. | — |
| | | aq | 310.7290 | — | -351.5 | — | — | — | — |
| $TlCl_3{\cdot}4H_2O$ | | cr | 382.7906 | — | -1503.7 | — | — | — | — |
| $TlCl_4^-$ | | ao | 346.1820 | — | -519.2 | -421.7 | — | 243. | — |
| $Tl_2Cl_2$ | | g | 479.6460 | — | -206.7 | — | — | — | — |
| $TlClO_3$ | | ai | 287.8212 | — | -98.62 | -40.35 | — | 287.9 | — |
| | | ao | 287.8212 | — | — | -43.20 | — | — | — |
| TlBr | | cr | 284.2790 | — | -173.2 | -167.36 | — | 120.5 | — |
| | | g | 284.2790 | — | -37.7 | — | — | — | — |

**FIGURE 13.15**
Page 2-135 from the NBS Tables of Chemical Thermodynamic Properties [*J. Phys. Chem. Ref. Data*, vol. 11, suppl.2 (1982)].

and so

$$\begin{aligned}\Delta_{\text{sol}}G^\circ &= \Delta_f G^\circ[\text{Tl}^+(\text{aq})] + \Delta_f G^\circ[\text{Cl}^-(\text{aq})] - \Delta_f G^\circ[\text{TlCl(s)}]\\ &= -32.40\ \text{kJ}\cdot\text{mol}^{-1} - 131.228\ \text{kJ}\cdot\text{mol}^{-1} - (-184.92\ \text{kJ}\cdot\text{mol}^{-1})\\ &= 21.29\ \text{kJ}\cdot\text{mol}^{-1}\end{aligned}$$

and so

$$\begin{aligned}K_{\text{sp}} &= e^{-\Delta_{\text{sol}}G^\circ/RT} = \exp\left[-\frac{21.29\times 10^3\ \text{J}\cdot\text{mol}^{-1}}{(8.3145\ \text{J}\cdot\text{mol}^{-1}\cdot\text{K}^{-1})(298.15\ \text{K})}\right]\\ &= 1.86\times 10^{-4}\end{aligned}$$

**EXAMPLE 13–11**

Using the values of $\Delta_f G^\circ[\text{Ag}^+(\text{aq})] = 77.107\ \text{kJ}\cdot\text{mol}^{-1}$, $\Delta_f G^\circ[\text{Br}^-(\text{aq})] = -103.96\ \text{kJ}\cdot\text{mol}^{-1}$, and $\Delta_f G^\circ[\text{AgBr(s)}] = -96.90\ \text{kJ}\cdot\text{mol}^{-1}$ at 298.15 K, calculate the value of $K_{\text{sp}}$ of AgBr(s) at 298.15 K.

SOLUTION: The equation for the dissolution of AgBr(s) is

$$\text{AgBr(s)} \longrightarrow \text{Ag}^+(\text{aq}) + \text{Br}^-(\text{aq})$$

and

$$\begin{aligned}\Delta_r G^\circ &= \Delta_f G^\circ[\text{Ag}^+(\text{aq})] + \Delta_f G^\circ[\text{Br}^-(\text{aq})] - \Delta_f G^\circ[\text{AgBr(s)}]\\ &= 70.047\ \text{kJ}\cdot\text{mol}^{-1}\end{aligned}$$

and

$$\begin{aligned}K_{\text{sp}} &= e^{-\Delta_r G^\circ/RT} = \exp\frac{-70.047\times 10^3\ \text{J}\cdot\text{mol}^{-1}}{(8.3145\ \text{J}\cdot\text{K}^{-1}\cdot\text{mol}^{-1})(298.15\ \text{K})}\\ &= 5.35\times 10^{-13}\end{aligned}$$

We can also use tabulated values of Gibbs energies of formation of ions to calculate acid-dissociation constants. For example, the *NBS Tables of Chemical Thermodynamic Properties* give

$$\Delta_f G^\circ[\text{HCOO}^-(\text{aq})] = -351.0\ \text{kJ}\cdot\text{mol}^{-1}$$

and

$$\Delta_f G^\circ[\text{HCOOH(aq)}] = -372.3\ \text{kJ}\cdot\text{mol}^{-1}$$

at 298.15 K. The equation for the dissociation of formic acid can be written as

$$\text{HCOOH(aq)} \rightleftharpoons \text{H}^+(\text{aq}) + \text{HCOO}^-(\text{aq})$$

and so

$$\begin{aligned}\Delta_r G^\circ &= \Delta_f G^\circ[\mathrm{H^+(aq)}] + \Delta_f G^\circ[\mathrm{HCOO^-(aq)}] - \Delta_f G^\circ[\mathrm{HCOOH(aq)}] \\ &= 0 - 351.0\ \mathrm{kJ\cdot mol^{-1}} - (-372.3\ \mathrm{kJ\cdot mol^{-1}}) \\ &= 21.3\ \mathrm{kJ\cdot mol^{-1}}\end{aligned}$$

and

$$\begin{aligned}K = e^{-\Delta_r G^\circ/RT} &= \exp\left[-\frac{21.3\times 10^3\ \mathrm{J\cdot mol^{-1}}}{(8.3145\ \mathrm{J\cdot mol^{-1}\cdot K^{-1}})(298.15\ \mathrm{K^{-1}})}\right] \\ &= 1.86\times 10^{-4}\end{aligned}$$

**EXAMPLE 13–12**

Given the following data from the *NBS Tables of Chemical Thermodynamic Properties*, calculate the value of the protonation constant of aminomethane at 298.15 K.

$$\begin{aligned}\Delta_f G^\circ[\mathrm{CH_3NH_2(aq)}] &= 20.77\ \mathrm{kJ\cdot mol^{-1}} \\ \Delta_f G^\circ[\mathrm{CH_3NH_3^+(aq)}] &= -39.86\ \mathrm{kJ\cdot mol^{-1}} \\ \Delta_f G^\circ[\mathrm{OH^-(aq)}] &= -157.244\ \mathrm{kJ\cdot mol^{-1}} \\ \Delta_f G^\circ[\mathrm{H_2O(l)}] &= -237.129\ \mathrm{kJ\cdot mol^{-1}}\end{aligned}$$

SOLUTION: The equation for the protonation of $CH_3NH_2(aq)$ is

$$\mathrm{CH_3NH_2(aq) + H_2O(l) \rightleftharpoons CH_3NH_3^+(aq) + OH^-(aq)}$$

and so

$$\begin{aligned}\Delta_r G^\circ &= \Delta_f G^\circ[\mathrm{CH_3NH_3^+(aq)}] + \Delta_f G^\circ[\mathrm{OH^-(aq)}] \\ &\qquad - \Delta_f G^\circ[\mathrm{CH_3NH_2(aq)}] - \Delta_f G^\circ[\mathrm{H_2O(l)}] \\ &= -39.86\ \mathrm{kJ\cdot mol^{-1}} - 157.244\ \mathrm{kJ\cdot mol^{-1}} \\ &\qquad - (20.77\ \mathrm{kJ\cdot mol^{-1}}) - (-237.129\ \mathrm{kJ\cdot mol^{-1}}) \\ &= 19.26\ \mathrm{kJ\cdot mol^{-1}}\end{aligned}$$

and

$$\begin{aligned}K = e^{-\Delta_r G^\circ/RT} &= \exp\left[-\frac{19.26\times 10^3\ \mathrm{J\cdot mol^{-1}}}{(8.3145\ \mathrm{J\cdot mol^{-1}\cdot K^{-1}})(298.15\ \mathrm{K^{-1}})}\right] \\ &= 4.23\times 10^{-4}\end{aligned}$$

## 13–12. Batteries and Fuel Cells Are Devices That Use Chemical Reactions to Produce Electric Currents

Batteries are devices that are commonly used to provide energy for numerous applications. Batteries are classified as primary if they are not rechargeable and as secondary if they are chargeable. Examples of primary batteries are alkaline batteries, commonly used in flashlights and toys; lithium batteries, used in wristwatches; and mercury batteries, used in heart pacemakers, hearing aids, and computers, all of which require the constant voltage this type of battery provides. Examples of secondary batteries are lead storage batteries used in automobiles, and nickel-cadmium (NiCad) batteries used in cordless tools, shavers, toothbrushes, and many other devices.

The cell diagram of a mercury battery can be described by

$$\mathrm{Zn(s)|ZnO(s)|KOH(paste)|HgO(s)|Hg(l)|steel}$$

with a simplified version of the corresponding cell reaction being given by

$$\mathrm{Zn(s) + HgO(s) \longrightarrow ZnO(s) + Hg(l)}$$

Notice that all the species in the above cell reaction are in condensed phases. Consequently, there is no change in reaction concentration during discharge, which means that the voltage (1.35 V) stays constant during discharge. This constant voltage, along with a long shelf life, is one of the important distinguishing features of a mercury battery.

Its low atomic mass and strong reducing potential give lithium many advantages as a material to use in batteries. For a given size, lithium batteries have the highest power-to-mass ratio of all batteries. They can also be designed to have long drain rates and long lifetimes, excellent characteristics for an application such as cardiac pacemakers and wristwatches. A lithium battery can power a wristwatch for more than five years.

The classic example of a secondary battery is a lead storage battery, whose cell diagram is

$$\mathrm{Pb(s)|PbSO_4(s)|H_2SO_4(aq)|PbO_2(s),\ PbSO_4(s)|Pb(s)}$$

where $\mathrm{PbO_2(s),\ PbSO_4(s)}$ denotes a heterogeneous mixture of the two solids. When fully charged, this cell produces about two volts. The 12-V batteries commonly used in automobiles consist of six cells in series. The overall cell reaction upon discharge is

$$\mathrm{Pb(s) + PbO_2(s) + 2\ H_2SO_4(aq)} \underset{\text{charge}}{\overset{\text{discharge}}{\rightleftharpoons}} \mathrm{2\ PbSO_4(s) + 2\ H_2O(l)}$$

During discharge, lead sulfate is formed at both electrodes and sulfuric acid is used up. In the process of being charged, the above reaction is reversed.

Another commonly used rechargeable battery is a nickel-cadmium (nicad) battery. A sealed nickel-cadmium battery is more stable than a lead storage battery and can be left inactive for long periods. The overall charge-discharge reaction can be written as

$$2\,\mathrm{NiOOH(s)} + \mathrm{Cd(s)} + 2\,\mathrm{H_2O(l)} \underset{\text{charge}}{\overset{\text{discharge}}{\rightleftharpoons}} 2\,\mathrm{Ni(OH)_2(s)} + \mathrm{Cd(OH)_2(s)}$$

In almost all batteries, a metal in the form of an electrode is oxidized by an oxidizing agent. Although batteries have the convenience of being portable and self-contained, they are not suitable for very large scale energy production. A promising alternative is a fuel cell, which is an electrochemical device that utilizes the oxidation of a fuel to produce electricity. Fuel cells differ from batteries in that the fuel and oxidizer are fed continuously into the cell, so that its operating life is essentially unlimited. The $H_2(g)$-$O_2(g)$ fuel cell, which has been a power source on many manned space flights, is shown schematically in Figure 13.16. Hydrogen is fed into a porous, metallic (often nickel) electrode, where it is oxidized to form water according to

$$\mathrm{H_2(g)} + 2\,\mathrm{OH^-(aq)} \longrightarrow 2\,\mathrm{H_2O(l)} + 2\,e^- \tag{13.56}$$

The $OH^-(aq)$ is supplied by the concentrated NaOH(aq) or KOH(aq) solution bathing the two electrodes. The oxygen is reduced at the other electrode according to

$$\mathrm{O_2(g)} + 2\,\mathrm{H_2O(l)} + 4\,e^- \longrightarrow 4\,\mathrm{OH^-(aq)} \tag{13.57}$$

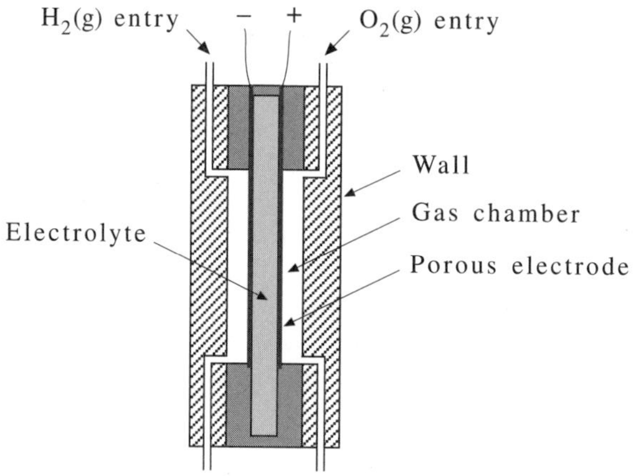


**FIGURE 13.16**
A schematic illustration of a $H_2(g)$-$O_2(g)$ fuel cell. The $H_2(g)$ is passed over a porous electrode, where oxidation takes place according to Equation 13.56. The resulting electrons pass through the external circuit, and the oxidizer [using $O_2(g)$] is reduced according to Equation 13.57. The current is carried through the cell by $OH^-(aq)$.

Note that the overall reaction is the combustion of hydrogen and oxygen to form water

$$2\,H_2(g) + O_2(g) \longrightarrow 2\,H_2O(l)$$

The current through the cell electrolyte is carried by the $Na^+(aq)$ and $OH^-(aq)$, so high concentrations are used to favor a low internal cell resistance.

The major appeal of fuel cells is that they are not limited by the inherent efficiency of heat engines. Recall from Section 6–7 that the maximum efficiency that we could obtain if we burned hydrogen to run a heat engine is given by (Equation 6.35)

$$\text{maximum efficiency} = \frac{T_h - T_c}{T_h}$$

where $T_h$ is the temperature of the hot reservior and $T_c$ is the temperature of the cold reservoir. Thus, if we were to run the reaction in a heat engine at 900 K with a low-temperature reservior at 300 K, the efficiency would be

$$\text{maximum efficiency} = \frac{900\text{ K} - 300\text{ K}}{900\text{ K}} = \frac{2}{3}$$

or 67%. Using the data in Table 5.2, the maximum amount of energy available is $(2/3)\Delta_r H^\circ = (2/3)(2)(285.83\text{ kJ}\cdot\text{mol}^{-1}) = 381.07$ kJ if two moles of $H_2(g)$ are burned in one mole of $O_2(g)$ under standard conditions. The value of $E^\circ$ for a half cell whose reaction is described by Equation 13.56 is $E^\circ = 0.8277$ V, and that for a half cell whose reaction is described by Equation 13.57 is 0.401 V, and so $E^\circ$ for the cell is 1.229 V. For a cell operating reversibly under standard conditions at constant $T$ and $P$, the maximum work that can be done by the cell is $-w_{rev}$, which is equal to $\Delta_r G^\circ$, and so

$$\begin{aligned} -w_{rev} &= nFE^\circ = (4\text{ mol})(96485\text{ C}\cdot\text{mol}^{-1})(1.229\text{ V}) \\ &= 474\text{ kJ} \end{aligned}$$

The above calculation is done for a system under ideal conditions, but the message is clear. In practice, fuel cells can realize efficiencies close to 90%, whereas heat engines have efficiencies up to about 30%.

**EXAMPLE 13–13**

Compare the idealized amount of work available when one mole of $CH_4(g)$ is burned in a fuel cell at 298 K and in a heat engine, with $T_h = 900$ K and $T_c = 300$ K. Calculate the value of $E^\circ$ for this fuel cell.

SOLUTION: The combustion reaction of one mole of methane is given by

$$CH_4(g) + 2\,O_2(g) \longrightarrow CO_2(g) + 2\,H_2O(l) \tag{13.58}$$

In Equation 13.58, the oxidation state of the carbon atom goes from −4 to +4, and that of each of the four oxygen atoms goes from 0 to −2. Thus, $n = 8$ in Equation 13.58.

The maximum work available when one mole of methane is burned in a fuel cell (under standard conditions) is given by (see Table 12.1)

$$\begin{aligned} -w_{\text{rev}} &= -\Delta_r G^\circ \\ &= -(2)(-237.141\ \text{kJ}\cdot\text{mol}^{-1}) - (-394.389\ \text{kJ}\cdot\text{mol}^{-1}) \\ &\qquad + (-50.768\ \text{kJ}\cdot\text{mol}^{-1}) \\ &= 817.903\ \text{kJ} \end{aligned}$$

The value of $\Delta_r H^\circ$ is given by (see Table 5.2)

$$\begin{aligned} \Delta_r H^\circ &= (2)(-285.83\ \text{kJ}\cdot\text{mol}^{-1}) + (-393.509\ \text{kJ}\cdot\text{mol}^{-1}) \\ &\qquad - (-74.81\ \text{kJ}\cdot\text{mol}^{-1}) \\ &= -890.26\ \text{kJ} \end{aligned}$$

and so the work available when one mole of methane is burned in a heat engine (with $T_h = 900$ K and $T_c = 300$ K) is

$$w = (-\Delta_r H^\circ)\left(\frac{900\ \text{K} - 300\ \text{K}}{900\ \text{K}}\right) = 593.51\ \text{kJ}\cdot\text{mol}^{-1}$$

The value of $E^\circ$ for the fuel cell is

$$E^\circ = -\frac{\Delta_r G^\circ}{nF} = -\frac{-817.903\ \text{kJ}\cdot\text{mol}^{-1}}{(8)(96485\ \text{C}\cdot\text{mol}^{-1})} = 1.06\ \text{V}$$

Because of their high efficiency and clean operation, fuel cells offer an attractive means for future energy production. Their high cost, however, prevents their widespread use. A major factor of this high cost is the catalyst (often platinum) needed to promote the reactions taking place at the electrodes. The development of electrodes that catalyze the oxidation of hydrocarbon fuels will most likely lead to the extensive use of fuel cells.

## Problems

**13-1.** Write the equations for the electrode reactions and the overall cell reaction for the electrochemical cells whose cell diagrams are

**a.** $Pb(s)|PbI_2(s)|HI(aq)|H_2(g)|Pt(s)$
**b.** $Cu(s)|Cu(ClO_4)_2(aq)||AgClO_4(aq)|Ag(s)$
**c.** $In(s)|In(NO_3)_3(aq)||CdCl_2(aq)|Cd(s)$
**d.** $Sn(s)|SnCl_2(aq)||AgNO_3(aq)|Ag(s)$

**13-2.** Write the equations for the electrode reactions and the overall cell reaction for the electrochemical cells whose cell diagram are

**a.** $Pb(s)|PbSO_4(s)|K_2SO_4(aq)|Hg_2SO_4(s)|Hg(l)$
**b.** $Pt(s)|H_2(g)|HCl(aq)|Hg_2Cl_2(s)|Hg(l)$

c. $Zn(s)|ZnO(s)|NaOH(aq)|HgO(s)|Hg(l)$
d. $Cd(s)|CdSO_4(aq)|Hg_2SO_4(s)|Hg(l)$

**13-3.** Consider an electrochemical cell in which the reaction is described by the equation

$$2\ HCl(aq) + Ca(s) \rightleftharpoons CaCl_2(aq) + H_2(g)$$

Predict the effect of the following changes on the cell voltage:

a. decrease in the amount of Ca(s)
b. increase in the pressure of $H_2(g)$
c. increase in the concentration of HCl(aq)
d. dissolution of $Ca(NO_3)_2(s)$ in the $CaCl_2(aq)$ solution

**13-4.** Given the following equation for an electrochemical cell reaction

$$H_2(g) + PbSO_4(s) \rightleftharpoons 2\ H^+(aq) + SO_4^{2-}(aq) + Pb(s)$$

predict the effect of the following changes on the cell voltage:

a. increase in the pressure of $H_2(g)$
b. increase in the size of the lead electrode
c. decrease in the pH of the cell electrolyte
d. dissolution of $Na_2SO_4(s)$ in the cell electrolyte
e. decrease in the amount of $PbSO_4(s)$
f. dissolution of a small amount of NaOH(s) in the cell eletrolyte

**13-5.** Determine the value of $n$ in the Nernst equation for the following equations:

a. $CH_4(g) + 2\ O_2(g) \longrightarrow CO_2(g) + 2\ H_2O(l)$
b. $2\ Zn(s) + Ag_2O_2(s) + 2\ H_2O(l) + 4\ OH^-(aq) \longrightarrow 2\ Ag(s) + 2\ Zn(OH)_4^{2-}(aq)$
c. $Cd(s) + Hg_2SO_4(s) \longrightarrow 2\ Hg(l) + Cd^{2+}(aq) + SO_4^{2-}(aq)$
d. $C_3H_8(g) + 5\ O_2(g) \longrightarrow 3\ CO_2(g) + 4\ H_2O(l)$

**13-6.** Consider the electrochemical cell whose cell diagram is $Pb(s)|PbBr_2(s)|HBr(aq)|H_2(1\text{ bar})|Pt(s)$. Given emf versus molality of HBr(aq) data for this cell, describe how you would determine the value of $E^\circ$.

**13-7.** Consider the electrochemical cell whose cell diagram is $Pt(s)|H_2(g)|HCl(aq)|Hg_2Cl_2(s)|Hg(l)$. Given emf versus molality of HCl(aq) data for this cell, describe how you would determine the value of $E^\circ$.

**13-8.** After having determined the value of $E^\circ$ in Problem 13–6, describe how you would determine the mean ionic activity coefficient of HBr(aq) as a function of molality.

**13-9.** After having determined the value of $E^\circ$ in Problem 13–7, describe how you would determine the mean ionic activity coefficient of HCl(aq) as a function of molality.

**13-10.** Without consulting Table 11.3, derive an expression for the activity of each of the following electrolytes in terms of the mean ionic activity coefficient and the molality:

a. KCl(aq) b. $CaCl_2(aq)$ c. $K_2SO_4(aq)$ d. $ZnSO_4(aq)$

**13-11.** Use the following data at 298.15 K for a cell whose cell diagram is $Pt(s)|H_2(1\ bar)|HCl(m)|AgCl(s)|Ag(s)$ to determine the value of $E^\circ$.

| $m/mol\cdot kg^{-1}$ | $E/V$ | $m/mol\cdot kg^{-1}$ | $E/V$ |
|---|---|---|---|
| 0.003215 | 0.52042 | 0.011195 | 0.45860 |
| 0.004488 | 0.50380 | 0.013407 | 0.44974 |
| 0.005619 | 0.49262 | 0.01710 | 0.43783 |
| 0.007311 | 0.47957 | 0.02563 | 0.41824 |
| 0.009138 | 0.46859 | 0.05391 | 0.38222 |

**13-12.** Use the following data at 298.15 K for a cell whose cell diagram is $Zn(s)|ZnCl_2(aq)|AgCl(s)|Ag(s)$ to determine the value of $E^\circ$.

| $m/mol\cdot kg^{-1}$ | $E/V$ | $m/mol\cdot kg^{-1}$ | $E/V$ |
|---|---|---|---|
| 0.002941 | 1.1983 | 0.04242 | 1.1090 |
| 0.007814 | 1.1650 | 0.09048 | 1.0844 |
| 0.01236 | 1.1495 | 0.2211 | 1.0556 |
| 0.02144 | 1.1310 | 0.4499 | 1.0328 |

**13-13.** Using the data in Table 13.2, calculate the emf at 298.15 K of the electrochemical cell whose cell diagram is $Pb(s)|PbI_2(s)|HI(a=0.800)|H_2(1\ bar)|Pt(s)$.

**13-14.** Use the data in Problem 13–12 and the resulting value of $E^\circ$ (0.9845 V) to determine the value of the mean ionic activity coefficient of $ZnCl_2(aq)$ at each given molality.

**13-15.** Use the data in Table 13.1 and the resulting value of $E^\circ$ (0.2224 V) to determine the value of the mean ionic activity coefficient of HCl(aq) at each given molality.

**13-16.** Use the data in Example 13–4 and the resulting value of $E^\circ$ (0.073 V) to determine the value of the mean ionic activity coefficient of HBr(aq) at each given molality.

**13-17.** Using the data in Table 13.2, calculate the value of the standard emf at 298.15 K of a cell whose cell diagram is $Hg(l)|Hg_2Cl_2(s)|HCl(aq)|H_2(1\ bar)|Pt(s)$. Which electrode is positive?

**13-18.** Using the data in Table 13.2, calculate the value of the standard emf at 298.15 K of a cell whose cell diagram is $Pb(s)|PbSO_4(s)|K_2SO_4(aq)|Hg_2SO_4(s)|Hg(l)$. Which electrode is positive?

**13-19.** Using the data in Table 13.2, calculate the value of the emf at 298.15 K of a cell whose cell diagram is $Ni(s)|Ni^{2+}(a=0.0250)||Zn^{2+}(a=0.300)|Zn(s)$. Which electrode is positive?

**13-20.** Write the cell diagram for an electrochemical cell whose cell reaction is $H_2(1\ bar) + I_2(s) \longrightarrow 2\ HI(aq)$. Using the data in Table 13.2, calculate the value of $E^\circ$ at 298.15 K for the cell.

**13-21.** Write the cell diagram for an electrochemical cell whose cell reaction is $PbCl_2(s) \longrightarrow Pb^{2+}(aq) + 2\ Cl^-(aq)$. Using the data in Table 13.2, calculate the value of $E^\circ$ at 298.15 K for the cell.

**13-22.** Write the cell diagram for an electrochemical cell whose cell reaction is $2\ HCl(aq) + Zn(s) \longrightarrow ZnCl_2(aq) + H_2(g)$. Using the data in Table 13.2, calculate the value of $E^\circ$ at 298.15 K for the cell.

**13-23.** Calculate the value of the emf at 298.15 K of a cell whose cell diagram is $Pt(s)|Sn^{2+}(a=0.200), Sn^{4+}(a=0.400)||Fe^{3+}(a=0.300)|Fe(s)$.

**13-24.** Use the data in Table 13.2 to calculate the value of $E$ at 298.15 K for the cell whose cell reaction is

$$Cu(s) + 2\ AgNO_3(0.100\ m) \rightleftharpoons 2\ Ag(s) + Cu(NO_3)_2(1.00\ m)$$

Take $\gamma_\pm = 0.734$ for $AgNO_3(0.100\ m)$ and 0.338 for $Cu(NO_3)_2(1.00\ m)$. Will the reaction proceed spontaneously?

**13-25.** Use the data in Table 13.2 to calculate the value of $E$ at 298.15 K for the cell whose cell reaction is

$$Cd(s) + ZnCl_2(1.00\ m) \rightleftharpoons Zn(s) + CdCl_2(1.00\ m)$$

Take $\gamma_\pm = 0.341$ for $ZnCl_2(1.00\ m)$ and 0.0669 for $CdCl_2(1.00\ m)$. Will the reaction proceed spontaneously?

**13-26.** Consider the electrochemical cell whose cell diagram is $Zn(s)|ZnI_2(m)|I_2(s)|Pt(s)$. Given that $\gamma_\pm = 0.605$ for 0.500 molal $ZnI_2(aq)$, calculate the value of the emf of the cell at 298.15 K.

**13-27.** Consider the electrochemical cell whose cell diagram is $Cd(s)|CdCl_2(m)|AgCl(s)|Ag(s)$. Given that $\gamma_\pm = 0.0669$ for 1.00 molal $CdCl_2(aq)$, calculate the value of the emf of the cell at 298.15 K.

**13-28.** The standard emf for the reaction described by

$$HClO(aq) + H^+(aq) + 2\ Cr^{2+}(aq) \longrightarrow 2\ Cr^{3+}(aq) + Cl^-(aq) + H_2O(l)$$

is $E^\circ = 1.90$ V. Use the data in Table 13.2 to calculate the value of $E^\circ_{red}$ at 298.15 K for the electrode reaction

$$HClO(aq) + H^+(aq) + 2\ e^- \longrightarrow Cl^-(aq) + H_2O(l)$$

**13-29.** Design an electrochemical cell that can be used to determine the activity coefficients of $H_2SO_4(aq)$ solutions. Write its cell diagram and the corresponding Nernst equation, relating $E$ to $\gamma_\pm$. Determine the value of $E^\circ$ at 298.15 K from Table 13.2.

**13-30.** Data are given below for the emf versus concentration at 298.15 K for the cell whose cell diagram is $Pt(s)|H_2(1\ bar)|HBr(aq)|AgBr(s)|Ag(s)$. Determine the value of $E^\circ$ and calculate the value of $\gamma_\pm$ of HBr(aq) for each concentration.

| $m/10^{-4}\ mol\cdot kg^{-1}$ | 3.198 | 4.042 | 8.444 | 13.55 | 18.50 |
|---|---|---|---|---|---|
| $E/V$ | 0.48775 | 0.47604 | 0.43850 | 0.41450 | 0.39891 |

**13-31.** The temperature dependence of the standard emf of a cell whose cell diagram is Pt(s)|$H_2$(1 bar)|HCl($a$ = 1.00)|AgCl(s)|Ag(s) is given by

$$E^\circ/\text{V} = 0.23659 - (4.8564 \times 10^{-4}\ {}^\circ\text{C}^{-1})t - (3.4205 \times 10^{-6}\ {}^\circ\text{C}^{-2})t^2$$
$$+ (5.869 \times 10^{-9}\ {}^\circ\text{C}^{-3})t^3 \qquad 0^\circ\text{C} \le t \le 50^\circ\text{C}$$

where the temperature is in degrees Celsius. Write the associated cell reaction, and determine the standard change in Gibbs energy, the enthalpy, and the entropy at 298.15 K.

**13-32.** The temperature dependence of the standard emf of a cell whose cell diagram is Pt(s)|$H_2$(1 bar)|HBr($a$ = 1.00)|$Hg_2Br_2$(s)|Hg(l) is given by

$$E^\circ/\text{V} = 0.13970 - (1.54 \times 10^{-4}\ {}^\circ\text{C}^{-1})(t - 25.0^\circ\text{C}) - (3.6 \times 10^{-6}\ {}^\circ\text{C}^{-2})(t - 25.0^\circ\text{C})^2$$

where $t$ is the Celsius temperature. Determine the change in the standard Gibbs energy, the standard enthalpy, and the standard entropy at 298.15 K. Given that $S^\circ[H_2(g)] =$ 130.684 $\text{J}\cdot\text{K}^{-1}\cdot\text{mol}^{-1}$, $S^\circ[Hg_2Br_2(s)] = 218.0$ $\text{J}\cdot\text{K}^{-1}\cdot\text{mol}^{-1}$, and $S^\circ[Hg(l)] =$ 76.02$\text{J}\cdot\text{K}^{-1}\cdot\text{mol}^{-1}$ at 298.15 K, calculate the value of $S^\circ$ for $Br^-$(aq). Compare your answer with the value in the *NBS Tables of Chemical Thermodynamic Properties*.

**13-33.** Use the data in Table 13.2 to calculate the value of the solubility product of $Hg_2Cl_2$(s) at 298.15 K.

**13-34.** Use the data in Table 13.2 to calculate the value of the solubility product of $Ag_2SO_4$(s) at 298.15 K.

**13-35.** Use the data in Table 13.2 to calculate the value of the solubility product of $PbSO_4$(s) at 298.15 K.

**13-36.** Given that the standard reduction potential of 2 $D^+$(aq) + 2 $e^-$ $\longrightarrow$ $D_2$(g) is −0.0034 V, calculate the value of the equilibrium constant of

$$2\ \text{H}^+(\text{aq}) + \text{D}_2(\text{g}) \longrightarrow 2\ \text{D}^+(\text{aq}) + \text{H}_2(\text{g})$$

at 298.15 K.

**13-37.** Given the reduction electrode-reaction data at 298.15 K

$$\text{HClO(aq)} + \text{H}^+(\text{aq}) + 2\,e^- \longrightarrow \text{Cl}^-(\text{aq}) + \text{H}_2\text{O(l)} \qquad E^\circ = 1.49\ \text{V}$$

$$\text{ClO}^-(\text{aq}) + \text{H}_2\text{O(l)} + 2\,e^- \longrightarrow \text{Cl}^-(\text{aq}) + 2\ \text{OH}^-(\text{aq}) \qquad E^\circ = 0.90\ \text{V}$$

calculate the value of the acid-dissociation constant of HClO(aq). *Hint*: You need to use the fact that $K_w = 1.00 \times 10^{-14}$.

**13-38.** Given the reduction electrode-reaction data at 298.15 K

$$\text{ClO}_2(\text{aq}) + e^- \longrightarrow \text{ClO}_2^-(\text{aq}) \qquad E^\circ = 1.15\ \text{V}$$

$$\text{ClO}_2(\text{aq}) + \text{H}^+(\text{aq}) + e^- \longrightarrow \text{HClO}_2(\text{aq}) \qquad E^\circ = 1.27\ \text{V}$$

calculate the value of the acid-dissociation constant of $HClO_2$(aq).

**13-39.** The value of the acid-dissociation constant of propanoic acid can be determined using the cell

$$\mathrm{Pt(s)|H_2(g)|HP(aq),\ NaP(aq),\ NaCl(aq)|AgCl(s)|Ag(s)}$$

Use the following data at 10°C to determine the value of $K_a$ for propanoic acid at 10°C. Take $E^\circ$ for the cell to be 0.23142 V at 10°C.

| $m_{\mathrm{HP}} = m_{\mathrm{NaP}} = m_{\mathrm{NaCl}}$ | $E/\mathrm{V}$ |
|---|---|
| 0.006442 | 0.62854 |
| 0.007055 | 0.62622 |
| 0.009225 | 0.61973 |
| 0.010812 | 0.61590 |
| 0.01660 | 0.60544 |
| 0.02274 | 0.59757 |
| 0.026833 | 0.59360 |
| 0.03067 | 0.59037 |
| 0.03108 | 0.59001 |

**13-40.** The value of the acid-dissociation constant of acetic acid can be determined using the cell

$$\mathrm{Pt(s)|H_2(g)|HAc(aq),\ NaAc(aq),\ NaCl(aq)|AgCl(s)|Ag(s)}$$

Use the following data at 0°C to determine the value of $K_a$ for acetic acid at 0°C. Take $E^\circ$ for the cell to be 0.23655 V at 0°C.

| $m_{\mathrm{HAc}}/\mathrm{mol\cdot kg^{-1}}$ | $m_{\mathrm{NaAc}}/\mathrm{mol\cdot kg^{-1}}$ | $m_{\mathrm{NaCl}}/\mathrm{mol\cdot kg^{-1}}$ | $E/\mathrm{V}$ |
|---|---|---|---|
| 0.0047790 | 0.0045990 | 0.0048960 | 0.61995 |
| 0.012035 | 0.011582 | 0.012426 | 0.59826 |
| 0.021006 | 0.020216 | 0.021516 | 0.58528 |
| 0.049220 | 0.047370 | 0.050420 | 0.56545 |
| 0.081010 | 0.077960 | 0.082970 | 0.55388 |
| 0.090560 | 0.087160 | 0.092760 | 0.55128 |

**13-41.** The value of the acid-dissociation constant of formic acid can be determined using the cell

$$\mathrm{Pt(s)|H_2(g)|HFo(aq),\ NaFo(aq),\ NaCl(aq)|AgCl(s)|Ag(s)}$$

Determine the equation for the overall reaction of this cell and write the corresponding Nernst equation. Now use the relation

$$K_a = \frac{a_{\mathrm{H^+}} a_{\mathrm{Fo^-}}}{a_{\mathrm{HFo}}}$$

to show that

$$E = E^\circ - \frac{RT}{F} \ln \frac{m_{\mathrm{HFo}} \gamma_{\mathrm{HFo}} m_{\mathrm{Cl^-}} \gamma_{\mathrm{Cl^-}}}{m_{\mathrm{Fo^-}} \gamma_{\mathrm{Fo^-}}} - \frac{RT}{F} \ln K_a$$

Show that this equation can be written as

$$\frac{F}{RT}(E - E^{\circ}) + \ln \frac{m_{\text{HFo}} m_{\text{Cl}^-}}{m_{\text{Fo}^-}} = -\ln K_a - \ln \frac{\gamma_{\text{HFo}} \gamma_{\text{Cl}^-}}{\gamma_{\text{Fo}^-}}$$

$$= -\ln K_a'$$

Describe how the value of $K_a$ can be determined by plotting

$$\frac{F}{RT}(E - E^{\circ}) + \ln \frac{m_{\text{HFo}} m_{\text{Cl}^-}}{m_{\text{Fo}^-}} = -\ln K_a'$$

against ionic strength and extrapolating to zero. Unlike the case of propanoic acid presented in Example 13–10, the $H^+(aq)$ and $Fo^-(aq)$ from the dissociation of HFo(aq) cannot be ignored. The way to proceed is as follows. First, calculate the values of $\ln K_a'$ and $I_m$ neglecting the $H^+(aq)$ and $Fo^-(aq)$ from the dissociation of formic acid. Plot $\ln K_a'$ against $I_m$, and obtain a preliminary value of $K_a$ by extrapolating to zero ionic strength. Now use the preliminary value of $K_a$ to calculate $m_{H^+}$ using

$$K_a = \frac{m_{\text{H}^+} m_{\text{Fo}^-}}{m_{\text{HFo}}} \frac{\gamma_{\text{H}^+} \gamma_{\text{Fo}^-}}{\gamma_{\text{HFo}}} = \frac{m_{\text{H}^+}(m_{\text{NaFo}} + m_{\text{H}^+})}{m_{\text{HFo}} - m_{\text{H}^+}} \gamma_{\pm}^2$$

where $m_{\text{HFo}}$ and $m_{\text{NaFo}}$ are the stoichiometric concentrations of HFo(aq) and NaFo(aq). Realizing that $m_{H^+}$ will be fairly small, neglect $m_{H^+}$ with respect to $m_{\text{HFo}}$ and $m_{\text{NaFo}}$ and write

$$m_{\text{H}^+} = \frac{m_{\text{HFo}}}{m_{\text{NaFo}}} \frac{K_a}{\gamma_{\pm}^2}$$

The value of $\gamma_{\pm}$ can be estimated using Equation 11.56 on a molality scale

$$\ln \gamma_{\pm} = -\frac{1.171(I_m/\text{ mol}\cdot\text{kg}^{-1})^{1/2}}{1 + (I_m/\text{ mol}\cdot\text{kg}^{-1})^{1/2}}$$

Using the above procedure and the following data at 25°C, calculate the value of $K_a$ for formic acid at 25°C.

| $m_{\text{HFo}}/\text{mol}\cdot\text{kg}^{-1}$ | $m_{\text{NaFo}}/\text{mol}\cdot\text{kg}^{-1}$ | $m_{\text{NaCl}}/\text{mol}\cdot\text{kg}^{-1}$ | $E/\text{V}$ |
|---|---|---|---|
| 0.0065380 | 0.0081509 | 0.0070544 | 0.57842 |
| 0.011760 | 0.014661 | 0.012689 | 0.56294 |
| 0.024450 | 0.030482 | 0.026381 | 0.54398 |
| 0.035750 | 0.044570 | 0.038574 | 0.53427 |
| 0.048630 | 0.060627 | 0.052471 | 0.52647 |
| 0.098760 | 0.12312 | 0.10656 | 0.50882 |

**13-42.** The value of the dissociation constant of water, $K_w$, can be determined using the cell

$$\text{Pt(s)}|\text{H}_2\text{(g)}|\text{KOH(aq), KCl(aq)}|\text{AgCl(s)}|\text{Ag(s)} \qquad (1)$$

Show that the emf of this cell is given by

$$E = E^{\circ} - \frac{RT}{F} \ln m_{\text{H}^+} m_{\text{Cl}^-} \gamma_{\text{H}^+} \gamma_{\text{Cl}^-}$$

where $E^\circ$ is the standard emf of the cell

$$\mathrm{Pt(s)|H_2(g)|HCl(aq)|AgCl(s)|Ag(s)}$$

Using the relation

$$K_\mathrm{w} = m_{\mathrm{H^+}} m_{\mathrm{OH^-}} \frac{\gamma_{\mathrm{H^+}}\gamma_{\mathrm{OH^-}}}{a_{\mathrm{H_2O}}}$$

show that

$$\frac{F}{RT}(E - E^\circ) + \ln\frac{m_{\mathrm{KCl}}}{m_{\mathrm{KOH}}} = \ln\frac{\gamma_{\mathrm{OH^-}}}{\gamma_{\mathrm{Cl^-}} a_{\mathrm{H_2O}}} - \ln K_\mathrm{w}$$
$$= -\ln K'_\mathrm{w}$$

Using the following data for the cell given by Equation 1 with $m_{\mathrm{KOH}} = 0.0100\ \mathrm{mol \cdot kg^{-1}}$, determine the value of $K_\mathrm{w}$ at each temperature by plotting the left side of the above equation against ionic strength and extrapolating to zero. Take $E^\circ$ to be given by

$$E^\circ/\mathrm{V} = 0.23634 - 0.00047396(t/^\circ\mathrm{C}) - 3.2331 \times 10^{-6}(t/^\circ\mathrm{C})^2$$
$$- 9.1128 \times 10^{-9}(t/^\circ\mathrm{C})^3 + 1.477 \times 10^{-10}(t/^\circ\mathrm{C})^4$$

| $m_{\mathrm{KCl}}/\mathrm{mol \cdot kg^{-1}}$ | $E/\mathrm{V}$ at 0°C | $E/\mathrm{V}$ at 10°C | $E/\mathrm{V}$ at 20°C | $E/\mathrm{V}$ at 30°C | $E/\mathrm{V}$ at 40°C | $E/\mathrm{V}$ at 50°C | $E/\mathrm{V}$ at 60°C |
|---|---|---|---|---|---|---|---|
| 0.0100000 | 1.0462 | 1.0478 | 1.04947 | 1.0512 | 1.0530 | 1.0548 | 1.0567 |
| 0.020000 | 1.0299 | 1.0309 | 1.03203 | 1.0332 | 1.0344 | 1.0356 | 1.0370 |
| 0.030000 | 1.0204 | 1.0211 | 1.02183 | 1.0226 | 1.0235 | 1.0244 | 1.0253 |
| 0.040000 | 1.0136 | 1.0141 | 1.01458 | 1.0151 | 1.0158 | 1.0164 | 1.0171 |
| 0.050000 | 1.0084 | 1.0087 | 1.00903 | 1.0094 | 1.0098 | 1.0102 | 1.0107 |
| 0.070000 | 1.0005 | 1.0005 | 1.00057 | 1.0007 | 1.0008 | 1.0010 | 1.0012 |
| 0.10000 | 0.99217 | 0.99192 | 0.991700 | 0.99151 | 0.99136 | 0.99123 | 0.99114 |
| 0.20000 | 0.97605 | 0.97529 | 0.974550 | 0.97380 | 0.97309 | 0.97239 | 0.97170 |
| 0.30000 | 0.96680 | 0.96571 | 0.964640 | 0.96357 | 0.96251 | 0.96145 | 0.96041 |
| 0.40000 | 0.96029 | 0.95901 | 0.957730 | 0.95643 | 0.95513 | 0.95383 | 0.95250 |
| 0.50000 | 0.95537 | 0.95391 | 0.952440 | 0.95098 | 0.94948 | 0.94798 | 0.94648 |
| 0.60000 | 0.95138 | 0.94978 | 0.948170 | 0.94655 | 0.94492 | 0.94328 | 0.94164 |
| 0.70000 | 0.94805 | 0.94634 | 0.944620 | 0.94288 | 0.94113 | 0.93936 | 0.93758 |
| 0.80000 | 0.94517 | 0.94337 | 0.941540 | 0.93970 | 0.93784 | 0.93596 | 0.93406 |
| 0.90000 | 0.94271 | 0.94082 | 0.938910 | 0.93462 | 0.93257 | 0.93049 | 0.92839 |
| 1.2500 | 0.93604 | 0.93389 | 0.931710 | 0.92950 | 0.92726 | 0.92500 | 0.92271 |
| 1.5000 | 0.93234 | 0.93005 | 0.927740 | 0.92539 | 0.92301 | 0.92060 | 0.91817 |
| 1.7500 | 0.92955 | 0.92714 | 0.924690 | 0.92204 | 0.91971 | 0.91717 | 0.91460 |
| 2.0000 | 0.92703 | 0.92454 | 0.922010 | 0.91934 | 0.91435 | 0.91159 | 0.90880 |
| 2.5000 | 0.92309 | 0.92042 | 0.917720 | 0.91497 | 0.91218 | 0.90935 | 0.90648 |
| 3.0000 | 0.92004 | 0.91723 | 0.914370 | 0.91146 | 0.90851 | 0.90555 | 0.90249 |
| 3.2500 | 0.91875 | 0.91587 | 0.912940 | 0.90997 | 0.90675 | 0.90388 | 0.90077 |
| 3.5000 | 0.91751 | 0.91458 | 0.911600 | 0.90857 | 0.90550 | 0.90237 | 0.89920 |

Now plot $\ln K_w$ against $1/T$, curve fit your result to

$$\ln K_w = -\frac{a}{T} - b\ln T + cT + d$$

and use Equation 12.29 to determine the value of $\Delta_r H^\circ$ for the dissociation of water as a quadratic polynomial in $T$. The experimental value of $\Delta_r H^\circ$ at 25°C is 55.9 $kJ\cdot mol^{-1}$.

**13-43.** Use the data in Table 13.2 to calculate the value of the solubility product of $PbCl_2(s)$ at 298.15 K, and compare your result with the value that you obtain using the data in the *NBS Tables of Chemical Thermodynamic Properties* [*J. Phys. Chem. Data*, vol. II, suppl. 2 (1982)].

**13-44.** Use the data in Table 13.2 to calculate the value of the solubility product of $Hg_2SO_4(s)$ at 298.15 K, and compare your result with the value that you obtain using the data in the *NBS Tables of Chemical Thermodynamic Properties* [*J. Phys. Chem. Data*, vol. II, suppl. 2 (1982)].

**13-45.** Given that $\Delta_f G^\circ[Pb^{2+}(aq)] = -24.43\ kJ\cdot mol^{-1}$, $\Delta_f G^\circ[SO_4^{2-}(aq)] = -744.53\ kJ\cdot mol^{-1}$, and $\Delta_f G^\circ[PbSO_4(s)] = -813.4\ kJ\cdot mol^{-1}$ at 298.15 K, calculate the value of the solubility product of $PbSO_4(s)$. Compare your result with the one obtained in Problem 13–35.

**13-46.** Use the data in Figure 13.15 along with $\Delta_f G^\circ[Br^-(aq)] = -103.96\ kJ\cdot mol^{-1}$ to calculate the value of the solubility product of thallium(I) bromide at 298.15 K.

**13-47.** Use the data in *NBS Tables of Chemical Thermodynamic Properties* [*J. Phys. Chem. Data*, vol. II, suppl. 2 (1982)] to calculate the value of the solubility product of barium sulfate at 298.15 K.

**13-48.** Given that $\Delta_f G^\circ[HSO_4^-(aq)] = -755.91\ kJ\cdot mol^{-1}$ and that $\Delta_f G^\circ[SO_4^{2-}(aq)] = -744.53\ kJ\cdot mol^{-1}$, calculate the value of the second acid-dissociation constant of sulfuric acid at 298.15 K.

**13-49.** Use the data in *NBS Tables of Chemical Thermodynamic Properties* [*J. Phys. Chem. Data*, vol. II, suppl. 2 (1982)] to calculate the value of the first acid-dissociation constant of arsenic acid, $H_3AsO_4(aq)$ at 298.15 K.

**13-50.** Use the data in *NBS Tables of Chemical Thermodynamic Properties* [*J. Phys. Chem. Data*, vol. II, suppl. 2 (1982)] to calculate the value of the first base-protonation constant of hydrazine at 298.15 K.

**13-51.** Compare the idealized amount of work available when one mole of propane is burned in a fuel cell at 298 K and in a heat engine with $T_h = 900$ K and $T_c = 300$ K. Calculate the value of $E^\circ$ for this fuel cell.

**13-52.** Derive a general relation for the pressure dependence of the emf of an electrochemical cell at constant temperature. Show that for the cell whose diagram is $Pt(s)|H_2(g)|HCl(aq)|AgCl(s)|Ag(s)$ that

$$\left(\frac{\partial E}{\partial P}\right)_T = -\frac{2(\overline{V}_{HCl} + \overline{V}_{Ag}) - (\overline{V}_{H_2} + 2\overline{V}_{AgCl})}{F}$$

where $\overline{V}_{\mathrm{HCl}}$ is the partial molar volume of the HCl(aq) at molality $m$, $\overline{V}_{\mathrm{H_2}}$ is the partial molar volume of $H_2$(g) at pressure $P$, and $\overline{V}_{\mathrm{Ag}}$ and $\overline{V}_{\mathrm{AgCl}}$ are the molar volumes of Ag(s) and AgCl(s). Argue that at relatively low pressures

$$E_2 - E_1 \approx \frac{RT}{2F} \ln \frac{P_2}{P_1}$$

where $E_j$ is the emf of the cell at pressure $P_j$.

**13-53.** Consider the cell whose cell diagram is

$$\mathrm{Pt(s)|H_2(g)|HCl(0.100\ m)|Hg_2Cl_2(s)|Hg(l)}$$

Show that $\Delta_r V \approx -\overline{V}_{\mathrm{H_2}}$. Using a virial equation of state through the second virial coefficient (Equation 2.23), show that (see Problem 13–52)

$$\left(\frac{\partial E}{\partial P}\right)_T = \frac{RT}{2F}\left[\frac{1}{P} + B_{2P}(T) + \cdots\right]$$

Now use the data in Table 2.7 and Figure 2.15 to estimate the value of $B_{2P}(T) = B_{2V}(T)/RT$ for $H_2$(g) at 25°C. [The value of $B_{2V}^*(T^*)$ from numerical tables is 0.416.] Finally, use the following data at 25°C for the above cell to calculate the value of $E$ as a function of pressure, and compare the result of your calculation with the experimental results graphically.

| $P$/atm | $E$/V |
|---|---|
| 1.0000 | 0.39900 |
| 37.800 | 0.44560 |
| 51.600 | 0.44960 |
| 110.20 | 0.45960 |
| 204.70 | 0.46830 |
| 439.30 | 0.48040 |
| 568.80 | 0.48500 |
| 701.80 | 0.48910 |
| 754.40 | 0.49030 |
| 893.90 | 0.49380 |
| 1035.2 | 0.49750 |

*The next five problems involve calculating the value of $E°$ for an electrode reaction that is a combination of other electrode reactions.*

**13-54.** Sometimes it might be necessary to calculate the value of $E°$ for a reduction electrode reaction that is a combination of other reduction electrode reactions. For example, consider the two reduction electrode reactions

**1.** $8\ \mathrm{H^+(aq)} + \mathrm{MnO_4^-(aq)} + 5\ e^- \longrightarrow 4\ \mathrm{H_2O(l)} + \mathrm{Mn^{2+}(aq)}$ $\qquad E_1^\circ = 1.491\ \mathrm{V}$

**2.** $4\ \mathrm{H^+(aq)} + \mathrm{MnO_2(s)} + 2\ e^- \longrightarrow 2\ \mathrm{H_2O(l)} + \mathrm{Mn^{2+}(aq)}$ $\qquad E_2^\circ = 1.208\ \mathrm{V}$

We can use these data to calculate the value of $E°$ for

**3.** $4\ \mathrm{H^+(aq)} + \mathrm{MnO_4^-(aq)} + 3\ e^- \longrightarrow \mathrm{MnO_2(s)} + 2\ \mathrm{H_2O(l)}$

We first note that Equation 3 results from subtracting Equation 2 from Equation 1. It is important to realize that $E_3^\circ$ is *not* equal to $E_1^\circ - E_2^\circ$, however, because $E^\circ$ is an *intensive* property. The standard Gibbs energy change for Equation 3 *is* given by $\Delta G_3^\circ = \Delta G_1^\circ - \Delta G_2^\circ$, however. Using this fact, show that

$$E_3^\circ = \frac{5E_1^\circ - 2E_2^\circ}{3} = 1.680 \text{ V}$$

**13-55.** Given that

$$Hg_2^{2+}(aq) + 2\,e^- \longrightarrow 2\,Hg(l) \qquad E^\circ = +0.796 \text{ V}$$

$$2\,Hg^{2+}(aq) + 2\,e^- \longrightarrow Hg_2^{2+}(aq) \qquad E^\circ = +0.907 \text{ V}$$

calculate the value of $E^\circ$ for

$$Hg^{2+}(aq) + 2\,e^- \longrightarrow Hg(l)$$

**13-56.** Given that

$$Cr^{2+}(aq) + 2\,e^- \longrightarrow Cr(s) \qquad E^\circ = -0.91 \text{ V}$$

$$Cr^{3+}(aq) + e^- \longrightarrow Cr^{2+}(aq) \qquad E^\circ = -0.41 \text{ V}$$

calculate the value of $E^\circ$ for

$$Cr^{3+}(aq) + 3\,e^- \longrightarrow Cr(s)$$

**13-57.** The previous three problems develop the idea that you must use values of $\Delta G^\circ$ to calculate values of $E^\circ$ of reduction electrode reactions from the values of $E^\circ$ from other electrode reactions. When the combination is such that the electrons on each side of the equation cancel, however, we do not have to use $\Delta G^\circ$ as an intermediate quantity and can write $E_{rxn}^\circ = E_{ox}^\circ + E_{red}^\circ$ directly as we did in Equation 13.22. In this problem, we derive Equation 13.22.

Consider the two reduction electrode reactions

**1.** $A + n_1\,e^- \longrightarrow X \qquad E_{1,red}^\circ$

**2.** $B + n_2\,e^- \longrightarrow Y \qquad E_{2,red}^\circ$

Show that the combination in which the electrons cancel is

**3.** $n_2\,A + n_1\,Y \longrightarrow n_1\,B + n_2\,X$

Show that $\Delta G_3^\circ$ for Equation 3 is

$$\Delta G_3^\circ = n_2 \Delta G_1^\circ - n_1 \Delta G_2^\circ$$
$$= -n_1 n_2 F E_{1,red}^\circ + n_1 n_2 F E_{2,red}^\circ$$

and that

$$E_3^\circ = E_{1,red}^\circ - E_{2,red}^\circ = E_{1,red}^\circ + E_{2,ox}^\circ$$

which is Equation 13.22

**13-58.** Consider the reduction electrode reaction

$$4\,H^+(aq) + MnO_4^-(aq) + 3\,e^- \longrightarrow MnO_2(s) + 2\,H_2O(l) \qquad E^\circ = 1.679\ V$$

Using the fact that the dissociation constant for $H_2O(l)$ is $K_w = 1.008 \times 10^{-14}$ at 298.15 K, calculate the value of $E^\circ$ for

$$2\,H_2O(l) + MnO_4^-(aq) + 3\,e^- \longrightarrow MnO_2(s) + 4\,OH^-(aq)$$

**Lars Onsager** was born in Oslo, Norway on November 27, 1903, and died in 1976. He received a degree in chemical engineering from the Norwegian Institute of Technology in Trondheim in 1925, after which he went to Zurich to discuss with Peter Debye a flaw he had discovered in the Debye-Hückel theory. Onsager was a graduate student with Debye from 1926 to 1928, during which time he developed the Onsager limiting law for the conductivity of dilute solutions of electrolytes. In 1928, he emigrated to the United States and spent five years as an associate in chemistry at Brown University, where he developed the Onsager reciprocal relations, which the Norwegian Institute of Technology rejected as his doctoral thesis. In 1933, he became a Sterling and Gibbs postdoctoral fellow at Yale University. When it was dicovered that he had never received his doctorate, the department persuaded him to submit a thesis for a Yale doctorate. Rather than submit a thesis on the reciprocal relations, he chose instead a mathematical topic. In 1945, he was appointed as the J. Willard Gibbs Professor of Theoretical Chemistry and remained at Yale until his retirement in 1972. He became a US citizen in 1947. He taught only graduate courses at Yale because of his reputation as a difficult lecturer. Students often jokingly referred to his statistical mechanics courses as "Advanced Norwegian I and II." He spent his remaining years at the Center for Theoretical Studies at the University of Miami at Coral Gables, Florida, working in biophysics. Onsager had a wide range of interests in theoretical chemistry and physics including the development of the thermodynamic theory of irreversible processes. Onsager received the Nobel Prize for chemistry in 1968 for the discovery of the reciprocal relations bearing his name, which are fundamental for the thermodynamics of irreversible processes.

CHAPTER 14

# Nonequilibrium Thermodynamics

To this point, we have discussed the thermodynamics only of systems in equilibrium. Realize that thermodynamics allows us to derive relationships between measurable properties such as temperature, pressure, and heat capacity. Note that these are properties of systems in equilibrium. There are many measurable properties of nonequilibrium systems such as thermal conductivity, electrical conductivity, and diffusion coefficients. Note that these properties involve processes that are the result of differences in temperature, electrical potential, and concentration, respectively, so that such systems are not in equilibrium.

We want to investigate whether we can extend our treatment of thermodynamics so that we can derive relationships between various nonequilibrium properties. It so happens that we can do this with the introduction of a few reasonable, additional postulates. This part of thermodynamics is called *nonequilibrium thermodynamics* or the *thermodynamics of irreversible processes*.

In the first two sections of this chapter, we will reexamine the Second Law of Thermodynamics. We begin there because the entropy, although strictly an equilibrium thermodynamic property, clearly has a "directionality" to it because $dS \geq 0$ for any natural process in an isolated system. In the subsequent two sections, we will introduce time into our thermodynamic equations and discuss the linear flux-force relations that are central to nonequilibrium thermodynamics for systems that are not too far from equilibrium. The most celebrated result of these two sections are the Onsager reciprocal relations, which we will see are the keys for deriving the relationships between various nonequilibrium properties. We present our first practical application of nonequilibrium thermodynamics in Section 14–5 and show how various apparently disparate electrokinetic quantities are actually related to each other. Then, in Section 14–6, we use a simple chemical kinetic scheme to show the fundamental basis of the Onsager reciprocal relations. In the next few sections, we use nonequilibrium thermodynamics to derive equations for the magnitude of the liquid junction potential, which we stated in Chapter 13 is the electrical potential that results whenever two different electrolyte solutions are brought into contact. All the systems we discuss in the first nine sections are discontinuous systems, in the sense that they consist of two subsystems at different temperatures, chemical potentials, or electrical potentials in contact with each other. In Section 14–10, we extend our approach to continuous systems, which are those in which temperature, chemical potential, or electrical potential vary smoothly from one region to another. In particular, we will study diffusion in binary and ternary solutions. We discuss a few general principles of nonequilibrium thermodynamics in the last two sections. In Section 14–11, we show that the entropy production is a minimum when a system is in a steady state near equilibrium as compared with an equilibrium state, at which point the entropy production is zero. In the last section, we briefly discuss extensions of nonequilibrium thermodynamics to systems that are far from equilibrium.

## 14–1. Entropy Is Always Produced in a Spontaneous Process

In our discussion of the Second Law of Thermodynamics in Section 6–4, we showed that (Equation 6.17)

$$\begin{aligned} dS &> 0 \quad \text{(spontaneous process in an isolated system)} \\ dS &= 0 \quad \text{(reversible process in an isolated system)} \end{aligned} \tag{14.1}$$

Equations 14.1 apply only to processes that take place in isolated systems. For other types of systems, we found it convenient to view $dS$ as consisting of two parts. One part of $dS$ is the entropy created in the system by any spontaneous processes occurring within it, and the other part is the change in entropy resulting from the exchange of energy as heat between the system and its surroundings. These two contributions account for the entire change in the entropy. We denoted the part of $dS$ that is created by any spontaneous process by $dS_{\text{prod}}$. This quantity is always positive. We denoted the part of $dS$ that results from the exchange of energy as heat with the surroundings by $dS_{\text{exch}}$. This quantity is given by $\delta q/T$, where $T$ is the temperature of the surroundings; it can be positive, negative, or zero. The quantity $\delta q$ will be $\delta q_{\text{rev}}$ if the exchange is carried out reversibly and $\delta q_{\text{irr}}$ if it is carried out irreversibly. Thus, we can write for *any* process (Equation 6.18)

$$\begin{aligned} dS &= dS_{\text{prod}} + dS_{\text{exch}} \\ &= dS_{\text{prod}} + \frac{\delta q}{T} \end{aligned} \tag{14.2}$$

For a reversible process, $dS_{\text{prod}} = 0$ and $\delta q = \delta q_{\text{rev}}$, so that we have

$$dS = \frac{\delta q_{\text{rev}}}{T} \tag{14.3}$$

For an irreversible or spontaneous process, $dS_{\text{prod}} > 0$ and $\delta q = \delta q_{\text{irr}}$, so that

$$dS > \frac{\delta q_{\text{irr}}}{T} \tag{14.4}$$

Combining Equations 14.3 and 14.4 gives

$$dS \geq \frac{\delta q}{T} \tag{14.5}$$

which is Clausius' statement of the Second Law of Thermodynamics.

We can illustrate these ideas concretely with the following example. Consider a two-compartment system as shown in Figure 14.1. Each compartment is in equilibrium with a heat reservoir at different temperatures $T_1$ and $T_2$, and the two compartments are separated by a rigid, heat-conducting wall. The total change of energy of compartment 1 is

$$\delta q_1 = \delta_e q_1 + \delta_i q_1 \tag{14.6}$$

where $\delta_e q_1$ is the energy as heat exchanged with its reservoir and $\delta_i q_1$ is the energy as heat exchanged with compartment 2. Similarly,

$$\delta q_2 = \delta_e q_2 + \delta_i q_2 \tag{14.7}$$

Essentially by definition

$$\delta_i q_1 = -\delta_i q_2 \tag{14.8}$$

Now each compartment is in equilibrium with its heat reservoir, so

$$dS_1 = \left(\frac{\partial S}{\partial U_1}\right)_{V_1} dU_1 = \frac{dU_1}{T_1} \qquad \text{(constant } V\text{)}$$

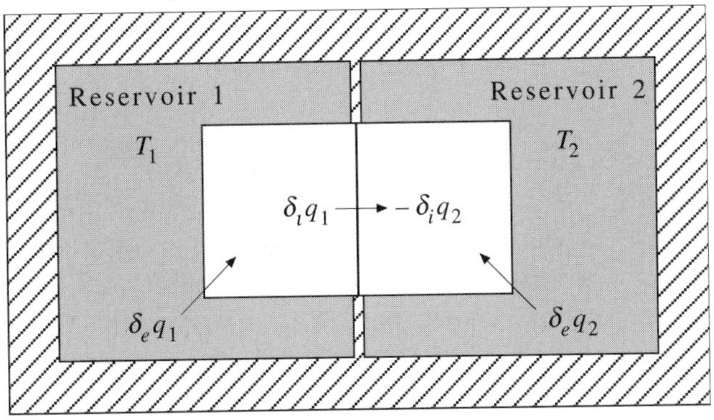


**FIGURE 14.1**
A two-compartment system with each compartment in contact with an (essentially infinite) heat reservoir, one at temperature $T_1$ and the other at temperature $T_2$. The two compartments are separated by a rigid, heat-conducting wall.

and

$$dS_2 = \left(\frac{\partial S}{\partial U_2}\right)_{V_2} dU_2 = \frac{dU_2}{T_2} \qquad \text{(constant } V\text{)}$$

Using the fact that $S = S_1 + S_2$, we obtain

$$dS = \frac{dU_1}{T_1} + \frac{dU_2}{T_2} \tag{14.9}$$

(See Equation 6.15.) But $dU_1 = \delta q_1 = \delta_e q_1 + \delta_i q_1$ (Equation 14.6) and $dU_2 = \delta_e q_2 + \delta_i q_2$ (Equation 14.7), so Equation 14.9 becomes

$$dS = \frac{\delta_e q_1}{T_1} + \frac{\delta_e q_2}{T_2} + \frac{\delta_i q_1}{T_1} + \frac{\delta_i q_2}{T_2}$$

Finally, Equation 14.8 allows us to write

$$\begin{aligned} dS &= dS_{\text{exch}} + dS_{\text{prod}} \\ &= \frac{\delta_e q_1}{T_1} + \frac{\delta_e q_2}{T_2} + \delta_i q_1 \left(\frac{1}{T_1} - \frac{1}{T_2}\right) \end{aligned} \tag{14.10}$$

where

$$dS_{\text{exch}} = \frac{\delta_e q_1}{T_1} + \frac{\delta_e q_2}{T_2} \tag{14.11}$$

is the entropy *exchanged* with the reservoirs (surroundings) and

$$dS_{\text{prod}} = \delta_i q_1 \left(\frac{1}{T_1} - \frac{1}{T_2}\right) = \delta_i q_1 \left(\frac{T_2 - T_1}{T_1 T_2}\right) \tag{14.12}$$

is the entropy *produced* within the two-compartment system.

The condition $dS_{\text{prod}} \geq 0$ implies that $\delta_i q_1$ must be positive if $T_2 > T_1$. Physcially, this means that energy as heat flows from compartment 2 to compartment 1 if $T_2 > T_1$, as we should expect. Conversely, $\delta_i q_1$ must be negative if $T_2 < T_1$, or energy as heat flows from compartment 1 to compartment 2 if $T_1 > T_2$. Note that the directionality of the heat flow is dictated by $dS_{\text{prod}} \geq 0$. The value of $dS_{\text{exch}}$ is arbitrary in the sense that it can be positive, negative, or zero and has nothing to do with the directionality of the heat flow.

The separation of $dS$ into the sum of $dS_{\text{exch}}$ and $dS_{\text{prod}}$ is fundamental. We will see that nonequilibrium thermodynamics is based upon the fact that $dS_{\text{prod}} > 0$ for any spontaneous process.

## 14–2. Entropy Always Increases When There Is a Material Flow from a Region of Higher Chemical Potential to a Region of Lower Chemical Potential

Consider the isolated two-compartment system shown in Figure 14.2. Each of the two compartments is in equilibrium individually, but the two compartments are not in equilibrium with each other. The two compartments are separated by a diathermal (heat-conducting), flexible, permeable wall; therefore, energy as heat, volume, and matter can flow from one compartment to the other. Each compartment must be large enough that the transport of energy or matter from one compartment to the other does not appreciably disturb the equilibrium state of each compartment. The total entropy of the two-compartment system is $S = S_1 + S_2$, with $S_1 = S_1(U_1, V_1, n_1)$ and $S_2 = S_2(U_2, V_2, n_2)$. Because each compartment is in equilibrium individually, we can write

$$dS_1 = \left(\frac{\partial S_1}{\partial U_1}\right)_{V_1,n_1} dU_1 + \left(\frac{\partial S_1}{\partial V_1}\right)_{U_1,n_1} dV_1 + \left(\frac{\partial S_1}{\partial n_1}\right)_{U_1,V_1} dn_1$$

$$= \frac{dU_1}{T_1} + \frac{P_1 dV_1}{T_1} - \frac{\mu_1 dn_1}{T_1} \tag{14.13}$$

with a similar equation for $dS_2$. The evaluation of the derivatives in Equation 14.13 follows from the equation $dU = TdS - PdV + \mu dn$. Therefore,

$$dS_{\text{prod}} = dS_1 + dS_2$$

$$= \frac{dU_1}{T_1} + \frac{dU_2}{T_2} + \frac{P_1 dV_1}{T_1} + \frac{P_2 dV_2}{T_2} - \frac{\mu_1 dn_1}{T_1} - \frac{\mu_2 dn_2}{T_2} \geq 0 \tag{14.14}$$

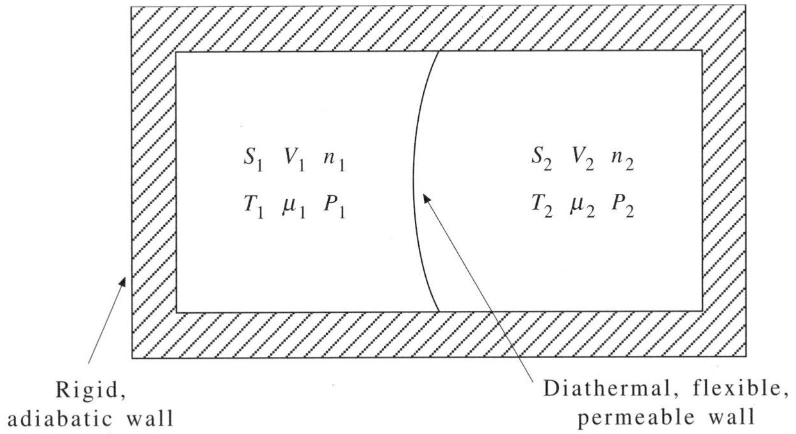


**FIGURE 14.2**
An isolated two-compartment system in which the two compartments are separated by a diathermal (heat-conducting), flexible, permeable wall. Each compartment is at equilibrium individually, but the two compartments are not in equilibrium with each other.

where we have written $dS_{\text{prod}}$ instead of $dS$ because the two-compartment system is isolated.

The two-compartment system in Figure 14.2 is isolated, so

$$U_1 + U_2 = \text{constant} \qquad V_1 + V_2 = \text{constant} \qquad n_1 + n_2 = \text{constant}$$

or

$$dU_1 = -dU_2 \qquad dV_1 = -dV_2 \qquad dn_1 = -dn_2 \tag{14.15}$$

Substituting Equation 14.15 into Equation 14.14 gives

$$dS_{\text{prod}} = dU_1\left(\frac{1}{T_1} - \frac{1}{T_2}\right) + dV_1\left(\frac{P_1}{T_1} - \frac{P_2}{T_2}\right) - dn_1\left(\frac{\mu_1}{T_1} - \frac{\mu_2}{T_2}\right) \geq 0 \tag{14.16}$$

The first term in Equation 14.16 is essentially Equation 14.12. Let's look at the second term. Equation 14.16 says that

$$TdS_{\text{prod}} = dV_1(P_1 - P_2) \geq 0 \tag{14.17}$$

if $T_1 = T_2$ and $\mu_1 = \mu_2$. Thus, $dV_1 > 0$ if $P_1 > P_2$, which means that the flexible wall will move to the right in Figure 14.2 if $P_1 > P_2$, which is what we should expect.

The third term in Equation 14.16 has to do with material flow, or diffusion. If we take $T_1 = T_2 = T$ and $P_1 = P_2$ for simplicity, then Equation 14.16 is

$$TdS_{\text{prod}} = -dn_1(\mu_1 - \mu_2) \geq 0 \tag{14.18}$$

Recall that (replacing activity by concentration)

$$\mu(T, c) = \mu^\circ(T) + RT \ln c$$

for a solution of concentration $c$, or that $\mu$ depends logarithmically on the concentration. Thus, Equation 14.18 says that $dn_1 < 0$ if $\mu_1 > \mu_2$, which means that matter will flow from compartment 1 to compartment 2 if the concentration of the diffusing species is greater in compartment 1 than in compartment 2, as we know from experience.

So far we have been treating an isolated (two-compartment) system in this section. Suppose instead that the two-compartment system is in contact with a heat bath at temperature $T$, so that $T_1 = T_2 = T$. Furthermore, for simplicity, let the wall separating the two compartments be rigid so that $V_1$ and $V_2$ are fixed. In this case, the two-compartment system is at constant $T$ and constant $V_1$ and $V_2$. The appropriate thermodynamic state function under these conditions is the Helmholtz energy, $A = A(T, V = V_1 + V_2, n)$, and the governing equation is $dA = \mu dn$ at constant $T$ and $V$. Thus, we have for the two-compartment system,

$$\begin{aligned} dA &= dA_1 + dA_2 = \mu_1 dn_1 + \mu_2 dn_2 \\ &= dn_1(\mu_1 - \mu_2) \leq 0 \end{aligned} \tag{14.19}$$

Note that $dn_1$ must be negative if $\mu_1 > \mu_2$, meaning that matter will flow from compartment 1 to compartment 2, as expected.

We can relate Equation 14.19 to Equation 14.18 in the following way. Because the $V = V_1 + V_2$ is fixed, we can write

$$dU = \delta_e q = T dS_{\text{exch}} = T(dS - dS_{\text{prod}}) \tag{14.20}$$

for the two-compartment system, and so

$$dS_{\text{prod}} = \frac{T dS - dU}{T} = -\frac{dA}{T} \qquad \text{(constant } T \text{ and } V\text{)}$$

To make the complete connection between Equations 14.18 and 14.19, substitute Equation 14.19 into the above equation to obtain

$$T dS_{\text{prod}} = -dA = -dn_1(\mu_1 - \mu_2) \geq 0 \tag{14.21}$$

in agreement with Equation 14.18.

**EXAMPLE 14–1**

Suppose a two-compartment system is enclosed by flexible, diathermal walls and is immersed in a heat bath at temperature $T$ and pressure $P$. Derive the analog of Equation 14.21 for this system. Assume that the wall separating the two compartments allows only matter to flow from one compartment to the other.

SOLUTION: We start with

$$\begin{aligned} dG &= -SdT + VdP + \mu dn \\ &= \mu dn \qquad \text{(constant } T \text{ and } P\text{)} \end{aligned}$$

Because the pressure is held constant, we can write (see Equation 14.20)

$$dH = \delta_e q = T dS_{\text{exch}} = T(dS - dS_{\text{prod}})$$

or

$$dS_{\text{prod}} = \frac{T dS - dH}{T} = -\frac{dG}{T} \qquad \text{(constant } T \text{ and } P\text{)}$$

Therefore, the analog of Equation 14.21 is

$$T dS_{\text{prod}} = -dG = -dn_1(\mu_1 - \mu_2) \geq 0$$

## 14–3. Many Flux–Force Relations Are Linear

To this point, we have really discussed nothing new. The contents of Sections 14–1 and 14–2 are contained in Problems 6–16 and 9–19. To introduce the basic approach of nonequilibrium thermodynamics, let's consider an isolated two-compartment system in which the two compartments are separated by a rigid, diathermal, permeable wall, so that matter and energy as heat can flow from one compartment to the other (Figure 14.3). The thermodynamic equation describing this system is Equation 14.16 with $dV_1 = 0$

$$dS_{\text{prod}} = dU_1\left(\frac{1}{T_1} - \frac{1}{T_2}\right) - dn_1\left(\frac{\mu_1}{T_1} - \frac{\mu_2}{T_2}\right) \geq 0$$

or

$$dS_{\text{prod}} = dU_1\left(\frac{1}{T_1} - \frac{1}{T_2}\right) + dn_1\left(\frac{\mu_2}{T_2} - \frac{\mu_1}{T_1}\right) \geq 0 \qquad (14.22)$$

where we have chosen to write the two terms on the right such that the sign of each term is positive. We now enter the realm of nonequilibrium thermodynamics by dividing Equation 14.22 by $dt$ to obtain

$$\frac{dS_{\text{prod}}}{dt} = \frac{dU_1}{dt}\left(\frac{1}{T_1} - \frac{1}{T_2}\right) + \frac{dn_1}{dt}\left(\frac{\mu_2}{T_2} - \frac{\mu_1}{T_1}\right) \geq 0 \qquad (14.23)$$

We will now write Equation 14.23 in standard notation:

$$\frac{dS_{\text{prod}}}{dt} = \dot{S}_{\text{prod}} = J_U X_U + J_n X_n \geq 0 \qquad (14.24)$$

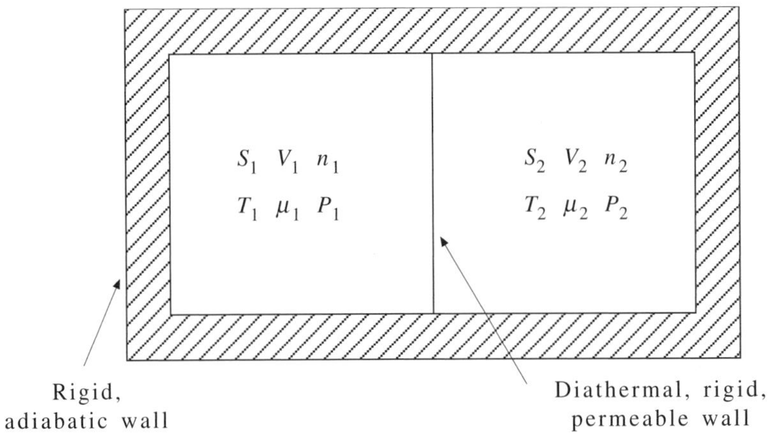


**FIGURE 14.3**
An isolated two-compartment system where the two compartments are separated by a rigid, diathermal, permeable wall. Each compartment is in equilibrium individually, but the two compartments are not in equilibrium with each other.

where

$$J_U = \frac{dU_1}{dt} = -\frac{dU_2}{dt}; \qquad J_n = \frac{dn_1}{dt} = -\frac{dn_2}{dt} \tag{14.25}$$

and

$$X_U = \frac{1}{T_1} - \frac{1}{T_2}; \qquad X_n = \frac{\mu_2}{T_2} - \frac{\mu_1}{T_1} \tag{14.26}$$

The quantities $J_U$ and $J_n$ are called fluxes; $J_U$ is the flux of energy and $J_n$ is the flux of matter. The quantities $X_U$ and $X_n$ are called thermodynamic forces; their nonzero values are responsible for the fluxes of energy and matter. Note that fluxes are the time derivatives of extensive thermodynamic properties and that forces are differences in intensive thermodynamic properties. The equality in Equation 14.24 holds only for the equilibrium state. From here on, we will consider only nonequilibrium systems, so we will use only the inequality in Equation 14.24.

Equations 14.22 and 14.23 are very different in character. Equation 14.22 is an equation of classical thermodynamics and the quantities appearing in that equation depend only upon the properties of each compartment. In Equation 14.23, on the other hand, $dU_1/dt$ and $dn_1/dt$ are kinetic quantities, and their values depend not only upon the properties of each compartment but also upon the nature of the wall separating the two compartments and the mechanism of the transport of energy and matter through the wall.

The fluxes and forces in Equation 14.24 are related to each other. Experimentally, the relation is often linear. For example, Fourier's law of heat conduction says that the flux of energy as heat is directly proportional to the temperature difference,

$$J_U \propto T_2 - T_1$$

But

$$T_2 - T_1 \propto \frac{T_2 - T_1}{T_1 T_2} = \frac{1}{T_1} - \frac{1}{T_2} = X_U$$

so we write

$$J_U \propto \frac{1}{T_1} - \frac{1}{T_2} = X_U$$

Similarly, Fick's law of (isothermal) diffusion says that the flux of matter is directly proportional to the concentration difference,

$$J_n \propto c_2 - c_1$$

which can be expressed in the form (Problem 14–5)

$$J_n \propto \mu_2 - \mu_1 \qquad \text{(constant } T\text{)}$$

Perhaps more familiar than Fourier's law or Fick's law is Ohm's law, which says that an electric current, $I$, is directly proportional to a voltage difference, $E_2 - E_1$,

$$I \propto E_2 - E_1$$

The point is that $J_i \propto X_i$ in all three cases; the fluxes are linearly dependent on the forces.

Because of the classical linear flux-force relations introduced above, you might expect to say that $J_U = aX_U$ and $J_n = bX_n$ ($a$ and $b$ are proportionality constants) in Equation 14.24. It's not quite that simple, however. Each of the so-called classical flux-force laws applies to experimental conditions in which there is a difference in only one quantity, temperature in the case of Fourier's law, concentration (or chemical potential) in the case of Fick's law, or voltage in the case of Ohm's law. In the more general case, in which there are differences in more than one quantity as in Equation 14.24, it turns out experimentally that

$$\begin{aligned} J_n &= L_{nn}X_n + L_{nU}X_U \\ J_U &= L_{Un}X_n + L_{UU}X_U \end{aligned} \tag{14.27}$$

where the $L$s are called *phenomenological coefficients*, which are to be determined experimentally. The physical significance of the diagonal phenomenological coefficients ($L_{UU}$ and $L_{nn}$) in Equation 14.27 is that they relate a flux to its "own" force. The fluxes and forces in Equation 14.24 are defined such that $L_{UU}$ and $L_{nn}$ are positive. The off-diagonal phenomenological coefficients ($L_{nU}$ and $L_{Un}$) in Equation 14.27 are called *coupling* or *cross coefficients* and relate an indirect dependence of a flux on the "direct" force of some other flux. Cross coefficients may be positive or negative.

**EXAMPLE 14–2**

Show that $L_{nU}$ and $L_{Un}$ in Equation 14.27 have the same units.

SOLUTION: The two terms in Equation 14.24 must have the same units (those of $\dot{S}_{\text{prod}}$), so

$$J_U X_U \longleftarrow \text{units} \longrightarrow J_n X_n$$

Equations 14.27 shows that $J_n$ and $L_{nU}X_U$ have the same units and that $J_U$ and $L_{Un}X_n$ have the same units. If we substitute these results into the above "unit" equation, we find that

$$L_{Un}X_n X_U \longleftarrow \text{units} \longrightarrow L_{nU}X_U X_n$$

or simply that

$$L_{Un} \longleftarrow \text{units} \longrightarrow L_{nU}$$

The physical interpretation of the coupling terms in relations such as Equation 14.27 can sometimes be difficult to see. But a nice, simple example is given by the isothermal diffusion of a solution of sucrose in water. Fick's law says that there will be a flux of water as a result of a difference in the chemical potential of water from one region to another and that there will be a flux of sucrose as a result of a chemical potential difference of sucrose. The sucrose molecules, however, are partially hydrated, so the flux of sucrose molecules contributes to a flux of water molecules as well. Thus, there is a flux of water molecules as a result of a chemical potential difference of the sucrose in addition to a flux of water molecules as a result of a chemical potential difference of water. The flux-force relation describing this coupling is

$$J_{\mathrm{w}} = L_{\mathrm{ww}} X_{\mathrm{w}} + L_{\mathrm{ws}} X_{\mathrm{s}}$$

where $J_{\mathrm{w}}$ is the flux of water molecules, $X_{\mathrm{w}}$ is the chemical potential difference of water, and $X_{\mathrm{s}}$ is the chemical potential difference of sucrose. $J_{\mathrm{s}}$ has a similar equation:

$$J_{\mathrm{s}} = L_{\mathrm{ss}} X_{\mathrm{s}} + L_{\mathrm{sw}} X_{\mathrm{w}}$$

If the water molecules and sucrose molecules diffused completely independently of each other, then the coupling coefficients $L_{\mathrm{ws}}$ and $L_{\mathrm{sw}}$ would be zero.

## 14–4. The Onsager Reciprocal Relations Say That $L_{ij} = L_{ji}$

In this section, we summarize and generalize the results of the previous section and then introduce the Onsager reciprocal relations, which constitute a fundamental principle of nonequilibrium thermodynamics. For any isolated two-compartment system, we define a set of thermodynamic forces, $X_1, X_2, \ldots, X_N$ and a set of corresponding fluxes $J_1, J_2, \ldots, J_N$ in such a way that $\dot{S}_{\mathrm{prod}} = dS_{\mathrm{prod}}/dt$, the rate of entropy production, is given by

$$\dot{S}_{\mathrm{prod}} = J_1 X_1 + J_2 X_2 + \cdots + J_N X_N > 0 \tag{14.28}$$

We then express the fluxes as linear combinations of the forces according to

$$\begin{aligned} J_1 &= L_{11} X_1 + L_{12} X_2 + \cdots + L_{1N} X_N \\ J_2 &= L_{21} X_1 + L_{22} X_2 + \cdots + L_{2N} X_N \\ &\vdots \\ J_N &= L_{N1} X_1 + L_{N2} X_2 + \cdots + L_{NN} X_N \end{aligned} \tag{14.29}$$

The formalism that we are developing here is called *linear nonequilibrium thermodynamics* because the flux-force relations are restricted to being linear. Although a number of flux-force relationships are linear over a wide range of conditions, not all are, and we must bear in mind that all our results will be restricted to the linear regime. We can think of Equations 14.29 as the leading terms of expansions of $J_1$, $J_2$, ... as power series (see MathChapter C) in $X_1, X_2, \ldots$. There are no constant terms because the fluxes all vanish if all the forces vanish. We will see in Section 14–6 that chemical reactions provide a case in which the linear flux-force relations restrict us to situations very close to chemical equilibrium.

We now come to a fundamental relation of nonequilibrium thermodynamics that was discovered by Lars Onsager in the early 1930s. Onsager showed that if the fluxes and forces are defined according to Equations 14.28 and 14.29, then the phenomenological coefficients satsify the condition

$$L_{ij} = L_{ji} \tag{14.30}$$

Equations 14.30 are called the *Onsager reciprocal relations*. Just as equilibrium thermodynamics gives us relations between various equilibrium thermodynamic properties, we will see that the Onsager reciprocal relations give us relations between various properties of systems that are not in equilibrium. Apparently Onsager first encountered reciprocal relations while investigating the properties of some simple model systems, and he then was able to prove them in general using the principles of statistical mechanics.

The positive-definite nature of $\dot{S}_{\text{prod}}$ given by Equation 14.28 and the reciprocal relations lead to a number of relations between the phenomenological coefficients. Let's take $N = 2$ in Equations 14.28 and 14.29 for simplicity. If we substitute Equation 14.29 (with $N = 2$) into Equation 14.28 (with $N = 2$) and use the fact that $L_{12} = L_{21}$, we obtain

$$\dot{S}_{\text{prod}} = L_{11}X_1^2 + 2L_{12}X_1X_2 + L_{22}X_2^2 > 0 \tag{14.31}$$

Equation 14.31 must be satisfied for any values of $X_1$ and $X_2$. If we let $X_2 = 0$, we obtain $\dot{S}_{\text{prod}} = L_{11}X_1^2 > 0$, which implies that $L_{11} > 0$. Similarly, we prove that $L_{22} > 0$ by letting $X_1 = 0$ in Equation 14.31. A less transparent condition that Equation 14.31 be satisfied is that (Problem 14–8)

$$L_{11}L_{22} - L_{12}^2 > 0 \tag{14.32}$$

Physically, Equation 14.32 means that the off-diagonal terms cannot dominate the flux-force relations. The diagonal terms in Equations 14.29 are the dominant ones, as you might expect.

The above results are easy to generalize. If we simply take

$$\dot{S}_{\text{prod}} = \sum_{i=1}^{N}\sum_{j=1}^{N} L_{ij}X_iX_j$$

then we obtain the general results

$$\begin{aligned} &L_{ii} > 0 \\ &L_{ii}L_{jj} - L_{ij}^2 > 0 \end{aligned} \tag{14.33}$$

In the next section, we will apply the formalism of nonequilibrium thermodynamics to a specific example called electrokinetics and show how the Onsager reciprocal relations connect various electrokinetic quantities.

## 14–5. Various Electrokinetic Quantities Are Related by the Onsager Reciprocal Relations

There are a number of various so-called electrokinetic phenomena and they all play important roles in colloid stability and the flow properties of colloids. For years, certain electrokinetic quantities were known to be experimentally related to each other, but despite intense effort, no satisfactory theoretical explanation was found for these relations. We will see in this section that the Onsager reciprocal relations provide such an explanation. Before deriving the basic nonequilibrium thermodynamic equations describing these electrokinetic phenomena, we will describe two types of experiments that are carried out.

Consider the schematic diagram of the appartus shown in Figure 14.4. The two arms, or compartments, are separated by a porous diaphragm that allows material transport. An electrode is immersed in each compartment so that an electrical potential difference $\Delta\psi$ can be imposed across the diaphragm. Each compartment contains a solution, which, for simplicity, we take as an aqueous solution of a 1–1 electrolyte such as potassium chloride. As you might expect, an electric current results when a potential difference is imposed across the diaphragm. Perhaps less obvious is that there is a volume flow through the diaphragm, which causes the liquid level in the two arms in Figure 14.4 to be unequal. The volume flow continues until finally a steady state is reached. At this point, the pressure difference, $\Delta P$, resulting from the unequal liquid levels in the two arms just balances the imposed potential difference, $\Delta\psi$, and the volume flow stops. The resulting value of $\Delta P$ depends upon the solutions in the two compartments and upon the nature of the diaphragm, and it is proportional to $\Delta\psi$. The flow of a fluid volume through a diaphragm that is caused by an applied electrical

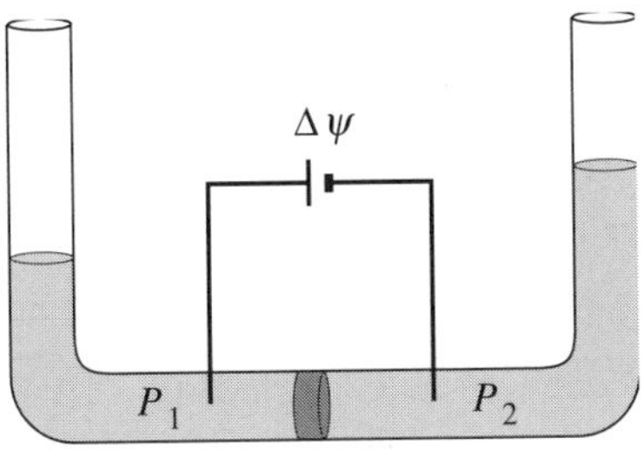


**FIGURE 14.4**
A schematic illustration of an apparatus to measure electroosmotic pressure. Each compartment contains identical solutions of an electrolyte, and the two compartments are separated by a permeable diaphragm. Electrodes are immersed in each solution so that an electrical potential can be applied across the diaphragm.

potential difference is called *electroosmosis*. For a given system and diaphragm, the electroosmotic pressure (EOP) is defined as

$$\text{EOP} = \left(\frac{\Delta P}{\Delta \psi}\right)_{J_V=0} \tag{14.34}$$

where $J_V$ is the volume flow across the diaphragm. The electroosmotic pressure is the proportionality constant relating $\Delta P$ to $\Delta \psi$ at the point when the volume flow has ceased.

Now consider the opposite experiment with the same system in which the electrodes are short circuited so that $\Delta \psi = 0$ and the solution is forced through the diaphragm by a piston (Figure 14.5). In this experiment, there is an electric current through the diaphragm as well as a volume flow, $J_V$. It turns out that the magnitude of the electric current is directly proportional to the volume flow. We define a quantity called the *streaming current*, SC, by the relation

$$\text{SC} = \left(\frac{I}{J_V}\right)_{\Delta \psi=0} \tag{14.35}$$

The streaming current is the proportionality constant relating the current $I$ to the volume flow $J_V$.

It is found empirically that the magnitude of the electroosmotic pressure differs only in sign from the streaming current, so that

$$\text{EOP} = -\text{SC} \tag{14.36}$$

We will see below that Equation 14.36 results from the Onsager reciprocal relations.

**EXAMPLE 14–3**

Show that the electroosmotic pressure and the streaming current have the same units.

SOLUTION: The SI units of $\Delta P$ are $\text{N}\cdot\text{m}^{-2} = \text{J}\cdot\text{m}^{-3}$ and those of $\Delta \psi$ are $\text{V} = \text{J}\cdot\text{C}^{-1}$, so that the units of EOP are

$$\text{EOP} = \left(\frac{\Delta P}{\Delta \psi}\right)_{J_V=0} \overset{\text{units}}{\longleftrightarrow} \frac{\text{N}\cdot\text{m}^{-2}}{\text{J}\cdot\text{C}^{-1}} = \frac{\text{J}\cdot\text{m}^{-3}}{\text{J}\cdot\text{C}^{-1}} = \text{C}\cdot\text{m}^{-3}$$

The SI units of $I$ are $\text{C}\cdot\text{s}^{-1}$, and those of $J_V$ are $\text{m}^3\cdot\text{s}^{-1}$, so that the units of SC are

$$\text{SC} = \left(\frac{I}{J_V}\right)_{\Delta \psi=0} \overset{\text{units}}{\longleftrightarrow} \frac{\text{C}\cdot\text{s}^{-1}}{\text{m}^3\cdot\text{s}^{-1}} = \text{C}\cdot\text{m}^{-3}$$

Thus the units of EOP and SC are the same.

There are other pairs of electrokinetic quantities that are related to each other. For example, if the fluid is forced through the diaphragm shown in Figure 14.4 or 14.5

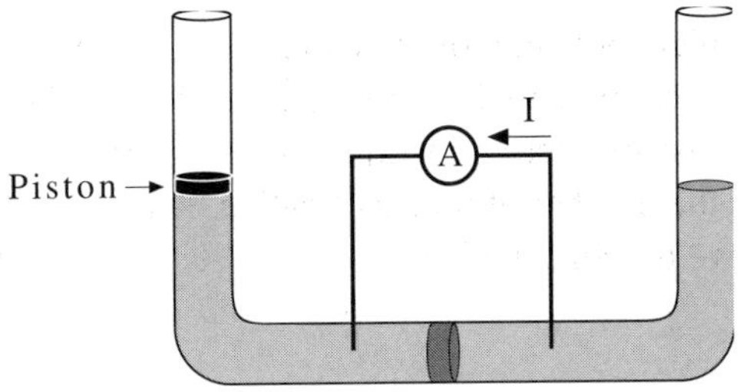


**FIGURE 14.5**
A schematic diagram of an apparatus to measure streaming current. Each compartment contains identical solutions of an electrolyte and the two compartments are separated by a porous diaphragm. An ammeter measures the electric current that accompanies the fluid flow caused by the piston.

with a pressure difference $\Delta P$, a voltage difference results. If this voltage difference is measured with a potentiometer so that $I = 0$, then $\Delta\psi$ turns out to be proportional to $\Delta P$. This fact allows us to define a *streaming potential*, SP, by

$$\text{SP} = \left(\frac{\Delta\psi}{\Delta P}\right)_{I=0} \tag{14.37}$$

Similarly, when a potential difference is applied across the diaphragm under conditions in which $\Delta P = 0$ (horizontal tube), there is a volume flow, $J_V$, as well as a current flow, $I$. It turns out that $J_V$ is proportional to $I$ under these conditions, and we define a quantity *electroosmotic flow*, EOF, by

$$\text{EOF} = \left(\frac{J_V}{I}\right)_{\Delta P=0} \tag{14.38}$$

The streaming potential and electroosmotic flow are related to each other by

$$\text{SP} = -\text{EOF} \tag{14.39}$$

Equations 14.36 and 14.39 are just two of a number of relations between various electrokinetic quantities. We will now show how Equations 14.36 and 14.39 result from the Onsager reciprocal relations.

If the apparatus illustrated in Figure 14.4 is at a fixed temperature and pressure, then the appropriate thermodynamic equations are

$$G = G_1 + G_2$$

with

$$dG_1 = \mu_{1+}dn_{1+} + \mu_{1-}dn_{1-} + \mu_{1\text{w}}dn_{1\text{w}} + e\psi_1 dn_{1+} - e\psi_1 dn_{1-} \tag{14.40}$$

where $\mu_{1+}$ is the chemical potential of the cation, $\mu_{1-}$ is the chemical potential of the anion, $\mu_{1\mathrm{w}}$ is the chemical potential of the water, and $\psi_1$ is the electrical potential in compartment 1. The last two terms in Equation 14.40 represent the electrical work involved in changing the number of moles of cation by $dn_{1+}$ and the number of moles of anion by $dn_{1-}$. If we use an equation similar to Equation 14.40 for compartment 2, then

$$\begin{aligned} dG &= \mu_{1+}dn_{1+} + \mu_{2+}dn_{2+} + \mu_{1-}dn_{1-} + \mu_{2-}dn_{2-} + \mu_{1\mathrm{w}}dn_{1\mathrm{w}} + \mu_{2\mathrm{w}}dn_{2\mathrm{w}} \\ &\quad + e\psi_1 dn_{1+} + e\psi_2 dn_{2+} - e\psi_1 dn_{1-} - e\psi_2 dn_{2-} \end{aligned}$$

Using the facts that $dn_{1+} = -dn_{2+}$, $dn_{1-} = -dn_{2-}$, and $dn_{1\mathrm{w}} = -dn_{2\mathrm{w}}$ gives

$$\begin{aligned} dG &= (\mu_{1+} - \mu_{2+})dn_{1+} + (\mu_{1-} - \mu_{2-})dn_{1-} + (\mu_{1\mathrm{w}} - \mu_{2\mathrm{w}})dn_{1\mathrm{w}} \\ &\quad + e(\psi_1 - \psi_2)dn_{1+} - e(\psi_1 - \psi_2)dn_{1-} \end{aligned} \tag{14.41}$$

Example 14–1 showed that $dS_{\mathrm{prod}} = -dG/T$ for a system at constant $T$ and $P$, so Equation 14.41 can be written as

$$T dS_{\mathrm{prod}} = \Delta\mu_+ dn_{1+} + \Delta\mu_- dn_{1-} + \Delta\mu_{\mathrm{w}} dn_{1\mathrm{w}} + e\Delta\psi dn_{1+} - e\Delta\psi dn_{1-} \tag{14.42}$$

where $\Delta\mu_+ = \mu_{2+} - \mu_{1+}$, $\Delta\mu_- = \mu_{2-} - \mu_{1-}$, $\Delta\mu_{\mathrm{w}} = \mu_{2\mathrm{w}} - \mu_{1\mathrm{w}}$, and $\Delta\psi = \psi_2 - \psi_1$. Now divide Equation 14.42 by $dt$ to get

$$T\dot{S}_{\mathrm{prod}} = J_+\Delta\mu_+ + J_-\Delta\mu_- + J_{\mathrm{w}}\Delta\mu_{\mathrm{w}} + eJ_+\Delta\psi - eJ_-\Delta\psi \tag{14.43}$$

where $J_+ = dn_{1+}/dt$, $J_- = dn_{1-}/dt$, and $J_{\mathrm{w}} = dn_{1\mathrm{w}}/dt$. The flow of the neutral 1–1 electrolyte across the diaphragm is $J_{\mathrm{s}} = J_+ = J_-$, the electric current across it is $I = eJ_+ - eJ_-$, and the chemical potential of a 1–1 electrolyte is $\mu_{\mathrm{s}} = \mu_+ + \mu_-$. Substituting these three relations into Equation 14.43 gives

$$T\dot{S}_{\mathrm{prod}} = J_{\mathrm{s}}\Delta\mu_{\mathrm{s}} + J_{\mathrm{w}}\Delta\mu_{\mathrm{w}} + I\Delta\psi \tag{14.44}$$

Most electrokinetic experiments are carried out such that the concentration of the solution is the same in both compartments. In this case, there is no concentration dependence of $\Delta\mu_{\mathrm{s}}$ or $\Delta\mu_{\mathrm{w}}$, so (Equation 9.9)

$$\Delta\mu_i = \overline{V}_i\Delta P$$

where $i$ stands for s or w. Substituting $\Delta\mu_i = \overline{V}_i\Delta P$ into Equation 14.44 gives

$$T\dot{S}_{\mathrm{prod}} = (J_{\mathrm{s}}\overline{V}_{\mathrm{s}} + J_{\mathrm{w}}\overline{V}_{\mathrm{w}})\Delta P + I\Delta\psi$$

The term in parentheses is simply the total volume flow, $J_V$, across the diaphragm, so we finally have our basic entropy production equation for electrokinetic phenomena:

$$T\dot{S}_{\mathrm{prod}} = J_V\Delta P + I\Delta\psi \tag{14.45}$$

The corresponding linear flux-force relations are

$$J_V = L_{VV}\Delta P + L_{VI}\Delta\psi \tag{14.46a}$$

$$J_I = I = L_{IV}\Delta P + L_{II}\Delta\psi \tag{14.46b}$$

We can now use Equations 14.46 to prove Equations 14.36 and 14.39. The electroosmotic pressure (Equation 14.34) is obtained by letting $J_V = 0$ in Equation 14.46*a* and solving for $\Delta P/\Delta\psi$ to obtain

$$\text{EOP} = \left(\frac{\Delta P}{\Delta\psi}\right)_{J_V=0} = -\frac{L_{VI}}{L_{VV}} \tag{14.47}$$

The streaming current (Equation 14.35) is obtained by setting $\Delta\psi = 0$ in Equations 14.46 and dividing Equation 14.46*b* by 14.46*a* to obtain

$$\text{SC} = \frac{L_{IV}}{L_{VV}} \tag{14.48}$$

Equations 14.47 and 14.48 show that EOP $= -$SC because $L_{VI} = L_{IV}$; thus proving the validity of Equation 14.36.

**EXAMPLE 14–4**

Prove that the streaming potential, SP, is the negative of electroosmotic flow, EOF.

SOLUTION: According to Equation 14.37, SP $= (\Delta\psi/\Delta P)_{I=0}$. Let $I = 0$ in Equation 14.46*b* to get SP $= -L_{IV}/L_{II}$. Electroosmotic flow is given by EOF $= (J_V/I)_{\Delta P=0}$ (Equation 14.38). Let $\Delta P = 0$ in Equations 14.46 and divide one by the other to obtain EOF $= (J_V/I)_{\Delta P=0} = L_{VI}/L_{II}$. Therefore, SP $= -$EOF because $L_{IV} = L_{VI}$.

Problems 14–13 to 14–15 have you derive relations between other electrokinetic quantities.

## 14–6. The Onsager Reciprocal Relations Are Based on the Principle of Detailed Balance

In this section, we will use a chemical kinetic scheme to gain some insight to the origin of the Onsager reciprocal relations. Of the many applications of nonequilibrium thermodynamics to various processes, the application to chemical kinetics is exceptionally limited in practice because of the requirement that the flux-force relations be linear. This linearity requirement means that chemically reacting systems must be close to equilibrium for the formalism of nonequilibrium thermodynamics to be applicable. Nevertheless, we will see in this section how the Onsager reciprocal relations arise by studying a simple chemical reaction.

Let's start with a reversible first-order elementary chemical reaction described by

$$\mathrm{X} \underset{k_{\mathrm{YX}}}{\overset{k_{\mathrm{XY}}}{\rightleftharpoons}} \mathrm{Y}$$

This reaction is too simple to involve the Onsager reciprocal relations, but we can use it to introduce some notation. The rate of this reaction is given by

$$\frac{d[\mathrm{X}]}{dt} = -k_{\mathrm{XY}}[\mathrm{X}] + k_{\mathrm{YX}}[\mathrm{Y}]$$

where [X] and [Y] are concentrations. We can define the flow of this reaction from left to right by

$$J = -\frac{d[\mathrm{X}]}{dt} = \frac{d[\mathrm{Y}]}{dt} = k_{\mathrm{XY}}[\mathrm{X}] - k_{\mathrm{YX}}[\mathrm{Y}] \tag{14.49}$$

At equilibrium, $J = 0$ and Equation 14.49 gives

$$k_{\mathrm{XY}}[\mathrm{X}]_{\mathrm{eq}} = k_{\mathrm{YX}}[\mathrm{Y}]_{\mathrm{eq}}$$

Now define

$$\alpha_{\mathrm{X}} = [\mathrm{X}] - [\mathrm{X}]_{\mathrm{eq}} \quad \text{and} \quad \alpha_{\mathrm{Y}} = [\mathrm{Y}] - [\mathrm{Y}]_{\mathrm{eq}} \tag{14.50}$$

Note that $\alpha_{\mathrm{X}} + \alpha_{\mathrm{Y}} = 0$ because $[\mathrm{X}] + [\mathrm{Y}] = [\mathrm{X}]_{\mathrm{eq}} + [\mathrm{Y}]_{\mathrm{eq}}$. Using this result, Equation 14.49 can be expressed in terms of $\alpha_{\mathrm{X}}$ by (Problem 14–16)

$$J = \alpha_{\mathrm{X}}(k_{\mathrm{XY}} + k_{\mathrm{YX}}) \tag{14.51}$$

Now let's apply nonequilibrium thermodynamics to this reaction. For convenience only (see Problem 14–17), we take the reaction system to be isolated. From the equation

$$dU = TdS - PdV + \mu_{\mathrm{X}}dn_{\mathrm{X}} + \mu_{\mathrm{Y}}dn_{\mathrm{Y}}$$

we see that

$$dS_{\mathrm{prod}} = -\frac{\mu_{\mathrm{X}}}{T}dn_{\mathrm{X}} - \frac{\mu_{\mathrm{Y}}}{T}dn_{\mathrm{Y}} \qquad (\text{constant } U \text{ and } V)$$

or, upon dividing by $dt$

$$\dot{S}_{\mathrm{prod}} = -\frac{\mu_{\mathrm{X}}}{T}\frac{dn_{\mathrm{X}}}{dt} - \frac{\mu_{\mathrm{Y}}}{T}\frac{dn_{\mathrm{Y}}}{dt} \tag{14.52}$$

If we divide Equation 14.52 by $V$, we obtain

$$\begin{aligned}\frac{\dot{S}_{\mathrm{prod}}}{V} &= -\frac{\mu_{\mathrm{X}}}{T}\frac{d[\mathrm{X}]}{dt} - \frac{\mu_{\mathrm{Y}}}{T}\frac{d[\mathrm{Y}]}{dt} \\ &= -\frac{d[\mathrm{X}]}{dt}\left(\frac{\mu_{\mathrm{X}}}{T} - \frac{\mu_{\mathrm{Y}}}{T}\right) \end{aligned} \tag{14.53}$$

where we have used the fact that $[X] + [Y] = \text{constant}$.

Equation 14.53 is our fundamental entropy production equation. We can write Equation 14.53 as

$$\frac{\dot{S}_{\text{prod}}}{V} = J\mathcal{A} \tag{14.54}$$

where $J$ is given by Equation 14.49 and where

$$\mathcal{A} = \frac{\mu_X}{T} - \frac{\mu_Y}{T} \tag{14.55}$$

is called the *affinity*. The linear flux-force relation is

$$J = L\mathcal{A} = L\left(\frac{\mu_X - \mu_Y}{T}\right) \tag{14.56}$$

Experimentally, the linear relation between $J$ and $\mathcal{A}$ given in Equation 14.56 is valid only for small values of $\mathcal{A}/R$ where $R$ is the molar gas constant. This means that [X] and [Y] must be close to their equilibrium values because $\mu_X = \mu_Y$ at equilibrium. Problem 14–18 shows that

$$\mathcal{A} = R\left(\frac{\alpha_X}{[X]_{eq}} - \frac{\alpha_Y}{[Y]_{eq}}\right) = \frac{R\alpha_X}{k_{YX}[Y]_{eq}}(k_{XY} + k_{YX})$$

if $\alpha_X/[X]_{eq} \ll 1$ and $\alpha_Y/[Y]_{eq} \ll 1$. If we substitute this expression for $\mathcal{A}$ into Equation 14.56, we obtain

$$J = \frac{LR}{k_{YX}[Y]_{eq}}\alpha_X(k_{XY} + k_{YX}) \tag{14.57}$$

If we compare Equation 14.57 with Equation 14.51, we see that $L = k_{YX}[Y]_{eq}/R = k_{XY}[X]_{eq}/R$.

---

**EXAMPLE 14–5**

Consider the elementary chemical reaction described by

$$X + Y \rightleftharpoons Z$$

Show that the affinity in this case is given by

$$\mathcal{A} = \frac{\mu_X + \mu_Y - \mu_Z}{T}$$

SOLUTION: The flow of this reaction from left to right is given by

$$J = -\frac{d[X]}{dt} = -\frac{d[Y]}{dt} = +\frac{d[Z]}{dt}$$

The basic thermodynamic equation for $dS$ is

$$dS = \frac{dU}{T} + \frac{P}{T}dV - \frac{\mu_X}{T}dn_X - \frac{\mu_Y}{T}dn_Y - \frac{\mu_Z}{T}dn_Z$$

and so

$$\dot{S}_{prod} = -\frac{\mu_X}{T}\frac{dn_X}{dt} - \frac{\mu_Y}{T}\frac{dn_Y}{dt} - \frac{\mu_Z}{T}\frac{dn_Z}{dt}$$

Divide by $V$ to obtain

$$\begin{aligned}\frac{\dot{S}_{prod}}{V} &= -\frac{\mu_X}{T}\frac{d[X]}{dt} - \frac{\mu_Y}{T}\frac{d[Y]}{dt} - \frac{\mu_Z}{T}\frac{d[Z]}{dt} \\ &= -\frac{d[X]}{dt}\left(\frac{\mu_X}{T} + \frac{\mu_Y}{T} - \frac{\mu_Z}{T}\right) \\ &= J\left(\frac{\mu_X}{T} + \frac{\mu_Y}{T} - \frac{\mu_Z}{T}\right) = J\mathcal{A}\end{aligned}$$

Now let's look at the elementary reaction scheme described by

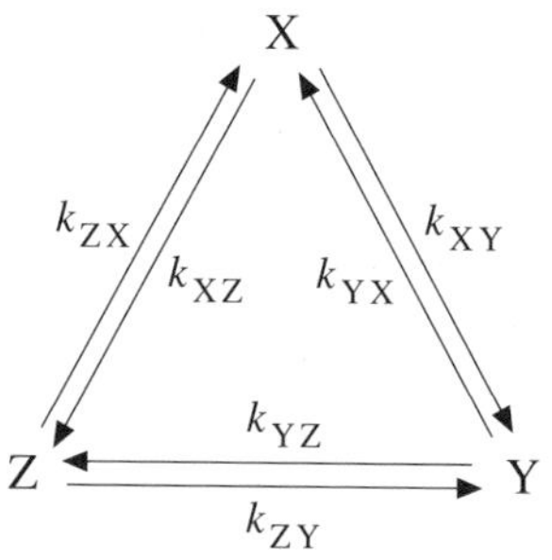


(14.58)

This is the reaction scheme used by Onsager to glean the basis of the reciprocal relations. We can write the kinetic equations as

$$\begin{aligned}J_1 &= k_{XY}[X] - k_{YX}[Y] \\ J_2 &= k_{YZ}[Y] - k_{ZY}[Z] \\ J_3 &= k_{ZX}[Z] - k_{XZ}[X]\end{aligned} \tag{14.59}$$

Only two of these kinetic equations are independent because $[X] + [Y] + [Z] =$ constant. The affinities associated with these three flows are

$$\begin{aligned}\mathcal{A}_1 &= \frac{\mu_X}{T} - \frac{\mu_Y}{T} \\ \mathcal{A}_2 &= \frac{\mu_Y}{T} - \frac{\mu_Z}{T} \\ \mathcal{A}_3 &= \frac{\mu_Z}{T} - \frac{\mu_X}{T}\end{aligned} \tag{14.60}$$

These three affinities are not independent, however, because $\mathcal{A}_3 = -(\mathcal{A}_1 + \mathcal{A}_2)$. Consequently, the entropy production equation consists of only two terms

$$\frac{\dot{S}_{\text{prod}}}{V} = J_1\mathcal{A}_1 + J_2\mathcal{A}_2 + J_3\mathcal{A}_3 \\ = (J_1 - J_3)\mathcal{A}_1 + (J_2 - J_3)\mathcal{A}_2 \tag{14.61}$$

The associated linear flux-force equations are

$$J_1 - J_3 = L_{11}\mathcal{A}_1 + L_{12}\mathcal{A}_2 \\ J_2 - J_3 = L_{21}\mathcal{A}_1 + L_{22}\mathcal{A}_2 \tag{14.62}$$

At equilibrium, $\mathcal{A}_1 = \mathcal{A}_2 = 0$ and so $J_1 - J_3 = 0$ and $J_2 - J_3 = 0$, or

$$J_1 = J_2 = J_3 \tag{14.63}$$

Equation 14.63 is an interesting result. According to this equation, the condition of thermodynamic equilibrium does not require that all the flows vanish, only that they be equal. Therefore, according to Equation 14.63, the reaction may circulate in one direction or the other indefinitely. This notion, however, violates a fundamental principle of nature, which says that under equilibrium conditions, any molecular process and its reverse process must take place, on the average, at the same rate. The consequences of this principle to the reaction scheme in Equation 14.58 is that each of the individual flows in Equation 14.59 must be zero. Thus, not only must these flows be equal to each other at equilibrium, but they individually must be equal to zero. This principle is called the *principle of detailed balance*. The condition of detailed balance

$$J_1 = J_2 = J_3 = 0 \tag{14.64}$$

requires that

$$k_{\text{XY}}[\text{X}]_{\text{eq}} = k_{\text{YX}}[\text{Y}]_{\text{eq}} \\ k_{\text{YZ}}[\text{Y}]_{\text{eq}} = k_{\text{ZY}}[\text{Z}]_{\text{eq}} \\ k_{\text{ZX}}[\text{Z}]_{\text{eq}} = k_{\text{XZ}}[\text{X}]_{\text{eq}} \tag{14.65}$$

**EXAMPLE 14–6**

Show that the principle of detailed balance requires that $k_{\text{XY}}k_{\text{YZ}}k_{\text{ZX}} = k_{\text{XZ}}k_{\text{ZY}}k_{\text{YX}}$. Interpret this result physically.

SOLUTION: Solve Equation 14.65*a* for $[\text{Y}]_{\text{eq}}/[\text{X}]_{\text{eq}}$, Equation 14.65*b* for $[\text{Z}]_{\text{eq}}/[\text{Y}]_{\text{eq}}$, and Equation 14.65*c* for $[\text{X}]_{\text{eq}}/[\text{Z}]_{\text{eq}}$ and multiply the three results together to get $k_{\text{XY}}k_{\text{YZ}}k_{\text{ZX}} = k_{\text{XZ}}k_{\text{ZY}}k_{\text{YX}}$. Thus, the product of the three rate constants in one direction is equal to the product of the three rate constants in the other direction. Note that the principle of detailed balance says that the six rate constants are not independent.

Just as we did for the case X $\rightleftharpoons$ Y, we consider the kinetic system described by Equation 14.58 to be near equilibrium. In this case

$$\begin{aligned} J_1 &= k_{XY}\alpha_X - k_{YX}\alpha_Y \\ J_2 &= k_{YZ}\alpha_Y - k_{ZY}\alpha_Z \end{aligned} \tag{14.66}$$

The third flux, $J_3$, is a linear combination of $J_1$ and $J_2$ and is given by (Problem 14–20)

$$J_3 = -\frac{k_{XZ}}{k_{XY}} J_1 - \frac{k_{ZX}}{k_{ZY}} J_2 \tag{14.67}$$

The two independent affinities are (Problem 14–21)

$$\begin{aligned} \mathcal{A}_1 &= \frac{R}{k_{XY}[X]_{eq}} J_1 \\ \mathcal{A}_2 &= \frac{R}{k_{YZ}[Y]_{eq}} J_2 \end{aligned} \tag{14.68}$$

Combining Equations 14.67 and 14.68 gives

$$\begin{aligned} J_1 - J_3 &= \left(\frac{k_{XY} + k_{XZ}}{k_{XY}}\right) \frac{k_{XY}[X]_{eq}}{R} \mathcal{A}_1 + \frac{k_{ZX}}{k_{ZY}} \frac{k_{YZ}[Y]_{eq}}{R} \mathcal{A}_2 \\ J_2 - J_3 &= \frac{k_{XZ}k_{XY}}{k_{XY}} \frac{[X]_{eq}}{R} \mathcal{A}_1 + \left(\frac{k_{ZY} + k_{ZX}}{k_{ZY}}\right) \frac{k_{YZ}[Y]_{eq}}{R} \mathcal{A}_2 \end{aligned} \tag{14.69}$$

If we compare Equations 14.69 with Equations 14.62, we see that

$$L_{12} = \frac{k_{ZX}k_{YZ}[Y]_{eq}}{k_{ZY}R} \qquad \text{and} \qquad L_{21} = \frac{k_{XZ}[X]_{eq}}{R}$$

If we use the detailed balance conditions $k_{YZ}[Y]_{eq} = k_{ZY}[Z]_{eq}$ and $k_{XZ}[X]_{eq} = k_{ZX}[Z]_{eq}$ (Equations 14.65*b* and *c*), then $L_{12}$ and $L_{21}$ become

$$L_{12} = \frac{k_{ZX}[Z]_{eq}}{R} \qquad \text{and} \qquad L_{21} = \frac{k_{ZX}[Z]_{eq}}{R}$$

Thus, we see that the principle of detailed balance is the basis of the Onsager reciprocal relations. Onsager later proved the validity of reciprocal relations more generally using statistical mechanics.

## 14–7. Electrochemical Potentials Play the Role of the Chemical Potentials for Charged Systems in Different Phases

Before we discuss electrochemical systems in the next few sections, it is convenient to introduce a quantity called the *electrochemical potential*. If a region has an electrostatic potential $\psi$, changing the charge by an amount $dQ$ will result in an amount of work

$\psi dQ$ being performed. The First Law of Thermodynamics for an electrochemical system becomes

$$dU = TdS - PdV + \sum_j \mu_j dn_j + \psi dQ \tag{14.70}$$

In Equation 14.70, the second term on the right represents pressure-volume work, the third term represents the work involved in changing the composition of the system, and the fourth term represents electrical work. The total charge of the system is given by

$$Q = \sum_j z_j F n_j \tag{14.71}$$

where $z_j$ is the valence, $n_j$ is the number of moles of the $j$th species, and $F$ is the Faraday constant. Substituting Equation 14.71 into Equation 14.70 gives

$$dU = TdS - PdV + \sum_j (\mu_j + z_j F\psi) dn_j \tag{14.72}$$

Equation 14.72 shows that the work involved in changing the composition of an electrochemical system is composed of two parts, a chemical part, $\mu_j dn_j$, and an electrical part, $z_j F\psi dn_j$. We define a quantity, $\tilde{\mu}_j$, called the *electrochemical potential of component j*, by

$$\tilde{\mu}_j = \mu_j + z_j F\psi \tag{14.73}$$

Using this definition of $\tilde{\mu}_j$, Equation 14.72 becomes

$$dU = TdS - PdV + \sum_j \tilde{\mu}_j dn_j \tag{14.74}$$

The electrochemical potential plays the same role in electrochemical systems that the chemical potential plays in systems composed of neutral species.

**EXAMPLE 14–7**

Show that the condition for chemical equilibrium for the general reaction described by

$$\nu_A \text{ A} + \nu_B \text{ B} \rightleftharpoons \nu_Y \text{ Y} + \nu_Z \text{ Z}$$

is

$$\nu_Y \tilde{\mu}_Y + \nu_Z \tilde{\mu}_Z = \nu_A \tilde{\mu}_A + \nu_B \tilde{\mu}_B$$

where each species may be charged and in different phases at different electric potentials.

SOLUTION: First subtract $d(TS)$ and add $d(PV)$ to Equation 14.74 to obtain

$$dG = -SdT + VdP + \sum_j \tilde{\mu}_j dn_j \tag{14.75}$$

Note that

$$\left(\frac{\partial G}{\partial n_j}\right)_{T,P,n_{k \neq j}} = \tilde{\mu}_j \tag{14.76}$$

which is a generalization of Equation 9.26. Just as we did in the beginning of Chapter 12, we introduce the extent of reaction $\xi$ to obtain Equations 12.2, and then simply follow the argument to Equation 12.6 to obtain

$$\left(\frac{\partial G}{\partial \xi}\right)_{T,P} = \nu_{\text{Y}}\tilde{\mu}_{\text{Y}} - \nu_{\text{Z}}\tilde{\mu}_{\text{Z}} - \nu_{\text{A}}\tilde{\mu}_{\text{A}} - \nu_{\text{B}}\tilde{\mu}_{\text{B}}$$

The condition for equilibrium is that $(\partial G/\partial \xi)_{T,P} = 0$, and so we have

$$\nu_{\text{A}}\tilde{\mu}_{\text{A}} + \nu_{\text{B}}\tilde{\mu}_{\text{B}} = \nu_{\text{Y}}\tilde{\mu}_{\text{Y}} + \nu_{\text{Z}}\tilde{\mu}_{\text{Z}} \tag{14.77}$$

It turns out that all the equations involving $\mu$ that we derived in earlier chapters can be rewritten in terms of $\tilde{\mu}$ to include electrochemical systems.

We can apply Equation 14.77 to the case in which two aqueous electrolyte solutions are separated by a membrane that is permeable to one of the ions. If we denote the two solutions by $\alpha$ and $\beta$ and the permeable ion by $i$, then Equation 14.77 becomes

$$\tilde{\mu}_i^{\alpha} = \tilde{\mu}_i^{\beta}$$

or

$$\mu_i^{\alpha} + z_i F\psi^{\alpha} = \mu_i^{\beta} + z_i F\psi^{\beta}$$

Using the fact that $\mu_i = \mu_i^{\circ} + RT \ln a_j$, where $a_j$ is the activity of the ion $i$, we have

$$\Delta\psi = \psi^{\beta} - \psi^{\alpha} = \frac{RT}{z_i F} \ln \frac{a_i^{\alpha}}{a_j^{\beta}} \tag{14.78}$$

Thus, we see that an electric potential across the membrane results. This potential difference counteracts the tendency of the $i$ ions to equalize their concentrations on the two sides of the membrane. This potential difference is often called the *Nernst potential*, especially in biophysics. A number of biological membranes are much more permeable to one particular ion than to others. For example, squid axon nerve cell membranes in their resting state are almost exclusively permeable to potassium ions. The concentrations of the potassium ions on the two sides of the membrane are about

400 mmol·L$^{-1}$ and 20 mmol·L$^{-1}$. If we assume that the activity coefficients of these solutions are about the same, then the Nernst potential at 25°C is

$$\Delta\psi = \frac{(8.3145\ \mathrm{J\cdot mol^{-1}\cdot K^{-1}})(298.15\ \mathrm{K})}{96\,485\ \mathrm{C\cdot mol^{-1}}} \ln \frac{400\ \mathrm{mmol\cdot L^{-1}}}{20\ \mathrm{mmol\cdot L^{-1}}}$$
$$= 0.077\ \mathrm{V} = 77\ \mathrm{mV}$$

which is typical of resting-state potentials in biological membranes.

## 14–8. The Magnitude of the Liquid Junction Potential Depends Upon Transport Numbers

Consider the electrochemical cell whose cell diagram is (Figure 14.6)

$$\mathrm{Ag(s)|AgCl(s)|NaCl}(a_1) \vdots \mathrm{NaCl}(a_2)\mathrm{|AgCl(s)|Ag(s)} \qquad (14.79)$$

The vertical dotted line between the NaCl($a_1$) and the NaCl($a_2$) represents the junction between the two solutions. The silver-silver chloride electrode reacts reversibly with the chloride ions, and the reaction that takes place at the right electrode (reduction) is described by

$$\mathrm{AgCl(s)} + e^- \longrightarrow \mathrm{Ag(s)} + \mathrm{Cl^-}(a_2)$$

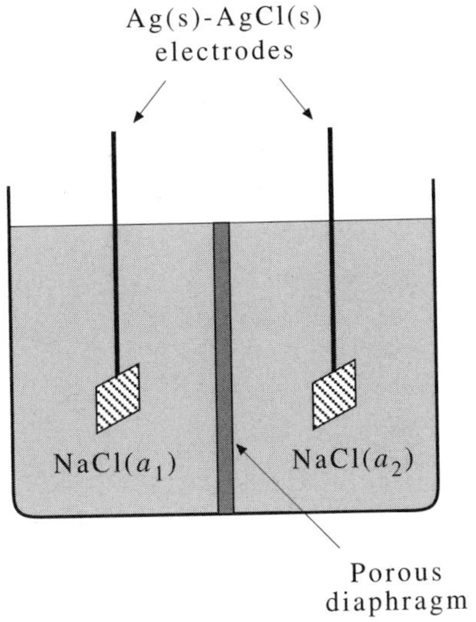


**FIGURE 14.6**
A schematic illustration of the cell whose cell diagram is
$\mathrm{Ag(s)|AgCl(s)|NaCl}(a_1) \vdots \mathrm{NaCl}(a_2)\mathrm{|AgCl(s)|Ag(s)}$.
The silver-silver chloride electrode reacts reversibly with the chloride ions. The two NaCl(aq) solutions may be separated by a porous diaphragm to minimize their mixing. The net reaction of this cell is NaCl($a_1$) → NaCl($a_2$), and the cell is called a concentration cell.

where the electron, $e^-$, is in the Ag(s) electrode. The reaction that takes place at the left electrode (oxidation) is described by

$$\text{Ag(s)} + \text{Cl}^-(a_1) \longrightarrow \text{AgCl(s)} + e^-$$

The equation for the overall cell reaction is given by the sum of these two equations

$$\text{Cl}^-(a_1) \longrightarrow \text{Cl}^-(a_2)$$

Thus, the driving force of the cell is due only to an activity change, or a concentration change. Such a cell is called a *concentration cell.*

If we apply Equation 14.77 to the equations for the two electrode reactions given by above, we have

$$\mu^\alpha_{\text{Ag}} + \tilde{\mu}^\alpha_{\text{Cl}^-} = \mu^\alpha_{\text{AgCl}} + \tilde{\mu}^\alpha_{e^-} \tag{14.80}$$

and

$$\mu^\beta_{\text{Ag}} + \tilde{\mu}^\beta_{\text{Cl}^-} = \mu^\beta_{\text{AgCl}} + \tilde{\mu}^\beta_{e^-} \tag{14.81}$$

where $\alpha$ and $\beta$ denote the solution electrode compartments. Using the fact that the electrodes are similar, we can write $\mu^\alpha_{\text{Ag}} = \mu^\beta_{\text{Ag}}$ and $\mu^\alpha_{\text{AgCl}} = \mu^\beta_{\text{AgCl}}$, so if we subtract Equation 14.81 from Equation 14.80, we have

$$\tilde{\mu}^\alpha_{e^-} - \tilde{\mu}^\beta_{e^-} = \tilde{\mu}^\alpha_{\text{Cl}^-} - \tilde{\mu}^\beta_{\text{Cl}^-} = \Delta\tilde{\mu}_{\text{Cl}^-} \tag{14.82}$$

Because the electrons are in similar (silver electrode) phases in the two electrode compartments, the chemical part of the electrochemical potential cancels in $\tilde{\mu}^\alpha_{e^-} - \tilde{\mu}^\beta_{e^-}$, and so

$$\tilde{\mu}^\alpha_{e^-} - \tilde{\mu}^\beta_{e^-} = -F(\phi^\alpha - \phi^\beta) \tag{14.83}$$

where $\phi^\alpha$ and $\phi^\beta$ are the electrical potentials of the electrons in the silver electrodes. The difference in these potentials is the electromotive force of the cell, $E$. Thus, we see from Equations 14.82 and 14.83 that the emf of the cell shown in Figure 14.6 is given by

$$E = \Delta\phi = -\frac{\Delta\tilde{\mu}_{\text{Cl}^-}}{F} \tag{14.84}$$

Problem 14–31 helps you show that

$$E = \Delta\phi = \frac{\Delta\tilde{\mu}_j}{z_j F} \tag{14.85}$$

if the electrodes react reversibly with a $z_j$-valent ion.

Now if we let the electrical potential that an ion experiences in electrode compartment $\alpha$ or $\beta$ be $\psi^{\alpha}$ or $\psi^{\beta}$, then (Equation 14.73)

$$\tilde{\mu}_j^{\alpha} - \tilde{\mu}_j^{\beta} = \Delta\tilde{\mu}_j = \mu_j^{\alpha} - \mu_j^{\beta} + z_j F(\psi^{\alpha} - \psi^{\beta}) = \Delta\mu_j + z_j F \Delta\psi$$

Solving for $\Delta\psi$ and using Equation 14.85 gives us

$$\Delta\psi = E - \frac{\Delta\mu_j}{z_j F} \tag{14.86}$$

In Equation 14.86, $\Delta\psi$ is the liquid junction potential and $E$ is the emf of the cell.

We will now develop the nonequilibrium thermodynamic equations for this (isothermal) system. Following the reasoning that we used in previous sections, we may write

$$T\dot{S}_{\text{prod}} = J_+\Delta\tilde{\mu}_+ + J_-\Delta\tilde{\mu}_- + J_{\text{w}}\Delta\mu_{\text{w}} \tag{14.87}$$

where $J_+$ ($J_-$) is the flow of cations (anions) and $J_{\text{w}}$ is the flow of water molecules. We will first transform Equation 14.87 into one involving the flow of neutral salt, $J_{\text{s}}$, rather than the individual ions. For concreteness, we will assume that the electrodes react reversibly with the anion (as in Figure 14.6). In this case, the flow of the cation is the same as the flow of neutral salt because the cation is not produced or removed at the electrodes. For a 1–1 salt such as NaCl, $J_{\text{s}} = J_+$ when the electrodes react reversibly with the $Cl^-$(aq) ions. But for an arbitrary salt, $M_{\nu_+}A_{\nu_-}$, $J_+ = \nu_+ J_{\text{s}}$ (Problem 14–32). According to Equation 11.32,

$$\Delta\mu_{\text{s}} = \nu_+\Delta\tilde{\mu}_+ + \nu_-\Delta\tilde{\mu}_-$$

or

$$\Delta\tilde{\mu}_+ = \frac{\Delta\mu_{\text{s}} - \nu_-\Delta\tilde{\mu}_-}{\nu_+} \tag{14.88}$$

Substituting $J_+ = \nu_+ J_{\text{s}}$ and Equation 14.88 into Equation 14.87 gives us

$$\begin{aligned} T\dot{S}_{\text{prod}} &= \nu_+ J_{\text{s}}\left(\frac{\Delta\mu_{\text{s}} - \nu_-\Delta\tilde{\mu}_-}{\nu_+}\right) + J_-\Delta\tilde{\mu}_- + J_{\text{w}}\Delta\mu_{\text{w}} \\ &= J_{\text{s}}\Delta\mu_{\text{s}} + (J_- - \nu_- J_{\text{s}})\Delta\tilde{\mu}_- + J_{\text{w}}\Delta\mu_{\text{w}} \end{aligned} \tag{14.89}$$

Introducing the electroneutrality condition $z_+\nu_+ + z_-\nu_- = 0$ and Equation 14.85 into Equation 14.89 gives us (Problem 14–33)

$$T\dot{S}_{\text{prod}} = J_{\text{s}}\Delta\mu_{\text{s}} + IE + J_{\text{w}}\Delta\mu_{\text{w}} \tag{14.90}$$

where $I$, the electric current passing across the liquid junction, is given by

$$I = z_+ F J_+ + z_- F J_- = \nu_+ z_+ F J_{\text{s}} + z_- F J_- \tag{14.91}$$

Note that the force that drives the electric current across the liquid junction is the emf of the cell, and not the liquid junction potential.

Equation 14.90 isn't quite our desired form for $T\dot{S}_{\text{prod}}$. The total volume flow is given by

$$J_V = J_s\overline{V}_s + J_w\overline{V}_w \tag{14.92}$$

But $\overline{V}_w \gg \overline{V}_s$, especially in a dilute solution, so $J_V \approx J_w\overline{V}_w$. Thus, we can write Equation 14.90 in the form

$$T\dot{S}_{\text{prod}} = J_s\Delta\mu_s + IE + J_V\frac{\Delta\mu_w}{\overline{V}_w}$$

Experimentally, a concentration cell such as that depicted in Figure 14.6 is run such that $J_V$, the volume flow from one compartment to the other, is zero, so our basic entropy production equation is

$$T\dot{S}_{\text{prod}} = J_s\Delta\mu_s + IE \tag{14.93}$$

The associated phenomenological equations are

$$\begin{aligned} I &= L_{11}E + L_{12}\Delta\mu_s \\ J_s &= L_{21}E + L_{22}\Delta\mu_s \end{aligned} \tag{14.94}$$

The emf of the cell is determined under conditions of zero current flow, so if we set $I = 0$ in Equations 14.94, we find that

$$E = -\frac{L_{12}}{L_{11}}\Delta\mu_s \qquad (I = 0,\ J_V = 0) \tag{14.95}$$

The ratio $L_{12}/L_{11}$ has a nice physical interpretation. To see what this is, first divide $J_s$ by $I$ in Equations 14.94 with $\Delta\mu_s = 0$ to get

$$\left(\frac{J_s}{I}\right)_{\Delta\mu_s=0} = \frac{L_{21}}{L_{11}} = \frac{L_{12}}{L_{11}} \tag{14.96}$$

where we have used the Onsager reciprocal relation in going from the second to the third ratio. But using the relation $J_+ = \nu_+ J_s$ (the electrodes react reversibly with the anion), we see that

$$\frac{J_s}{I} = \frac{J_+}{\nu_+ I} = \frac{z_+FJ_+}{\nu_+z_+FI} = \frac{1}{\nu_+z_+F}\frac{I_+}{I} \qquad (J_V = 0,\ \Delta\mu_s = 0) \tag{14.97}$$

The ratio $I_+/I$ in Equation 14.97 is the fraction of the total ionic current that is carried by the cation. Contrary to what you might have thought, the cations and anions do not necessarily carry the same amount of current because they do not necessarily move at the same speed. For example, the chloride ions in a NaCl(aq) solution carry about 60% of the ionic current. The fraction of the electric current carried by an ion of type $j$, $I_j/I$, is called the *transport number* and is denoted by

$$t_j = \frac{I_j}{I} \tag{14.98}$$

For a binary salt, $t_+ + t_- = 1$. Transport numbers can be determined by a variety of experimental methods (see Problem 14–28).

We can write Equation 14.97 in terms of $t_+ = I_+/I$.

$$\frac{J_s}{I} = \frac{t_+}{\nu_+ z_+ F} \qquad (J_V = 0,\ \Delta\mu_s = 0) \tag{14.99}$$

Solving Equation 14.99 for $t_+$ gives

$$t_+ = \nu_+ z_+ F\left(\frac{J_s}{I}\right) = \nu_+ z_+ F\left(\frac{L_{12}}{L_{11}}\right) \qquad (\Delta\mu_s = 0)$$

or

$$\frac{L_{12}}{L_{11}} = \frac{t_+}{\nu_+ z_+ F} \tag{14.100}$$

If we substitute Equation 14.100 into Equation 14.95 and substitute that result into Equation 14.86, we obtain

$$\Delta\psi = -\frac{t_+}{\nu_+ z_+ F}\Delta\mu_s - \frac{\Delta\mu_-}{z_- F}$$

Last, we use $\Delta\mu_s = \nu_+\Delta\tilde{\mu}_+ + \nu_-\Delta\tilde{\mu}_-$, electroneutrality ($\nu_+ z_+ + \nu_- z_- = 0$), and the fact that $t_+ + t_- = 1$ ($t_+$ and $t_-$ are the fractions of current carried by the cations and anions) to get (Problem 14–35)

$$F\Delta\psi = -\frac{t_+}{z_+}\Delta\mu_+ - \frac{t_-}{z_-}\Delta\mu_- \tag{14.101}$$

If we use the relations $\Delta\mu_+ = RT\ln(a_{+,2}/a_{+,1})$ and $\Delta\mu_- = RT\ln(a_{-,2}/a_{-,1})$, then Equation 14.101 becomes

$$\Delta\psi = -t_+\frac{RT}{F}\ln\frac{a_{+,2}}{a_{+,1}} + t_-\frac{RT}{F}\ln\frac{a_{-,2}}{a_{-,1}} \tag{14.102}$$

Let's apply Equation 14.102 to a 1–1 electrolyte. If we replace the activities by concentrations in Equation 14.102, we have $a_{+,2} = c_{+,2} = c_2$, $a_{-,2} = c_{-,2} = c_2$, $a_{+,1} = c_{+,1} = c_1$, and $a_{-,1} = c_{-,1} = c_1$, and so

$$\Delta\psi = (t_- - t_+)\frac{RT}{F}\ln\frac{c_2}{c_1} \tag{14.103}$$

Note that Equation 14.103 says that the magnitude of the liquid junction potential depends upon $t_- - t_+$ in this case. It turns out that $t_{K^+} \approx t_{Cl^-}$, so we expect that the liquid junction potential should be quite small for a concentration cell whose cell diagram is

$$\text{Ag(s)}|\text{AgCl(s)}|\text{KCl}(c_1) \vdots \text{KCl}(c_2)|\text{AgCl(s)}|\text{Ag(s)}$$

**EXAMPLE 14–8**

Use Equation 14.103 and the data below to estimate the liquid junction potential at 25°C for the cells

$\text{Ag(s)}|\text{AgCl(s)}|\text{MCl}(c_1) \vdots \text{MCl}(c_2)|\text{AgCl(s)}|\text{Ag(s)}$.

| | $t_{M^+}$ | $c_1/\text{mol·L}^{-1}$ | $c_2/\text{mol·L}^{-1}$ |
|---|---|---|---|
| $M^+ = Na^+$ | 0.392 | 0.010 | 0.0050 |
| $M^+ = K^+$ | 0.490 | 0.010 | 0.0050 |
| $M^+ = H^+$ | 0.825 | 0.010 | 0.0050 |

SOLUTION: For $Na^+$, $t_{Na^+} = 0.392$, so $t_{Cl^-} = 0.608$. Therefore,

$$\Delta\psi = (0.608 - 0.392)\frac{(8.314\ \text{K·mol}^{-1}\text{·K}^{-1})(298.15\ \text{K})}{96\,485\ \text{C·mol}^{-1}} \ln\frac{0.0050}{0.010}$$

$$= -3.85 \times 10^{-3}\ \text{V} = -3.85\ \text{mV}$$

The values for $K^+$ and $H^+$ are $-0.36$ mV and $+11.6$ mV, respectively. The experimental values are estimated to be $-3.68$ mV, $-0.33$ mV, and $+11.13$ mV, respectively. Note that the magnitude of $\Delta\psi$ for KCl(aq) is about a factor of 10 smaller than the others.

Equation 14.101 is often written in the differential form

$$-Fd\psi = \sum_j \frac{t_j}{z_j} d\mu_j \qquad (14.104)$$

Equation 14.104 is a fundamental equation for the liquid junction potential. We will apply Equation 14.104 to a number of special cases in the next section.

## 14–9. The Liquid Junction Potential Is Well Approximated by the Henderson Equation

Once again, we consider a cell whose cell diagram is

$$\text{Ag(s)}|\text{AgCl(s)}|\text{MCl}(a_1) \vdots \text{MCl}(a_2)|\text{AgCl(s)}|\text{Ag(s)}$$

Equation 14.104 for this cell is

$$\begin{aligned} -F d\psi &= t_{\mathrm{M}^+} d\mu_{\mathrm{M}^+} - t_{\mathrm{Cl}^-} d\mu_{\mathrm{Cl}^-} \\ &= t_{\mathrm{M}^+} RT\, d \ln a_{\mathrm{M}^+} - t_{\mathrm{Cl}^-} RT\, d \ln a_{\mathrm{Cl}^-} \\ &= t_{\mathrm{M}^+} RT\, d \ln a_{\mathrm{M}^+} a_{\mathrm{Cl}^-} - RT\, d \ln a_{\mathrm{Cl}^-} \end{aligned} \tag{14.105}$$

where we have used the fact that $t_{\mathrm{M}^+} + t_{\mathrm{Cl}^-} = 1$ in going from the second line to the third line. If we assume that $t_{\mathrm{M}^+}$ is constant and integrate Equation 14.105 from electrode compartment 1 to electrode compartment 2, we get

$$\begin{aligned} -\Delta\psi &= t_{\mathrm{M}^+} \frac{RT}{F} \ln \frac{(a_{\mathrm{M}^+} a_{\mathrm{Cl}^-})_2}{(a_{\mathrm{M}^+} a_{\mathrm{Cl}^-})_1} - \frac{RT}{F} \ln \frac{a_{\mathrm{Cl}^-,2}}{a_{\mathrm{Cl}^-,1}} \\ &= 2t_{\mathrm{M}^+} \frac{RT}{F} \ln \frac{a_{\pm,2}}{a_{\pm,1}} - \frac{RT}{F} \ln \frac{a_{\mathrm{Cl}^-,2}}{a_{\mathrm{Cl}^-,1}} \end{aligned} \tag{14.106}$$

where we have used the relation $a_{\pm}^2 = a_{\mathrm{M}^+} a_{\mathrm{Cl}^-}$ in going from the first line to the second line. Before we can use Equation 14.106, we must deal with the (nonexperimental) ratio $a_{\mathrm{Cl}^-,2}/a_{\mathrm{Cl}^-,1}$. According to the Debye-Hückel theory, or its simple extension (see Equation 11–57),

$$\ln \gamma_j = -\frac{1.173 z_j^2 (I_c/\,\mathrm{mol{\cdot}L^{-1}})^{1/2}}{1 + (I_c/\,\mathrm{mol{\cdot}L^{-1}})^{1/2}}$$

Note that this equation says that $\gamma_{\mathrm{M}^+} = \gamma_{\mathrm{Cl}^-}$, so $a_{\mathrm{M}^+} = a_{\mathrm{Cl}^-}$ in each compartment (Problem 14–29). If we accept this result, then we can substitute $a_{\mathrm{Cl}^-} = (a_{\mathrm{M}^+} a_{\mathrm{Cl}^-})^{1/2} = (a_{\pm}^2)^{1/2} = a_{\pm}$ into the second term on the right side of Equation 14.106 to obtain

$$\begin{aligned} \Delta\psi &= (1 - 2t_{\mathrm{M}^+}) \frac{RT}{F} \ln \frac{a_{\pm,2}}{a_{\pm,1}} \\ &= (t_{\mathrm{Cl}^-} - t_{\mathrm{M}^+}) \frac{RT}{F} \ln \frac{a_{\pm,2}}{a_{\pm,1}} \end{aligned} \tag{14.107}$$

We used Equation 14.107 (with activities replaced by concentrations) in Example 14–8 to calculate $\Delta\psi$ for several concentration cells. We can also use Equation 14.104 to calculate the values of the emfs of the concentration cells in Example 14–8. We start with the first line of Equation 14.105 and use $t_{\mathrm{Cl}^-} = 1 - t_{\mathrm{M}^+}$ to obtain

$$\begin{aligned} -F d\psi &= t_{\mathrm{M}^+} (d\mu_{\mathrm{M}^+} + d\mu_{\mathrm{Cl}^-}) - d\mu_{\mathrm{Cl}^-} \\ &= t_{\mathrm{M}^+} d\mu_{\mathrm{MCl}} - d\mu_{\mathrm{Cl}^-} \end{aligned} \tag{14.108}$$

where we have used Equation 11.32, which says that $\mu_s = \nu_+\mu_+ + \nu_-\mu_-$. Because the electrodes react reversibly with the $Cl^-$(aq) ions, we now use Equation 14.84 in the differential form

$$FdE = -d\mu_{Cl^-} + Fd\psi$$

to write Equation 14.108 as

$$FdE = -t_{M^+}d\mu_{MCl} = -t_{M^+}RTd\ln a_{MCl}$$

If we assume that $t_{M^+}$ is constant and integrate from electrode compartment 1 to electrode compartment 2, we obtain

$$E = -t_{M^+}\frac{RT}{F}\ln\frac{a_{MCl,2}}{a_{MCl,1}} = -2t_{M^+}\frac{RT}{F}\ln\frac{a_{\pm,2}}{a_{\pm,1}} \tag{14.109}$$

where we have used the fact that $a = a_\pm^2$. Recall that the electrodes are reversible with respect to $Cl^-$(aq); if they were reversible with respect to the cation, then $t_{M^+}$ would be replaced by $-t_{Cl^-}$ in Equation 14.109 (Problem 14–36).

**EXAMPLE 14–9**

Use Equation 14.109 and the data in Example 14–8 to calculate the value of $E$ for each cell in Example 14–8.

SOLUTION: For $Na^+$ we have

$$\begin{aligned} E &= -(2)(0.392)\frac{(8.314\ \text{K}\cdot\text{mol}^{-1}\cdot\text{K}^{-1})(298.15\ \text{K})}{96\,485\ \text{C}\cdot\text{mol}^{-1}}\ln\frac{0.0050}{0.010} \\ &= 13.96\ \text{mV} \end{aligned}$$

The values for $K^+$ and $H^+$ are 17.45 mV and 29.38 mV, respectively. The experimental values are estimated to be 13.5 mV, 16.81 mV, and 28.3 mV, respectively. The small discrepancies between the calculated and the experimental results come from the assumption that $t_{M^+}$ is constant.

To this point, we have considered only cells in which the two solutions in contact contained the same salt. Let's now consider cells of the type

$$\text{Ag(s)}|\text{AgCl(s)}|\text{MCl}(c_1) \vdots \text{M}'\text{Cl}(c_2)|\text{AgCl(s)}|\text{Ag(s)}$$

Before we can integrate Equation 14.104 for any case other than two solutions of the same salt in contact, we must either know or assume the relative concentrations of the ions with respect to each other, in addition to making some assumption concerning single-ion activities. There are a number of integration schemes for Equation 14.104, but we will discuss only one of them, which is due to P. Henderson. Before we can present Henderson's integration scheme, we must introduce the *mobility* of an ion. An ion in a dilute gas will be accelerated by an electric field, but an ion in a solution such as an aqueous solution will quickly (of the order of $10^{-12}$ s, see Problem 14–39) come to a constant velocity as the viscous drag on the ion balances the force arising from the electric field. If we let $v_j$ be the (constant) drift velocity of the $j$th ion, then the mobility of the $j$th ion, $u_j$, is defined by the equation

$$v_j = u_j \mathcal{E} \tag{14.110}$$

where $\mathcal{E}$ is the electric field strength. Thus, the mobility is the drift velocity of an ion in a unit electric field. Because the units of $v_j$ are $\text{m}\cdot\text{s}^{-1}$ and those of $\mathcal{E}$ are $\text{V}\cdot\text{m}^{-1}$, the SI units of mobility, $u_j$, are $\text{m}^2\cdot\text{V}^{-1}\cdot\text{s}^{-1}$.

For a 1–1 electrolyte such as NaCl, the current density, $j$, is given by

$$\begin{aligned} j &= c_+ z_+ F v_+ + c_- z_- F v_- \\ &= (c_+ z_+ F u_+ + c_- z_- F u_-)\mathcal{E} \end{aligned} \tag{14.111}$$

Note that the units of $j$ are $(\text{mol}\cdot\text{dm}^{-3})(\text{C}\cdot\text{mol}^{-1})(\text{m}\cdot\text{s}^{-1}) = \text{A}\cdot\text{m}^{-2}$, where $A$ stands for amperes. The first term of the right side of Equation 14.111 denotes a current flux of cations, and the second term denotes a current flux of anions. The cations and anions are moving in opposite directions under the influence of $\mathcal{E}$, but they have opposite signs, so the electric current flux is in the same direction. Mobilities are customarily taken to be positive quantities and Equation 14.111 is written as

$$j = (c_+ z_+ F u_+ + c_- |z_-| F u_-)\mathcal{E} \tag{14.112}$$

Note that $j_+ = c_+ z_+ F u_+ \mathcal{E}$ and $j_- = c_- |z_-| F u_- \mathcal{E}$, so the cation transport number $t_+$ is given by

$$t_+ = \frac{j_+}{j} = \frac{c_+ z_+ F u_+ \mathcal{E}}{c_+ z_+ F u_+ \mathcal{E} + c_- |z_-| F u_- \mathcal{E}} = \frac{u_+}{u_+ + u_-}$$

for a single 1–1 electrolyte. For a mixture of 1–1 electrolytes, $t_i$ becomes (Problem 14–40)

$$t_i = \frac{c_i u_i}{\sum_j c_j u_j} \tag{14.113}$$

for the $i$th ion.

We are now ready to derive the so-called Henderson equation for the liquid junction potential. To carry out the integrations in Equation 14.104, we must make assumptions about how the $t_j$ and the $a_j$ vary across the liquid junction. We will first assume that we can use concentrations for activities. The basic assumption of the Henderson integration scheme is that the liquid junction consists of a narrow region with a continuous variation of solution composition from one electrode compartment to the other. For example, consider the liquid junction $\text{HCl}(c_1) \vdots \text{NaCl}(c_2)$. We will assume that the solution within the liquid junction is a mixture of HCl(aq) and NaCl(aq) and that the composition of the solution varies as

$$c_{\mathrm{HCl}} = (1 - x)c_1 \qquad \text{and} \qquad c_{\mathrm{NaCl}} = xc_2$$

where $x$ varies from 0 to 1 from electrode compartment 1 to electrode compartment 2. Note that $c_{\mathrm{HCl}} = c_1$ and $c_{\mathrm{NaCl}} = 0$ when $x = 0$ and that $c_{\mathrm{HCl}} = 0$ and $c_{\mathrm{NaCl}} = c_2$ when $x = 1$. The variable $x$ is the proportion of the solution in electrode compartment 2 that makes up the solution within the liquid junction.

Generally, at a position within the liquid junction where the mixing fraction is $x$, the concentration of ion $j$ will be

$$c_j(x) = c_{j,1} + (c_{j,2} - c_{j,1})x \qquad 0 \le x \le 1 \tag{14.114}$$

where $c_{j,1}$ and $c_{j,2}$ are the compositions of ion $j$ in electrode compartments 1 and 2, respectively. For the $\mathrm{HCl}(c_1) \vdots \mathrm{NaCl}(c_2)$ liquid junction that we discussed above,

$$\begin{aligned} c_{\mathrm{H^+}}(x) &= c_{\mathrm{H^+},1}(1 - x) = c_1(1 - x) \\ c_{\mathrm{Na^+}}(x) &= c_{\mathrm{Na^+},2}x = c_2 x \\ c_{\mathrm{Cl^-}}(x) &= c_{\mathrm{Cl^-},1} + (c_{\mathrm{Cl^-},2} - c_{\mathrm{Cl^-},1})x = c_1 + (c_2 - c_1)x \end{aligned} \tag{14.115}$$

We must also make assumptions about how the transport numbers vary with solution composition. To do this, we simply substitute Equation 14.114 into Equation 14.113 to obtain

$$t_j(x) = \frac{c_j(x)u_j}{\sum_i c_i(x)u_i} = \frac{c_{j,1}u_j + (c_{j,2} - c_{j,1})u_j x}{\sum_i c_{i,1}u_i + \sum_i (c_{i,2} - c_{i,1})u_i x} \tag{14.116}$$

where we assume that the mobilities are constants. When these approximations are used in Equation 14.104, the resulting equation for $\Delta\psi$ is called the *Henderson equation.*

---

**EXAMPLE 14–10**

Use Equations 14.115 and 14.116 and the mobilities listed in Table 14.1 to plot the transport numbers of $H^+$(aq), $Na^+$(aq), and $Cl^-$(aq) as a function of mixing fraction $x$ for the liquid junction $\mathrm{HCl(0.050\ M)} \vdots \mathrm{NaCl(0.100\ M)}$.

SOLUTION: The denominator of Equation 14.116 is

$$\sum_i c_i(x)u_i/10^{-8}\ \mathrm{M{\cdot}m^2{\cdot}V^{-1}{\cdot}s^{-1}} =$$

$$(0.050)(36.3)(1 - x) + (0.100)(5.19)x + (0.050)(7.91) + (0.100 - 0.050)(7.91)x$$

$$= 2.211 - 0.901x$$

and so

$$t_{H^+}(x) = \frac{(0.050)(36.3) - (0.050)(36.3)x}{2.211 - 0.901x} = \frac{1.815 - 1.815x}{2.211 - 0.901x}$$

$$t_{Na^+}(x) = \frac{(0.100)(5.19)x}{2.211 - 0.901x} = \frac{0.519x}{2.211 - 0.901x}$$

and $t_{H^+} + t_{Na^+} + t_{Cl^-} = 1$. The numerical results are plotted in Figure 14.7.

**TABLE 14.1**
The mobilities of various ions at 298.15 K at infinite dilution.

| Ion | $u/10^{-8}$ m$^2$·V$^{-1}$·s$^{-1}$ | Ion | $u/10^{-8}$ m$^2$·V$^{-1}$·s$^{-1}$ |
|---|---|---|---|
| $Ag^+$(aq) | 6.42 | $Br^-$(aq) | 8.09 |
| $Cs^+$(aq) | 8.01 | $CH_3COO^-$(aq) | 4.24 |
| $H^+$(aq) | 36.3 | $Cl^-$(aq) | 7.91 |
| $K^+$(aq) | 7.62 | $I^-$(aq) | 7.96 |
| $NH_4^+$(aq) | 7.62 | $NO_3^-$(aq) | 7.40 |
| $Na^+$(aq) | 5.19 | $OH^-$(aq) | 20.6 |
| $Rb^+$(aq) | 8.06 | | |

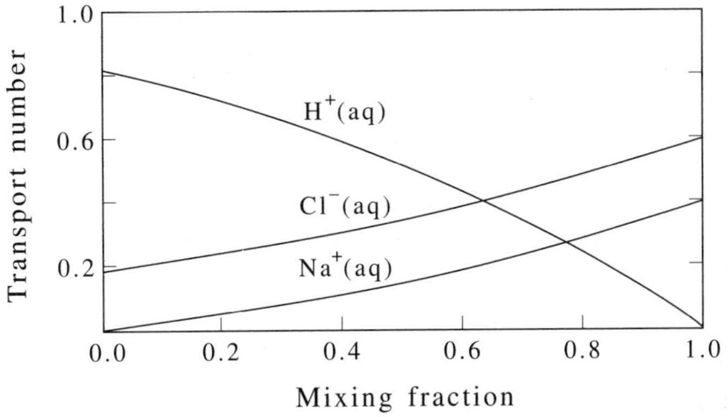


**FIGURE 14.7**
The transport numbers of $H^+$(aq), $Na^+$(aq), and $Cl^-$(aq) within the liquid junction HCl(0.050 M) $\vdots$ NaCl(0.100 M) plotted against the mixing fraction according to the Henderson integration scheme.

Let's apply these approximations to a cell such as

$$\text{Ag(s)|AgCl(s)|MCl}(c_1)\,\vdots\,\text{MCl}(c_2)\text{|AgCl(s)|Ag(s)}$$

In this case,

$$t_+ = \frac{c_{\text{M}^+}(x)u_{\text{M}^+}}{c_{\text{M}^+}(x)u_{\text{M}^+} + c_{\text{Cl}^-}(x)u_{\text{Cl}^-}} = \frac{u_{\text{M}^+}}{u_{\text{M}^+} + u_{\text{Cl}^-}}$$

because $c_{\text{M}^+}(x) = c_{\text{Cl}^-}(x)$. If we substitute Equation 14.114 into Equation 14.104 and integrate, then we obtain (Problem 14–41)

$$\begin{aligned}\Delta\psi &= \frac{u_{\text{Cl}^-} - u_{\text{M}^+}}{u_{\text{M}^+} + u_{\text{Cl}^-}}\frac{RT}{F}\ln\frac{c_{\text{MCl},2}}{c_{\text{MCl},1}} \\ &= (t_{\text{Cl}^-} - t_{\text{M}^+})\frac{RT}{F}\ln\frac{c_{\text{MCl},2}}{c_{\text{MCl},1}}\end{aligned}$$

in agreement with Equation 14.107.

Now let's consider a liquid junction such as $\text{HCl}(c)\,\vdots\,\text{KCl}(c)$. In this case, the Henderson equation yields (Problem 14–42)

$$\Delta\psi = \frac{RT}{F}\ln\frac{u_{\text{H}^+} + u_{\text{Cl}^-}}{u_{\text{K}^+} + u_{\text{Cl}^-}} \tag{14.117}$$

**EXAMPLE 14–11**

Use the data in Table 14.1 to calculate the value of $\Delta\psi$ at 25°C for the liquid junction $\text{HCl(0.010 M)}\,\vdots\,\text{NaCl(0.010 M)}$. The experimental value is 31.2 mV.

SOLUTION: Using the mobilities in Table 14.1, Equation 14.117 gives us

$$\begin{aligned}\Delta\psi &= \frac{(8.314\ \text{K}\cdot\text{mol}^{-1}\cdot\text{K}^{-1})(298.15\ \text{K})}{96\,485\ \text{C}\cdot\text{mol}^{-1}}\ln\frac{36.3\times10^{-8} + 7.91\times10^{-8}}{5.19\times10^{-8} + 7.91\times10^{-8}} \\ &= 31.3\ \text{mV}\end{aligned}$$

in good agreement with the experimental value of 31.2 V.

There are other integration schemes for Equation 14.104, but Henderson's is relatively simple and gives satisfactory agreement with experimental values.

## 14–10. The Flux–Force Relations for Continuous Systems Involve Gradients of Thermodynamic Quantities Rather Than Differences

All the systems we have discussed in this chapter consist of two compartments, in which some thermodynamic property such as temperature or concentration changes rather abruptly from one compartment to the other. In the vocabulary of nonequilibrium thermodynamics, we say that such systems are discontinuous. We can imagine systems such as a metallic bar in which the temperature varies smoothly from one end to the other, or a solution in which the concentration varies smoothly from one region to another. We say that systems like these are continuous systems. In this section, we will show how nonequilibrium thermodynamics can be formulated to treat continuous systems. For simplicity only, we will consider systems in which the properties vary in only one direction, which we take to be along the $x$ axis. We now assume that we can subdivide the system into slices of thickness $\Delta x = x_2 - x_1$ (Figure 14.8), which is large enough that thermodynamic variables such as $T$, $P$, and $\mu$ have well-defined local values but small enough that these local values are the same everywhere within the region of thickness $\Delta x$. We furthermore assume that these local thermodynamic quantities satisfy the same thermodynamic equations that we have derived for equilibrium systems. These two assumptions are formalized by the *Postulate of Local Equilibrium.* This postulate obviously places restrictions on the systems we can treat, but experience shows that it is a good approximation for many systems. Most systems of chemical interest vary rather smoothly on a molecular scale, with notable exceptions being systems involving phenomena such as shock waves or explosions.

Let's consider Equation 14.23,

$$\dot{S}_{\text{prod}} = \frac{dU_1}{dt}\left(\frac{1}{T_1} - \frac{1}{T_2}\right) + \frac{dn_1}{dt}\left(\frac{\mu_2}{T_2} - \frac{\mu_1}{T_1}\right) \geq 0$$

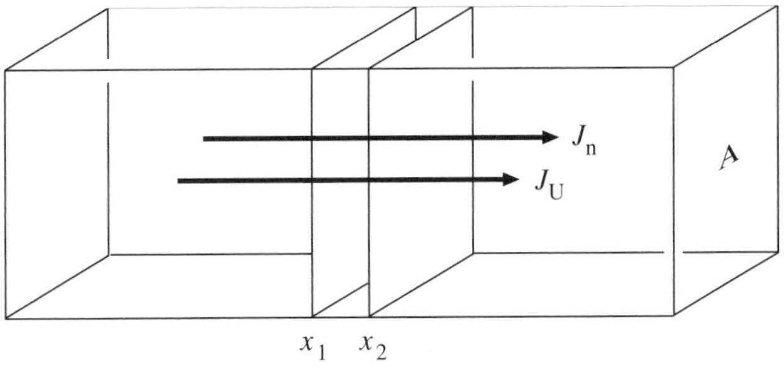


**FIGURE 14.8**
The geometry used to derive Equation 14.119. The positively directed fluxes $J_U$ and $J_n$ are shown pointing in the positive $x$ direction.

which applies to the discontinuous system shown in Figure 14.3. We now apply this equation to any subregion of volume $A\Delta x$ and divide $\dot{S}_{\text{prod}}$ by $V = A\Delta x$ to write

$$\sigma = \frac{\dot{S}_{\text{prod}}}{V} = \left(-\frac{1}{A}\frac{dU_1}{dt}\right)\frac{\Delta(1/T)}{\Delta x} + \frac{1}{A}\left(\frac{dn_1}{dt}\right)\frac{\Delta(\mu/T)}{\Delta x} \geq 0 \tag{14.118}$$

where $\Delta(1/T) = 1/T_2 - 1/T_1$ and $\Delta(\mu/T) = \mu_2/T_2 - \mu_1/T_1$. Realizing that $\Delta x$ is small, we write Equation 14.118 as

$$\sigma = J_U\frac{\partial(1/T)}{\partial x} + J_n\frac{\partial(-\mu/T)}{\partial x} \geq 0 \tag{14.119}$$

where

$$J_U = -\frac{1}{A}\frac{dU_1}{dt} \qquad \text{and} \qquad J_n = -\frac{1}{A}\frac{dn_1}{dt} \tag{14.120}$$

The quantities $J_U$ and $J_n$ are called *fluxes*. Note that $J_U$ and $J_n$ in Equation 14.120 differ from $J_U$ and $J_n$ in Equation 14.24 by a factor of $1/A$. The derivatives in Equation 14.119 are called gradients of $1/T$ and $-\mu/T$, respectively.

As an example of the application of nonequilibrium thermodynamics to continuous systems, we will consider diffusion in an isothermal system (isothermal diffusion). In this case, we have

$$\Phi = T\sigma = \sum_{i=1}^{n} J_i\left(-\frac{\partial\mu_i}{\partial x}\right)_{P,T} \tag{14.121}$$

where $\Phi$ simply stands for $T\sigma$ and $n$ is the number of components in the system. The chemical potentials in Equation 14.121 are not independent because of the Gibbs-Duhem equation (Equation 10.10), which we can write as

$$\sum_{i=1}^{n} c_i\left(-\frac{\partial\mu_i}{\partial x}\right)_{P,T} = 0 \tag{14.122}$$

We will use Equation 14.122 to eliminate the chemical potential of the solvent in favor of the others. As usual, we let the chemical potential of the solvent be $\mu_1$, so

$$\frac{\partial\mu_1}{\partial x} = -\frac{1}{c_1}\sum_{i=2}^{n} c_i\left(\frac{\partial\mu_i}{\partial x}\right)_{P,T}$$

Substitute this result into Equation 14.122 to obtain

$$\begin{aligned}\Phi &= \sum_{i=2}^{n}\left(J_i - \frac{c_i}{c_1}J_1\right)\left(-\frac{\partial\mu_i}{\partial x}\right)_{P,T} \\ &= \sum_{i=2}^{n} J_i^d\left(-\frac{\partial\mu_i}{\partial x}\right)_{P,T}\end{aligned} \tag{14.123}$$

**EXAMPLE 14–12**
Discuss the units in Equation 14.123.

SOLUTION: The units of $\sigma = \dot{S}_{\text{prod}}/V$ are J·K$^{-1}$·m$^{-3}$·s$^{-1}$, so the units of $\Phi = T\sigma$ are J·m$^{-3}$·s$^{-1}$. The units of the $J_i^d$ are mol·m$^{-2}$·s$^{-1}$ and of the $\partial\mu_i/\partial x$ are J·mol$^{-1}$·m$^{-1}$, so the units on the right side of Equation 14.123 are
(mol·m$^{-2}$·s$^{-1}$)(J·mol$^{-1}$·m$^{-1}$) = J·m$^{-3}$·s$^{-1}$, in agreement with the units of $\Phi = T\sigma$.

We can give a physical interpretation to the $J_i^d$ in Equation 14.123. If we use the fact that a flux, $J_i$, is equal to $c_i v_i$, where $v_i$ is the velocity of component $i$ (Problem 14–44), then the $J_i^d$ can be expressed as

$$J_i^d = c_i\left(\frac{J_i}{c_i} - \frac{J_1}{c_1}\right) = c_i(v_i - v_1)$$

Thus, we see that $J_i^d$ is the flux of solute $i$ relative to that of the solvent (component 1).

Let's consider a binary solution, such as sucrose in water or sodium chloride in water. In such cases, there is only one independent flow and Equation 14.123 is

$$\Phi = J_2^d\left(-\frac{\partial\mu_2}{\partial x}\right)_{T,P} \tag{14.124}$$

with just one flux-force equation

$$J_2^d = L_{22}\left(-\frac{\partial\mu_2}{\partial x}\right)_{T,P} = -L_{22}\left(\frac{\partial\mu_2}{\partial x}\right)_{T,P} \tag{14.125}$$

Equation 14.125 can be written in terms of a gradient in the concentration rather than a gradient in the chemical potential by using the relation $\mu_2 = \mu_2^\circ(T, P) + RT\ln a_2$ and assuming that $a_2 = c_2$. This gives

$$\left(\frac{\partial\mu_2}{\partial x}\right)_{T,P} = \left(\frac{\partial\mu_2}{\partial c_2}\right)_{T,P}\left(\frac{\partial c_2}{\partial x}\right)_{T,P} = \frac{RT}{c_2}\left(\frac{\partial c_2}{\partial x}\right)_{T,P}$$

Equation 14.125 now becomes

$$J_2^d = -L_{22}\frac{RT}{c_2}\frac{\partial c_2}{\partial x} \tag{14.126}$$

Equation 14.126 has been known experimentally since the 19th century and is called *Fick's law of diffusion*. It is usually written as

$$\text{flux} = -D\frac{\partial c}{\partial x} \tag{14.127}$$

where $D$ is the *diffusion coefficient*. The negative sign in Equations 14.126 and 14.127 simply means that the direction of the diffusional flux is opposite the concentration gradient. By comparing Equations 14.126 and 14.127, we see that

$$D = \frac{L_{22}RT}{c_2} \tag{14.128}$$

Note that because $L_{22} > 0$ (Equation 14.33), $D$ is an intrinsically positive quantity.

The phenomenological coefficients $L_{22}$ vary approximately linearly with concentration, so Equation 14.128 says that diffusion coefficients should be fairly independent of concentration. The plots of the diffusion coefficients of LiCl(aq), NaCl(aq), and KCl(aq) at 25°C against the square root of the concentration in Figure 14.9 show that the diffusion coefficients have only a slight dependence on concentration.

Let's now discuss a solution of two solutes in water (a ternary system). In this case, Equation 14.123 becomes

$$\Phi = J_2^d\left(-\frac{\partial \mu_2}{\partial x}\right) + J_3^d\left(-\frac{\partial \mu_3}{\partial x}\right) \tag{14.129}$$

with the two flux-force relations

$$\begin{aligned} J_2^d &= L_{22}\left(-\frac{\partial \mu_2}{\partial x}\right) + L_{23}\left(-\frac{\partial \mu_3}{\partial x}\right) \\ J_3^d &= L_{32}\left(-\frac{\partial \mu_2}{\partial x}\right) + L_{33}\left(-\frac{\partial \mu_3}{\partial x}\right) \end{aligned} \tag{14.130}$$

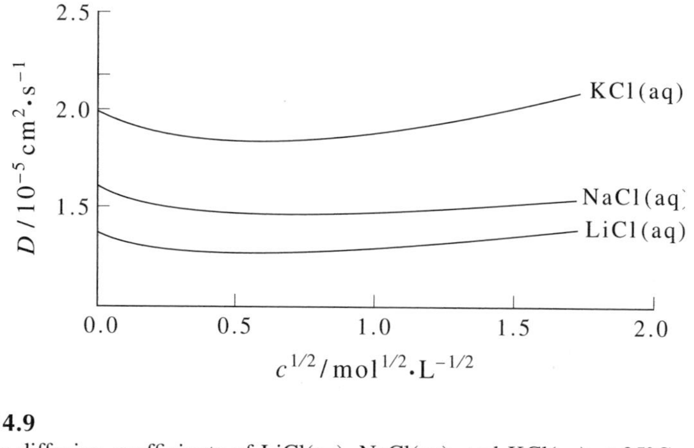


**FIGURE 14.9**
A plot of the diffusion coefficients of LiCl(aq), NaCl(aq), and KCl(aq) at 25°C against the square root of the concentration, showing that the diffusion coefficients are fairly independent of concentration.

with $L_{23} = L_{32}$. In an ideal solution, $\mu_2$ depends only on $c_2$ and $\mu_3$ depends only on $c_3$, but in a general nonideal ternary solution, $\mu_2$ and $\mu_3$ depend on both $c_2$ and $c_3$ (Problem 14–45). Therefore, we have

$$\frac{\partial \mu_2}{\partial x} = \frac{\partial \mu_2}{\partial c_2}\frac{\partial c_2}{\partial x} + \frac{\partial \mu_2}{\partial c_3}\frac{\partial c_3}{\partial x} = \mu_{22}\frac{\partial c_2}{\partial x} + \mu_{23}\frac{\partial c_3}{\partial x}$$

$$\frac{\partial \mu_3}{\partial x} = \frac{\partial \mu_3}{\partial c_2}\frac{\partial c_2}{\partial x} + \frac{\partial \mu_3}{\partial c_3}\frac{\partial c_3}{\partial x} = \mu_{32}\frac{\partial c_2}{\partial x} + \mu_{33}\frac{\partial c_3}{\partial x}$$

where $\mu_{ij} = \partial \mu_i / \partial c_j$. Equations 14.130 become

$$\begin{aligned} J_2^d &= (L_{22}\mu_{22} + L_{23}\mu_{32})\left(-\frac{\partial c_2}{\partial x}\right) + (L_{22}\mu_{23} + L_{23}\mu_{33})\left(-\frac{\partial c_3}{\partial x}\right) \\ J_3^d &= (L_{32}\mu_{22} + L_{33}\mu_{32})\left(-\frac{\partial c_2}{\partial x}\right) + (L_{32}\mu_{23} + L_{33}\mu_{33})\left(-\frac{\partial c_3}{\partial x}\right) \end{aligned} \tag{14.131}$$

or

$$\begin{aligned} J_2^d &= -D_{22}\frac{\partial c_2}{\partial x} - D_{23}\frac{\partial c_3}{\partial x} \\ J_3^d &= -D_{32}\frac{\partial c_2}{\partial x} - D_{33}\frac{\partial c_3}{\partial x} \end{aligned} \tag{14.132}$$

where the $D_{ij}$ are defined by comparing Equations 14.133 and 14.134.

Note that the reciprocal relations $L_{32} = L_{32}$ do *not* lead to $D_{23} = D_{32}$, nor does $L_{23} = 0$ lead to $D_{23} = 0$. Only when both $\mu_{23} = \mu_{32} = 0$ and $L_{23} = 0$ will $D_{23} = 0$.

**EXAMPLE 14–13**

Use the following experimental data for a NaCl(0.250 M)–KCl(0.250 M) solution at 25°C to verify the Onsager reciprocal relations.

$D_{22} = 1.38 \times 10^{-5}\ \text{cm}^2\cdot\text{s}^{-1}$ $\quad D_{32} = 0.15 \times 10^{-5}\ \text{cm}^2\cdot\text{s}^{-1}$
$D_{23} = 0.010 \times 10^{-5}\ \text{cm}^2\cdot\text{s}^{-1}$ $\quad D_{33} = 1.83 \times 10^{-5}\ \text{cm}^2\cdot\text{s}^{-1}$
$\mu_{22}/RT = 5.762\ \text{M}^{-1}$ $\quad \mu_{32}/RT = 1.633\ \text{M}^{-1}$
$\mu_{23}/RT = 1.652\ \text{M}^{-1}$ $\quad \mu_{33}/RT = 5.601\ \text{M}^{-1}$

SOLUTION: If we solve Equations 14.133 and 14.134 for $L_{23}$ and $L_{32}$, we find that

$$L_{23} = \frac{D_{23}\mu_{22} - D_{22}\mu_{23}}{\mu_{22}\mu_{33} - \mu_{23}\mu_{32}} \quad \text{and} \quad L_{32} = \frac{D_{32}\mu_{33} - D_{33}\mu_{32}}{\mu_{22}\mu_{33} - \mu_{23}\mu_{32}}$$

Substituting the experimental values for the various quantities gives

$$RTL_{23}/\text{M}\cdot\text{cm}^2\cdot\text{s}^{-1} = -7.5 \times 10^{-7} \quad \text{and} \quad RTL_{32}/\text{M}\cdot\text{cm}^2\cdot\text{s}^{-1} = -7.3 \times 10^{-7}$$

which agree with each other within experimental error.

## 14–11. A Steady State Is a State of Minimum Entropy Production

In these final two sections, we will prove some general results of nonequilibrium thermodynamics. First, we will prove a result involving steady-state systems. We will see that the steady state plays a similar role in nonequilibrium thermodynamics to that which the equilibrium state plays in equilibrium thermodynamics. For simplicity, let's consider the simple system shown in Figure 14.3. This system is described by Equations 14.24, 14.25, 14.26, and 14.27,

$$\dot{S}_{\text{prod}} = J_U X_U + J_n X_n > 0 \tag{14.133}$$

with

$$\begin{aligned} J_n &= L_{nn} X_n + L_{nU} X_U \\ J_U &= L_{Un} X_n + L_{UU} X_U \end{aligned} \tag{14.134}$$

Substituting Equation 14.134 into Equation 14.133 and using $L_{Un} = L_{nU}$ gives

$$\dot{S}_{\text{prod}} = L_{UU} X_U^2 + 2L_{nU} X_n X_U + L_{nn} X_n^2 \tag{14.135}$$

At equilibrium, $\dot{S}_{\text{prod}} = 0$ and all the fluxes and forces are equal to zero; the properties of the system are uniform and there are no fluxes. At a steady state, the fluxes and forces in Equation 14.133 do not change with time and $\dot{S}_{\text{prod}} > 0$. One way to maintain a steady state in Figure 14.3 is to use very large compartments (strictly speaking, infinitely large) so that the flux of energy and matter from one compartment to the other does not alter the temperature or the chemical potential in either compartment.

Now let's consider the case in which only $X_U$ is held constant. This case can be achieved experimentally by using a system of two relatively small compartments in each of which the chemical potential changes as molecules flow from one to the other, but where one compartment is in contact with a heat bath at temperature $T_1$ and the other compartment is in contact with a heat bath at temperature $T_2$. Contrary to what you might think, the concentrations (or chemical potentials) of the two compartments do not equalize, even though molecules can pass from one compartment to the other. Instead, the system comes to a steady state in which the material flux, $J_n$, is equal to zero but a (now constant) difference in concentration (or chemical potential) still remains between the two compartments. Thus, even though $J_n = 0$, $X_n \neq 0$ at the steady state. The following example derives this result in terms of equations.

---

**EXAMPLE 14–14**

Determine the values of $X_n$, $J_U$, and $\dot{S}_{\text{prod}}$ for the system described above, when $X_U$ is held fixed but $X_n$ is allowed to adjust to its steady-state value.

SOLUTION: We set $J_n = 0$ in Equation 14.134$a$ and solve for $X_n$ to get

$$X_n = -\frac{L_{nU}}{L_{nn}} X_U$$

Thus, we see that the steady-state force, $X_n$, is not equal to zero even though its associated flux, $J_n$, is. Substitute this result into Equation 14.134$b$ to get

$$J_U = \left( L_{UU} - \frac{L_{nU}^2}{L_{nn}} \right) X_U$$

If we substitute this result into Equation 14.133, we obtain

$$\dot{S}_{\text{prod}} = \left( L_{UU} - \frac{L_{nU}^2}{L_{nn}} \right) X_U^2 > 0$$

The inequality here follows from Equation 14.33.

Let's see how $\dot{S}_{\text{prod}}$ varies with $X_n$ keeping $X_U$ fixed. To do this mathematically, we differentiate $\dot{S}_{\text{prod}}$ given by Equation 14.135 with respect to $X_n$ to obtain

$$\left( \frac{\partial \dot{S}_{\text{prod}}}{\partial X_n} \right)_{X_U} = 2L_{nn}X_n + 2L_{nU}X_U$$

But the right side here is equal to $2J_n$ by Equation 14.134$a$, so we have

$$\left( \frac{\partial \dot{S}_{\text{prod}}}{\partial X_n} \right)_{X_U} = 2J_n = 0 \qquad (14.136)$$

where we have used the fact that $J_n = 0$ at the steady state. A second derivative of $\dot{S}_{\text{prod}}$ with respect to $X_n$ gives $(\partial^2 \dot{S}_{\text{prod}}/\partial X_n^2) = 2L_{nn} > 0$, so we see that the rate of entropy production is a minimum at a steady state.

Equation 14.136 has a nice physical interpretation. At steady state, the unrestrained force will adjust itself so that the rate of entropy production is a minimum. Equation 14.136 is an example of the Principle of Minimum Entropy Production. An equilibrium state is a state of zero entropy production; a steady state is a state of minimum entropy production. In a sense, a steady state plays the same role in nonequilibrium thermodynamics as an equilibrium state plays in classical (equilibrium) thermodynamics.

For the case treated in Example 14–14, the rate of entropy production is a minimum. Thus, if we hold $T_1$ and $T_2$ constant and let energy flux and material flux take place until a steady state is reached, then $\dot{S}_{\text{prod}}$ will decrease to its minimum value consistent with the fixed (nonzero) thermodynamic force $X_U$. If we were then to let $X_U$ vanish by removing the contact of the compartments with their respective heat baths, the system would come to equilibrium, when $\dot{S}_{\text{prod}} = 0$.

We can illustrate these results pictorially in Figure 14.10, which shows $\dot{S}_{\text{prod}}$ given by Equation 14.135 plotted against $X_U$ and $X_n$. As Equation 14.135 indicates, $\dot{S}_{\text{prod}}$ is a quadratic function of $X_U$ and $X_n$, so $\dot{S}_{\text{prod}}$ appears as a quadratic surface in Figure 14.10. Because $\dot{S}_{\text{prod}} \geq 0$, the quadratic surface has only positive values and is centered at the origin of the $X_U$ and $X_n$ axes. Let's start at some arbitrary point, $P_1$, with coordinates $X'_U$ and $X'_n$ on the surface. Now if we keep $X'_U$ fixed and let $X'_n$ vary, the point will move along the parabola formed by the intersection of the $\dot{S}_{\text{prod}}$ surface with the plane $X_U = X'_U =$ constant until it reaches a minimum point designated by $P_2$ with coordinates $X_n = 0$ and $X_U = X'_U$. If both $X_U$ and $X_n$ are allowed to vary, then $\dot{S}_{\text{prod}}$ will continue to decrease until the point $\dot{S}_{\text{prod}} = 0$, where the system is now in a state of equilibrium. Problem 14–50 shows that a steady state, once established, is stable with respect to small fluctuations in the nonfixed forces.

In closing this section, we will prove the Principle of Minimum Entropy Production more generally. In a steady state, certain (thermodynamic) forces are held fixed, and the properties of the system do not change with time. It is observed experimentally that if the forces $X_1, X_2, \ldots, X_k$ are held fixed (by contact with external baths, say), then the other forces $X_{k+1}, \ldots, X_n$ will eventually attain constant values that may or may not be zero. The fluxes $J_1, J_2, \ldots, J_k$ will attain constant nonzero values, and the fluxes $J_{k+1}, \ldots, J_n$ will vanish. This is essentially the definition of a steady state. Now $\dot{S}_{\text{prod}}$ is given by

$$\dot{S}_{\text{prod}} = \sum_{i=1}^{n}\sum_{j=1}^{n} L_{ij} X_i X_j \tag{14.137}$$

with $X_1, X_2, \ldots, X_k$ fixed and $X_{k+1}, \ldots, X_n$ variable. Problem 14–52 helps you prove that $\dot{S}_{\text{prod}}$ attains a minimum value with respect to each of the nonfixed forces, $X_{k+1}, \ldots, X_n$. Realize that the Principle of Minimum Entropy Production is limited to cases in which all the fluxes are linearly related to the forces.

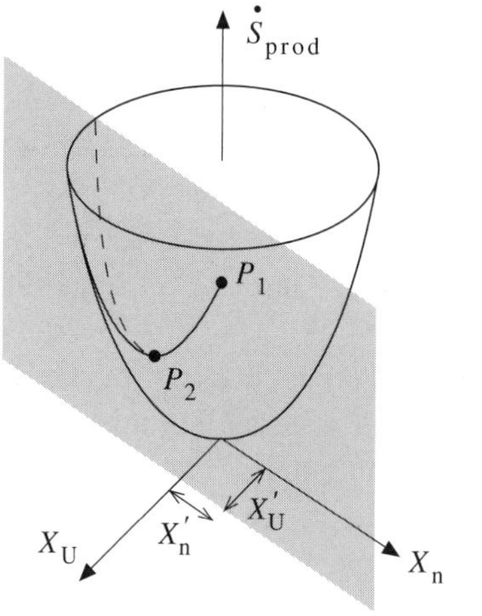


**FIGURE 14.10**
A pictorial illustration of the principle of minimum entropy production. The quadratic surface is $\dot{S}_{\text{prod}}$ given by Equation 14.135 plotted against $X_U$ and $X_n$. We start at the point $P_1$ with the coordinates $X'_U$ and $X'_n$ and then allow $X_n$ to vary. The result is that the system will move to the point $P_2$ at the minimum of the parabola formed by the intersection of the quadratic surface and the plane $X'_U =$ constant.

## 14–12. The Glansdorff–Prigogine Inequality Applies to Systems That Do Not Necessarily Have Linear Flux–Force Relations

In Section 14–11, we showed that the entropy production in a steady state is a minimum with respect to the set of forces that are not fixed. We proved it, however, only under the conditions that (1) the flux-force relations are linear, (2) the Onsager reciprocal relations are valid, and (3) the phenomenological coefficients are independent of time. Thus, our principle of minimum entropy production in steady states is not universally valid. In this final section, we will discuss a more general condition for the nature of a steady state.

The total entropy production in a continuous system is given by

$$\mathcal{P} = \dot{S}_{\text{prod}} = \int \sigma dV = \int \sum_i J_i X_i dV \tag{14.138}$$

where the $J_i$ and $X_i$ are the thermodynamic fluxes and forces, respectively. The time derivative of $\mathcal{P}$ is

$$\frac{\partial \mathcal{P}}{\partial t} = \int \sum_{i=1}^{n} J_i \frac{\partial X_i}{\partial t} dV + \int \sum_{i=1}^{n} X_i \frac{\partial J_i}{\partial t} dV \tag{14.139}$$

which we write as

$$\frac{\partial \mathcal{P}}{\partial t} = \frac{\partial_X \mathcal{P}}{\partial t} + \frac{\partial_J \mathcal{P}}{\partial t} \tag{14.140}$$

The Belgian chemists Peter Glansdorff and Ilya Prigogine have shown that

$$\frac{\partial_X \mathcal{P}}{\partial t} = \int \sum_{i=1}^{n} J_i \frac{\partial X_i}{\partial t} dV \le 0 \tag{14.141}$$

whereas nothing can be said about the sign of $\partial_J \mathcal{P}/\partial t$. This means that for systems that are far from equilibrium, a stationary state does not necessarily correspond to a state of minimum entropy production.

We will prove Equation 14.141 for the example of the one-dimensional heat flow in a narrow bar of length $l$. In this case, $n = 1$ and $J_i = J_U$, the flux of energy as heat, and $X_i = \partial(1/T)/\partial x$, the gradient of the reciprocal temperature, and so Equation 14.141 becomes

$$\frac{\partial_X \mathcal{P}}{\partial t} = \int_0^l J_U \frac{\partial}{\partial t}\left[\frac{\partial(1/T)}{\partial x}\right] dx = \int_0^l J_U \frac{\partial}{\partial x}\left[\frac{\partial(1/T)}{\partial t}\right] dx$$

To get from the second term to the third term, we have used the fact that cross partial second derivatives are equal (MathChapter D). Now integrate by parts to obtain

$$\frac{\partial_X \mathcal{P}}{\partial t} = \left| J_U \left[\frac{\partial(1/T)}{\partial t}\right]\right|_0^l - \int_0^l \left[\frac{\partial(1/T)}{\partial t}\right] \frac{\partial J_U}{\partial x} dx \tag{14.142}$$

If the temperature is fixed at the two ends of the bar so that $\partial(1/T)/\partial t = 0$ at $x = 0$ and $x = l$, then the first term on the right side of Equation 14.142 vanishes and we have

$$\frac{\partial_X \mathcal{P}}{\partial t} = -\int_0^l \left[\frac{\partial(1/T)}{\partial t}\right]\frac{\partial J_U}{\partial x}dx$$

$$= \int_0^l \frac{1}{T^2}\left(\frac{\partial T}{\partial t}\right)\left(\frac{\partial J_U}{\partial x}\right)dx \tag{14.143}$$

We can write $\partial J_U/\partial x$ in terms of $\partial T/\partial t$ by referring to Figure 14.11. The difference between the flux of energy at $x + \Delta x$ and that at $x$, $J_U(x + \Delta x) - J_U(x)$, is equal to the change in energy in the volume $A\Delta x$. If we let $u$ be the energy density, then we have

$$[J_U(x + \Delta x) - J_U(x)]A = -\frac{\partial u}{\partial t}A\Delta x \tag{14.144}$$

The negative sign occurs because the energy within the volume $A\Delta x$ decreases with time if $J_U(x + \Delta x)$ is larger than $J_U(x)$. Equation 14.144 is simply a conservation of energy condition. If we divide both sides of Equation 14.144 by $\Delta x$ and let $\Delta x$ be small, then we have

$$\frac{\partial J_U}{\partial x} = -\frac{\partial u}{\partial t} \tag{14.145}$$

We can write $\Delta u$ as $\rho\overline{C}_V\Delta T$ where $\rho$ is the molar density, and so Equation 14.145 becomes

$$\frac{\partial J_U}{\partial x} = -\rho\overline{C}_V\frac{\partial T}{\partial t} \tag{14.146}$$

If we substitute this result into Equation 14.143, we find that

$$\frac{\partial_X \mathcal{P}}{\partial t} = -\int_0^l \frac{\rho\overline{C}_V}{T^2}\left(\frac{\partial T}{\partial t}\right)^2 dx \le 0 \tag{14.147}$$

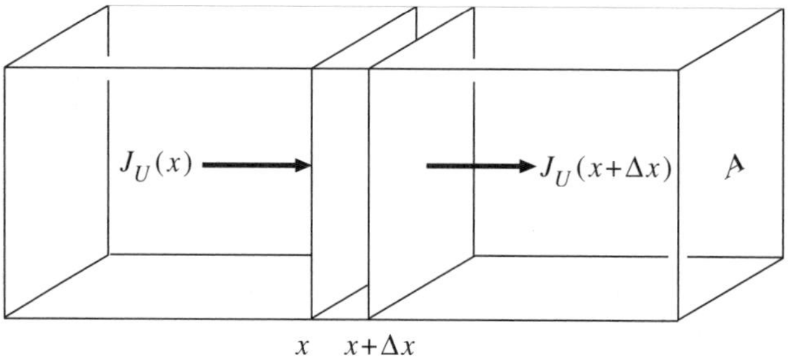


**FIGURE 14.11**
The geometry used to derive Equation 14.145, which represents the conservation of energy.

We obtain the inequality because the integrand in Equation 14.147 is always positive (see Problem 3–40 or 8–58).

**EXAMPLE 14–15**

Prove the inequality in Equation 14.141 for the case of one-dimensional isothermal diffusion in a binary solution.

SOLUTION: In this case, we see from Equation 14.124 that

$$T\dot{S}_{\text{prod}} = J_2^d \left(-\frac{\partial \mu_2}{\partial x}\right)_{T,P}$$

and so there is only one flux and force, with $J_1 = J_2^d$ and $X_1 = \partial(-\mu_2/T)/\partial x = (-1/T)(\partial \mu_2/\partial x)$. Substitute $J_1$ and $X_1$ into Equation 14.141 and integrate by parts to get

$$\frac{\partial_X \mathcal{P}}{\partial t} = \left| -\frac{J_2^d}{T}\frac{\partial \mu_2}{\partial t}\right|_0^l + \int_0^l \frac{1}{T}\left(\frac{\partial \mu_2}{\partial t}\right)\frac{\partial J_2^d}{\partial x}dx$$

The first term on the right vanishes if we maintain the chemical potentials (concentrations) fixed at each end, so that

$$\frac{\partial_X \mathcal{P}}{\partial t} = \int_0^l \frac{1}{T}\left(\frac{\partial \mu_2}{\partial t}\right)\frac{\partial J_2^d}{\partial x}dx \tag{14.148}$$

By the same argument that we used to derive Equation 14.145 (Problem 14–53),

$$\frac{\partial J_2^d}{\partial x} = -\frac{\partial c_2}{\partial t} \tag{14.149}$$

Equation 14.149 simply expresses the conservation of the number of diffusing molecules. Substitute Equation 14.149 into Equation 14.148 to obtain

$$\frac{\partial_X \mathcal{P}}{\partial t} = -\int_0^l \frac{1}{T}\left(\frac{\partial \mu_2}{\partial t}\right)\frac{\partial c_2}{\partial t}dx \tag{14.150}$$

But

$$\frac{\partial c_2}{\partial t} = \left(\frac{\partial c_2}{\partial \mu_2}\right)_{V,T,n_1}\left(\frac{\partial \mu_2}{\partial t}\right)$$

and so Equation 14.150 becomes

$$\frac{\partial_X \mathcal{P}}{\partial t} = -\int_0^l \frac{1}{T}\left(\frac{\partial c_2}{\partial \mu_2}\right)_{V,T,n_1}\left(\frac{\partial \mu_2}{\partial t}\right)^2 dx \le 0$$

The inequality occurs because $T$ and $(\partial c_2/\partial \mu_2)_{V,T,n_1}$ are intrinsically positive quantities (Problem 14–54).

Note that in both proofs of Equation 14.141, we did not assume a linear flux-force relation, and so we never assumed the validity of the Onsager reciprocal relations. Problems 14–55 and 14–56 have you prove the Glansdorff-Prigogine inequality for the chemical reaction scheme X $\rightleftharpoons$ Y and the triangular scheme given by Equation 14.58, respectively.

Nonequilibrium thermodynamics as we have presented in this chapter is restricted to processes that are near equilibrium, have linear flux-force relations, and obey the Onsager reciprocal relations. Although these are not serious restrictions for many commonly studied processes such as heat flow and diffusion, there are a number of important cases for which the restrictions are unacceptable. Noteworthy examples abound in biophysical and physiological systems, in which rapid metabolic processes and other biochemical reactions play central roles. Considerable research has focussed on extending nonequilibrium thermodynamics to systems far from equilibrium, particularly in biophysics.

# Problems

**14-1.** Consider a two-compartment system in contact with a heat bath at temperature $T$, so that $T_1 = T_2 = T$. Let the two-compartment system be surrounded by rigid, impermeable walls, but let the wall separating the two compartments be permeable and flexible. Show that $dA = dn_1(\mu_1 - \mu_2) - dV_1(P_1 - P_2) \leq 0$. Now show that

$$dS_{\text{prod}} = -\frac{dA}{T} = -dn_1(\mu_1 - \mu_2) + dV_1(P_1 - P_2) \geq 0$$

**14-2.** Discuss the physical meaning of each term on the right side of Equation 14.22.

**14-3.** Extend Equation 14.23 to include a flexible wall between the two compartments.

**14-4.** Show that the two terms on the right side of $\dot{S}_{\text{prod}} = J_U X_U + J_n X_n$ have the same units as $\dot{S}_{\text{prod}}$.

**14-5.** Show that Fick's law in the form $J_n \propto \mu_2 - \mu_1$ can be written as $J_n \propto c_2 - c_1$, when $\mu_2 - \mu_1$ is small.

**14-6.** Extend Equations 14.24 to 14.26 to include a flexible wall between the two compartments.

**14-7.** If the system in Figure 14.3 has a flexible wall between the two compartments, then

$$\dot{S}_{\text{prod}} = J_U X_U + J_n X_n + J_V X_V > 0$$

where $J_V = dV_1/dt$ and $X_V = (P_1/T_1) - (P_2/T_2)$. Write out the linear flux-force relations for this system. Show that $L_{UV}$ and $L_{VU}$ have the same units.

**14-8.** Prove that $L_{11}L_{22} > L_{12}^2$.

**14-9.** Prove that $L_{ii} > 0$ and that $L_{ii}L_{jj} > L_{ij}^2$ if $\dot{S}_{\text{prod}} = \sum_i \sum_j L_{ij} X_i X_j$.

**14-10.** Prove that the largest phenomenological coefficient must be one of the diagonal ones.

**14-11.** In this problem, we will show that we can use various linear combinations of fluxes and forces and still preserve the Onsager reciprocal relations. In many applications of nonequilibrium thermodynamics, certain linear flux-force relations are more convenient than others, and the result of this problem says that we can use any convenient linear combinations that we want. We will prove this result for only a special case, but the result is general.

First start with $\dot{S}_{\text{prod}} = J_1 X_1 + J_2 X_2$ with

$$\begin{aligned} J_1 &= L_{11} X_1 + L_{12} X_2 \\ J_2 &= L_{21} X_1 + L_{22} X_2 \end{aligned} \tag{1}$$

To keep the algebra to a minimum, define new fluxes by

$$J_1' = aJ_1 + bJ_2 \qquad J_2' = J_2$$

where $a$ and $b$ are constants. Now solve these two equations for $J_1$ and $J_2$ and substitute them into $\dot{S}_{\text{prod}}$ to obtain

$$\dot{S}_{\text{prod}} = J_1' \frac{X_1}{a} + J_2' \left( X_2 - \frac{b}{a} X_1 \right) = J_1' X_1' + J_2' X_2'$$

which serves to define $X_1'$ and $X_2'$. Now define the phenomenological coefficients $M_{ij}$ by

$$\begin{aligned} J_1' &= M_{11} X_1' + M_{12} X_2' \\ J_2' &= M_{21} X_1' + M_{22} X_2' \end{aligned}$$

Convert these equations into the form of Equations 1, and show that $M_{12} = M_{21}$ follows from $L_{12} = L_{21}$.

**14-12.** The diagonal terms in Equation 14.46 can be directly related to experimentally measurable quantities. Show that the *mechanical conductance*, $(J_V/\Delta P)_{\Delta\psi=0}$ is equal to $L_{VV}$. Show that the *electrical conductance*, $(I/\Delta\psi)_{\Delta P=0}$ is equal to $L_{II}$.

**14-13.** The *second electroosmotic flow* is defined as $(J_V/\Delta\psi)_{\Delta P=0}$, and the *second streaming current* is defined as $(I/\Delta P)_{\Delta\psi=0}$. Show that these quantities are equal.

**14-14.** Instead of writing the fluxes as linear combinations of the forces, we can write the forces as linear combinations of the fluxes. For two forces and two fluxes, we have

$$\begin{aligned} X_1 &= R_{11} J_1 + R_{12} J_2 \\ X_2 &= R_{21} J_1 + R_{22} J_2 \end{aligned}$$

Show that

$$R_{11} = \frac{L_{22}}{\Delta} \qquad R_{12} = -\frac{L_{12}}{\Delta} \qquad R_{21} = -\frac{L_{21}}{\Delta} \qquad R_{22} = \frac{L_{11}}{\Delta}$$

where $\Delta = L_{11}L_{22} - L_{12}L_{21}$. Note that $R_{12} = R_{21}$ as a consequence of the Onsager reciprocal relations, $L_{12} = L_{21}$.

**14-15.** Show that the two electrokinetic quantities $(\Delta\psi/J_V)_{I=0}$ (*second streaming potential*) and $(\Delta P/I)_{J_V=0}$ (*second electroosmotic pressure*) are equal to each other. *Hint*: See the previous problem.

**14-16.** Derive Equation 14.51.

**14-17.** At the beginning of Section 14–6, we discussed the chemical equation $X \rightleftharpoons Y$ and took the reaction system to be isolated. Derive the same final result considering the reaction system to be held at a fixed temperature and volume.

**14-18.** Show that $\mathcal{A} = (\mu_X - \mu_Y)/T$ is equal to $R\{(\alpha_X/[X]_{eq}) - (\alpha_Y/[Y]_{eq})\} = R\alpha_X(k_{XY} + k_{YX})/k_{YX}[Y]_{eq}$ where $\alpha_X = [X] - [X]_{eq}$ and when the reaction system is near equilibrium.

**14-19.** Consider the elementary chemical reaction described by

$$\nu_A\ A + \nu_B\ B \rightleftharpoons \nu_Y\ Y + \nu_Z\ Z$$

Show that the affinity in this case is given by $\mathcal{A} = (\nu_A\mu_A + \nu_B\mu_B - \nu_Y\mu_Y - \nu_Z\mu_Z)/T$.

**14-20.** Derive Equation 14.67.

**14-21.** Show that $\mathcal{A}_1 = RJ_1/k_{XY}[X]_{eq}$ and that $\mathcal{A}_2 = RJ_2/k_{YZ}[Y]_{eq}$ for the triangular reaction scheme discussed in Section 14–6 when it is near equilibrium.

**14-22.** Discuss why the reaction scheme

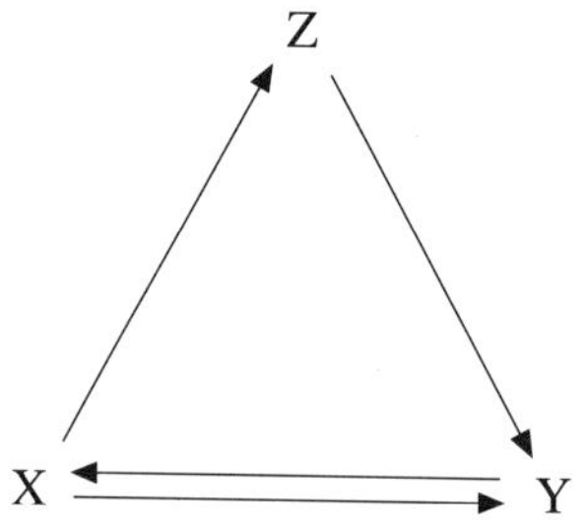


is not allowed.

**14-23.** This problem illustrates an alternate derivation of the Onsager reciprocal relations for the triangular kinetic scheme discussed in Section 14–6. First show that $\mathcal{A}_3$ given by Equation 14.60 can be written as

$$\mathcal{A}_3 = \frac{R}{k_{ZX}[Z]_{eq}} J_3$$

Now use Equations 14.68 and the fact that $\mathcal{A}_3 = -(\mathcal{A}_1 + \mathcal{A}_2)$ to derive

$$J_1 - J_3 = \frac{k_{XY}[X]_{eq} + k_{ZX}[Z]_{eq}}{R}\mathcal{A}_1 + \frac{k_{ZX}[Z]_{eq}}{R}\mathcal{A}_2$$

$$J_2 - J_3 = \frac{k_{ZX}[Z]_{eq}}{R}\mathcal{A}_1 + \frac{k_{YZ}[Y]_{eq} + k_{ZX}[Z]_{eq}}{R}\mathcal{A}_2$$

so that $L_{12} = L_{21}$.

**14-24.** Calculate the value of the transmembrane potential at 298.15 K of a membrane that is permeable only to potassium ions if the solution on the two sides of the membrane are

0.200 M and 0.020 M in KCl(aq). Assume that the activity coefficients in the two solutions are the same.

**14-25.** Consider a membrane that is permeable to sodium ions. Show that the Gibbs energy required to transport one mole of sodium ions across the membrane is given by

$$\Delta G = RT \ln \frac{a_{\mathrm{Na^+},2}}{a_{\mathrm{Na^+},1}} + F\Delta\psi$$

where $\Delta\psi = \psi_2 - \psi_1$. Take $\Delta\psi$ to be 70 mV, $a_{\mathrm{Na^+},2}/a_{\mathrm{Na^+},1}$ to be 10, and $T$ to be 37°C, and calculate the value of $\Delta G$.

**14-26.** In this problem, we will discuss the *Donnan effect*, which occurs for systems like the one shown in Figure 14.3. Two solutions are separated by a membrane that is permeable to small ions but not to polymers or protein molecules.

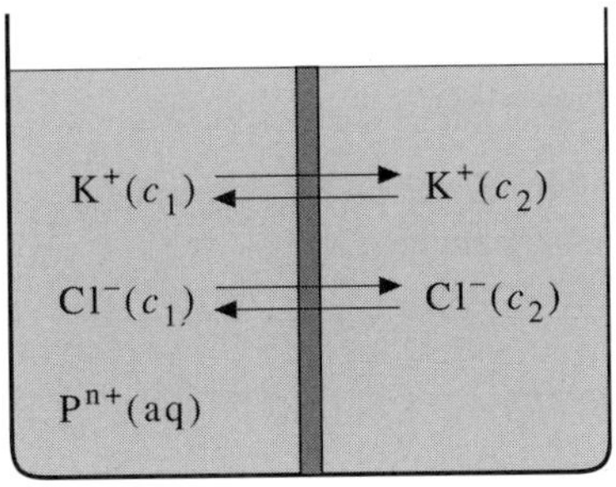


If we ignore any small transport of water molecules, the condition for equilibrium across the membrane is

$$\tilde{\mu}_{\mathrm{K^+},1} = \tilde{\mu}_{\mathrm{K^+},2} \qquad \text{and} \qquad \tilde{\mu}_{\mathrm{Cl^-},1} = \tilde{\mu}_{\mathrm{Cl^-},2}$$

Show that if we replace activities by concentrations, then these two conditions give us

$$\Delta\psi = \psi_2 - \psi_1 = \frac{RT}{F} \ln \frac{c_{\mathrm{K^+},1}}{c_{\mathrm{K^+},2}} = \frac{RT}{F} \ln \frac{c_{\mathrm{Cl^-},2}}{c_{\mathrm{Cl^-},1}}$$

The potential, $\Delta\psi$, is called the *Donnan potential*. Show that $c_{\mathrm{K^+},1}c_{\mathrm{Cl^-},1} = c_{\mathrm{K^+},2}c_{\mathrm{Cl^-},2} = c^2$. Now show that electroneutrality gives

$$c_{\mathrm{K^+},1} + nc_{\mathrm{P^{n+}}} = c_{\mathrm{Cl^-},1} \qquad \text{and} \qquad c_{\mathrm{K^+},2} = c_{\mathrm{Cl^-},2} = c$$

and that

$$c^2_{\mathrm{K^+},2} = c_{\mathrm{K^+},1}(c_{\mathrm{K^+},1} + nc_{\mathrm{P^{n+}}})$$

Use these equations to verfiy the entries in the last four columns in the following table (all concentrations are mol·L$^{-1}$ and the temperature is 298.15 K).

| $nc_{\text{P}^{n+}}$ | $c_{\text{K}^+,2} = c_{\text{Cl}^-,2}$ | $c_{\text{K}^+,1}$ | $c_{\text{Cl}^-,1}$ | $c_{\text{K}^+,1}/c_{\text{K}^+,2}$ | $-\Delta\psi/\text{mV}$ |
|---|---|---|---|---|---|
| 0.0020 | 0.0010 | 0.00041 | 0.0024 | 0.41 | 23 |
| | 0.010 | 0.0091 | 0.0111 | 0.91 | 2.4 |
| | 0.100 | 0.099 | 0.101 | 0.99 | 0.26 |
| 0.020 | 0.0010 | 0.000050 | 0.020 | 0.050 | 77 |
| | 0.010 | 0.0041 | 0.024 | 0.41 | 23 |
| | 0.100 | 0.091 | 0.111 | 0.91 | 2.4 |

It is often desirable to suppress the Donnan effect and the above table shows that this can be done by adding relatively high concentrations of salt.

**14-27.** Consider the cell whose cell diagram is

$$\text{Pt(s)}|\text{H}_2(\text{g}, P_1)|\text{HCl(aq)}|\text{H}_2(\text{g}, P_2)|\text{Pt(s)}$$

Write the equations for the electrode reactions and the overall reaction of this cell. Would you call this a concentration cell?

**14-28.** This problem illustrates a method for determining transport numbers experimentally. Consider the schematic diagram below.

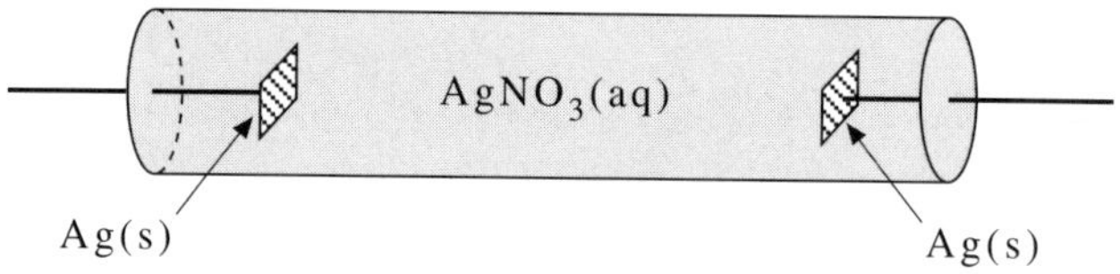


The tube is filled with $AgNO_3$(aq) and there are silver electrodes at each end of the tube. When a current is passed through the cell (electrolysis), the reactions at the two electrodes are described by $\text{Ag(s)} \rightarrow \text{Ag}^+(\text{aq}) + e^-$ and $\text{Ag}^+(\text{aq}) + e^- \rightarrow \text{Ag(s)}$. Thus, if one mole of charge (one faraday) is passed through the solution, then one mole of $Ag^+$(aq) ions will form at one electrode and one mole will be removed at the other. The current between the electrodes is carried by $Ag^+$(aq) ions moving in one direction and $NO_3^-$(aq) ions moving in the other direction. Because the $Ag^+$(aq) ions do not carry all the current, they do not move away from the electrode at which they are formed as fast as they form. Similarly, they do not arrive at the other electrode as fast as they are removed. Therefore, $Ag^+$(aq) ions accumulate around the electrode at which they are produced and are depleted around the electrode at which they are removed.

Consider the electrolysis cell pictured above to be divided into two electrodes compartments, and suppose that each electrode compartment initially contains 0.1000 moles of $AgNO_3$(aq). Now suppose that 0.0100 faradays are passed through the $AgNO_3$(aq) solution, and we find that there are 0.1053 moles of $AgNO_3$(aq) in the electrode compartment in which $Ag^+$(aq) is produced and 0.0947 moles in the compartment in which $Ag^+$(aq) is removed. If there were no migration of $Ag^+$(aq) ions from one electrode compartment to the other, the passage of 0.0100 faradays would result in an increase of 0.0100 moles $Ag^+$(aq) in the electrode compartment in which $Ag^+$(aq) is produced. The observed increase, however, is only 0.0053 moles; therefore, (0.0100 – 0.0053) moles = 0.0047 moles of $Ag^+(aq)$ must have migrated out of the electrode compartment. The fraction of the total current flow (0.0100 faradays) carried by the $Ag^+(aq)$ ions is

$$t_+ = \frac{0.0047 \text{ mol}}{0.0100 \text{ mol}} = 0.47$$

Similarly, at the other electrode, the decrease of the amount of $Ag^+(aq)$ is (0.1000 – 0.0947) moles = 0.0053 moles. But in the absence of migration, the passage of 0.0100 faradays would result in a decrease of 0.0100 moles, so 0.0047 moles must have migrated into the electrode compartment. So we see once again that

$$t_+ = \frac{0.0047 \text{ mol}}{0.0100 \text{ mol}} = 0.47$$

Suppose now that 638 C are passed through a $AgNO_3(aq)$ solution and that 3.412 g of $AgNO_3(aq)$ are found in one electrode compartment and 4.602 g in the other. Calculate the transport number of $Ag^+(aq)$ in $AgNO_3(aq)$. Assume that the initial amount of $AgNO_3(aq)$ in the two compartment is the same.

**14-29.** Show that the Debye-Hückel theory says that $a_{M^+} = a_{A^-}$ for a 1–1 electrolyte. *Hint*: See Equation 11.48.

**14-30.** Use Equation 14.103 and the following data to calculate the values of the liquid junction potentials at 298.15 K and compare your results with the experimental values given below. Use concentrations in place of activities.

| Electrolyte | $c_1/\text{mol}\cdot\text{L}^{-1}$ | $c_2/\text{mol}\cdot\text{L}^{-1}$ | $t_+$ | $\Delta\psi/\text{mV}$ |
|---|---|---|---|---|
| NaCl(aq) | 0.040 | 0.020 | 0.389 | −3.70 |
| | 0.030 | 0.020 | 0.389 | −2.17 |
| KCl(aq) | 0.040 | 0.020 | 0.490 | −0.34 |
| | 0.030 | 0.020 | 0.490 | −0.20 |
| HCl(aq) | 0.040 | 0.020 | 0.827 | +11.01 |

**14-31.** Show that $E = \Delta\tilde{\mu}_j/z_j F$ for an electrode that reacts reversibly with an ion of valence $z_j$.

**14-32.** Show that the flux of an ion of a salt $M_{\nu^+}A_{\nu^-}$ is given by $J_+ = \nu_+ J_s$ or $J_- = \nu_- J_s$, where $J_s$ is the flux of neutral salt, as long as neither ion is produced or removed in an electrode reaction.

**14-33.** Derive Equation 14.90 from Equation 14.89.

**14-34.** Show that the volume flow defined by Equation 14.92 has units of $\text{L}\cdot\text{s}^{-1}$.

**14-35.** Derive Equation 14.101.

**14-36.** Show that Equation 14.109 becomes

$$E = 2t_- \frac{RT}{F} \ln \frac{a_{\pm,2}}{a_{\pm,1}}$$

if the electrodes react reversibly with the cation instead of the anion.

**14-37.** Use Equation 14.109 and the data below to calculate the value of the emf at 298.15 K of each cell given. Compare your results with the experimental values. Assume that the activity coefficients are unity in each case.

| Electrolyte | $c_1/\text{mol}\cdot\text{L}^{-1}$ | $c_2/\text{mol}\cdot\text{L}^{-1}$ | $t_+$ | $E_{\text{exptl}}/\text{mV}$ |
|---|---|---|---|---|
| NaCl(aq) | 0.020 | 0.010 | 0.391 | 13.41 |
| | 0.040 | 0.0050 | 0.391 | 39.63 |
| KCl(aq) | 0.020 | 0.010 | 0.490 | 16.56 |
| | 0.040 | 0.0050 | 0.490 | 49.63 |
| HCl(aq) | 0.020 | 0.010 | 0.826 | 28.05 |
| | 0.040 | 0.0050 | 0.826 | 84.16 |

**14-38.** We claimed in Section 14–9 that an ion in solution in an electric field will quickly come to a constant velocity because of the viscous drag on the ion from the solvent molecules. Let the viscous drag be linearly proportional to the velocity of the ion but in the opposite direction. Then Newton's equation for the motion of the ion is

$$m\frac{dv}{dt} = -fv + zeE$$

where $f$ is the *friction constant*. Show that if the ion is at rest initially, then the solution to this equation is

$$v_j(t) = \frac{zeE}{f}(1 - e^{-ft/m})$$

Note that when $ft/m \gg 1$, then $v$ has a constant value $v = zeE/f$. We can obtain this result by letting $v$ be a constant in Newton's equation above. The mobility is defined as $u = v/E$, so we see that $u = Ze/f$.

**14-39.** A famous expression for the friction constant is $f = 6\pi\eta a$, where $\eta$ is the viscosity of the solvent and $a$ is the radius of the ion. This expression for $f$ is called *Stokes's law*. Given that the viscosity of water is $8.9 \times 10^{-4}\ \text{kg}\cdot\text{m}^{-1}\cdot\text{s}^{-1}$ at 25°C, estimate the magnitude of $f$ for an ion such as $Na^+$(aq). Now estimate $f/m$ and use the result of the previous problem to show that $v$ attains its steady value in about $10^{-12}$ s to $10^{-13}$ s.

**14-40.** Show that the transport number of ion $i$ in a mixture of ions is given by

$$t_i = \frac{c_i u_i}{\sum_j c_j u_j}$$

**14-41.** Use the Henderson integration scheme to derive Equation 14.107 (with activities replaced by concentrations) for a liquid junction $\text{MCl}(c_1) \vdots \text{MCl}(c_2)$.

**14-42.** Use the Henderson integration scheme to derive Equation 14.117 for a liquid junction $\text{HCl}(c) \vdots \text{KCl}(c)$.

**14-43.** Use the data in Table 14.1 to calculate the value of $\Delta\psi$ at 25°C for the following liquid junctions. Compare your results with the given experimental values. All concentrations are $0.0100\ \text{mol}\cdot\text{L}^{-1}$.

| Junction | $\Delta\psi/\text{mV}$ | Junction | $\Delta\psi/\text{mV}$ |
|---|---|---|---|
| $\text{HCl}(c)\vdots\text{KCl}(c)$ | 25.73 | $\text{NaCl}(c)\vdots\text{NH}_4\text{Cl}(c)$ | −4.26 |
| $\text{HCl}(c)\vdots\text{NH}_4\text{Cl}(c)$ | 27.02 | $\text{NaCl}(c)\vdots\text{KCl}(c)$ | −5.65 |
| $\text{HCl}(c)\vdots\text{NaCl}(c)$ | 31.16 | $\text{NaCl}(c)\vdots\text{CsCl}(c)$ | −5.39 |

**14-44.** Prove to yourself that a material flux, $J$, is equal to $cv$, where $c$ is concentration and $v$ is velocity.

**14-45.** Show for a mixture of two 1–1 electrolytes, 2 and 3, that the Debye-Hückel theory gives

$$\begin{aligned}\mu_2 &= \mu_2^\circ + 2RT\ln\gamma_{\pm,2}c_2\\ &= \mu_2^\circ + 2RT\ln c_2 - 1.659RT(c_2+c_3)^{1/2}\end{aligned}$$

Evaluate and compare the magnitude of $\mu_{22} = (\partial\mu_2/\partial c_2)_T$ and $\mu_{23} = (\partial\mu_2/\partial c_3)_T$ at $c_2 = c_3 = 0.0010$ M, 0.010 M, and 0.10 M.

**14-46.** Under what conditions do $(\partial\mu_2/\partial c_3)_T$ and $(\partial\mu_3/\partial c_2)_T$ equal zero?

**14-47.** Use the following experimental data for a NaCl(0.250 M)–KCl(0.500 M) solution at 25°C to verify the Onsager reciprocal relations.

$$\begin{aligned}
&D_{22} = 1.35\times10^{-5}\ \text{cm}^2\cdot\text{s}^{-1} && D_{32} = 0.22\times10^{-5}\ \text{cm}^2\cdot\text{s}^{-1}\\
&D_{23} = 0.013\times10^{-5}\ \text{cm}^2\cdot\text{s}^{-1} && D_{33} = 1.86\times10^{-5}\ \text{cm}^2\cdot\text{s}^{-1}\\
&\mu_{22}/RT = 5.251\ \text{M}^{-1} && \mu_{32}/RT = 1.129\ \text{M}^{-1}\\
&\mu_{23}/RT = 1.149\ \text{M}^{-1} && \mu_{33}/RT = 3.105\ \text{M}^{-1}
\end{aligned}$$

**14-48.** Use the following experimental data for a NaCl(0.500 M)–KCl(0.500 M) solution at 25°C to verify the Onsager reciprocal relations.

$$\begin{aligned}
&D_{22} = 1.40\times10^{-5}\ \text{cm}^2\cdot\text{s}^{-1} && D_{32} = 0.17\times10^{-5}\ \text{cm}^2\cdot\text{s}^{-1}\\
&D_{23} = 0.018\times10^{-5}\ \text{cm}^2\cdot\text{s}^{-1} && D_{33} = 1.87\times10^{-5}\ \text{cm}^2\cdot\text{s}^{-1}\\
&\mu_{22}/RT = 2.989\ \text{M}^{-1} && \mu_{32}/RT = 0.870\ \text{M}^{-1}\\
&\mu_{23}/RT = 0.892\ \text{M}^{-1} && \mu_{33}/RT = 2.851\ \text{M}^{-1}
\end{aligned}$$

**14-49.** Prove that $\dot{S}_{\text{prod}}$ given by Equation 14.135 is necessarily a minimum at $X_U = X_n = 0$.

**14-50.** In this problem, we will prove that a steady state, once established, is stable with respect to small fluctuations in the variable forces. We do this here only for a system with two fluxes and two forces. (The following problem gives a more general treatment.) Let $X_1$ be fixed and $X_2$ be variable, so that $J_1 \neq 0$ but $J_2 = 0$. Now let the variable force $X_2$ vary from its steady-state value by an amount $\delta X_2$. Show that $\delta J_2 = L_{22}\delta X_2$. Because $L_{22} > 0$, however, $\delta J_2$ must have the same sign as $\delta X_2$. Now argue that if $\delta X_2 > 0$, then $\delta J_2 > 0$ means that $X_2$ is made smaller, thus causing $X_2$ to go back toward its steady-state value. Now argue that the same is true if $\delta X_2 < 0$.

**14-51.** In this problem, we will develop a general proof about the stability of the steady state for a system near equilibrium. Let $X_1, X_2, \ldots, X_k$ be fixed forces and $X_{k+1}, \ldots, X_n$ be variable forces, so that $J_{k+1} = J_{k+2} = \cdots = J_n = 0$. Now let one of the variable forces, $X_m$, change slightly by $\delta X_m$, so that $X_m^{\text{ss}} \to X_m^{\text{ss}} + \delta X_m$, with $k+1 \le m \le n$. Show that $\delta J_m = L_{mm}\delta X_m$. Argue as we did in the previous problem that the sign of $\delta J_m$ will always be such that the flux $J_m$ reduces $X_m$ back to its steady-state value.

**14-52.** In this problem, we prove that

$$\dot{S}_{\text{prod}} = \sum_{i=1}^{n}\sum_{j=1}^{n} L_{ij} X_i X_j$$

attains a minimum value with respect to each of the nonfixed forces, $X_{k+1}, \ldots, X_n$, in a steady state. The mathematical condition that $\dot{S}_{\text{prod}}$ be a minimum (and not a maximum or an inflection point) is that

$$\frac{\partial \dot{S}_{\text{prod}}}{\partial X_\alpha} = 0 \qquad \frac{\partial \dot{S}_{\text{prod}}}{\partial X_\beta} = 0 \qquad \begin{array}{l} k+1 \le \alpha \le n \\ k+1 \le \beta \le n \end{array}$$

and

$$\left(\frac{\partial^2 \dot{S}_{\text{prod}}}{\partial X_\alpha^2}\right)\left(\frac{\partial^2 \dot{S}_{\text{prod}}}{\partial X_\beta^2}\right) > \left(\frac{\partial^2 \dot{S}_{\text{prod}}}{\partial X_\alpha \partial X_\beta}\right)^2 \qquad \begin{array}{l} k+1 \le \alpha \le n \\ k+1 \le \beta \le n \end{array}$$

Now show that

$$\frac{\partial \dot{S}_{\text{prod}}}{\partial X_\alpha} = \sum_{j=1}^{n} L_{\alpha j} X_j + \sum_{i=1}^{n} L_{i\alpha} X_i = 2J_\alpha = 0 \qquad k+1 \le \alpha \le n$$

and

$$\frac{\partial \dot{S}_{\text{prod}}}{\partial X_\beta} = \sum_{j=1}^{n} L_{\beta j} X_j + \sum_{i=1}^{n} L_{i\beta} X_i = 2J_\beta = 0 \qquad k+1 \le \beta \le n$$

Now show that

$$\frac{\partial^2 \dot{S}_{\text{prod}}}{\partial X_\alpha^2} = 2L_{\alpha\alpha} \qquad \frac{\partial^2 \dot{S}_{\text{prod}}}{\partial X_\beta^2} = 2L_{\beta\beta} \qquad \begin{array}{l} k+1 \le \alpha \le n \\ k+1 \le \beta \le n \end{array}$$

and

$$\frac{\partial^2 \dot{S}_{\text{prod}}}{\partial X_\alpha \partial X_\beta} = \frac{\partial^2 \dot{S}_{\text{prod}}}{\partial X_\beta \partial X_\alpha} = 2L_{\alpha\beta} \qquad \begin{array}{l} k+1 \le \alpha \le n \\ k+1 \le \beta \le n \end{array}$$

The condition that $\dot{S}_{\text{prod}}$ be a minimum is that $L_{\alpha\alpha}L_{\beta\beta} > L_{\alpha\beta}^2$, which is exactly the same as Equation 14.32, which is a consequence of the positive definite nature of $\dot{S}_{\text{prod}}$ (see Problem 14–8).

**14-53.** Derive Equation 14.149.

**14-54.** Use the approach of Problem 8–57 to repartition $n_2 = n_{\text{solute}}$ between two initially identical compartments to show that $(\partial^2 A/\partial n_2^2)_{V,T,n_1} > 0$ and that $(\partial \mu_2/\partial n_2)_{V,T,n_1} = V^{-1}(\partial \mu_2/\partial c_2)_{V,T,n_1} > 0$.

**14-55.** Prove the Glansdorff-Prigogine inequality for the chemical scheme $X \rightleftharpoons Y$. Do not assume that the system is near equilibrium.

**14-56.** Prove the Glansdorff-Prigogine inequality for the chemical scheme given by Equation 14.58. Do not assume that the reaction is near equilibrium.

**14-57.** Combine Equations 14.127 and 14.149 to derive the so-called diffusion equation

$$\frac{\partial c}{\partial t} = D\frac{\partial^2 c}{\partial x^2}$$

The solutions to this equation give the concentration of a diffusing substance at any point and time, starting from some intitial distribution of concentration. The solution to the diffusion equation if all the diffusing substance is initially located at the origin is

$$c(x,t) = \frac{c_0}{(4\pi Dt)^{1/2}}e^{-x^2/4Dt}$$

where $c_0$ is the total concentration, initially located at the origin. First show that $c(x,t)$ does indeed satisfy the diffusion equation. Plot $c(x,t)/c_0$ versus $x$ for increasing values of $Dt$ and interpret the result physically. What do you think the follow integral equals?

$$\int_{-\infty}^{\infty} c(x,t)dx = \frac{c_0}{(4\pi Dt)^{1/2}}\int_{-\infty}^{\infty} e^{-x^2/4Dt}dx = ?$$

**14-58.** We can interpret $c(x,t)dx/c_0$ in the previous problem as the probability that a diffusing particle is located between $x$ and $x + dx$ at time $t$ (MathChapter B). First show that $c(x,t)/c_0$ is normalized. Now calculate the average distance that the particle will be found from the origin. Interpret your result physically. Now show that

$$\langle x^2 \rangle = \int_{-\infty}^{\infty} \frac{c(x,t)dx}{c_0} = \frac{1}{(4\pi Dt)^{1/2}}\int_{-\infty}^{\infty} e^{-x^2/4Dt}dx = 2Dt$$

The root-mean-square distance traveled by a diffusing particle is

$$x_{\text{rms}} = \langle x^2 \rangle^{1/2} = (2Dt)^{1/2}$$

Given that $D \approx 10^{-5}$ $\text{cm}^2\cdot\text{s}^{-1}$ for an ion diffusing in water, calculate the value of $x_{\text{rms}}$ at $10^{-12}$ s, $10^{-9}$ s, and $10^{-6}$ s. Discuss your results.

**14-59.** Consider a closed circuit consisting of two different metallic conductors where the junctions between the dissimilar metals are maintained at different temperatures, $T_1$ and $T_2$. It turns out that a voltage with a resulting electric current will be generated in this circuit. This effect was first observed by T.J.Seebeck in 1822 and is now known as the *Seebeck effect*. The corresponding emf in the circuit (measured with a potentiometer) depends upon the temperatures at the two junctions and upon the two metals. Thus, if one junction is fixed at a known temperature, then the temperature at the other junction can be determined by measuring the emf. Such a device is called a *thermocouple* and is used extensively to measure temperature.

Let's now consider the inverse effect, where the two junctions are at the same temperature and an electric current is maintained across the junctions between the two metals. In this case, the temperature at one junction will increase and the temperature at the other will decrease, unless heat is supplied or removed at the junctions. The flux of energy as heat that must be supplied to a junction to maintain its initial temperature is directly proportional to the electric current and changes sign when the direction of the current is reversed. The evolution or absorption of energy as heat when an electric current passes across the junction of two dissimilar metals is called the *Peltier effect*, after its discoverer, J.C.A.Peltier. We define the Peltier coefficient $\pi$ by

$$J_U = \pi I \qquad\qquad (\Delta T = 0)$$

where $J_U$ is the flux of energy as heat $(\mathrm{J\cdot m^{-2}\cdot s^{-1}})$ and $I$ is the flux of the electric current $(\mathrm{C\cdot m^{-2}\cdot s^{-1}})$.

The entropy production equation for a thermocouple can be written as

$$\sigma = J_U\left[\frac{d(1/T)}{dx}\right] + I\left(\frac{1}{T}\frac{d\phi}{dx}\right)$$

where $\phi$ is the electric potential in the circuit. Show that both sides of this equation have the same units. Write out the two linear flux-force relations and show that

$$\pi = T\frac{dE}{dT}$$

(Recall that the emf, $E$, is measured under the condition of no current flow.) This equation relating the Seebeck effect and the Peltier effect has been known experimentally since the middle of the 1800s. Not until Onsager's formulation of the reciprocal relations was it derived correctly, however. There were several previous derivations, none of which was correct, and some of which were quite bizarre.

# Answers to the Numerical Problems

## Chapter 1

**1-1.** $1.50 \times 10^{15}$ s$^{-1}$; $5.00 \times 10^4$ cm$^{-1}$; $9.93 \times 10^{-19}$ J

**1-2.** $3 \times 10^{13}$ s$^{-1}$; $1 \times 10^{-5}$ m; $2 \times 10^{-20}$ J

**1-3.** 0.67 cm$^{-1}$; 0.015 m; $1.3 \times 10^{-23}$ J

**1-4.** $2 \times 10^{-15}$ J

**1-5.** (a) $1.07 \times 10^{16}$; (b) $5.41 \times 10^{15}$ photons; (c) $2.68 \times 10^{15}$ photons

**1-6.** $4.738 \times 10^{14}$ s$^{-1}$; $3.139 \times 10^{-19}$ J

**1-7.** $1.70 \times 10^{15}$ photon·s$^{-1}$

**1-8.** $2.178\,69 \times 10^{-18}$ J

**1-9.** 3

**1-10.** 2

**1-12.** 13.598 eV

**1-14.** $1.282 \times 10^{-6}$ m

**1-15.** 52.917 72 pm

**1-18.** The final equation illustrates conservation of energy.

**1-21.** $9.104\,432 \times 10^{-31}$ kg

**1-22.** $2.178\,688 \times 10^{-18}$ J

**1-24.** $k = 2D_e\beta^2$; 479 N·m$^{-1}$ = 479 kg·s$^{-2}$

**1-25.** 0.0181 pm$^{-1}$

**1-26.** $\gamma = -6D_e\beta^3$

**1-27.** $\tilde{x} = 0.01961$; $\tilde{x}\tilde{\nu} = 56.59$ cm$^{-1}$

**1-28.** 406 N·m$^{-1}$; $1.27 \times 10^{-14}$ s

**1-29.** $\nu = 9.63 \times 10^{12}$ s$^{-1}$; $\varepsilon_0 = 3.19 \times 10^{-21}$ J

**1-30.** 83.9 N·m$^{-1}$; $1.20 \times 10^{-13}$ s

**1-31.** $\tilde{\nu}_{obs}/\text{cm}^{-1} = 2988.7v - 51.565v(v+1)$

**1-32.** 1051.8 cm$^{-1}$; 1567.5 cm$^{-1}$

**1-33.** $\tilde{\nu} = 2169.0$ cm$^{-1}$; $\tilde{x}\tilde{\nu} = 13.0$ cm$^{-1}$

**1-36.** $I = 3.348 \times 10^{-47}$ kg·m$^2$; $R_e = 142.3$ pm

**1-37.** 79 880 cm$^{-1}$ = 955.6 kJ·mol$^{-1}$

**1-38.** 113.0 pm

**1-39.** 305.5 pm

**1-40.** $1.36 \times 10^{11}$ revolution·s$^{-1}$

**1-41.** $\tilde{\nu}/\text{cm}^{-1} = 20.806(J+1) - 0.0018(J+1)^3$; $\tilde{B} = 10.403$ cm$^{-1}$; $\tilde{D} = 0.00044$ cm$^{-1}$

**1-42.** $\tilde{\nu}/\text{cm}^{-1} = 3.8454(J+1) - 2.555 \times 10^{-5}(J+1)^3$; $\tilde{B} = 1.9227$ cm$^{-1}$; $\tilde{D} = 6.39 \times 10^{-6}$ cm$^{-1}$

**1-43.** 16.929 cm$^{-1}$; 33.849 cm$^{-1}$; 50.753 cm$^{-1}$; 67.631 cm$^{-1}$; 84.477 cm$^{-1}$; 101.28 cm$^{-1}$; 118.04 cm$^{-1}$

**1-44.** $N_2$: 3, 2, 1; $C_2H_2$: 3, 2, 7; $C_2H_4$: 3, 3, 12; $C_2H_6$: 3, 3, 18; $C_6H_6$: 3, 3, 30

**1-45.** HCN: 3, 2, 4; $CD_4$: 3, 3, 9; $SO_3$: 3, 3, 6; $SF_6$: 3, 3, 15; $(CH_3)_2CO$: 3, 3, 24

**1-46.** $8.845 \times 10^{-20}$ J = 53.27 kJ·mol$^{-1}$

**1-47.** $1.871 \times 10^{-19}$ J = 112.6 kJ·mol$^{-1}$

**1-48.** The center of mass lies 7.2 pm beyond the central nitrogen atom. $I = 6.645 \times 10^{-46}$ kg·m$^2$; $\tilde{B} = 0.4213$ cm$^{-1}$

## MathChapter A

**A-1.** 0.8596

**A-2.** 1.4142

**A-3.** 4.965

**A-4.** 0.615 atm

**A-5.** 0.077 780

**A-6.** 0.3473, 1.532, −1.879

**A-7.** 0.0750

**A-10.** $\ln 2 = 0.693\,147$; $n = 10$

**A-11.** 0.886 2269

**A-12.** 6.493 94

## Chapter 2

**2-1.** $2.98 \times 10^6$ atm, $3.02 \times 10^6$ bar

**2-2.** $7.39 \times 10^2$ torr, 0.972 atm

**2-3.** 1.00 atm

**2-4.** −40°

**2-6.** $3.24 \times 10^4$ molecules, $1.85 \times 10^{19}$ $cm^3 \cdot mol^{-1}$
**2-7.** 44.10
**2-9.** $y_{H_2} = 0.77$, $y_{N_2} = 0.23$
**2-10.** 2.2 bar, 2.2 bar
**2-11.** $Cl_2$
**2-12.** 62.3639 $dm^3 \cdot torr \cdot K^{-1} \cdot mol^{-1}$
**2-15.** 0.04998 $dm^3 \cdot mol^{-1}$; 0.03865 $dm^3 \cdot mol^{-1}$; 0.01663 $dm^3 \cdot mol^{-1}$
**2-16.** 353 bar; 8008 bar; 438 bar; 284 bar
**2-17.** 1031 bar; 411 bar
**2-18.** 10.00 $mol \cdot L^{-1}$; 10.28 $mol \cdot L^{-1}$
**2-19.** 1570 bar; −4250 bar
**2-21.** R-K: 345 bar; P-R: 129 bar
**2-22.** 0.07073 $L \cdot mol^{-1}$, 0.07897 $L \cdot mol^{-1}$, 0.2167 $L \cdot mol^{-1}$; 14.14 $mol \cdot L^{-1}$; 4.615 $mol \cdot L^{-1}$.
**2-23.** RK: 20.13 $mol \cdot L^{-1}$, 5.148 $mol \cdot L^{-1}$; PR: 23.61 $mol \cdot L^{-1}$, 5.564 $mol \cdot L^{-1}$
**2-24.** vdW: 4.786 $mol \cdot L^{-1}$, 0.5741 $mol \cdot L^{-1}$; RK: 6.823 $mol \cdot L^{-1}$, 0.6078 $mol \cdot L^{-1}$; PR: 8.116 $mol \cdot L^{-1}$, 0.6321 $mol \cdot L^{-1}$
**2-35.** $\overline{V} \approx 78.5$ $cm^3 \cdot mol^{-1}$
**2-37.** 0.00150 $bar^{-1}$
**2-38.** $-5.33 \times 10^{-3}$ $dm^3 \cdot mol^{-1}$, −0.0213 $dm^3 \cdot mol^{-1}$
**2-40.** 1 kJ vs. 100 kJ
**2-43.** Yes
**2-44.** −15.15 $cm^3 \cdot mol^{-1}$
**2-45.** −60 $cm^3 \cdot mol^{-1}$
**2-52.** $(C \cdot m)^2(m^3)/(C^2 \cdot s^2 \cdot kg^{-1} \cdot m^{-3})(m^6) = kg \cdot m^2 \cdot s^{-1} = J$
**2-54.** $5.86 \times 10^{-78}$ $J \cdot m^6$; $1.35 \times 10^{-77}$ $J \cdot m^6$

## MathChapter B

**B-1.** $a/2$
**B-2.** $\dfrac{a^2}{12} - \dfrac{a^2}{2n^2\pi^2}$
**B-3.** 1/2
**B-6.** $\left(\dfrac{8k_BT}{\pi m}\right)^{1/2}$
**B-7.** $\frac{3}{2}k_BT$

## Chapter 3

**3-7.** $-\hbar\gamma B_z$, 0
**3-8.** $\exp(-0.010\ K/T)$
**3-10.** $\langle E \rangle = Nk_BT$
**3-11.** $\langle E \rangle = \dfrac{3}{2}Nk_BT - \dfrac{aN^2}{V}$
**3-12.** $\frac{3}{2}Nk_BT$, $P = \dfrac{Nk_BT}{V-b}$
**3-13.** $Nk_B$
**3-14.** $\frac{3}{2}Nk_B$
**3-18.** $\overline{C}_V$ is a function of $T^* = T/\Theta_E$, where $\Theta_E = h\nu/k_B$
**3-23.** (a) 6 (b) 9 (c) 12
**3-24.** 9 total, 3 allowed
**3-25.** 6 allowed terms
**3-26.** 27 total, 1 allowed
**3-27.** 10 allowed terms
**3-28.** $1.94 \times 10^{-6}$
**3-29.** 0.092
**3-30.** 1420
**3-31.** 0.286
**3-35.** $f_\upsilon = e^{-h\upsilon\nu/k_BT}(1 - e^{-h\nu/k_BT})$; 0.9999, $1.01 \times 10^{-6}$, $1.03 \times 10^{-12}$
**3-36.** 1.000, 0.9962, 0.9650
**3-41.**

| Fraction of atoms, 1000 K | Fraction of atoms, 2500 K |
|---|---|
| 1.00 | 1.00 |
| $2.55 \times 10^{-11}$ | $5.79 \times 10^{-5}$ |
| $4.97 \times 10^{-11}$ | $1.15 \times 10^{-4}$ |
| $8.27 \times 10^{-17}$ | $3.69 \times 10^{-7}$ |

**3-42.** 29.06 $J \cdot mol^{-1} \cdot K^{-1}$
**3-43.** 3420 K

## MathChapter C

**C-1.** $1.25 \times 10^{-3}\%$, $4.97 \times 10^{-3}\%$, . . . , 0.467%
**C-2.** 0.249%, 0.499%, . . . , 4.92%
**C-3.** $1 + \dfrac{x}{2} - \dfrac{x^2}{8} + O(x^3)$
**C-4.** $\dfrac{e^{-\frac{1}{2}\beta h\nu}}{1 - e^{-\beta h\nu}}$
**C-6.** 1
**C-10.** 1
**C-11.** $\dfrac{a^3}{3} - \dfrac{a^4}{4} + \cdots$

## Chapter 4

**4-4.** $f_2(T=300\text{K}) = 4.8 \times 10^{-36}$; $f_2(T=1000\text{K}) = 2.5 \times 10^{-11}$; $f_2(T=2000\text{K}) = 5.0 \times 10^{-6}$

**4-5.** $f_2(T=300\text{K}) = 9.0 \times 10^{-32}$; $f_2(T=1000\text{K}) = 4.9 \times 10^{-10}$; $f_2(T=2000\text{K}) = 2.2 \times 10^{-5}$

**4-7.** $D_e = D_0 + \frac{1}{2}R\Theta_{\text{vib}}$ CO: 1083 kJ·mol$^{-1}$; NO: 638.1 kJ·mol$^{-1}$; $K_2$: 54.1 kJ·mol$^{-1}$

**4-8.** 6332 K, 4462 K

**4-10.** $f_{v>0} = 7.6 \times 10^{-7}$ at 300 K; $f_{v>0} = 1.46 \times 10^{-2}$ at 1000 K

**4-11.** $f_{v>0} = e^{-\Theta_{\text{vib}}/T} = 1.01 \times 10^{-9}$ for $H_2$; 0.0683 for $Cl_2$; 0.358 for $I_2$; etc.

**4-12.** 87.4 K, 43.7 K

**4-13.** 9 or 10

**4-14.** $N_2$: 0.32%; $H_2$: 9.45%

**4-16.** $\approx 20\%$

**4-18.** NO(g) at 300 K, $J_{\text{max}} = 7$; at 1000 K $J_{\text{max}} = 14$

**4-21.** $\Theta_{\text{vib},j} = 5360$ K: $(\overline{C}_{Vj}/R) = 1.05 \times 10^{-2}$;

$\Theta_{\text{vib},j} = 5160$ K: $(\overline{C}_{Vj}/R) = 1.36 \times 10^{-2}$;

$\Theta_{\text{vib},j} = 2290$ K: $(\overline{C}_{Vj}/R) = 3.35 \times 10^{-1}$

**4-22.** $\Theta_{\text{elec},1} = 227.6$ K; $\Theta_{\text{elec},2} = 325.9$ K; $q_{\text{elec}} = 5 + 3e^{-227.6\text{ K}/T} + e^{-325.9\text{ K}/T} =$ 8.803 at 5000 K

**4-23.** 2, 1, 12, 24, 2, 4

**4-24.** $\Theta_{\text{rot}} = 2.141$ K; $\Theta_{\text{vib},1} = 3016$ K; $\Theta_{\text{vib},2} = 1026$ K; $\Theta_{\text{vib},3} = 4765$ K; $\overline{C}_V/R = 6.21$

**4-25.** 2; $2.360 \times 10^{-46}$ kg·m$^2$; 1.702 K; 2842 K; 4849 K; 4715 K; 1049 K; 863.3 K; $4.34R$

**4-28.** $I = 6.746 \times 10^{-46}$ kg·m$^2$, $\Theta_{\text{rot}} = 0.597$ K

**4-29.** see Table 18.4; $5.30R$

**4-34.** $\Theta_{\text{vib},D_2} = 4480$ K; $\Theta_{\text{rot},D_2} = 42.6$ K; $\Theta_{\text{vib,HD}} = 5484$K; $\Theta_{\text{rot,HD}} = 64.4$ K

**4-35.** $\ln q_{\text{rot}}(T) = \ln \frac{T}{\Theta_{\text{rot}}} + \frac{1}{3}\left(\frac{\Theta_{\text{rot}}}{T}\right) + \frac{1}{90}\left(\frac{\Theta_{\text{rot}}}{T}\right)^2 + \cdots$

**4-37.** yes; 2140 K; no

**4-38.** 4 degrees of freedom;

$$U = \frac{3}{2}RT + \frac{R\Theta_{\text{vib}}}{2} + \frac{R\Theta_{\text{vib}}e^{-\Theta_{\text{vib}}/T}}{1 - e^{-\Theta_{\text{vib}}/T}}$$

**4-39.** (a) $3R/2$ (b) $7R/2$ (c) $6R$ (d) $13R/2$ (e) $12R$

## MathChapter D

**D-1.** $\kappa = 1/P$

**D-2.** $\alpha = 1/T$

**D-4.** $\frac{3}{2}Nk_BT = \frac{3}{2}nRT$

**D-7.** 0; $a/V^2$; $3A/2T^{1/2}V(V+B)$

**D-9.** 0; 0; $-3A/4T^{3/2}V(V+B)$

**D-10.** exact

**D-11.** inexact; exact

## Chapter 5

**5-1.** KE = 9.80 kJ; $u = 44.3$ m·s$^{-1}$; 22.2°C

**5-2.** 15 bar; 3000 J

**5-3.** 28.8 bar; 3.60 J

**5-4.** 4.01 kJ

**5-5.** −1.73 kJ

**5-6.** 11.4 kJ

**5-7.** +325 J; +309 J; they differ because $w$ is a path function

**5-9.** −3.93 kJ·mol$^{-1}$

**5-10.** −3.92 kJ·mol$^{-1}$

**5-12.** $V_1 = 11.35$ L; $V_2 = 22.70$ L; $T_2 =$ 1090 K; $\Delta U = 10.2$ kJ·mol$^{-1}$; $\Delta H =$ 17.0 kJ·mol$^{-1}$; $q = 13.6$ kJ; $w = -3.40$ kJ

**5-13.** 418 J

**5-19.** $T_2 = 226$ K, $w = -898$ J

**5-20.** 519 K

**5-21.** 421 K

**5-22.** $q_P = 122.9$ kJ·mol$^{-1}$, $\Delta H = 122.9$ kJ·mol$^{-1}$, $\Delta U = 113.2$ kJ·mol$^{-1}$, $w = +9.77$ kJ·mol$^{-1}$; $q_V = 113.2$ kJ·mol$^{-1}$, $\Delta H = 122.9$ kJ·mol$^{-1}$, $\Delta U = 113.2$ kJ·mol$^{-1}$, $w = 0$

**5-23.** $\Delta_r U^\circ = 288.3\ \text{kJ}\cdot\text{mol}^{-1}$
**5-24.** 74.6 kg
**5-25.** 295 K
**5-26.** 3340 kJ
**5-35.** $\Delta_r H = 416$ kJ
**5-36.** $\Delta_r H = -521.6$ kJ
**5-37.** $\Delta_r H = +2.9$ kJ
**5-38.** $\Delta_r H^\circ[\text{fructose}] = +1249.3\ \text{kJ}\cdot\text{mol}^{-1}$
**5-39.** methanol: $-22.7\ \text{kJ}\cdot\text{g}^{-1}$; $N_2H_4(l) = -19.4\ \text{kJ}\cdot\text{g}^{-1}$
**5-40.** 32.5 kJ
**5-41.** (a) $-44.14$ kJ, exothermic (b) $-429.87$ kJ, exothermic
**5-42.** $43.8\ \text{kJ}\cdot\text{mol}^{-1}$; $44.0\ \text{kJ}\cdot\text{mol}^{-1}$ from Table 19.2
**5-43.** $136.964\ \text{kJ}\cdot\text{mol}^{-1}$
**5-44.** $-394.478\ \text{kJ}\cdot\text{mol}^{-1}$
**5-46.** 4040 K
**5-48.** $64.780\ \text{kJ}\cdot\text{mol}^{-1}$
**5-49.** $1.50R$
**5-50.** $-13.3$ kJ; $-15.7$ kJ
**5-53.** Drops by 30 K
**5-56.** $0.027\ \text{J}\cdot\text{mol}^{-1}\cdot\text{K}^{-1}\cdot\text{atm}^{-1}$

## MathChapter E

**E-1.** $x^5 + 5x^4 + 10x^3 + 10x^2 + 5x + 1$
**E-2.** $x^2 + 2xy + 2xz + y^2 + 2yz + z^2$
**E-3.** $x^4 + 4x^3y + 4x^3z + 6x^2y^2 + 12x^2yz + 6x^2z^2 + 4xy^3 + 12xy^2z + 12xyz^2 + 4yz^3 + 6y^2z^2 + 4y^3z + y^4 + 4xz^3 + z^4$
**E-4.** 6
**E-5.** Each number in a row is the sum of the two numbers above it.
**E-6.** 84
**E-7.** $1.12 \times 10^{-5}$ vs. 0.0194 in Table J.1

## Chapter 6

**6-2.** $dz/y$
**6-6.** $q_{\text{rev}} = \int_{T_1}^{T_4} \overline{C}_V(T)dT + \int_{T_4}^{T_1} \overline{C}_V(T)dT - \int_{\overline{V}_1}^{\overline{V}_2} P_2 d\overline{V}$; $\Delta\overline{S} = R \ln \frac{V_2}{V_1}$
**6-8.** $5.76\ \text{J}\cdot\text{K}^{-1}$; positive because the gas is expanding
**6-9.** $19.1\ \text{J}\cdot\text{K}^{-1}$; positive because the gas is expanding
**6-10.** $q_{\text{rev}} = P_1(\overline{V}_2 - \overline{V}_1)$; $\Delta S = R \ln \left(\frac{\overline{V}_2 - b}{\overline{V}_1 - b}\right)$
**6-12.** $q_{\text{rev}} = -P_2(\overline{V}_2 - \overline{V}_1)$; $\Delta S = R \ln \left(\frac{\overline{V}_2 - b}{\overline{V}_1 - b}\right)$
**6-13.** $\Delta S = 37.4\ \text{J}\cdot\text{K}^{-1}$
**6-14.** $\Delta\overline{S} = 30.6\ \text{J}\cdot\text{K}^{-1}\cdot\text{mol}^{-1}$
**6-17.** $\Delta S$ can be positive or negative for an isothermal process; $\Delta S = -5.76\ \text{J}\cdot\text{K}^{-1}$
**6-18.** $\Delta S = 217.9\ \text{J}\cdot\text{K}^{-1}$
**6-19.** $\Delta S = 44.0\ \text{J}\cdot\text{K}^{-1}$
**6-25.** $\Delta S_{\text{sys}} = 13.4\ \text{J}\cdot\text{K}^{-1}$; $\Delta S_{\text{surr}} = -13.4\ \text{J}\cdot\text{K}^{-1}$; $\Delta S_{\text{tot}} = 0$
**6-26.** $\Delta S_{\text{surr}} = 0$; $\Delta S_{\text{sys}} = 13.4\ \text{J}\cdot\text{K}^{-1}$; $\Delta S_{\text{tot}} = 13.4\ \text{J}\cdot\text{K}^{-1}$
**6-27.** $\Delta\overline{S} = 192.78\ \text{J}\cdot\text{K}^{-1}\cdot\text{mol}^{-1}$
**6-28.** $y_1 = 0.5$
**6-29.** $\Delta_{\text{mix}}\overline{S} = 5.29\text{J}\cdot\text{K}^{-1}$
**6-33.** $\Delta S = 95.60\ \text{J}\cdot\text{K}^{-1}\cdot\text{mol}^{-1}$
**6-37.** $\exp(-1.5 \times 10^{17})$
**6-38.** $(1/2)^{N_A}$
**6-40.** $164.1\ \text{J}\cdot\text{K}^{-1}\cdot\text{mol}^{-1}$
**6-41.** $191.6\ \text{J}\cdot\text{K}^{-1}\cdot\text{mol}^{-1}$
**6-42.** $213.8\ \text{J}\cdot\text{K}^{-1}\cdot\text{mol}^{-1}$
**6-43.** $193.1\ \text{J}\cdot\text{K}^{-1}\cdot\text{mol}^{-1}$
**6-45.** 21% at 1 atm; 41% at 25 atm

## Chapter 7

**7-2.** $37.5\ \text{J}\cdot\text{K}^{-1}$
**7-3.** $192.6\ \text{J}\cdot\text{K}^{-1}$
**7-4.** $38.75\ \text{J}\cdot\text{K}^{-1}$
**7-5.** $44.51\ \text{J}\cdot\text{K}^{-1}$

**7-10.**

| Substance | $\Delta_{vap}\overline{S}/\ J\cdot K^{-1}\cdot mol^{-1}$ |
|---|---|
| Pentane | 83.41 |
| Hexane | 84.39 |
| Heptane | 85.5 |
| Ethylene oxide | 89.9 |
| Benzene | 86.97 |
| Diethyl ether | 86.2 |
| Tetrachloromethane | 85.2 |
| Mercury | 93.85 |
| Bromine | 90.3 |

**7-11.**

| Substance | $\Delta_{fus}\overline{S}/\ J\cdot K^{-1}\cdot mol^{-1}$ |
|---|---|
| Pentane | 58.7 |
| Hexane | 73.5 |
| Heptane | 77.6 |
| Ethylene oxide | 32.0 |
| Benzene | 35.7 |
| Diethyl ether | 46.3 |
| Tetrachloromethane | 13 |
| Mercury | 9.77 |
| Bromine | 40 |

**7-14.** $192.05\ J\cdot K^{-1}\cdot mol^{-1}$

**7-16.** $223.2\ J\cdot K^{-1}\cdot mol^{-1}$ compared to $223.1\ J\cdot K^{-1}\cdot mol^{-1}$

**7-18.** $237.8\ J\cdot K^{-1}\cdot mol^{-1}$

**7-20.** $196.7\ J\cdot K^{-1}\cdot mol^{-1}$

**7-21.** $139.3\ J\cdot K^{-1}\cdot mol^{-1}$

**7-22.** $272.6\ J\cdot K^{-1}\cdot mol^{-1}$

**7-23.** $274.3\ J\cdot K^{-1}\cdot mol^{-1}$

**7-24.** $154.7\ J\cdot K^{-1}\cdot mol^{-1}$; residual entropy

**7-25.** $185.6\ J\cdot K^{-1}\cdot mol^{-1}$

**7-30.** $222.8\ J\cdot K^{-1}\cdot mol^{-1}$

**7-31.** $159.9\ J\cdot K^{-1}\cdot mol^{-1}$; residual entropy

**7-32.** $193.1\ J\cdot K^{-1}\cdot mol^{-1}$

**7-33.** $245.4\ J\cdot K^{-1}\cdot mol^{-1}$

**7-34.** $173.7\ J\cdot K^{-1}\cdot mol^{-1}$

**7-35.**
$S^\circ(H_2, 298.15K) = 130.7\ J\cdot K^{-1}\cdot mol^{-1}$;
$S^\circ(D_2, 298.15K) = 144.9\ J\cdot K^{-1}\cdot mol^{-1}$;
$S^\circ(HD, 298K) = 143.8\ J\cdot K^{-1}\cdot mol^{-1}$

**7-36.** $253.6\ J\cdot K^{-1}\cdot mol^{-1}$. The experimental value is $253.7\ J\cdot K^{-1}\cdot mol^{-1}$.

**7-37.** $234.3\ J\cdot K^{-1}\cdot mol^{-1}$. The experimental value is $240.1\ J\cdot K^{-1}\cdot mol^{-1}$. The difference is due to residual entropy.

**7-38.** $-172.7\ J\cdot K^{-1}\cdot mol^{-1}$

**7-39.** $-49.6\ J\cdot K^{-1}\cdot mol^{-1}$

**7-40.** (a) $CO_2$ (b) $CH_3CH_2CH_3$ (c) $CH_3CH_2CH_2CH_2CH_3$

**7-41.** (a) $D_2O$ (b) $CH_3CH_2OH$ (c) $CH_3CH_2CH_2CH_2NH_2$

**7-42.** (d) > (a) > (b) > (c)

**7-43.** (c) > (b) $\approx$ (d) > (a)

**7-44.** translational for both

**7-45.** $239.5\ J\cdot K^{-1}\cdot mol^{-1}$

**7-46.** $188.2\ J\cdot K^{-1}\cdot mol^{-1}$

**7-47.** (a) $2.86\ J\cdot K^{-1}\cdot mol^{-1}$
(b) $-242.9\ J\cdot K^{-1}\cdot mol^{-1}$
(c) $-112.0\ J\cdot K^{-1}\cdot mol^{-1}$

**7-48.** (a) $-332.3\ J\cdot K^{-1}\cdot mol^{-1}$
(b) $252.66\ J\cdot K^{-1}\cdot mol^{-1}$
(c) $-173.0\ J\cdot K^{-1}\cdot mol^{-1}$

## Chapter 8

**8-1.** $\Delta_{vap}\overline{G}(80.09^\circ C) = 0$;
$\Delta_{vap}\overline{G}(75.0^\circ C) = 0.43\ kJ\cdot mol^{-1}$;
$\Delta_{vap}\overline{G}(85.0^\circ C) = -0.44\ kJ\cdot mol^{-1}$

**8-2.** $\Delta_{vap}\overline{G}(80.09^\circ C) = 0$;
$\Delta_{vap}\overline{G}(75.0^\circ C) = 435.9\ J\cdot mol^{-1}$;
$\Delta_{vap}\overline{G}(85.0^\circ C) = -429.8\ J\cdot mol^{-1}$; no

**8-5.** $P\overline{V} = RT$ and $P(\overline{V} - b) = RT$

**8-7.** $-0.0508\ kJ\cdot mol^{-1}$

**8-8.** $R$

**8-9.** $7.87 \times 10^{-3}\ dm^3\cdot bar^{-1}\cdot K^{-1} = 0.787\ J\cdot K^{-1}\cdot mol^{-1}$

**8-12.** from data: 155.6 bar; from van der Waals: -4.80 bar; from Redlich-Kwong: 161.1 bar

**8-13.** $-0.0552\ kJ\cdot mol^{-1}$

**8-16.** $(\partial\overline{C}_P/\partial P)_T = 4.47 \times 10^{-4}\ dm^3\cdot mol^{-1}\cdot K^{-1}$;
$\overline{C}_P = 25.27\ J\cdot K^{-1}\cdot mol^{-1}$

**8-17.** $138.8\ J\cdot mol^{-1}$

**8-19.** $V$ and $U$

**8-20.** $0.0156\ J\cdot K^{-1}\cdot mol^{-1}$

**8-21.** $0.866\ J\cdot K^{-1}\cdot mol^{-1}$
**8-22.** $0.466\ J\cdot K^{-1}\cdot mol^{-1}$
**8-30.** $\gamma \approx 0.63$
**8-51.**

| Gas | Ar | $N_2$ | $CO_2$ |
|---|---|---|---|
| $\mu_{JT}$(theor.)/K·atm$^{-1}$ | 0.44 | 0.24 | 1.38 |
| $\mu_{JT}$(exp.)/K·atm$^{-1}$ | 0.43 | 0.26 | 1.3 |
| Percent Difference | 3.4 | 6.6 | 6.6 |

**8-52.**

| Gas | Ar | $N_2$ | $CO_2$ |
|---|---|---|---|
| $T_i$(theor.)/K | 791 | 634 | 1310 |
| $T_i$(exp.)/K | 794 | 621 | 1500 |
| Percent Difference | 0.378 | 2.09 | 12.7 |

**8-53.** Ar: 42.6 K, $N_2$: 25.7 K, $CO_2$: 129 K.
**8-55.** $-19.7\ J\cdot K^{-1}\cdot mol^{-1}$ versus $-19.1\ J\cdot K^{-1}\cdot mol^{-1}$ for an ideal gas.
**8-56.** $-20.6\ J\cdot K^{-1}\cdot mol^{-1}$ versus $-19.1\ J\cdot K^{-1}\cdot mol^{-1}$ for an ideal gas.

## Chapter 9

**9-1.** No. Its normal melting point is higher than the triple point temperature.
**9-4.** 11.1 torr, 172.4 K
**9-6.** 1556 bar
**9-9.** 352.8 K
**9-10.** $T_c = 305.4$ K
**9-16.** $T_c = 152$ K
**9-17.** $\Delta_{vap}\overline{H} = 35.26\ kJ\cdot mol^{-1}$
**9-20.** $dT/dP = 27.9\ K\cdot atm^{-1}$; at 2 atm, the boiling point is about 127.9°C
**9-21.** $29.5\ kJ\cdot mol^{-1}$
**9-22.** $59.62\ kJ\cdot mol^{-1}$
**9-23.** $383\ cm^3\cdot mol^{-1}$
**9-27.** 1070 torr
**9-28.** $41.2\ kJ\cdot mol^{-1}$
**9-29.** $T_{vap} = 2010$ K; $\Delta_{vap}\overline{H} = 179.6\ kJ\cdot mol^{-1}$
**9-30.** $T_{sub} = 386$ K; $\Delta_{sub}\overline{H} = 62.3\ kJ\cdot mol^{-1}$
**9-31.** $51.6\ kJ\cdot mol^{-1}$
**9-32.** $410.8\ kJ\cdot mol^{-1}$
**9-33.** $\Delta_{sub}\overline{H} = 27.6\ kJ\cdot mol^{-1}$
**9-34.** 1.12
**9-36.** $\Delta_r H^\circ = 1895\ J\cdot mol^{-1}$; $\Delta_r S^\circ = -3.363\ J\cdot mol^{-1}$; $P = 15\,000$ bar
**9-37.** $-42.72\ kJ\cdot mol^{-1}$
**9-39.** $-48.43\ kJ\cdot mol^{-1}$
**9-40.** $-50.25\ kJ\cdot mol^{-1}$
**9-41.** $-45.53\ kJ\cdot mol^{-1}$
**9-42.** 0.0315 atm
**9-43.** 0.0315 atm
**9-45.** 0.0313 atm

## Chapter 10

**10-4.** $G = \mu n = U - TS + PV$
**10-5.** $G = \mu n = A + PV$
**10-16.** $n^l/n^{vap} = 0.58$
**10-18.** The vapor phase is richer in the more volatile component.
**10-20.** $x_1 = 0.463$; $y_1 = 0.542$
**10-26.** $P_{total} = 140$ torr; $y_1 = 0.26$
**10-27.** $y_1 < x_1$ because $P_1^* < P_2^*$
**10-29.** $P_1^* = 120$ torr; $P_2^* = 140$ torr; $k_{H,1} = 162$ torr; $k_{H,2} = 180$ torr
**10-44.** Yes
**10-45.** No
**10-47.** $\overline{G}^E/R = 0.8149x_1x_2(1 + 0.4183x_1)$ is not symmetric about $x_1 = x_2 = 1/2$.
**10-48.** $a_{tri}^{(R)} = 0.181$; $\gamma_{tri}^{(R)} = 0.631$
**10-49.** $P_1^* = 78.8$ torr; $P_1 = 30.6$ torr; $a_1^{(R)} = 0.39$; $\gamma_1^{(R)} = 1.6$; $k_{H,1} = 180.7$ torr; $a_1^{(H)} = 0.17$; $\gamma_1^{(H)} = 0.68$
**10-52.** No
**10-57.** $\overline{G}^E/RT = x_1x_2[\alpha + \beta(1 - x_1/2)]$

## Chapter 11

**11-1.** $4.78\ mol\cdot L^{-1}$; $7.24\ mol\cdot kg^{-1}$; molality is independent of temperature
**11-2.** $18\ mol\cdot L^{-1}$
**11-3.** $1.7\ g\cdot mL^{-1}$
**11-4.** 0.00893
**11-6.** $0.060\ mol\cdot kg^{-1}$; $0.313\ mol\cdot kg^{-1}$; $0.660\ mol\cdot kg^{-1}$; $1.484\ mol\cdot kg^{-1}$; $3.960\ mol\cdot kg^{-1}$
**11-7.** $0.73\ mol\cdot kg^{-1}$
**11-9.** $2.83\ mol\cdot L^{-1}$
**11-15.** $x_1 = 0.9487$; $\gamma_1 = 0.983$
**11-18.** $\gamma_{2m} = 1.186$

**11-19.** $\phi - 1 = 0.2879$; the integral $= 0.272$; $\ln \gamma_{2m} = 0.560$; $\gamma_{2m} = 1.75$

**11-21.** 0.958

**11-22.** 0.902

**11-24.** $6.87\ \mathrm{K \cdot kg \cdot mol^{-1}}$

**11-26.** $2.93\ \mathrm{K \cdot kg \cdot mol^{-1}}$

**11-27.** $K_b = 2.53\ \mathrm{K \cdot kg \cdot mol^{-1}}$; 147

**11-29.** 72 000

**11-31.** 58.0 atm

**11-40.** $10\ \mathrm{mol \cdot kg^{-1}}$

**11-41.** The value of $\nu$ comes out to be 1.02, which means that $HgCl_2(aq)$ is undissociated at the given conditions.

**11-42.** $\nu = 3$;
$K_2HgI_4(aq) \longrightarrow 2\,K^+(aq) + HgI_4^{2-}(aq)$

**11-43.** $Pt(NH_3)_4^{2+}(aq),\ 2\,Cl^-(aq)$
$Pt(NH_3)_3Cl^+(aq),\ Cl^-(aq)$
$Pt(NH_3)_2Cl_2(aq)$
$K^+(aq),\ Pt(NH_3)Cl_3^-(aq)$
$2\,K^+(aq),\ PtCl_4^{2-}(aq)$

**11-44.** One third of $0.315\ \mathrm{mol \cdot L^{-1}}$, or $0.105\ \mathrm{mol \cdot L^{-1}}$

**11-48.** 0.889

**11-51.** electroneutrality

**11-55.** The thickness of the ionic atmosphere of a 1–1 electrolyte is twice as large as that of a 2–2 electrolyte.

**11-62.** The pressure is proportional to the molality squared because HCl(aq) dissociates into $H^+(aq)$ and $Cl^-(aq)$.

## Chapter 12

**12-1.** (a) (1); $n_0 - \xi, \xi, \xi$; (2); $n_0 - \xi$, $n_1 + \xi, \xi$
(b) (1); $n_0 - 2\xi, 2\xi, \xi$ (2); $n_0 - 2\xi, 2\xi$, $n_1 + \xi$
(c) (1); $n_0 - \xi, 2n_0 - 2\xi, \xi$ (2); $n_0 - \xi$, $n_0 - 2\xi, \xi$

**12-2.** The equilibrium-constant expression for the second reaction is the square root of that of the first reaction.

**12-3.** Yes, $\xi_{eq}$ decreases with increasing $P$.

**12-6.** $\xi_{eq}$ increases as $P$ increases, in accord with Le Châtelier's principle

**12-7.** $\xi_{eq}/n_0 = 0.0783$ at $P = 0.080$ bar; $\xi_{eq}/n_0 = 0.0.633$ at $P = 0.160$ bar; $\xi_{eq}$ decreases as $P$ increases, in accord with Le Châtelier's principle

**12-8.** $K_P = 17.4$; the values of equilibrium constants depend upon the reference state

**12-10.** (a) $\Delta_r G^\circ = 4.729\ \mathrm{kJ \cdot mol^{-1}}$; $K_P = 0.148$
(b) $\Delta_r G^\circ = -16.205\ \mathrm{kJ \cdot mol^{-1}}$; $K_P = 692$
(c) $\Delta_r G^\circ = -32.734\ \mathrm{kJ \cdot mol^{-1}}$; $K_P = 6.80 \times 10^5$

**12-11.** (a) $K_c = 5.97 \times 10^{-3}$; (b) $K_c = 692$; (c) $K_c = 4.17 \times 10^8$

**12-13.** $K_P = 2.94 \times 10^{-3}$; $K_c = 2.11 \times 10^{-5}$

**12-15.** $\xi_{eq} = 0.31$

**12-17.** $K_3 = 3.37$

**12-21.** The reaction as written will proceed to the right.

**12-22.** The reaction as written will proceed to the left.

**12-23.** $\Delta_r H^\circ = 6.91\ \mathrm{kJ \cdot mol^{-1}}$

**12-24.** $K_P = 1.35$

**12-25.** $K_P = 14.9$

**12-26.** $\Delta_r H^\circ = 99.6\ \mathrm{kJ \cdot mol^{-1}}$

**12-27.** $\Delta_r H^\circ = 12.02\ \mathrm{kJ \cdot mol^{-1}}$

**12-29.** $\Delta_r G^\circ = -23.78\ \mathrm{kJ \cdot mol^{-1}}$; $\Delta_r H^\circ = -89.30\ \mathrm{kJ \cdot mol^{-1}}$; $\Delta_r S^\circ = -124.8\ \mathrm{J \cdot mol^{-1} \cdot K^{-1}}$

**12-30.** $\Delta_r H^\circ = 266.5\ \mathrm{kJ \cdot mol^{-1}}$

**12-31.** $K_P = 3.889 \times 10^{-4}, 0.7367, 9.554$; $\Delta_r H^\circ = 125.9\ \mathrm{kJ \cdot mol^{-1}}$; $\Delta_r S^\circ =$ $60.61\ \mathrm{J \cdot K^{-1} \cdot mol^{-1}}$, $60.41\ \mathrm{J \cdot K^{-1} \cdot mol^{-1}}$, $60.73\ \mathrm{J \cdot K^{-1} \cdot mol^{-1}}$

**12-32.** At 1000 K, $\Delta_r G^\circ = 4.914\ \mathrm{kJ \cdot mol^{-1}}$, $\Delta_r H^\circ = 90.2\ \mathrm{kJ \cdot mol^{-1}}$, $\Delta_r S^\circ = 85.3\ \mathrm{J \cdot K^{-1} \cdot mol^{-1}}$

**12-34.** $K = 52.29$

**12-35.** $K_P(900\ \mathrm{K}) = 1.47$, $K_P(1000\ \mathrm{K}) = 0.52$, $K_P(1100\ \mathrm{K}) = 0.22$, $K_P(1200\ \mathrm{K}) = 0.11$,; $\Delta_r H^\circ = -76.8\ \mathrm{kJ \cdot mol^{-1}}$

**12-36.** $1.46 \times 10^{-3}$

**12-37.** $K_P(900\ \mathrm{K}) = 0.56$; $K_P(1200\ \mathrm{K}) = 1.66$

**12-38.** $K_P = 12.3 \times 10^{-5}$

**12-39.** $\Delta_r H^\circ = 153.8\ \text{kJ}\cdot\text{mol}^{-1}$

**12-41.** $\Delta_r H^\circ = 98.8\ \text{kJ}\cdot\text{mol}^{-1}$

**12-43.** $K = 3.37$

**12-44.** At 900 K, $K_P(\text{JANAF}) = 1.28$, $K_P(\text{Problem 24-35}) = 1.47$

At 1000 K, $K_P(\text{JANAF}) = 0.472$, $K_P(\text{Problem 24-35}) = 0.52$

At 1100 K, $K_P(\text{JANAF}) = 0.208$, $K_P(\text{Problem 24-35}) = 0.22$

**12-45.** $K_P(\text{JANAF}) = 1.32 \times 10^{-3}$, $K_P(\text{Problem 24-36}) = 1.46 \times 10^{-3}$

**12-46.** $K_P(\text{JANAF}) = 8.75 \times 10^{-5}$, $K_P(\text{Problem 24-38}) = 12.3 \times 10^{-5}$

**12-47.**

| $T$/K | $K_P$(JANAF) | $K_P$(Problem 24-39) |
|---|---|---|
| 800 | $3.05 \times 10^{-5}$ | $3.14 \times 10^{-5}$ |
| 900 | $4.26 \times 10^{-4}$ | $4.08 \times 10^{-4}$ |
| 1000 | $3.08 \times 10^{-3}$ | $3.19 \times 10^{-3}$ |
| 1100 | $1.66 \times 10^{-2}$ | $1.72 \times 10^{-2}$ |
| 1200 | $6.78 \times 10^{-2}$ | $7.07 \times 10^{-2}$ |

**12-49.** $2.443 \times 10^{32}\ \text{m}^{-3}$

**12-50.** $3.84 \times 10^{35}\ \text{m}^{-3}$ versus $3.86 \times 10^{35}\ \text{m}^{-3}$

**12-51.** $1.87 \times 10^{35}\ \text{m}^{-3}$ versus $1.91 \times 10^{35}\ \text{m}^{-3}$

**12-52.** $5.66 \times 10^{35}\ \text{m}^{-3}$ versus $5.51 \times 10^{35}\ \text{m}^{-3}$

**12-53.** $D_0 = 427.8\ \text{kJ}\cdot\text{mol}^{-1}$

**12-54.** $D_0 = 1642\ \text{kJ}\cdot\text{mol}^{-1}$

**12-55.** $D_0 = 1598\ \text{kJ}\cdot\text{mol}^{-1}$

**12-56.** $K_\gamma \approx 0.53$

**12-57.** $K_\gamma \approx 1.1$; therefore $K_P$ at 500 bar must be smaller than $K_P$ at one bar

**12-58.** $\ln a = 1.08$ at 100 bar

**12-59.** $\Delta_r H^\circ = 159.2\text{kJ}\cdot\text{mol}^{-1}$; $\Delta_r S^\circ = 217.7\text{J}\cdot\text{mol}^{-1}\cdot\text{K}^{-1}$; $\Delta_r G^\circ = 94.3\text{kJ}\cdot\text{mol}^{-1}$

**12-60.** $\Delta_r H^\circ = 31.67\text{kJ}\cdot\text{mol}^{-1}$; $\Delta_r S^\circ = 96.51\text{J}\cdot\text{mol}^{-1}\cdot\text{K}^{-1}$; $\Delta_r G^\circ = 2.910\text{kJ}\cdot\text{mol}^{-1}$

**12-61.** 3800 bar

**12-63.** $0.051\text{mol}\cdot\text{L}^{-1}$

**12-64.** $3.4 \times 10^{-4}\ \text{mol}\cdot\text{L}^{-1}$

**12-65.** $3.3 \times 10^{-4}\ \text{mol}\cdot\text{L}^{-1}$

## Chapter 13

**13-1.** (a)

$\text{Pb(s)} + 2\,\text{HI(aq)} \longrightarrow \text{PbI}_2\text{(s)} + \text{H}_2\text{(g)}$

(b) $\text{Cu(s)} + 2\,\text{AgClO}_4\text{(aq)} \longrightarrow 2\,\text{Ag(s)} + \text{Cu(ClO}_4)_2\text{(aq)}$

(c) $2\,\text{In(s)} + 3\,\text{Cd}^{2+}\text{(aq)} \longrightarrow 3\,\text{Cd(s)} + 2\,\text{In}^{3+}\text{(aq)}$

(d)

$\text{Sn(s)} + 2\,\text{Ag}^+\text{(aq)} \longrightarrow 2\,\text{Ag(s)} + \text{Sn}^{2+}\text{(aq)}$

**13-2.** (a) $\text{Pb(s)} + \text{Hg}_2\text{SO}_4\text{(s)} \longrightarrow 2\,\text{Hg(l)} + \text{PbSO}_4\text{(s)}$

(b)

$\text{H}_2\text{(g)} + \text{Hg}_2\text{Cl}_2\text{(s)} \longrightarrow 2\,\text{Hg(l)} + 2\,\text{HCl(aq)}$

(c) $\text{Zn(s)} + \text{HgO(s)} \longrightarrow \text{Hg(l)} + \text{ZnO(s)}$

(d) $\text{Cd(s)} + \text{Hg}_2\text{SO}_4\text{(s)} \longrightarrow 2\,\text{Hg(l)} + \text{CdSO}_4\text{(aq)}$

**13-3.** (a) no effect; (b) decrease; (c) increase; (d) decrease

**13-4.** (a) increase; (b) no effect; (c) decrease; (d) decrease; (e) no effect; (f) increase

**13-5.** (a) 8; (b) 4; (c) 2; (d) 20

**13-10.** (a) $\gamma_\pm^2 m^2$; (b) $4\gamma_\pm^3 m^3$; (c) $4\gamma_\pm^3 m^3$; (d) $\gamma_\pm^2 m^2$

**13-11.** 0.2223 V

**13-12.** 0.984 V

**13-13.** 0.358 V

**13-14.** 0.834; 0.745; 0.704; 0.656; 0.587; 0.521; 0.450; 0.400

**13-15.** 0.788; 0.827; 0.863; 0.894; 0.909; 0.927; 0.940

**13-16.** 0.966; 0.951; 0.930; 0.906; 0.878; 0.838

**13-17.** −0.268 V, The $\text{H}_2\text{(g)}$ electrode is positive.

**13-18.** 0.9738 V, The Pb(s) electrode is positive.

**13-19.** −0.481 V, The Zn(s) electrode is positive.

**13-20.** $\text{H}_2\text{(g)}|\text{HI(aq)}|\text{I}_2\text{(s)}|\text{Pt(s)}$; 0.5345 V

**13-21.** $\text{Pb(s)}|\text{PbCl}_2\text{(aq)}|\text{PbCl}_2\text{(s)}|\text{Pb(s)}$; −0.140 V

**13-22.** $Zn(s)|ZnCl_2(aq)||HCl(aq)|H_2(g)$; $+0.763$ V

**13-23.** $-0.21$ V

**13-24.** 0.419 V; yes

**13-25.** $-0.297$ V; no

**13-26.** 1.326 V

**13-27.** 0.712 V

**13-28.** 1.49 V

**13-29.** $H_2(g)|H_2SO_4(aq)|PbSO_4(s)|Pb(s)$; $-0.3583$ V

**13-30.** 0.0732 V: 0.981, 0.975, 0.969, 0.963, 0.955

**13-31.** $H_2(g) + 2\,AgCl(s) \longrightarrow 2\,Ag(s) + 2\,HCl(aq)$; $E^\circ = 0.2224$ V; $\Delta_r G^\circ = -42.92\ \text{kJ}\cdot\text{mol}^{-1}$; $\Delta_r S^\circ = -124.6\ \text{J}\cdot\text{mol}^{-1}\cdot\text{K}^{-1}$; $\Delta_r H^\circ = -80.07\ \text{kJ}\cdot\text{mol}^{-1}$

**13-32.** $H_2(g) + Hg_2Br_2(s) \longrightarrow 2\,Hg(l) + 2\,HBr(aq)$; $\Delta_r G^\circ = -26.96\ \text{kJ}\cdot\text{mol}^{-1}$; $\Delta_r S^\circ = -29.7\ \text{J}\cdot\text{mol}^{-1}\cdot\text{K}^{-1}$; $\Delta_r H^\circ = -35.8\ \text{kJ}\cdot\text{mol}^{-1}$; $S^\circ[Br^-(aq)] = 83.5\ \text{J}\cdot\text{mol}^{-1}\cdot\text{K}^{-1}$

**13-33.** $1.41 \times 10^{-18}$

**13-34.** $1.16 \times 10^{-5}$

**13-35.** $1.40 \times 10^{-8}$

**13-36.** 1.30

**13-37.** $8.8 \times 10^{-9}$

**13-38.** $9.4 \times 10^{-3}$

**13-39.** $1.33 \times 10^{-5}$

**13-40.** $1.65 \times 10^{-5}$

**13-41.** $1.78 \times 10^{-4}$

**13-42.** The values of $K_w$ at 0°C, 10°C, 20°C, 30°C, 40°C, 50°C, and 60°C are $1.15 \times 10^{-15}$, $2.94 \times 10^{-15}$, $6.85 \times 10^{-15}$, $1.45 \times 10^{-14}$, $2.92 \times 10^{-14}$, $5.45 \times 10^{-14}$, and $9.60 \times 10^{-14}$, respectively. The corresponding values of $\Delta_r H^\circ$ are (in $\text{kJ}\cdot\text{mol}^{-1}$) are 63.88, 60.33, 57.26, 54.67, 52.58, 50.97, and 49.84, The value of $K_w$ at 25°C comes out to be $1.01 \times 10^{-4}$ and the value of $\Delta_r H^\circ$ comes out to be $55.90\ \text{kJ}\cdot\text{mol}^{-1}$.

**13-43.** $1.85 \times 10^{-5}$ and $1.71 \times 10^{-5}$

**13-44.** $7.90 \times 10^{-7}$ and $7.99 \times 10^{-7}$

**13-45.** $1.64 \times 10^{-8}$

**13-46.** $3.70 \times 10^{-6}$

**13-47.** $1.08 \times 10^{-10}$

**13-48.** $1.01 \times 10^{-2}$

**13-49.** $5.65 \times 10^{-3}$

**13-50.** $9.85 \times 10^{-7}$

**13-51.** 1480 kJ versus 2108 kJ; $E^\circ = 1.09$ V

**13-53.** $B_{2P} = 6.1 \times 10^{-4}\ \text{atm}^{-1}$ (estimated), or $5.39 \times 10^{-4}\ \text{atm}^{-1}$ (using tabluated values)

**13-55.** 0.852 V

**13-56.** $-0.74$ V

## Chapter 14

**14-24.** 59 mV

**14-25.** $12.7\ \text{kJ}\cdot\text{mol}^{-1}$

**14-27.** $H_2(P_1) \rightarrow H_2(P_2)$; the cell is a concentration cell

**14-28.** 0.47

**14-30.** $-3.95$ mV; $-2.31$ mV; $-0.356$ mV; $-0.208$ mV; $+11.6$ mV

**14-37.** 13.93 mV; 41.78 mV; 17.45 mV; 52.36 mV; 29.42 mV; 88.26 mV

**14-39.** Take the radius of the $Na^+(aq)$ ion to be 100 pm. Then $f = 1.7 \times 10^{-12}\ \text{kg}\cdot\text{s}^{-1}$. Take the mass of the $Na^+(aq)$ ion to be 100 amu. Then $f/m = 1 \times 10^{13}\ \text{s}^{-1}$.

**14-43.** 26.9 mV; 26.9 mV; 31.2 mV; $-4.37$ mV; $-4.37$ mV; $-5.01$ mV

**14-45.** $\mu_{23}/\mu_{22} = -0.013$ for $c_2 = c_3 = 0.0010$ M; $-0.043$ for $c_2 = c_3 = 0.010$ M; $-0.15$ for $c_2 = c_3 = 0.10$ M

**14-46.** For an ideal solution.

**14-47.** $RTL_{23}/\text{M}\cdot\text{cm}^2\cdot\text{s}^{-1} = -9.9 \times 10^{-7}$ and $RTL_{32}/\text{M}\cdot\text{cm}^2\cdot\text{s}^{-1} = -9.5 \times 10^{-7}$

**14-48.** $RTL_{23}/\text{M}\cdot\text{cm}^2\cdot\text{s}^{-1} = -1.5 \times 10^{-6}$ and $RTL_{32}/\text{M}\cdot\text{cm}^2\cdot\text{s}^{-1} = -1.5 \times 10^{-6}$

**14-57.** The integral is equal to zero.

**14-58.** $\langle x \rangle = 0$; $x_{\text{rms}}/\text{cm} = 4 \times 10^{-9},\ 1 \times 10^{-7},\ 4 \times 10^{-6}$

# Illustration Credits

## CHAPTER-OPENING PHOTOS

Chapter 1: Courtesy of AIP Emilio Segrè Visual Archives, W.F. Meggers Collection.

Chapter 2: Kamerlingh Onnes Laboratory, Leiden. Courtesy of the Caltech Archives, Earnest Watson Collection.

Chapter 3: Courtesy of John Simon (memorial stone) and University of Vienna, Courtesy of AIP Niels Bohr Library.

Chapter 4: Courtesy of UC Berkeley.

Chapter 5: Courtesy of AIP Emilio Segrè Visual Archives, *Physics Today* Collection.

Chapter 6: Courtesy of AIP Emilio Segrè Visual Archives, Landè Collection.

Chapter 7: Rijksmuseum voor de Geschiedenis van de Natuurwetenschappen, Leiden. Courtesy of AIP Emilio Segrè Visual Archives.

Chapter 8: Photograph by E. Lange, courtesy of AIP Emilio Segrè Visual Archives.

Chapter 9: Courtesy of AIP Emilio Segrè Visual Archives.

Chapter 10: Reprinted courtesy of The Bancroft Library, University of California, Berkeley.

Chapter 11: (Debye) Photograph by Francis Simon, courtesy AIP Emilio Segrè Visual Archives.

Chapter 12: Courtesy of AIP Emilio Segrè Visual Archives, W. Cady Collection.

Chapter 13: Reprinted with permission from the *Journal of Chemical Education*, Vol. 61, No. 1, (1984). Copyright © 1984 by Division of Chemical Education, Inc., American Chemical Society.

Chapter 14: Courtesy of Yale University Library, Manuscripts and Archives.

## TEXT FIGURES

Fig. 2.10: With permission from Goug-jen Su, *Industrial and Engineering Chemistry Research* 38, 803 (1946). Copyright 1946 American Chemical Society.

Chapter 5 quote: With permission of Open Court Trade & Academic Books, a division of Carus Publishing Company, Peru, IL, from *Albert Einstein: Philosopher Scientist*, edited by P.A. Schilpp, copyright 1973.

Fig. 8.11: Reprinted with permission from R.H. Newton, *Industrial and Engineering Chemistry Research* 27, 302 (1935). Copyright 1935 American Chemical Society.

Fig. 9.14: With permission from *Journal of Physical and Chemical Reference Data* 17, 1541 (1988). Copyright 1988 American Chemical Society and the American Institute of Physics for the National Institute of Science and Technology.

Table 12.4: With permission from *Journal of Physical and Chemical Reference Data* 14, supplement 1 (1985). Copyright 1985 American Chemical Society and the American Institute of Physics for the National Institute of Science and Technology.

Fig. 13.15: With permission from *Journal of Physical and Chemical Reference Data* 11, supplement 2 (1982). Copyright 1982 American Chemical Society and the American Institute of Physics for the National Institute of Science and Technology.

# Index

## Some Mathematical Formulas

$$\sin(x \pm y) = \sin x \cos y \pm \cos x \sin y$$
$$\cos(x \pm y) = \cos x \cos y \mp \sin x \sin y$$
$$\sin x \sin y = \tfrac{1}{2}\cos(x-y) - \tfrac{1}{2}\cos(x+y)$$
$$\cos x \cos y = \tfrac{1}{2}\cos(x-y) + \tfrac{1}{2}\cos(x+y)$$
$$\sin x \cos y = \tfrac{1}{2}\sin(x+y) + \tfrac{1}{2}\sin(x-y)$$
$$\sin^2 x + \cos^2 x = 1$$

$$\cosh x = \frac{e^x + e^{-x}}{2} \qquad \sinh x = \frac{e^x - e^{-x}}{2} \qquad \tanh x = \frac{e^x - e^{-x}}{e^x + e^{-x}}$$

$$f(x) = f(a) + f'(a)(x-a) + \frac{1}{2!}f''(a)(x-a)^2 + \frac{1}{3!}f'''(a)(x-a)^3 + \cdots$$

$$e^x = 1 + x + \frac{x^2}{2!} + \frac{x^3}{3!} + \frac{x^4}{4!} + \cdots$$

$$\cos x = 1 - \frac{x^2}{2!} + \frac{x^4}{4!} - \frac{x^6}{6!} + \cdots$$

$$\sin x = x - \frac{x^3}{3!} + \frac{x^5}{5!} - \frac{x^7}{7!} + \cdots$$

$$\ln(1+x) = x - \frac{x^2}{2} + \frac{x^3}{3} - \frac{x^4}{4} + \cdots \qquad -1 < x \le 1$$

$$\frac{1}{1-x} = 1 + x + x^2 + x^3 + x^4 + \cdots \qquad x^2 < 1$$

$$(1 \pm x)^n = 1 \pm nx \pm \frac{n(n-1)}{2!}x^2 \pm \frac{n(n-1)(n-2)}{3!}x^3 + \cdots \qquad x^2 < 1$$

$$\int_0^\infty x^n e^{-ax}\,dx = \frac{n!}{a^{n+1}} \qquad (n \text{ positive integer})$$

$$\int_0^\infty e^{-ax^2}\,dx = \left(\frac{\pi}{4a}\right)^{1/2}$$

$$\int_0^\infty x^{2n} e^{-ax^2}\,dx = \frac{1 \cdot 3 \cdot 5 \cdots (2n-1)}{2^{n+1}a^n}\left(\frac{\pi}{a}\right)^{1/2} \qquad (n \text{ positive integer})$$

$$\int_0^\infty x^{2n+1} e^{-ax^2}\,dx = \frac{n!}{2a^{n+1}} \qquad (n \text{ positive integer})$$

$$\int \frac{dx}{ax+b} = \frac{1}{a}\ln(ax+b)$$

$$\int \frac{dx}{ax^2+bx+c} = \frac{1}{\sqrt{-q}}\ln\frac{2ax+b-\sqrt{-q}}{2ax+b+\sqrt{-q}} \qquad q = 4ac - b^2$$

$$\int \frac{dx}{x(ax+b)} = -\frac{1}{b}\ln\frac{ax+b}{x}$$